XIII Mexican Workshop on Particles and Fields 2011

(MWPF)

Journal of Physics: Conference Series Volume 378

Leon, Guanajuato, Mexico
20-26 October 2011

Editors:

Juan Barranco
David Delepine
Guillermo Contreras
Mauro Napsuciale

ISBN: 978-1-62276-812-7
ISSN: 1742-6588

Printed from e-media with permission by:

Curran Associates, Inc.
57 Morehouse Lane
Red Hook, NY 12571

Some format issues inherent in the e-media version may also appear in this print version.

Printed by Curran Associates, Inc. (2013)

For permission requests, please contact the Institute of Physics
at the address below.

Institute of Physics
Dirac House, Temple Back
Bristol BS1 6BE UK

Phone: 44 1 17 929 7481
Fax: 44 1 17 920 0979

techtracking@iop.org

Additional copies of this publication are available from:

Curran Associates, Inc.
57 Morehouse Lane
Red Hook, NY 12571 USA
Phone: 845-758-0400
Fax: 845-758-2634
Email: curran@proceedings.com
Web: www.proceedings.com

Astroparticles, Cosmology, Strings and Beyond Standard Model

Electroweak and Flavour Physics

Hadronic Matters

Editorial

Juan Barranco

Physics Department, Guanajuato University, Loma del Bosque 103, col. Loma del Campestre, 37150, Leon (Mexico)

E-mail: jbarranc@fisica.ugto.mx

Guillermo Contreras

Departamento de Fisica Aplicada Centro de Investigacion y de Estudios Avanzados del Instituto Politecnico Nacional, Merida (Mexico)

E-mail: jgcn@mda.cinvestav.mx

David Delepine

Physics Department, Guanajuato University, Loma del Bosque 103, col. Loma del Campestre, 37150, Leon (Mexico)

E-mail: delepine@fisica.ugto.mx

Mauro Napsuciale

Physics Department, Guanajuato University, Loma del Bosque 103, col. Loma del Campestre, 37150, Leon (Mexico)

E-mail: mauro@fisica.ugto.mx

The XIII Mexican Workshop on Particles and Fields (MWPF) took place from 20-26 October, 2011, in the city of León, Guanajuato, México. This is a biennial meeting organized by the Division of Particles and Fields of the Mexican Physical Society designed to gather specialists in different areas of high energy physics to discuss the latest developments in the field. The thirteenth edition of this meeting was hosted by the Department of Cultural Studies of Guanajuato University in a nice surrounding environment dedicated to the Arts and Culture.

The XIII MWPF was organized by three working groups who organized the corresponding sessions around three topics. The first one was Strings, Cosmology, Astroparticles and Physics Beyond the Standard Model. In this category we included: Cosmic Rays, Gamma Ray Bursts, Physics Beyond the Standard Model (theory and experimental searches), Strings and Cosmology. The working group for this topic was formed by Arnulfo Zepeda, Oscar Loaiza, Axel de la Macorra and Myriam Mondragón. The second topic was Hadronic Matter which included Perturbative QCD, Jets and Diffractive Physics, Hadronic Structure, Soft QCD, Hadron Spectroscopy, Heavy Ion Collisions and Soft Physics at Hadron Colliders, Lattice Results and Instrumentation. The working group for this topic was integrated by Wolfgang Bietenholz and Mariana Kirchbach. The third topic was Electroweak and Flavor Physics in whose classification we had Quark masses and mixing, Electroweak Symmetry Breaking, Heavy Quark Physics, Neutrinos, CP Violation, CKM and Rare Decays. The working group for this topic was formed by David Delepine, Lorenzo Díaz, Gerardo Herrera and Olga Félix Beltrán.

The three topics included invited talks dedicated to review theoretical and experimental aspects of the corresponding topic. We had also plenary topical sessions during the mornings and early afternoons and parallel thematic sessions were held late in the afternoons. The reviews sessions were delivered by invited speakers of international prestige on the corresponding subjects and having an active collaboration with the Mexican scientific community. The plenary topical sessions and parallel thematic sessions were given by active researchers both from abroad and working in México. The program included also a permanent poster session and the organizing committee awarded the best poster - a decision made by the three working groups- with a silver medal.

On Sunday 23th of October,we had an excursion to the ancient archeological site of Plazuelas near Penjamo (Guanajuato). Plazuelas is a prehispanic archaeological site located just north of San Juan el Alto, some 2.7 kilometers north of federal highway 90 (Penjamo-Guadalajara), and about 11 kilometers west of the city of Penjamo in the state of Guanajuato, Mexico. The site has been recently open to the public. This site is a very complex city, it was edified following the natural landscape surrounding it. This city was occupied between the 600 and 900 a.C.
After spending some time around this very interesting archaeological site, we went to Hacienda Corralejo , an hacienda where Tequila Corralejo is produced and we had a guided visit of the "Tequilera" where the famous Corralejo blue bottle of Tequila is done. Hacienda Corralejo is very close to the orginal Hacienda where Miguel Hidalgo (a Mexican independence hero) was born which now is just a set of foundations and a portal, considered national monuments. There, we enjoyed a delicious meal in the cellars of the Corralejo Hacienda before returning to Leon.

The XIII MWPF was sponsored by several institutions: Consejo Nacional de Ciencia y Tecnología (CONACyT) through the Red Nacional de Física de Altas Energías and individual research projects, Consejo de Ciencia y Tecnológico del Estado de Guanajuato (CONCyTEG), Universidad de Guanajuato, Universidad Nacional Autónoma de México, Universidad de Guanajuato, Centro de Investigaciones de Estudios Avanzados del IPN (CINVESTAV), Universidad Michoacana de San Nicolás de Hidalgo, Benemérita Universidad Autónoma de Puebla and Universidad Autónoma de San Luis Potosí. Also, we wish to thak to those who helped in the process of getting financial support for the meeting, specially Dr. Juan Carlos D'Olivo, President of the Red Nacional de Física de Altas Energías and Dr. José Luis Lucio Martínez, Rector of Universidad de Guanajuato, Campus León. These proceedings have been published thanks to the support of PIFI 2011.

This meeting was possible due to the commitment of the working groups and we wish to thank to their members for the decisive collaboration with the organizing committee. At the local level, we thank our graduate students: Carolina Luján, Vannia González, Selim Gomez and Carlos Alberto Vaquera for their invaluable contribution in the organization of the large amount of little but important things around the meeting.

Finally, we would like to thank all the speakers for delivering excellent talks which contributed to the success of the event. We are also grateful to all the participants for the nice academic and human atmosphere during the meeting and for providing their write-ups on time.

The National organizing committee was formed by

(i) Arnulfo Zepeda (CINVESTAV-DF).

(ii) Guillermo Contreras (CINVESTAV-Mrida).

(iii) David Delepine (DF-UG).

(iv) Axel de la Macorra (IAC/IF-UNAM).

(v) Lorenzo Díaz (BUAP).
(vi) Gerardo Herrera (CINVESTAV-DF).
(vii) Mariana Kirchbach (IF-UASLP).
(viii) Oscar Loaiza (DF-UG).
(ix) Myriam Mondragón (IF-UNAM).
(x) Mauro Napsuciale (DF-UG).
(xi) Humberto Salazar Ibarguen (BUAP).
(xii) Andrés Sandoval (IF-UNAM).
(xiii) Juan Carlos DOlivo (ICN-UNAM)).
(xiv) Wolfgang Bietenholz (ICN-UNAM).
(xv) Olga Félix Beltrán (BUAP).

Journal of Physics: Conference Series **378** (2012) 011001 doi:10.1088/1742-6596/378/1/011001

Dinner in cellars of Corralejo Hacienda

Journal of Physics: Conference Series **378** (2012) 011001
doi:10.1088/1742-6596/378/1/011001

Visit at Corralejo Hacienda

XIII Mexican Workshop on Particles and Fields

Thursday 20 October 2011 - Wednesday 26 October 2011

León, Guanajuato

Programme

Table of contents

Revision

Revision

Hadronic Matter

Hadronic Matter

Electroweak and Flavor Physics

Astroparticles, Cosmology, Strings and Beyond the Standard Model

Electroweak and Flavor Physics

Hadronic Matter

Astroparticles, Cosmology, Strings and Beyond the Standard Model

Wednesday 26 October 2011

Electroweak and Flavor Physics

Summaries I,II and III

Thursday 20 October 2011

Revision: Electroweak and Flavor, R1 - Auditorio Jorge Ibargüengoitia (09:00-10:00)

- Conveners: Dr. Díaz, Lorenzo

time	[id] title	presenter
09:00	[81] Searches for the higgs boson	DE LA CRUZ BURELO, Eduard

Revision: Astroparticles, Strings, Cosmology and Beyond the Standard Model, R2 - Auditorio Jorge Ibargüengoitia (10:00-11:00)

- Conveners: Dr. Díaz, Lorenzo

time	[id] title	presenter
10:00	[7] The Status of the Dark Matter Paradigm	Prof. MATOS, Tonatiuh

Astroparticles, Cosmology, Strings and Beyond the Standard Model - Auditorio Jorge Ibargüengoitia (11:30-13:30)

- Conveners: Dr. Aranda, Alfredo

time	[id] title	presenter
11:30	[77] Whish you were here (The Higgs)	Dr. DIAZ-CRUZ, Lorenzo
12:00	[34] Possible texture zeros for mass matrices of the quarks and leptons	Mr. GONZALEZ CANALES, Felix
12:30	[59] Synthesis of plastic scintillation and applications	Dr. LEÓN MONZÓN, Ildefonso
13:00	[78] Ultrahigh Cosmic Rays: The highest energy frontier	Dr. SALAZAR, Humberto

Astroparticles, Cosmology, Strings and Beyond the Standard Model - Auditorio Jorge Ibargüengoitia (16:00-16:30)

- Conveners: Dr. Sabido, Miguel

time	[id] title	presenter
16:00	[69] QM lessons for standard cosmology	Dr. REYES SANTOS, Marco A.

Revision: Astroparticles, String, Cosmology and Beyond the Standard Model, R3 - Auditorio Jorge Ibargüengoitia (16:30-17:30)

- Conveners: Dr. Sabido, Miguel

time	[id] title	presenter
16:30	[9] Challenges of string theory: particle physics and cosmology	Dr. RAMOS-SANCHEZ, Saul

Astroparticles, Cosmology, Strings and Beyond the Standard Model - Auditorio Jorge Ibargüengoitia (17:30-18:00)

***** Cosmic Rays***

***** Gamma Ray Bursts***

***** Beyond the Standard Model (theory and experimental searches)***

***** Strings***

***** Cosmology***

- Conveners: Dr. Sabido, Miguel

time	[id] title	presenter
17:30	[39] Diffuse supernova neutrino background and scalar field dark matter	Dr. BARRANCO, Juan

Poster session + Brindis - Auditorio Jorge Ibargüengoitia (18:00-20:00)

Friday 21 October 2011

Revision: Hadronic Matter, R4 - Auditorio Jorge Ibargüengoitia (09:00-10:00)

- Conveners: Dr. Kirchbach, Mariana

time	[id] title	presenter
09:00	[52] ALICE first physics results	Dr. HERRERA, Gerardo

Revision: Hadronic Matter, R5 - Auditorio Jorge Ibargüengoitia (10:00-11:00)

- Conveners: Dr. Kirchbach, Mariana

time	[id] title	presenter
10:00	[35] Twenty-First Century Lattice Gauge Theory: Consequences of the QCD Lagrangian	Dr. KRONFELD, Andreas Kronfeld

Hadronic Matter - Auditorio Jorge Ibargüengoitia (11:30-13:30)

Perturbative QCD, Jets and Diffractive Physics.

Hadronic Structure, soft QCD, Spectroscopy.

Heavy Ion Collisions and Soft Physics at Hadron Colliders.

Lattice Results.

Instrumentation.

- Conveners: Dr. De la Cruz, Eduard

time	[id] title	presenter
11:30	[3] Ultraviolet divergences in classical field equations with noise: the dynamics of the chiral phase transition in QCD	Prof. KREIN, Gastao
12:30	[15] The infrared fixed point of Yang-Mills theories: A renormalization group analysis	Prof. WEBER, Axel
13:00	[40] Heavy quarks in a QCD plasma: energy loss or more baryons than mesons	Dr. CUAUTLE, Eleazar CUAUTLE, Eleazar

Hadronic Matter - Auditorio Jorge Ibargüengoitia (16:00-17:30)

Perturbative QCD, Jets and Diffractive Physics.

Hadronic Structure, soft QCD, Spectroscopy.

Heavy Ion Collisions and Soft Physics at Hadron Colliders.

Lattice Results.

Instrumentation.

- **Conveners: Dr. Bietenholz, Wolfgang**

time	[id] title	presenter
16:00	[60] QCD flavour-blindness on the lattice and in the real world	Dr. RAKOW, Paul
17:00	[64] B hadrons spectroscopy at the D0 experiment in RunII of the Tevatron	Dr. DE LA CRUZ BURELO, Eduard

Hadronic Matter - Audiovisual-2 (18:00-18:30)

Perturbative QCD, Jets and Diffractive Physics.

Hadronic Structure, soft QCD, Spectroscopy.

Heavy Ion Collisions and Soft Physics at Hadron Colliders.

Lattice Results.

Instrumentation.

- **Conveners: Dr. Barranco, Juan**

time	[id] title	presenter
18:00	[50] Search for Z rare decays with CDF data.	Ms. PSIHAS, Fernanda

Hadronic Matter - Audiovisual-1 (18:00-19:30)

Perturbative QCD, Jets and Diffractive Physics.

Hadronic Structure, soft QCD, Spectroscopy.

Heavy Ion Collisions and Soft Physics at Hadron Colliders.

Lattice Results.

Instrumentation.

- **Conveners: Prof. Krein, Gastao**

time	[id] title	presenter
18:00	[16] Conformal symmetry patterns in high-lying unflavored meson spectra	Dr. KIRCHBACH, Mariana
18:30	[23] Electromagnetic moments of elementary particles from a second order formalism	Dr. DELGADO-ACOSTA, Ernesto Germán
19:00	[70] Renormalization of Second Order Formalism for Spin 1/2 Fermions	Mr. RENÉ, Ángeles-Martínez

Electroweak and Flavor Physics - Audiovisual-2 (18:30-19:30)

***Quark masses and mixing**

***Electroweak Symmetry Breaking**

***Heavy Quark Physics**

***Neutrinos**

***CP Violation, CKM and Rare Decays**

- Conveners: Dr. Barranco, Juan

time	[id] title	presenter
18:30	[68] Measurement of the CP-violating phase $\phi_s^{J/\psi \phi}$ using the flavor-tagged decay $B_s^0 \rightarrow J/\psi \phi$ in 8 fb^{-1} of $p \overline p$ collisions	Mr. MAGAÑA VILLALBA, Ricardo
19:00	[67] Signal of FCNC and CPV through $h \rightarrow c \overline{b} W^-$ in two Higgs doublet model	Dr. HALIM, José

Saturday 22 October 2011

Revision: Electroweak and Flavor, R6 - Auditorio Jorge Ibargüengoitia (09:00-10:00)

- Conveners: Dr. Delepine, David

time	[id] title	presenter
09:00	[25] Probing new physics at SuperKEKB	Dr. YAMAUCHI, Masa

Revision: Electroweak and Flavor, R7 - Auditorio Jorge Ibargüengoitia (10:00-11:00)

- Conveners: Dr. Delepine, David

time	[id] title	presenter
10:00	[32] Flavor now and then	Dr. ARANDA, Alfredo

Electroweak and Flavor Physics - Auditorio Jorge Ibargüengoitia (11:30-13:30)

**** Quark masses and mixing***

**** Electroweak Symmetry Breaking***

**** Heavy Quark Physics***

**** Neutrinos***

**** CP Violation, CKM and Rare Decays***

- Conveners: Dr. López Castro, Gabriel

time	[id] title	presenter
11:30	[66] The SuperB project : physics potential and current status	Dr. GUY, Wormser
12:30	[42] The flavour permutational symmetry S(3)	Prof. MONDRAGON BALLESTEROS, Alfonso
13:00	[28] Effects of heavy Majorana neutrinos in semileptonic heavy quark decays	Mr. QUINTERO POVEDA, Néstor

Electroweak and Flavor Physics - Auditorio Jorge Ibargüengoitia (16:00-16:30)

**** Quark masses and mixing***

**** Electroweak Symmetry Breaking***

**** Heavy Quark Physics***

**** Neutrinos***

**** CP Violation, CKM and Rare Decays***

- Conveners: Prof. Mondragón, Alfonso

time	[id] title	presenter
16:00	[18] Single Top quark production at lepton colliders	Dr. LARIOS, Francisco

Revision: Electroweak and Flavor R8 - Auditorio Jorge Ibargüengoitia (16:30-17:30)

- Conveners: Dr. Mondragón, Alfonso

time	[id] title	presenter
16:30	[57] Neutrino oscillations and non standard interactions	Dr. OMAR, Miranda

Hadronic Matter - Auditorio Jorge Ibargüengoitia (18:00-19:00)

Perturbative QCD, Jets and Diffractive Physics.

Hadronic Structure, soft QCD, Spectroscopy.

Heavy Ion Collisions and Soft Physics at Hadron Colliders.

Lattice Results.

Instrumentation.

- **Conveners: Dr. Weber, Axel**

time	[id] title	presenter
18:00	[37] The omega-rho-pi coupling and the influence of heavier resonances	Dr. TOLEDO, Genaro
18:30	[80] An approach for the Quantization of Spin- 3/2 beyond the Rarita-Schwinger Framework.	Dr. VILLANUEVA, Victor

Astroparticles, Cosmology, Strings and Beyond the Standard Model - Auditorio Jorge Ibargüengoitia (19:00-19:30)

* ***Cosmic Rays***
* ***Gamma Ray Bursts***
* ***Beyond the Standard Model (theory and experimental searches)***
* ***Strings***
* ***Cosmology***

- **Conveners: Dr. Weber, Axel**

time	[id] title	presenter
19:00	[82] The Lorentz violation in a 5D susy model.	Mr. GARCÍA, José Didino

Monday 24 October 2011

Revision: Astroparticles, Cosmology, Strings and Beyond the Standard Model, R9 - Auditorio Jorge Ibargüengoitia (09:00-10:00)

- **Conveners: Dr. Mondragón, Myriam**

time	[id] title	presenter
09:00	[63] Probing the small-scale structure of spacetime	LEHNERT, Ralf

Revision: Astroparticles, Cosmology, Strings and Beyond the Standard Model, R10 - Auditorio Jorge Ibargüengoitia (10:00-11:00)

- **Conveners: Dr. Mondragón, Myriam**

time	[id] title	presenter
10:00	[14] The Cosmic Frontier: Cosmic Rays, Gamma Rays, and Neutrinos	Dr. SINNIS, Gus

Astroparticles, Cosmology, Strings and Beyond the Standard Model - Auditorio Jorge Ibargüengoitia (11:30-13:30)

Cosmic Rays

Gamma Ray Bursts

Beyond the Standard Model (theory and experimental searches)

Strings

Cosmology

- **Conveners: Dr. Loaiza, Oscar**

time	[id] title	presenter
11:30	[53] E-instantons in Two Dimensional Models	GARCÍA COMPEÁN, Hugo
12:00	[56] SUSY breaking in the baryonic branch	Dr. CÁCERES, Elena
12:30	[65] Noncommutative Minisuperspace and Lambda	Dr. SABIDO, Miguel
13:00	[10] Towards flavor physics from the heterotic string	Dr. RAMOS-SANCHEZ, Saul

Astroparticles, Cosmology, Strings and Beyond the Standard Model - Auditorio Jorge Ibargüengoitia (16:00-17:30)

- **Conveners: Dr. Zepeda, Arnulfo**

time	[id] title	presenter
16:00	[17] Simulation of Fano Factor in HAWC-30 Array	CASTILLO, Mario
16:30	[30] Ultra-high energy extragalactic neutrinos interacting with ultra-light dark matter	Dr. MOURA, Celio

time	[id] title	presenter
17:00	[24] Exploring the light and heavy composition of cosmic rays with KASCADE-Grande	Dr. ARTEAGA VELAZQUEZ, Juan Carlos

Astroparticles, Cosmology, Strings and Beyond the Standard Model - Audiovisual-1 (18:00-19:30)

*** Cosmic Rays**

*** Gamma Ray Bursts**

*** Beyond the Standard Model (theory and experimental searches)**

*** Strings**

*** Cosmology**

- Conveners: Dr. Arteaga, Juan Carlos

time	[id] title	presenter
18:00	[41] Study of cosmic rays in the range of 10^17-10^18 eV with extentions of Pierre Auger Observatory.	Mr. VARELA, Enrique
18:30	[13] Cosmological model by flux compactification in generalized varieties	DAMIAN, Cesar
19:00	[21] Near-horizon geometry from flux compactification	VAZQUEZ MERCADO, Liliana

Hadronic Matter - Audiovisual-2 (18:00-19:30)

Perturbative QCD, Jets and Diffractive Physics.

Hadronic Structure, soft QCD, Spectroscopy.

Heavy Ion Collisions and Soft Physics at Hadron Colliders.

Lattice Results.

Instrumentation.

- Conveners: Dr. Hofmann, Christoph

time	[id] title	presenter
18:00	[20] Dyson-Schwinger equations for infrared Yang-Mills theories in different dimensions	Mr. DALL'OLIO, Pietro
18:30	[26] Search for dilepton and lepton+ MET resonances at high mass with ATLAS Experiment at LHC	Ms. CASTANEDA-MIRANDA, Elizabeth
19:00	[38] A Bethe-Salpeter equation for unstable states.	Mr. MALDONADO VILLAMIZAR, Felix

Tuesday 25 October 2011

Revision: Hadronic Matter, R11 - Auditorio Jorge Ibargüengoitia (09:00-10:00)

- Conveners: Dr. Bietenholz, Wolfgang

time	[id] title	presenter
09:00	[5] Review on Unitary ChPT	Dr. OLLER, José Antonio

Revision: Hadronic Matter, R12 - Auditorio Jorge Ibargüengoitia (10:00-11:00)

- Conveners: Dr. Bietenholz, Wolfgang

time	[id] title	presenter
10:00	[62] Quarkonia and heavy-flavour production in heavy-ion collisions	Dr. TAPIA TAKAKI, Daniel

Hadronic Matter - Auditorio Jorge Ibargüengoitia (11:30-13:30)

Perturbative QCD, Jets and Diffractive Physics.

Hadronic Structure, soft QCD, Spectroscopy.

Heavy Ion Collisions and Soft Physics at Hadron Colliders.

Lattice Results.

Instrumentation.

- Conveners: Dr. Oller, José Antonio

time	[id] title	presenter
11:30	[51] Results from the LHCb Experiment	POLYCARPO, Érica
12:30	[2] Baryon properties in large-N chiral perturbation theory	Prof. FLORES, Ruben
13:00	[49] Baryon chiral perturbation theory transferred to hole-doped antiferromagnets on the honeycomb lattice	Dr. HOFMANN, Christoph

Hadronic Matter - Auditorio Jorge Ibargüengoitia (16:00-16:30)

Perturbative QCD, Jets and Diffractive Physics.

Hadronic Structure, soft QCD, Spectroscopy.

Heavy Ion Collisions and Soft Physics at Hadron Colliders.

Lattice Results.

Instrumentation.

- Conveners: Dr. Moura, Celio

time	[id] title	presenter
16:00	[4] Spin and flavor content of octet baryons	Dr. BIJKER, Roelof

Electroweak and Flavor Physics - Auditorio Jorge Ibargüengoitia (16:30-17:30)

*** Quark masses and mixing**

*** Electroweak Symmetry Breaking**

*** Heavy Quark Physics**

*** Neutrinos**

*** CP Violation, CKM and Rare Decays**

- **Conveners: Dr. Moura, Celio**

time	[id] title	presenter
16:30	[54] The decay D+ -> K- pi+ pi+	Dr. RAMÍREZ, Carlos
17:00	[79] R for the Mu2e Experiment	CARMINE, Pagliarone

Astroparticles, Cosmology, Strings and Beyond the Standard Model - Audiovisual-1 (18:00-18:30)

*** Cosmic Rays**

*** Gamma Ray Bursts**

*** Beyond the Standard Model (theory and experimental searches)**

*** Strings**

*** Cosmology**

- **Conveners: Dr. Delgado, Germán**

time	[id] title	presenter
18:00	[19] High-energy neutrino production in active galactic nuclei	Dr. ARTEAGA VELAZQUEZ, Juan Carlos

Electroweak and Flavor Physics - Audiovisual-2 (18:00-19:00)

*** Quark masses and mixing**

*** Electroweak Symmetry Breaking**

*** Heavy Quark Physics**

*** Neutrinos**

*** CP Violation, CKM and Rare Decays**

- **Conveners: Dr. Gómez Bock, Melina**

time	[id] title	presenter
18:00	[33] Tensorial Non Standard and Unparticle interactions in neutrino-electron Scattering.	GARCES, Estela
18:30	[31] On the Higgs potential in the MS3IESM	Mr. SALDAÑA SALAZAR, Ulises Jesús

Hadronic Matter - Audiovisual-1 (18:30-19:30)

Perturbative QCD, Jets and Diffractive Physics.

Hadronic Structure, soft QCD, Spectroscopy.

Heavy Ion Collisions and Soft Physics at Hadron Colliders.

Lattice Results.

Instrumentation.

- Conveners: Dr. Delgado, Germán

time	[id] title	presenter
18:30	[27] Topological summation in lattice gauge theory	Dr. BIETENHOLZ, Wolfgang
19:00	[76] Searches for TeV-gravity signatures in final states with lepton and Jets with the ATLAS detector	CASTAÑEDA HDZ., Alfedro

Astroparticles, Cosmology, Strings and Beyond the Standard Model - Audiovisual-2 (19:00-19:30)

**** Cosmic Rays***

**** Gamma Ray Bursts***

**** Beyond the Standard Model (theory and experimental searches)***

**** Strings***

**** Cosmology***

- Conveners: Dr. Gómez Bock, Melina

time	[id] title	presenter
19:00	[44] A WWW Interface to generate Aires & Corsika Steer Files.	Mr. DE LEÓN ACUÑA, Cederik León

Wednesday 26 October 2011

Electroweak and Flavor Physics - Auditorio Jorge Ibargüengoitia (09:00-11:00)

Quark masses and mixing

Electroweak Symmetry Breaking

Heavy Quark Physics

Neutrinos

CP Violation, CKM and Rare Decays

- Conveners: Dr. Félix Beltrán, Olga

time [id] title	presenter
09:00 [36] The I-tracker proposal for MU2E Experiment at Fermilab	Prof. PAGLIARONE, Carmine Elvezio Prof. GRANCAGNOLO, Francesco
10:00 [55] Higgs renormalization in the NMSSM	Dr. GÓMEZ BOCK, Melina
10:30 [29] Flavour Parametrization of couplings with fermions in the 2HDM: the case with the 4-zero texture mass matrix	Dr. LÓPEZ, Lao Tse

Summaries I,II and III - Auditorio Jorge Ibargüengoitia (11:30-13:30)

- Conveners: Dr. Napsuciale, Mauro

Peer review statement

Abstract Metrics

All papers published in this volume of *Journal of Physics: Conference Series* have been peer reviewed through processes administered by the proceedings Editors. Reviews were conducted by expert referees to the professional and scientific standards expected of a proceedings journal published by IOP Publishing.

Dates

Issue 1 (2012)

Journal of Physics: Conference Series **378** (2012) 012001 doi:10.1088/1742-6596/378/1/012001

Ultra-high energy extragalactic neutrinos interacting with ultra-light dark matter

J. Barranco[1], O. G. Miranda[2], C. A. Moura[3], T. I. Rashba[4], and F. Rossi-Torres[5]

[1] Instituto de Astronomía, Universidad Nacional Autonoma de México, Mexico, DF 04510, Mexico

[2] Departamento de Física, Centro de Investigación y de Estudios Avanzados del IPN, Apdo. Postal 14-740 07000 México, D.F., Mexico

[3] Centro de Ciências Naturais e Humanas, Universidade Federal do ABC (UFABC), Rua Santa Adélia 166, 09210-170 Santo André, SP, Brazil

[4] IZMIRAN, Institute of Terrestrial Magnetism, Ionosphere and Radio Wave Propagation of the Russian Academy of Sciences, 142190, Troitsk, Moscow region, Russia

[5] Instituto de Física Teórica, Universidade Estadual Paulista, Rua Dr. Bento Teobaldo Ferraz, 271 - Bl. II, 01140-070, São Paulo, SP, Brazil

E-mail: celio.moura@ufabc.edu.br

Abstract. We show the results and discussions of the study of a possible suppression of the extragalactic neutrino flux during its propagation due to a nonstandard interaction with a candidate field to dark matter. In particular, we show the study of neutrino interaction with an ultra-light scalar field. It is shown that the extragalactic neutrino flux may be suppressed by such an interaction, leading to a mechanism to reduce the ultra-high energy neutrino flux. We calculate both the cases of non-self-conjugate as well as self-conjugate ultra-light dark matter. In the first case, the suppression is independent of the neutrino and dark matter masses. We conclude that care must be taken when explaining limits on the neutrino flux through source acceleration mechanisms only, since there could be other mechanisms, as absorption during propagation, for the reduction of the neutrino flux [1].

1. Introduction

It is almost a general assumption that Ultra-high Energy (UHE) neutrinos should arrive at the Earth coming from very distant sources like Active Galactic Nuclei (AGN) and Gamma Ray Bursts (GRB) following straight trajectories. Neutrinos interact only through weak interactions, as they are not charged and if they have a nonzero magnetic moment it must be very small. Hence, there would be no interaction preventing them to travel cosmological distances [2]. The efforts to improve sensitivity to the UHE neutrino flux may test the existence of the Berezinsky-Zatsepin (BZ) neutrinos [3]. Those neutrinos are generated through the same process that predicts the Greisen-Zatsepin-Kuzmin (GZK) cutoff [4] of Cosmic Ray (CR) flux, which has been observed by HiRes [5] and the Auger Observatory [6]. The ultra energetic CRs, most probably composed by protons and nuclei, may have an origin at relatively close sources, of the order of 100 Mpc for protons and even less for nuclei [7]. In fact, UHE CRs are most probably arriving from nearby AGNs [8]. But we observe that despite all the efforts to detect high energy extragalactic neutrinos, i.e., neutrinos with energies higher than 10^3 TeV, more and more restricted limits on their flux have been reported by several experiments due to the non observation of these neutrinos [9, 10, 11, 12, 13]. In figure 1, the results for Auger [14], IceCube-40 [15], Anita-II [16], Rice [17], and two flux predictions are shown [18]. Both the different energy bins of width 0.5 in $\log_{10} E_\nu$ and the integrated limit are shown in the case of Auger.

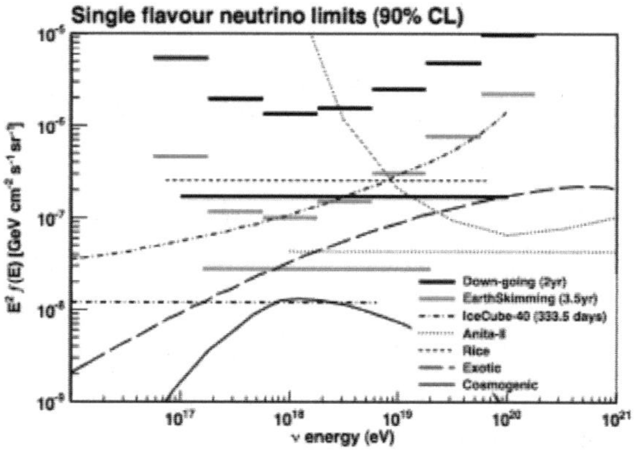

Figure 1. Limits on the neutrino flux from [14]. Differential and integrated upper limits (90% C.L.) from the Pierre Auger Observatory for a diffuse flux of down-going ν (2 yr of full Auger) and Earth skimming ν_τ (3.5 yr of full Auger). Other experiments are also included. For more explanations see text and [14].

In addition, based on astronomical and cosmological observations, it is difficult to deny the existence of Dark Matter (DM) and Dark Energy (DE). Cosmological observations of clusters of galaxies indicate that the density fraction of DM is $\Omega_{\rm DM} \approx 0.227 \pm 0.014$ [19]. The two most popular solution to the DM problem are *i)* a slight modification on the Newtonian dynamics [20] and *ii)* relic particles [21]. A DM relic particle candidate must be non-relativistic, must be stable on the cosmological time scale, and must interact weakly with other particles. Some of the possible candidates are: WIMPS [22], axions [23, 24, 25, 26, 27], MeV-scalar fields [28, 29], technicolor candidates [30, 31, 32], and ultra-light scalar fields [33, 34, 35, 36, 37, 38]. An ultra-light scalar field is motivated as DM because it can alleviate some of the problems that arise at galactic scale in the standard paradigm of cold DM, namely, the origin of cusped halos [39] and the overproduction of substructure [40].

Neutrino interaction with DM, ν-DM for short, could have strong implications at cosmological scales. Interactions of neutrinos with light scalar fields have been studied with some interesting implications noticed, such as a reduction of the relic neutrino density, leading to a neutrinoless universe [41], or a modification on the CMB spectra [42, 43], or even a connection between the smallness of neutrino mass and a MeV-mass scalar field DM [44]. Furthermore, ν-DM interaction might affect the flux of UHE neutrinos. In particular, such interaction may suppress the neutrino flux resulting in a kind of GZK cutoff for neutrinos. Many DM candidates were analyzed in this context: heavy neutrinos as dark matter [45, 46, 47], lightest supersymmetric particles (LSP) discussed in Ref. [46] and updated in Ref. [48], and MeV-mass scalar field [28, 42, 44]. In all these cases the suppression is small, not interfering in the propagation of UHE neutrinos.

Nevertheless, we show here that a coupling between relic ultra-light scalar fields and neutrinos may imply a suppression of the UHE neutrino flux, in which case there may be a confusion between the flux limit at the source and a reduction of the UHE neutrino flux during propagation. Previous analysis have put constraints on the ν-DM interaction couplings for those models by using, for instance, SN1987A neutrino data or possible imprints on the angular power spectra of CMB anisotropies [42]. Nevertheless, those limits do not apply to our case, since they were obtained by assuming a mass of the scalar field $m_\phi > 10$ MeV while we explore the possibility that ultra-light scalar field with $m_\phi \ll 1$ eV can couple to neutrinos. Such a small mass for the scalar particle gives a cross section with different behavior compared the one previously reported in other works [28, 29], allowing for new phenomena like a flux suppression for reasonable values of the coupling constants. Consequences of this type of ultra-light scalar fields have been studied in other astrophysical contexts like in the equilibrium of degenerate stars [49]. In this work we use the ν-DM coupling as described in [29].

In the next Section we show how the neutrino flux may vary due to a nonstandard interaction of neutrinos with DM. We also introduce the space of parameters, namely, cross section and DM mass, that lead to an important effect on the neutrino propagation through cosmological distances. In Section 3, we

Journal of Physics: Conference Series **378** (2012) 012001 doi:10.1088/1742-6596/378/1/012001

show the elastic scattering cross section for self and non-self-conjugate scalar fields while Section 4 is devoted to study the suppression to the neutrino flux due to the proposed ν-DM interaction. Finally, in Section 5, we discuss our results and conclude.

2. Neutrino Flux and Dark Matter Density

Once the neutrinos are produced at extragalactic sources, they have to propagate through distances of the order of 100-1000 Mpc to arrive at the Earth. For this scale of distances it is a good approximation to consider the distribution of sources and DM as homogeneous and isotropic. Then, considering ν-DM interaction, one can calculate the neutrino mean free path $\lambda = (n\sigma)^{-1}$, where n is the DM particle density ($\rho_{\rm DM}/m_{\rm DM}$) and σ is the ν-DM cross section. Therefore, at a distance L from the source the total flux expected is given by:

$$F(L) = F_0 e^{-L/\lambda}\,, \tag{1}$$

where F_0 is the flux when no interaction with the dark matter medium is considered. From Eq. (1) we learn that, e.g., a mean free path of approximately one third of the mean distance to the sources ($\lambda \sim L/3$) gives a 95% suppression of the initial flux. We do not show the result considering the evolution of sources because it does not change the result considerably and do not affect our conclusions.

Considering, for instance, $\lambda = L/3 \sim 33$ Mpc, we can compute the cross section σ as a function of the DM particle's mass $m_{\rm DM}$. We consider for the mean DM density in the universe, $\rho_{\rm DM} = 1.2 \times 10^{-6}$ GeV/cm^3 [19]. In figure 2 we show the space of parameters that results in a 95% neutrino flux suppression or more.

We observe that there is a flux suppression even for a very small cross section provided that $m_{\rm DM}$ is also very small. So we may conclude that even very weakly interacting particles may have an effect on the neutrino flux if the mass of the background DM particle is extremely small, giving a high DM number density n.

Such an ultra-light particle has already been considered. DM scalar field candidates with mass in the range $10^{-22} - 10^{-24}$ eV have been proposed as viable DM particles [33, 34, 35, 36, 37, 38]. Even lighter scalar fields, with masses lower than 10^{-33} eV, have been postulated in order to explain DE [50, 51, 52, 53].

Concerning ultra-light scalar fields as DM, they have been studied for a number of self-interaction potentials, like quadratic ones [33]. The main idea behind these models is that scalar fields were unified fields at a very early time after the origin of the universe. As the universe expands, the scalar fields cool together with the rest of the particles and finally they decouple from the rest of the matter. They condensate once they reach a critical temperature T_C. For the case of a complex scalar field the critical temperature is given by [54]

$$T_C = \sqrt{\frac{3q}{m_\phi}}\,, \tag{2}$$

where q is the charge density, defined as the excess of particles n over antiparticles $\bar{n}$, $q = n - \bar{n}$, and m_ϕ is the mass of the scalar field. From this formula one sees that an asymmetry of scalar particles, n, over antiparticles, $\bar{n}$, is required in order to have a high T_C. One can make an estimate [55] for the value of T_C considering $n \gg \bar{n}$, in which case the antiparticle contribution to the dark matter density is negligible and, therefore,

$$\rho_{\rm DM} \simeq n m_\phi\,. \tag{3}$$

If we consider that the present dark matter density is $\rho_{\rm DM} \simeq 0.23\rho_c$ and $\rho_c \simeq 4.19 \times 10^{-11}$ eV^4 we conclude that $n \simeq 10^{12}$ eV^3. For a scalar field mass of $m_\phi = 10^{-23}$ eV we get a critical temperature of condensation $T_c \simeq 10^{17}$ eV.

In this example, condensation occurs at very early stages of the evolution of the universe. After the scalar field condenses, most of these bosons lie in the ground state and one coherent field is appropriate to describe its evolution as the universe expands. It was shown in [56] that, for a coherent scalar field

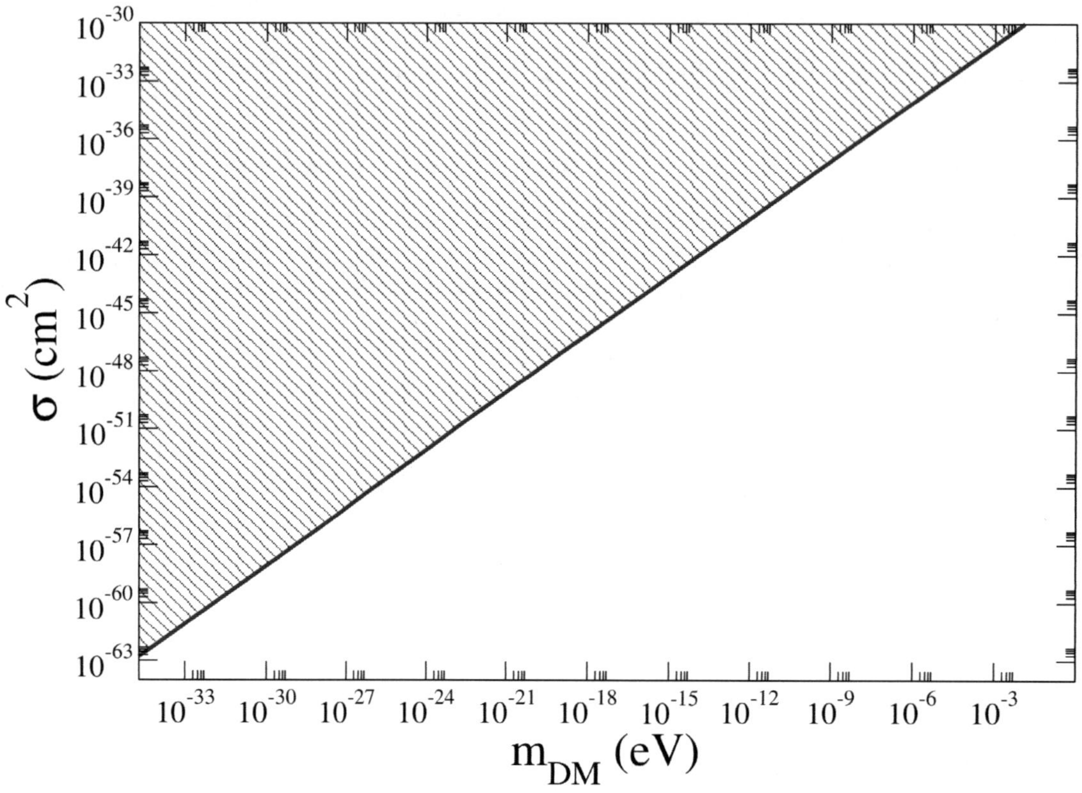

Figure 2. Neutrino cross section inducing a 95% neutrino flux suppression, as a function of the DM particle's mass $m_{\rm DM}$. The shaded area shows the suppression parameters' space. We consider a density $\rho_{\rm DM} = 1.2 \times 10^{-6}$ GeV/cm^3 and a mean source distance of $L = 100$ Mpc.

with a potential $V(\phi) \simeq \phi^k$, the energy density decreases as $\rho_\phi \sim a^{-6k/(k+2)}$, with a the cosmic scale factor. For our case of interest of scalar field DM [33], the potential is $V(\phi) \sim \phi^2$ and then, the energy density decreases as $\rho_\phi \sim a^{-3}$, i.e., it evolves as dust and hence as cold dark matter.

We show our results for two different situations: when the total amount of DM in the universe is composed of the ultra-light scalar field ($\rho_\phi = \rho_{\rm DM}$) and for the case of a multi-component DM in the universe [35], where the ultra-light scalar field density is a ten percent fraction of the total density in the ΛCDM model. Therefore, in this case

$$n = 0.1 \frac{\rho_{\rm DM}}{m_\phi}, \tag{4}$$

where m_ϕ is the scalar field mass and the ultra-light DM density is given by $\rho_\phi = 0.1\rho_{\rm DM}$. Any other DM fraction can be obtained by simply rescaling the results.

3. The cross section

The kind of interaction we are assuming is the elastic scattering $\nu + \phi \to \nu + \phi$, where ϕ is the ultra-light scalar field. To compute an expression for the cross section in this process, we assume one of the models proposed in ref. [29], where the DM candidate can be either self-conjugate ($\phi = \phi^*$) or non-self-conjugate ($\phi \neq \phi^*$).

The Lagrangian for the interaction is given by

$$\mathcal{L} = g_{\nu\phi}\, \bar{\nu}\, \phi\, P_R F + H.c. , \tag{5}$$

where $g_{\nu\phi}$ is the ν-DM coupling, P_R denotes the chiral projector $(1+\gamma^5)/2$ and F denotes a new spin one half fermion that mediates the interaction.

3.1. Non-self-conjugate scalar field dark matter ($\phi \neq \phi^*$)

For this case, only the u-channel contributes to the cross section amplitude [29], and it is given by:

$$\sigma \simeq \frac{g_{\nu\phi}^4}{32\pi} \frac{s}{(u - M_I^2)^2} , \tag{6}$$

where the center of mass energy is $\sqrt{s} = \sqrt{2m_\phi E_\nu}$, E_ν is the neutrino energy, and M_I is the mass of the intermediate particle for the ν-DM interaction. Since we are considering an ultra-light scalar field, $u << M_I^2$ for all the considered energy range. In that limit the cross section can be written in the form:

$$\sigma \simeq \left(\frac{g_{\nu\phi}}{M_I}\right)^4 \frac{m_\phi E_\nu}{16\pi} . \tag{7}$$

3.2. Self-conjugate scalar field dark matter ($\phi = \phi^*$)

In this case there is a contribution from the s-channel [29]. Neglecting the neutrino mass, in the local limit approximation ($u, s \ll M_I^2$), both contributions from u and s channel cancel each other. But considering the neutrino mass the cross section, although small, is not exactly zero and it is given by the expression:

$$\begin{aligned} \frac{d\sigma}{d\Omega} &= \frac{g_{\nu\phi}^4}{32\pi^2}\Big[\frac{m_\nu^2}{4}\left(\frac{1-\cos\theta}{(s-M_I^2)^2} + \frac{1-\cos\theta}{(u-M_I^2)^2}\right) \\ &+ \left(\frac{1}{s-M_I^2} - \frac{1}{u-M_I^2}\right)^2 \left(\frac{s}{4}(1+\cos\theta) - \frac{m_\phi^2}{4}(1-\cos\theta)\right) \\ &+ \left(\frac{1}{s-M_I^2} - \frac{1}{u-M_I^2}\right) m_\nu^2 \left(\frac{1+\cos\theta}{2(s-M_I^2)} + \frac{1}{u-M_I^2}\right) \\ &+ \frac{2m_\nu^4}{s(u-M_I^2)(s-M_I^2)}\Big] . \end{aligned} \tag{8}$$

In the limit where $s, u \ll M_I$ the cross section, after integration in solid angle, is reduced to

$$\sigma \simeq \left(\frac{g_{\nu\phi}}{M_I}\right)^4 \frac{m_\nu^2}{16\pi} . \tag{9}$$

In our computation of Eqs. (6) and (8) we consider, for simplicity, the limit $m_\nu^2 << s$, which for a neutrino energy of the order of $E_\nu \sim 10^{18}$ eV, and neutrino mass around 1 eV, translates into a restriction for the mass of the scalar field DM $m_\phi >> \mathcal{O}\ (10^{-18})$ eV. We remark that, however, it is possible to consider smaller masses for the scalar field DM, recalculating the cross section without working in this limit.

Journal of Physics: Conference Series **378** (2012) 012001 doi:10.1088/1742-6596/378/1/012001

4. Neutrino flux suppression

4.1. Non-self-conjugate scalar field dark matter

Remembering that $\lambda = (n\sigma)^{-1}$ and using Eq. (7) for σ, we have the neutrino mean free path given by

$$\begin{aligned} \lambda &= 16\pi \left(\frac{M_I/g_{\nu\phi}}{\text{GeV}}\right)^4 \left(\frac{\text{GeV}}{E_\nu}\right) \left(\frac{\text{GeV/cm}^3}{\rho_\phi}\right) \text{GeV}^2\text{cm}^3 \\ &\simeq L_0 \left(\frac{M_I/g_{\nu\phi}}{\text{GeV}}\right)^4 \left(\frac{10^{18}\text{eV}}{E_\nu}\right) \left(\frac{\text{GeV/cm}^3}{\rho_\phi}\right), \end{aligned} \tag{10}$$

where $L_0 \simeq 42$ pc.

Depending specially on the ν-DM coupling strength, a neutrino flux suppression is possible. In this case, the non observation of UHE neutrino events by the experiments do not mean that there is a limit for the neutrino production in the extragalactic sources, but rather a reduction of the neutrino flux due to the ν-DM interaction. We can estimate the necessary strength of the coupling $g_{\nu\phi}$ to produce an important effect on the neutrino flux for propagation through distances of the order of 100 Mpc or more. It is also important to notice that, in this case, the mean free path is independent of the mass of the scalar field DM.

From Eqs. (1) and (10) we conclude that if the interaction have at least a strength given by the coupling

$$\frac{g_{\nu\phi}}{M_I} \gtrsim \left[\ln\left(\frac{F_0}{F}\right) \frac{L_0}{\rho_\phi E_\nu L}\right]^{\frac{1}{4}}, \tag{11}$$

the source flux limit is loose due to the above argument. In Eq. (11) L is given in Mpc, E_ν in GeV, ρ_ϕ in GeV/cm^3, and M_I in GeV.

We show this result graphically in figure 3. Considering a non suppressed neutrino flux F_0, equal to the one calculated by Mannheim, Protheroe, and Rachen (MPR) [57], we show in this figure the value of the ratio $\frac{g_{\nu\phi}}{M_I}$ that leads to a suppression giving a neutrino flux, F, equal to the Waxman and Bahcall (WB) bound [58, 59]. The solid line is for a single ultra-light field forming the DM in the universe. The dashed one is for a 10% ultra-light DM component.

According to the WB bound, the maximum allowed neutrino flux is of the order of $\mathcal{O}(10^{-8})\varepsilon_Z$ GeV cm^{-2}s^{-1}sr^{-1}, where ε_Z is of order unity and includes possible contribution of so far unobserved high redshift sources and the effect of redshift in neutrino energy. In this model protons are confined in the astrophysical sources and undergo photoproduction of mesons and neutrons. The mesons' decay generates the neutrino flux, while the neutrons escape from the acceleration site, decay, and produce the observed UHE cosmic rays.

On the other hand, the MPR flux model consider neutron optically thick sources ($\tau_{n\gamma} >> 1$), i.e., in the photoproduction process neutrinos escape from the sources, but not the cosmic rays. Therefore, the relation between the neutrino and the cosmic ray fluxes is not direct and the neutrino flux limit is relaxed up to $\mathcal{O}(10^{-6})$ GeV cm^{-2}s^{-1}sr^{-1}. For these models it corresponds to a total neutrino flux ratio at the Earth of the order of $F_{\text{WB}}/F_{\text{MPR}} \approx \mathcal{O}(0.01)$, which is the ratio we use in Eqs. (11) and (13). For a recent discussion of the neutrino flux limits from various experiments see [60].

The value of L is fixed to be $L = 5 \times 10^2$ Mpc. One can see that an ultra-light scalar field, if it exists as DM in the universe, may generate an extragalactic neutrino flux suppression effect. For instance, for a 10% scalar field DM component, if the ratio $\frac{g_{\nu\phi}}{M_I}$ is of the order 0.1, neutrinos with energies of the order of 10^{18} eV would present a suppression reducing the flux from the MPR to the WB value.

4.2. Self-conjugate scalar field dark matter

A similar analysis can be made for the cross section given in Eq. (9), valid for the case of self-conjugate scalar field dark matter. The resulting neutrino mean free path is

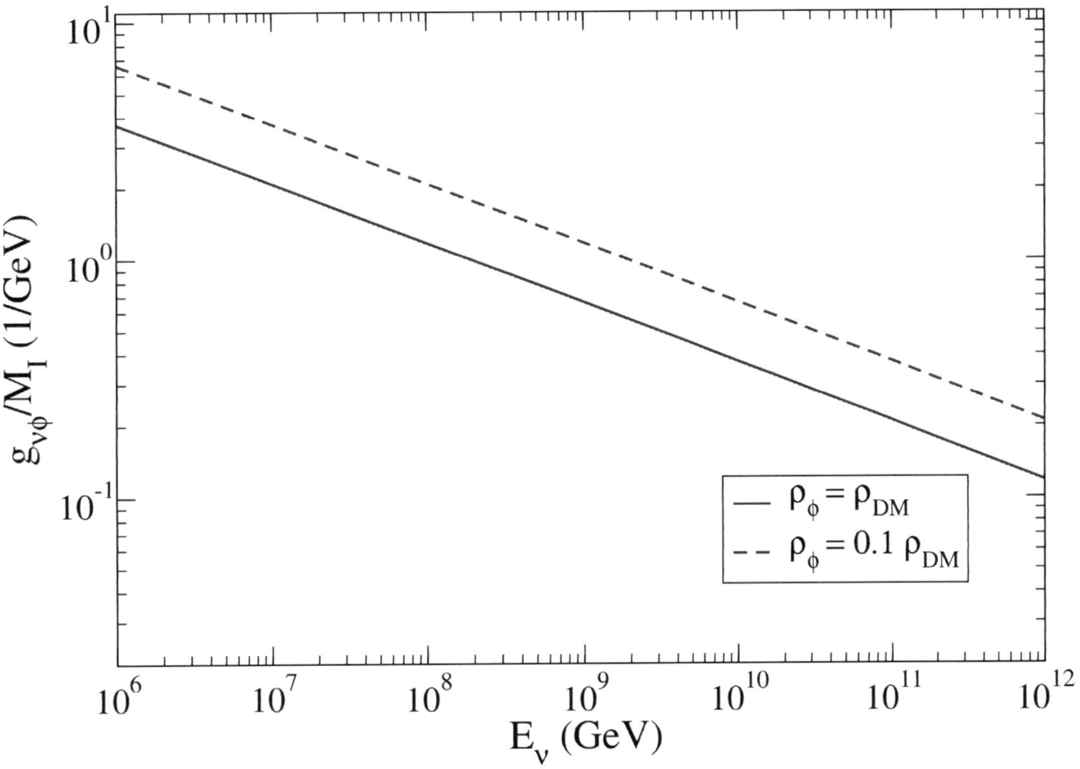

Figure 3. Ratio $\frac{g_{\nu\phi}}{M_I}$ as a function of the neutrino energy E_ν that induces a neutrino flux suppression, for the non-self-conjugate dark matter forming the total amount of DM (solid line) or a 10% fraction (dashed line). The regions above the curves predict stronger suppression. The considered mean distance to the sources is $L = 5 \times 10^2$ Mpc and the neutrino flux with no suppression is taken according to the MPR limit prediction.

$$\begin{aligned}
\lambda &= 16\pi \times 10^{-6} \left(\frac{M_I/g_{\nu\phi}}{\text{GeV}}\right)^4 \left(\frac{\text{eV}}{m_\nu}\right)^2 \left(\frac{\text{GeV/cm}^3}{\rho_\phi}\right) \left(\frac{m_\phi}{10^{-15}\text{eV}}\right) \text{GeV}^2\text{cm}^3 \\
&\simeq L_0 \left(\frac{M_I/g_{\nu\phi}}{\text{GeV}}\right)^4 \left(\frac{\text{eV}}{m_\nu}\right)^2 \left(\frac{\text{GeV/cm}^3}{\rho_\phi}\right) \left(\frac{m_\phi}{10^{-18}\text{eV}}\right), \qquad (12)
\end{aligned}$$

where $L_0 \simeq 42$ pc. The required ratio $\frac{g_{\nu\phi}}{M_I}$ in order to have a reduction of the flux F_0, with no suppression mechanism, to a flux F, or lower, due to the ν-DM interaction should be

$$\frac{g_{\nu\phi}}{M_I} \gtrsim \left[\ln\left(\frac{F_0}{F}\right)\frac{L_0 m_\phi}{\rho_\phi m_\nu^2 L}\right]^{\frac{1}{4}}. \qquad (13)$$

The above ratio is shown in figure 4 as a function of the mass of the scalar field DM candidate, using $L = 5 \times 10^2$ Mpc and $m_\nu = 1$ eV. We assume again F_0 equal to the MPR prediction, while the suppressed flux, F, is considered to be equal or below the WB limit. The solid line is for a single ultra-light field forming the DM in the universe. The dashed one is for a 10% ultra-light DM component.

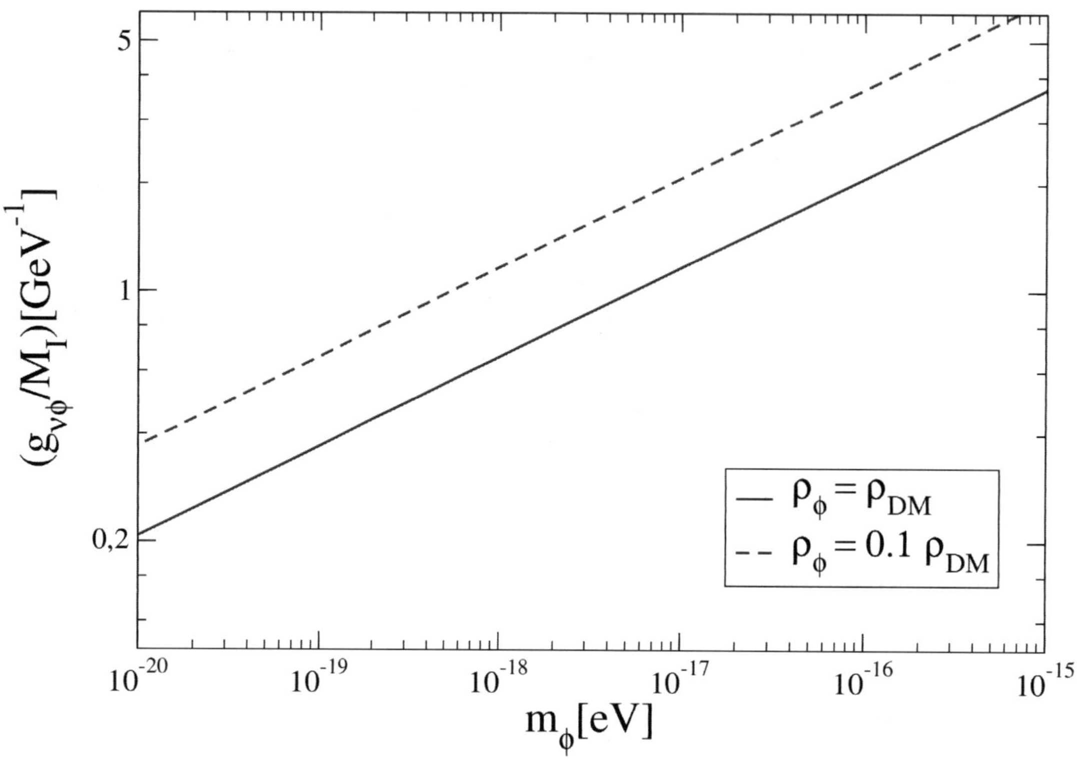

Figure 4. Ratio $\frac{g_{\nu\phi}}{M_I}$ as a function of the mass of the scalar field DM m_ϕ that induces a neutrino flux suppression (regions above the curves), for the self-conjugate dark matter forming the total amount of DM (solid line) or a 10% fraction (dashed line). We consider the mean distance to the sources $L = 5 \times 10^2$ Mpc, a neutrino mass of 1 eV and the neutrino flux with no suppression is taken according to the MPR limit prediction.

5. Discussion and Conclusion

There are several experiments, like IceCube and the Pierre Auger Observatory, expecting to detect extragalactic neutrinos. However, neutrinos with energies above 10^{15} eV, coming from extragalactic sources, have not been detected yet. We study the possibility that UHE neutrinos could be absorbed while traveling from their sources to the Earth. In particular we illustrate this idea by considering an ultra-light particle as a component of the Dark Matter in the Universe.

We consider a mechanism for the neutrino interaction based on a scalar field dark matter model and we show that in this case the propagation of extragalactic neutrinos from sources 100 Mpc or farther from the Earth may be affected. This would give negative results on neutrino telescopes or UHE neutrino detectors. On the other hand, despite neutrinos from a nearby supernova could interact with the DM halo around the collapsing star, the scale involved would not be sufficient for an absorption like the one proposed here.

Although nonstandard interactions of neutrinos with Dark Matter particles had been considered in the literature before, no important effect on neutrino propagation had been predicted. In most of the literature, light scalar field DM had been considered to be relativistic and the coupling to neutrinos was constrained due to measurable effects on the CMB spectra. In this work we have considered non-relativistic ultra-light scalar fields, proposed in the literature, that besides their gravitational effects, may not have other measurable astrophysical consequences.

To our knowledge, this is the first example of a possible suppression of the extragalactic neutrino flux due to propagation effects. Therefore, care must be taken when using the limits obtained by such experiments, since those limits can be due to two factors: source limit and/or absorption due to UHE neutrino-ultra light scalar field Dark Matter interaction during neutrino propagation from the source to the Earth. On the other hand, a positive signal of UHE neutrinos could be useful to put restrictions on models that contains a light scalar field DM candidate.

Similar arguments could be applied for particles other than neutrinos. For instance, in [26] it was analyzed the suppression of charged particles due to the interaction with a pseudoscalar and it was shown that the axion can play the role of a shield for high energy cosmic rays.

Acknowledgments

C.A.M. would like to thank his collaborators for the productive discussions during the time of work that gave rise to the results presented here, and the organizing committee of the XIII MWPF for their hospitality and financial support. F.R.T. thanks CNPq.

References

[1] J. Barranco, O. G. Miranda, C. A. Moura, T. I. Rashba and F. Rossi-Torres, JCAP **1110** (2011) 007 [arXiv:1012.2476 [astro-ph.CO]].
[2] See, e.g., F. Halzen & D. Hooper, *High-energy neutrino astronomy: The cosmic ray connection*, Rept. Prog. Phys. **65**, (2002) 1025. [arXiv:astro-ph/0204527].
[3] V. S. Beresinsky & G. T. Zatsepin, Phys. Lett. B **28**, 423 (1969).
[4] K. Greisen, *End To The Cosmic Ray Spectrum?*, Phys. Rev. Lett. **16**, (1966) 748;
G. T. Zatsepin and V. A. Kuzmin, *Upper limit of the spectrum of cosmic rays*, JETP Lett. **4** (1966) 78.
[5] R. U. Abbasi *et al.* [HiRes Coll.], *Observation of the GZK cutoff by the HiRes experiment*, Phys. Rev. Lett. **100** (2008) 101101 [arXiv:astro-ph/0703099].
[6] J. Abraham *et al.* [Pierre Auger Coll.], *Observation of the suppression of the flux of cosmic rays above* $4 \times 10^{19} eV$, Phys. Rev. Lett. **101** (2008) 061101 (2008) [arXiv:0806.4302 [astro-ph]].
[7] D. Hooper, S. Sarkar and A. M. Taylor, *The intergalactic propagation of ultra-high energy cosmic ray nuclei*, Astropart. Phys. **27** (2007) 199. [arXiv:astro-ph/0608085].
[8] J. Abraham *et al.* [Pierre Auger Coll.], *Correlation of the highest energy cosmic rays with nearby extragalactic objects*, Science **318** (2007) 938 [arXiv:0711.2256 [astro-ph]].
J. Abraham *et al.* [Pierre Auger Coll.], *Correlation of the highest-energy cosmic rays with the positions of nearby active galactic nuclei*, Astropart. Phys. **29** (2008) 188 [Erratum-ibid. **30**, 45 (2008)]. [arXiv:0712.2843 [astro-ph]].
[9] M. Ackermann *et al.*, *Flux Limits On Ultra High Energy Neutrinos With Amanda-B10*, Astropart. Phys. **22** (2005) 339.
[10] S. Desai *et al.* [Super-Kamiokande Coll.], *Study of TeV Neutrinos with Upward Showering Muons in Super-Kamiokande*, Astropart. Phys. **29** (2008) 42 [arXiv:0711.0053 [hep-ex]].
[11] J. Abraham *et al.* [Pierre Auger Coll.], *Upper limit on the diffuse flux of UHE tau neutrinos from the Pierre Auger Observatory*, Phys. Rev. Lett. **100** (2008) 211101; M. Ackermann *et al.* [IceCube Coll.], *Search for Ultra High-Energy Neutrinos with AMANDA-II*, Astrophys. J. **675** (2008) 1014; [arXiv:0711.3022 [astro-ph]]; K. Martens [HiRes Coll.], *HiRes Estimates and Limits for Neutrino Fluxes at the Highest Energies*, [arXiv:0707.4417 [astro-ph]]; S. W. Barwick *et al.* [ANITA Coll.], *Constraints on cosmic neutrino fluxes from the ANITA experiment*, Phys. Rev. Lett. **96** (2006) 171101 [arXiv:astro-ph/0512265]; V. Aynutdinov *et al.* [BAIKAL Coll.], *Search for a diffuse flux of high-energy extraterrestrial neutrinos with the NT200 neutrino telescope*, Astropart. Phys. **25** (2006) 140 [arXiv:astro-ph/0508675]; P. W. Gorham, C. L. Hebert, K. M. Liewer, C. J. Naudet, D. Saltzberg and D. Williams, *Experimental limit on the cosmic diffuse ultrahigh-energy neutrino flux*, Phys. Rev. Lett. **93** (2004) 041101 [arXiv:astro-ph/0310232]; N. G. Lehtinen, P. W. Gorham, A. R. Jacobson and R. A. Roussel-Dupre, *FORTE satellite constraints on ultra-high energy cosmic particle fluxes*, Phys. Rev. D **69** (2004) 013008 [arXiv:astro-ph/0309656]; I. Kravchenko *et al.*, *RICE limits on the diffuse ultra-high energy neutrino flux*, Phys. Rev. D **73** (2006) 082002 [arXiv:astro-ph/0601148]; J. A. Aguilar *et al.*, Nucl. Instrum. Methods Phys. Res., Sect. A **570**, 107 (2007).
[12] J. Abraham *et al.* [Pierre Auger Coll.], *Limit on the diffuse flux of ultrahigh energy tau neutrinos with the surface detector of the Pierre Auger Observatory*, Phys. Rev. D **79** (2009) 102001; O. Blanch Bigas [Pierre Auger Collaboration], *Upper Limit On The Diffuse Flux Of Uhe Tau Neutrinos*, Nucl. Phys. Proc. Suppl. **190** (2009) 156.
[13] R. Abbasi *et al.* [IceCube Collaboration], *Limits on Neutrino Emission from Gamma-Ray Bursts with the 40 String IceCube Detector*, Phys. Rev. Lett. **106** (2011) 141101. [arXiv:1101.1448 [astro-ph.HE]].
[14] P. Abreu *et al.* [The Pierre Auger Collaboration], arXiv:1107.4805 [astro-ph.HE].
[15] The IceCube Collaboration, Phys. Rev. D, 2011, **83**:092003.

[16] The ANITA Collaboration, Phys. Rev. D, 2010, **82**:022004, Erratum arXiv:1011.5004v1 [astro-ph].
[17] I. Kravchenko et al., Phys. Rev. D, 2006, **73**: 082002.
[18] M. Ahlers et al., Astropart. Phys., 2010, **34**: 106-115; D. Semikoz, G. Sigl, JCAP, 2004, **0404**:003.
[19] N. Jarosik *et al.*, *Seven-Year Wilkinson Microwave Anisotropy Probe (WMAP) Observations: Sky Maps, Systematic Errors, and Basic Results*, [arXiv:1001.4744 [astro-ph.CO]].
[20] M. Milgron, *A modification of newtonian dynamics as a possible alternative to the hidden mass hypothesis*, Astrophys. J. **270** (1983) 365.
[21] G. Bertone, D. Hooper & J. Silk, *Particle dark matter: Evidence, candidates and constraints*, Phys. Rept. **405** (2005) 279 [arXiv:hep-ph/0404175].
[22] G. Jungman, M. Kamionkowski and K. Griest, *Supersymmetric dark matter*, Phys. Rept. **267** (1996) 195 [arXiv:hep-ph/9506380].
[23] R. D. Peccei & H. R. Quinn, *Constraints Imposed By CP Conservation In The Presence Of Instantons*, Phys. Rev. D **16** (1977) 1791.
[24] R. D. Peccei & H. R. Quinn, *CP Conservation In The Presence Of Instantons*, Phys. Rev. Lett. **38** (1977) 1440.
[25] N. V. Mikheev and E. N. Narynskaya, *Axion-medium effect on fermion dispersion*, Phys. Atom. Nucl. **71** (2008) 1487.
[26] A. A. Andrianov, D. Espriu, F. Mescia and A. Renau, *The axion shield*, Phys. Lett. B **684** (2010) 101 [arXiv:0912.3151 [hep-ph]].
[27] M. Dine, W. Fischler and M. Srednicki, *A Simple Solution To The Strong CP Problem With A Harmless Axion*, Phys. Lett. B **104** (1981) 199; A. R. Zhitnitsky, *On Possible Suppression Of The Axion Hadron Interactions. (In Russian)*, Sov. J. Nucl. Phys. **31** (1980) 260.
[28] C. Boehm, T. A. Ensslin and J. Silk, *Are light annihilating dark matter particles possible?*, J. Phys. G **30** (2004) 279 [arXiv:astro-ph/0208458]; D. Hooper, F. Ferrer, C. Boehm, J. Silk, J. Paul, N. W. Evans and M. Casse, *MeV dark matter in dwarf spheroidals: A smoking gun?*, Phys. Rev. Lett. **93** (2004) 161302 [arXiv:astro-ph/0311150]; C. Boehm, D. Hooper, J. Silk, M. Casse and J. Paul, *MeV Dark Matter: Has It Been Detected?*, Phys. Rev. Lett. **92** (2004) 101301 [arXiv:astro-ph/0309686]; C. Boehm, P. Fayet and J. Silk, *Light and heavy dark matter particles*, Phys. Rev. D **69** (2004) 101302 [arXiv:hep-ph/0311143].
[29] C. Boehm and P. Fayet, *Scalar dark matter candidates*, Nucl. Phys. B **683** (2004) 219 [arXiv:hep-ph/0305261];
[30] R. Foadi, M. T. Frandsen and F. Sannino, *Technicolor Dark Matter*, Phys. Rev. D **80** (2009) 037702 [arXiv:0812.3406 [hep-ph]].
[31] S. B. Gudnason, C. Kouvaris and F. Sannino, *Towards working technicolor: Effective theories and dark matter*, Phys. Rev. D **73** (2006) 115003 [arXiv:hep-ph/0603014].
[32] S. B. Gudnason, C. Kouvaris and F. Sannino, *Dark Matter from new Technicolor Theories*, Phys. Rev. D **74** (2006) 095008 [arXiv:hep-ph/0608055].
[33] T. Matos, J. A. Vazquez and J. Magana, *ϕ^2 as Dark Matter*, [arXiv:0806.0683]
[34] A. Arbey, J. Lesgourgues and P. Salati, *Quintessential haloes around galaxies*, Phys. Rev. D **64** (2001) 123528 [arXiv:astro-ph/0105564].
[35] L. Amendola and R. Barbieri, *Dark matter from an ultra-light pseudo-Goldsone-boson*, Phys. Lett. B **642** (2006) 192 [arXiv:hep-ph/0509257].
[36] A. Bernal, F. Siddhartha Guzman, *Scalar Field Dark Matter: Head-on interaction between two structures*, Phys. Rev. **D74** (2006) 103002 [astro-ph/0610682]; A. Bernal, F. S. Guzman, *Scalar Field Dark Matter: non-spherical collapse and late time behavior*, Phys. Rev. **D74** (2006) 063504 [astro-ph/0608523]; F. S. Guzman, L. A. Urena-Lopez, *Gravitational cooling of self-gravitating Bose-Condensates*, Astrophys. J. **645** (2006) 814-819 [astro-ph/0603613]; F. S. Guzman, L. A. Urena-Lopez, *Newtonian collapse of scalar field dark matter*, Phys. Rev. **D68** (2003) 024023 [astro-ph/0303440]; T. Matos, F. S. Guzman, L. A. Urena-Lopez, *Scalar field as dark matter in the universe*, Class. Quant. Grav. **17** (2000) 1707-1712 [astro-ph/9908152]; S. J. Sin, *Late time cosmological phase transition and galactic halo as Bose liquid*, Phys. Rev. **D50** (1994) 3650-3654. [hep-ph/9205208]; S. U. Ji, S. J. Sin, *Late time phase transition and the galactic halo as a bose liquid: 2. The Effect of visible matter*, Phys. Rev. **D50** (1994) 3655-3659. [hep-ph/9409267].
[37] J. Lesgourgues, A. Arbey and P. Salati, *A Light Scalar Field At The Origin Of Galaxy Rotation Curves*, New Astron. Rev. **46** (2002) 791.
[38] W. Hu, R. Barkana and A. Gruzinov, *Cold and fuzzy dark matter*, Phys. Rev. Lett. **85** (2000) 1158 [arXiv:astro-ph/0003365].
[39] B. Moore *et al.*, *Dark Matter Substructure Within Galactic Halos*, Astrophys. J. **524** (1999) L19.
[40] W. B. Lin, D. H. Huang, X. Zhang and R. H. Brandenberger, *Non-thermal production of WIMPs and the sub-galactic structure of the universe*, Phys. Rev. Lett. **86** (2001) 954 [arXiv:astro-ph/0009003].
[41] J. F. Beacom, N. F. Bell and S. Dodelson, *Neutrinoless Universe*, Phys. Rev. Lett. **93** (2004) 121302 [arXiv:astro-ph/0404585].
[42] G. Mangano *et al.*, *Cosmological bounds on dark matter - neutrino interactions*, Phys. Rev. D **74** (2006) 043517 [arXiv:astro-ph/0606190].

[43] P. Serra *et al.*, *Constraints on neutrino – dark matter interactions from cosmic microwave background and large scale structure data*, arXiv:0911.4411 [astro-ph.CO].
[44] C. Boehm *et al.*, *Are small neutrino masses unveiling the missing mass problem of the universe?*, Phys. Rev. D **77** (2008) 043516 [arXiv:hep-ph/0612228].
[45] T. J. Weiler, *Big Bang Cosmology, Relic Neutrinos, And Absorption Of Neutrino Cosmic Rays*, Astrophys. J. **285** (1984) 495.
[46] T. J. Weiler, *Dips in the UHE cosmic ray neutrino spectrum from absorption by dark matter, Given at High-energy Neutrino Astrophysics Workshop, on Astrophysics of High-energy Neutrinos: Particle Physics, Sources, Production Mechanisms and Detection Prospects, Honolulu, Hawaii, 23-26 Mar 1992.*
[47] E. Roulet, *Ultrahigh-Energy Neutrino Absorption By Neutrino Dark Matter*, Phys. Rev. D **47** (1993) 5247.
[48] G. Barenboim, O. Mena Requejo and C. Quigg, *Neutrino coannihilation on dark-matter relics?*, Phys. Rev. D **74** (2006) 023006 [arXiv:astro-ph/0604215].
[49] J. A. Grifols, *Scalar Dark Matter and Cold Stars*, Astropart. Phys. **25** (2006) 98-105. [astro-ph/0505602].
[50] N. Weiss, *Possible origins of a small nonzero cosmological constant* Phys. Lett. B **197** (1987) 42.
[51] J. A. Frieman, C. T. Hill, A. Stebbins and I. Waga, *Cosmology with ultralight pseudo Nambu-Goldstone bosons*, Phys. Rev. Lett. **75** (1995) 2077 [arXiv:astro-ph/9505060].
[52] J. E. Kim, *Axion and almost massless quark as ingredients of quintessence*, JHEP **9905** (1999) 022 [arXiv:hep-ph/9811509].
[53] L. J. Hall, Y. Nomura and S. J. Oliver, *Evolving dark energy with w not equal -1*, Phys. Rev. Lett. **95** (2005) 141302 [arXiv:astro-ph/0503706].
[54] L. A. Urena-Lopez, JCAP **0901** (2009) 014. [arXiv:0806.3093 [gr-qc]].
[55] A. P. Lundgren, M. Bondarescu, R. Bondarescu, J. Balakrishna, Astrophys. J. **715**, L35 (2010). [arXiv:1001.0051 [astro-ph.CO]].
[56] M. S. Turner, Phys. Rev. **D28** (1983) 1243.
[57] K. Mannheim, R. J. Protheroe and J. P. Rachen, *On the cosmic ray bound for models of extragalactic neutrino production*, Phys. Rev. D **63** (2001) 023003 [arXiv:astro-ph/9812398].
[58] E. Waxman & J. N. Bahcall, *High energy neutrinos from astrophysical sources: An upper bound*, Phys. Rev. D **59** (1999) 023002 [arXiv:hep-ph/9807282].
[59] J. N. Bahcall & E. Waxman, *High energy astrophysical neutrinos: The upper bound is robust*, Phys. Rev. D **64** (2001) 023002 [arXiv:hep-ph/9902383].
[60] J. Brunner, *Diffuse Neutrino Flux*, [arXiv:1107.1593 [astro-ph.HE]].

Journal of Physics: Conference Series **378** (2012) 012002 doi:10.1088/1742-6596/378/1/012002

Towards Cosmological Models by Compactification on a Non-geometric Twisted Torus

C. Damian and O. Loaiza-Brito

Departamento de Física, DCI, Campus León, Universidad de Guanajuato, C.P. 37150, Guanajuato, Mexico

E-mail: cesaredas@fisica.ugto.mx, oloaiza@fisica.ugto.mx

Abstract. We study a type IIB superstring compactification in the presence of nongeometric fluxes. We study the non supersymmetric solutions to the equations of motion. The corresponding vacuum is found by employing a genetic algorithm constrained by the Bianchi identities. We explore the space of the non supersymmetric vacua and look for possible De Sitter solutions. We find that such vacua are possible if supersymmetry is broken by all the three moduli.

1. Introduction

After performing a string compactification from ten dimensions to four, the effective theory corresponds to a gauge supergravity theory coupled to light scalar fields [1]. These scalar fields parametrize the size and shape of the compact space and define the coupling constants and masses of the remaining fields of the effective theory. So in order to construct a phenomenological attractive theory they must acquire values which are compatible with the observed values. In the last years several ways have been proposed in order to stabilize all the moduli, mainly by turning on fluxes on the compact space [3, 4] and by considering the presence of exotic ingredients such as orientifolds, and generalized structures on the internal space. Some of them breake supersymmetry in a localized way [5]. Some other examples of moduli stabilization are by mean of non perturbative effects on the Kähler potential or on the superpotential [5, 6, 7]. One interesting approach is to consider the possibility to obtain a classical de Sitter vacuum in order to explain the accelerated universe we observe [2].

The parameters of the effective gauged supergravity theory are associated either to the fluxes or to features of the background in which the compactification is performed. However in some cases fluxes which have a natural interpretation in a theory may not have a obvious interpretation in its T-dual part [8, 9, 10, 11, 12].

In the context of moduli compactification by fluxes it was argued that a compactification on space manifolds with negative curvature [13] leads to a lift in the vacua and suggests that De Sitter space can be constructed through these manifolds. However the stability of the vacua is in many cases only metastable [14]. In particular the twisted torus is an example of a negatively curvature manifold. The twist on the torus can be interpreted as a geometric flux in the context of Type IIB string theory which is T dual to a Type IIA in which the geometric fluxes are mapped to a kind of flux which cannot be interpreted globally as a geometrical feature of the space time manifold. So this compactification which cannot be interpreted as a space time manifold leads to the notion of non geometric fluxes.

In order to get an intuition of the non-geometric fluxes consider for example a compactification on a six dimensional torus T^6 threaded with NS-NS 3-form flux H_{abc} where $a, b, ... \in 1, 2, ..., 6$ with nonzero values in the compact dimensions.The corresponding dimensional reduction is understood as a class of generalized Scherk-Schawrz dimensional reduction [15, 16, 17, 18, 19]. Under a T-duality, for example in direction a, the $H-$flux is mapped to a geometric flux associated with a twist in the torus topology. In the presence of this geometric flux, the metric on the twisted torus acquires a contribution which can be written as $(dx^a - f^a_{bc}x^c dx^b)^2$, where f^a_{bc} is integrally quantized and characterizes the geometric flux of the compactification.

Even in the presence of geometric flux f^a_{bc}, however, we can perform another T-duality on direction b, since the metric can be chosen to be independent of the coordinate x^b. Carrying out this T-duality explicitly leads to a dual torus, which is locally geometric, but it cannot be described globally in terms of a fixed geometry, due to the appearance of non-geometric duality transformation in the boundary conditions which patch together local descriptions of the compactified space. This non-geometric flux which arise after a second T-duality transformation, is labeled as Q^{ab}_c. It is known that the related superpotential depends on all the moduli of the compactification, implying that it is possible to look at classical level, for De sitter solutions. Therefore, we shall study type IIB string compactifications on a six dimensional torus threaded with non-geometric fluxes, besides the usual 3-form fluxes, in order to look for positive valued vacua.

Our work is organized as follows: in section 2, a type IIB compactification on a non-geometric torus is presented by preserving only one of the supersymmetries. In section 3 the constraints that the fluxes must satisfy are presented together with a set of solutions of these constraints. In section 4 we show a systematic analysis of the equations of motion by breaking supersymmetry through one or several moduli. In section 5, the effective scalar potential is constructed for each solution previously found. Finally in section 6 we give some final comments and perspectives of the work.

2. Flux compactification

We consider a Type IIB compactification with O3-planes transverse to the compact directions in order to avoid tadpole conditions in the presence of 3-form fluxes [20, 21]. The final $N = 1, D = 4$ contents include three complex moduli, namely: axio-dilaton S, a complex structure moduli τ and a Kähler moduli U.

We consider the standard tree-level Kähler potential given by

$$K = -3ln\left(-i(\tau - \bar{\tau})\right) - 3ln\left(-i(U - \bar{U})\right) - ln\left(-i(S - \bar{S})\right). \tag{1}$$

The contribution of the non-geometric fluxes to the superpotential is given in [8], and it is obtained by adding a term to the the well known Gukov-Vafa-Witten [22]

$$W = \int (F_3 - SH_3 - QU) \wedge \Omega. \tag{2}$$

By applying the Buscher rules to a type IIB torus compactification in the presence of a NS-NS 3-form flux, the metric of the non-geometric torus is derived and it is given by

$$ds^2 = \frac{1}{1+Q^2z^2}(dx^2+dy^2)+dz^2.$$

This background is characterized by the non-geometric flux Q_a^{bc}.

The $N=1$ four dimensional effective gauged supergravity scalar potential is given by the standard scalar potential [23]. The equations for a supersymmetric vacuum are $D_iW=0$ which correspond to the F terms. For simplicity, we shall consider a six-dimensional torus made of three identical copies. This reduces the number of moduli which, as recently discussed, increases the chances to obtain De Sitter vacua [13]. Hence, by defining the complex coordinates on each torus as $z^a = x^\alpha + \tau x^i$ where $\alpha = 1,3,5$ and $i = 2,4,6$, the T-dual invariant superpotential is given by

$$\begin{aligned} W &= a_0 - 3a_1\tau + 3a_2\tau^2 - a_3\tau^3 + S(b_0 - 3b_1\tau + 3b_2\tau^2 - b_3\tau^3) \\ &+ 3U(c_0 + (\check{c}_1 + \hat{c}_1 + \check{c}_1)\tau + (\check{c}_2 + \hat{c}_2 + \check{c}_2)\tau^2 - c_3\tau^3) \end{aligned} \quad (3)$$

where the coefficients are the components of each fluxes over integrated over a cycle. The coefficients are related with a flux as is indicated in the Table 1.

Table 1. Fluxes in the duality invariant superpotential.

Term	IIB Flux integer	Integer flux
1	$\bar{F}_{ijk}$	a_0
τ	$\bar{F}_{ij\gamma}$	a_1
τ^2	$\bar{F}_{i\beta\gamma}$	a_2
τ^3	$\bar{F}_{\alpha\beta\gamma}$	a_3
S	$\bar{H}_{ijk}$	b_0
U	$\bar{Q}_k^{\alpha\beta}$	c_0
$S\tau$	$\bar{H}_{\alpha jk}$	b_1
$U\tau$	$\bar{Q}_k^{\alpha j}, \bar{Q}_k^{i\beta}, \bar{Q}_\alpha^{\beta\gamma}$	$\check{c}_1, \hat{c}_1, \tilde{c}_1$
$S\tau^2$	$\bar{H}_{i\beta\gamma}$	b_2
$U\tau^2$	$\bar{Q}_\gamma^{i\beta}, \bar{Q}_\beta^{\gamma i}, \bar{Q}_k^{ij}$	$\check{c}_2, \hat{c}_2, \tilde{c}_2$
$S\tau^3$	$\bar{H}_{\alpha\beta\gamma}$	b_3
$U\tau^3$	$\bar{Q}_\gamma^{ij}$	c_3

Here we have adopted the notation given in [10] in which $\bar{F}_{abc}$ is the integrated flux over the cycle abc. Taking the isotropic case where $\bar{F}_{ij\gamma} = \bar{F}_{ki\beta} = \bar{F}_{jk\alpha}$ there is only one representative flux for any given combination of indices. The same notation holds for the rest of fluxes.

As mentioned above, the presence of fluxes contributes with an amount of internal energy which must by cancelled by the addition of a negative tensioned object, in order to cancel the tadpole. These requirements lead to a set of constraints that the fluxes must satisfy. The R-R constraints for the fluxes are;

$$a_0b_3 - 3a_1b_2 + 3a_2b_1 - a_3b_0 = 16 \tag{4}$$

$$a_0c_3 + a_1(\check{c}_2 + \hat{c}_2 - \tilde{c}_2) - a_2((\check{c}_1 + \hat{c}_1 - \tilde{c}_1) - a_3c_0 = 0 \tag{5}$$

There are also constraints of the same nature on the NS fluxes which are obtained starting by the Bianchi identity $dH = 0$ and its T-dual constraints. There are a total of 8 constraints given by

$$c_0b_2 - \tilde{c}_1b_1 + \hat{c}_1b_1 - \check{c}_2b_0 = 0, \tag{6}$$

$$\check{c}_1b_3 - \hat{c}_2b_2 + \tilde{c}_2b_2 - c_3b_1 = 0, \tag{7}$$

$$c_0b_3 - \tilde{c}_1b_2 + \hat{c}_1b_2 - \check{c}_2b_1 = 0, \tag{8}$$

$$\check{c}_1b_2 - \hat{c}_2b_1 + \tilde{c}_2b_1 - c_3b_0 = 0, \tag{9}$$

$$c_0\tilde{c}_2 - \check{c}_1^2 + \tilde{c}_1\hat{c}_1 - \hat{c}_2c_0 = 0, \tag{10}$$

$$c_3\tilde{c}_1 - \check{c}_2^2 + \tilde{c}_2\hat{c}_2 - \hat{c}_1c_3 = 0, \tag{11}$$

$$c_3c_0 - \check{c}_2\hat{c}_1 + \tilde{c}_2\check{c}_1 - \hat{c}_1\tilde{c}_2 = 0, \tag{12}$$

$$\hat{c}_2\tilde{c}_1 - \tilde{c}_1\check{c}_2 + \check{c}_1\hat{c}_2 - c_0c_3 = 0. \tag{13}$$

For a formal derivation of these constraints, see [9].

3. Solutions of the equations of motion

Since we are not considering the presence of D-branes, there are no D-terms in the superpotential. and breaking SUSY is achieved only through the F terms. Our model consists only on three moduli, namely the complex structure τ, the Kähler modulus U and the axio-dilaton S. A complete solution of the supersymmetric cases is presented in [11] with AdS vacua as expected. However by breaking SUSY we shall explore the possibility to obtain classical De Sitter vacua.

Breaking SUSY through only one of the moduli can be achieved in three forms. First, breaking SUSY through the axio-dilaton implies that the F terms of the complex structure and Kähler moduli must be zero. However the condition of a zero Kähler derivate through S and U directions violates the tadpole condition. Second, breaking SUSY through the Kähler modulus leads to zero volume of the compact space or an infinite coupling constant e^{ϕ} where our perturbative approach is not valid. Finally SUSY breaking through the complex structure modulus implies that the physical solutions must satisfy that $Im(P_2\bar{P}_3) = 0$. Solving this equation for the complex structure provides a way to a stabilize this modulus. The SUSY equations of motion for U and S can be solved, providing a stabilization of the axio-dilaton S and leaving unfixed the Kähler modulus. So the stabilization of the Kähler modulus depends on the minima that the scalar potential acquires. This possibility is studied in the following sections.

On the other hand, breaking SUSY through U and S leads to a violation of the Bianchi identity. This follows from the fact that the solution to the equation of motion implies that $P_1(\tau) = P_2(\tau) = P_3(\tau)$, so the fluxes associated with similar terms in τ must have the same value. However this condition leads to $a_1 = b_1 = 2c_1 - \tilde{c}_1$ and $a_2 = b_2 = \tilde{c}_2 - 2c_2$ which implies that the first Bianchi identity must be equal to zero which clearly violates the tadpole condition. SUSY breakdown through the Kähler moduli and the axio-dilaton leads to unphysical conditions which include a zero volume of the compact space and an infinite coupling constant as well.

Finally breaking SUSY through all the moduli implies that all the moduli must be stabilized dynamically through reaching the minima of the scalar potential. The values of the corresponding fluxes must satisfy the Bianchi identities and the moduli must acquire positive values.

In order to explore the space of solutions, we employ a genetic algorithm, and we look for the minimization of the scalar potential. The solutions of the equations of motion, without taking into account the case in which the scalar potential vanishes at the minimum, are summarized in the Table 2. It is important to stress out that a configuration of fluxes which satisfy the solutions does not satisfy automatically the tadpole conditions. Table 2 only remarks the fact that it is possible to find a consistent flux configuration.

Table 2. Solutions to the equations of motion.

F term	Tadpole condition	Physical solution
D_τ=0, D_S=0, $D_U\neq$0	No satisfied	$Im(S) = 0, Im(U) = 0$
D_τ=0, $D_S\neq$0, D_U=0	No satisfied	$Im(S) = 0, Im(U) = 0$
$D_\tau\neq$0, D_S=0, D_U=0	Satisfied	$Im(P_2\bar{P}_3) = 0$
$D_\tau\neq$0, $D_S\neq$0, D_U=0	Satisfied	$Im(S), Im(U) = 0$
$D_\tau\neq$0, D_S=0, $D_U\neq$0	Satisfied	$Im(S), Im(U) = 0$
D_τ=0, $D_S\neq$0, $D_U\neq$0	No satisfied	$a_i = b_i = c_i$

4. Discussion of the vacua

In order to get a set of solutions to the equations of motion which stabilize all the moduli, a genetic algorithm is employed.

Genetic algorithms are probabilistic global search and optimization methods that mimic the metaphor of natural biological evolution. The computational procedure operates on a set of initial values for the moduli (population) which gives a potential solution by applying the principle of survival of the fittest to produce successively better approximations to a minima of the scalar potential. At each generation of the numerical procedure, a new set of moduli is created by the process of selecting individuals according to their level of fitness in the moduli space and

Journal of Physics: Conference Series **378** (2012) 012002 doi:10.1088/1742-6596/378/1/012002

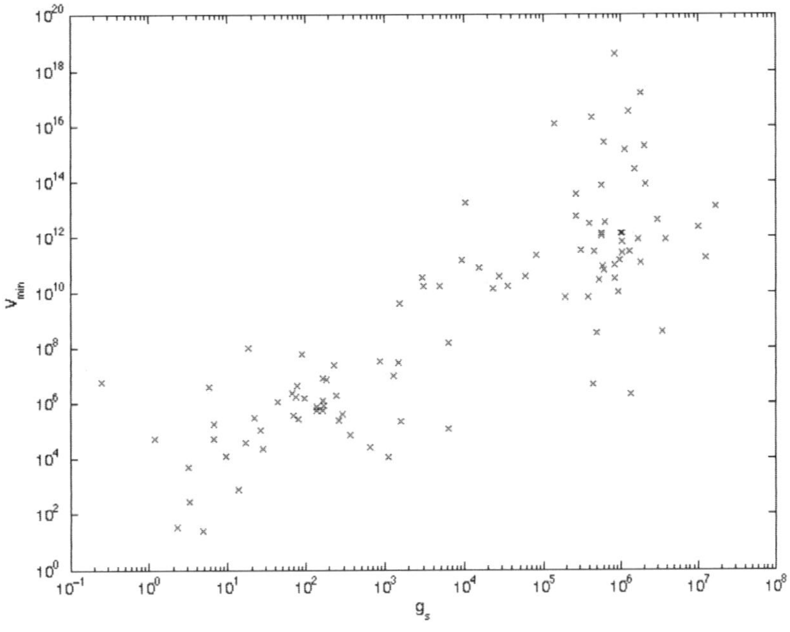

Figure 1. Logarithmic plot of g vs Λ for breaking SUSY through complex structure moduli.

reproducing them using feasible adaptation. This process leads to the evolution of moduli that are better suited to the minimum value of the scalar potential than the set of moduli from which they were created, so the algorithm successively improve the solution.

In addition we have introduced the SUSY constraints to the genetic algorithm . The numerical procedure begins with a solution to the RR and NS constraints. The solution is restricted to positive even fluxes. Once a solution to the constraints is obtained the genetic algorithm initializes with random values on the moduli. The iterative process proceeds to find a set of moduli which gives the minima value of the scalar potential. Then the constraints are check after each iteration until the maxima tolerance is reached.

In general the algebraic solution to the constraints is only valid if some of the fluxes is non zero. For example, solving the Equation (6) for a_3 is only valid if b_0 is different of zero. So the set of solutions generated turn off randomly fluxes only when the constraints equations does not yield to unphysical solutions.

In contrast to the supersymmetric solution studied in [9], where all the moduli are stabilized in a supersymmetric vacuum, we consider only two cases: the stabilization of four real moduli using the supersymmetric conditions for U and S while the remaining two real moduli, related to the complex structure τ are stabilized by minimizing the scalar potential, and the stabilization of all moduli by minimizing the scalar potential where supersymmetry is broken through all moduli.

Breaking SUSY through the complex structure τ implies the existence of De Sitter solutions. The corresponding$g - \Lambda$ space distribution is shown in Figure 1. We find that small cosmological

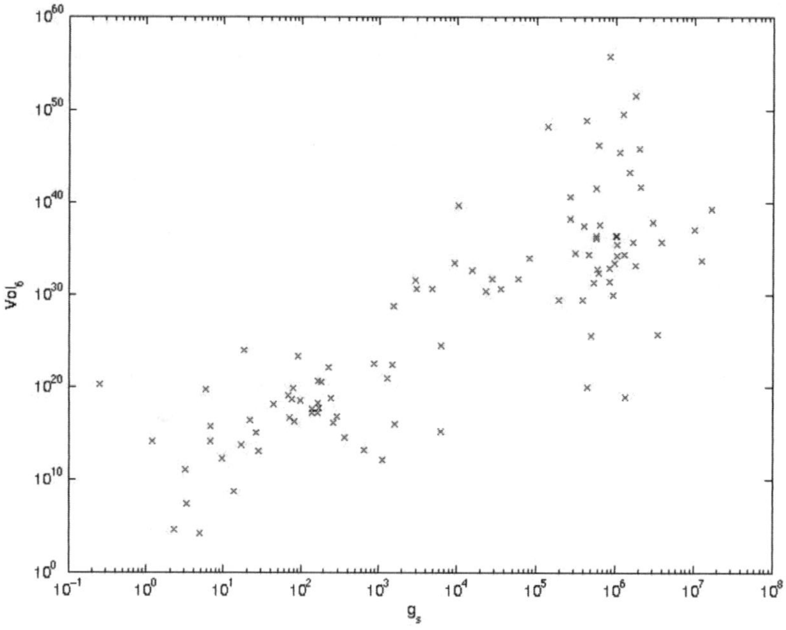

Figure 2. Logarithmic plot of g vs Vol_6 for breaking SUSY through complex structure moduli.

constants are very unfavored since we see that almost all these solutions are related with a large string coupling constant (fixed by the stabilization of the axio-dilaton), rendering our solutions unviable since they are far away from the perturbative regime where our approach is valid.

On the other hand, the vacua distribution in terms of the internal volume, lies in a region of small volume (small Kähler modulus) suggesting as well that the supergravity approximation is not valid.

In contrast to the above models, if SUSY is broken through all the moduli the $g - \Lambda$ distribution (Figure 3) seems to lie on a region of small coupling constant. The corresponding vacua have a positive value for all solutions. Finally as observed in Figure 4, the set of solutions in this case is constraint in a region in which the internal volume is big (an order of 10^4) implying that the supergravity approach we have taken is indeed valid. Therefore, we have found a rich region of De Sitter vacua in which all our approximations can be thrust. An extended analysis about their stability and the study of the existence of inflation conditions are under research [24].

5. Final comments

The present work shows a landscape of classical De Sitter vacua constructed by a Type IIB compactification on a six-dimensional torus threaded with NS-NS, RR and non-geometric fluxes and in the presence of an orientifold 3-plane. This constructions leads to a $\mathcal{N} = 1$ supersymmetric four-dimensional space-time. In effective theory there are three moduli related to the complex structure, Kähler modulus and the axio-dilaton field S of the corresponding three identical

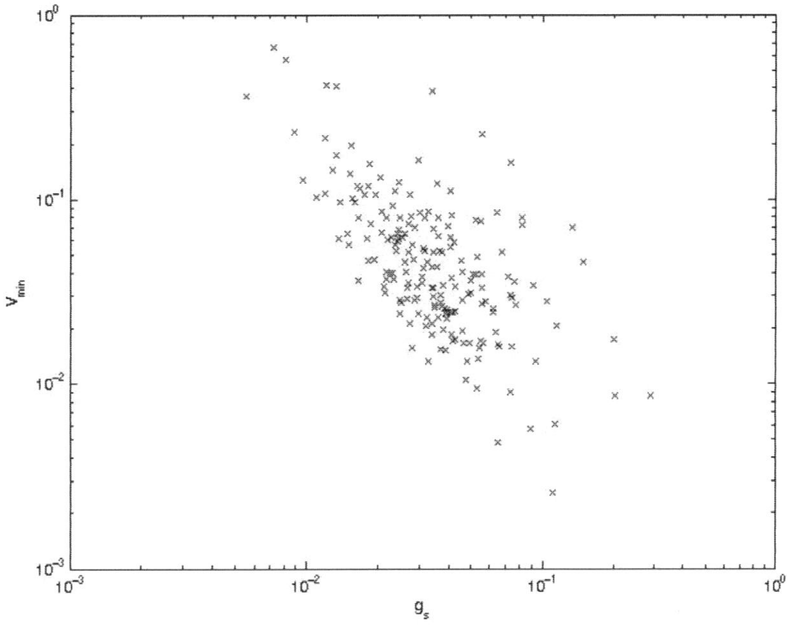

Figure 3. Logarithmic plot of g vs Λ for breaking SUSY through all the moduli.

two-dimensional torus.

We find that the viability of the solutions depends on which supersymmetric conditions are satisfied. There are only two cases in which physical solutions can be achieved and they correspond to the cases in which supersymmetry is broken through the complex structure or thorugh all moduli.

The first case leads to vacua in which the string coupling constant is large and the internal volume is very small. In such scenario our perturbative approach and use of supergravity is not valid.

However, for the second case we find suitable physical conditions in which our approximations are valid. Although the cosmological constant is positive and small, still it seems that fine-tuning is required. For these cases, we consider is worth to study whether they are compatible with inflation scenarios. For that it is necessary to study some extra features as their stability. We shall explore cosmological models within these vacua in a forthcoming work.

Acknowledgments

O.L.-B. is partially supported by a CONACyT grant with contract number 132166, by PROMEP and by DAIP. C. D.-A. is supported by a PhD CONACyT grant.

References

[1] M. R. Douglas and S. Kachru, 2007, [hep-th/0610102v3].
[2] A. Riess and others, 1998, *Atron. J.*.116.

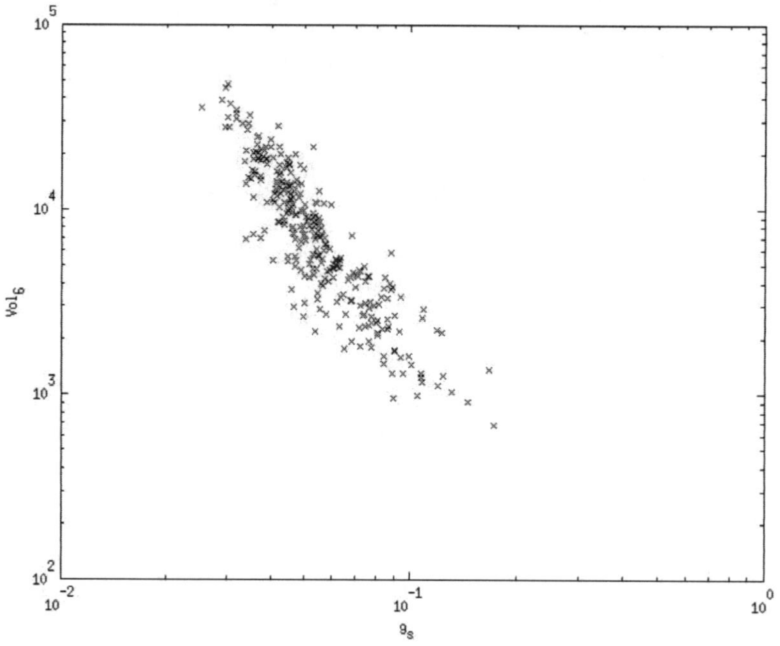

Figure 4. Logarithmic plot of g vs Vol_6 for breaking SUSY through all the moduli.

[3] S. Kachru, M. B. Shulz and Trivedi, 2003, *Phys. Rev. D*, **0310**, [hep-th/0201028].
[4] S. B. Gibbdings, S. Kachru and J. Polchinsky, 2001, *Phys. Rev. D*, **66**, [hep-th/0105097].
[5] S. Kachru, R. Kallosh, A. Linde, and S. Trivedi, 2003, *Phys. Rev. D*, **68**, [hep-th/0301240].
[6] V. Balasubramanian, P. Berglund, J. Conlon, and F. Quevedo, 2005, *JHEP*, **03**, [hep-th/0502058].
[7] J. Conlon, F. Quevedo, and K. Suruliz, 2005, *JHEP*, **08**, [hep-th/0505076].
[8] G. Aldazabal, P.G. Camara, A. Font, and L. E. Ibaez, 2006, [hep-th/0602089].
[9] J. Shelton, W. Taylor, and B. Wecht, 2006, [hep-th/0607015v2].
[10] B. Wetch, 2007, [hep-th/0708.3984v1].
[11] J. Shelton, W. Taylor, and B. Wecht, 2005, *JHEP*, **10**. [jhep/102005085].
[12] J. Shelton, W. Taylor, and B. Wecht, 2006, [hep-th/0508133v4].
[13] G. Shiu, Y. Sumitomo, 2011, [hep-th/1107.2925].
[14] T. Wrase and M. Zagermann, 2010, [hep-th/1003.0029v1].
[15] J. Schrek and J.H. Shwarz, 1979, *Nucl. Phys.* B, **153**.
[16] N. Kaloper and R. C. Myers, 1999, *JHEP*, **9905**, [jhep/9901045].
[17] G. Dall'Agata and S. Ferrara, 2005, *Nucl. Phys.* B, **717**, [hep-th/0502066].
[18] L. Andrianopoli, M. A. Lled and M. Trigiante, *JHEP*, **0505**, [hep-th/0502083].
[19] C. M. Hull and R. A. Reid-Edwards, [hep-th/0503114].
[20] B. de Wit, D.J. Smit adn N.D. Hari Dass, 1987, *Nucl. Phys. B*, **283**.
[21] J. M. Maldacena and C. Nuez, *Int. J. Mod. Phys. A*, **16**, [hep-th/0007018].
[22] J. Wess and J. Bagger, 1992, *Princeton* USA: *Univ. Pr.*
[23] S. Gukov, C. Vafa, E. Witten, *Nucl. Phys. B*, **584**, [hep-th/9906070].
[24] C. Damian, O. Loaiza-Brito, work in progress.

Journal of Physics: Conference Series **378** (2012) 012003 doi:10.1088/1742-6596/378/1/012003

Simulation of Fano factor at HAWC-30 array

M. A. Castillo[1],H. Salazar[1],A. Tepe[2],L.M. Villaseñor[3] on behalf of the HAWC Collaboration

[1]FCFM-BUAP, Avenida San Claudio y 18 Sur, Colonia San Manuel, Ciudad Universitaria, Puebla, Mexico.

[2]Center for Relativistic Astrophysics, School of Physics, Georgia Institute of Technology, 837 State St., Atlanta, GA 30332-0430, USA.

[3]IFM-UMSNH, Edificio C-3, Ciudad Universitaria, Morelia, Mxico.

E-mail: mcastillomaldonado@yahoo.com

Abstract. The High Altitude Water Cherenkov detector is a gamma-ray observatory which is able to scan the sky in energy from 100 GeV to 100 TeV and will be localized at Sierra La Negra volcano at 4100 m a.s.l. near to Puebla, México. In 2011 an engineering array called VAMOS was installed and in 2012, it will start the deployment of HAWC first step, the HAWC-30 array, with 30 water Cherenkov detectors. In this work it is presented the results of simulations where the goal is to get Fano factor in order to simulate the HAWC-30 array sensibility to gamma-ray bursts using the single particle technique.

1. Introduction

Gamma-ray bursts are short and intense pulses of gamma rays. The bursts last from a fraction of a second to hundreds of seconds. The overall observed fluences range from 10^{-4}erg/cm^2 to 10^{-7}erg/cm^2 (the lower limit depends, of course, on the characteristic of the detectors an not on the bursts themselves). This corresponds to an isotropic luminosity of $10^{51} - 10^{52}$erg/sec, making GRBs the most luminous objects in the sky [1]. The Large Area Telescope (LAT) is the principal scientific instrument on the Fermi Gamma-ray Space Telescope Spacecraft and it is an imaging high-energy gamma-ray telescope covering a energy range from about 20 MeV to more than 300 GeV. The Fermi observations also show the most intense GeV emissions occur promptly and extend longer than the emission at low energies. A wide field of view, high duty factor observatory, such as HAWC, is required to observe these prompt emissions and determine their extension at high energy. In this work it is studied the possibility to detect high energy GRBs assuming their spectra is a power law form as Fermi has already detected, not with HAWC, but with HAWC-30 array, a first step in the whole detector construction.

2. The HAWC observatory

The HAWC observatory, a second generation water Cherenkov detector built on the success of the first generation, wide field-of-view, high duty cycle, TeV gamma-ray observatory, Milagro. HAWC is located in Sierra Negra volcano near to Puebla, México at 4100 m a.s.l. (N 18°59'48' and W 97°18'34"). The observatory will consist of 300 water Cherenkov detectors (WCDs) with 7.3 m on diameter and 4.5 m depth each with $\sim$200,000 liters of ultra clean water, densely

Figure 1. HAWC observatory site. Artistic conception

packed to cover an area of 25,000 m^2 instrumented with three 20 cm and one 25 cm upward-looking photomultiplier tubes mounted on the bottom of the WCD. The expanded detector area (~10 times the Milagro bottom layer), increased altitude, and optical isolation of the detector elements lead to a 15-fold increase in sensitivity to Milagro, which performed the deepest wide-field survey of TeV sky to date. This is achieved by applying the experience with Milagro to the design of HAWC and reusing the Milagro's PMTs and front-end electronics.

2.1. HAWC construction

The HAWC observatory will be constructed in three stages where each is an array of tanks that includes the previous one. The first stage is the HAWC-30 array (to be finished by Summer, 2012), the next one to be constructed is HAWC-100 array (to be finished by Spring, 2013) and finally HAWC-300 (to be finished by Summer, 2014). In Figure 2 it is shown a layout with each array.

2.2. Engineering array

It is worth to mention that it has been constructed an engineering array (7-tanks array with the same dimensions to be used for HAWC) where all subsystems would be tested. This array is called VAMOS (Verification And Measuring Observatory Systems) and it has been taking data since October, 2011.

Journal of Physics: Conference Series **378** (2012) 012003 doi:10.1088/1742-6596/378/1/012003

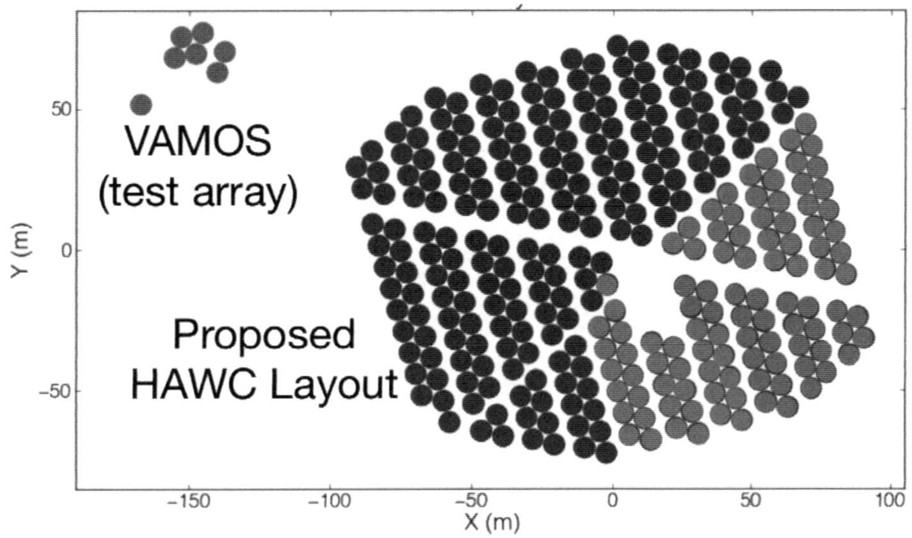

Figure 2. Layout of HAWC observatory. In red are the WCDs that form a 30 tank array, in magenta and red the ones for HAWC-100 and finally those in red, magenta and blue are HAWC-300

2.3. HAWC Scientific goals

Among the HAWC observatory scientific goals are: to map the galactic diffuse gamma-ray emission from 1 TeV to 100 TeV, to measure the spectrum of galactic sources to highest energies, to study transient emission from AGN, to study the local anisotropy in the cosmic radiation observed by Milagro and to measure its energy spectrum with better precision, monitor the sky for VHE emission for gamma-ray bursts and search for signals from fundamental particles. Although most of these goals will require the full HAWC array, some studies can be accomplished with a smaller (partially built) detector. In particular the single particle technique will provide the opportunity to detect high energy gamma-ray bursts already with HAWC-30.

3. Single particle technique

HAWC is going to observe gamma-rays by detecting, at ground level, the secondary particles that compose the extensive air showers. It is taking advantage of the Cherenkov light produced in water when a charged particle pass through, with high speed greater than the speed of light in this medium. The Cherenkov radiation consist of photons with wavelength in the range from 350 nm to 500 nm. 90% of the particles at ground level are photons [4]. High energy photons may produce electron-positron pairs, which then can be detected by Water Cherenkov Detectors.

HAWC will use two kinds of data acquisition (DAQ) systems: the main DAQ system and the scaler DAQ system. The present work is dedicated to the scaler DAQ system. The technique applied below consists of monitoring all PMTs in fixed periods of 10 ms with a threshold of $\frac{1}{4}$ photo-electron, waiting for an increase in the background rate in coincidence with a GRB detected e.g. by Fermi. The single particle technique does not provide a reconstruction information on single showers but it provides better sensitivity than main DAQ system at low energies, where full event reconstruction may not be possible.

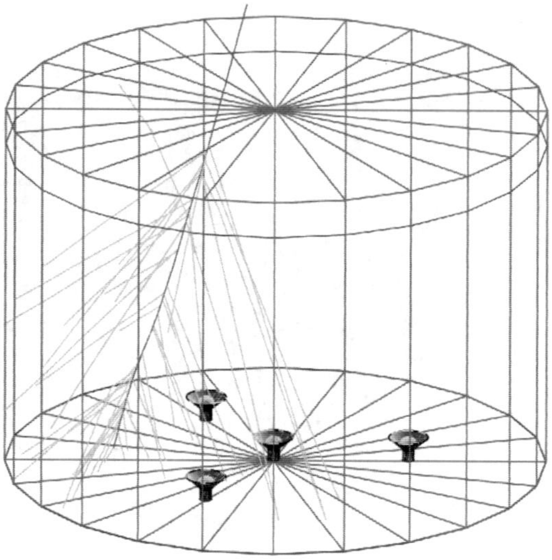

Figure 3. A HAWC water Cherenkov detector. A muon (in red) is passing through WCD and Cherenkov photons are created in a cone with a semi angle of $\sim$ 42°.

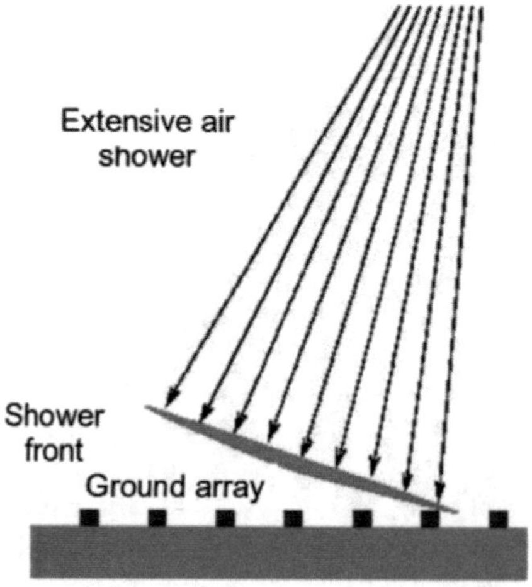

Figure 4. Detection of a shower as usually. For this operation mode HAWC will use main DAQ system.

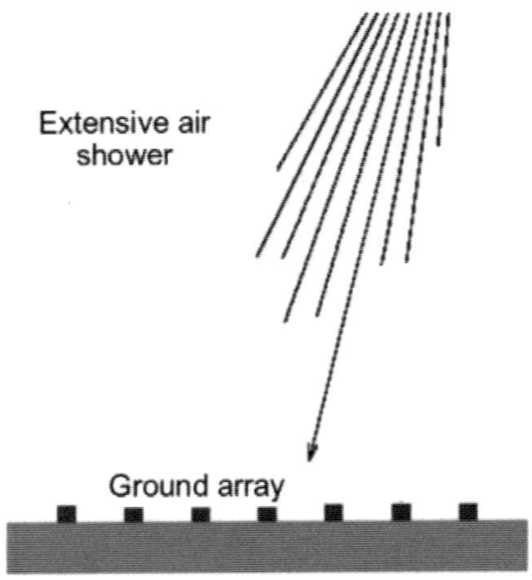

Figure 5. Single particle technique. Detection of some of the secondary particles that belong to the shower. For this operation mode HAWC will use scaler DAQ system.

4. The fano factor

The Fano factor describes how much wider the real distribution is with respect to a Poisson distribution with the same mean value but without correlation. It is defined as:

$$F = \frac{\sigma^2}{\mu}, \tag{1}$$

In formula (1), σ is the spread of data and μ is the variance of a Poisson distribution with the same mean of the real distribution but without correlation. The Fano factor is a measure for the correlation in the random data. For scaler DAQ system, the random variable would be the number of hits in all PMTs in a given time window. For a Poisson process, the variance in the count equals the mean count, so $F = 1$. The most important sources of correlated noise are (hierarchically): a) air showers resulting in more than one PMT registering a signal, b) PMT after pulses and c) Michel's electrons due to muons that stop inside WCDs. The correlation result in a distribution that is wider than a Poisson distribution [3] . It is important to simulate Fano factor in order to know the HAWC capabilities to detect GRBs.

5. Simulation features

5.1. Fano factor simulation features

In order to get the simulated Fano factor of HAWC 30 it is mandatory to simulate the background expected at site level. For HAWC detector the main background is caused by secondary cosmic rays and extensive air showers (EASs), which are initiated by primary particle interactions at the top of the atmosphere. So, the first step was to simulate the EASs and for this purpose we used CORSIKA. All runs are done by using protons as primary particles, with energies in the range from 5 GeV to 50 TeV, zenith angles between 0° and 60°, the slope in the energy spectra of ~ 2.7 and the ground level at 4100 m a.s.l. In total, 2×10^{10} primary interactions were simulated. Every event contains all sub-particles created at ground with their respective momentum, position and arrival time. The second step was to simulate the detector response to the ground level particles. In order to pass all particles through the detector, we used a GEANT4 based code. For this step it is worth to mention that each CORSIKA file has been oversampled 10 times in the GEANT4 code. Besides, every shower core was limited to a rectangular zone

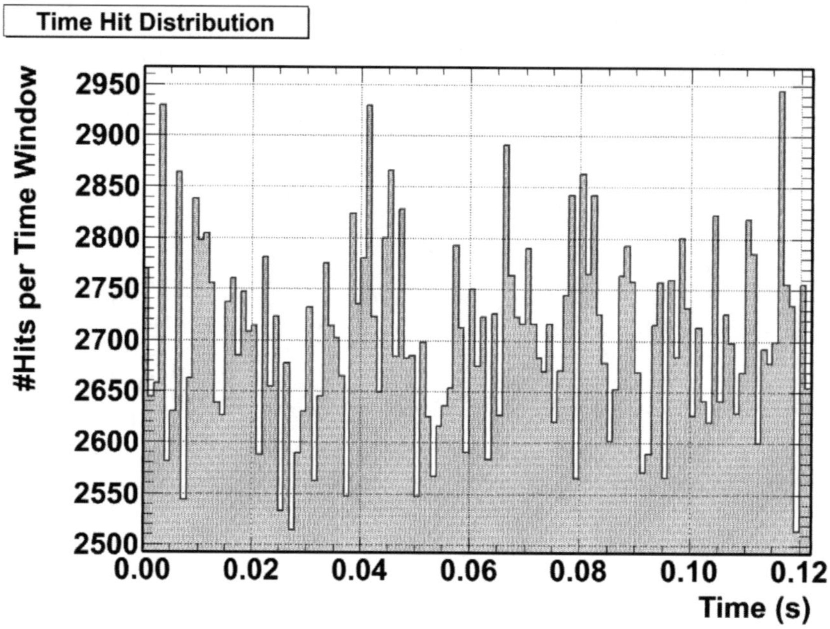

Figure 6. Rate behavior in time for HAWC-30 array assuming 3 PMTs per WCD (MC simulation).

of 1500 m each side which includes the array. What this code gives us is the number of photo-electrons created by the secondary particles in each PMT. The WCDs localization is showed at Figure 2. At the last step the PMT afterpulses and electronics response are simulated, and finally the Fano factor is calculated.

5.2. Scaler effective area simulation features

For extensive air showers was used CORSIKA, this time with gammas as primaries with energies in the range from 5 GeV to 10 TeV, zenith angles between 0° and 65°, the slope of the energy spectra of -2 and ground level of 4100 m a.s.l. After that we used the GEANT4 based code to get detector response to ground level particles, and finally the official analysis software was used.

6. Results of Fano factor simulations

In the simulation presented in this work it has been considered 3 PMTs per tank (20 cm photomultiplier tubes) and the tanks localization is presented in red in Figure 2. This is because of the relatively new decision taken by the collaboration to increase the number of PMTs per tank from three to four. As it was mentioned in section 5, we only have 125 ms of background simulated so we chose a time window of 1 ms in order to calculate the rate and Fano factor for HAWC-30 array.

In Figure 6, the rate behavior in time is shown, y-axis is the number of hits in all PMTs each millisecond and x-axis is the time with 1 millisecond in each bin, and in Figure 7 it is shown the rate distribution, i.e. y-axis is the frequency of occurrence and x-axis is the number of hits in all PMTs each millisecond.

Journal of Physics: Conference Series **378** (2012) 012003 doi:10.1088/1742-6596/378/1/012003

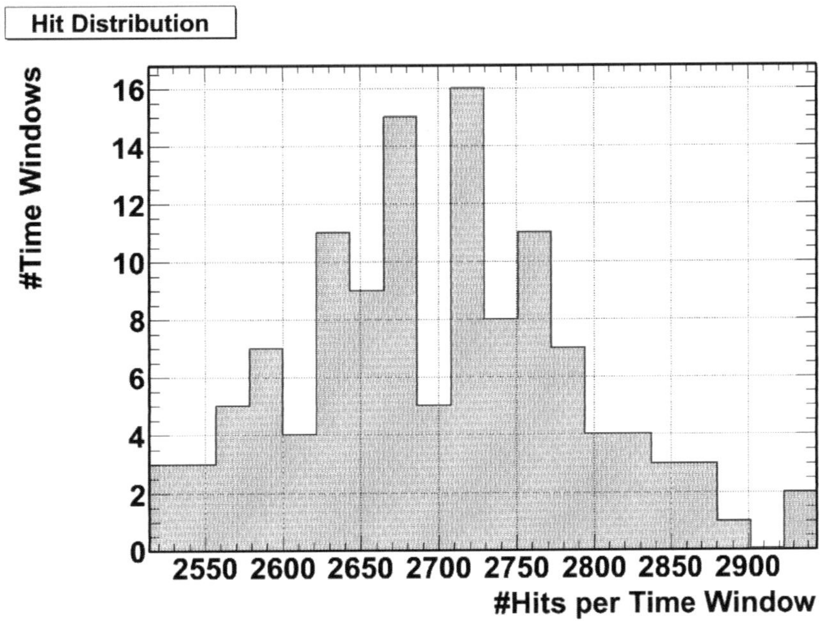

Figure 7. An example of the rate distribution for HAWC-30 array assuming 3 PMTs per WCD (MC simulation).

We found that simulated background rate per PMT is 30 kHz and the simulated Fano factor for HAWC-30 array is 2.9.

7. Results of scaler effective area simulations

The effective area $A_{eff}(E, \theta)$ of HAWC-30 array describes its efficiency of converting an incoming gamma-ray flux at the top of the atmosphere in detected hits. The effective area is calculated by simulating the response of the detector to extensive air showers generated by incoming gamma-rays [3]. The calculation uses formula (2), where A_{trhown} is the area over which the simulated EAS are thrown, $N_{obs}(E, \theta)$ is the number of showers that result in at least one PMT being hit, N_{thrown} is the number of simulated EASs and η_{PMT} is the average number of PMTs hits for this energy E and Zenith θ,

$$A_{eff}^{Scaler} = A_{thrown}\eta_{PMT}\frac{N_{obs}(E, \theta)}{N_{thrown}(E, \theta)} \,. \tag{2}$$

The obtained scaler effective area is shown in Fig 8. Note that the meaning of the scaler effective area differs from the traditional concept of effective area used e.g. for air shower arrays. Because of the number of PMTs hits is the signal (and not the individual showers) the scaler effective area is not restricted to the detector area, so it never saturates to a value. It is interesting to compare this effective area with the effective area of Fermi LAT. For instance, for a primary photon energy of 30 GeV and a zenith angle between 0° and 25.8^o the Fermi LAT effective area differs from the traditional concept of effective area used e.g. for air shower arrays is $\sim 1\text{m}^2$ and the scaler effective area of HAWC-30 array is $\sim 4900\text{m}^2$. The scaler effective area decreases as the zenith angle increases.

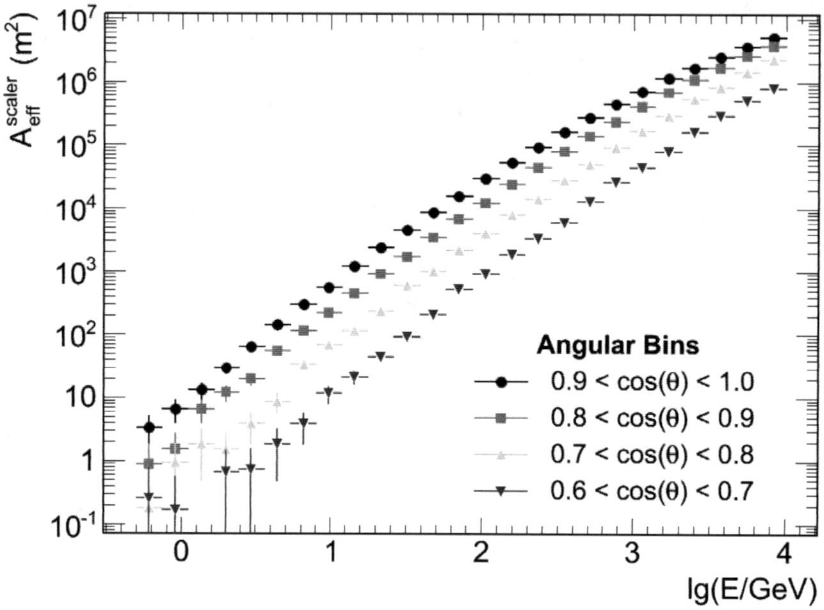

Figure 8. Scaler effective area of HAWC-30 array for 4 zenith angle bands and different primary photon energies.

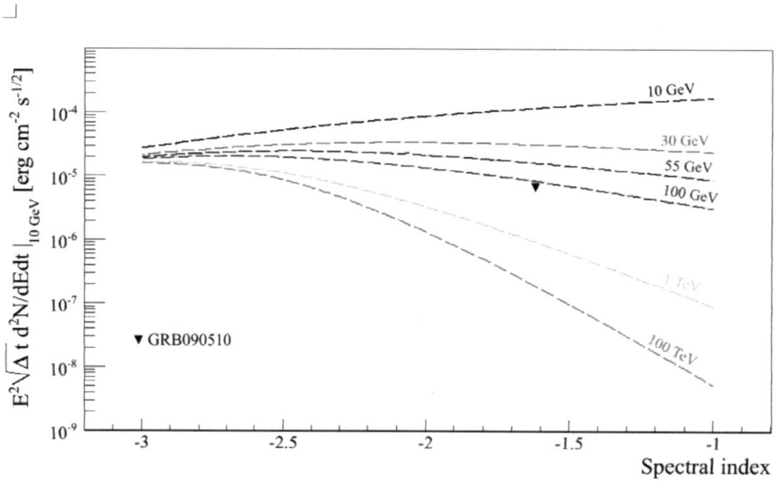

Figure 9. Scaler DAQ system sensitivity to GRBs for HAWC-30 array taking into account 3 PMTs/tank. Necessary flux at 10 GeV multiplied by square root of the GRB duration to produce a 5 σ signal. The range of zenith angle is from 25.8^o to 36.8^o.

8. Sensitivity to GRBs

In this section we describe the HAWC-30 array sensitivity to GRBs, as obtained using the single particle technique (scaler DAQ system) described above. The background for HAWC-30 array

Journal of Physics: Conference Series **378** (2012) 012003 doi:10.1088/1742-6596/378/1/012003

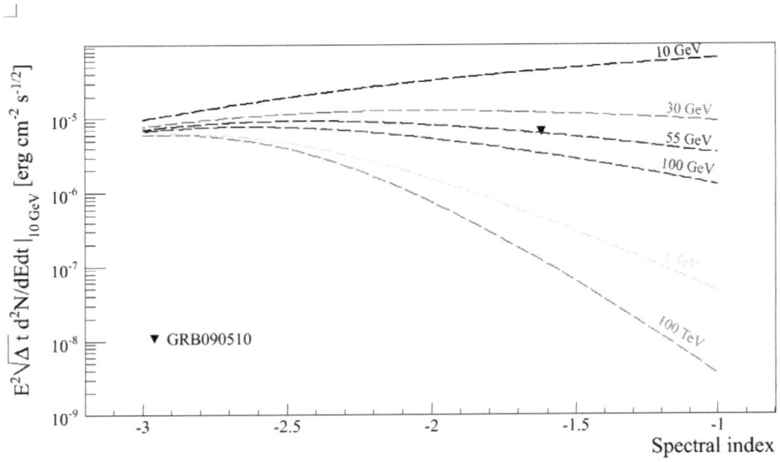

Figure 10. Scaler DAQ system sensitivity to GRBs for HAWC-30 array taking into account 3 PMTs/tank). Necessary flux at 10 GeV multiplied by square root of t he GRB duration to produce a 5 σ signal. The range of zenith angle is from 0° to 25.8^o.

is 30 WCDs $\times$ 3 PMTs $\times$ 30 kHz = 2.7 MHz.

The signal to background ratio or significance of an observation with HAWC is defined as

$$Sigf = \sqrt{\frac{\Delta T}{F\eta_{PMT}R_{PMT}}} \int_{E_{min}}^{E_{max}} dE \frac{dN}{dE} A_{eff}^{scaler}, \tag{3}$$

in formula (3), ΔT is the observation time window, F is the Fano factor, η_{PMT} the number of PMTs in the detector, R_{PMT} is the average PMT rate, θ is the observed GRB zenith angle, $A_{eff}^{scaler}(\theta)$ is the scaler effective area and $\frac{dN}{dE}$ is the observed GRB flux.

For these calculations we assumed that GRB energy spectrum, in the high energy zone, is a power law form with a high energy cutoff. The calculations are done in two zenith angle bands: from 0° to 25.8^o (Figure 10) and from 25.8^o to 36.87^o (Figure 9) and the following values for the high energy cutoff in the GRB espectrum: 10 GeV, 30 GeV, 55 GeV, 100 GeV, 1 TeV and 100 TeV and a spectral index in the range from -3 to -1. Effects of extra galactic background light (EBL) have been ignored. The computed sensitivity is shown in Figures 9 and 10, in terms of differential flux at 10 GeV multiplied by square root of the GRB duration. The sensitivity corresponds to a 5 σ detection of the GRB. The flux of the GRB 090510 measured by Fermi LAT is shown for reference [5].

In Figures 9 and 10 one can see as the zenith angle for GRBs decreases the necessary flux to detect a GRB at 5 σ decreases and, as the energy cutoff increases the necessary flux to detect GRBs at 5 σ decreases and as the spectral index increases the necessary flux to detect GRBs decreases. The GRB090510 could be seen if its energy spectrum would extend to 55 GeV.

9. Conclusions

The HAWC observatory is a new generation wide field of view gamma-ray telescope currently under construction in Mexico. The wide field of view and high duty cycle of HAWC offer exciting prospects for the first detection of a gamma-ray burst by a ground-based experiment. In this paper the capabilities of the partially build array (HAWC-30) to detect high energy GRBs using single particle technique are analyzed. As a comparison GRB090510 flux could be seen with

HAWC-30 array if its energy spectrum would extend to 55 GeV. Thus, although much smaller than complete HAWC, HAWC-30 provides a realistic opportunity to observe powerful GRBs. Measurements could constrain high energy cutoffs in GRB spectra, thus providing information on intrinsic spectral cutoff of the GRB or EBL absorption.

References

[1] T. Piran, Rev. Mod. Phys. **76** (2004) 1143 [astro-ph/0405503].
[2] Pennini, F. and Plastino, A. J. Phys. Conf. Ser. **246** (2010) 012030
[3] A. U. Abeysekara *et al.* [HAWC Collaboration], Astropart. Phys. **35**, 641 (2012) [arXiv:1108.6034 [astro-ph.HE]].
[4] D. Allard, I. Allekotte, C. Alvarez, H. Asorey, H. Barros, X. Bertou, O. Burgoa and M. Gomez Berisso *et al.*, Nucl. Instrum. Meth. A **595** (2008) 70.
[5] M. Ackermann *et al.* [Fermi LAT and GBM Collaborations], Astrophys. J. **716** (2010) 1178 [arXiv:1005.2141 [astro-ph.HE]].

Journal of Physics: Conference Series **378** (2012) 012004 doi:10.1088/1742-6596/378/1/012004

Superfield formalism for 5D Lorentz violating models

J.D. García-Aguilar[1], **A. Pérez-Lorenzana**[2]

Departamento de Física, Centro de Investigación y de Estudios Avanzados del I.P.N., Apdo. Post. 14-740 07000 México D.F., México.

E-mail: [1]jdgarcia@fis.cinvestav.mx, [2]aplorenz@fis.cinvestav.mx

Abstract. We construct a supersymmetric higher dimensional theory which includes 5D Lorentz violation terms in a superfield formalism. We show that in order to obtain a consistent construction which incorporates the most general terms that explicitly breaks the 5D Lorentz invariance, it is only necessary to redefine the supersymmetric auxiliary fields. We also analyze the meaning of these violation terms and point out the possibility of resizing the radius of compactification to obtain an equivalent 4D Lorentz invariant effective theory.

1. Introduction

On December 2011, CMS and ATLAS groups presented their latest results on the Higgs search, where they showed the possibility of finding this particle with a mass between 119 and 130 GeV [1]. A Higgs particle with such a mass leads to the famous hierarchy problem in the Standard Model. As it is well known, supersymmetry [2] (SUSY) and Extra Dimensional theories (XD) [3] offer some suggestions to solve this problem, unfortunately we do not have any experimental evidence of their existence yet.

SUSY is introduced as a global symmetry of the Lagrangian which transforms fermions into boson and viceversa, that implies the existence at least of one superpartner for each particle of the SM. On an exact SUSY theory, a particle and its superpartner must have same charge, spin and mass. There is not any indication for the existence of superpartners, and for this reason global supersymmetry, if present, must be broken, but a convincing mechanism to obtain this rupture is yet unknown. [5]. Most supersymmetric theories can be rewritten in superspace formalism which was proposed by Salam and Strathdee in 1974 [6]. This formalism involves the consideration of superfield $\Phi(x,\theta,\bar{\theta})$ defined on an 8-dimensional space which is the product of ordinary space-time with a 4-dimensional space whose points are labeled by the anticommuting Grassman variables θ_α.

Extra dimensional models, on the other hand, introduce additional space-like coordinates in the usual theories, trying to solve some conceptual problems of the Standard Model (SM). However, the required compactification of such dimensions leads to an infinite number of particles arranged on the so called Kaluza Klein (KK) towers, where different levels incorporate copies of SM particle representations, but with larger masses. Lowest levels corresponding to SM particle spectrum. Mass gaps and the KK spectrum characteristics do depend on the geometry of the compact space. [4]. One of the common features of such constructions, is that compactification of the extra space implies the breaking of the higher dimensional space-time symmetries. In particular, in flat extra space models, higher dimensional Lorentz invariance cannot be advocated in general as a true symmetry. Nevertheless, the way this symmetry is broken and the physical

Journal of Physics: Conference Series **378** (2012) 012004 doi:10.1088/1742-6596/378/1/012004

implications of the breaking at the level of the effective theories had not yet been studied in detail so far. This is the issue we are dealing with, and here we present some technical advances on our studies. In a previous work we build a SUSY model with one extra dimension in which the 5D Lorentz symmetry is broken in a explicit way, whereas the 4D Lorentz symmetry remains after the compactification process [7]. The goal of this work is to present the model discussed above in a superfield formalism which explicitly depict the degrees of freedom that are relevant at the level of the 4D effective theory, and further discuss the physical meaning of the terms that break Lorentz invariance.

The outline for this work is as follow: In the section 2 we briefly present a review of the superfield formalism for standard SUSY theories which should serve as the reference frame for the rest of the discussion. In the section 3 we show the convention for Dirac matrices in five dimensions for the model, and rewrite the Dirac Lagrangian in terms of two Weyl spinors. In the section 4 and 5 we present the 5D SUSY model on our superfield formalism. In section 6 we discuss how a modification on compactification radius provides a mechanism to hide the terms of Lorentz violation. Finally, our conclusions are presented in section 7.

2. Superfield formalism.

The SUSY algebra is generated by the momentum operators P_μ and the Weyl spinor supersymmetry generators Q_α and $\bar{Q}_{\dot{\alpha}}$

$$\begin{aligned}
&[Q_\alpha, P_\mu] = [\bar{Q}_{\dot{\alpha}}, P_\mu] = [P_\mu, P_\nu] = 0 \\
&\{Q_\alpha, Q_\beta\} = \{\bar{Q}_{\dot{\alpha}}, \bar{Q}_{\dot{\beta}}\} = 0 \\
&\{Q_\alpha, \bar{Q}_{\dot{\beta}}\} = 2\sigma^\mu_{\alpha\dot{\beta}} P_\mu \ ,
\end{aligned} \tag{1}$$

and, a finite element of the corresponding group is given by

$$G\left(x^\mu, \theta, \bar{\theta}\right) = \exp\left(i\left(\theta Q + \bar{\theta}\bar{Q} - x^\mu P_\mu\right)\right) \ .$$

Hausdorff formula leads to express the group product as

$$G\left(a^\mu, \xi, \bar{\xi}\right) G\left(x^\mu, \theta, \bar{\theta}\right) = G\left(x^\mu + a^\mu + i\xi\sigma^\mu\bar{\theta} - i\theta\sigma^\mu\bar{\xi}, \theta + \xi, \bar{\theta} + \bar{\xi}\right) \ .$$

Therefore, the SUSY generators induce a translation in group parameter space

$$\exp\left(i\left(\xi Q + \bar{\xi}\bar{Q} - a^\mu P_\mu\right)\right) : \left(x^\mu, \theta, \bar{\theta}\right) \longrightarrow \left(x^\mu + a^\mu + i\xi\sigma^\mu\bar{\theta} - i\theta\sigma^\mu\bar{\xi}, \theta + \xi, \bar{\theta} + \bar{\xi}\right) .$$

For a superspace function $S(x^\mu, \theta, \bar{\theta})$ (referred to as a superfield) we have

$$\begin{aligned}
S\left(x^\mu + a^\mu - i\xi\sigma^\mu\bar{\theta} + i\theta\sigma^\mu\bar{\xi}, \theta + \xi, \bar{\theta} + \bar{\xi}\right) &= S\left(x^\mu, \theta, \bar{\theta}\right) + \left(a^\mu + i\xi\sigma^\mu\bar{\theta} - i\theta\sigma^\mu\bar{\xi}\right)\frac{\partial S}{\partial x^\mu} \\
&\quad + \xi^\alpha \frac{\partial S}{\partial \theta^\alpha} + \bar{\xi}_{\dot{\alpha}} \frac{\partial S}{\partial \bar{\theta}_{\dot{\alpha}}} + \ldots
\end{aligned}$$

from which it follows that the action of the SUSY algebra on superfields is generated by

$$\begin{aligned}
P_\mu &= i\partial_\mu \\
iQ_\alpha &= \frac{\partial}{\partial \theta^\alpha} + i\sigma^\mu_{\alpha\dot{\alpha}}\bar{\theta}^{\dot{\alpha}}\partial_\mu \\
i\bar{Q}_{\dot{\alpha}} &= -\frac{\partial}{\partial \bar{\theta}^{\dot{\alpha}}} - i\theta^\alpha\sigma^\mu_{\alpha\dot{\alpha}}\partial_\mu .
\end{aligned} \tag{2}$$

Journal of Physics: Conference Series **378** (2012) 012004 doi:10.1088/1742-6596/378/1/012004

At this level we can define fermionic derivatives which anticommute with the generators of the supersymmetry algebra, an example for they is given by

$$\begin{aligned} D_\alpha &= \frac{\partial}{\partial\theta^\alpha} - i\sigma^\mu_{\alpha\dot\alpha}\bar\theta^{\dot\alpha}\partial_\mu \\ \bar D_{\dot\alpha} &= -\frac{\partial}{\partial\bar\theta^{\dot\alpha}} + i\theta^\alpha\sigma^{\alpha\dot\alpha}\mu\partial_\mu \,. \end{aligned}$$

and these covariant derivatives obey the algebra

$$\begin{aligned} \{D_\alpha, \bar D_{\dot\alpha}\} &= 2\sigma^\mu\alpha\dot\alpha P_\mu \\ \{D_\alpha, D_\beta\} &= \{\bar D_{\dot\alpha}, \bar D_{\dot\beta}\} = 0 \end{aligned} \tag{3}$$

A general superfield $S(x^\mu, \theta, \bar\theta)$ may be expanded as a power series in θ and $\bar\theta$ involving not more than two powers of θ and θ because they are two-component Grassmann variables. The coefficients of the various powers of θ and $\bar\theta$ in this expansion are ordinary fields (functions of x^μ). The superfield on which has been imposed the constraint $\bar D_{\dot\alpha}S = 0$ is denoted by Φ and we refer to it as a chiral superfield. Furthermore, in this work we are going to define the chiral superfield as

$$\begin{aligned} \Phi(x^\mu, \theta, \bar\theta) = A(x) + \sqrt{2}\theta\chi(x) + \theta^2F(x) - i\left(\theta\sigma^\mu\bar\theta\right)\partial_\mu A(x) \\ + \frac{i}{\sqrt{2}}\theta^2\left(\partial_\mu\chi(x)\sigma^\mu\bar\theta\right) - \frac{1}{4}\theta^2\bar\theta^2\partial_\mu\partial^\mu A(x) \end{aligned}$$

here $\mu = 0, \ldots, 3$, A is a complex scalar field, χ is a Weyl spinor, F is an auxiliary field and $\sigma^\mu = (\mathbb{I}, \vec\sigma)$. To obtain the supersymmetric transformations for the fields, lets consider an infinitesimal supersymmetry transformation on the chiral superfield

$$\Phi \to \Phi + \delta\Phi \,,$$

where

$$\delta\Phi = i\left(\xi Q + \bar\xi\bar Q\right)\Phi \,.$$

Thus, this leads to

$$\begin{aligned} \delta\Phi &= \xi^\alpha\left(\frac{\partial}{\partial\theta^\alpha} + i\sigma^\mu_{\alpha\dot\alpha}\bar\theta^{\dot\alpha}\partial_\mu\right)\Phi + \bar\xi^{\dot\alpha}\left(\frac{\partial}{\partial\bar\theta^{\dot\alpha}} + i\theta^\alpha\sigma^\mu_{\alpha\dot\alpha}\partial_\mu\right)\Phi \\ &= \sqrt{2}\xi\chi + 2\xi\theta F - 2i\theta\sigma^\mu\bar\xi\partial_\mu A + \frac{i}{\sqrt{2}}\theta^2\partial_\mu\left(\chi\sigma^\mu\bar\xi\right) + \ldots \\ &= \delta A + \sqrt{2}\theta\delta\chi + \theta^2\delta F + \ldots \,, \end{aligned} \tag{4}$$

which implies

$$\begin{aligned} \delta A &= \sqrt{2}\xi\chi \\ \delta\chi &= \sqrt{2}\xi F - i\sqrt{2}\sigma^\mu\bar\xi\partial_\mu A \\ \delta F &= \sqrt{2}\partial_\mu\chi\sigma^\mu\bar\xi \,. \end{aligned} \tag{5}$$

In order to find an object $\mathcal{L}(\Phi)$ such that $\delta\mathcal{L}$ is a total derivative under SUSY transformation, we can exploit that

- For a general scalar superfield $S = \ldots + \theta^2\bar\theta^2 D(x)$, the D-term transforms as a total derivative.

Journal of Physics: Conference Series **378** (2012) 012004 doi:10.1088/1742-6596/378/1/012004

- For a chiral superfield $\Phi = \ldots + \theta^2 F(x)$, the F-term transforms as a total-derivative (see (5)).

That means, the most general Lagrangian for a chiral superfield can be written as

$$\mathcal{L} = K\left(\Phi^\dagger, \Phi\right)\Big|_D + \left(W(\Phi)|_F + h.c.\right) .$$

The function K is known as the Kähler potential. This is a real function of Φ and $\Phi^\dagger$. $W(\Phi)$ is known as the superpotential, and it is a holomorphic function of the chiral superfield Φ. For example, to the *Wess Zumino model* Lagrangian,

$$\mathcal{L} = \partial_\mu A^\dagger \partial^\mu A + i\chi\sigma^\mu\partial_\mu\bar{\chi} + F^\dagger F , \tag{6}$$

corresponds the Kähler potential

$$K = \Phi^\dagger\Phi.$$

3. Spinors in five dimensions.

Without loss of generality, a massless fermion field can be defined as the solution of the equation of motion for the Dirac Lagrangian

$$\mathcal{L} = i\bar{\Psi}\Gamma^M\partial_M\Psi ,$$

where $M = 0\ldots, 3, 5$, Gamma matrices (Γ^M) satisfy the Clifford algebra relation

$$\{\Gamma^M, \Gamma^N\} = 2\eta^{MN}\mathbb{I}$$

where $\mathbb{I}$ is the identity matrix and η^{MN} is the Minkowski metric. In this work we use the convention $\eta^{MN} = diag(1, -1, \ldots, -1)$.

In five dimension we use the Weyl basis

$$\Gamma^\mu = \begin{pmatrix} 0 & \sigma^\mu \\ \bar{\sigma}^\mu & 0 \end{pmatrix}, \qquad \Gamma^5 = i\begin{pmatrix} -\mathbb{I} & 0 \\ 0 & \mathbb{I} \end{pmatrix}. \tag{7}$$

Therefore, a fermion on five space-time dimensions is necessarily a four component spinor, nevertheless for our proposes we can always arrange it in terms of two two-component spinors as

$$\Psi = \begin{pmatrix} \psi_\alpha \\ \bar{\chi}^{\dot{\alpha}} \end{pmatrix} .$$

Under this decomposition, the corresponding 5D Dirac Lagrangian can be rewritten as

$$\mathcal{L} = i\chi\sigma^\mu\partial_\mu\bar{\chi} + i\psi\sigma^\mu\partial_\mu\bar{\psi} - \psi\partial_5\chi + \bar{\chi}\partial_5\bar{\psi} + \text{total derivatives} . \tag{8}$$

4. Superfield formalism for a 5D SUSY model.

In order to proceed with the development of our formulation, lets us consider the general 5D action

$$S = \int d^5x \left(\partial_M A^\dagger\partial^M A + \partial_M B^\dagger\partial^M B + i\bar{\Psi}\Gamma^M\partial_M\Psi + P^\dagger P + Q^\dagger Q\right) , \tag{9}$$

where A and B are complex scalar fields; Ψ is given as above, and P and Q are general auxiliary fields. This is a straightforward generalization of the action in Eq. (6) which contains twice the number of degrees of freedom. The goal next would be to conveniently rewrite the above action in terms of a 4D-like superfield formalism which would explicitly exhibit the degrees of freedom

Journal of Physics: Conference Series **378** (2012) 012004 doi:10.1088/1742-6596/378/1/012004

that are relevant at the 4D effective level of the theory (upon compactification). This formalism, as we will see later, will also allow us for a straightforward introduction of Lorentz violation 5D terms on the theory. Notice that this is actually possible, since the 5D construction turns out to be a $N = 2$ SUSY model, for which the hyperepresentation can always be decomposed (at the level of 4D) as made out of a couple of $N = 1$ chiral fields. As a matter of fact, by using the 5D fermion represented in terms of two Weyl spinor we introduced in previous section, it is easy to see that for a SUSY transformation each Weyl spinor would require its own complex scalar field and its own auxiliary field. Therefore, the field content so far would be right what it is required to have two chiral superfields.

For our proposes, let us define the required chiral superfields as

$$\Phi = \Phi\left(A(x^M), \psi(x^M), F(x^M)\right) ,$$
$$\Omega = \Omega\left(B(x^M), \chi(x^M), G(x^M)\right) ,$$

where to keep things as general as possible we named the corresponding auxiliary fields as F and G. As it is clear, with these superfields one could straightforwardly write down two Wess-Zumino like actions, as

$$S' = \int d^5x \left(\Phi^\dagger\Phi\Big|_D + \Omega^\dagger\Omega\Big|_D \right) ,$$

nonetheless, these terms do not reproduce the total action as given in Eq. (9) As a matter of fact, the difference among both expressions gives

$$S - S' = \int d^5x \left(\partial_5 A^\dagger \partial^5 A + \partial_5 B^\dagger \partial^5 B + i\bar{\Psi}\Gamma^5\partial_5\Psi + P^\dagger P + Q^\dagger Q - F^\dagger F - G^\dagger G \right) . \quad (10)$$

Notice that last essentially contains all terms with fifth derivatives. Thus, the simple Wess-Zumino-like action misses the actual terms depicting the fact that the construction lives in five dimension, and, furthermore, it implicitly breaks 5D Lorentz invariance, although, of course, it maintains 4D Lorentz symmetry. However, considering the following interaction-like term that involves both superfields

$$S_{int} = \int d^5x \left(\Phi\partial_5\Omega|_F + h.c. \right) \quad (11)$$

where

$$\Phi\partial_5\Omega \supset \theta^2 \left(F\partial_5 B - G\partial_5 A - \psi\partial_5\chi \right) ,$$
$$\Phi^\dagger\partial_5\Omega^\dagger \supset \bar{\theta}^2 \left(F^\dagger\partial_5 B^\dagger - G^\dagger\partial_5 A^\dagger + \bar{\chi}\partial_5\bar{\psi} \right) ,$$

one could require that $S_{int} = S - S'$, which leads to the relationship

$$\partial_5 A^\dagger \partial^5 A + \partial_5 B^\dagger \partial^5 B + P^\dagger P + Q^\dagger Q - F^\dagger F - G^\dagger G = F\partial_5 B - G\partial_5 A + F^\dagger\partial_5 B^\dagger - G^\dagger\partial_5 A^\dagger .$$

Last expression is easily satisfied, provided that one takes

$$F = P - \partial_5 B^\dagger$$
$$G = Q + \partial_5 A^\dagger . \quad (12)$$

Therefore, the action (9) can be rewritten in terms of superfields considering two Kähler potentials, one superpotential and the redefinition of the auxiliary fields, such that $S = S' + S_{int}$. Here, the fifth derivative components get explicitly exposed, which would be particularly useful in what follows.

Journal of Physics: Conference Series **378** (2012) 012004 doi:10.1088/1742-6596/378/1/012004

5. 5D Lorentz violation terms.

Here we are interested only in those modifications to the above theory which explicitly break 5D Lorenz symmetry, but leaving 4D Lorenz invariance intact. This problem was already considered in general in a previous work, where those terms were identified (see Ref. [7]). Here we will focus on presenting the deformations that we use to incorporate 5D Lorentz violation to the scheme.

Key observation for our goal is that by introducing a single parameter deformation change in the expressions given by Eqs. (12), and in the superpotential (11), such that,

$$\begin{aligned} F' &= P - a\partial_5 B^\dagger \\ G' &= Q + a\partial_5 A^\dagger \\ S'_{int} &= a \int d^5x \left(\Phi \partial_5 \Omega|_F + h.c. \right) \end{aligned}$$

then, the action we obtain from $S'' = S' + S'_{int}$ becomes

$$S'' = \int d^5x \left(\partial_\mu A^\dagger \partial^\mu A + \partial_\mu B^\dagger \partial^\mu B + i\bar{\Psi}\Gamma^\mu \partial_\mu \Psi + P^\dagger P + Q^\dagger Q \right) \tag{13}$$

$$+ \int dx^5 \left(a^2 \partial_5 A^\dagger \partial^5 A + a^2 \partial_5 B^\dagger \partial^5 B + ia\bar{\Psi}\Gamma^5 \partial_5 \Psi \right) . \tag{14}$$

This expression can be rewritten, by taking the reparameterization $a = 1 + \kappa$, as

$$\begin{aligned} S'' &= \int d^5x \left(\partial_M A^\dagger \partial^M A + \partial_M B^\dagger \partial^M B + i\bar{\Psi}\Gamma^M \partial_M \Psi + P^\dagger P + Q^\dagger Q \right) \\ &+ \int d^5x \left((2\kappa + \kappa^2) \left(\partial_5 A^\dagger \partial^5 A + \partial_5 B^\dagger \partial^5 B \right) + i\kappa \bar{\Psi}\Gamma^5 \partial_5 \Psi \right) , \end{aligned} \tag{15}$$

which is the very same action discussed in our previous work [7], which was found to contain all possible terms that violate 5D Lorentz symmetry, but remained invariant under 4D Lorentz transformations and under Supersymmetry, just as desired. We should stress that this procedure of reconstructing the last action neatly suggest that 5D Lorentz violation can be expressed as a simple deformation of the fifth derivatives. This will have immediate implications at the level of the effective 4D theory, in particular on the KK mass spectrum.

6. Parameter interpretation.

To have a more clear physical idea of the effects, and the meaning of the Lorentz violation parameter we have introduced above, let us consider the aftercome of the compactification process on the theory. As an example, considering only the Lagrangian of the scalar like field,

$$\mathcal{L} = \partial_M A^\dagger \partial^M A + \omega^2 \partial_5 A^\dagger \partial^5 A = -A^\dagger \left(\partial_M \partial^M + \omega^2 \partial_5 \partial^5 \right) A . \tag{16}$$

A solution of the equation of motion for the A field (on compact space) can be in general separated as

$$A(x^\mu, y) = \sum_n \phi_n(x^\mu) \zeta_n(x^5) ,$$

where ϕ and ζ correspond to the four and fifth wave function components, respectively. Thus, to open the mode spectrum, one can always require that

$$\partial_\mu \partial^\mu \phi_n = -m_n^2 \phi_n ,$$

which leads to

$$\zeta_n(x^5) = C_1 \cos\left(\alpha_n x^5\right) + C_2 \sin\left(\alpha_n x^5\right)$$

Journal of Physics: Conference Series **378** (2012) 012004 doi:10.1088/1742-6596/378/1/012004

where $\alpha_n = \left(m_n/\sqrt{1+\omega^2}\right)$. Assuming that fifth dimension has being compacted into a circle of radius $2\pi R$, last expression implies that eigen masses are given as

$$m_n = \left(\frac{n}{R}\right)\sqrt{1+\omega^2}\ . \tag{17}$$

That means, the Lorentz violation parameter corresponds to a shift on the mass for each Kaluza-Klein mode, except for the zero mode.

Interestingly, if one rather suppose to have a usual Lorentz invariant theory (that is where $\omega = 0$), but instead consider a compactification on a circle with a radius $2\pi R'$, it is known that KK mass would simply be

$$m_n = \frac{n}{R'}\ . \tag{18}$$

A comparison between Eq. (17) and Eq. (18) straightforwardly shows that a simple redefinition of the former compactification radius, such that

$$R = R'\sqrt{1+\omega^2},$$

will give the same spectrum as that of 5D Lorentz violating theory. Therefore, at least at this level, it is clear that a non 5D Lorentz invariant theory can be seen, at the effective level, as derived from a totally Lorentz invariant one, just by a simple redefinition on the compactification radius

7. Conclusions

In this work we have presented a superfield formulation for a five dimensional SUSY and Lorentz invariant theory, which clearly shows how 5D Lorentz violation can be incorporated in a consistent way, respecting supersymmetry, through a simple parameter deformation on the fifth derivative involved in a superpotential-like term, and, in simple field redefinition for the SUSY auxiliary fields. The construction explicitly requires two chiral superfields, which means the possibility of studying, under certain conditions, the rupture of $N = 1$ to $N = 2$ supersymmetry. This is yet a problem we are still working on. Furthermore, we presented how the Lorentz violation terms can be hidden in a redefinition of the compactification radius of the original theory, making the effective theory equivalent to that obtained from a rather Lorentz invariant formulation. Nevertheless, it is important to remark that this redefinition do depend on the particular Lorentz violation parameter, and thus, only in the case where this were universal for all involved fields the effect of 5D Lorentz breaking could be scaled away. Otherwise, asymmetries on the KK mass gaps for different fields would be expected.

References.

[1] http://cms.web.cern.ch/org/cms-papers-and-results
[2] S. P. Martin, In *Kane, G.L. (ed.): Perspectives on supersymmetry II* 1-153 [hep-ph/9709356].
[3] N. Arkani-Hamed, S. Dimopoulos and G. R. Dvali, Phys. Lett. B **429** (1998) 263 [hep-ph/9803315].
[4] A. Perez-Lorenzana, AIP Conf. Proc. **562**, 53 (2001) [arXiv:hep-ph/0008333].
[5] D. Bailin and A. Love, *Bristol, UK: IOP (1994) 322 p. (Graduate student series in physics)*
[6] A. Salam and J. A. Strathdee, Phys. Rev. D **11**, 1521 (1975).
[7] J. D. Garcia-Aguilar, A. Perez-Lorenzana and O. Pedraza-Ortega, AIP Conf. Proc. **1396**, 140 (2011).

XIII Mexican Workshop on Particles and Fields | IOP Publishing
Journal of Physics: Conference Series **378** (2012) 012005 | doi:10.1088/1742-6596/378/1/012005

ν production in Centaurus A and M87 from γ ray interactions with the gas and dust at the sources

J.C. Arteaga-Velázquez

Instituto de Física y Matemáticas, Universidad Michoacana,
Edificio C3, Cd. Universitaria, 58040 Morelia, Michoacan, Mexico

E-mail: arteaga@ifm.umich.mx

Abstract.
Centaurus A and M87 are the closest galaxies with active galactic nuclei and TeV γ-ray emission. The existence of such TeV radiation suggests the production of a neutrino flux from the photo-hadronic interactions of the γ-photons and the gas and dust content of the active galaxies. Using a simple model of Centaurus A and M87, the corresponding ν luminosities at source and their fluxes at Earth were calculated. The neutrino fluxes associated with the aforementioned process resulted to be $E^2\Phi_{\nu+\bar{\nu}} \lesssim 10^{-13}\ \mathrm{s^{-1}\ GeV\,cm^{-2}}$, more than 6 orders of magnitude below the modern upper limits from neutrino telescopes. It will be shown that, at high-energies, relevant for neutrino astronomy, these ν fluxes are not competitive with those fluxes derived from astrophysical scenarios involving cosmic ray acceleration.

PACS numbers: 95.85.Ry, 95.85.Pw, 98.54.Cm, 98.62.-g

Keywords: Centaurus A, M87, Neutrinos, TeV γ-rays

1. Introduction

Centaurus A and M87 are two giant elliptical radio galaxies, which are classified as Fanaroff-Riley I (FR I) type sources [1]. Centaurus A, also known as NGC 5128, is located at a distance of 3.4 Mpc from the Earth [2]. Meanwhile, M87 is some 16.7 Mpc away [3]. These galaxies host the closest active galactic nuclei to the Earth and are the sources of high-energy radiation in the form of TeV γ-rays [4]-[11]. Recently, the Pierre Auger observatory (PAO) detected some ultrahigh-energy cosmic ray (UHECR) events within 3.1° around the direction of Centaurus A [12]. Besides, the PAO showed a weak correlation between the arrival direction of cosmic rays with energies above 6×10^{19} eV and the distribution of AGN's [12]. Both PAO's results suggest, in this way, that Centaurus A and M87 could also be the production place of UHECR.

The presence of high-energy gamma- and cosmic rays in Centaurus A and M87 would imply the existence of an associated flux of high-energy neutrinos produced from the interactions of both types of high-energy radiation with the ambient gas

Journal of Physics: Conference Series **378** (2012) 012005 doi:10.1088/1742-6596/378/1/012005

and photons of the corresponding galaxies. For the case of UHECR, ν production mediated by pion photoproduction has been widely studied in the literature (see, for example, [13, 14, 15, 16]) and its contribution is though to be the main mechanism of ν emission. However, it remains to be proved beyond any reasonable doubt that UHECR are produced in Centaurus A and M87 to guarantee the production of neutrinos from the above process. On the other hand, although it is known that several AGN's are γ-ray emitters, the mechanisms of neutrino production induced by γ-ray collisions [17, 18, 19] in that context are scarcely studied, mainly because they are not expected to produce important ν fluxes due to their low cross sections. In any case, these secondary processes should be also contributing to the production of high-energy neutrinos in active galaxies. In the present work, one of such ν production mechanisms is studied: the photo-hadronic interaction of gamma-rays with the gas and dust content of active galaxies, in particular, for NGC5128 and M87, assuming that the gamma sources are located at the core of their AGN's ‡. Accordingly, the corresponding ν fluxes are estimated and compared with those calculated in the framework of some astrophysical models involving UHECR.

2. The neutrino induced flux

Let's be $L^*_\gamma(\epsilon_\gamma, r) = dN_\gamma/dt\, d\epsilon_\gamma$ the photon spectral luminosity of the AGN measured in the source frame for a photon energy ϵ_γ at a distance r from the core. Let's also assume that the gas and dust of the host galaxy are found in the form of protons and that their kinetic energies are negligible in comparison with the energies of the gamma-rays (supposition that can work for the hot, warm and cold gasses of the interstellar medium §).

In this way, the neutrino spectral luminosity, $L_\nu(\epsilon_\nu)$ induced by the interaction of the gamma-photons with the intergalactic material of the galaxy on their way out along the direction of the observer is given by

$$L_\nu(\epsilon_\nu)d\epsilon_\nu = \int_0^{r_f}\int_{\epsilon_{\gamma,i}}^{\epsilon_{\gamma,f}} \rho_H(r)\sigma_{\gamma P}(\epsilon_\gamma)Y^{\gamma P\to\nu}(\epsilon_\gamma,\epsilon_\nu)L^*_\gamma(\epsilon_\gamma, r)d\epsilon_\gamma dr, \qquad (1)$$

where ρ_H is the density of target protons at the distance r, $\sigma_{\gamma P}$ is the γP cross section at a photon energy ϵ_γ and $Y^{\gamma P\to\nu}(\epsilon_\gamma,\epsilon_\nu)$ is the neutrino yield, which is defined as the number of ν's with energy in the interval $d\epsilon_\nu$ around ϵ_ν produced by a photon with energy in the range $[\epsilon_\gamma, \epsilon_\gamma + d\epsilon_\gamma]$ after a collision with a proton at rest. Energy losses of the parent particles in the ν-production chain have been neglected.

Since, $L^*_\gamma(\epsilon_\gamma, r)$ is unknown at the core of the AGN, the gamma-ray luminosity at source derived from Earth observations, $L_\gamma(\epsilon_\gamma)$, will be used instead in equation (1).

‡ The exact position of the high-energy source inside active galaxies is still an open problem. One candidate is the core of AGN's. In fact, multiwave-length observations of M87 point out that the TeV gamma-ray production site could be located at the nucleus of its AGN [3].

§ For example, the temperature of the gas supply for the central engine of some AGN's has been estimated to be of the order of keV [20, 21, 22], in such a medium, the mean kinetic energy expected for protons is roughly of the same order of magnitude, which results to be small in comparison with the gamma energies considered in this work ($\epsilon_\gamma > 10^{-0.8}$ GeV).

Note that by doing so the estimated ν flux could be lower than the real one, since the actual γ-ray luminosity could be bigger than $L_\gamma(\epsilon_\gamma)$ at the interior of the AGN [23]. Now, by using the above approximation, equation (1) is reduced to the following expression:

$$L_\nu(\epsilon_\nu)d\epsilon_\nu = \Sigma_H \int_{\epsilon_{\gamma,i}}^{\epsilon_{\gamma,f}} \sigma_{\gamma P}(\epsilon_\gamma) Y^{\gamma P \to \nu}(\epsilon_\gamma, \epsilon_\nu) L_\gamma(\epsilon_\gamma) d\epsilon_\gamma, \tag{2}$$

with

$$\Sigma_H = \int_0^{r_f} \rho_H(r) dr. \tag{3}$$

Here, Σ_H is the column density of target protons along the line of sight from $r = 0$ up to r_f, which corresponds to the size of the halo of the active galaxy. Integration limits for the energy parameters are defined as follows: $(\epsilon_{\gamma,i}, \epsilon_{\gamma,f}) = (10^{-0.8}\,\mathrm{GeV}, 10^6\,\mathrm{GeV})$ and $(\epsilon_{\nu,i}, \epsilon_{\nu,f}) = 10^{-5}\,\mathrm{GeV}, 10^6\,\mathrm{GeV})$. The lower limit on the gamma-energies is set just at the threshold for pion photoproduction, $\epsilon_{\gamma,th} = m_\pi(m_\pi c^2 + 2m_p c^2)/2m_p \approx 10^{-0.8}\,\mathrm{GeV}$ [24]. Meanwhile, the upper limit is fixed at $10^6\,\mathrm{GeV}$, considering that the gamma-ray emission could be of leptonic origin and therefore that it could have a cut at the high-energy regime, somewhere around $10^2\,\mathrm{TeV}$ (see, for example, [25]).

The flux of neutrinos at Earth is estimated from the ν luminosity of the active galaxy by means of the formula

$$\Phi_{\nu+\bar{\nu}}(\epsilon'_\nu) = \frac{(1+z)^2}{4\pi D_L^2} \cdot L_{\nu+\bar{\nu}}(\epsilon_\nu) \tag{4}$$

where z is the redshift of the galaxy and $\epsilon'_\nu = \epsilon_\nu/(1+z)$ is the neutrino energy after redshift energy losses‖. D_L is the luminosity distance to the source, which is defined as

$$D_L(z) = (1+z) \int_0^z c\,dz/H(z). \tag{5}$$

Here, c is the speed of light in vacuum and $H(z)$ is the Hubble parameter at redshift z. For a cosmological scenario dominated by vacuum energy, with $\Omega_\Lambda = 1 - \Omega_m$,

$$c/H(z) = (c/H_0)[1 - \Omega_m + \Omega_m(1+z)^3]^{-1/2}. \tag{6}$$

Along this paper, the following parameters will be used: $\Omega_\Lambda = 0.74$ and $c/H_0 = 1.28 \times 10^{28}\,\mathrm{cm}$ [26].

For the final fluxes of individual ν flavors, neutrino oscillations will be taken into account [27]. That implies that after a long way to the Earth, neutrinos of different types will be present in the total flux in a $(\nu_e + \bar{\nu}_e)$:$(\nu_\mu + \bar{\nu}_\mu)$: $(\nu_\tau + \bar{\nu}_\tau) = 1:1:1$ ratio.

3. Gamma ray luminosities

Several gamma-ray observatories have measured the differential photon spectra of Centaurus A and M87 in the MeV-TeV regime. For Centaurus A, measurements of the γ-ray flux in the interval $0.1 - 1\,\mathrm{GeV}$ have been performed with the EGRET detector on

‖ Primed quantities are measured at Earth.

Journal of Physics: Conference Series **378** (2012) 012005 doi:10.1088/1742-6596/378/1/012005

board of the Compton Gamma-Ray Observatory [4, 5], while in the $0.1 - 30\,\mathrm{GeV}$ energy range, data has been collected with the *Fermi*-LAT instrument [6]. Besides in the very high-energy region ($\epsilon'_\gamma > 100\,\mathrm{GeV}$), data has been registered with the telescopes of the H.E.S.S. experiment [7]. In case of M87, its gamma-ray flux has been measured with the *Fermi*-LAT telescope in the range from 0.2 to 32 GeV and with HEGRA [8], HESS [9] and VERITAS [10] in the energy regime of $0.1 - 30\,\mathrm{TeV}$.

Using the above observational data, the γ luminosities of the sources can be easily estimated. Let's be $\Phi_\gamma(\epsilon'_\gamma) = dN_\gamma/dt'dA'd\epsilon'_\gamma$ the differential photon spectrum (in units of photons per unit time, unit area and interval of energy) detected at Earth from the active galaxy. Assuming a steady and isotropic emission, the photon spectral luminosity of the AGN is calculated by using

$$L_\gamma(\epsilon_\gamma) = 4\pi D_L^2 \cdot \frac{\Phi_\gamma(\epsilon_\gamma)}{(1+z)^2} \cdot \mathrm{e}^{\tau(\epsilon_\gamma, z)}, \tag{7}$$

where corrections for both adiabatic energy losses due to the redshift and attenuation due to interactions with the background radiation are considered. The latter is done introducing the exponential term, where the parameter $\tau(\epsilon_\gamma, z)$ appears. Here, τ represents the $\gamma\gamma$ optical depth for a photon traveling from the source with initial energy ϵ_γ. The values used for τ were taken from reference [28].

Photon spectral luminosities of Centaurus A and M87 at source, multiplied by ϵ_γ, as derived from experimental data are presented in figure 1. Along the plots, individual fits using a power-law function with a cut-off at high-energies,

$$L_\gamma^{\mathrm{fit}}(\epsilon_\gamma) = b \cdot \left[\frac{\epsilon_\gamma}{\mathrm{TeV}}\right]^a \cdot e^{-(\epsilon_\gamma/10^2\,\mathrm{TeV})}, \tag{8}$$

are presented. The cut in expression (8) is introduced to model a γ-ray flux of leptonic origin. In the above equation, a and b stand for the fit parameters. The results of the fit are shown in table 1. Formula (8) will be employed in expression (2), when calculating the ν luminosities.

By integrating $L_\gamma(\epsilon_\gamma)$, the integral luminosity, $\mathcal{L}_\gamma$, is obtained. Using equation (8) along with the parameters of table 1, a value of $\mathcal{L}_\gamma(\epsilon_\gamma > 100\mathrm{MeV}) = 7.2 \times 10^{40}\,\mathrm{erg}\cdot\mathrm{s}^{-1}$ is derived for NGC 5128 and, $\mathcal{L}_\gamma(\epsilon_\gamma > 100\,\mathrm{MeV}) = 1.1 \times 10^{42}\,\mathrm{erg}\cdot\mathrm{s}^{-1}$, for M87. The integral luminosity of M87 results to be almost one order of magnitude bigger than that of Centaurus A.

Table 1: Values of the fit parameters a and b of formula (8) for Centaurus A and M87.

AGN	$\log_{10}[b/\mathbf{s}^{-1}\cdot\mathbf{TeV}^{-1}]$	a
Centaurus A	38.53 ± 0.09	-2.44 ± 0.03
M87	40.53 ± 0.03	-2.08 ± 0.02

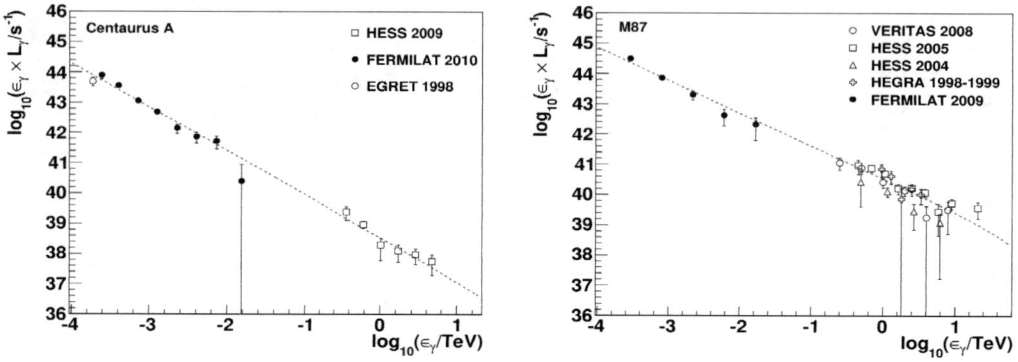

Figure 1: Photon spectral luminosities for Centaurus A (left) and M87 (right) derived from γ-ray measurements (data points) performed with different experiments (see text). The results of the fits with formula (8) are shown in each case (segmented lines).

4. Gas and dust distributions of the galaxies

4.1. Centaurus A

Due to its proximity to Earth, Centaurus A has been well studied in the literature (see for example, the reviews [29, 30]). To calculate Σ_H along the line of sight a simple model is built based on the observational data. The main parameters are presented in table 2.

The most prominent features of Centaurus A are the nuclear region, the circumnuclear disk, the dust lane, the jets and the halo [29, 30]. All of them, but the jet structures, are included in the model, since the jets do not lie along the line of sight.

In this model of Centaurus A, the VLBI radio core [31] is assumed to be the source of the γ radiation. This supposition is not yet confirmed, but it is still consistent with observations [7, 6]. A source with radius $R = 0.01\,\mathrm{pc}$ (corresponding to the upper limit on the size of the radio source established by VLBI observations [31]) and density $n_H \approx 10^6\,\mathrm{cm}^{-3}$ (as derived in [32] assuming free-free absorption in the core of Centaurus A and using data from mm and X-ray observations) will be considered in this paper.

X-ray observations of the nucleus of Centaurus A reveal the presence of two spectra with different degrees of nuclear absorption [33]. The spectra could be associated either with the emission of a single X-ray source attenuated by different components of a strong absorber [33, 34, 35] or with the combined emission from the accretion disk and the pc-scale VLBI jet [22]. Here, the first scenario will be considered. In particular, it will be assumed that the absorber is in the form of a cold cloud entirely surrounding the core [36] characterized by a column density with weighted-mean of $2.4 \times 10^{23}\,\mathrm{cm}^{-2}$ (best-fit model from [33]). For simplicity, a cloud with uniform density and radius $R = 0.1\,\mathrm{pc}$ (equivalent to the emission radius of the Fe Kα lines observed in the X-ray spectra, which seem to belong to the absorber [22]) will be employed in the calculations.

Journal of Physics: Conference Series **378** (2012) 012005 doi:10.1088/1742-6596/378/1/012005

Table 2: Extension and density distribution of the gas structures in the model of Centaurus A. Here, R represents the cylindrical radius and r, the spherical one.

Structure	R **[pc]**	n_H **[cm^{-3}]**
Core	0.01	10^6
X-ray absorber	$0.01 - 0.1$	8.6×10^5
Nuclear disk	$0.1 - 40$	3.4×10^2
Circumnuclear disk	$40 - 200$	$(5.5 \times 10^5)/r^2$
Dust lane	$(0.8 - 7) \times 10^3$	$(3.6 \times 10^{13})/r^4$
ISM and halo	35×10^3	$4 \times 10^{-2}[1 + (r/500\,\text{pc})^2]^{-0.6}$

Along the central region of Centaurus A, gas and dust seem to be organized in the form of a warped disk [2, 37], which can be modeled by means of tilted rings with different position (PA) and inclination (i) angles (see [38] and references therein). These tilted rings will be incorporated into the models of the nuclear disk [39] and the dust lane [29]. However, for the circumnuclear disk, only a single ring will be used [40].

The nuclear disk will be considered as the result of individual tilted rings distributed between $r = 0.1$ and 40 pc. The position (PA) and inclination (i) angles of the rings are taken from the SINFONI data [39]. Since observations in [39] are reported only for $r = 0.824 - 31.3$ pc, for those nuclear rings lying outside the previous range the angular parameters will be kept constant. An aspect ratio $k(R) = h(R)/R = 0.5$ is assumed for the rings, where $h(R)$ is the thickness as a function of the cylindrical radius, R.

For the circumnuclear disk a single ring with external radius $R = 200$ pc, thickness of 80 pc and internal radius $R = 40$ pc will be assumed, in accordance with the studies of $CO(1-0)$ absorption and H_2 emission against the nucleus [40]. The disk with $PA = 155°$ and $i = 70°$ is considered to have a gas mass of $\sim 10^7 M_\odot$ distributed in the form $n_H \sim r^{-2}$, where r is the spherical radius [40]. This expression will be extrapolated up to the region $r < 40$ pc in order to find the density inside the nuclear disk.

The dust lane, extended from $r = 800$ pc [29] up to 7 kpc [30], will be also described with the tilted ring model [29, 41, 42]. The position and inclination angles of the rings are taken from references [2, 38, 43] and the disk aspect ratio ($k(R) = h(R)/R = 0.1 \cdot [R/824\,\text{pc}]^{0.9}$), from [2]. A dusty disk with a mass of the order $1.3 \times 10^9 M_\odot$ [30] and a density distribution of the form r^{-4} [2, 42] will be assumed in this work.

Finally, the last structure to include is the halo, which has a radius of at least $r_f = 35$ kpc according to XMM-Newton observations [44]. It has been found from studies of NGC 5128 performed with the Chandra and XMM-Newton, that the density profile of the halo for $r \lesssim 10$ kpc can be very well represented by a beta model for distances, $n_H = n_0[1 + (r/r_0)^{-1.5\xi}]$, with $n_0 = 4 \times 10^{-2}\,\text{cm}^{-3}$, $r_0 = 500$ pc and $\xi = 0.40 \pm 0.04$ [44]. Along the paper, this expression will be used to infer the density distribution of the interstellar medium (ISM) and the material in the halo up to a distance of 35 kpc

Journal of Physics: Conference Series **378** (2012) 012005 doi:10.1088/1742-6596/378/1/012005

form the AGN's core.

By integrating numerically the density of target protons along the line of sight, the corresponding column density is found and amounts to $4.4 \times 10^{23}\,\mathrm{cm}^{-2}$.

4.2. M87

Astronomical observations show the presence of a core [45], a nuclear disk [46], a kpc jet [47, 48] and a halo [49, 50, 51] in M87. As it was the case for Centaurus A, only the jets will be left out of the present model of M87. The principal structures involved in the model, as well as their densities and sizes, are summarized in table 3.

The source of gamma rays will be located located at the core of the radio galaxy in the model. In fact, results of the 2008 multi-wavelength campaign on M87 show evidence in favor of a correlation between the radio core and the TeV emissions [3]. Following the model proposed in [11] to explain the *Fermi*-LAT MeV/GeV observations, a gamma source with a radius of $r_1 = 4.5\,\mathrm{mpc}$ is assumed. Since the density of the source is uncertain, the value $n_H = 10^6\,\mathrm{cm}^{-3}$ will be used, which is in agreement with limits presented in [52] and estimations from [53] (based on observations of UV and optical emission lines measured with instruments of the Hubble Space Telescope).

The nuclear disk of M87 will be modeled using a thin disk with a radius $r = 100\,\mathrm{pc}$ [46, 54] and thickness of 10 pc (as in reference [55]). This is just a simplified picture, since the nuclear disk has a more complex structure. Observations with the Hubble Space Telescope show spiral arms in the nuclear disk, which are apparently connected with 1 kpc-scale filaments of gas [46, 54]. The position and inclination angles of the nuclear disk are $PA = 6°$ and $i = 35°$, respectively [56]. The mean surface density of the disk in the model will be $\Sigma_H = 10^{20}\,\mathrm{cm}^{-2}$, which is within the range of values derived from the results of different studies [53, 55, 57].

The density profile of the ISM and halo of M87 will be described with the beta model of reference [58], $n_H = n_0[1 + (r/r_0)^{-1.5\xi}]$, where $n_0 = 4.2 \times 10^{-2}\,\mathrm{cm}^{-3}$, $r_0 = (7.87 \pm 1.36) \times 10^3\,\mathrm{pc}$ and $\xi = 0.436 \pm 0.008$. The parameters of the above function were obtained from fits to the X-ray data from the Einstein observatory of the hot halo of M87 [58]. The radius of the halo will be set at $r_f = 260\,\mathrm{kpc}$, distance at which the measured gas is still associated with M87 [58].

Using the model described in this section, the column density of target protons in M87 along the line of sight of the active galaxy is estimated. The result is $\Sigma_H = 2.90 \times 10^{21}\,\mathrm{cm}^{-2}$, two orders of magnitude lower than that for Centaurus A.

5. The γP hadronic cross section and the neutrino yields

The total γP hadronic cross section above $\sqrt{s} = 5\,\mathrm{GeV}$ is evaluated with the parameterized formula of references [26, 59]:

$$\sigma_{\gamma P} = \delta \cdot Z + \delta \cdot B \log^2(s/s_0) + Y(s_1/s)^{\eta_1}, \tag{9}$$

Journal of Physics: Conference Series **378** (2012) 012005 doi:10.1088/1742-6596/378/1/012005

Table 3: Extension and density distribution of main gas structures in the model of M87. Here, R represents the cylindrical radius and r, the spherical one.

Structure	R **[pc]**	n_H **[cm^{-3}]**
Core	0.0045	10^6
Nuclear disk	100	3.24
Halo	260×10^3	$4.2 \times 10^{-2}[1 + (r/7.87\,\text{kpc})^2]^{-0.654}$

where $\sqrt{s}$ is the center-of-mass energy and $\sqrt{s_1}$ is an energy scale fixed at 1 GeV. $\delta = 0.00308$, $Z = 35.45\,\text{mb}$, $B = 0.308\,\text{mb}$, $Y = 0.0320\,\text{mb}$ and $\eta_1 = 0.458$, while $\sqrt{s_0} = 5.38$ GeV [26]. These parameters were obtained by the COMPETE Collaboration by means of a global fit to current accelerator data with $\sqrt{s} \geq 5\,\text{GeV}$ [59].

For center-of-mass energies inside the interval $\sqrt{s} = 1 - 5\,\text{GeV}$ the total hadronic cross section is calculated by interpolating the $\sigma_{\gamma P}$ data presented in [26].

On the other hand, neutrino yields are obtained using the Monte Carlo program SOPHIA 2.01 [60]¶. The procedure is as follows: Gamma-photons and neutrino energy ranges are divided in several intervals (in logarithmic scale). Then collisions of gamma-photons with energy $\log_{10} \epsilon_{\gamma,i}$ against protons at rest are simulated with SOPHIA and the number of neutrinos, n_{ij}, produced with energy in the bin $\log_{10} \epsilon_{\nu,j}$ is counted. Neutrino yields, $Y_{ij}^{\gamma P \to \nu}$, are estimated averaging n_{ij} over all events induced by photons with energy $\log_{10} \epsilon_{\gamma,i}$. The ν yields were calculated only for electron and muon neutrinos, since simulations show that the tau neutrino production in γP collisions is negligible in the energy range under consideration.

6. Results

Electron and muon neutrino luminosities at source for NGC 5128 and M87 are shown in figure 2. They are restricted at low- and high-energies by the pion photoproduction energy threshold and the exponential energy cut imposed on the γ-ray luminosities, respectively. Inbetween, ν luminosities follow a simple power-law only interrupted at low-energies ($\epsilon_\nu \sim 10^{-4}\,\text{TeV}$) by a small bump that emerges as a result of the enhancement of the cross section due to the production of hadronic resonances in γP interactions.

From data presented in figure 2, integrated neutrino luminosities can be derived. They are found to be quite low. Integrating $L_{\nu+\bar{\nu}}$ for energies above 100 MeV and summing for electron a muon neutrinos, in case of Centaurus A, the following value results $\mathcal{L}_{\nu+\bar{\nu}}(\epsilon_\nu > 100\,\text{MeV}) = 1.04 \times 10^{36}\,\text{erg} \cdot \text{s}^{-1}$. Meanwhile, for M87, the value $\mathcal{L}_{\nu+\bar{\nu}}(\epsilon_\nu > 100\,\text{MeV}) = 1.28 \times 10^{35}, \text{erg} \cdot \text{s}^{-1}$ is obtained. These integrated luminosities

¶ Simulations with PYTHIA 6.4 [61] were also used to check out the results for the neutrino yields at high-energies. They gave also values of the same order of magnitude than those found with SOPHIA.

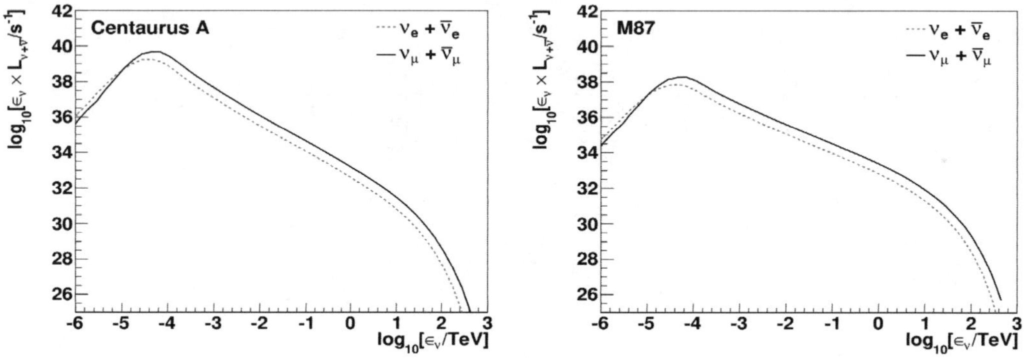

Figure 2: Electron (dotted line) and muon (solid line) neutrino luminosities expected from Centaurus A (left) and M87 (right) at source and produced by interactions of their own γ radiation with the gas and dust in the corresponding galaxies.

are from 5 to 7 orders of magnitude smaller than the ones corresponding to γ-rays.

In figure 3, the final neutrino fluxes (per flavor) arriving at Earth are presented and are compared with the 90% confidence level upper bounds obtained with the Antares[62], Amanda[63] and ICECUBE[64] detectors for the ν fluxes from Centaurus A and M87. The difference between experimental bounds and expected values are too big, the latter ones are more than 6 orders of magnitude smaller than the experimental limits. Therefore, neutrinos from Centaurus A and M87 coming from interactions of their own γ-ray flux with the respective gas and dust could escape detection at the modern neutrino observatories.

In figure 3 (left panel), the expected ν fluxes from Centaurus A from three different models involving cosmic rays are also shown for comparison. In all cases, the cosmic ray fluxes (composed by protons) are normalized using the data from the Pierre Auger observatory [12]. One model, due to Hylke et al., assumes that acceleration takes place at the jets [14]. The second model, worked out by Kachelriess et al., invokes cosmic ray acceleration close to the core (with spectral index $\alpha = 2.7$) [13]. The last model, proposed in [15], uses also the γ-ray data from the HESS detector [65] to put an upper limit on the neutrino flux assuming a pionic origin of the TeV radiation. This limit is also applied in [16] to M87 (see right panel of figure 3) by noting that luminosities from Centaurus A and M87 are similar at TeV energies. It is clear from the above figure that at high-energies, relevant for neutrino astronomy ($\epsilon_\nu \gtrsim 1\,\mathrm{TeV}$), neutrinos from the cosmic ray channel dominate over the modest contribution from γP interactions studied in this paper.

7. Conclusions

The GeV-TeV gamma-ray flux detected from Centaurus A and M87 should be producing a feeble ν flux arising from its photo-hadronic interactions with the material of the

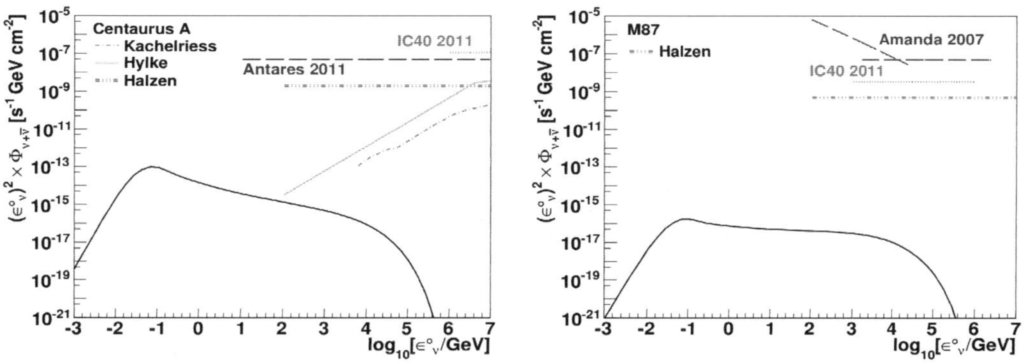

Figure 3: Neutrino and antineutrino fluxes expected from Centaurus A (left) and M87 (right) at Earth for each neutrino type. The flux is produced by interactions of their own γ radiation with the gas and dust at the sources. Neutrino oscillations are taken into account. Individual 90% C.L upper limits on the neutrino fluxes established with the Amanda [63], ICECUBE [64] and Antares [62] detectors for each galaxy are also shown. For M87, two different ν limits from Amanda are presented (segmented lines). They were estimated assuming different spectral indices for the differential ν flux: $\gamma = -2$ and -3 (horizontal and inclined lines, respectively) [63]. Finally, predictions for Centaurus A (Kachelriess [13], Hylke [14] and Halzen [15, 16]) and M87 (Halzen [15, 16]) based on different models involving cosmic rays are plotted.

sources. Estimated ν luminosities are from 10^{-5} to 10^{-7} times the corresponding luminosities for γ-rays, in case of Centaurus A and M87, respectively. For the simple models here analyzed, derived ν fluxes are found to be $E^2\Phi_{\nu+\bar{\nu}} \lesssim 10^{-13}$ s^{-1} GeV cm^{-2}, i.e. more than 6 orders of magnitude below the upper experimental limits. That result implies that these fluxes could escape experimental detection at modern neutrino telescopes like ANTARES and ICECUBE. Estimations are still dependent on the real position of the γ ray source at the active galaxy and on the shape and magnitude of the γ-ray spectrum at the interior of the source, which are at the moment unknown.

Acknowledgments

The author thanks R. Engel for his suggestions on the Monte-Carlo programs for photo-hadronic interactions. This work has been partially supported by the Coordinación de la Investigación Científica de la Universidad Michoacana.

References

[1] B.L. Fanaroff and J.M. Riley, MNRAS 167 (1974) 31P.
[2] A. Quillen, ApJ 645 (2006) 1092.
[3] The VERITAS Collaboration, the VLBA 43 GHz M87 Monitoring Team, the H.E.S.S. Collaboration and the MAGIC Collaboration, Science 325 (2009) 444, arXiv: 0908.0511.

[4] H. Steinle et al., A&A 33 (1998) 97.
[5] P. Sreekumar et al., Astrop. Phys. 11 (1999) 221.
[6] A. A. Abdo at al., ApJ 719 (2010) 1433, arXiv: 1006.5463.
[7] F. Aharonian et al., ApJ 695 (2009) L40, arXiv: 0903.1582.
[8] Aharonian F. et al., ApJ 614 (2004) 897.
[9] Aharonian F. et al., A&A 457 (2006) 899.
[10] V.A. Acciari et al., ApJ 679 (2008) 397, arXiv: 0802.1951.
[11] A. A. Abdo et al., ApJ 707 (2009) 55, arXiv: 0910.3565.
[12] J. Abraham et al. (PAO Collaboration), Science 318 (2007) 939, arXiv: 0711.2256.
[13] M. Kachelriess et al., New Journal of Physics 11 (2009) 065017.
[14] B. Hylke et al., Phys. Rev., D78 (2008) 083009.
[15] F. Halzen and A. O'Murchadha, arXiv: 0802.0887v2.
[16] F. Halzen, arXiv: 1111.1131v1.
[17] S. Razzaque et al., Phys. Rev. D 73 (2006) 103005.
[18] C. D. Dermer and G. Menon, High Energy Radiation from Black Holes: Gamma Rays, Cosmic Rays, and Neutrinos, Princeton University Press, 2009.
[19] H.Y. Chiu, R.C. Stabler, Phys. Rev. 122 (1961) 1317. V.I. Ritus, JETP 14 (1962) 915. G. Beaudet et al., Phys. Rev. 154 (1966) 1445.
[20] S. W. Allen et al., Mon. Not. R. Astron. Soc. 372 (2006) 21.
[21] D. A. Evans et al., ApJ 642 (2006) 96, arXiv:0512600.
[22] D. A. Evans et al., ApJ 612 (2004) 786.
[23] M. Kachelriess et al., PASA 27 (2010) 482-489.
[24] R.J. Protheroe and P. A. Johnson, Astrop. Phys. 4 (1996) 253, arXiv: 9506119.
[25] M. M. Reynoso, M. C. Medina, G. E. Romero, A&A 531 A30 (2011).
[26] K. Nakamura et al. (Particle Data Group), Review of particle physics, J. Phys. G 37 (2010) 075021.
[27] K. Nakamura and S.T Petcov, Neutrino Mass, mixing and oscillations, in: reference [26].
[28] A. Franceschini et al., A&A 487, 837 (2008), arXiv: 0805.1841.
[29] R. Morganti, PASA 27 (2010) 463, arXiv:1003.5568.
[30] F.P. Israel, Astron. Astrophys. Rev. 8 (1998) 237.
[31] K.I. Kellermann et al., ApJ 475 (1997) L93.
[32] Z. Abraham et al., Mon. Not. R. Astron. Soc. 375 (2007) 171.
[33] A. Markowitz et al., ApJ 665 (2007) 209.
[34] T.J. Turner et al., ApJ 475 (1997) 118.
[35] P.R. Wozniak, et al., MNRAS 299 (1998) 449.
[36] S. Miyazaki et al., Pbl. Astron. Soc. Japan 48 (1996) 801.
[37] A.D. Tubbs, ApJ 241 (1980) 969.
[38] A.C. Quillen, PASA 27 (2010) 396, arXiv:0912.0632.
[39] N. Neumayer et al., ApJ 671 (2007) 1329.
[40] F.P. Israel et al., A&A 227 (1990) 342.
[41] R.A. Nicholson at al., ApJ 387 (1992) 503.
[42] A.C. Quillen et al., ApJ 391 (1992) 121.
[43] D. Espada et al., ApJ 695 (2009) 116.
[44] R. P. Kraft et al., ApJ 698 (2009) 2036.
[45] M.H. Cohen et al., ApJ 158 (1969) L83.
[46] H.C. Ford et al., ApJ 435 (1994) L27.
[47] H.D. Curtis, Pub. Lick. Obs. 13 (1918) 31.
[48] J. Biretta in: D. Burgarella et al. (Eds.), Astrophysical Jets, Cambridge University Press, (1994) 263. J.A. Biretta & W. Junor, Proc. Natl. Acad. Sci. 92 (1995) 11364.
[49] R. Malina et al., ApJ 209 (1976) 678.
[50] E.J. Schreier et al., ApJ 261 (1982) 42.
[51] C.L. Sarazin, X-ray Emission from Clusters of Galaxies, Cambridge Astrophysics Series, Cambridge

University Press, 1988.
[52] A. Neronov & F.A. Aharonian, ApJ 671 (2007) 85.
[53] B.M. Sabra et al., ApJ 584 (2003) 164.
[54] H. Ford & Z. Tsvetanov, Lecture Notes in Physics 530 (1999) 278.
[55] J.C. Tan et al., ApJ 689 (2008) 775.
[56] Z.I. Tsvetanov et al., Proceedings of the M87 Workshop, Ringberg castle, Germany (1998) 15 arXiv: 9803178.
[57] M.A. Dopita et al., ApJ 490 (1997) 202.
[58] D. Fabricant & P. Gorenstein, ApJ 267 (1983) 535.
[59] J.R. Crudell et al. (COMPETE Collab.), Phys. Rev. D65 (2002) 074024.
[60] A. Mücke, R. Engel, J.P. Rachen, R.J. Protheroe, and T. Stanev, Comp. Phys. Commun. 124 (2000) 290, arXiv: 9903478.
[61] T. Sjöstrand et al., JHEP05(2006)026.
[62] S. Adrián-Martínez et al., ApJL 743 (2011) L14, arXiv:1108.0292v1.
[63] M. Ackermann, Astrophys. Space. Sci. 309 (2007) 421.
[64] R. Abbasi et al., ApJ 732 (2011) 18.
[65] F. Aharonian et al., A&A. 441 (2005) 465.

Journal of Physics: Conference Series **378** (2012) 012006 doi:10.1088/1742-6596/378/1/012006

Robinson-Bertotti solution from flux compactification

O. Loaiza-Brito

Departamento de Física, DCI, Campus León, Universidad de Guanajuato,C.P. 37150, Guanajuato, México

E-mail: oloaiza@fisica.ugto.mx

L. Vazquez-Mercado

Departamento de Física, DCI, Campus León, Universidad de Guanajuato,C.P. 37150, Guanajuato, México

E-mail: livazquezm@fisica.ugto.mx

Abstract. We present a 10 dimensional supergravity compactification threaded with a simple flux configuration of Neveu-Schwarz-Neveu-Schwarz (NS-NS) and Ramond-Ramond (RR) 3-form fluxes to obtain the near-horizon geometry of an extremal black hole, namely, the Robinson-Bertotti solution. The constraints and conditions that fluxes must fulfill to attain an $AdS_2 \times S^2$ symmetry in 4d space-time, were found by Einsteins equations, Bianchi identities and by the integrability conditions of 10d chiral spinors.

1. Introduction

The $AdS_2 \times S^2$ geometry of extremal black holes is a solution that has been studied under different approaches [1]. In the context of $\mathcal{N} = 2, d = 4$ ungauged supergravity, this geometry emerges from the compactification of type IIB superstring theory on a Calabi-Yau (CY) threefold. The presence of branes wrapping internal cycles of the CY are considered. The effective theory in 4d space-time is interpreted as an extremal supersymmetric black hole. The $AdS_2 \times S^2$ geometry appears when the near-horizon limit is taken.

We are interested in $AdS_2 \times S^2$ solutions of supergravity theories within a flux compactification scenario. This construction is studied in detail in a previous work by the authors [2]. The aim of it is to study the conditions under which, a compactification in presence of fluxes, yields a 4d space-time of the type $AdS_2 \times S^2$. The solution we present, considers a constant dilaton and hence, a constant warping factor. It consists of type IIB string theory compactification on a $SU(3)$-structure manifold, Ricci-flat and containing torsion terms. The compactification is performed in the presence of RR and NS-NS 3-form fluxes. The flux configuration we use, satisfies the Einstein equations, Bianchi identities and the integrability conditions of the corresponding ten dimensional spinors. Besides, it yields a null contribution to the 4d scalar curvature. Therefore, we provide a new way to obtain the $AdS_2 \times S^2$ geometry as a near-horizon solution. We might mention that a well known solution [3], which includes 5-form RR fluxes is also considered as a solution in our work.

The content of the manuscript is as follows. In section 2, we study the contribution of some configuration of fluxes, within the context of supergravity compactification, which yield a Ricci-

Journal of Physics: Conference Series **378** (2012) 012006 doi:10.1088/1742-6596/378/1/012006

flat space-time formed by the product $AdS_2 \times S^2$. We see how the presence of fluxes modifies (and enriches) the scalar curvatures. So we can have positive, null or negative curvature. While, in the absence of fluxes, the only allowed curvatures are null or negative [4]. The analysis is performed, in spirit of [4], by computing the contribution of the energy-momentum tensor of the n-form fluxes to the scalar curvature. Then, we study the compatibility of such fluxes with Einstein equations and Biachi identities. We find a simple flux configuration, consisting of 3-form RR and NS-NS fluxes, which fulfills these constraints. Section 3 contains the scalar curvatures of spaces AdS_2 and S^2, computed as function of flux numbers from the integrability conditions on the 10d spinors on type IIB supergravity. For this purpose, we choose the simplest flux configuration with the 3-forms from section 2. We look for solutions preserving $\mathcal{N} = 2$ supergravity in 4d. This condition in addition to the assumption of maximally symmetric 2d spaces, imposes the space-time $AdS_2 \times S^2$ to be Ricci-flat. At the end of the manuscript we present our conclusions.

2. Flux supergravity compactification

We begin our study by defining what kind of fluxes are compatible with a compactification of the type IIB string theory on a 6-dimensional manifold into a space formed by the product $AdS_2 \times S^2$. The 6d manifold possesses an $SU(3)$-structure. The symmetry associated to this compactification setup is $SO(1,1) \times SO(2) \times SU(3)$.

Let us start by considering the most generic 10d metric which is invariant under Poincaré transformations in 4 dimensions [5, 6]

$$ds^2 = e^{2A(y)}(\widetilde{g}_{ij}dx^i dx^j + \widetilde{g}_{ab}dx^a dx^b) + h_{mn}dy^m dy^n. \tag{1}$$

We have assigned indices i,j to the coordinates of AdS_2 space, with $i = 0, j = 1$ and indices a,b to the coordinates of S^2 space, with $a = 2, b = 3$. We also have the coordinates of the internal space, whose indices are the letters m,n, with $m, n = 4, ..., 9$. Finally, for a generic $4d$ coordinate, we use Greek letters $\mu, \nu = 0, ..., 4$.

What follows next is to calculate the curvature of the maximally symmetric 2d spaces. With this purpose, by taking the Einstein trace-reverse equations, the Ricci scalar $R(\widetilde{g}_{ij}) \equiv \widetilde{R}_1$ for AdS_2 satisfies

$$\widetilde{R}_1 + e^{2A}(-T^i_i + \frac{1}{4}T^L_L) = e^{-2A}\nabla^2 e^{2A}, \tag{2}$$

where T_{MN} is the energy-momentum tensor in 10d. Integration over the internal manifold fixes the right-hand side of this equation to vanish. Then, a relation between curvature and the field content contribution from the second term of left-hand side, is established. On the other hand, we have the expression of the energy-momentum tensor for a general n-form [4, 7]

$$\mathcal{T}_1 \equiv -T^i_i + \frac{1}{4}T^L_L = -\mathcal{F}_{iM_1...M_{n-1}}\mathcal{F}^{iM_1...M_{n-1}} + \frac{n-1}{4n}\mathcal{F}^2. \tag{3}$$

A similar result is obtained for $\tilde{R}_2$ and $\mathcal{T}_2$ for S^2. Now, it is necessary to consider specific flux configurations which preserve the $SO(1,1) \times SO(2)$ symmetry in 4d. At this point we can have different configurations of n-forms fulfilling this requirement. We can consider *internal fluxes*, with all their indices filling the internal space, or fluxes with indices in one of the 2d spaces, i.e., fluxes of the general form $\mathcal{F}_n = \omega_2 \wedge f_{n-2}$. The contribution of the later to the curvature can be positive or negative, as we shall see.

2.1. Ricci flat space

We study the contribution to 4d scalar curvature of different fluxes by following [4, 7]. We start with those which have 2 indices in one of the 2d subspaces, then, we continue our analysis with internal fluxes and finally, we take into account fluxes with four indices in the external 4d space. Remember that we are considering fluxes compatible with the symmetry $SO(1,1) \times SO(2)$.

Case I

Fluxes with general form $\mathcal{F}_n = \omega_2 \wedge f_{n-2}$, with 2 indices filling one of the 2d spaces, namely the AdS_2 space. Let ω_2, which is the volume 2-form of the space-time AdS_2, be

$$\omega_2 = \frac{1}{2}\omega_{ij} dx^i \wedge dx^j \tag{4}$$

The first term of the right-hand side of Eq.(3) yields

$$\mathcal{F}_{jL_1 \ldots L_{n-1}} \mathcal{F}^{jL_1 \ldots L_{n-1}} = \frac{2}{n} F^2 \tag{5}$$

From the above expressions, (4) and (5), the contribution to $\widetilde{R}_1$ for $\mathcal{T}_1$ in terms of a general form flux $\mathcal{F}_n$ is

$$\mathcal{T}_1 = \frac{n-9}{4n} F^2. \tag{6}$$

Since $F^2 \leq 0$, it follows that all n-forms in type IIB theory contribute positively to $\mathcal{T}_1$ and negatively to $\widetilde{R}_1$, except for 9-forms which contribution make the curvature vanishes.

Case II

Fluxes whose general form is $\mathcal{G}_n = \tilde{\omega}_2 \wedge g_{n-2}$, with

$$\tilde{\omega}_2 = \frac{1}{2}\omega_{ab} dx^a \wedge dx^b, \tag{7}$$

where the contribution to $\widetilde{R}_{(2)}$ by $\mathcal{T}_{(2)}$ is given by

$$\mathcal{T}_2 = \frac{n-9}{4n} G_n^2. \tag{8}$$

The contribution to $\mathcal{T}_2$ by the field strengths n-forms in type IIB theory is negative. Hence, the contribution to the S^2 curvature, turns to be positive. For values of $n = 9$, curvature vanishes.

Case III

Finally, we consider internal fluxes and fluxes of the form $\mathcal{G}_n = Vol_4 \wedge h_{n-4}$. The contribution of internal fluxes to $\mathcal{T}$ is,

$$\mathcal{T} = \frac{n-1}{2n} F_n^2 \tag{9}$$

From this expression we see that 1-forms make curvature vanishes while all n-forms of type IIB theory yield a negative curvature. In addition, the contribution to curvature from fluxes $\mathcal{G}$ is given by

$$\mathcal{T} = -\frac{9-n}{2n} \mathcal{G}_n^2. \tag{10}$$

The above cases illustrate the possibility of having positive, negative or null scalar curvatures. Since our goal is to obtain a Ricci-flat 4d space-time, based on

$$\widetilde{R}_4 = \widetilde{R}_1 + \widetilde{R}_2 = -e^{2A}\mathcal{T} = -e^{2A}(\mathcal{T}_1 + \mathcal{T}_2)\,, \tag{11}$$

Journal of Physics: Conference Series **378** (2012) 012006 doi:10.1088/1742-6596/378/1/012006

the natural step is to choose fluxes such that, their contribution to $\mathcal{T}$ leads to a vanishing curvature, i.e. fluxes satisfying $\mathcal{T}_1 + \mathcal{T}_2 = 0$. This constraint is not just satisfied by RR 5-form fluxes $F_5 = \omega_2 \wedge F_3$ [3], but by a wider number of flux configurations. An important remark is the fact that, the fluxes $\mathcal{F}_n$ and $\mathcal{G}_n$, can be chosen such that no tadpole is generated in the internal space X_6, avoiding in this way, the presence of orientifolds. This fact allows to keep two supersymmetries in four dimensions.

2.2. Examples

Some examples will illustrate the obtention of a null curvature by the incorporation of specific fluxes configurations. We will consider the presence of fluxes from type IIB supergravity theory. Let us take 5-form fluxes which general formula is $f_2 \wedge F_3$. The coefficients are F_{ijmnp} for the AdS_2 and F_{abmnp} for S^2. Recall that we are labeling 2d coordinates x on AdS_2 with letters $\{i, j, k, l\}$ and on S^2 with letters $\{a, b, c, d\}$, while internal coordinate are denoted by indices $\{m, n, p, ...\}$. The corresponding curvatures are

$$\widetilde{R}_1 = -\frac{e^{2A(y)}}{5}|F_5|^2, \qquad \widetilde{R}_2 = \frac{e^{2A(y)}}{5}|F_5|^2. \tag{12}$$

Hence, the 4d scalar curvature vanishes. On the other hand, we can choose a more general flux configuration, let H_n be an n-form with two legs on AdS_2 (and the rest of them on the internal space X_6) and an internal flux $\mathcal{F}_m$. Then, the corresponding 4d scalar curvature vanishes if

$$|H_n^2| = \frac{2n}{9-n}\frac{m-1}{m}\mathcal{F}_m^2. \tag{13}$$

Now, we shall focus on a simpler case in which NS-NS 3-form fluxes are turned on. Let NS-NS flux H_3 and RR flux F_3 be defined by

$$\begin{aligned} H_3 &= \left(N dx^0 \wedge dx^1 + M dx^2 \wedge dx^3\right) \wedge d\alpha, \\ F_3 &= \left(P dx^0 \wedge dx^1 + Q dx^2 \wedge dx^3\right) \wedge d\alpha, \end{aligned} \tag{14}$$

with α being a function of internal coordinates. The next step is to obtain the scalar curvatures, which according to (6) are,

$$\begin{aligned} \widetilde{R}_1 &= -2e^{2A(y)}(N^2 + P^2)(\nabla\alpha)^2, \\ \widetilde{R}_2 &= 2e^{2A(y)}(M^2 + Q^2)(\nabla\alpha)^2. \end{aligned} \tag{15}$$

From these expressions we can find that total 4d curvature vanishes if $M^2 + Q^2 = N^2 + P^2$. Under these conditions the 4d space-time with an $AdS_2 \times S^2$ geometry becomes the near-horizon limit of an extremal Reisner-Nordström black-hole.

Some remarks are important. First, because of the $SU(3)$-structure of internal manifold we have assume, it is not possible to expand a flux in terms of internal vector components. For this reason, we take a smeared internal index of the 3-form fluxes as the most general case. Second, we have $H_3 \wedge F_3 = 0$, then our choice is consistent with the absence of RR five-form fluxes. This fact does not force us to have a CY as internal space, because, as we shall further see, some torsion terms arise. In the next section we will study and verify the compatibility of our flux configuration with 10d Einstein equations and Bianchi identities.

2.3. Einstein equations

In this section we will compute the 2d Ricci tensors for each 2d subspace following [7]. First, let us consider the bosonic part of the type IIB superstring action,

$$S = \frac{1}{2\kappa_{10}^2} \int d^{10}x \, \sqrt{-G} e^{-2\phi} \left(R - \frac{1}{12 \, Im \, \tau} |G_3|^2 \right), \tag{16}$$

where G_{MN} is the 10d metric and $G_3 = F_3 - \tau H_3$. This action is written in the string frame. It follows now, that the 10d component of the Ricci tensor is given by,

$$R_{MN} = -\frac{1}{Im \, \tau} \left(\frac{G_3^2}{48} G_{MN} - \frac{1}{4} G_{MQR} \bar{G}_N{}^{QR} \right). \tag{17}$$

Recall that we want to preserve the symmetries of the compactification setup, and it is necessary to consider the adequate metric. The most general one we choose is

$$ds^2 = e^{2A(y)} \tilde{g}_{ij} dx^i dx^j + e^{2B(y)} \tilde{g}_{ab} dx^a dx^b + e^{-2A(y)} \tilde{h}_{mn} dy^m dy^n. \tag{18}$$

Then, the corresponding Ricci tensors are

$$R_{ij} = \tilde{R}_{ij} - e^{4A} \left(\tilde{\nabla}^2 A + 2\tilde{\nabla} A \cdot \tilde{\nabla} B - 2(\tilde{\nabla} A)^2 \right) \tilde{g}_{ij}, \tag{19}$$

$$R_{ab} = \tilde{R}_{ab} - e^{2(A+B)} \left(\tilde{\nabla}^2 B - 2\tilde{\nabla} A \cdot \tilde{\nabla} B + 2(\tilde{\nabla} B)^2 \right) \tilde{g}_{ab}, \tag{20}$$

where $\tilde{\nabla}$ is the covariant derivative with respect to the non-warped metric $\tilde{h}_{mn}$. From these expressions and the corresponding components of Eq. (17), the Ricci tensors of subspaces AdS_2 and S^2 are, respectively

$$\begin{aligned} \tilde{R}_{ij} &= e^{4A} \left(\tilde{\nabla}^2 A + 2\tilde{\nabla} A \cdot \tilde{\nabla} B - 2(\tilde{\nabla} A)^2 \right) \tilde{g}_{ij} - \frac{1}{Im \, \tau} \left(\frac{G_3^2}{48} G_{ij} - \frac{1}{4} G_{iQR} \bar{G}_j{}^{QR} \right), \\ \tilde{R}_{ab} &= e^{2(A+B)} \left(\tilde{\nabla}^2 B - 2\tilde{\nabla} A \cdot \tilde{\nabla} B + 2(\tilde{\nabla} B)^2 \right) \tilde{g}_{ab} \\ & \quad - \frac{1}{Im \, \tau} \left(\frac{G_3^2}{48} G_{ab} - \frac{1}{4} G_{aQR} \bar{G}_b{}^{QR} \right). \end{aligned} \tag{21}$$

These expressions are written in terms of the nonwarped metric. The next step is to compute the Bianchi identities in order to look for possible constraints that fluxes must fulfill.

2.4. Biachi identities

Bianchi identities are trivially satisfied because we are working with 3-form fluxes having a smeared index on the internal space. Hence, it is necessary to look for more astringent conditions provided by the dual Bianchi identities, $d * F_3 = d * H_3 = 0$. From these we get

$$\begin{aligned} (Q + \tau M) \left(-2\partial_m (2A + B) \tilde{\partial}^m \alpha + \tilde{\partial}^2 \alpha \right) &= 0 \\ (P + \tau N) \left(+2\partial_m (-4A + B) \tilde{\partial}^m \alpha + \tilde{\partial}^2 \alpha \right) &= 0. \end{aligned} \tag{22}$$

Journal of Physics: Conference Series **378** (2012) 012006 doi:10.1088/1742-6596/378/1/012006

It follows that $A = B$ for nonvanishing $Q + \tau M$ and $P + \tau N$. On the other hand, it seems that even though M_2 and $\tilde{M}_2$ are independent, they share the same warping factor in a background threaded with fluxes (14). Then, for $A = B$ both expressions reduce to

$$\tilde{\partial}^2 \alpha = 6e^{-2A}\partial_m A \partial^m \alpha = \frac{3}{2} e^{-6A}(\partial_m e^{4A})(\partial^m \alpha). \tag{23}$$

Comparing (21) with the Bianchi identities, we have, for the AdS_2 space

$$\begin{aligned}\tilde{\nabla}^2(e^{4A} - \alpha) &= 2\tilde{R}_1 + \frac{1}{2} e^{-6A}(\partial_m e^{4A} \partial^m e^{4A}) - \frac{3}{2}\partial_m e^{4A}\partial^m \alpha \\ &+ \frac{1}{4Im\ \tau}\left(-P^2 - \tau\bar{\tau}N^2 + 2(Im\ \tau)PN\right) e^{-2A}\partial_m \alpha \partial^m \alpha.\end{aligned} \tag{24}$$

The configuration (14) has been taken to compute the flux components. Similarly, by making the comparison of R_{ab} with Bianchi identities, the result is

$$\begin{aligned}\tilde{\nabla}^2(e^{4A} - \alpha) = 2\tilde{R}_2 + \frac{1}{2} e^{-6A}(\partial_m e^{4A} \partial^m e^{4A}) - \frac{3}{2}\partial_m e^{4A}\partial^m \alpha \\ + \frac{1}{4Im\ \tau}\left(Q^2 + \tau\bar{\tau}M^2 + 2(Im\ \tau)QM\right) e^{-2A}\partial_m \alpha \partial^m \alpha,\end{aligned} \tag{25}$$

for the S^2 space. Since $\widetilde{R}_4 = \widetilde{R}_{(1)} + \widetilde{R}_{(2)}$ must vanish for $AdS_2 \times S^2$, we add (24) and (25). The result is

$$\begin{aligned}\tilde{\nabla}^2(e^{4A} - \alpha) &= e^{-6A}(\partial_m e^{4A} \partial^m e^{4A}) - 3\partial_m e^{4A}\partial^m \alpha + \frac{1}{4Im\ \tau}\left((Q^2 - P^2) - \tau\bar{\tau}(M^2 - N^2)\right. \\ &\left. + 2(Im\ \tau)(QM + PN)\right) e^{-2A}\partial_m \alpha \partial^m \alpha.\end{aligned} \tag{26}$$

The left side of (26) vanishes if we integrate over a compact manifold. Consequently, the values for the flux numbers and the warping factor are constrained. There are several possibilities. Among all these, we can choose a nonconstant warping factor but the solution would be more complex and it is beyond the scope of this study. Then, we will concentrate on the simplest solution involving a constant warping factor. We are looking for the minimal conditions under which we can construct the Robinson-Bertotti solution in 4d. There are some cases to consider (all of them satisfying the general condition for $\hat{\mathcal{T}} = 0$, $M^2 + Q^2 = N^2 + P^2$) in which Eq. (26) vanishes:

(i) $(M, N, P, Q) \neq 0$ and $M = -N, Q = P$.
(ii) $P = 0$ and $\frac{M}{Q} = \frac{\tau\bar{\tau}-1}{2\ Im\ \tau}$
(iii) $Q = 0$ and $\frac{N}{P} = \frac{\tau\bar{\tau}+1}{2\ Im\ \tau}$
(iv) $N = 0$ and $\frac{Q}{M} = \frac{\tau\bar{\tau}-1}{2\ Im\ \tau}$
(v) $M = 0$ and $\frac{P}{N} = \frac{\tau\bar{\tau}+1}{2\ Im\ \tau}$

For the above expressions, $H_3 \wedge F_3$ is satisfied as expected for a configuration without a 5-form RR flux.

We proceed to compute the curvature scalars from the integrability conditions on the 2d components of 10d spinors. We shall concentrate on a configuration of RR and NS-NS 3-form fluxes given by (14) satisfying one of these constraints, it turns out that conditions (2) to (4) are easier to be considered in this procedure.

3. Near-Horizon geometry from 3-form fluxes

Our next task, is to compute the curvature scalars of the 2d spaces from the integrability conditions on the 2d components of 10d spinors within a type IIB supergravity flux compactification scenario. We are working with flux configuration (14). The form of the 10d space-time is $AdS_2 \times S^2 \times X^6$, with an $SU(3)$-structure for X^6 and arbitrary curvatures for AdS_2 and S^2. In principle, we are not assuming any relation among the flux coefficients but we look for solutions preserving $\mathcal{N} = 2$ 4d supergravity. Under this condition, the AdS_2 and S^2 curvatures are restricted to be equal in magnitude. Therefore the 4d space-time will be Ricci-flat.

When supersymmetry is preserved, the gravitino variation is

$$\delta\Psi_M = \nabla_M\epsilon - \frac{1}{4}\,\not{H}_M\sigma^3\epsilon + \frac{1}{16}e^{\phi}\,\not{F}_3\Gamma_M\sigma^1\epsilon = 0 \tag{27}$$

where σ are Pauli matrices, and $\epsilon = \begin{pmatrix} \epsilon^1 \\ \epsilon^2 \end{pmatrix}$ are the chiral spinors. Since the ten-dimensional chiral spinors $\epsilon^{1,2}$ are correlated by the vanishing of the gravitino variation (27), we shall obtain an independent equation for each one by using the vanishing supersymmetric variation of the fermi fields.

We start by taking the 4d component of $\delta\Psi^1$ in a background threaded with the flux content (14). Using the fact that

$$[\Gamma_i, \Gamma_{jkm}] = \{\Gamma_i, \Gamma_{abm}\} = 0, \tag{28}$$

the M_2-component of gravitino variation is

$$\nabla_i\epsilon - \frac{1}{4}\not{H}_i\sigma^3\epsilon + \frac{1}{16}e^{\phi}\Gamma_i(\not{F}_3^{(1)} - \not{F}_3^{(2)})\sigma^1\epsilon = 0, \tag{29}$$

where $\not{F}_3^{(1)} = F^{01m}\Gamma_{01m}$ and $\not{F}_3^{(2)} = F^{23m}\Gamma_{23m}$. The strategy we will follow is commuting Γ_i with $\not{F}_3$ in the last term of the above equation by using the dilatino variation for a constant dilaton given by

$$\delta\lambda = -\frac{1}{2}\not{H}_3\sigma^3\epsilon - \frac{1}{4}e^{\phi}\not{F}_3\sigma^1\epsilon = 0, \tag{30}$$

this will allow to decouple the ten-dimensional spinors $\epsilon^{1,2}$. However, this seems difficult to perform unless $\not{F}^{(1)}{}_3$ or $\not{F}^{(2)}{}_3$ vanishes, but according to section 3, this is an available condition on the fluxes. Then, we choose $P = 0$ corresponding to $F^{01m} = 0$. The i-component of the gravitino variation yields

$$\left(\nabla_i - \frac{1}{4}\not{H}_i + \frac{1}{8}\Gamma_i\not{H}_3\right)\epsilon^1 = 0, \tag{31}$$

a similar expression for the a-component is computed.

From now on we are going to concentrate on the equations for ϵ^1, while the expressions for ϵ^2 are similar. It is important to notice that both spinors are decoupled just in the presence of non-trivial fluxes H_3 and F_3.

Then, we can express (31) as

$$\nabla_i^{(T)}\epsilon^1 = (\nabla_i + \kappa_i)\epsilon^1 = 0, \tag{32}$$

with $\kappa_i = -\frac{1}{4}\not{H}_i + \frac{1}{8}\Gamma_i\not{H}_3$.

The next step is to compute the corresponding components of the connection by using the metric $ds^2 = e^{2A(y)}\widetilde{g}_{\mu\nu}dx^\mu dx^\nu + h_{mn}dy^m dy^n$. The i-component of the covariant derivative of the spinor ϵ^1 is

$$\nabla_i^T\epsilon^1 = \left(\widetilde{\nabla}_i - \frac{1}{2}\gamma_i\widetilde{\gamma}\otimes\widetilde{\sigma}\otimes\not{\partial}A + \kappa_i\right)\epsilon^1 = 0, \tag{33}$$

we have denoted the covariant derivative with respect to the metric g_{ij} as $\widetilde{\nabla}$ and $\nabla_i = \widetilde{\nabla}_i - \frac{1}{2}\gamma_i\widetilde{\gamma}\otimes\widetilde{\sigma}\otimes\not{\partial}A$. On the other hand, we have an expression for the Riemann tensor in terms of the connection,

$$[\widetilde{\nabla}_i, \widetilde{\nabla}_j]\epsilon^1 = \frac{1}{4}\widetilde{R}_{ij}{}^{kl}\gamma_{kl}\epsilon^1, \tag{34}$$

and together with the variation of gravitino, we will have,

$$[\widetilde{\nabla}_i, \widetilde{\nabla}_j]\epsilon^1 = -(\partial_n A\partial^n A)\gamma_{ij} + \frac{1}{2}[\gamma_i\widetilde{\gamma}\otimes\widetilde{\sigma}\otimes\not{\partial}A, \kappa_j] + \frac{1}{2}[\kappa_i, \gamma_j\widetilde{\gamma}\otimes\widetilde{\sigma}\otimes\not{\partial}A] + [\kappa_i, \kappa_j]\epsilon^1. \tag{35}$$

If we take the constant warping factor as a solution, the last expression reduces to $[\kappa_i, \kappa_j]\epsilon^1$. Therefore, the new expression for the Riemann tensor is

$$\frac{1}{4}\widetilde{R}_{ij}{}^{kl}\gamma_{kl} - [\kappa_i, \kappa_j] = 0. \tag{36}$$

On this point it is worthy to emphasize some facts. The first one is to notice that in the fluxless case, the Riemann tensor and the contorsion would be zero, leading us to Minkowskian space-time. As a second fact, let us consider the flux case, then, even for a constant warping factor, there are contributions to the curvature because of the contorsion term. Let us compute explicitly the $i = 0, j = 1$-components of $[\kappa_i, \kappa_j]$ which yields

$$\begin{aligned}[\kappa_0, \kappa_1] &= \frac{1}{16}[\not{H}_0, \not{H}_1] + \frac{1}{64}[\Gamma_0\not{H}_3, \Gamma_1\not{H}_3] - \frac{1}{32}[\not{H}_0, \Gamma_1\not{H}_3] - \frac{1}{32}[\Gamma_0\not{H}_3, \not{H}_1] \\ &= -\frac{1}{32}(N^2 + M^2)(\nabla\alpha^2)\gamma_{01}. \end{aligned} \tag{37}$$

Finally, the curvature of the AdS_2 subspace is given by

$$R_1 = -\frac{1}{8}(N^2 + M^2)(\nabla\alpha)^2, \tag{38}$$

where expressions (36) and (37) were used and then, there is the maximal symmetry on AdS_2 with symmetry $SO(1,1)$. A similar analysis reveals, that the scalar curvature for the S^2 subspace with symmetry $SO(2)$, is exactly the same that R_1 but with opposite sign, i.e. $R_2 = -R_1$. The total curvature of the 4d space formed by the product $AdS_2 \times S^2$ vanishes. Hence, the result is a unique solution called the near-horizon geometry.

There is a remarkable fact in the analysis carried in sections 2 and 3. In the former, the relation among fluxes was established by hand in order to reproduce the near-horizon geometry. While, in the later, this relation is established by requiring $\mathcal{N} = 2$ supergravity in 4d, which, in consequence lead us to decoupling both spinors, ϵ^1 and ϵ^2. Besides, we have found an alternative way for constructing a 4d space-time with an $AdS_2 \times S^2$ symmetry by turning on 3-form fluxes, including NS-NS, rather than the inclusion of only 5-form fluxes.

Before we continue, there are some remarks which summarize the results so far obtained:

(a) In the fluxless case, i.e. for $M = N = 0$ the result is a Minkowski space-time.

(b) Accordingly with (38), RR fluxes seem not to play a role in the curvature, however they can not vanish because in their absence, the contribution to $\widetilde{R}_4$ from Einstein equations by $\mathcal{T}$ would not be zero.

(c) There exists a metric, $\tilde{g}_{\mu\nu}$, which can be induced from R_1 and R_2. The form of the metric is

$$ds_4^2 = -\frac{x_1^2}{h}dx_0^2 + \frac{h}{x_1^2}dx_1^2 + hdx_2^2 + h\ sin^2 x_2 dx_3^2\,, \tag{39}$$

with $h = 2/|\tilde{R}_1|$. In addition to reproduce the curvature of the space-time in $4d$, this metric must satisfy the $4d$ effective theory. If the presence of moduli scalar fields, as a result of the compactification procedure is ignored, the effective theory contains gravity and fields with two antisymmetric indices induced by the presence of RR and NS-NS fields, which are effectively interpreted as electromagnetic tensor fields with no sources. The NS-NS and RR fluxes have both a smeared leg on the internal manifold. Hence, in the presence of an homogeneous electromagnetic field of the form

$$F_{t\rho} = 2|\widetilde{R}_1| = \frac{1}{4}(N^2 + M^2)(\nabla\alpha)^2, \tag{40}$$

there exists a unique solution for the Einstein-Maxwell equations known as the Robinson-Bertotti solution, which is the near-horizon metric of an extremal black hole. Another noticeable point is that coordinates of the F_3 and H_3 fluxes, are related to the homogeneous electromagnetic field.

(d) The curvatures of the subspaces are proportional to the flux numbers $(N^2 + M^2)$. If the numbers are big, the spaces will be highly curved and its horizon area will be small.

(e) The integrability conditions for the second spinor ϵ^2 yield the same result already discussed. The reason lies in the contorsion term

$$\kappa_i = \frac{1}{4}\not{H}_i - \frac{1}{8}\Gamma_i\not{H}. \tag{41}$$

Then, the curvatures of both subspaces remain unchanged with respect to those computed by the equations of ϵ^1.

(f) From the cases reviewed at the end of section 2, another possibility is to consider a RR flux configuration in which $Q = 0$. This choice will yield the near-horizon metric as a solution in 4d space-time too.

4. Conclusions

We present a simple way to obtain the Robinson-Bertotti solution by compactifying a type IIB supergravity theory in the presence of only NS-NS, H_3 and RR, F_3 3-form fluxes. We assume a 10d space-time conformed by the product of two maximally symmetric 2d spaces and a 6d space with $SU(3)$-structure, i.e. $AdS_2 \times S^2 \times X^6$. By requiring a 4d Ricci flat space, we find the conditions that flux configuration must fulfill in order to satisfy this symmetry. To achieve the goal we use Einstein equations and dual Bianchi identities. As a result of this analysis we find that one viable solution involves a constant warping factor and the condition $M^2+Q^2 = P^2+N^2$ for the flux numbers.

On the other hand, we use the integrability conditions on the 2d components of the chiral 10d spinors ϵ^1 and ϵ^2, with the purpose of finding the expressions for the 2d subspaces curvatures. An interesting feature extracted from this approach is the fact that, by preserving $\mathcal{N} = 2$ in 4d (which implies the decoupling of spinors in Eq.(27)), the conditions that fluxes must fulfill that renders a Ricci-flat space-time arises naturally, provided by supersymmetry.

Another issue is the possibility to obtain positive and negative curvatures since there are many other different flux configurations that can be considered. It is important to notice that the possibility to achieve different values for the curvature increases by not requesting a maximally symmetric 4d space-time.

As an alternative to the well known case of five forms associated to D3-branes and considered in literature so far, we show a way to construct a 4d space-time with an $AdS_2 \times S^2$ symmetry by turning on 3-form fluxes including NS-NS fluxes.

Acknowledgments

We thank the organizers of the XIII Mexican Workshop on Particles and Fields for accepting this talk. O Loaiza-Brito is supported by PROMEP, CONACyT-132166 and DAIP grants. L Vazquez-Mercado is supported by a CONACyT doctoral fellowship.

References

[1] V. Balasubramanian and F. Larsen, Nucl. Phys. B **528**, 229 (1998) [arXiv:hep-th/9802198].
[2] O. Loaiza-Brito and L. Vazquez-Mercado, Phys. Rev. D **84**, 066010 (2011) [arXiv:1012.2338 [hep-th]].
[3] H. Ooguri, C. Vafa and E. P. Verlinde, Lett. Math. Phys. **74**, 311 (2005) [hep-th/0502211].
[4] J. M. Maldacena and C. Nunez, Int. J. Mod. Phys. A **16**, 822 (2001) [hep-th/0007018].
[5] B. de Wit, D. J. Smit and N. D. Hari Dass, Nucl. Phys. B **283**, 165 (1987).
[6] T. Ortin, Cambridge Unversity, Cambridge University Press, 2004
[7] S. B. Giddings, S. Kachru and J. Polchinski, Phys. Rev. D **66**, 106006 (2002) [hep-th/0105097].

Branching ratio of the invisible decays of the lightest Higgs boson h^0 in the MSSM

M. del R. Aparicio Méndez, J. E. Barradas Guevara, L. A. Cordero Cid and O. Félix Beltrán

Fac. de Cs. Físico Matemáticas, Benemérita Universidad Autónoma de Puebla, Apdo. Postal 1152, 72570 Puebla, Pue., M éxico.
Fac. de Cs. de la Electrónica, Benemérita Universidad Autónoma de Puebla, Apdo. Postal 542, C. P. 72570, Puebla, Pue., México.

E-mail: ma_rocio_a_m@yahoo.com.mx

Abstract. The Supersymmetric models predicts the existence of light neutralinos with masses around 50 GeV as good candidates for detection of physics beyond the Standard Model, as well as other supersymmetric particles for different channels of Higgs boson decays. It this work, we presented the branching ratios of the lightest Higgs boson to neutralinos and charginos in the context of the Minimal Supersymmetric Standard Model. Although the invisible modes have been studied previously, is not known in detail the region of parameter space for supersymmetric sector where the mode reaches a significant branching ratio. We presented the analysis the parameters space looking for these regions by sampling techniques.

1. Introduction

Supersymmetry (SUSY) is a generalization of the space-time symmetries of quantum field theory that transforms fermions into bosons and vice versa [1]. At present, there is no experimental evidence that nature is supersymmetric. If supersymmetry were an exact symmetry in nature, then particles and their superpartners (which differ in spin by half a unit) would be degenerate in mass. Since this is not observed in data, supersymmetry cannot be an exact symmetry and must be broken. If we have supersymmetric theories at low energies, the electroweak scale is linked to the effective scale of the breaking of supersymmetry, both scales are characterized by the vacuum expectation value of the Higgs field (246 GeV). It is possible that supersymmetry will ultimately explain the origin of the large hierarchy of energy scales from the W and Z masses to the Planck scale.

2. Basic Concepts of the MSSM

The most economical candidate for a realistic SUSY model with a gauge group SU(3) $\times$ SU(2) $\times$ U(1), and a minimal content of particles is the Minimal Supersymmetric Standard Model (MSSM) [1]. This model consists of taking the Standard Model and adding the corresponding supersymmetric partners [1]. In addition, the MSSM contains two hypercharge $Y = \pm 1$ Higgs doublets, which is the minimal structure for the Higgs sector of an anomaly-free supersymmetric extension of the Standard Model. The supersymmetric structure of the theory also requires (at least) two Higgs doublets to generate mass for both "up"-type and "down"-type quarks (and charged leptons) [6]. All renormalizable supersymmetric interactions consistent with (global) B-L conservation (B=baryon number and L=lepton number) are included [2].

As a consequence of B-L invariance, the MSSM possesses a multiplicative R-parity invariance, where $R = (-1)^{3(B-L)} + 2S$ for a particle of spin S. Note that this implies that all the ordinary Standard Model particles have even R parity, whereas the corresponding supersymmetric partners have odd R parity. The conservation of R parity in scattering and decay processes has a crucial impact on supersymmetric phenomenology. For example, starting from an initial state involving ordinary (R- even) particles, it follows that supersymmetric particles must be produced in pairs. In general, these particles are highly unstable and decay into lighter states. However, R parity invariance also implies that the lightest supersymmetric particle (LSP) is absolutely stable, and must eventually be produced at the end of a decay chain initiated by the decay of heavy unstable supersymmetric particle [3].

3. Neutralino sector

The higgsinos and electroweak gauginos mix with each other to produce the mass eigenstates that could be detected experimentally. The neutral higgsinos ($\widetilde{H}_1^0$ and

$\widetilde{H}_2^0$) and the neutral gauginos ($\widetilde{B}^0, \widetilde{W}^0$) combine to form four mass eigenstates called *neutralinos* [4]. The charged higgsinos ($\widetilde{H}_1^+$ and $\widetilde{H}_2^-$) and winos ($\widetilde{W}^+$ and $\widetilde{W}^-$) mix to form two mass eigenstates with charge ± 1 called *charginos*. We will denote the neutralino and chargino mass eigenstates by $\widetilde{\chi}_i^0$ $(i = 1, 2, 3, 4)$ and $\widetilde{\chi}_i^\mp$ $(i = 1, 2)$. By convention, these are labeled in ascending order, so that $m_{\widetilde{\chi}_1^0} < m_{\widetilde{\chi}_2^0} < m_{\widetilde{\chi}_3^0} < m_{\widetilde{\chi}_4^0}$ and $m_{\widetilde{\chi}_1^\pm} < m_{\widetilde{\chi}_2^\pm}$. The lightest neutralino, $\widetilde{\chi}_1^0$, is usually assumed to be the LSP, unless there is a lighter gravitino or unless R-parity is not conserved, through it is the only a MSSM particle that can make a good dark matter candidate.

The neutralinos are determined by diagonalizing the corresponding mass matrix. In the minimal supersymmetric model the corresponding mass matrix depends on three unknown mass scales, namely, μ, M_2 and M_1, and on the ratio $\tan\beta \equiv v_2/v_1$, where $v_2(v_1)$ is the vacuum expectation value of the Higgs field which couples to the up-type quarks(down-type quarks); μ is the supersymmetric Higgs-boson-mass parameter; and M_2 and M_1 are the gaugino mass parameters associated with the SU(2) and U(1) subgroups of the standard model. It is common to reduce the parameter freedom by assuming that M_2 and M_1 are related to the gaugino mass M_3 of the SU(3) [5].

4. Higgs sector

Supersymmetry imposes new requirements on the theory regarding the structure of the Higgs boson. In order to supersymmetric theories don't show anomalies, two Higgs doublets are required and so the theory is renormalizable. It is known that within the standard model only one Higgs doublet is required to give mass to quarks and leptons, while in the MSSM two Higgs doublets ($\widetilde{H}_1^0, \widetilde{H}_1^-$) and ($\widetilde{H}_2^0, \widetilde{H}_2^+$) are required to give mass to up-type quarks and down-type quarks and the corresponding leptons. Therefore, this model contains two Higgs doublets [6]. The MSSM considers five physical Higgs particles: a pair of charged scalar Higgs bosons $H^\pm$, two CP-even neutral scalar Higgs bosons h^0 and H^0 (by convention is taken $m_{H^0} \geq m_{h^0}$) and finally a CP- odd pseudoscalar neutral Higgs boson A^0 (in this paper its mass is denoted as m_A). The three neutral Higgs bosons are determined by the diagonalization of the mass matrix of neutral Higgs bosons, which depends on the Higgs potential.

5. Results and conclusions

Decays of neutral Higgs bosons to charginos neutralinos are important in order to determine the branching ratios of the different supersymmetric decay channels. There exists a considerable probability for the Higgs boson decays into a pair of lighter supersymmetric particles $h^0 \to \widetilde{\chi}_i\widetilde{\chi}_j$ (LSP lightest neutralino). The LSP could be almost degenerate with other charged states (charginos), which the Higgs boson can decay, and thus the product of the charged states are too weak to be detected. If R-parity is conserved, the LSP $\widetilde{\chi}_i$ (neutralino or chargino lightest) doesn't decay and escapes the detection. Therefore, the Higgs boson is invisible [7].

In the following, we present the results obtained at loop-level for the lightest neutralino masses $\tilde{\chi}_1^0$ ($m_{\tilde{\chi}_1^0}$), the masses of neutral Higgs boson lightest h^0 (m_{h^0}) and the *branching ratio* of the possible decays of h^0 in the MSSM. These results are obtained using the program HDECAY [8].

Taking into account the bounds of mass of 90 GeV for chargino and the lightest Higgs boson ($\tilde{\chi}_1^{\pm}$ and h^0), a search of data was carried out where the branching ratios for the decay $h^0 \to \tilde{\chi}_1^0\tilde{\chi}_1^0$ (BRANEU) may be more dominant.

With the mass bounds imposed for h^0 and $\tilde{\chi}_1^{\pm}$, and the parameters range 50 GeV $\leq m_A \leq$ 1000 GeV, $2 \leq \tan\beta \leq 50$, and 100 GeV $\leq \mu \leq$ 1000 GeV, we are maintaining the parameter M = 100 GeV fixed; and two graphs were made. It was find the masses of $\tilde{\chi}_1^0$ as function of the parameters μ and $\tan\beta$, and also the branching ratios of the possible decay channels of h^0, which are as function of the masses of h^0. These plots are located in Figure 1. With these same parameters in the same range except 0.5 $\leq \tan\beta \leq$ 50, another search was made of the branching ratios of the different decay channels of h^0.

The results obtained are contained in Figure 2 and values for the different parameters are reported in Table 1.

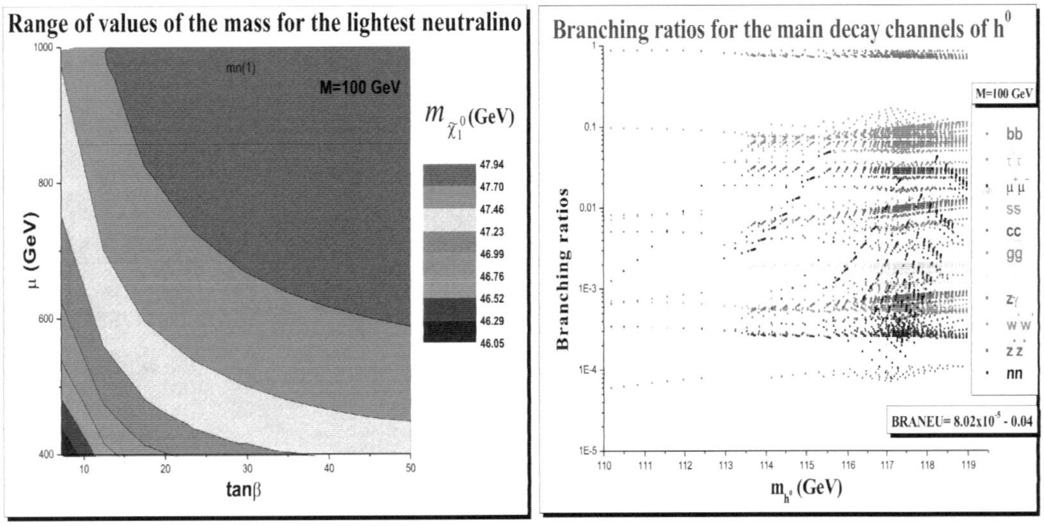

Figure 1. Range of masses values for $\tilde{\chi}_1^0$ and The Branching Ratios for different decay channels of the lightest Higgs boson h^0, with M = 100 GeV fixed, and 100 GeV $\leq \mu \leq$1000 GeV, 50 GeV $\leq m_A \leq$1000 GeV, 2$\leq \tan\beta \leq$50.

The Figure 2 shows the behavior of the branching ratios for different decay channels of the lightest Higgs boson h^0 as function of m_{h^0}, where the maximum value of the BRANEU is 0.06 and the mass obtained for the neutralinos are in a range of 45.38 to 47.94 GeV that falls within the bounds of mass for the neutralinos.

After having carried out a study of possible decay channels for the decay of the lightest neutral Higgs boson in the MSSM, it is concluded that the decay of the Higgs

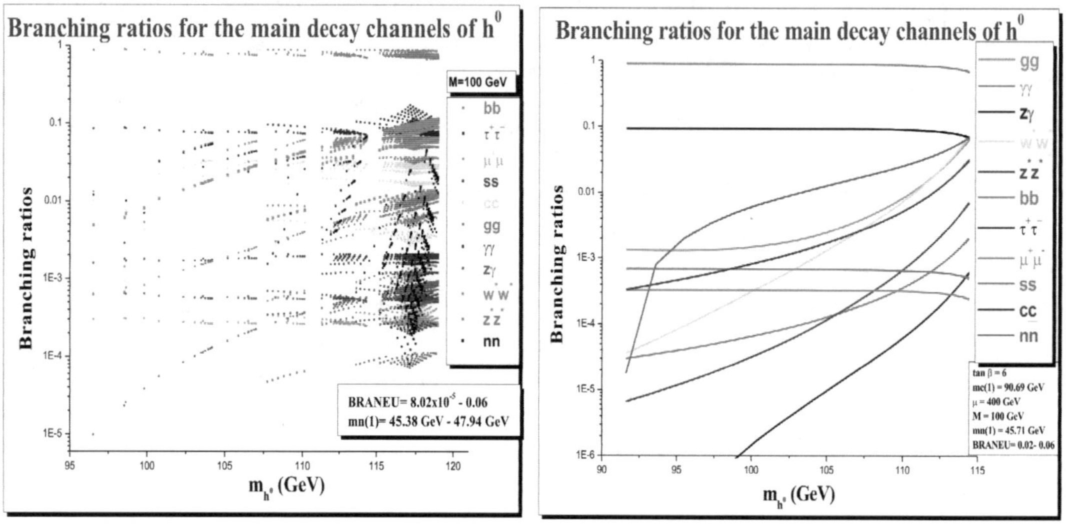

Figure 2. Branching ratios of different decay channels of the lightest Higgs boson h^0, with M = 100 GeV fixed, and 100 GeV $\leq \mu \leq$1000 GeV; 50 GeV$\leq m_A \leq$1000 GeV; $0.5 \leq \tan\beta \leq$50.

$\tan\beta$	μ [GeV]	M [GeV]	$m_{\widetilde{\chi}_1^0}$ [GeV]	$m_{\widetilde{\chi}_1^+}$ [GeV]	$BRA-$ NEU	m_A [GeV]	M_{h^0} [GeV]
6	400	100	45.71	90.69	0.06	1000	114.4

Table 1. Mass values for lightest neutralino $\widetilde{\chi}_1^0$, chargino $\widetilde{\chi}_1^+$ and Higgs boson h^0 corresponding to values in which is reached the best branching ratio for the decay $h^0 \to \widetilde{\chi}_1^0\widetilde{\chi}_1^0$.

bosons are dominant in invisible modes in different regions of the parameter space, hence the importance in continuing their study to better understand the supersymmetric spectrum.

References

[1] H. E. Haber and G.L. Kane, Phys. Rep. **117**, 75 (1985).
[2] H. E. Haber, hep-ph/9709450v1.
[3] K. Hagiwara *et al.* Particle Data Group, Phys. Rev. D **66**, 010001 (2002).
[4] S. P. Martin, hep-ph/9709356v5.
[5] M. M. El Kheishen, A. A. Shafik and A. A. Aboshousha, Phys. Rev. D **45**, 4345 (1992).
[6] J. F. Gunion, H. E. Haber, G. Kane and S. Dawson, *The Higgs Hunter's Guide*, Perseus Books Group, Cambridge, Massachussetts, 1990.
[7] S. P. Martin and J. D. Wells, hep-ph/9903259v1.
[8] A. Djouadi, J. Kalinowski and M. Spira, hep-ph/9704448v1.

Journal of Physics: Conference Series **378** (2012) 012008 doi:10.1088/1742-6596/378/1/012008

Quantum mechanics lessons for standard cosmology

Marco A. Reyes

Department of Physics, DCI-CLE, Universidad de Guanajuato, Apdo. Postal E143, 37150, León, Gto., México

E-mail: marco@fisica.ugto.mx

Abstract. By recalling the relevance of the Sturm-Liouville theory has had on the solutions of quantum mechanics problems, here it is explored the possibility of getting some insight to the solutions for a standard cosmology model for inflation, from a time independent Schrödinger type equation derived from the equations of motion for a single scalar field in a flat space time with a FRW metric and a cosmological constant.

1. Introduction

In 1925 Erwin Schrödinger proposed his model for wave mechanics; his first task was to describe the dynamics of an electron in a hydrogen atom. Ever since his model was proposed, it was found that the second order differential equations appearing in some of the most interesting time independent problems of quantum mechanics (QM) had been studied in the century before this model was born, and that they pertain to the set of equations studied in the Sturm-Liouville (SL) theory. Such is the case of the time independent equation for the hydrogen atom, whose radial solutions are the generalized Laguerre polynomials, and where the angular part, as in any central problem, is solved in terms of the associated Legendre polynomials, as part of the solution of Laplace's equation; another case is the simple harmonic oscillator, whose solutions come in terms of the Hermite polynomials. In fact, the boundary conditions imposed in the QM problems is what implies the appearance of discrete solutions, an aspect which is usually cited as the quantization of the problem, or the quantization of the parameter in question; the energy, for example.

Some of the special functions found on QM problems were initially used to solve non-mechanical problems: the Laguerre polynomials, named after Edmond Laguerre (1834 1886), had been used to numerically compute the Gaussian quadrature of some integrals, while the Hermite polynomials, named after Charles Hermite (1822-1901), conform a sequence that arises in various probability problems, numerical computations of Gaussian quadratures, and can also be used to describe Dirac's Delta function. While the hydrogen atom is the only QM problem analytically solved, there are very few second order ordinary differential equations with known analytical solutions, the most important for physics being those pertaining to the SL theory, where hermiticity is the most important property.

On the other hand, the general relativity (GR) equations are in some cases reduced to second order ordinary differential equation, and the special functions of the SL theory again immediately become the candidates for analytical solutions. As QM inherited some properties from the SL theory with regards to the analytic solutions, one may ask what could be learned from the QM solutions that could give us some insight into the GR problems.

Consider for example the case of a single scalar field (SF) as the source for inflation, a useful mechanism to understand the inflationary period that provided the initial conditions of the early Universe from which the large scale structure that we see can be obtained. In this scenario, the functional form of the field and the potential driving its dynamics is rather arbitrarily inserted into the equations, since there is no unique prescription or phenomenology to determine them. But, for a given potential function, it may be possible to consider the equivalent QM problem and from its dynamics gain some insight to the problem.

2. Bifurcations in a QM phase space portrait

Let us first consider the classical picture of a one dimensional time independent QM problem,[1] where the particle is subjected to a potential well, with $U(x) < 0$ inside the well region, $a < x < b$, and $U = 0$ elsewhere, and where we are interested in the bounded states solutions of the problem, with $E < 0$. The Schrödinger equation in this case is

$$\left[-\frac{1}{2}\frac{d^2}{dx^2} + U(x)\right] \psi(x) = E\ \psi(x). \tag{1}$$

where we set $\hbar = m = 1$. Now let us redefine the problem by changing the coordinate x by t and call it *time*, and let the function $\psi(t)$ be the new coordinate $q(t)$ and its *time* derivative its associated momentum $p(t) = \frac{d\psi}{dt}$. Then, the following equations hold

$$\begin{aligned} \frac{dq}{dt} &= p \\ \frac{dp}{dt} &= 2\left[U(t) - E\right]. \end{aligned} \tag{2}$$

Therefore, for this QM problem we have an associated classical picture, where the dynamics are determined by the Hamiltonian $H(p,q) = p^2/2 + (E - U(t))q^2$, with the boundary conditions $q(\pm\infty) = 0$. The dynamics could be determined from the following matrix relation

$$\frac{d\vec{q}}{dt} = \Lambda(t)\vec{q}(t) \tag{3}$$

where

$$\Lambda(t) = \begin{vmatrix} 0 & 1 \\ 2\,(U(t) - E) & 0 \end{vmatrix} \tag{4}$$

is the dynamics generator matrix and

$$\vec{q}(t) = \begin{vmatrix} q \\ p \end{vmatrix}. \tag{5}$$

Outside of the potential well, where $U(t) \equiv 0$, the generator becomes time independent

$$\Lambda = \begin{vmatrix} 0 & 1 \\ 2|E| & 0 \end{vmatrix} \text{for } t \le a \text{ or } t \ge b \tag{6}$$

leading to the explicit solutions for eq.(3)

$$\vec{q}(t) = \begin{cases} e^{\Lambda(t-a)}\vec{q}(a) & \text{for } t \le a \\ e^{\Lambda(t-b)}\vec{q}(b) & \text{for } t \ge b \end{cases} \tag{7}$$

Moreover, since $\Lambda^2 = 2|E|\mathbf{1}$, it possesses two eigenvalues, $\lambda_\pm = \pm\sqrt{2|E|}$, with the corresponding eigenvectors

$$\vec{e}_+ = \left| \begin{matrix} 1 \\ +\sqrt{2|E|} \end{matrix} \right|, \quad \vec{e}_- = \left| \begin{matrix} 1 \\ -\sqrt{2|E|} \end{matrix} \right| \tag{8}$$

Therefore, for any given value of the parameter E the time evolution preserve the fixed directions $\vec{e}_\pm$. Hereafter we assume that the phase space trajectory began in the direction $\vec{e}_+$. In the QM case, E is called an eigenvalue of the Schrödinger equation (1) *iff* there exists a non-trivial trajectory that vanishes both at $t \to -\infty$ and $t \to +\infty$, that is, if after traversing the potential well the phase space trajectory that originated in the direction $\vec{e}_+$ at $t < a$ is driven to a trajectory that moves in the $\vec{e}_-$ direction for $t > b$. However, this is the most unusual case, since there exist only a countable set of values for which this occurs, while there is an infinite number of values of the parameter E for which the trajectories diverge in the direction $\vec{e}_+$ after traversing the potential well. The phase space trajectories are depicted in Fig.1.

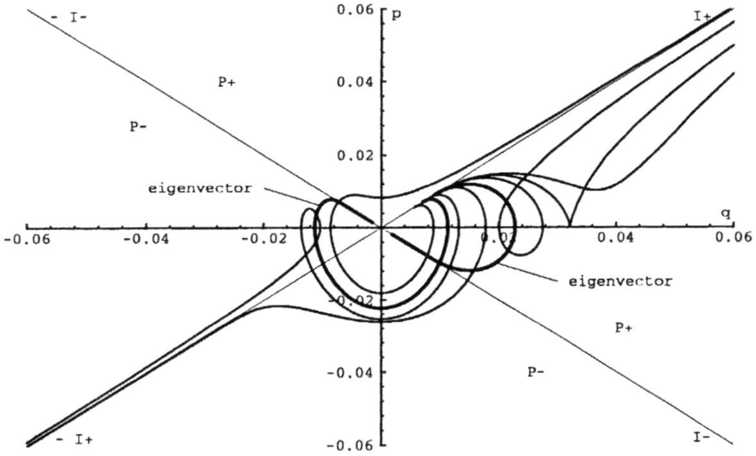

Figure 1. The bifurcations of phase space trajectories. The fixed directions $\vec{e}_\pm$ determine the eigenvalue problem.

An interesting aspect of this picture is that the asymptotes directions determined by $\vec{e}_\pm$ are fixed, no matter what the form of the potential well be, and that they determine a constant ratio between the values of the phase space coordinates, q and p for large $|t|$. This is also true if the potential function has two different constant values outside of the well.

We shall now consider a QM analogue of standard cosmology, and apply there some of the notions introduced here.

3. The QM equivalent of standard cosmology

Consider now an homogeneous and isotropic universe, i.e., a FRW background with a scalar homogeneous field $\phi(t)$ minimally coupled to gravity and nonzero cosmological constant

$$\int d^4x\sqrt{-g}\left[R + \Lambda + \frac{1}{2}(\nabla\phi)^2 + V(\phi)\right], \tag{9}$$

where $(\nabla\phi)^2 = g^{\mu\nu}\partial_\mu\partial_\nu$ and $V(\phi)$ is the potential energy of the field. In order to describe the dynamics of the scalar field during inflation the usual treatment is performed [2]-[7], leading to the pair of equations

$$3H^2 = \frac{1}{2}\dot{\phi}^2 + V(\phi) + \Lambda \tag{10}$$

$$\ddot{\phi} + 3H\dot{\phi} = -\frac{dV(\phi)}{d\phi}, \tag{11}$$

where dot means derivative with respect to time, and we set $M_{Pl} = c = 1$. The time derivative of eq.(10) is related to eq.(11) through the momentum equation

$$\dot{H} = -\frac{1}{2}\dot{\phi}^2. \tag{12}$$

With the use of eqs. (11) and (12) the dynamics of the model may be described by the single equation:

$$3H^2 + \dot{H} = V(\phi) + \Lambda. \tag{13}$$

which can be recognized as a Riccati equation for the Hubble parameter $H(t)$. Now, since the time independent one dimensional Schrödinger equation can be transformed into a Riccati equation, its appearance here suggests the introduction of a QM analogue to inflationary cosmology. This has been proposed earlier, leading however to different approaches: a nonlinear Schrödinger equation [8], or a case of cosmology with a perfect fluid but in the context of classical mechanics ([9, 10].) We have developed our own treatment before [11], and we shall relate it here to the previous section.

The Riccati equation (13) is transformed into a Schrödinger equation when the Hubble parameter $H(t)$ is related to a wave function $\psi(t)$ through

$$H \equiv \frac{1}{3}\frac{\dot{\psi}(t)}{\psi(t)}, \tag{14}$$

taking the form

$$\left[-\frac{d^2}{dt^2} + 3\ V(t)\right]\psi(t) = -3\Lambda\ \psi(t). \tag{15}$$

This transformation immediately brings some analogies between the two models, QM and inflationary cosmology, in particular when the Hubble parameter is not singular, implying that $\psi(t)$ has to be an at least C^1 class function. In that case, we have to consider the ground state solutions of all known exactly solvable bound state problems of QM, where the SF potential $V(\phi)$ and the cosmological constant Λ become the QM potential $U(x)$ and ground state energy eigenvalue E_g, respectively,

$$3V(\phi(t)) + 3\Lambda \leftrightarrow 2U(x) - 2E_g \tag{16}$$

We have realized this analogies before [11]. The general algebraic procedure is very simple; let us recall here the case of the QM simple harmonic oscillator. In this case, when we substitute the ground state eigenfunction $\psi_g(x) = c_0\exp\{-x^2/2\}$ into eq.(14) we find that $H(t) = -\frac{w}{3}t$; then, use eq.(12) we find that $\phi(t) = \sqrt{\frac{2w}{3}}\,t$, which together with eq.(16) gives us that $V(\phi) = \frac{w}{2}\phi^2$. If we simply replace $w = 2\lambda$ we have finally that $V(\phi) = \lambda\phi^2$. [1] It is amazing that the same type of functional dependence appears in both models. Other combinations of QM typical potentials

[1] In this case $U(x) = \frac{1}{2}w^2x^2 = 2\lambda^2x^2$.

Journal of Physics: Conference Series **378** (2012) 012008 doi:10.1088/1742-6596/378/1/012008

and inflationary cosmology potentials can be found elsewhere, for one dimensional and radial problems.[11]

It is interesting that the exact analytic solution for the $\lambda\phi^2$ potential differs very much from the required behavior for an inflationary model, where the main reason would be that typically this potential is used together with the slow roll approximation, which imposes special conditions on the dynamics of the field ($\ddot{\phi} = 0$, for example.) To begin with, a cosmological constant is needed, or the exact potential would need to be $\lambda\phi^2 + const.$[12] Also, the QM analogue cited here for $V(\phi) = \lambda\phi^2$, gives a value of $H(t) = -\frac{2\lambda}{3}t$ for the Hubble parameter, implying that the equation for the dynamics of the field, eq.(11), be

$$\ddot{\phi} - 2\lambda t\,\dot{\phi} + 2\lambda\phi = 0 \qquad (17)$$

whose solution clearly is $\phi(t) = const. \times t$, whereas a constant Hubble parameter, the slow roll case, would derive in a equation where oscillations are still possible.

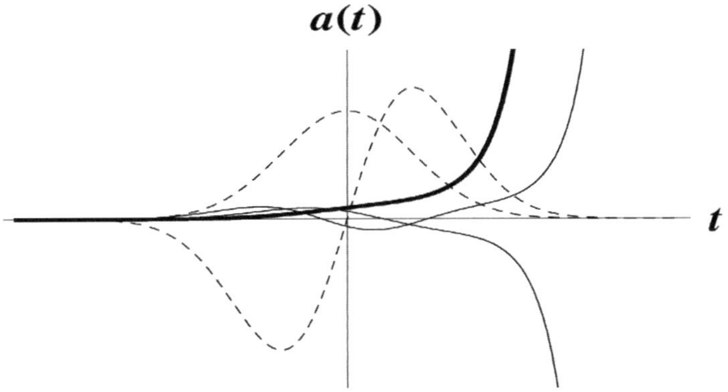

Figure 2. Different forms of $a(t)$ obtained from one QM problem.

In Fig.(2) different curves for the scale factor obtained from the QM analogy of this problem are depicted. The dashed curves correspond to $a(t)$ obtained from the ground state and first excited state eigenfunctions, $\psi_0(x)$ and $\psi_1(x)$, of the harmonic oscillator problem. The solid curves draw the scale factor for three different cosmological constants, derived from the energy eigenvalue E, with $E < E_0$, $E_0 < E < E_1$ and $E_1 < E < E_2$, whose wave functions diverge towards $+\infty$, $-\infty$ and to $+\infty$ again, the first one without nodes and the other with increasing number of nodes. Therefore, only the wave function with $E < E_0$ could lead to a solution with $a(t) > 0$ for all t, describing a cosmological solution for an ever expanding universe (bold solid line.) There is an continuum of solutions with $E < E_0$ that diverge to $+\infty$ as t grows, and for all of them the slope $\frac{\dot{a}}{a}$ is determined by the cosmological constant Λ, the "energy" parameter of the corresponding QM problem. Besides the analogies found between the two models [11], this may be the most interesting aspect of the relation between both models, since there are implications for the form of the Hubble parameter, as is stated in eq.(14).

4. Conclusion

In this work it is shown that two completely separate models, one being a cosmological model for inflation and the other a one dimensional quantum mechanics time independent model, may have more in common than what is expected, and that their relations could give us some insight to the description of some of the most relevant parameters in standard cosmology; in the present case, the Hubble parameter is being studied with respect to the way bifurcations are produced in a QM bounded problem, showing that these parameter is highly related to the cosmological constant.

References

[1] Mielnik B and Reyes M A 1996 *Jour. Phys* A **29** 6009-6026.
[2] Bassett B A, Tsujikawa S and Wands D 2006 *Rev. Mod. Phys.* **78** 536.
[3] Copeland E J, Sami M and Tsujikawa S 2006 *Int. J. Mod. Phys. D.* **15** 1753-1936.
[4] Liddle A R and Lyth D H 2000 *Cosmological Inflation and Large Scale Structure* (Cambridge University Press).
[5] Linde A 1990 *Particle Physics and Inflationary Cosmology* (Harwood Academic Publishers).
[6] Copeland E J, Kolb E W, Liddle A R and Lidsey J E 1993 *Phys. Rev. D* **48** 2529-1993.
[7] Andrew R. Liddle, Paul Parsons and John D. Barrow, *Phys. Rev. D.* **50** (1994) 7222-7232.
[8] Hawkins R M and Lidsey J E 2002 *Phys. Rev.* D **66** 023523.
[9] Lima J A S, Moreira J A M and Santos J 1998 *Gen. Rel. Grav.* **30** 425.
[10] Lima J A S 2001 *Am. J. Phys.* **69** 1245
[11] Barbosa-Cendejas N and Reyes M A 2010 The Schrodinger picture of standard cosmology *Preprint* gr-qc/1001.0084.
[12] Reyes M A 2008 On exact solutions to the scalar field equations in standard cosmology *Preprint* gr-qc/0806.2292.

Journal of Physics: Conference Series **378** (2012) 012009 doi:10.1088/1742-6596/378/1/012009

Noncommutativity, Cosmology and Λ

M. Sabido[1], S. Pérez-Payán[2]
Departamento de Física de la Universidad de Guanajuato,
A.P. E-143, C.P. 37150, León, Guanajuato, México
E-mail: [1]msabid@fisica.ugto.mx
E-mail: [2]sinuhe@fisica.ugto.mx

E. Mena
Centro Universitario de la Ciénega, Universidad de Guadalajara
Ave. Universidad 1115 Edif. B, C.P. 47820 Ocotlán, Jalisco, México
E-mail: emena@cuci.udg.mx

C. Yee-Romero
Departamento de Matemáticas, Facultad de Ciencias
Universidad Autónoma de Baja California, Ensenada, Baja California, México
E-mail: carlos.yee@uabc.edu.mx

Abstract. The effects of phase space deformations in standard scalar field cosmology are studied. The deformation is introduced by modifying the symplectic structure of the minisuperspace variables to have a deformed Poisson algebra among the coordinates and the canonical momenta. The resulting theory is knowns as noncomutative cosmology. We show the effects of this deformation in scalar field cosmology and establish a relationship between the noncommutative parameters and Λ. This work is based on the papers, arXiv:1111.6137 and arXiv:1111.6136.

1. Introduction

The initial interest in noncommutative field theory [1] slowly but steadily permeated in the realm of gravity, from which several approaches to noncommutative gravity [2] where proposed. All of these formulations showed that the end result of a noncommutative theory of gravity, is a highly nonlinear theory. In order to study the effects of noncommutativity on different aspects of the universe, noncommutative cosmology was presented in [3]. The authors noticed that the noncommutative deformations modify the noncommutative fields, and conjectured that the effects of the full noncummutative theory of gravity should be reflected in the minisuperspace variables. This was achieved by introducing the Moyal product of functions in the Wheeler-DeWitt equation, in the same manner as is done in noncommutative quantum mechanics [4, 5].

Although the noncommutative deformations of the minisuperspace where originally analyzed at the quantum level, by an effective noncommutativity on the minisuperspace, classical noncommutative formulations have been proposed. The main idea of this classical noncommutativity is based on the assumption that modifying the Poisson brackets of the classical theory gives the noncommutative equations of motion [3, 5, 4, 6].

We will work with an FRW universe and a scalar field. This model has been used to explain several aspects of our universe, like inflation, dark energy and dark matter. Our approach to noncommutativity is trough its introduction in a phase space constructed in the minisuperspace variables, and is achieved by modifying the symplectic structure, in the same manner as in [4, 5, 6]. It will be showed that in the absence of a cosmological constant, the behavior of the scale factor can account for a late time acceleration, furthermore it can be seen from the solution that there is no initial singularity. We follow this line of reasoning and give a proposal for the origin of Λ, by considering noncommutative deformation of a 5 dimensional theory of gravity with a compact extra dimension.

We will start in Section 2, by introducing the commutative model. In Section 3, the noncommutative model is presented, as well as the dynamics of the cosmological model. We will show that with our approach late time acceleration can be accounted, contrary to the commutative case. In. section 4, we present a model where the cosmological constant arises from the phase space deformation. Finally, the last section is devoted for conclusions and final remarks.

2. The Commutative Model

As already suggested, cosmology presents an attractive arena for noncommutative models, both in the quantum as well as classical level. One of the features of noncommutative field theories is UV/IR mixing, this effectively mixes short scales with long scales, from this fact one may expect that even if noncommutativity is present at a really small scale, by this UV/IR mixing, the effects might be present at an older time of the universe.

We start with a homogeneous and isotropic universe endowed with a flat Friedmann-Robertson-Walker metric

$$ds^2 = -N^2(t)dt^2 + a^2(t)\left[dr^2 + r^2 d\Omega\right], \qquad (1)$$

where as usual $a(t)$ is the scale factor of the universe and $N(t)$ is the lapse function. The action we will be working with is the Einstein-Hilbert action and a scalar field ϕ as the matter content for the model. In units $8\pi G = 1$, the action takes the form

$$S = \int dt \left\{ -\frac{3a\dot{a}^2}{N} + a^3\left(\frac{\dot{\phi}^2}{2N} + NV(\phi)\right)\right\}, \qquad (2)$$

here $V(\phi)$ is the scalar potential. From now on we will restrict to the case of constant potential ($V(\phi) = -\Lambda$).

For the purposes of introducing the deformation to the minisuperspace variables an appropriate redefinition needs to be made, we make the following change of variables

$$x = m^{-1}a^{3/2}\sinh(m\phi), \; y = m^{-1}a^{3/2}\cosh(m\phi). \qquad (3)$$

where $m^{-1} = 2\sqrt{2/3}$. In these new variables we calculate the Hamiltonian and is given by

$$H_c = N\left(\frac{1}{2}P_x^2 + \frac{\omega^2}{2}x^2\right) - N\left(\frac{1}{2}P_y^2 + \frac{\omega^2}{2}y^2\right), \qquad (4)$$

where $\omega^2 = -\frac{3}{4}\Lambda$. This is the canonical Hamiltonian which is a first-class constraint as is usual in general relativity. Since we do not have second class constraints in the model we will continue to work with the usual Poisson brackets and the relations of commutation between the phase space variables

$$\{x_i, x_j\} = 0, \qquad \{P_{x_i}, P_{y_j}\} = 0, \qquad \{x_i, P_{x_j}\} = \delta_{ij}. \tag{5}$$

At this point, we have a minisuperspace spawned by the new variables (x, y). To find the dynamics of this model, we need to solve the equations of motions, these are derived as usual by using Hamiltons equations. For the particular model the equations are

$$\begin{aligned} \dot{x} &= -P_x, \qquad \dot{y} = P_y, \\ \dot{P}_x &= \omega^2 x, \qquad \dot{P}_y = -\omega^2 y, \end{aligned} \tag{6}$$

these equations can easily be integrated

$$x(t) = X_0 Cos(|\omega| t + \delta_x), \quad y(t) = Y_0 Cos(|\omega| t + \delta_y). \tag{7}$$

In order to satisfy the Hamiltonian constraint we introduce the solutions to the Hamiltonian, this gives a relationship between the integration constants, it easy to verify that $X_0 = \pm Y_0$. Finally we write the solution in the original variables

$$a^3(t) = V_0 \left[\cos\left(\delta_x - \delta_y\right) + \cos\left(2|\omega| t + \delta_x + \delta_y\right)\right], \tag{8}$$

where $\delta_x - \delta_y = 2n\pi$ in order to have a positive volume for the universe, due to the periodic nature of the functions we can choose $n = 0$. From the equation for the volume we can see, that you get a periodic universe, furthermore we get zero volume when $t = \frac{(2n+1)}{4|\omega|}\pi$.

3. Deformed Space Model

The original ideas of deformed phase space, or more precisely deformed minisuperspace, where done in conection with noncommutative cosmology [3]. As already mentioned, in order to avoid the complications of a noncommutative theory of gravity, they introduce a deformation to the minisuperspace in order to incorporate noncommutativity. Usually the deformation is introduced by the Moyal brackets, which is based in the Moyal product. To study the behavior of the model in a deformed phase space framework such that the minisuperspace variables do not conmmute with each other; noncommutativity between the phase space variables can be understood by replacing the usual product with the star product, also know as the Moyal product law between two arbitrary functions of position and momentum, as

$$(f \star g)(x) = \exp\left[\frac{1}{2}\alpha^{ab}\partial_a^{(1)}\partial_b^{(2)}\right] f(x_1)g(x_2)|_{x_1=x_2=x}, \tag{9}$$

such that

$$\alpha = \begin{pmatrix} \theta_{ij} & \delta_{ij} + \sigma_{ij} \\ -\delta_{ij} - \sigma_{ij} & \beta_{ij} \end{pmatrix}, \tag{10}$$

where the 2×2 matrices θ and β are assumed to be constant, antisymmetric and represent the noncommutativity in the coordinates and momenta respectively and σ is a product of θ and β. With this product law, the α-deformed Poisson brackets can be written as

$$\{f, g\}_\alpha = f \star_\alpha g - g \star_\alpha f. \tag{11}$$

An alternative an far more useful construction, is based on symplectic manifolds [8]. Once the deformation has been done one arrives to a modified Poisson algebra

$$\{x_i, x_j\}_\alpha = \theta_{ij}, \quad \{x_i, p_j\}_\alpha = \delta_{ij} + \sigma_{ij}, \quad \{p_i, p_j\}_\alpha = \beta_{ij}. \tag{12}$$

Making the following transformation on the classical phase space variables $\{x, y, p_x, p_y\}$

$$\hat{x} = x + \frac{\theta}{2}p_y, \qquad \hat{y} = y - \frac{\theta}{2}p_x,$$
$$\hat{p}_x = p_x - \frac{\beta}{2}y, \qquad \hat{p}_y = p_y + \frac{\beta}{2}x, \tag{13}$$

now the algebra reads

$$\{\hat{y}, \hat{x}\} = \theta, \quad \{\hat{u}, \hat{p}_x\} = \{\hat{y}, \hat{p}_y\} = 1 + \sigma, \quad \{\hat{p}_y, \hat{p}_x\} = \beta, \tag{14}$$

where $\sigma = \theta\beta/4$. Now that we have constructed the modified phase space, we apply the transformation to the Hamiltonian, where after defining

$$\omega_1^2 = \frac{4(\beta - \omega^2\theta)}{4 - \omega^2\theta^2}, \quad \omega_2^2 = \frac{4(\omega^2 - \beta^2/4)}{4 - \omega^2\theta^2}, \tag{15}$$

we get

$$\widehat{H} = \frac{1}{2}\left\{\hat{p}_x^2 - \hat{p}_y^2 + \omega_1^2(\hat{x}\hat{p}_y + \hat{y}\hat{p}_x) + \omega_2^2(\hat{x}^2 - \hat{y}^2)\right\}. \tag{16}$$

We can construct a bidimensional vector potential $\hat{A}_x = -\frac{\omega_1^2}{2}\hat{y}$, $\hat{A}_y = \frac{\omega_1^2}{2}\hat{x}$ from were a magnetic field $B = \omega_1^2$ is calculated, this result allow us to write the effects of the noncommutative deformation as minimal coupling on the Hamiltonian, $\widehat{H} = [(p_x - \hat{A}_x)^2 + \omega_3^2\hat{x}^2] - [(p_y - \hat{A}_y)^2 + \omega_3^2\hat{y}^2]$, this expression can be rewritten in terms of the magnetic B-field as

$$\begin{aligned} \widehat{H} &= [(\hat{p}_x^2 + (\omega_3^2 - B^2/4)\hat{x}^2] - [(\hat{p}_y^2 + (\omega_3^2 - B^2/4)\hat{y}^2] \\ &+ B(\hat{y}\hat{p}_x + \hat{x}\hat{p}_y), \end{aligned} \tag{17}$$

where $\omega_3^2 = \omega_1^2 + \omega_2^2$. This is a typical result in 2-dimensional noncommutativity, where the effects of the noncommutative deformation can be encoded in a perpendicular constant magnetic field.

To obtain the dymanics for our model, we can derive the equations of motion from the Hamiltonian (29)

$$\begin{aligned} \dot{\hat{x}} &= \{x, \widehat{H}\} = \frac{1}{2}[2\hat{p}_x + \omega_1^2 y], \\ \dot{\hat{y}} &= \{y, \widehat{H}\} = \frac{1}{2}[-2\hat{p}_y + \omega_1^2 x], \\ \dot{\hat{p}}_x &= \{\hat{p}_x, \widehat{H}\} = \frac{1}{2}[-\omega_1^2\hat{p}_y - 2\omega_2^2 x], \\ \dot{\hat{p}}_y &= \{\hat{p}_y, \widehat{H}\} = \frac{1}{2}[-\omega_1^2\hat{p}_x + 2\omega_2^2 y], \end{aligned} \tag{18}$$

defining new variables η and ζ as $\eta = \hat{x} + \hat{y}$, $\zeta = \hat{y} - \hat{x}$, this set of equations reduce to

$$\begin{aligned} \ddot{\eta} - \omega_1^2\dot{\eta} + \frac{1}{4}(4\omega_2^2 + \omega_1^4)\eta = 0, \\ \ddot{\zeta} + \omega_1^2\dot{\zeta} + \frac{1}{4}(4\omega_2^2 + \omega_1^4)\zeta = 0, \end{aligned} \tag{19}$$

solving this equations we can get the solutions in terms of the noncomutative variables $\hat{x}(t)$ and $\hat{y}(t)$

$$\hat{x}(t) = \frac{1}{2}\left[e^{\frac{\omega_1^4}{4}t}\cosh\left(\omega'^2 t\right) - e^{-\frac{\omega_1^4}{4}t}\cosh\left(\omega'^2 t\right)\right], \tag{20}$$

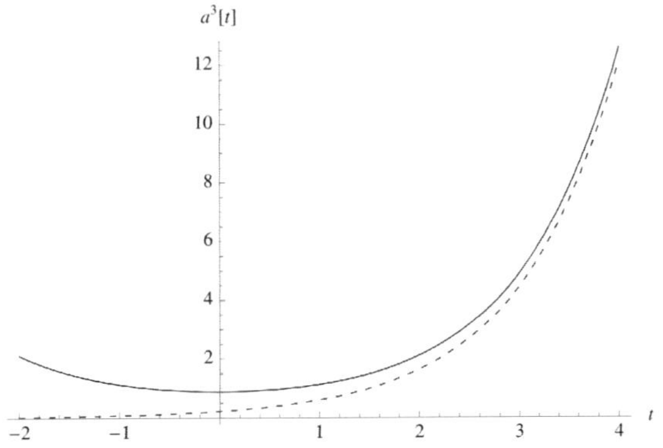

Figure 1. Dynamics of the phase space deformed model for the values $X_0 = Y_0 = 1, \delta_2 = \delta 1 = 0$, $\omega = 0$ and $\beta = 1$. The solid line corresponds to the volume of the universe, calculated with the noncommutative model. The dotted line corresponds to the volume of the de Sitter spacetime. For large values of t the behavior is the same.

$$\hat{y}(t) = \frac{1}{2}\left[e^{\frac{\omega_1^4}{4}t}\cosh\left(\omega'^2 t\right) + e^{-\frac{\omega_1^4}{4}t}\cosh\left(\omega'^2 t\right)\right], \tag{21}$$

where $\omega'^2 = (3\omega_1^4 + 16\omega_2^2)/4$.

Up to this point we have obtained the equations of motion using the deformed Poisson algebra (14). In order to find a solution we define the new variables η and ζ which decouple equations (19) into two differential equations. In the original viariables the volume of the universe is given by

$$a^3(t) = V_0 \cosh^2\left(\frac{1}{4}t\beta\right), \tag{22}$$

where we have taken the $\lim \omega' \to 0$. From the definition of ω, this limit means that $\Lambda = 0$, there is no cosmological constant. From Figure 1 we can notice two things: first we can see that with this construction the volume of the universe is not zero, there is no initial singularity for this cosmological model. Secondly, for large values of t our cosmological model behaves like a de Sitter cosmology. Comparing the models in the late time limit enables us to show that for late times, the noncommutative parameter β can play the role of the cosmological constant.

4. A model for Λ

Following the results of the previous section we will present a model for Λ. We will work with an empty (4+1) theory of gravity with cosmological constant Λ

$$I = \int \sqrt{-g}\,(R - \Lambda)\,dtd^3x d\rho, \tag{23}$$

where $\{t, x^i\}$ are the coordinates of the 4-dimensional space time and ρ represents the coordinate of the fifth dimension. We are interested in Kaluza-Klein cosmology, so an FRW type metric is assumed, which is of the form

$$ds^2 = -dt^2 + \frac{a^2(t)dr^i dr^i}{\left(1 + \frac{\kappa r^2}{4}\right)^2} + \phi^2(t)d\rho^2, \tag{24}$$

where $\kappa = 0, \pm 1$ and $a(t)$, $\phi(t)$ are the scale factors of the universe and the compact dimension. Substituting this metric into the action (23) and integrating over the spatial dimensions, we obtain an 4-dimensional effective lagrangian

$$L = \frac{1}{2}\left(a\phi\dot{a}^2 + a^2\dot{a}\phi - \kappa a\phi + \frac{1}{3}\Lambda a^3\phi\right). \tag{25}$$

For the purposes of simplicity and calculations, we can rewrite this lagrangian in a more convenient way

$$L = \frac{1}{2}\left[\left(\dot{x}^2 - \omega^2 x^2\right) - \left(\dot{y}^2 - \omega^2 y^2\right)\right], \tag{26}$$

where the new variables where defined as

$$x = \frac{1}{\sqrt{8}}\left(a^2 + a\phi - \frac{3\kappa}{\Lambda}\right),\ y = \frac{1}{\sqrt{8}}\left(a^2 - a\phi - \frac{3\kappa}{\Lambda}\right), \tag{27}$$

and $\omega = -\frac{2\Lambda}{3}$. The equations of motion are those derived from varying the action action (26). The Hamiltonian constraint for the model is calculated from the action and is given by

$$H = \frac{1}{2}\left[\left(\dot{x}^2 + \omega^2 x^2\right) - \left(\dot{y}^2 + \omega^2 y^2\right)\right] = 0, \tag{28}$$

which describes an isotropic oscillator-ghost-oscillator system. Now that we have defined the deformed phase space, we can see the effects on the proposed cosmological model. From the action (26) we can obtain the hamiltonian constraint (28), inserting relations (14) into the constraint we get

$$\begin{aligned} H\Psi(\hat{u},\hat{v}) = &\left\{\left(\hat{p}_u - \frac{2(\beta-\theta\omega^2)}{4-\omega^2\theta^2}\hat{v}\right)^2 - \left(\hat{p}_v + \frac{2(\beta-\theta\omega^2)}{4-\omega^2\theta^2}\hat{u}\right)^2\right. \\ &+ \left.\left(\frac{4(\beta-\theta\omega^2)^2}{(4-\omega^2\theta^2)^2} + \frac{4(\omega^2-\beta^2/4)}{4-\omega^2\theta^2}\right)\hat{u}^2 - \left(\frac{4(\beta-\theta\omega^2)^2}{(4-\omega^2\theta^2)^2} + \frac{4(\omega^2-\beta^2/4)}{4-\omega^2\theta^2}\right)\hat{v}^2\right\}\Psi(\hat{u},\hat{v}) = 0, \end{aligned} \tag{29}$$

making the following identifications

$$\begin{aligned} \omega'^2 &\equiv \frac{4(\beta-\theta\omega^2)^2}{(4-\omega^2\theta^2)^2} + \frac{4(\omega^2-\beta^2/4)}{4-\omega^2\theta^2}, \\ A_{\hat{u}} &\equiv \frac{-2(\beta-\theta\omega^2)}{4-\omega^2\theta^2}\hat{v}, \quad A_{\hat{v}} \equiv \frac{2(\beta-\theta\omega^2)}{4-\omega^2\theta^2}\hat{u}, \end{aligned} \tag{30}$$

we can rewrite (29) in a much simpler and suggestive form

$$H = \left\{\left[(\hat{p}_u - A_{\hat{u}})^2 + \omega'^2\hat{u}^2\right] - \left[(\hat{p}_v - A_{\hat{v}})^2 + \omega'^2\hat{v}^2\right]\right\}, \tag{31}$$

which is a two dimensional anisotropic ghost-oscillator [7]. From (31) we can see that the terms $(p_i - A_i)$ can be associated to a minimal coupling term. We found that ω is defined in terms of the cosmological constant, then modifications to the oscillator frequency will imply modifications to the effective cosmological constant. In our case we get

$$\tilde{\omega}^2 = \frac{4(\omega^2-\beta^2/4)}{4-\omega^2\theta^2}. \tag{32}$$

This $\tilde{\omega}$ was obtained by a definition of the effective cosmological constant $\tilde{\Lambda}_{eff} = -\frac{3}{2}\tilde{\omega}^2$ and finally get a redefinition of the effective cosmological constant due to noncommutative parameters

$$\tilde{\Lambda}_{eff} = \frac{4(\Lambda_{eff} + \frac{3}{8}\beta^2)}{4 - \frac{2}{3}\theta^2 \mid \Lambda_{eff} \mid}. \tag{33}$$

In the case where there is no deformation on the coordinates, the noncommutative paramenter $\theta = 0$ and we have that the frequency and the effective cosmological constant are given by

$$\tilde{\omega}^2 = \omega^2 - \frac{\beta^2}{4}, \qquad \text{and} \qquad \tilde{\Lambda}_{eff} = \Lambda_{eff} + \frac{3\beta^2}{8}. \tag{34}$$

We can see that noncommutative parameter β and Λ_{eff} compete to give the effective cosmological constant $\tilde{\Lambda}_{eff}$. If we consider the case of a flat universe with a vanishing Λ_{eff} we see, that $\tilde{\Lambda}_{eff} = \frac{3\beta^2}{8}$. This gives the origin of the cosmological constant in connection with the deformation parameter. As already argued the presence of the deformation parameters θ and β gives an alternative approach to the origin of the cosmological constant, in this approach the cosmological constant is given by the parameter β and is considered a new fundamental constant. Then in this case, one can trace the origin of Λ to the noncommutative deformation between the canonical momenta associated to the 4 dimensional scale factor and the canonical momenta associated to the compact extra dimension.

Discussion and Outlook

In this letter we have constructed a deformed phase space model of scalar field cosmology. The deformation is introduced by modifying the symplectic structure of the minisuperspace variables, in order to have a deformed Poisson algebra among the coordinates and momenta. This construction is consistent with the assumption taken in noncommutative quantum cosmology [3, 4, 5, 6], and enable us to study the effects of phase space deformations in scalar field cosmology.

The deformed phase space model is obtained making the transformation (13) on the canonical Hamiltonian (4) which allow us to work out a Hamiltonian that depends on the deformed variables $\hat{x}_i$ and $\hat{p}_i$. To obtain the noncommutative equations of motion for the model, an in order to find solutions, a convenient change of variables was made. Finally, with the solutions, we where able to return to the original variables, and find that the volume of the universe is given by equation (22).

We found that with our model the volume of the universe behaves like a de Sitter cosmology for large values of t even when $\Lambda = 0$. This allows us to get a relation between the cosmological constant and the deformed parameter for the momenta by comparing the late time evolution of the volume of the noncommutative model with the volume of a de Sitter universe. Evidence of a possible relationship between the late time acceleration of the universe and the noncommutative parameters has been accumulating [7, 9, 10], our results also point in this direction, based on this observation in a model that gives some insight of the origin of Λ.

Finally, we constructed 5 dimensional model in where it is conjectured that the cosmological constant has its origin in the noncommutative deformation between the canonical momenta associated to the 4 dimensional scale factor and the canonical momenta associated to the compact extra dimension.

Acknowledgments

This work is supported by DAIP grant 125/11 and CONACYT grants 62253, 84798, 135023. S.P.P is supported by CONACyT PhD. grant.

References

[1] N. Seiberg and E. Witten, JHEP **9909**, 032 (1999); A. Connes, M. R. Douglas, and A. Schwarz, JHEP **9802:003** (1998); M. R. Douglas and N. A. Nekrasov, Rev. Mod. Phys. **73**, 977 (2001).

[2] H. Garcia-Compean, O. Obregon, C. Ramirez and M. Sabido, Phys. Rev. D **68**, 044015 (2003); H. Garcia-Compean, O. Obregon, C. Ramirez and M. Sabido, Phys. Rev. D **68**, 045010 (2003); P. Aschieri, M. Dimitrijevic, F. Meyer and J. Wess, Class. Quant. Grav. **23**, 1883 (2006); X. Calmet and A. Kobakhidze, Phys. Rev. D **72**, 045010 (2005); L. Alvarez-Gaume, F. Meyer and M. A. Vazquez-Mozo, Nucl. Phys. B **753**, 92 (2006); S. Estrada-Jimenez, H. Garcia-Compean, O. Obregon and C. Ramirez, Phys. Rev. D **78**,124008 (2008).

[3] H. Garcia-Compean, O. Obregon and C. Ramirez, Phys. Rev. Lett. **88**, 161301 (2002).

[4] G. D. Barbosa and N. Pinto-Neto, Phys. Rev. D **70**, 103512 (2004).

[5] L. O. Pimentel and C. Mora, Gen. Rel. Grav. **37** (2005) 817; L. O. Pimentel and O. Obregon, Gen. Rel. Grav. **38**, 553 (2006); M. Aguero, J. A. Aguilar S., C. Ortiz, M. Sabido and J. Socorro, Int. J. Theor. Phys. **46**, 2928 (2007); W. Guzman, C. Ortiz, M. Sabido, J. Socorro and M. A. Aguero, Int. J. Mod. Phys. D **16**, 1625 (2007);B. Vakili, N. Khosravi and H. R. Sepangi, Class. Quant. Grav. **24** (2007) 931

[6] W. Guzman, M. Sabido and J. Socorro, Phys. Rev. D **76**, 087302 (2007).

[7] B. Vakili, P. Pedram, S. Jalalzadeh, Phys. Lett. **B687**, 119-123 (2010).

[8] W. Guzman, M. Sabido, J. Socorro, Phys. Lett. **B697**, 271-274 (2011);V. Guillemin and S. Sternberg, *Cambridge, UK: Univ. Pr. (1990) 468 p.*

[9] O. Obregon, M. Sabido, E. Mena, Mod. Phys. Lett. **A24**, 1907-1914 (2009).

[10] O. Obregon, I. Quiros, Phys. Rev. **D84**, 044005 (2011).

Journal of Physics: Conference Series **378** (2012) 012010 doi:10.1088/1742-6596/378/1/012010

Probing the small-scale structure of spacetime

Ralf Lehnert

Indiana University Center for Spacetime Symmetries, Bloomington, IN 47405, USA

E-mail: ralehner@indiana.edu

Abstract. Many theoretical attempts at understanding the Planck-scale structure of spacetime can accommodate minute departures from Lorentz and CPT invariance. At presently attainable energies, such effects are expected to be accurately described within effective field theory. Such a field theory, in turn, can be employed to identify currently feasible experimental tests. This work presents an overview of the theoretical motivations, the low-energy framework, and the phenomenological implications within this context.

1. Introduction

In the last two decades, there has been a renewed interest in experimental and phenomenological studies of the structure of spacetime at microscopic distance scales. The two primary reasons for this development are the following. First, novel ultrahigh-sensitivity experimental techniques have been established, and new and better observational data have become available [1]. Second, theoretical approaches to more fundamental physics have suggested the possibility of small deviations from our conventional spacetime picture. In particular, the symmetries underlying the spacetime-manifold structure—translation, Lorentz, and CPT invariance—may be violated by minuscule amounts [2]. The present work is primarily concerned with Lorentz- and CPT-symmetry breakdown.

For the identification and interpretation of suitable Lorentz and CPT tests, a theoretical framework allowing for deviations from these symmetries is needed. In the present context, the establishment of such a framework is hampered by the absence of fully realistic and complete theories underlying the Standard Model of particle physics (SM) and general relativity (GR). With these consideration in mind, a framework that captures a wide variety of Lorentz- and CPT-violating effects at presently attainable energies would be desirable and would have to be constructed by hand.

Over the last two decades, this goal has indeed been achieved: a framework has been developed that is based on effective field theory [3] and has become known as the Standard-Model Extension (SME) [4, 5, 6]. The SME contains the SM and GR as limiting cases; it can therefore be applied in all presently feasible experimental situations permitting a theoretical description of practically all current and near-future Lorentz and CPT tests. To date, the SME has provided the basis for over 100 phenomenological studies [1] across many physical systems, such as cosmic radiation [7], meson factories [8] and other particle colliders [9], resonance cavities [10], neutrinos [11], precision spectroscopy [12], and gravity [13].

In addition to experimental and phenomenological applications, the SME has provided a framework for numerous theoretical analyses of violations of Lorentz and CPT invariance. These analyses were concerned with the mathematical structure of effective field theories

Journal of Physics: Conference Series 378 (2012) 012010 doi:10.1088/1742-6596/378/1/012010

with spacetime-symmetry breaking [14], spontaneous Lorentz violation and Nambu–Goldstone modes [15], classical limits of Lorentz- and CPT-breaking physics [16], loop corrections and renormalizability [17, 18], as well as certain aspects of non-renormalizable contributions [19]. The above studies have added to the strength of the theoretical basis of the SME.

The outline of this work is as follows. Section 2 reviews the basic ideas behind the philosophy and the construction of the SME. A few remarks about possible mechanisms for Lorentz and CPT violation in more fundamental physics are contained in Sec. 3. Section 4 concludes this work with the discussion of some classes of experimental searches for departures from Lorentz and CPT symmetry.

2. SME basics

As mentioned in the introduction, the SME has to be constructed by hand, so some prior considerations are necessary. These include the physical framework, the scope, and the actual implementation of Lorentz and CPT breaking.

The physical framework for the description of deviations from spacetime symmetries at small distance scales should, for example, be microscopic; it should also be appropriate for the description of potential quantum-gravity features, such as the low-energy limit of an underlying Planck-level discreteness of spacetime. Effective field theory appears to be suitable in this context: it has been successfully used to describe non-relativistic many-body effects in condensed-matter systems. Note that such systems also exhibit a discrete structure at the atomic level, but the long wave-length limit is nevertheless governed by field theory. From a more general perspective, effective field theory is also a versatile tool in various other subfields of physics, such as nuclear and elementary-particle physics as well as hydrodynamics, and this tool is well understood from a theoretical perspective. For these reasons, a low-energy description of Lorentz and CPT violation in terms of effective field theory seems to be a general and well-founded assumption, and the starting point for the construction of the SME will be a field-theory Lagrangian.

The scope of the model is another important consideration. In the absence of a unique and fully realistic quantum-gravity theory, it is practical to perform broad experimental searches for Lorentz and CPT breakdown in a wide variety of physical systems. For this reason, the SME needs to allow for a general description various types of Lorentz and CPT violation, and it should do so for all known particle species. This can be achieved if the SME Lagrangian $\mathcal{L}_{\rm SME}$ contains both the usual SM Lagrangian $\mathcal{L}_{\rm SM}$ as well as the Einstein–Hilbert Lagrangian $\mathcal{L}_{\rm GR}$ for GR. This yields

$$\mathcal{L}_{\rm SME} = \mathcal{L}_{\rm SM} + \mathcal{L}_{\rm GR} + \delta\mathcal{L}_{\rm LV}\,, \tag{1}$$

as a sensible ansatz for the SME model. Here, $\delta\mathcal{L}_{\rm LV}$ represents Lorentz- and CPT-violating departures from established physics across all presently accessible physical systems.

The Lagrangians $\mathcal{L}_{\rm SM}$ and $\mathcal{L}_{\rm GR}$ are known explicitly, so it remains to determine the Lorentz- and CPT-breaking piece $\delta\mathcal{L}_{\rm LV}$. Various ways can be envisaged to construct this term. Whatever this construction is, it should maintain coordinate independence of physics: coordinates are just a convenient mathematical tool for the description of physical quantities, but their choice should not affect the actual physics. With coordinate independence in mind, one might consider implementing Lorentz and CPT breaking via six spacetime transformations that are more general than the usual Lorentz transformations $\Lambda^{\mu\nu}$, but contain $\Lambda^{\mu\nu}$ as limiting cases at, say, low energies. We won't follow this approach here because from a mathematical viewpoint, the Lorentz group is known to be stable. This essentially means that small group deformations are isomorphic to the original Lorentz group, so that it would be difficult to imagine observable effects. Another possibility for introducing Lorentz and CPT violation while maintaining coordinate independence is through non-dynamical background vectors and tensors

Journal of Physics: Conference Series **378** (2012) 012010 doi:10.1088/1742-6596/378/1/012010

$b^\mu, c^{\mu\nu}, (k_{AF})^\mu, \ldots$ that select preferred directions. This type of Lorentz and CPT breaking provides, in fact, the basis for the construction of the SME.

With the above considerations, we are finally in the position to write down explicitly sample terms contained in $\delta\mathcal{L}_{\text{LV}}$ in the flat-spacetime limit:

$$\delta\mathcal{L}_{\text{LV}} \supset b^\mu\overline{\psi}\gamma_5\gamma_\mu\psi,\ c^{\mu\nu}\overline{\psi}\gamma_\mu\partial_\nu\psi,\ (k_{AF})^\mu A^\nu\tilde{F}_{\mu\nu},\ \ldots, \tag{2}$$

where ψ is a SM fermion, and $\tilde{F}_{\mu\nu}$ denotes the dual field strength corresponding to the SM photon. The non-dynamical background vectors and tensors $b^\mu, c^{\mu\nu}, (k_{AF})^\mu, \ldots$ were introduced earlier; they control the type and extent of deviations from Lorentz symmetry. Both b^μ and $(k_{AF})^\mu$ have dimensions of mass. In addition to Lorentz invariance, they also violate CPT symmetry. The $c^{\mu\nu}$ coefficient is dimensionless and preserves CPT. These coefficients lead to physical effects that can be measured or constrained in experiments.

It is evident that the above construction possesses a large degree of generality. In fact, infinitely many terms of the form (2) can be added to $\delta\mathcal{L}_{\text{LV}}$. From a phenomenological perspective, the dominant terms are most interesting. These terms would be expected to be given by the subset of relevant and marginal operators one can construct. This yields a set of finitely many contributions to $\delta\mathcal{L}_{\text{LV}}$ with a manageable number of Lorentz-violating coefficients. Other physically desirable features, like translation invariance and the usual gauge structure of the SM, can be imposed. This determines the most popular limit of the SME called the minimal Standard-Model Extension (mSME).

We remark that in curved-spacetime contexts involving non-negligible gravitational effects, it is generally difficult to impose spacetime independence on the Lorentz-breaking background vectors and tensors. Moreover, dynamical features must be added to the background Lorentz-violating quantities (such as *spontaneous* Lorentz breakdown) to ensure compatibility with the underlying geometric manifold structure.

3. Possible underlying sources for Lorentz breaking

Thus far, we have merely parametrized possible low-energy remnants of Lorentz violations within effective field theory. While such an approach was forced on us from a phenomenological perspective, it leaves unanswered the question concerning the origin of Lorentz breaking at a more fundamental level. As mentioned in the introduction, this question is presently difficult to address because of the absence of a completely satisfactory theory underlying the SM and GR. However, one can make some general remarks regarding possible mechanisms that can produce Lorentz-violating background fields, such as $b^\mu, c^{\mu\nu}, (k_{AF})^\mu, \ldots$ introduced above. This section provides brief overviews of some of the mechanisms cited in Ref. [2].

Spontaneous Lorentz and CPT violation.—From a theoretical point of view, spontaneous symmetry breaking (SSB) is an attractive mechanism for creating the Lorentz-violating $b^\mu, c^{\mu\nu}, (k_{AF})^\mu, \ldots$ coefficients in the SME. SSB is well established in condensed-matter physics. Moreover, in elementary-particle physics SSB is presently the preferred way for a theoretically consistent description of heavy spin-1 bosons, and it is therefore a major ingredient in the SM.

The basic idea behind SSB is that the lowest-energy state of the system, which is commonly associated with the vacuum, requires the field value to be non-zero. Such non-vanishing vacuum expectation values (VEVs) typically do not reflect the dynamical symmetries of the system and are the key observable effect associated with SSB. In the case of Lorentz violation, these VEVs correspond to the $b^\mu, c^{\mu\nu}, (k_{AF})^\mu, \ldots$ coefficients contained in the SME. The interactions required for the spontaneous breakdown of Lorentz symmetry are difficult to fit into the framework of conventional renormalizable gauge theories. String field theory, on the other hand, provides a setting in which such interactions are natural.

Varying scalars.—Various theoretical approaches to more fundamental physics involve additional scalar fields beyond the SM Higgs. Moreover, some astrophysical observations, such

Journal of Physics: Conference Series **378** (2012) 012010 doi:10.1088/1742-6596/378/1/012010

as the flatness of the universe and its accelerated expansion, are often taken as experimental hints for novel scalars: the idea is that in such cosmological contexts, a scalar field can acquire a non-zero global value with a time evolution closely tied to that of the scale factor.

As an explicit example, let us consider $N = 4$ supergravity in four spacetime dimensions. Although unrealistic in detail (its particle content cannot be matched exactly to observations), it is contained in M-theory and may therefore yield insight into some generic features of a candidate quantum-gravity model. One such feature is the aforementioned presence of additional scalar fields: an axion a and dilation b, which are coupled via a known function $f(a, b)$ to the electromagnetic field strength, such that the term $f(a, b)F\tilde{F}$ is contained in the model's Lagrangian. Within a simple cosmological setting, it is then possible to determine the dependence of the axion and the dilaton on the comoving time t. This produces the effective lagrangian term $f(t)F\tilde{F}$. In a localized experiment, such a term would indeed be perceived as a varying coupling (in the present case as a time-dependent θ angle).

A scalar field with a global spacetime dependence, regardless of the mechanism that drives this effect, is typically associated with a violation of spacetime-translation symmetry. Note also that Lorentz invariance is normally broken in such cases: since the gradient of a varying scalar field must be non-zero (at least in some regions of spacetime), it selects a preferred direction. Mathematically, this result is intuitively reasonable because the definition of the Lorentz-transformation generators involves the energy–momentum tensor, which ceases to be conserved. It follows that the ordinary time-independent rotation and boost operators may not exist. Effective Lorentz violation can also be explicitly exposed in our toy supergravity model: an integration by parts at the level of the action yields

$$f(t)F\tilde{F} \to -2\,(\partial^\alpha f)\,A^\beta \tilde{F}_{\alpha\beta}\,, \tag{3}$$

which leaves unaffected the equations of motion under mild assumptions. The gradient $\partial^\alpha f$ of the cosmological background $f(t)$ is practically outside of experimental control for all local measurements. The gradient can then be taken as non-dynamical and—for time dependencies on cosmological scales—as approximately constant. Identifying $-2\,(\partial^\alpha f)$ with $(k_{AF})^\alpha$ in the SME context of Eq. (2), we explicitly see how this Lorentz- and CPT-violating Chern–Simons-type correction can arise in underlying physics.

Non-commutative field theory.—This popular approach to capture quantum aspects of the dynamics of spacetime is based on the idea that coordinates are no longer real numbers; they are promoted to operators that obey nontrivial commutation relations. A standard example often found in the literature is given by $[x^\mu, x^\nu] = i\theta^{\mu\nu}$. In this expression, the quantity $\theta^{\mu\nu} \neq 0$ is usually defined to be spacetime independent, and it therefore clearly exhibits the attributes of a coefficient in the SME, such as selecting preferred directions. For the physical interpretation of non-commutative models, one typically employs the Seiberg–Witten map, which transforms the model to a conventional field theory on Minkowski space. This Seiberg–Witten image of the non-commutative model still involves the non-dynamical and constant $\theta^{\mu\nu}$, which acts precisely like a coefficient in the SME.

Loop quantum gravity.—A further widely studied approach to a quantized version of GR is loop quantum gravity. In certain semiclassical investigations, a number of results have been obtained that are incompatible with the maintenance of Lorentz symmetry. For example, under certain reasonable assumptions both electrodynamics and fermions acquire loop-quantum-gravity corrections that correspond at low energies to various Lorentz-violating SME coefficients.

4. Experimental studies

Once a consistent test framework for Lorentz and CPT breaking has been established and convincingly motivated, it can be employed for its primary purpose: the identification, analysis, and comparison of Lorentz-symmetry tests. As mentioned in the introduction, the last decade or so has witnessed a revival of interest in Lorentz tests, and this section surveys some of the main ideas in this field.

Kinematical tests involving particle reactions.—One of the key effects of Lorentz and CPT violation is that one-particle dispersion relations are generally modified: they now contain the $b^\mu, c^{\mu\nu}, (k_{AF})^\mu, \ldots$ coefficients from the SME. Since these are taken as spacetime independent in non-gravitational mSME contexts, the corresponding modified 4-momenta remain conserved. This will normally affect the details of the energy–momentum kinematics in particle reactions. For instance, decay thresholds may be shifted, collision processes kinematically forbidden in the Lorentz-invariant case may now occur, and certain conventional decays or reactions may no longer be permitted.

As an example, let us look at the spontaneous emission of a photon from a free charge. In ordinary Lorentz-symmetric physics, 4-momentum conservation forbids this process. However, certain mSME coefficients can act like a crystal on electromagnetic radiation causing an "index of refraction" even in vacuum. This, in turn, can lead to a photon speed that is smaller than that of a charge. Paralleling the ordinary Cherenkov effect (when photons propagate slower inside a *macroscopic medium* with index of refraction $n > 1$), charges are then unstable against the emission of Cherenkov light in such a Lorentz- and CPT-breaking spacetime.

Depending on the type of Lorentz violation, this "vacuum Cherenkov effect" may or may not involve a threshold. Let us continue by focussing on those mSME coefficients that would lead to a threshold for vacuum Cherenkov radiation. In such a case, we may determine an experimental limit on the size of this type of Lorentz breaking in the following way. Charges propagating at speeds larger than the modified speed of photons would not be stable: they would decelerate to speeds below threshold via the emission of vacuum Cherenkov photons. It then becomes apparent that if high-energy stable charges are observed, they have to have energies below the threshold. Thus, the simple observation of highly energetic electrons, for example, determines a lower bound for the threshold energy, from which a limit on Lorentz violation can be extracted. Employing LEP electrons with energies up to 104.5 GeV in this context yields a (one-sided) constraint at the level of 1.2×10^{-11} on the appropriate mSME coefficients.

Another effect along these lines is photon splitting in vacuum. In ordinary Lorentz-invariant physics, such a decay process is forbidden kinematically by energy–momentum conservation. However, certain types of Lorentz breaking may actually lead to photon speeds greater than c instead of slowing light down. It is then possible for photons to propagate faster than the maximal attainable speed of charges. In analogy to the vacuum-Cherenkov case discussed above, in which charges beyond the threshold become unstable, one expects that highly energetic photons can now split into a charge–anticharge pair. A dispersion-relation analysis within the mSME test framework indeed confirms that this is the case. Like the vacuum Cherenkov effect, photon splitting in a Lorentz-breaking spacetime often occurs beyond a threshold, and it is then an excellent tool for the determination of observational constraints on this particular type of Lorentz breakdown. The line of reasoning parallels the Cherenkov case: if photons are observed to be stable, their energy must lie below the splitting threshold. One can then conclude that this threshold must be higher than the observed energy of these stable photons. This bound on the threshold energy can be used to establish a constraint on the value of the associated mSME coefficient. At the Tevatron collider, stable 300 GeV photons have been measured, which yields a (one-sided) limit of 5.8×10^{-12} on the appropriate Lorentz-violating coefficient.

We remark that the derivation of the above constraints requires not only the kinematical possibility of the decays, but also a sufficient efficiency of these processes. But for calculations of the decay rate, the dispersion relation alone is inadequate, and dynamical aspects need to be considered. Since the SME is an effective field theory, dynamical investigations, such as decay-rate and cross-section studies, can be performed. In the above context of vacuum Cherenkov radiation and photon splitting, the efficiency required for a conservative extraction of bounds

Journal of Physics: Conference Series **378** (2012) 012010 doi:10.1088/1742-6596/378/1/012010

has been confirmed validating our purely kinematical reasoning.

Spectropolarimetric studies of astrophysical sources.—The mSME's photon sector contains the aforementioned $(k_{AF})^\mu$ coefficient. It has mass dimension 1, and the corresponding operator breaks both Lorentz and CPT symmetry. The $(k_{AF})^\mu$ contribution to the Lagrangian is sometimes called a Chern–Simons term. This term would lead to various effects, such as birefringence of light, vacuum Cherenkov radiation, and cavity-frequency shifts. These Lorentz- and CPT-violating effects would lead to measurable signatures and can therefore be employed for experimental tests. Observational searches for birefringence in astrophysical photons are a particularly promising tool in this context because the cosmological scale of the propagation time translates directly into extremely high sensitivities to the $(k_{AF})^\mu$ coefficient. Spectropolarimetry of electromagnetic radiation from astrophysical objects has placed constraints better than 10^{-42} GeV on various components of $(k_{AF})^\mu$.

Investigations of cold antihydrogen.—Lorentz violation does not only affect free particles at high energies, but also bound-state energy levels. To gain some intuition, recall that departures from Lorentz symmetry are represented by background vectors and tensors in the SME, which is analogous to conventional physics in an external electromagnetic field. One therefore expects Lorentz breaking to affect bound systems via level shifts similar to the Zeeman and Stark effects, and this expectation can indeed be confirmed with SME calculations.

Lorentz breaking that is accompanied by CPT violation can be particularly well studied with matter–antimatter comparisons. In the present bound-state context, hydrogen (H) and antihydrogen ($\overline{\text{H}}$) spectroscopy provide an excellent probe for such CPT-violating effects. One of the transitions that can be considered is the unmixed 1S–2S transition: the expected experimental sensitivity for this measurement is at the 10^{-18} level, a promising reach in the context of a possible Planck suppression of effects related to the quantum structure of spacetime. Although a study within the mSME reveals that there are indeed Lorentz-violating level shifts in hydrogenlike systems, they are identical in the initial and final states at leading order, so that the transitions frequency remains unaffected. From this viewpoint, it appears that the unmixed 1S–2S transition is actually less suitable to search for unsuppressed Lorentz- and CPT-violating signatures governed by effective field theory. Within the mSME, one can show that the leading non-vanishing correction to this transition arises from relativistic effects, and it contains two additional factors of the fine-structure constant α. The transition-energy shift, which is expected to be tiny already at order zero in α, therefore comes with a further suppression factor of roughly ten thousand.

A further transition suitable for Lorentz- and CPT-violation searches is the spin-mixed 1S–2S transition. When H or $\overline{\text{H}}$ is confined with electromagnetic fields—such as in a Ioffe–Pritchard trap—both the 1S and the 2S states are split due to the conventional Zeeman effect. For this set-up, a study within the mSME establishes that the 1S–2S transition between the spin-mixed states is indeed altered at first order by Lorentz- and CPT-breaking coefficients. From a practical perspective, the magnetic-field dependence of this transition needs to be considered: it can limit the experimental sensitivity due to the inhomogeneity of the $\vec{B}$ field in the trap, which represents a drawback for studying this particular transition. However, the development of new experimental techniques may circumvent this drawback, and a frequency resolutions in the vicinity of the natural linewidth may then be achievable.

A third possibility for a transition suitable for Lorentz and CPT tests lies in the hyperfine Zeeman transitions within the 1S state itself. Even in the limit of a zero $\vec{B}$ field, the mSME establishes leading-order level shifts for two of the transitions between the Zeeman-split states. We note that this idea may also provide some experimental benefits: a variety of similar transitions, such as the usual H-maser line, can be well resolved in the laboratory.

Lorentz-violation searches involving Penning traps.—An analysis within the mSME demonstrates that not only energy levels in atoms can be shifted by departures from Lorentz and CPT symmetry, but also, for example, the levels of protons and antiprotons confined in a Penning trap. A perturbation-theory study establishes that only a single mSME coefficient (the previously mentioned CPT-violating b^μ-type background vector, which is contracted with the chiral current of a fermion) would lead to shifts in the transition frequency of the proton relative to the antiproton at first order. More concretely, the anomaly frequency of the proton is moved in a direction opposite to that of its antiparticle. This prediction can be employed to extract a clean experimental limit on the b^μ coefficient of the proton.

Tests at Neutral-meson factories.—Neutral-meson interferometry is widely appreciated to be an ultra-sensitive probe for CPT-symmetry: even minuscule differences between the meson and the antimeson mass would yield observable signatures suitable for interferometric experiments. Despite the fact that there is only a single mass parameter in the mSME for a each quark species and the associated antiquark species, these (anti)particles can nevertheless be affected differently by the Lorentz- and CPT-violating mSME coefficients. This can lead to a disparity between the dispersion relations for a meson and its antimeson, so that they can exhibit distinct energies although they have the same 3-momenta. Interferometric experiments are sensitive to such a difference in energy, and this idea provides the basis for measurements in this context. We note that in addition to the K-meson, also other neutral mesons can be studied with this idea. We remark in particular that not only CPT violation but also Lorentz breaking is present, so that boost- and rotation-dependent signals can be constrained.

Acknowledgments

The author wishes to thank Myriam Mondragón for the invitation to this stimulating meeting and for financial support to attend. This work has also been supported by the Indiana University Center for Spacetime Symmetries and by the Portuguese Fundação para a Ciência e a Tecnologia.

References

[1] V.A. Kostelecký and N. Russell, Rev. Mod. Phys. **83**, 11 (2011).
[2] See, e.g., V.A. Kostelecký and S. Samuel, Phys. Rev. D **39**, 683 (1989); V.A. Kostelecký and R. Potting, Nucl. Phys. B **359**, 545 (1991); J. Alfaro, H.A. Morales-Técotl, and L.F. Urrutia, Phys. Rev. Lett. **84**, 2318 (2000); S.M. Carroll *et al.*, Phys. Rev. Lett. **87**, 141601 (2001); J.D. Bjorken, Phys. Rev. D **67**, 043508 (2003); V.A. Kostelecký, R. Lehnert, and M.J. Perry, Phys. Rev. D **68**, 123511 (2003); O. Bertolami *et al.*, Phys. Rev. D **69**, 083513 (2004).
[3] V.A. Kostelecký and R. Potting, Phys. Rev. D **51**, 3923 (1995).
[4] D. Colladay and V.A. Kostelecký, Phys. Rev. D **55**, 6760 (1997); Phys. Rev. D **58**, 116002 (1998); V.A. Kostelecký, Phys. Rev. D **69**, 105009 (2004); V.A. Kostelecký and M. Mewes, arXiv:1112.6395.
[5] V.A. Kostelecký and R. Lehnert, Phys. Rev. D **63**, 065008 (2001).
[6] V.A. Kostelecký and M. Mewes, Phys. Rev. D **80**, 015020 (2009).
[7] See, e.g., V.A. Kostelecký and M. Mewes, Phys. Rev. Lett. **87**, 251304 (2001); Phys. Rev. D **66**, 056005 (2002); R. Lehnert, Phys. Rev. D **68**, 085003 (2003); R. Lehnert and R. Potting, Phys. Rev. Lett. **93**, 110402 (2004); Phys. Rev. D **70**, 125010 (2004); B. Altschul, Phys. Rev. D **78**, 085018 (2008).
[8] G. Amelino-Camelia *et al.*, Eur. Phys. J. C **68**, 619 (2010); KLOE Collaboration, A. Di Domenico, J. Phys. Conf. Ser. **171**, 012008 (2009); BaBar Collaboration, B. Aubert *et al.*, Phys. Rev. Lett. **100**, 131802 (2008); FOCUS Collaboration, J.M. Link *et al.*, Phys. Lett. B **556**, 7 (2003); KTeV Collaboration, H. Nguyen, hep-ex/0112046; V.A. Kostelecký and R.J. Van Kooten, Phys. Rev. D **82**, 101702(R) (2010); V.A. Kostelecký, Phys. Rev. D **64**, 076001 (2001); Phys. Rev. D **61**, 016002 (1999); Phys. Rev. Lett. **80**, 1818 (1998).
[9] See, e.g., M.A. Hohensee *et al.*, Phys. Rev. Lett. **102**, 170402 (2009); Phys. Rev. D **80**, 036010 (2009); J.-P. Bocquet *et al.* [GRAAL Collaboration], Phys. Rev. Lett. **104**, 241601 (2010); B. Altschul, Phys. Rev. D **80**, 091901 (2009); Phys. Rev. D **84**, 076006 (2011).
[10] H. Müller *et al.*, Phys. Rev. D **68**, 116006 (2003); Ch. Eisele *et al.*, Phys. Rev. Lett. **103**, 090401 (2009); M.E. Tobar *et al.*, Phys. Rev. D **80**, 125024 (2009).
[11] T. Katori, V.A. Kostelecký, and R. Tayloe,, Phys. Rev. D **74**, 105009 (2006); J.S. Díaz, V.A. Kostelecký, and M. Mewes, Phys. Rev. D **80**, 076007 (2009); V. Barger *et al.*, Phys. Rev. D **84**, 056014 (2011); J.S. Díaz and V.A. Kostelecký, arXiv:1108.1799.
[12] R. Bluhm, V.A. Kostelecký, and N. Russell, Phys. Rev. D **57**, 3932 (1998); H. Dehmelt *et al.*, Phys. Rev.

Lett. **83**, 4694 (1999); R. Mittleman *et al.*, Phys. Rev. Lett. **83**, 2116 (1999); G. Gabrielse *et al.*, Phys. Rev. Lett. **82**, 3198 (1999); D. Bear *et al.*, Phys. Rev. Lett. **85**, 5038 (2000); P. Wolf *et al.*, Phys. Rev. Lett. **96**, 060801 (2006); I. Altarev *et al.*, Phys. Rev. Lett. **103**, 081602 (2009); Europhys. Lett. **92**, 51001 (2010).

[13] Q.G. Bailey and V.A. Kostelecký, Phys. Rev. D **74**, 045001 (2006); J.B.R. Battat, J.F. Chandler, and C.W. Stubbs, Phys. Rev. Lett. **99**, 241103 (2007); H. Müller *et al.*, Phys. Rev. Lett. **100**, 031101 (2008); K.Y. Chung *et al.*, Phys. Rev. D **80**, 016002 (2009); V.A. Kostelecký and J.D. Tasson, Phys. Rev. Lett. **102**, 010402 (2009); Q.G. Bailey, Phys. Rev. D **80**, 044004 (2009); V.A. Kostelecký and J.D. Tasson, Phys. Rev. D **83**, 016013 (2011); R. Tso and Q.G. Bailey, Phys. Rev. D **84**, 085025 (2011).

[14] O.W. Greenberg, Phys. Rev. Lett. **89**, 231602 (2002); R. Lehnert, J. Math. Phys. **45**, 3399 (2004); Phys. Rev. D **74**, 125001 (2006); A.J. Hariton and R. Lehnert, Phys. Lett. A **367**, 11 (2007); V.A. Kostelecký, Phys. Lett. B **701**, 137 (2011).

[15] V.A. Kostelecký and S. Samuel, Phys. Rev. D **40**, 1886 (1989); R. Bluhm and V.A. Kostelecký, Phys. Rev. D **71**, 065008 (2005); R. Bluhm, S.-H. Fung, and V.A. Kostelecký, Phys. Rev. D **77**, 065020 (2008); R. Bluhm *et al.*, Phys. Rev. D **77**, 125007 (2008); V.A. Kostelecký and R. Potting, Phys. Rev. D **79**, 065018 (2009); J. Alfaro and L.F. Urrutia, Phys. Rev. D **81**, 025007 (2010); B. Altschul, Q. G. Bailey, and V.A. Kostelecký, Phys. Rev. D **81**, 065028 (2010); C. Armendáriz-Picón, A. Díez-Tejedor, and R. Penco, JHEP **1010**, 079 (2010).

[16] Q.G. Bailey and V.A. Kostelecký, Phys. Rev. D **70**, 076006 (2004); R. Lehnert, Rev. Mex. Fís. **56 (6)**, 469 (2010); Q.G. Bailey, Phys. Rev. D **82**, 065012 (2010); V.A. Kostelecký and N. Russell, Phys. Lett. B **693**, 443 (2010).

[17] R. Jackiw and V.A. Kostelecký, Phys. Rev. Lett. **82**, 3572 (1999); M. Pérez-Victoria, Phys. Rev. Lett. **83**, 2518 (1999); JHEP **0104**, 032 (2001); V.A. Kostelecký, C.D. Lane, and A.G.M. Pickering, Phys. Rev. D **65**, 056006 (2002); B. Altschul, Phys. Rev. D **69**, 125009 (2004); D.L. Anderson, M. Sher, and I. Turan, Phys. Rev. D **70**, 016001 (2004); D. Colladay and P. McDonald, Phys. Rev. D **77**, 085006 (2008); Phys. Rev. D **79**, 125019 (2009); M. Cambiaso, R. Lehnert, and R. Potting, in preparation.

[18] J. Alfaro *et al.*, Phys. Lett. B **639**, 586 (2006); Int. J. Mod. Phys. A **25**, 3271 (2010).

[19] R.C. Myers and M. Pospelov, Phys. Rev. Lett. **90**, 211601 (2003); B. Altschul and V.A. Kostelecký, Phys. Lett. B **628**, 106 (2005); C.M. Reyes, L.F. Urrutia, and J.D. Vergara, Phys. Rev. D **78**, 125011 (2008); Phys. Lett. B **675**, 336 (2009); C.M. Reyes, Phys. Rev. D **82**, 125036 (2010); M. Cambiaso and L.F. Urrutia, Phys. Rev. D **82**, 101502 (2010); M. Cambiaso, R. Lehnert, and R. Potting, arXiv:1201.3045 [hep-th], accepted for publication in Phys. Rev. D.

Journal of Physics: Conference Series **378** (2012) 012011 doi:10.1088/1742-6596/378/1/012011

Challenges of String Theory: Particle Physics and Cosmology

Saúl Ramos-Sánchez

Instituto de Física, Universidad Nacional Autónoma de México, Apdo. Postal 20-364, México 01000, México

E-mail: ramos@fisica.unam.mx

Abstract. We review some of the mechanisms that string theory offers to address phenomenological questions, such as the origin of our observable four-dimensional space-time, the standard model gauge symmetries and matter, as well as the possible source of the inflationary epoch. These notes are not comprehensive but attempt to provide a general view of the current search for physics cases within the context of string model building.

1. Preliminaries

Two of the major contributions of the 20th century to science are doubtless the standard model of particle physics (SM) and the cosmological model ΛCDM. The former combines the revolutionary ideas of special relativity and quantum mechanics to describe the smallest components of matter while the latter is based on general relativity and describes not only the large scale universe, but also its evolution in time.

Despite its enormous experimental success, the SM leaves some open questions. Among them, there is the origin of its 19 free parameters, the number and structure of the families of quarks and leptons, the neutrino masses (which add 7 more free parameters), and the reason of the apparent unification of the coupling constants α_i at a large energy scale. One urgent question is also the existence of the conjectured Higgs boson, for which recent data from the LHC suggest a mass around 125 GeV [1, 2, 3].

On the other hand, the ΛCDM model provides the best fit to the observational data about the CMB, supernovae and the structure of the universe at large scales. Curiously, this model indicates that the SM can only describe about 4% of the universe. The remaining 96% corresponds to a mixture of dark matter and dark energy, whose nature is still to be unveiled. In the ΛCDM model, the existence of a period of exponential expansion known as inflation is conjectured. This possibility is the strongest candidate as explanation of the structure of our universe due to the observations by COBE [4] and WMAP [5]. Unfortunately, the cosmological model does not clarify which mechanisms trigger and terminate inflation. Nor does it explain the nature of any of the dark components of the universe.

Aiming at solving the *aesthetical* problems of the SM and completing our cosmological understanding of the universe, several ideas have been concocted. Among the most relevant ones, we find: supersymmetry [6, 7, 8, 9] and extra dimensions [10, 11] as plausible keys to the stability of the electroweak scale and the origin of dark matter; and grand unification theories (GUTs) [12, 13] as possible explanation of the origin of the three microscopic fundamental

Journal of Physics: Conference Series **378** (2012) 012011 doi:10.1088/1742-6596/378/1/012011

forces and charge quantization. Although these proposals include some hurdles (such as the introduction of new free parameters and new exotic particles), they are considered to be promising solutions to the puzzles of contemporary physics. Nevertheless, they do not address why the major theories upon which our current understanding of physics rely are fully foreign to one another. Furthermore, none of these ideas deal with the quantization of gravity, which is fundamental to understand gravity at tiny scales and at the very early universe.

String theory is perhaps the best candidate to provide a quantized theory of gravity. Its only assumption is that all what surrounds us is composed of tiny vibrating strings instead of point-like particles. Surprisingly, this relatively simple assumption that originally attempted to address the physics of the strong interactions turned out to yield a quantum theory of gravity [14, 15, 16]. This breathtaking observation was further complemented by the fact that the emerging *string theory* also predicts the existence of extra dimensions and supersymmetry in our space-time (if the two-dimensional space-time swept by an evolving string –called *worldsheet*– is endowed with supersymmetry). Moreover, it leads to gauge symmetries as large as those proposed by GUTs, i.e. it has the potential of solving simultaneously several of our theoretical worries in both theories that are relevant to high energy physics [17, 18]. It is precisely this feature what encourages many physicists to expect string theory to be indeed the fundamental theory that Einstein longed for.

There exist 5 types of supersymmetric string theories in 10 dimensions: the heterotic (closed) strings which result from combining a supersymmetric string with a bosonic one and lead to $\mathcal{N}=1$ SUGRA in 10D with gauge group $E_8\times E_8$ or SO(32); the (open) strings type I with gauge group SO(32) which yields $\mathcal{N}=1$ SUGRA in 10D; and the (closed) strings type II A and B which lead to $\mathcal{N}=2$ supersymmetric fields without and with chirality, respectively. This means that, in contrast to field theory where an infinite number of actions are admissible, string theory is so restrictive that only 5 actions are allowed.

Since the "second revolution" in the 1990's it is known that all 5 string theories are different expressions of a more fundamental theory in 11 dimensions called M-theory [19, 20]. It was found that the low-energy limit of M-theory is 11-dimensional supergravity and that, from this effective theory, it is possible to arrive at the effective limits of the heterotic string with gauge group $E_8\times E_8$ and of the string theory type IIA. Further, compactifying the extra dimensions of the heterotic string with gauge group SO(32) on a tiny (small volume) manifold renders the same result as compactifying the extra space of the heterotic string with gauge group $E_8\times E_8$ on a huge (large volume) manifold. The same is true between the string theories type IIA and IIB. Further, the fifth theory, type I, studied at small coupling constant is equivalent to the heterotic string with gauge group SO(32) at large coupling constant. This set of relations between the various string theories that arise in string theory is called a series of dualities. One could be thus tempted to conclude that, if one wishes to characterize the 4-dimensional solutions emerging from string theory, it might suffice to fully understand one type of strings. However, it becomes rather difficult to fully understand a theory as its coupling constant becomes large, QCD being the best known example of the difficulties that emerge in this case.

The name *string* theory is rather antidemocratic, since the theory contains other dynamic structures beside strings. In the 1990's Joseph Polchinski discovered that the consistency of the theory (in presence of the cited dualities) requires the inclusion of objects that cover p spatial dimensions called Dirichlet p-branes or simply Dp-branes [21, 22], on which open strings with Dirichlet boundary conditions can end. These $p+1$-dimensional objects[1] can cover the whole 10-dimensional space-time predicted by string theory or only a subspace. Perhaps the most relevant aspects of D-banes are that they can yield gauge symmetries and that they appear in type II string theories, i.e. precisely where gauge groups cannot appear otherwise. The gauge

[1] A time-like dimension is included.

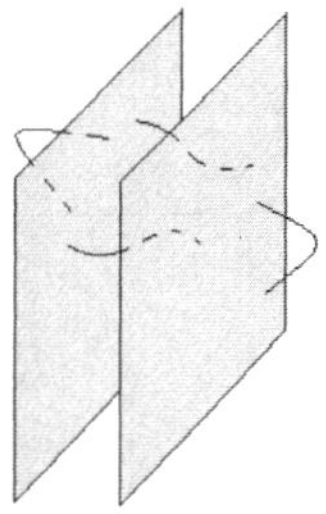

Figure 1: A stack of 2 D-branes. Independent open strings lead to the gauge bosons of the gauge group $U(2) = SU(2) \times U(1)$.

symmetries are born from considering all independent oriented open strings that can be attached to a D-brane or to a stack of superposed D-branes. These independent strings correspond to gauge fields in the low-energy effective theory confined to the space of the D-branes. For example, one single D-brane allows only one independent string and, therefore, one gauge boson in the emerging field theory, so that the corresponding gauge group is U(1). The situation becomes even more interesting when more than one D-brane are involved. For example, for a stack of 2 superposed D-branes there are four independent open strings (see fig. 1), leading to the gauge group U(2). In general, for stacks of N D-branes there are N^2 open strings which generate the gauge groups $U(N) = SU(N) \times U(1)$.[2]

It is important to mention that not every p is admissible. Consistency of the theory demands that only $p = 0 \mod 2$ $(0, 2, 4, 6, 8)$ and only $p = 1 \mod 2$ $(-1, 1, 3, 5, 7, 9)$ appear in the type IIA string theory and type IIB string theory, respectively. A D(-1) brane can be understood as an instanton whereas D0 branes are point-like particles. In type I string theory, also only odd values of p are allowed, although these are restricted to the values $p = 1, 5, 9$. This is, as we shall see below, due to the fact that type I can be seen as the one half of the type IIB theory left untouched by a special "surgery" called *orientifold*.

With all these ingredients at hand, one of the tasks of today's string theorists is to bring the enormous theoretical building that string theory represents in contact with the physics we know while giving answer to some (or all) of the persisting puzzles. This is the goal of a very active branch of string theory called *string phenomenology*, in which we are interested in these notes. A fast inspection of the aforementioned properties of string theory reveals that there are important differences between string theory and the successful theories we are trying to reproduce and improve. Those discrepancies are the fundamental challenges of string theory that we might enumerate as follows:

(i) 4D vs. 10D;
(ii) Matter states (quarks and leptons) from strings;
(iii) Exact (MS)SM matter spectrum and interactions (masses);
(iv) Stable cosmological constant.

The purpose of these notes is to provide some flavor of how string theory can give answer to these questions and what the prospects for observable physics are.

2. First challenge: 4D vs. 10D

The fact that our everyday life occurs in four dimensions is obvious to everybody. However, what would we expect if the additional dimensions were too small to be seen by the naked eye?

[2] If the branes wrap orbifold-invariant planes, the resulting group is symplectic Sp(N). Note that $SU(2) \simeq Sp(2)$.

a)

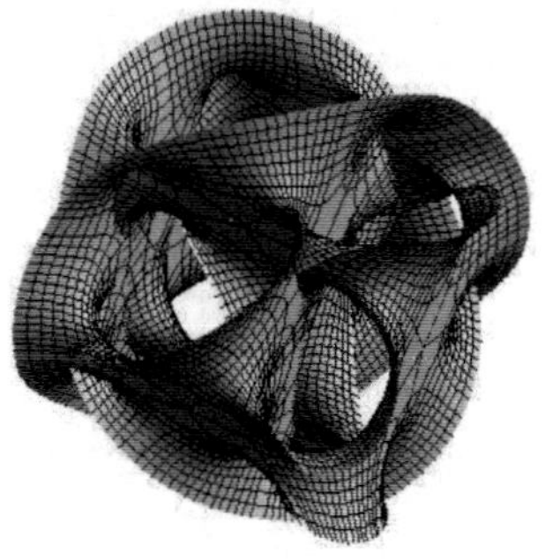

b)

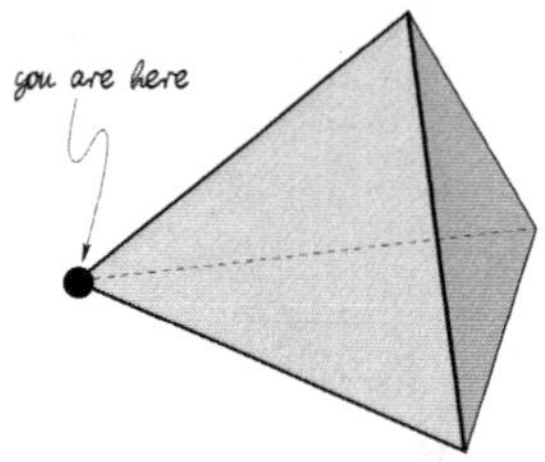

Figure 2: a) 3D projection of a quintic Calabi-Yau manifold. b) Orbifold space resulting from dividing a $\mathbb{Z}_2$ out of a 2D torus. The curvature singularities are the vertices.

A notable physical consequence of extra dimensions is that Newton's law would change at very small scales. For instance, if an extra dimension is replaced by a circle of radius R instead of an infinite line, the gravitational potential of a body with mass M valid within the circle would be [23]

$$V_5 = -G_5 \frac{M}{r_5^2}, \qquad \text{with} \quad r_5^2 = r_4^2 + \zeta^2 = x^2 + y^2 + z^2 + \zeta^2, \tag{1}$$

instead of the usual $V_4 = -G_4 M/r_4$, where G_4 and G_5 denote the Newton constants in 4D and 5D respectively, and ζ represents the coordinate of the additional spatial dimension. Very sensitive experiments have searched for this discrepancy at scales as small as $\sim$ 1mm without success [24]. But the extra dimension(s), if it (they) existed, could well be beyond the experimental threshold.

The conjecture of very small compact dimensions was introduced by Theodor Kaluza and Oskar Klein [25, 26]. Although in their pioneering works they failed to unify gravity with electromagnetism, as they attempted, with the advent of string theory their ideas have gained vigor.

2.1. Compactifications

Following the line of thought of Kaluza and Klein, a consistent technique to achieve models from string theory that at least approach to our apparently four-dimensional universe is to allow the six additional dimensions predicted by string theory to form Ricci flat compact spaces. That is, one conjectures that the whole space-time of strings X_{10} can be split into $X_6 \otimes \mathbb{M}^4$, where X_6 denotes a 6D compact manifold and $\mathbb{M}^4$ our observable Minkowski space-time. This method to "hide" the additional dimensions is called *compactification.* Since the consistent string theories are supersymmetric, one must impose further constraints to the shape of X_6 if one aims at a four-dimensional effective theory endowed with at most $\mathcal{N} = 1$ supersymmetry. To arrive at 4D $\mathcal{N} = 1$ supersymmetric field theories via compactification, the compact space must have SU(3) holonomy group (or a subgroup thereof larger than SU(2)).

These spaces can either be smooth like Calabi-Yau (CY) threefolds CY_3 [27] or contain a finite set of curvature singularities like an orbifold $\mathcal{O}_6$ [28, 29]. They are illustrated in fig. 2. Orbifolds are singular limits of CY manifolds, but the former are much simpler than the latter. The simplest orbifold is a circle S^1 divided by a $\mathbb{Z}_2$, i.e. where points on one side are identified with the points of the opposite side. By the identification, the two points at the poles become curvature singularities (also called orbifold singularities or fixed points - because they remain invariant under the $\mathbb{Z}_2$). On the other hand, one of the simplest CY_3 is the quintic, displayed in fig. 2a. In 6D, an Abelian toroidal orbifold is defined as the quotient of a T^6 by a discrete isometry $\mathbb{Z}_N$ or $\mathbb{Z}_N\times\mathbb{Z}_M$ of the torus which acts on the whole compact space. Since the holonomy group of the orbifold corresponds to the symmetry that is moded out, the isometry group must be embeddable within SU(3) but not in SU(2) to retain $\mathcal{N}=1$ supersymmetry in 4D. We shall be only interested in this kind of orbifolds.

Since the time of Kaluza and Klein, it is known that compactifying higher dimensional gravity leads to scalar fields in 4D called *moduli* that parametrize the size and shape of the compact space. These fields lack mass, becoming a source of extra massless matter that could be detected through their interactions. If there existed a mechanism to furnish them with potential (non-perturbatively or otherwise), these new fields would develop a mass and become important cosmologically. This can be considered both an advantage and a disadvantage: on the bright side, in the emerging effective field theory, these fields could account partly for the dark matter and for the conjectured inflaton. On the less appealing side, there might simply be too many moduli or they could acquire masses such that reheating or other cosmological events could be prevented. As we shall see, these worries, for which the term *(cosmological) moduli problem* was coined, are addressed with some success in different scenarios.

Independently of the personal taste, both CY and orbifold spaces can render some possibly useful 4D phenomenology. Originally, it was thought that this would only be true so long as the compact space be small enough to justify the experimental absence of evidence of deviations of Newton's law. However, as we will shortly see, today we know that D-branes open up a way out of this constraint. There is a second reason to avoid very large extra dimensions though. Kaluza and Klein also taught us that the 4D matter particles appearing in the effective theory have some heavier 'cousins' dubbed Kaluza-Klein (KK) modes. The KK modes arise from the components associated to the compact coordinates. Taking again the example of an additional dimension compactified on a circle of radius R, a 5D scalar field $\phi=\phi(x^\mu,\zeta)$ can be expressed as

$$\phi(x^\mu,\zeta) \;=\; \sum_{n\in\mathbb{Z}}\varphi_n(x^\mu)\,e^{i\frac{n}{R}\zeta}\,, \tag{2}$$

where the infinite set of 4D fields φ_n has mass (in natural units) $m_n=n/R$. Evidently, only φ_0 is massless. The rest of the φ_n are the massive KK modes and can be decoupled from the low-energy states if m_n is sufficiently large, i.e. if R is sufficiently small. E.g. for the KK modes to scape the LHC searches, R must be smaller than 10^{-18} cm in this example. This apparent issue is actually one of the most effective ways to prove that extra dimensions exist. In short: one should find a free massive photon or a much heavier electron. That would establish the existence and, if the number of extra dimensions is also figured out, also their size.

In the simplest string compactifications without D-branes, the theory itself dictates the size of the extra dimensions. Considering only closed strings, the heterotic strings are the best examples of theories that admit this compactification. In these scenarios, there are only two arbitrary constants: the string scale M_s (inversely proportional to the size of the strings) corresponding to the energy scale at which the interactions between strings are important, and the string coupling g_s. In the effective theory, i.e. for energies below M_s, one can compare the outcome from compactifying the effective 10D gravity plus the gauge interactions in X_6 against the expected 4D gravity plus SM interactions. Assuming that X_6 has volume V_6, we obtain

$$\begin{aligned} S_{het} &= \frac{M_s^8}{g_s^2}\int d^{10}x\, R_{10d} + \frac{M_s^6}{g_s^2}\int d^{10}x\,\left(\mathrm{tr}\,F_{MN}F^{MN}\right)_{10d} \\ &\approx \underbrace{\frac{M_s^8V_6}{g_s^2}}_{=M_{Pl}^2}\int d^4x\, R_{4d} + \underbrace{\frac{M_s^6V_6}{g_s^2}}_{=1/g_{SM}^2}\int d^4x\,\left(\mathrm{tr}\,F_{\mu\nu}F^{\mu\nu}\right)_{4d}\,, \end{aligned} \tag{3}$$

where R_{nd} is the Ricci scalar in n dimensions, $M,N=0,1,\ldots,9$, $M_{Pl}\sim 10^{18}$ GeV and

$g_{SM}^2 \sim 0.01$, assuming unification of all gauge interactions for simplicity. The second row is obtained by compactifying six dimensions and letting the moduli arising from the compact components of R_{10d} and F_{MN} not to yield observable interactions. Comparing the expressions before the integrals with those below the underbraces, we see that $M_s = M_{Pl} \cdot g_{SM} \sim 10^{17}$ GeV, which is very close to the proposed GUT scale $M_{GUT} \sim 10^{16}$ GeV. Further assuming that $g_s \approx g_{SM}$ close to the compactification scale, one finally finds that $V_6 \approx M_s^{-6}$ or, equivalently, that the size of the compact dimensions must be around 10^{-30} cm! This strong constraint (valid in heterotic constructions) can be considered a positive feature because it is a model independent *prediction*. However, it can also be a hindrance to get in touch with observations or to deal with precision coupling unification ($M_{GUT} \neq M_s$).

2.2. Brane-worlds

The might of D-branes is not only their capability to furnish type II string theories with gauge groups, but also the freedom they provide to design configurations in which the compact space can take the least beautiful shapes and still render 4D appealing physics. For one, with the aid of D-branes it is possible to escape the constraint on the size of the compact space. If the volume of the compact space is large, but our observable universe is confined to a D-brane that intersects only a region of the compact space with very small volume, the result is an acceptable 4D theory.

In string theory, *brane-worlds* are compactifications of type II (A or B) string theories, whose compact spaces contain some non-contractible loops wrapped by Dp-branes, with $p \geq 3$.[3] A natural consequence in these constructions is that certain kind of matter (particularly gauge bosons) is trapped within the intersection of the D-branes, isolated from some other physics occurring outside that intersection.

We can consider, for instance, a situation in which the gravity in 4D appears from gravity in 10D, i.e. from closed strings in a type II string theory, but the (unified) gauge forces are related to the open strings attached to a stack of D-branes. In this case, one arrives at

$$\begin{aligned} S_{II} &= \frac{M_s^8}{g_s^2}\int d^{10}x\, R_{10d} + \frac{M_s^{p-3}}{g_s}\int d^{p+1}x \left(\mathrm{tr}\, F_{\alpha\beta}F^{\alpha\beta}\right)_{(p+1)d} \\ &\approx \underbrace{\frac{M_s^8 V_6}{g_s^2}}_{=M_{Pl}^2}\int d^4x\, R_{4d} + \underbrace{\frac{M_s^{p-3}V_{p-3}}{g_s}}_{=1/g_{SM}^2}\int d^4x \left(\mathrm{tr}\, F_{\mu\nu}F^{\mu\nu}\right)_{4d} . \end{aligned} \tag{4}$$

The differences between this expression and eq. (3) appear in the second integral. First, the different power in the factor $1/g_s$ is because open strings couple with only one power of g_s. Second, the indices α, β run only over the $p+1$ dimensions of the D-branes. Let us assume that 3 spatial dimensions of the D-branes cover our $\mathbb{M}^4$ space-time; this implies that $V_6 = V_{p-3} \cdot V_\perp$, where V_{p-3} is the compact part of the volume of the D-branes and $V_\perp$, the compact volume of X_6 perpendicular to the dimensions of the D-branes. Comparing the coefficients of the integrals to the expected theory in 4D, we find that M_s is a function of the compact volume perpendicular to the D-brane $V_\perp$ and the total volume V_6 of X_6: $M_s^{7-p} \approx 10^{16}$ GeV $\cdot \sqrt{V_6}/V_\perp$. Whereas observations constrain the magnitude of V_6 (or equivalently V_{p-3}), one can adjust $V_\perp$ at will to get arbitrarily small values of M_s. This freedom allows one to set M_s as low as few TeV without affecting any experimental bounds.

[3] In a rather bottom-up approach, brane-worlds refer to higher dimensional field theories endowed with certain hypersurfaces (not precisely D-branes). The work by Lisa Randall and Raman Sundrum [10, 11] (see also [30]) is a good example of this case. We will not focus on this work.

Journal of Physics: Conference Series **378** (2012) 012011 doi:10.1088/1742-6596/378/1/012011

This way it turns out possible that some stringy effects due to the compactification process be already observable at LHC. In particular, the KK modes have masses that go as multiples of M_s in general compactifications. If M_s is accessible to current experiments, heavier states with the properties of the photon, neutrinos, electrons, etc. would display signatures in the accelerators.

3. Second challenge: matter fields from string theory

We call matter to all the fields that enter the SM, i.e. gauge bosons and chiral fermions. However, in a first step, we are interested in obtaining matter representations that transform as adjoints and as fundamentals of an arbitrary gauge group. We pursue here also a mechanism that yields small gauge groups, such as those of the SM or the smallest GUTs. The way how all kinds of matter appear depends on the particular scenario.

3.1. Compactifications

In the heterotic strings as well as in the string type I, the compactification of the additional dimensions must also affect their original gauge groups in order to retain the (modular) symmetries of the original stringy actions. The effect is an explicit breakdown of the original gauge groups ($\mathrm{E}_8\times\mathrm{E}_8$ or SO(32)) to a subgroup thereof. Compactifying on the simplest CY threefold, the so-called standard embedding, the gauge groups are broken according to

$$\begin{aligned} \mathrm{E}_8\times\mathrm{E}_8 &\longrightarrow \mathrm{E}_6\times\mathrm{E}_8\,, \\ \mathrm{SO}(32) &\longrightarrow \mathrm{SO}(26)\,. \end{aligned} \tag{5}$$

Based on this breaking, the branching rules of the original representations read

$$\begin{aligned} (\mathbf{248},\mathbf{248}) &\longrightarrow (\mathbf{78},\mathbf{1})\oplus(\mathbf{1},\mathbf{248})\oplus 3(\mathbf{27},\mathbf{1})\oplus 3(\overline{\mathbf{27}},\mathbf{1})\oplus 8(\mathbf{1},\mathbf{1})\,, \\ \mathbf{496} &\longrightarrow \mathbf{325}\oplus 6\times\mathbf{26}\oplus 16\times\mathbf{1}\,. \end{aligned} \tag{6}$$

Not all of these states survive the compactification due to further consistency conditions on the compactifications. However, the relevant feature to notice here is that, the splitting of the adjoint representations yields fundamental representations that can give rise to states of the (supersymmetric) SM in addition to the gauge bosons of the new unbroken gauge group. For example, a **27**-plet of E_6 contains all the gauge representations of a family of quarks and leptons of the SM.

Similarly, orbifold compactifications lead to the breakdown of the original gauge symmetry like in eq. (5). The difference is that, unlike in CY compactifications, the rank is never broken by the compactification. So, e.g. in the simplest case (orbifold standard embedding), E_8 is broken to $\mathrm{E}_6\times\mathrm{SU}(3)$ or $\mathrm{E}_6\times\mathrm{SU}(2)\times\mathrm{U}(1)$ or $\mathrm{E}_6\times\mathrm{U}(1)^2$. Further, invariance of the resulting states (what we call here the it spectrum of the model) under the action of the orbifold impose further constraints, yielding frequently less states than the branching rules might suggest. For example, compactifying the $\mathrm{E}_8\times\mathrm{E}_8$ heterotic string on a $T^6/\mathbb{Z}_6$–II orbifold leads to the unbroken gauge group $\mathrm{E}_6\times\mathrm{U}(1)^2\times\mathrm{E}_8$ and the non-adjoint matter stemming from the 10D vector multiplets is just $3\times(\mathbf{27},\mathbf{1})\oplus(\overline{\mathbf{27}},\mathbf{1})\oplus 3\times(\mathbf{1},\mathbf{1})$, where U(1) charges have been omitted.

In an orbifold, there is an additional source of matter fields. At the singularities, some strings that would be open in the string theory compactified on a smooth manifold are actually closed around the singularities due to the orbifold. To picture this let us take a $T^2/\mathbb{Z}_2$ orbifold. In this case, there are four singularities and the compact space of the torus is reduced to only half of it. Taking into account these ingredients, one arrives at the picture displayed in fig. 3. The original smooth space of the torus is deformed to a tetrahedron. As depicted in the figure, there are some closed strings around the vertices. If one unfolds the tetrahedron, it is clear

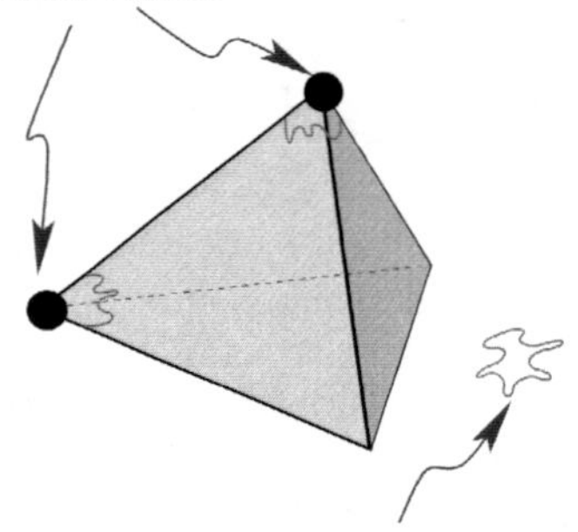

Figure 3: Orbifold $T^2/\mathbb{Z}_2$. The bulk states arise from the original 10D super Yang Mills and 10D gravity. The local or twisted states arise from the compactification itself and are attached to the singularities. These can be vector multiplets that form a local gauge group as well as chiral multiplets that form complete representations under the local gauge group.

that some edges have to be cut, rendering evident that the closed strings on the tetrahedron are open on a smooth (plane) manifold. The resulting states from those special strings are called *twisted* or simply *local*, as they attached to the singularities. In contrast, The states coming from the original vector and gravity multiplets in 10D (i.e. from the original string theory) are free to move everywhere. Retaking our previous $T^6/\mathbb{Z}_6$–II example, those local states yield the representations (omitting again the U(1) charges)

$$32 \times (\mathbf{27}, \mathbf{1}) \oplus 10 \times (\overline{\mathbf{27}}, \mathbf{1}) \oplus 217 \times (\mathbf{1}, \mathbf{1})\,.$$

Counting a **27** as a generation of quarks and leptons (as is done in E_6 GUTs), including the bulk or *untwisted* states, one obtains in this example a net number (generations minus anti-generations) of 24 generations.

3.2. Brane-worlds

In brane-worlds, we have learnt that vectorial multiplets accounting for the existence of gauge bosons arise from the open strings attached to a stack of N Dp branes. The dimensionality of the branes make no difference on the resulting gauge group (which can be SU(N) or Sp(N)). In order to obtain chiral supermultiplets, it is bit more complicated.

First, recall that the type II theories (where D-branes appear) yield $\mathcal{N} = 2$ supersymmetry in 10D. Thus, if one simply compactifies the additional dimensions onto a CY manifold or an admissible $\mathcal{O}_6$, the emerging 4D theory has $\mathcal{N} = 2$ supersymmetries too. To reduce the amount of remaining supersymmetry, it is needed to perform an orientifold (see e.g. [31] for an introduction to orientifolds). An orientifold is a generalization of an orbifold where the discrete isometry group G_1 is accompanied by a discrete operation of orientation-reversal in the worldsheet Ω, so that the final result is a theory of unoriented strings. In fact, the type I string theory can be obtained by letting the orientifold $G_1 = 1$ and $\Omega = \mathbb{Z}_2$ act on the type IIB string theory. A very important feature is that the invocation of an orientifold very often implies the introduction of open strings (absent in type IIB but present in type I strings). These new strings may be considered an analog of the twisted strings of the orbifold. Consistency of the theory (*tadpole cancellation*) restricts the amount of open-string states appearing after the orientifold and, in fact, this is the way how the gauge group SO(32) (and not any other) arises from type IIB.

Journal of Physics: Conference Series **378** (2012) 012011 doi:10.1088/1742-6596/378/1/012011

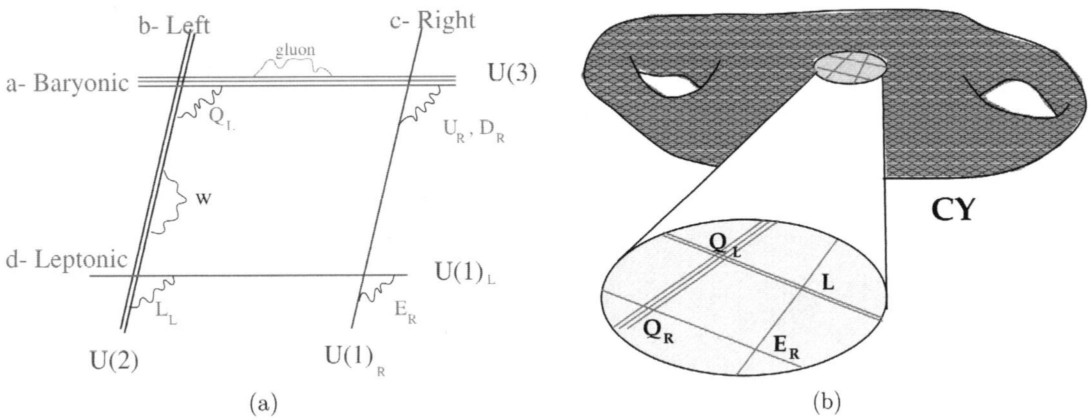

Figure 4: The Madrid model. (a) The SM matter arises from intersections of four stacks of D-branes. (b) The SM D-branes must wrap appropriate cycles of e.g. a CY threefold. Figure taken from [32].

The second source of chiral matter fields are intersecting branes (for a useful review of these constructions, see [33]). The starting point are D-branes that cover the 4D space that should correspond to ours and that wrap some of the non-contractible cycles of the compact space. For example, considering $X_6 = T^2 \times T^2 \times T^2$, there are 6 non-contractible cycles corresponding to the independent loops of the three 2-tori. If one introduces D6 branes in the type IIA string theory, three spatial brane dimensions can wrap three of those cycles. It is then clear that, if there are more than one D6 branes, one can accommodate the extra dimensions of the branes in different cycles, yielding points of intersection of the D-branes in the compact space.

As depicted in fig. 4a, at the intersection of two D-brane stacks there are open strings whose ends connect both stacks. It turns out that these strings yield physical states living in the intersection space (in this case, the 4D space we assume they both share) that are charged under the gauge groups associated to the stacks. In fact, the resulting states form (anti-)fundamental representations under those groups. For example, taking the upper left intersection of fig. 4a, one finds that the open strings would lead to the matter representations $(\mathbf{3},\mathbf{2})$ or $(\overline{\mathbf{3}},\mathbf{2})$ with some U(1) charges associated to the gauge groups of the parent D-branes $\mathrm{SU}(3)\times\mathrm{SU}(2)\times\mathrm{U}(1)^2$. Figure 4b depicts a situation in which the intersections proposed before occur in some region of a CY_3, where non-contractible cycles exist.

4. Third challenge: the (MS)SM spectrum & interactions

One of the final goals of string phenomenology is to reproduce the gauge forces and particle content of the SM. As we shall see shortly, this goal has been achieved in different scenarios at least at the supersymmetric level (the MSSM). The next step is then to address more serious questions like reproducing the matter interactions contained in the Lagrangian of the SM, providing thereby a plausible solution to the extant issues, such as the origin of baryon stability, the strong CP problem, the flavor problems, etc. Despite all the effort invested on the latter task, it is perhaps fair to say that this challenge is still open.

In this section, we shall discuss some of the methods employed in string theory to achieve models that approach at the (MS)SM and some qualities of the emerging models. We shall also mention how these constructions might lead to answers to persisting puzzles.

Journal of Physics: Conference Series **378** (2012) 012011 doi:10.1088/1742-6596/378/1/012011

4.1. Compactifications

Heterotic CY compactifications. Both kinds of compactifications, on smooth spaces (with and without fluxes) and on orbifolds, have been actively investigated since the 80's, although they were partly left aside when D-branes were discovered. During the last decade, perhaps due to the invention of more powerful computers, there have been intensive scans of the landscape of possible CY spaces leading to promising models, which have produced interesting results [34, 35, 36, 37, 38, 39, 40] (GUT models were derived from U(N) line bundles [41, 42]). Around 200 models with the exact spectrum of the MSSM have been found. The gauge group of these models is just $\mathrm{SU}(3) \times \mathrm{SU}(2) \times \mathrm{U}(1)_Y \times \mathrm{U}(1)_{B-L}$. The matter spectrum includes just three generations of quarks and leptons, some pairs of Higgs supermultiplets and no exotics of any kind.

Despite these appealing features, their advantages play also as challenges. One problem is that in the SM the symmetry $\mathrm{U}(1)_{B-L}$ appears only globally whereas in these constructions it is a gauge symmetry. The fact that there are no exotics particles of any kind implies that there are no additional particles that may be charged only under that symmetry, such that could trigger a spontaneous breakdown of $\mathrm{U}(1)_{B-L}$. Getting rid of this symmetry requires desperate measures. One method to solve it is by assuming that the sneutrinos (the bosonic partners of the neutrinos) develop somehow an expectation value [43].

Another complicated problem is related to the nature of CY manifolds. CY manifolds are smooth spaces, however even writing down the corresponding metric is a challenge. Therefore, this compactification scheme must abandon the calculability of (CFT in) string theory and apply low-energy geometric methods to arrive at e.g. the size of the interactions between the emerging matter fields. One requires then numerical methods combined with cohomology techniques that allow one to deal with the interactions of fields on a very complex geometric background. This effort has revealed some progress [44, 45], but further refinement must be achieved, to figure out e.g. what the explicit dependence of the coupling constants on the moduli is.

The qualitative picture of Yukawa couplings in CY compactifications is as follows. The Yukawa coupling is either order one or zero, however, it appears difficult to obtain hierarchical couplings at a *generic* point in moduli space. This does not mean that the CY is incapable of reproducing phenomenology, but it might mean that one would have to move close to a special point in moduli space. Whether this point can, at least to some extent, be identified with a configuration close to an orbifold point (or a partial blow-up of an orbifold) represents an interesting question. In fact, in the CY model [38, 45] there is no light Higgs pair at a generic point in moduli space: the mu-term only vanishes if the size of a certain cycle shrinks to zero. A μ term of order TeV forces us to move to a rather special point in moduli space.

Heterotic orbifold compactifications. Let us now focus on orbifold compactifications of the heterotic string, which have also gained much momentum in recent years. In fact, since the birth of string theory, successful models reproducing several features of the MSSM have been found [47, 48, 49, 50, 51, 52, 53]. The new ingredient that has contributed to the appearance of many phenomenologically appealing heterotic orbifold models is a feature which we would like to refer to as *gauge group topography*: different gauge groups can be realized at different points in internal space. In the interesting orbifolds that have been constructed, the 10D gauge group $\mathrm{E}_8 \times \mathrm{E}_8$ gets broken to different subgroups at different orbifold singularities. Figure 5 illustrates this situation in a two-dimensional $\mathbb{Z}_2$ orbifold. To understand fig. 5, we must only understand how the local gauge groups emerge. They are comprised out of the gauge bosons corresponding to generators that fulfill certain (local) orbifold invariance conditions, which can be recast schematically as

$$\text{generator} \cdot \text{local shift} \; = \; 0 \quad \text{mod } 1 \,, \tag{7}$$

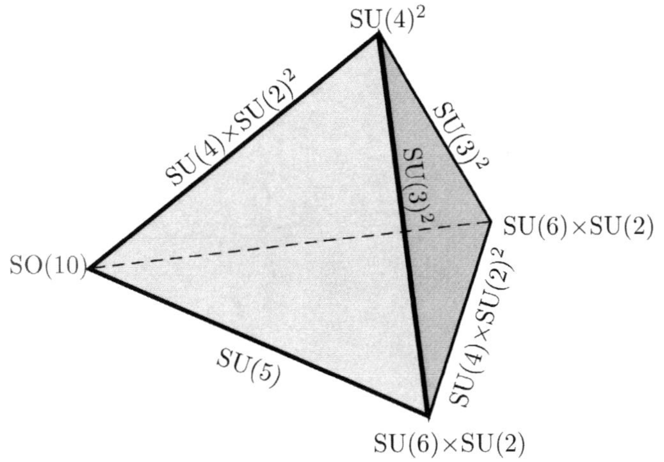

Figure 5: Gauge group topography. At different fixed points (corners of the tetrahedron), E_8 gets broken to different subgroups (U(1) factors are suppressed). At the edges we display the intersection of the two local gauge groups realized at the vertices. Figure taken from [46].

where "generator" is any of the generators of the 10D gauge group of the heterotic string, and the local shift [54, 51] introduced here depends on the fixed point [55]. Since these invariance conditions can be different at different fixed points, different gauge groups can live at the various fixed points.

In the field-theoretic description, one would say that the local gauge groups are made up from the generators for which the corresponding gauge boson has a non-vanishing profile at the singularity. This also means that the coset, $(E_8 \times E_8)/G_{\text{local}}$ with G_{local} denoting the local gauge group, comprises generators where the gauge bosons' profile has to vanish at the singularity. From this point of view it is also clear what the massless gauge interactions in 4D are: they are mediated by the bosons with flat profile, i.e. which fulfill the invariance conditions (7) everywhere. In other words, the 4D gauge group emerges as the intersection of local gauge groups in $E_8 \times E_8$.

Figure 5 displays not only the gauge groups at the singularities of the orbifold, but also the intersections between the different local groups. This is useful for two reasons. First, one notices that, even if two singularities are endowed with the same gauge group, their internal structure might be different and therefore lead to non-trivial intersections (see the vertices on the right). Second, focusing on the bottom left vertices, SO(10) and SU(6)×SU(2), one finds that local GUTs (like SO(10)) can lead to GUTs at the intersection. This is a hint that shall be used below.

Since the 4D gauge group is the intersection of the local gauge groups at the singularities, this implies that, when the gauge group topography is non-trivial, the local gauge groups at the orbifold singularities have to be *larger* than the SM, $G_{\text{local}} \supset G_{\text{SM}}$. This then leads to the picture of *local grand unification* [56].

From the requirement $G_{\text{local}} \supset G_{\text{SM}}$ we see immediately that the local gauge groups G_{local} can be groups that have been discussed in the context of GUTs, such as SU(5) and SO(10). It is important to note that fields confined to a region with certain gauge symmetry have to furnish complete representations of this symmetry. Thus, matter fields localized at a singularity with a GUT symmetry would appear as GUT multiplets in the low-energy theory *although the gauge symmetry of the low-energy theory was smaller.* This statement applies, in particular, to matter

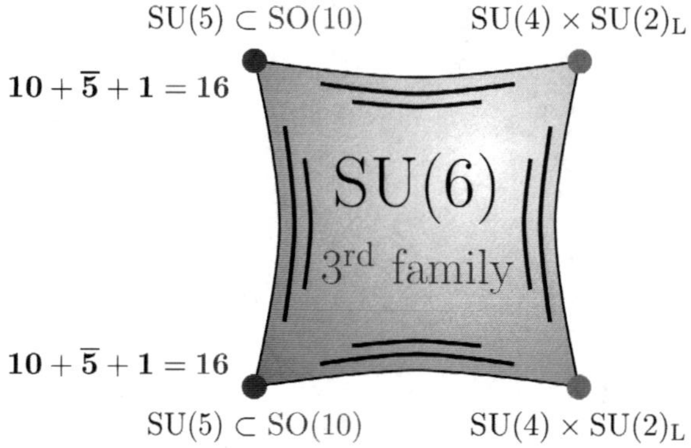

Figure 6: Orbifold GUT limit of model 1A of [52]. Figure taken from [46].

living in regions with SO(10) symmetry: it will furnish complete SO(10) representations, i.e. a **16**-plet will give rise to a complete family of quarks and leptons.

On the other hand, matter fields can also come from the bulk. Such fields do, like the gauge bosons, feel the symmetry breaking at every fixed point, and thus appear in split multiplets of the various local gauge groups. Together with what we have discussed in the previous paragraph, this provides a very simple scheme allowing to understand the simultaneous existence of complete and split GUT multiplets in nature (solving thereby the well-known doublet-triplet splitting problem of GUTs).

Based on these observations, a strategy to arrive at semi-realistic heterotic orbifold models (i.e. models with three chiral generations plus one pair of Higgs below the compactification scale) has been developed [57, 58]. The idea is to search for models containing **16**-plets living at singularities with SO(10) symmetry plus a pair of Higgs multiplets coming from the bulk. As a first shot, one would aim at all three generation arising from **16**-plets in order to explain the origin of the structure of the SM families. However, this situation is disfavored in heterotic orbifolds. The next simplest possibility is to demand the third generation to be different from the other two, i.e. two generations can rise from **16**-plets whereas the third one may come in split multiplets from the bulk or singularities without a GUT symmetry.

In fig. 6, we present the structure of one of the heterotic orbifolds constructed in [57, 58] and further studied in [52]. Focusing our attention on two of the six orbifolded dimensions, the resulting space is again a tetrahedron which is topologically equivalent to the pillow of fig. 6. The local symmetry of two (degenerate) singularities is SO(10), which contains SU(5). If we let these two dimensions be larger than the other four, the emerging theory is a 6D $\mathcal{N} = 1$ supersymmetric theory where the (now macroscopic) singularities of interest enjoy an SU(5) group with the representations $\mathbf{10} \oplus \overline{\mathbf{5}} \oplus \mathbf{1}$, i.e. with the complete **16**-plet split in SU(5) representations. If now the pillow is shrunk to submillimetric sizes, the emerging 4D theory has the gauge group of the SM and the two localized **16**-plets appear with the correct quantum numbers of the quarks and leptons. Furthermore, since the gauge group stems from a local SO(10), the quarks and fields are charged under a gauged $U(1)_{B-L}$ symmetry, pretty much as in the CY case discussed above. One useful remark at this point is that the two localized generations arising from the SO(10) local GUTs have exactly the same quantum numbers. In fact, they are identical. This characteristic originates the appearance of a discrete D_4 family symmetry [52, 59], which has several phenomenological applications, including the suppression of flavor changing neutral currents in gravity mediated supersymmetry breaking [60].

The origin of the third generation of quarks and leptons and the Higgs multiplets in these constructions is not so simple. We have mentioned that states living in the bulk form split

Table 1: Matter representations of the model 1A of [52]. It corresponds to the spectrum of the MSSM plus SM singlets and vectorlike exotics, which acquire masses bu a Higgs-like mechanism. The bulk moduli are uncharged under the gauge group. Quantum numbers w.r.t. the gauge group $\mathrm{SU}(3)_c \times \mathrm{SU}(2)_L \times [\mathrm{SU}(4) \times \mathrm{SU}(2)]_{hidden}$ are given in parenthesis. The hypercharge $\mathrm{U}(1)_Y$ and $\mathrm{U}(1)_{B-L}$ appear as subindices.

3 (net) generations					
3	$(\mathbf{3},\mathbf{2};\mathbf{1},\mathbf{1})_{1/6,1/3}$	q_i			
3	$(\overline{\mathbf{3}},\mathbf{1};\mathbf{1},\mathbf{1})_{-2/3,-1/3}$	$\bar{u}_i$			
3	$(\mathbf{1},\mathbf{1};\mathbf{1},\mathbf{1})_{1,1}$	$\bar{e}_i$			
3+1	$(\overline{\mathbf{3}},\mathbf{1};\mathbf{1},\mathbf{1})_{1/3,-1/3}$	$\bar{d}_i$	1	$(\mathbf{3},\mathbf{1};\mathbf{1},\mathbf{1})_{-1/3,1/3}$	d_i
3+1	$(\mathbf{1},\mathbf{2};\mathbf{1},\mathbf{1})_{-1/2,-1}$	ℓ_i	1	$(\mathbf{1},\mathbf{2};\mathbf{1},\mathbf{1})_{1/2,1}$	$\bar{\ell}_i$
3+13	$(\mathbf{1},\mathbf{1};\mathbf{1},\mathbf{1})_{0,1}$	$\bar{\nu}_i$	13	$(\mathbf{1},\mathbf{1};\mathbf{1},\mathbf{1})_{0,-1}$	ν_i
Higgses					
1	$(\mathbf{1},\mathbf{2};\mathbf{1},\mathbf{1})_{-1/2,0}$	h_d	1	$(\mathbf{1},\mathbf{2};\mathbf{1},\mathbf{1})_{1/2,0}$	h_u
Standard model singlets					
32	$(\mathbf{1},\mathbf{1};\mathbf{1},\mathbf{1})_{0,0}$	s_i^0	4	$(\mathbf{1},\mathbf{1};\mathbf{1},\mathbf{1})_{0,\pm 2}$	$\chi_i^\pm$
10	$(\mathbf{1},\mathbf{1};\mathbf{1},\mathbf{2})_{0,0}$	N_i			
Exotics					
5	$(\mathbf{1},\mathbf{1};\mathbf{1},\mathbf{2})_{0,-1}$	η_i	5	$(\mathbf{1},\mathbf{1};\mathbf{1},\mathbf{2})_{0,1}$	$\bar{\eta}_i$
6	$(\mathbf{3},\mathbf{1};\mathbf{1},\mathbf{1})_{-1/3,2/3}$	δ_i	6	$(\overline{\mathbf{3}},\mathbf{1};\mathbf{1},\mathbf{1})_{1/3,2/3}$	$\bar{\delta}_i$
2	$(\mathbf{3},\mathbf{1};\mathbf{1},\mathbf{1})_{1/6,-2/3}$	v_i	2	$(\overline{\mathbf{3}},\mathbf{1};\mathbf{1},\mathbf{1})_{-1/6,2/3}$	$\bar{v}_i$
14	$(\mathbf{1},\mathbf{1};\mathbf{1},\mathbf{1})_{1/2,*}$	s_i^+	14	$(\mathbf{1},\mathbf{1};\mathbf{1},\mathbf{1})_{-1/2,*}$	s_i^-
8	$(\mathbf{1},\mathbf{2};\mathbf{1},\mathbf{1})_{0,*}$	m_i	2	$(\mathbf{1},\mathbf{2};\mathbf{1},\mathbf{2})_{0,0}$	y_i
6	$(\mathbf{1},\mathbf{1};\mathbf{4},\mathbf{1})_{0,*}$	f_i	6	$(\mathbf{1},\mathbf{1};\overline{\mathbf{4}},\mathbf{1})_{0,*}$	$\bar{f}_i$
2	$(\mathbf{1},\mathbf{1};\mathbf{4},\mathbf{1})_{-1/2,-1}$	f_i^-	2	$(\mathbf{1},\mathbf{1};\overline{\mathbf{4}},\mathbf{1})_{1/2,1}$	f_i^+
Bulk moduli: S, T_1, T_2, T_3, U					

multiplets with respect to the local GUTs. In the model illustrated in fig. 6, one finds that the top quark and the Higgses arise from the bulk. This implies particularly that the magnitude of the interactions of these fields is of order unity. (This observation has been found to be relevant to explain the supersymmetric scenario known as *natural supersymmetry* in which the third generation sfermions and especially the stop are much lighter than all other superpartners [61].) The other quarks and leptons of the third generation are found at other singularities of the orbifold. As we shall see, interactions among fields located at singularities are exponentially suppressed by the distance that separate them.

In table 1 we show a summary of all matter representations of a semi-realistic orbifold model. We see that there is a net number of three SM generations plus a pair of supersymmetric Higgs multiplets. These are distinguished from the lepton doublets ℓ_i thanks to the gauged $\mathrm{U}(1)_{B-L}$ symmetry. There is also a set of exotic matter. The exotics are vectorlike because for each exotic X there is its antiparticle $\bar{X}$ too. A crucial point here is the existence of the SM singlet fields $S = \{s^0, \chi^\pm, N\}$. Taking as a given that supersymmetry – if it exists– must be unbroken well below the compactification scale that follows from eq. (3), many of the SM singlets must acquire develop an expectation value (VEV) of order $0.1M_{Pl}$. It turns out that there are two direct effects of these VEVs. Firstly, since the $\chi^\pm$ are charged under $\mathrm{U}(1)_{B-L}$ and some or all of them develop VEVs in a generic supersymmetric vacuum, $\mathrm{U}(1)_{B-L}$ is spontaneously broken down to a $\mathbb{Z}_2$ matter parity (useful to prevent rapid proton decay [62, 52]). Notice that this is an important advantage against the CY compactifications discussed earlier.

The second effect of the VEVs is that all exotics acquire masses very close to the compactification scale provided that superpotential couplings of the type $S^n X \bar{X}$, $n \in \mathbb{Z}$, are allowed by all the symmetries of the orbifold, i.e. provided that these couplings satisfy gauge invariance and certain stringy selection rules [63, 64, 65, 66]. It has been explicitly verified that those couplings exist and that indeed all vectorlike exotics attain masses through this Higgs-like mechanism.

Another important feature of the model under consideration is that the gauge group includes, besides $G_{\rm SM}$ and $U(1)_{B-L}$, a *hidden sector.* The hidden sector is understood here as a group under which the particles of the (MS)SM are uncharged. In this case, the hidden gauge group is SU(4) $\times$ SU(2), but SU(2) is spontaneously broken by the VEVs of some N_i, leaving only SU(4) as hidden group. This is relevant because a hidden sector with a pure Yang-Mills theory can trigger spontaneous breaking due to gaugino condensation. Roughly speaking, the hidden gauginos λ form bound states $\langle\lambda\lambda\rangle$ at the scale at which the interactions become strong,

$$\Lambda \sim M_{GUT}\, e^{-8\pi^2/b_0 g^2(M_{GUT})}\,, \qquad b_0 = \beta\text{-function coefficient of hidden group,} \tag{8}$$

(which follows from the corresponding renormalization group flow,) inducing thereby a supersymmetry-breaking F-term which is transmitted to the observable sector via gravitational interactions. The result is that the scale of supersymmetry breaking in the observable sector is given by [67]

$$m_{3/2} \sim \Lambda^3/M_{Pl}^2\,.$$

Assuming that there is unification of all the SM interactions at $M_{GUT} \sim 10^{16}$ GeV, implying $g^{-2}(M_{GUT}) \sim 2$, in the current case $m_{3/2} \sim few$ MeV. This scale is rather low given the current constraints on supersymmetric particles. However, the important message we learn from this is that these models induce naturally low-scale supersymmetry breaking [68].

With these ingredients in hand, one can investigate e.g. the structure of the Yukawa matrices. There are also superpotential couplings of the type $Y^u_{ij} q_i h_u \bar{u}_j$, where Y^u is a 3×3 matrix whose entries are of the form S^n/M_{Pl}^{n-3}. What is interesting is that, as announced before, the entry Y^u_{33} is order unity due to the fact that there is (almost) no obstruction that prevents interactions among bulk fields. However, since the other quarks are localized, the stringy selection rules [63, 64, 65, 66] do not allow trilinear couplings such as $q_{1/2} h_u \bar{u}_{1/2}$, so that powers of fields S are required. Since $\langle S\rangle \sim 0.1 M_{Pl}$, each power of S adds a suppression factor 1/10. So, the generic picture is that all quarks (up and down type) acquire masses once the Higgs mechanism takes place (i.e. once h_u and h_d acquire VEVs); however, there is a natural hierarchy between the top mass and the masses of all other quarks which depends on the specifics of each model. Similarly, also the Yukawa couplings of the leptons are non-trivial, implying that, as in the SM, all leptons and quarks acquire masses.

Nonetheless, to arrive at specific predictions from these constructions, such as the exact texture of the CKM matrix, the precise value of the quark and lepton masses, etc., one needs additionally perform a precise computation of the coefficients λ of couplings like $\lambda S^n q_i h_u \bar{u}_j$ directly from string theory. The advantage here is that, in contrast to CY compactification, we have the power of CFT tools that allow us to perform an analytic computations. In fact, couplings between three and four fields have already been computed [69, 70, 71, 72].

The starting point of the computation are the vertex operators V associated with the strings that yield the fields of interest. Vertex operators are operators in the 2D CFT of the worldsheet (the surface that the strings sweep as they evolve) that carry the whole information of the fields in the emerging 10D QFT (the so-called *target space*). The form of the vertex operators is known and strongly depends on two features: 1) whether the field is a boson or a fermion; and 2) whether the coupling is computed at tree or loop level in the worldsheet. For the first feature, there are vertex operators for bosons V^b and for fermions V^f. The second feature has to do with the fact that interactions between strings form different surfaces. For example, three closed strings interacting at tree level form the surface of a pair of trousers, which is topologically

Journal of Physics: Conference Series **378** (2012) 012011 doi:10.1088/1742-6596/378/1/012011

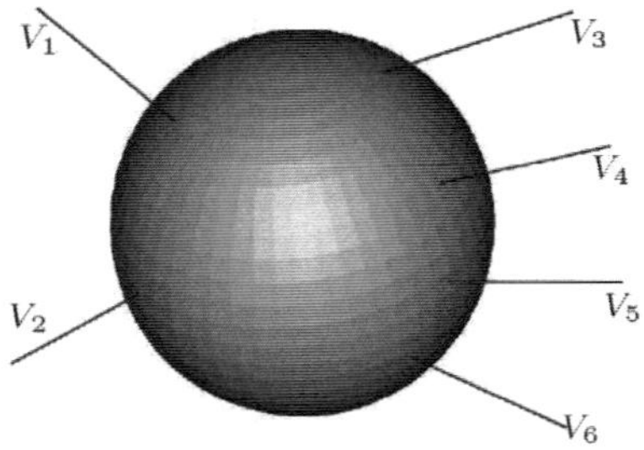

Figure 7: Computation of interaction strengths among six fields in heterotic orbifold models at tree level via vertex operators in CFT.

equivalent to a sphere when the 'wholes' of the trousers are pulled to infinity. Similarly, at one loop, the surface subtended by the string interaction is a 2D torus; at two loops, the surface has two handles, and so on. Focusing on tree-level interactions, the picture associated to the scattering amplitude of six strings is displayed in fig. 7.

Once the specifics of the vertex operators are known, we can compute the scattering amplitudes for the effective fields and deduce the coupling terms in the 4D effective field theory. A term Φ^L in the superpotential, with Φ a chiral superfield with components (ϕ, ψ), can be inferred most straightforwardly from an interaction of the form $\psi\,\psi\,\phi^{L-2}$. The computation required in this case is the following correlation function

$$\lambda = \langle V_1^f(z_1,\bar{z}_1)V_2^f(z_2,\bar{z}_2)V_3^b(z_3,\bar{z}_3)V_4^b(z_4,\bar{z}_4)\dots V_L^b(z_L,\bar{z}_L)\rangle\,, \tag{9}$$

where the vertex operators have to be adjusted to the form needed for interactions on a sphere. The points z_i on the complex plane are points of insertion on the sphere of interaction. The correlation function eq. (9) is a path integral, where the action is the string action of the heterotic string compactified on the orbifold, which is well known. (Further details can be found in ref. [66] or directly in the original literature [63, 64].)

In the simplest case of trilinear interactions, it has been found that the coupling strength goes like $\lambda \sim e^{-\alpha|u(T_i)|^2}$, where α is a numerical factor and $u(T_i)$ is the vector that parametrizes the distance between the fixed points at which the fields on interest are attached. This non-trivial interaction among distant strings is possible due to the existence of worldsheet instantons. Notice that u depends on the Kähler (also called volume) moduli T_i, whose values must be fixed (see sec. 5.1) prior to computing the actual value of λ. There is a second challenge. At non-renormalizable level, the integral expressions of eq. (9) become increasingly complicated. Therefore, it is necessary to implement computer algorithms that can deal with complicated integral in complex variables.

Despite all the complications, there is some progress in the computations and, some particular cases (where the value of the moduli has been fixed to order unity values and the VEVs of the SM singlets S has been tuned), one can arrive at the following eigenvalues of the Yukawa matrices.

$$|Y_u^{\text{diag}}| \sim \text{diag}(0.0003,\,0.06,\,1.1)\,, \qquad |Y_d^{\text{diag}}| \sim \text{diag}(0.0002,\,0.001,\,0.04)\,.$$

These values, although not quite realistic, are pretty close to the values of the Yukawa couplings that one expects in a scenario where $\tan\beta \sim 1$.

How generic are these results? Evidently, these results are quite encouraging. One might be tempted to claim that one of the goals of string phenomenology is about to become true. So it is

Journal of Physics: Conference Series **378** (2012) 012011 doi:10.1088/1742-6596/378/1/012011

natural to wonder whether these results are common to all or a large subset of compactifications. To answer this question, we have performed a scan based on the local GUT strategy depicted before. That is, we have started a search of models endowed with local GUTs and selected those among them that fulfill a series of priors such as couplings unification, absence of exotics and sizable Yukawa couplings for the top quark. The results were called the mini-landscape [57, 58] because they correspond to a set of about 300 models with properties similar to those explained before identified within a larger set of about 10 million (not so appealing) models. Due to the amount of data that has been studied, the huge task of finding those models would not have been possible without the orbifolder [73]. The orbifolder is a program design in C++ specifically for the computation of all the properties of Abelian heterotic orbifolds, which is now made public through the site `http://projects.hepforge.org/orbifolder/`. It can also be used on-line via our web interface: `http://stringpheno.fisica.unam.mx/orbifolder`. The idea of that publication is to allow one to overcome the technicalities of the heterotic compactifications and focus on a more phenomenological quest towards more appealing constructions and a better understanding of the models that are already available in the literature.

The message that one should draw from that fruitful search is that there exist patches in the landscape of string vacua which lead to phenomenologically promising models. It seems also sensible to say that the main property of those patches is that they are endowed with local GUTs at a certain level. The task to accomplish in the future is then to analyze thoroughly the models and to figure out whether they lead to any kind of predictions.

Unfortunately, as we mentioned earlier, contact between these orbifold constructions and particle phenomenology requires a full understanding of the details of the vacua that each model offers. These details are associated with the dynamics of the *moduli* fields that describe all geometrical features of the particular compactification. For instance, the masses of quarks and leptons as well as the dynamics of supersymmetry breakdown depend on the vacuum expectation values (VEVs) of the moduli, which are arbitrary at tree level in many string constructions. Therefore, in order to achieve a prediction from string theory and simultaneously to avoid severe cosmological constraints, a mechanism that fixes the VEVs of the moduli and gives them large masses is needed. This mechanism is usually called *moduli fixing* or *stabilization* and will be discussed below.

Before discussing some results on brane-worlds, a final comment is in order. Both CY and orbifold compactifications have been applied in some other string constructions, such as M-theory compactifications [74, 75, 76, 77] and type I phenomenology [78, 79]. Although some phenomenological prospects have been identified in these scenarios, it is fair to say that they do not compete with the results obtained in CY and orbifold compactifications.

4.2. Brane-worlds

Just as in the case of the promising constructions obtained by compactifying the heterotic strings, in the type II theories there has been a lot of progress in recent years. The main idea is encoded in the *Madrid model* [80, 81], which is displayed in fig. 4. In that picture, type II string theories (A or B) appear compactified on a CY and include an orientifold background, so that one ends up with a 4D $\mathcal{N}=1$ supersymmetric theory. Without paying much attention on the cancellation of tadpoles, the minimal model that allows for one generation of quarks and leptons of the (MS)SM contains four stacks of D6 or D5 branes, depending on whether one starts with type IIA or type IIB. There is a stack of 3 D-branes called *baryonic*, a stack of 2 D-branes called *left* (for left-chiral quarks and leptons are supposed to arise from this stack), and two "stacks" of one brane, one called *leptonic* (only leptons couple to this one) and the other called *right* (right-chiral particles are attached to this brane). The emerging gauge group in 4D (where all branes share the whole space) is, as expected, $\mathrm{U}(3)_a\times\mathrm{U}(2)_b\times\mathrm{U}(1)_c\times\mathrm{U}(1)_d$.

As we have learnt before, in these scenarios the gauge bosons are associated with those strings whose both ends are attached to the same stack. So e.g. we find the vector multiplets corresponding to the gluons in the baryonic stack and the weak vector bosons, in the left stack. Chiral matter states arise at the intersections of different stacks of branes. The direct consequence of the intersections proposed in fig. 4 is that the matter representations obtained at the four intersections are

$$(\mathbf{3},\mathbf{2})_{a_q,b_q,c_q,d_q} \oplus (\mathbf{1},\mathbf{2})_{a_\ell,b_\ell,c_\ell,d_\ell} \oplus (\mathbf{1},\mathbf{1})_{a_e,b_e,c_e,d_e} \oplus (\overline{\mathbf{3}},\mathbf{1})_{a_d,b_d,c_d,d_d}\,,$$

where the subindices denote the charges under the four different U(1) symmetries. Two observations must be made here. First, a linear combination of the U(1)s can yield the hypercharge, so that the resulting fields from this configuration are q_L, ℓ_L, e_R and $\bar{d}_R$. The second observation is that this is not a complete generation of quarks and leptons.

In order to rescue this model, one must rely on one unavoidable effect of the orientifold that we have not mentioned until here. Another way to interpret the worldsheet orientation reflection is by adding a D-brane parallel (not superposed) to the right stack, whose charges of the associated states be opposite to the c_i charges. We shall dub this D-brane *right**. This way, two new matter representations are induced:

$$(\overline{\mathbf{3}},\mathbf{1})_{a_u,b_u,c*_u,d_u} \oplus (\mathbf{1},\mathbf{1})_{a_\nu,b_\nu,c*_\nu,d_\nu} \to \bar{u}_R \text{ and } \bar{\nu}_R$$

that complete an (MS)SM matter generation.

Once the full matter content of one generation of quarks and leptons has been achieved, the next goal is to get the Higgs multiplets. This is achieved by proposing that, at some point, the right and right* brane stacks intersect the left stack. Each of these intersections will induce a matter representation $(\mathbf{1},\mathbf{2})$ with some U(1) charges that shall combine to provide the correct hypercharges of h_u and h_d.

This relatively simple setup provides all what one needs for the stable matter of nature, namely the first generation of quarks and leptons. For arriving at the second and third generations, one needs an additional ingredient. The key point is once more that chiral matter appear at the intersections of D-brane stacks. Thus, if e.g. there was a way to let the baryonic and the right (right*) stacks intersect more than once, one would obtain more than one copy of q_R. (From now on, we shall refer to $\bar{u}_R$ and $\bar{d}_R$ as q_R.)

In the original Madrid model, the six extra dimensions of the type II strings were not compactified on a CY, but on a factorizable 6D torus $T^6 = T^2 \times T^2 \times T^2$. Since D5 and D6 branes do not cover the whole compact space, it is reasonable not to expect that the stacks intersect in every single T^2. Another important element here is that they must not intersect

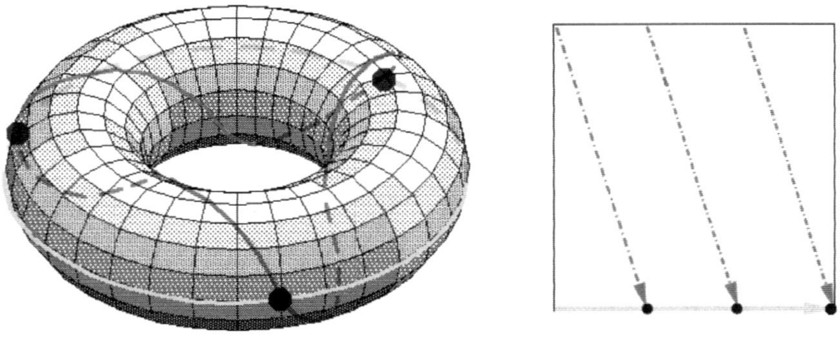

Figure 8: Three families of quarks q_R^i arising from intersections of D-brane stacks on a torus. The quarks arise from the strings attaching both D-brane stacks at the intersections. Both pictures are equivalent.

exactly perpendicularly; the can intersect at angles. Supposing e.g. that the baryonic and the right/right* stacks intersect on one of the tori at an angle $\sim 3\pi/10$, one achieves the picture displayed in fig. 8. This leads to three copies of q_R. In fig. 8, the horizontal/green line represents the baryonic stack where as the other line represents the right/right* stack. Notice that a slightly different angle leads to a different number of quark copies. The crucial point here is that branes have to wrap non-contractible compact cycles for this mechanism to operate. Therefore, this result is also valid for much more complicated spaces than a torus, like CY_3, so long as they exhibit such cycles and the D-brane stacks wrap them. This kind of arrangement has been successfully done for all brane stacks [80, 81], leading to exactly three (MS)SM generations plus a pair of Higgs multiplets.

Despite its elegant properties, the original Madrid has a very important problem. Tadpole cancellation ensures that all equations of motion of the theory have solutions. If this is not satisfied, the theory is not consistent *globally*, since tadpoles are a result of considering the whole compactification scheme. In fact, the original Madrid model is challenged by this constraint. The way to fix this problem is by adding further brane stacks and, therefore, further chiral and vector multiplets to the effective theory, so that, besides the matter content of the MSSM, one ends up with an extra bunch of exotic chiral matter. Another more subtle issue is anomaly cancellation. Generically, the U(1)s that appear from D-brane stacks are anomalous, i.e. are broken explicitly by quantum effects. It is then difficult (but not impossible) to come up with a linear combination of all the U(1)s involved that yields the correct hypercharge for the SM states and simultaneously does not suffer from anomalies.

As in heterotic compactifications, there is another more phenomenological question one must pose to a promising construction. What are the interactions between the emerging fields and how large are they? The interactions between strings originating at different brane intersections have been computed via worldsheet instantons that can connect strings that are disconnected in the compact space [32]. This will usually involve the presence of three different D-brane stacks, which determine the boundary conditions of the worldsheet instanton contributing to this Yukawa coupling, see figure 9.

Roughly speaking, the instanton contribution to the Yukawa coupling will be given by evaluating the classical action $e^{-S_{cl}}$ on the surface of minimal area connecting the three intersections. As a result, as before, Yukawa couplings will depend on several moduli of the

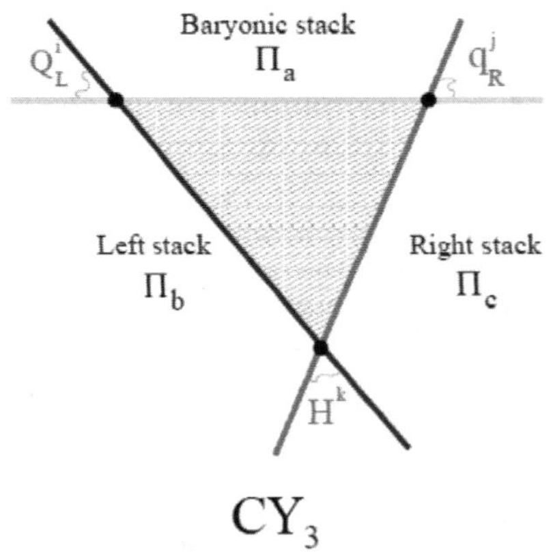

Figure 9: Yukawa couplings arise from instantonic interactions among states living at closely located intersections.

Journal of Physics: Conference Series **378** (2012) 012011 doi:10.1088/1742-6596/378/1/012011

theory, such as D-brane positions (open string moduli) and the compact manifold metric (closed string moduli).

In the Madrid model, since intersections such as the one exhibited in fig. 9 only appear for one generation of quarks [4], the result is that only the third generation of quarks and leptons can acquire a mass once the Higgs mechanism occurs. This situation can be alleviated by the introduction of non-perturbative effects. In particular, it was found that D-brane instantons (i.e. instantons in 10D space-time) can generate admissible couplings [82].

Unfortunately, this situation is in general much worse. Even though this kind of couplings can render phenomenologically useful interactions, in *globally* consistent models (i.e. those with tadpole cancellation) Yukawa couplings for all generations are forbidden by some U(1)s of the construction. Arguably, non-perturbative effects such as D-brane instantons can induce couplings of order unity (required for the top quark) while producing tiny couplings for the neutrinos, leading to semi-realistic Yukawa textures [82, 83] in type IIB brane-worlds with D7 branes.

There have been several models since the Madrid model getting closer to the properties of the SM. Recently, a thorough scan of type IIA models with intersecting branes has been performed. The main result is that there seem to be many (perhaps equivalent)[5] consistent models (i.e. with all tadpoles canceled) with the following features [84, 85, 86]:

- gauge group either $\mathrm{U}(3) \times Sp(2) \times \mathrm{U}(1)^n$ or $\mathrm{U}(3) \times \mathrm{U}(2) \times \mathrm{U}(1)^n$, $n \in \mathbb{Z}$, times a hidden sector gauge group;
- some models with and without $\mathrm{U}(1)_{B-L}$;
- hidden gauge group of rank smaller than four or no hidden gauge group;
- three (MS)SM matter generations;
- sizable Yukawa couplings for a generation of quarks and leptons;
- supersymmetric μ-term and proton-decay operators forbidden by gauge symmetries;
- large amount of vectorlike exotics; and
- large number of Higgs multiplet pairs.

All these properties are very promising. However, it still unclear how to generate non-perturbative couplings in a consistent manner, such that masses for the first generations can arise. In this case, it is not known whether the exotics can acquire large masses. Research along these lines is on course.

Another more subtle issue of this problem in these constructions is that there is no explanation to why the coupling constants of the known microscopic interactions seem to unify at large energy scales. The idea of GUTs is simply not consistent with this kind of setups.

5. Fourth challenge: towards cosmology

5.1. Moduli stabilization

As we have seen, many of the phenomenological properties of the string constructions we have discussed depend on some free parameters encoded in scalar fields called moduli. The moduli parametrize the metric of the compact space (size and shape) as well as the position and angles (or equivalently magnetization) of the D-branes. Since the moduli are truly free at perturbative level, i.e. they exhibit no perturbative potential that binds them, a 'good' vacuum with phenomenologically appealing properties is equivalent to any other arbitrary vacuum. In fact, if the moduli remained free, transitions between the different universes predicted by the

[4] To figure this out, one must place the left stack vertically and close to the left axis in fig. 8.

[5] It is still under debate whether the models in the classification are different nor whether all exotics are truly vectorlike in the most promising constructions.

Journal of Physics: Conference Series **378** (2012) 012011 doi:10.1088/1742-6596/378/1/012011

different vacua would be unavoidable. That definitively does not describe a universe like ours. Furthermore, even if those vacuum transitions could be avoided, free scalar fields can also be interpreted as mediators of as yet unknown forces and/or contribute in excess to the energy density of the universe, spoiling thereby cosmological processes like reheating, recombination and big bang nucleosynthesis. In short: moduli are not welcome!

There is only one way to solve this issue: *moduli stabilization.* This process consists in searching for all possible sources of vacuum energy related to the moduli at perturbative and non-perturbative level. There are some typical complications here. First, even after including non-perturbative elements –which are not as well understood as the perturbative ones–, frequently the resulting scalar potentials do not have a minimum at a finite value of the moduli; this is regularly the problem of the dilaton in heterotic string theory. Second, a normal problem of supergravity theories (like the effective limits of any string theory) is that, when there is a stable vacuum, it is usually at values for which $V(min) < 0$. This *anti-de Sitter* vacuum is bad news because the vacuum in which we live has $V(min) \sim 10^{-120}$, i.e. we live in a *de Sitter* vacuum which is fairly flat, almost Minkowski.

In string phenomenology, besides constructions that potentially explain particle physics, moduli stabilization has been independently studied in several scenarios. In type IIB string theory, it has been shown that conjugating fluxes, non-perturbative effects (gaugino condensation) and/or α' corrections could yield stable vacua [87, 88, 89]. There has also been some progress in fixing Kähler and complex structure moduli in heterotic compactifications on Calabi-Yau [90] and (generalized) half-flat [91] manifolds. In heterotic orbifold compactifications [28, 29], it is frequently argued that the absence of fluxes (other than the NS flux) can render much more difficult –if not impossible– the search for a stable vacuum. However, it is known that the inherent target space modular symmetries [92, 93, 94] together with nonperturbative effects such as gaugino condensation and string worldsheet instantons can provide a scalar potential with metastable minima, at least in toy models [95, 96, 97, 98, 99]. Actually, this question has been recently addressed in explicit heterotic orbifolds [100, 101].

Some of the essential features of moduli stabilization are captured in the so-called KKLT model [89]. Let us then describe how KKLT arrive to stabilized moduli in a de Sitter vacuum. The starting point is the type IIB theory compactified on a CY with strong warping on a particular region and that allows for fluxes to stabilize a large subset of moduli without breaking supersymmetry (all the complex structure moduli) [87]. After fixing these moduli, the fluxes render a constant contribution W_0 to the superpotential. In this ideal setup, the dilaton has acquired a mass too and all tadpoles have been canceled. The next assumption is that *only* the overall volume modulus, a Kähler modulus T, is left untouched and thus the string coupling depends only on this modulus, $g_s^{-2} = \mathrm{Re}\,T \equiv t$. Allowing for a strong interacting hidden sector whose gauginos condensate inducing a term proportional to Λ (see eq. (8)) in the superpotential. The result is the superpotential

$$W = W_0 + Ae^{aT}\,, \qquad |W_0| \ll 1\,,$$

which, together with the tree-level Kähler potential $K = -3\log\mathrm{Re}\,T = -3\log t$ yields the scalar potential

$$V(t) = -a^2A^2e^{-2at}/6t\,. \tag{10}$$

This scalar potential is always negative and, therefore, its minimum must render an anti-de Sitter vacuum. This, by itself, is not so beautiful, although it must be pointed out that the modulus t has been stabilized.

The next ingredient one must add in order to achieve a de Sitter vacuum is anti-D3 branes in the region of strong warping. These branes have the effect of breaking supersymmetry while introducing a new term in the scalar potential of the form D/t^3 (or D/t^2 if one introduces

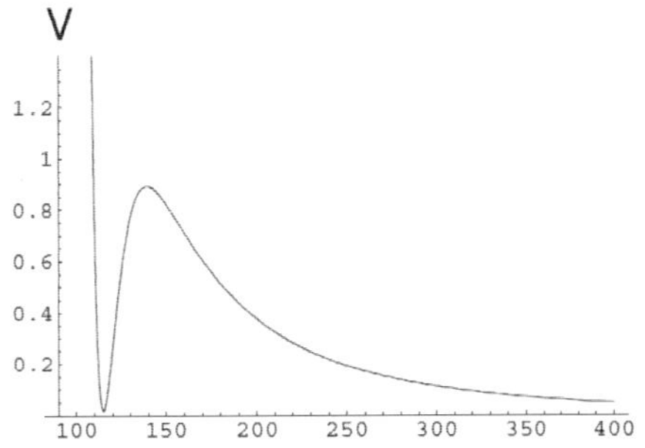

Figure 10: Moduli stabilization à la KKLT [89]. Including non-perturbative effects and an up-lifting sector, one arrives at stable De Sitter vacua.

D7 branes). In order to arrive at a de Sitter vacuum, D must be of order W_0^2, very small. Once all these effects are included, the scalar potential of the KKLT proposal takes the shape for V(t) displayed in fig. 10. In this figure, the values taken for the parameters are $A = 1, a = 0.1, W_0 = -10^{-4}$ and $D = 3 \times 10^{-9}$.

One of the unpleasant features of the KKLT proposal is that it requires an uplifting sector (the anti-D3 branes) which have no purpose other than pushing the minimum of V to positive values. It has been shown that this might actually not be needed at all, as long as T is not the only unfixed modulus [102]. In fact, if matter fields are available, it is very simple to arrive at a stabilized vacuum. Another unpleasant feature is the amount of fine-tuning needed to arrive at a de Sitter vacuum; in particular, the numbers W_0 and D must be adjusted very carefully not to lose the minimum of the potential. A very simple example of moduli stabilization without uplifting nor fine-tuning, based partly on the proposal of [102], is presented in Appendix A. Finally, the sector in which the SM matter should be located is not addressed. Hence, a realization of this model within string theory seems very challenging.

Despite its challenging features, the KKLT scenario has served as inspiration for many stringy attempts to arrive at realistic vacua and in type II theories (and M-theory), this model has become the standard method to stabilize moduli. The key generic ingredients are: 1) most moduli stabilized in a supersymmetric fashion, 2) non-perturbative effects in order to fix the remaining modulus (T in the case of type II theories, and the dilaton S for the heterotic strings), and 3) a uplifting sector which, as we have seen, can contain a set of branes or some matter fields. It seems thus clear that using this or a similar method, one can obtain a stable compactification scheme in which some cosmological problems are avoided.

5.2. Inflation

A successful string model should be able not only to reproduce the particle physics of our universe, but also to explain the structure and dynamics of the cosmos. One persisting problem is to explain the origin and dynamics of the rapid period of expansion of the observable dimensions, known as inflation (see [103] for a useful review). In field theory, it is assumed that there is some field, the inflaton, whose potential is almost flat and has a minimum for some value of the inflaton. Many single and multi-field inflationary models have been concocted and lead to acceptable e-folds, primordial fluctuations and non-gaussianities.

In the context of string theory, there has been an effort mostly in the type IIB string [104, 105, 106, 107] to derive inflation from first principles. Remarkably, it has been found that, in certain scenarios, admissible fluctuations and nongaussianities are more natural in a universe dominated by multi-field dynamics [108], as is always the case in string constructions.

In general, the inflaton must be an (MS)SM singlet scalar whose potential energy satisfies a

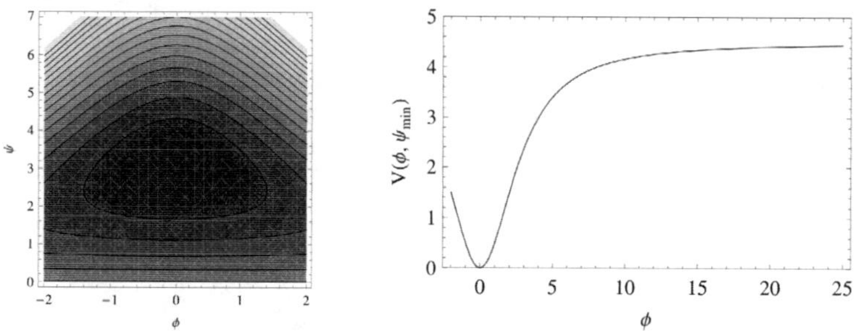

Figure 11: Modified hybrid inflation. The field ψ develops quickly a VEV inducing an inflationary potential for ϕ. We have taken $V = (\psi - 3)^2/2 + \phi^2\psi^2/16$.

series of constraints in order to produce inflation during an appropriate period of time. In the simplest case of slow-roll inflation, the scalar potential of the inflaton $V(\phi)$ must be very flat and the number of e-folds $-\int d\phi V/V'$ during the inflationary epoch must be something between 50 and 60. Since the moduli are scalar fields and, in some cases, their potential is very flat, it is natural to expect that some modulus play the role of the inflaton.

Interestingly, in the heterotic orbifolds with semi-realistic matter content, it has been found that some scalar fields whose squared masses lie between 10^{-100} and 10^{-30} in Planck units [100]. Since these quantities correspond to the curvature of the potential on the directions of those fields in moduli space, there are some almost flat directions which, nevertheless, lead to a minimum. The left panel of fig. 11 is a cartoon of this situation involving only two fields, where we have denoted them by ϕ and ψ.

Under these conditions, we can consider two different situations: when the coupling between ϕ and ψ cannot be neglected, and when $V(\phi, \psi) \approx V(\phi) + V(\psi)$. In the former case, we might arrive at the scenario described by hybrid inflation models. If the field ψ rapidly develops a VEV, the resulting scalar potential $V(\phi) = V(\phi, \psi_{\min})$ is just the ideal slow-roll potential, as displayed in the right panel of fig. 11. In the latter case, the fields with quasi-flat potential might be suitable to accommodate the multi-field curvaton-inflaton scenario investigated in [108], yielding more accessible values for cosmological observables, such as the non-linearity parameter f_{NL}. Of course, whether or not we get such behaviors in actual orbifold models is still a question that must be studied.

On the other hand, given that most of the moduli stabilization mechanism are based on the proposal of KKLT, it is natural to investigate what prospects there are for inflationary scenarios in that case. In KKLT-based models, the Hubble constant during the last stages of a string theory inflation model should be below the expected gravitino mass, $H \lesssim 1$ TeV [104, 109]. Therefore, one should find a realistic particle physics model where the non-perturbative string theory dynamics occurs at the LHC scale or even lower, and inflation occurs at a density at least 30 orders of magnitude below the Planck energy density [104, 109]. This represents a problem since the inflationary energy scales are close to the compactification scales, which are much higher than $m_{3/2}$.

This is not the first time that string theory and supergravity have encountered cosmological problems associated with the small value of the gravitino mass and of the moduli fields. The famous gravitino problem and the cosmological moduli problem are haunting us for more than two decades [110, 111]. Now we see that the smallness of the gravitino mass leads to an additional problem in the context of string cosmology [104, 109, 112].

There is an interesting idea introduced in [107] to solve this issue, which does not involve giving up the solution of the naturalness problem through supersymmetry (i.e. the value of

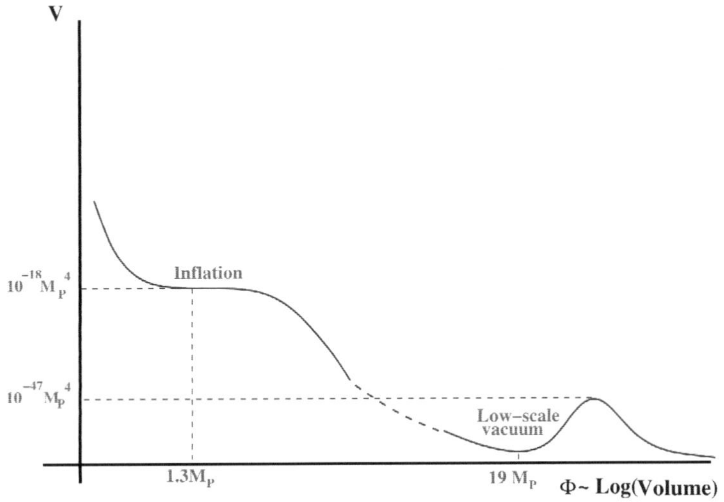

Figure 12: An illustration of the scenario proposed in [107]. At relatively small volume, high-scale inflation occurs due to fine-tuned quantum corrections. After inflation the volume modulus evolves over a long range of many Planck scales, eventually settling in the large volume minimum with TeV gravitino mass. Although the barrier protecting from decompactification is very small compared to the initial energies, an attractor solution guides the fields to the minimum and prevents overshooting.

$m_{3/2}$ is not increased radically) nor introducing new non-standard contributions to the Kähler potential plus huge fine-tuning [113]. The idea is that inflation should end with a runaway in field space, with the true minimum lying a very long distance in field space from the location where inflation occurs (e.g. about twenty Planckian distances, like in fig. 12). The problem to solve is that characteristic inflationary energy scales are much larger than those appropriate for supersymmetry breaking. The advantage of a runaway epoch is that evolution along runaway potentials like $e^{-\lambda\phi}$ is one of the few efficient ways of naturally dissipating large quantities of energy and reducing the scale of the potential by many orders of magnitude. If the true minimum of the scalar potential lies a long way along the runaway direction, it naturally has much lower characteristic energy scales than apply during inflation.

In this case, supergravity models of high-scale inflation consistent with low-scale supersymmetry breaking, $m_{3/2} \lesssim 1$ TeV, should have three stages. In the first, inflation occurs at high energy scales with $m_{3/2} \gg 1$ TeV during inflation. In the second, inflation ends with the fields fast-rolling towards a runaway direction. In the third stage, the presence of small initial quantities of radiation drives the fields to a minimum. During the evolution towards the minimum large amounts of energy are dissipated. Due to this dissipation of energy, overshooting (i.e. $\phi \longrightarrow \infty$) is avoided and the field falls into the global minimum of the potential in which $m_{3/2} \sim 1$ TeV. This scenario is illustrated in fig. 12.

The proposed scenarios are promising. However, a crucial step is missing. One must compute all details in explicit constructions and pursue actual predictions, which do not involve black boxes. In this sense, current efforts in heterotic constructions as those cited above are needed and hopefully shall render useful fruits.

6. Final remarks

During the last decades, string phenomenologists have worked in many different scopes in order to bring string theory *down to Earth*. Models that reproduce many properties of the MSSM exist in all different scenarios, but they all are challenged by the computation of the magnitude of the

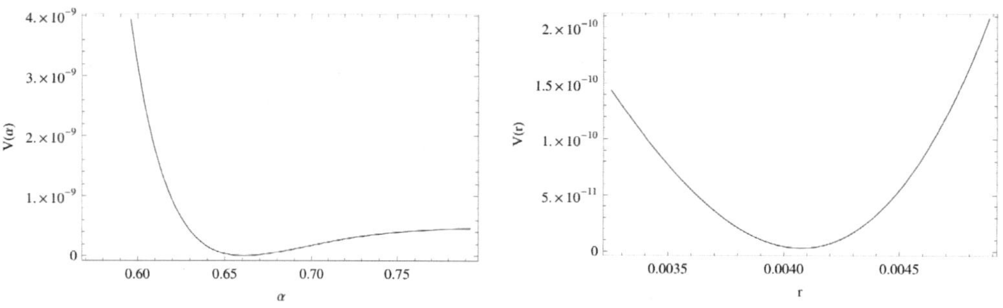

Figure A1: De Sitter vacuum from a simple no-scale model with a modulus T and a matter field Φ. No uplifting term nor fine-tuned parameters are needed.

couplings, which eventually might lead to predictions. The situation is even more complicated when cosmological features are addressed. It seems nevertheless that the current tools developed in string theories could serve to overcome those obstacles; one must *just* perform the crucial computations which, in some cases, will require new algorithms and mathematical methods.

Before concluding, it is worth discussing one recurring worry: the so-called landscape problem. The issue is that there is a plethora of CY, orbifold and other similar spaces with mathematically consistent properties to compactify the extra dimensions of string theories, and only very few or one (if at all) can reproduce our universe. How can one be wise enough to find out which one is the correct one? This formulation of the issue can lead to believe that it is impossible to circumvent it. However, it is useful to compare it with finding out both the correct metric describing the gravitation dynamics of our universe and the action that describes particle physics. The number of mathematically admissible metrics and actions is actually infinite. And one must make use of one's intuition to arrive at the correct results. In string theory, the situation is somewhat better. The mathematical consistency of the theory implies several constraints that yield a finite (but very large) set of possible compactification spaces and vacua. The task of a string phenomenologist is then to use (once again) her intuition and observable data to arrive at the compactification that describes our physics.

Acknowledgements
I am thankful to the organizers of the XIII Mexican School on Particles and Fields for their invitation. I am also indebted to O. Loaiza-Brito for the hospitality during the workshop and to G. Contreras for his support. This work was partially supported by CONACyT grants 82291 and 151234, and DGAPA-IACOD grant IA101811.

Appendix A. A simple exercise: de Sitter vacua with two moduli

Consider a hypothetical (type II) string compactification where all moduli excepting the overall volume T and a (bulk) matter field Φ have been fixed in a supersymmetric fashion. Including the Kähler potential at tree level and supposing a generic superpotential with gaugino condensation, the resulting model in the supergravity limit is defined by

$$W = \Phi^3 + A\Phi e^{-aT}\,, \qquad K = -3\log(T+\bar{T}) + \Phi\bar{\Phi}\,. \tag{A.1}$$

Let us take $T = \alpha + i\tau$ and $\Phi = re^{i\varphi}$ and suppose that $\tau = \varphi = 0$. The corresponding scalar potential with $A = -1$ and $a = 15$ is depicted in fig. A1. We find that the minimum is located at $r = 4 \times 10^{-3} M_{Pl}$ and $\alpha = 0.7 M_{Pl}$, which leads to the vacuum energy (cosmological constant) $V(min) \approx 3 \times 10^{-12} M_{Pl}^4$. In this vacuum, supersymmetry is broken because the gravitino mass is non-vanishing, $m_{3/2} = e^{K/2}|W| \approx 10^{-7} M_{Pl}$.

The main property of this model is the coupling between the gaugino condensate $\langle\lambda\lambda\rangle$ and the matter field giving rise to the superpotential given in eq. (A.1). The value chosen for a is quite natural as it can arise from a simple SU(3) Yang-Mills theory in the hidden sector. Surprisingly, the gravitino mass is quite suppressed and, although is still unrealistic, there might be variations of the current model that provide an acceptable $m_{3/2}$ as well as a proper cosmological constant. Despite its simplicity, this model seems much than the proposal by KKLT, since it avoids all naturalness issues present in previous attempts of stabilization.

References

[1] ATLAS-Collaboration 2011 Combination of higgs boson searches with up to 4.9 fb^{-1} of *pp* collision data taken at $\sqrt{s} = 7$ tev with the atlas experiment at the lhc ATLAS-CONF-2011-163
[2] CMS-Collaboration 2011 Combination of cms searches for a standard model higgs boson CMS PAS HIG-11-032
[3] Erler J 2012 (*Preprint* 1201.0695)
[4] Bennett C L *et al.* 1996 *Astrophys. J.* **464** L1–L4 (*Preprint* astro-ph/9601067)
[5] Komatsu E *et al.* (WMAP) 2011 *Astrophys. J. Supp.* **192** 18 (*Preprint* 1001.4538)
[6] Golfand Y A and Likhtman E P 1971 *JETP Lett.* **13** 323–326
[7] Volkov D V and Akulov V P 1972 *JETP Lett.* **16** 438–440
[8] Wess J and Zumino B 1974 *Nucl. Phys.* **B70** 39–50
[9] Nilles H P 1984 *Phys. Rept.* **110** 1–162
[10] Randall L and Sundrum R 1999 *Phys. Rev. Lett.* **83** 3370–3373 (*Preprint* hep-ph/9905221)
[11] Randall L and Sundrum R 1999 *Phys. Rev. Lett.* **83** 4690–4693 (*Preprint* hep-th/9906064)
[12] Georgi H and Glashow S L 1974 *Phys. Rev. Lett.* **32** 438–441
[13] Pati J C and Salam A 1974 *Phys. Rev.* **D10** 275–289
[14] Yoneya T 1973 *Nuovo Cim. Lett.* **8** 951–955
[15] Yoneya T 1974 *Prog. Theor. Phys.* **51** 1907–1920
[16] Scherk J and Schwarz J H 1974 *Nucl. Phys.* **B81** 118–144
[17] Gliozzi F, Scherk J and Olive D I 1977 *Nucl. Phys.* **B122** 253–290
[18] Brink L, Schwarz J H and Scherk J 1977 *Nucl. Phys.* **B121** 77
[19] Schwarz J H 1996 *Phys. Lett.* **B367** 97–103 (*Preprint* hep-th/9510086)
[20] Townsend P K 1996 (*Preprint* hep-th/9612121)
[21] Polchinski J Cambridge, UK: Univ. Pr. (1998) 402 p
[22] Polchinski J 1998 *String theory. Vol. 2: Superstring theory and beyond* (Cambridge, UK: Univ. Pr.) 531 P.
[23] Kehagias A and Sfetsos K 2000 *Phys. Lett.* **B472** 39–44 (*Preprint* hep-ph/9905417)
[24] Long J C, Chan H W and Price J C 1999 *Nucl. Phys.* **B539** 23–34 (*Preprint* hep-ph/9805217)
[25] Kaluza T 1921 *Sitzungsber. Preuss. Akad. Wiss. Berlin* **K1** 966
[26] Klein O 1926 *Z. Phys.* **37** 895
[27] Candelas P, Horowitz G T, Strominger A and Witten E 1985 *Nucl. Phys.* **B258** 46–74
[28] Dixon L J, Harvey J A, Vafa C and Witten E 1985 *Nucl. Phys.* **B261** 678–686
[29] Dixon L J, Harvey J A, Vafa C and Witten E 1986 *Nucl. Phys.* **B274** 285–314
[30] Huber S J and Shafi Q 2001 *Phys. Lett.* **B498** 256–262 (*Preprint* hep-ph/0010195)
[31] Dabholkar A 1997 (*Preprint* hep-th/9804208)
[32] Cremades D, Ibáñez L E and Marchesano F 2003 *JHEP* **07** 038 (*Preprint* hep-th/0302105)
[33] Marchesano Buznego F G 2003 (*Preprint* hep-th/0307252)
[34] Pokorski W and Ross G G 1999 *Nucl. Phys.* **B551** 515–548 (*Preprint* hep-ph/9809537)
[35] Donagi R, Ovrut B A, Pantev T and Waldram D 2002 *Adv. Theor. Math. Phys.* **5** 93–137 (*Preprint* hep-th/9912208)
[36] Donagi R, He Y H, Ovrut B A and Reinbacher R 2005 *Phys. Lett.* **B618** 259–264 (*Preprint* hep-th/0409291)
[37] Braun V, He Y H, Ovrut B A and Pantev T 2006 *JHEP* **05** 043 (*Preprint* hep-th/0512177)
[38] Bouchard V and Donagi R 2006 *Phys. Lett.* **B633** 783–791 (*Preprint* hep-th/0512149)
[39] Anderson L B, Gray J, He Y H and Lukas A 2010 *JHEP* **02** 054 (*Preprint* 0911.1569)
[40] Anderson L B, Gray J, Lukas A and Palti E 2011 *Phys. Rev.* **D84** 106005 (*Preprint* 1106.4804)
[41] Blumenhagen R, Moster S and Weigand T 2006 (*Preprint* hep-th/0603015)
[42] Blumenhagen R, Moster S, Reinbacher R and Weigand T 2007 *JHEP* **05** 041 (*Preprint* hep-th/0612039)

[43] Ambroso M and Ovrut B A 2010 *Int. J. Mod. Phys.* **A25** 2631–2677 (*Preprint* 0910.1129)
[44] Braun V, He Y H and Ovrut B A 2006 *JHEP* **04** 019 (*Preprint* hep-th/0601204)
[45] Bouchard V, Cvetič M and Donagi R 2006 *Nucl. Phys.* **B745** 62–83 (*Preprint* hep-th/0602096)
[46] Nilles H P, Ramos-Sánchez S, Ratz M and Vaudrevange P K S 2009 *Eur. Phys. J.* **C59** 249–267 (*Preprint* 0806.3905)
[47] Ibáñez L E, Kim J E, Nilles H P and Quevedo F 1987 *Phys. Lett.* **B191** 282–286
[48] Faraggi A E, Nanopoulos D V and Yuan K j 1990 *Nucl. Phys.* **B335** 347
[49] Kobayashi T, Raby S and Zhang R J 2005 *Nucl. Phys.* **B704** 3–55 (*Preprint* hep-ph/0409098)
[50] Buchmüller W, Hamaguchi K, Lebedev O and Ratz M 2006 *Phys. Rev. Lett.* **96** 121602 (*Preprint* hep-ph/0511035)
[51] Buchmüller W, Hamaguchi K, Lebedev O and Ratz M 2007 *Nucl. Phys.* **B785** 149–209 (*Preprint* hep-th/0606187)
[52] Lebedev O *et al.* 2008 *Phys. Rev.* **D77** 046013 (*Preprint* 0708.2691)
[53] Blaszczyk M *et al.* 2010 *Phys. Lett.* **B683** 340–348 (*Preprint* 0911.4905)
[54] Groot Nibbelink S, Hillenbach M, Kobayashi T and Walter M G A 2004 *Phys. Rev.* **D69** 046001 (*Preprint* hep-th/0308076)
[55] Ibáñez L E, Nilles H P and Quevedo F 1987 *Phys. Lett.* **B187** 25–32
[56] Buchmüller W, Hamaguchi K, Lebedev O and Ratz M 2005 (*Preprint* hep-ph/0512326)
[57] Lebedev O *et al.* 2007 *Phys. Lett.* **B645** 88–94 (*Preprint* hep-th/0611095)
[58] Lebedev O, Nilles H P, Ramos-Sánchez S, Ratz M and Vaudrevange P K S 2008 *Phys. Lett.* **B668** 331–335 (*Preprint* 0807.4384)
[59] Kobayashi T, Nilles H P, Plöger F, Raby S and Ratz M 2007 *Nucl. Phys.* **B768** 135–156 (*Preprint* hep-ph/0611020)
[60] Ko P, Kobayashi T, Park J h and Raby S 2007 *Phys. Rev.* **D76** 035005 (*Preprint* 0704.2807)
[61] Krippendorf S, Nilles H P, Ratz M and Winkler M W 2012 (*Preprint* 1201.4857)
[62] Dimopoulos S, Raby S and Wilczek F 1982 *Phys. Lett.* **B112** 133
[63] Hamidi S and Vafa C 1987 *Nucl. Phys.* **B279** 465
[64] Dixon L J, Friedan D, Martinec E J and Shenker S H 1987 *Nucl. Phys.* **B282** 13–73
[65] Font A, Ibáñez L E, Nilles H P and Quevedo F 1988 *Phys. Lett.* **B213** 274
[66] Kobayashi T, Parameswaran S L, Ramos-Sánchez S and Zavala I 2011 (*Preprint* 1107.2137)
[67] Ferrara S, Girardello L and Nilles H P 1983 *Phys. Lett.* **B125** 457
[68] Lebedev O *et al.* 2007 *Phys. Rev. Lett.* **98** 181602 (*Preprint* hep-th/0611203)
[69] Burwick T T, Kaiser R K and Müller H F 1991 *Nucl. Phys.* **B355** 689–711
[70] Erler J, Jungnickel D, Spalinski M and Stieberger S 1993 *Nucl. Phys.* **B397** 379–416 (*Preprint* hep-th/9207049)
[71] Kobayashi T and Ohtsubo N 1990 *Phys. Lett.* **B245** 441–446
[72] Kobayashi T and Ohtsubo N 1994 *Int. J. Mod. Phys.* **A9** 87–126
[73] Nilles H P, Ramos-Sánchez S, Vaudrevange P K S and Wingerter A 2012 *Comput. Phys. Commun.* **183** 1363–1380 (*Preprint* 1110.5229)
[74] Acharya B S and Witten E 2001 (*Preprint* hep-th/0109152)
[75] Acharya B S, Bobkov K, Kane G L, Kumar P and Shao J 2007 *Phys. Rev.* **D76** 126010 (*Preprint* hep-th/0701034)
[76] Acharya B S, Bobkov K, Kane G L, Shao J and Kumar P 2008 *Phys. Rev.* **D78** 065038 (*Preprint* 0801.0478)
[77] Acharya B S, Kane G, Kuflik E and Lu R 2011 *JHEP* **05** 033 (*Preprint* 1102.0556)
[78] Angelantonj C, Bianchi M, Pradisi G, Sagnotti A and Stanev Y S 1996 *Phys. Lett.* **B385** 96–102 (*Preprint* hep-th/9606169)
[79] Kakushadze Z 1998 *Nucl. Phys.* **B512** 221–236 (*Preprint* hep-th/9704059)
[80] Cremades D, Ibanez L E and Marchesano F 2002 *JHEP* **07** 022 (*Preprint* hep-th/0203160)
[81] Cremades D, Ibáñez L E and Marchesano F 2002 *Nucl. Phys.* **B643** 93–130 (*Preprint* hep-th/0205074)
[82] Blumenhagen R, Cvetič M, Kachru S and Weigand T 2009 *Ann. Rev. Nucl. Part. Sci.* **59** 269–296 (*Preprint* 0902.3251)
[83] Cvetič M, Halverson J and Richter R 2009 *JHEP* **12** 063 (*Preprint* 0905.3379)
[84] Gmeiner F, Blumenhagen R, Honecker G, Lüst D and Weigand T 2006 *JHEP* **01** 004 (*Preprint* hep-th/0510170)
[85] Gmeiner F and Honecker G 2007 *JHEP* **09** 128 (*Preprint* 0708.2285)
[86] Gmeiner F and Honecker G 2008 *JHEP* **07** 052 (*Preprint* 0806.3039)
[87] Giddings S B, Kachru S and Polchinski J 2002 *Phys. Rev.* **D66** 106006 (*Preprint* hep-th/0105097)
[88] Becker K, Becker M, Haack M and Louis J 2002 *JHEP* **06** 060 (*Preprint* hep-th/0204254)
[89] Kachru S, Kallosh R, Linde A D and Trivedi S P 2003 *Phys. Rev.* **D68** 046005 (*Preprint* hep-th/0301240)

[90] Anderson L B, Gray J, Lukas A and Ovrut B 2011 *JHEP* **02** 088 (*Preprint* 1010.0255)
[91] de Carlos B, Gurrieri S, Lukas A and Micu A 2006 *JHEP* **03** 005 (*Preprint* hep-th/0507173)
[92] Ferrara S, Lust D, Shapere A D and Theisen S 1989 *Phys. Lett.* **B225** 363
[93] Lauer J, Mas J and Nilles H P 1989 *Phys. Lett.* **B226** 251
[94] Lauer J, Mas J and Nilles H P 1991 *Nucl. Phys.* **B351** 353–424
[95] Cvetič M, Font A, Ibáñez L E, Lüst D and Quevedo F 1991 *Nucl. Phys.* **B361** 194–232
[96] Dundee B, Raby S and Westphal A 2010 *Phys. Rev.* **D82** 126002 (*Preprint* 1002.1081)
[97] de Carlos B, Casas J A and Muñoz C 1991 *Phys. Lett.* **B263** 248–254
[98] de Carlos B, Casas J A and Muñoz C 1993 *Nucl. Phys.* **B399** 623–653 (*Preprint* hep-th/9204012)
[99] Bailin D, Love A, Sabra W A and Thomas S 1994 *Mod. Phys. Lett.* **A9** 2543–2555 (*Preprint* hep-th/9405031)
[100] Parameswaran S L, Ramos-Sánchez S and Zavala I 2011 *JHEP* **01** 071 (*Preprint* 1009.3931)
[101] Kappl R, Petersen B, Raby S, Ratz M, Schieren R *et al.* 2011 *Nucl.Phys.* **B847** 325–349 26 pages (*Preprint* 1012.4574)
[102] Lebedev O, Nilles H P and Ratz M 2006 *Phys. Lett.* **B636** 126–131 (*Preprint* hep-th/0603047)
[103] Mazumdar A and Rocher J 2011 *Phys. Rept.* **497** 85–215 (*Preprint* 1001.0993)
[104] Kallosh R and Linde A D 2004 *JHEP* **0412** 004 (*Preprint* hep-th/0411011)
[105] Lalak Z, Ross G G and Sarkar S 2007 *Nucl. Phys.* **B766** 1–20 (*Preprint* hep-th/0503178)
[106] Covi L *et al.* 2008 *JHEP* **08** 055 (*Preprint* 0805.3290)
[107] Conlon J P, Kallosh R, Linde A D and Quevedo F 2008 *JCAP* **0809** 011 (*Preprint* 0806.0809)
[108] Burgess C P *et al.* 2010 *JHEP* **08** 045 (*Preprint* 1005.4840)
[109] Kallosh R and Linde A D 2007 *JCAP* **0704** 017 (*Preprint* 0704.0647)
[110] Weinberg S 1982 *Phys.Rev.Lett.* **48** 1303
[111] Coughlan G, Fischler W, Kolb E W, Raby S and Ross G G 1983 *Phys.Lett.* **B131** 59
[112] Buchmüller W, Hamaguchi K, Lebedev O and Ratz M 2005 *JCAP* **0501** 004 (*Preprint* hep-th/0411109)
[113] Badziak M and Olechowski M 2008 *JCAP* **0807** 021 (*Preprint* 0802.1014)

Journal of Physics: Conference Series **378** (2012) 012012 doi:10.1088/1742-6596/378/1/012012

A brief Review of the Scalar Field Dark Matter model

Juan Magaña[1]

Instituto de Astromía, Universiad Nacional Autónoma de México, 07000 México D.F., México.

E-mail: `jmagana@astroscu.unam.mx`

Tonatiuh Matos[1]

Departamento de Física, Centro de Investigación y de Estudios Avanzados del IPN, A.P. 14-740, 07000 México D.F., México.

E-mail: `tmatos@fis.cinvestav.mx`

Abstract. In the last time the cold dark matter (CDM) model has suggested more and more that it is not able to describe all the properties of nearby galaxies that can be observed in great detail as well as that it has some problems in the mechanism by which matter is more rapidly gathered into large-scale structure such as galaxies and clusters of galaxies. In this work we revisit an alternative model, the scalar field dark matter (SFDM) model, which proposes that the galactic haloes form by condensation of a scalar field (SF) very early in the Universe, i.e., in this model the haloes of galaxies are astronomical Bose-Einstein Condensate drops of SF. On the other hand, large-scale structures like clusters or superclusters of galaxies form similar to the ΛCDM model, by hierarchy, thus all the predictions of the ΛCDM model at cosmological scales are reproduced by SFDM. This model predicts that all galaxy haloes must be very similar and exist for higher redshifts than in the ΛCDM model. In the first part of this review we revisit the cosmological evolution of SFDM model with a scalar potential $m^2\Phi^2/2 + \lambda\Phi^4/4$ with two different frameworks: the field and fluid approach. We derive the evolution equations of the SF in the linear regime of perturbations as well. The scalar fluctuations have an oscillating growing mode and therefore, this kind of dark matter could lead to the early formation of gravitational structures in the Universe. We also revisit how BEC dark matter haloes exhibit a natural cut of the mass power spectrum. In the second part, we study the core central density profiles of BEC dark matter haloes and fit high-resolution rotation curves, we show a sample of some low surface brightness galaxies. The mean value of the logarithmic inner density slopes is $\alpha = -0.27 \pm 0.18$. Using a model independent new definition of the core in the BEC density profile, we show that the recent observation of the constant dark matter central surface density can be reproduced. We conclude that in light of the difficulties that the ΛCDM model is currently facing the SFDM model can be a worthy alternative to keep exploring further.

1. Introduction

Nowadays the most accepted model in cosmology which explains the evolution of the Universe is known as ΛCDM. In this model 4 per cent of the total content of the Universe is baryonic matter, 22 per cent is non-baryonic dark matter (DM) and the rest is in some form of cosmological

[1] Part of the Instituto Avanzado de Cosmología (IAC) collaboration http://www.iac.edu.mx/

constant. ΛCDM has achieved several observations with outstanding success, like for example the fact that the cosmic microwave background radiation can be explained in great detail and that it provides a framework within one can understand the large-scale isotropy of the Universe and important characteristics on the origin, nature and evolution of density fluctuations which are believed to give rise to galaxies and other cosmic structures. There remain, however, certain conflicts at galactic scales, like the cusp profile of central densities in galactic halos, the overpopulation of substructures predicted by N-body numerical simulations which are an order of magnitude larger than what has been observed, etc (see for example [18], [48], [72], and [78]).

There are other important issues of the ΛCDM model which require further considerations. Observations point out to a better understanding of the theory beginning with the Local Void, which contains far fewer galaxies than the expected. Another problem arises for the so called pure disk galaxies, which do not appear in numerical simulations of structure formation in the standard model. These problems would be solved, if the structure grew faster than it does in the standard paradigm [77].

On the other hand, [57] also found that the collision velocity of 3000km/s at R200 for the Bullet Cluster is very unlikely and might no be explained by the standard model. A final example of inconsistencies can be seen in a paper by [92], who found anomalies in the mass power spectrum obtained by SDSS and the one obtained with the CDM model. With these and other results it seems necessary to change the ΛCDM paradigm to try and explain the formation of structure in the Universe.

The incorporation of a new kind of DM, different from the one proposed by the ΛCDM model into the big bang theory holds out the possibility of resolving some of these issues. Several authors have proposed interesting alternatives in where they try to solve the latter explained difficulties that the CDM scenario seem not solve. Some of these models consider warm dark matter ([19], [97]), self interacting dark matter ([59],[90]), and other exotic scenarios as braneworld models [32]. In fact, there are models that do not include DM but instead modify the Newtonian force law [71, 87]. Other alternative scenario that has received much attention in the last years is the scalar field dark matter (SFDM) model. The main idea is simple (see [38]), the nature of the DM is completely determined by a fundamental scalar field Φ. The SFDM model proposes that galactic haloes form by Bose-Einstein condensation of a scalar field (SF) whose boson has an ultra-light mass of the order of $m \sim 10^{-22}$eV. From this mass it follows that the critical temperature of condensation $T_c \sim 1/m^{5/3}$ ~TeV is very high, therefore, they form Bose-Einstein Condensates (BEC) drops very early in the Universe. In addition, the Compton length $\lambda_c = 2\pi\hbar/m$ associated to this boson is about ~ kpc that corresponds to the dark halo-size of typical galaxies in the Universe. Thus, it has been proposed that these drops are the haloes of galaxies (see [67]), *i.e.*, that haloes are huge drops of SF. On the other hand, big structures form like in the ΛCDM model, by hierarchy [67, 91], thus, all successful predictions of the standard model at large scales are well reproduced by SFDM. In other words, in the SFDM model the haloes of galaxies do not form hierarchically, they are formed at the same time and in the same way when the Universe reaches the critical temperature of condensation of the SF. From this it follows that all galaxies must be very similar because they formed in the same manner and at the same moment [56]. Therefore, from this paradigm we have to expect that there exist well formed galaxy haloes at higher redshifts than in the ΛCDM model. In this model the scalar particles with that ultra light mass are such that their wave properties avoid the cusp problem and reduce the high number of small satellites by the quantum uncertainty principle [46, 61] which is another problem that is still present in the ΛCDM model [43, 48, 82, **?**]. Summarizing, it is remarkable that with only one free parameter, the ultra-light scalar field mass ($m \sim 10^{-22}$eV), the SFDM model fits:

(i) The evolution of the cosmological densities [68].

(ii) The acoustic peaks of the cosmic microwave background [83].

(iii) The scalar field has a natural cut off, thus the substructures in clusters of galaxies is avoided naturally. With a scalar field mass of $m_\phi \sim 10^{-22}$eV the amount of substructure is compatible with the observed one [67, 46, 91].

(iv) We expect that SFDM forms galaxies earlier than the CDM model, because they form BECs at a critical temperature $T_c >>$TeV. So if SFDM is right, we have to see big galaxies at high redshifts with similar features [68, 91].

(v) The rotation curves of big galaxies and LSB galaxies [58, 4, 42, 13, 82].

(vi) With this mass, the critical mass of collapse for a real scalar field is just $10^{12}\,M_\odot$, i.e., the one observed in galaxies haloes [1].

(vii) The observed properties of dwarf galaxies, i.e., the minimum length scale, the minimum mass scale, and their independence from the brightness [56].

(viii) And recently it has been demonstrated that if the scalar mass is $m \sim 10^{-22}$eV, the SFDM haloes would have cores large enough to explain the longevity of the cold clump in Ursa Minor and the wide distribution of globular clusters in Fornax (see [60]).

The idea was first considered by [88, 47] and independently by [64, 37, 38, 53] suggested bosonic dark matter as a model for galactic halos [28]. In the BEC model, DM haloes can be described, in the non-relativistic regime, as Newtonian gravitational condensates made up of ultra light bosons condensated in a single macroscopic wave function. Thus, BECs haloes can be described as a coherent scalar field Φ.

Several authors have introduced a dynamic scalar field with a certain potential $V(\Phi)$ as a candidate to dark matter, although there is not yet an agreement for the correct form of the potential of the field. One interesting work pointing this way was done by [65] and independently by [85] where they used a potential of the form $V(\Phi) = V_0\left[\cosh\left(\xi\Phi\right) - 1\right]$ to explain the core density problem for disc galaxy halos in the ΛCDM model (see also [98], [80] and [21]). [38] presented a model for the DM in spiral galaxies, in which they supposed that DM is an arbitrary SF endowed with a scalar potential. Other scalar potential widely used to describe DM is $V(\Phi) = m^2\Phi^2/2$ [93, 68]. This potential is very interesting because it can mimic the cosmological evolution of the Universe predicted by the ΛCDM model. If we consider a SF self-interaction, we need to add a quartic term to the SF potential [2, 5, 6, 7, 15], in this case the equation of state of the SF is that of a polytope of index n=1 (see [91, 42, 44, 3, 58]).

Different issues of the cosmological behavior of the SFDM/BEC model have been studied (see for example [67, 2, 46, 95, 61, 83, 84, 91, 44, 42, 68, 5, 69, 24, 99]). [46] proposed fuzzy dark matter that is composed of ultra light scalar particles who are initially in the form of a BEC. In their work [99] used a bosonic dark matter model to explain the structure formation via high-resolution simulations, [96, 95] reviewed the key properties that may arise from the bosonic nature of SFDM models. Recently [44, 42] developed a further analysis of the cosmological dynamics of SFDM/BEC as well as of the evolution of their fluctuations (see also [24]). In the same direction, [91] studied, the growth of scalar fluctuations and large-scale structure formation with the fluid approach of the SFDM/BEC model.

On the other hand, SFDM/BEC model has provided to be a good candidate for DM haloes of galaxies in the Universe because it can explain many aspects where the standard model fails ([13, 14, 55, 56, 54, 60, 43, 82]). In addition, many numerical simulations have been performed to study the gravitational collapse of SFDM/BEC model [11, 12, 20, 39, 40, 41, 35]. [22, 23] found approximate analytical expression and numerical solutions of the mass-radius relation of SFDM/BEC haloes. Recently, [81] give constraints on the boson mass to form and mantain more than one vortex in SFDM/BEC haloes. These constraints are in agreement with the ultra light mass founded in previous works (see also [100]). Lately, [60] performed N-body simulations

Journal of Physics: Conference Series **378** (2012) 012012 doi:10.1088/1742-6596/378/1/012012

to study the dynamics observed in Ursa Minor dwarf galaxy. They model the dark matter halo of Ursa Minor as a SFDM/BEC halo to establish constraints for the boson mass. Moreover, they introduce a dynamical friction analysis with the SFDM/BEC model to study the wide distribution of globular clusters in Fornax. An overall good agreement is found for the ultra light mass $\sim 10^{-22}$eV of bosonic dark matter.

The main objective of this review is to introduce the framework of the SFDM/BEC model that assumes that DM is a scalar field that involves a self-interacting potential of the form $V(\Phi) = m^2\Phi^2/2 + \lambda\Phi^4/4$, where $m \sim 10^{-22}$eV is the mass of the scalar field, [53], [67] and [46].

We solve the Friedmann equations for the SFDM/BEC model to show that it behaves just like the CDM model. Also, we show that the SFDM/BEC predicts galaxy formation earlier than the CDM model, because they form BEC at a critical temperature $T_c >> $ TeV. So, if SFDM is right, this would imply that we have to see big galaxies at high redshifts. In order to do this, we study the density fluctuations of the scalar field from a hydrodynamical point of view, this will give us some information about the energy density of dark matter halos necessary to obtain the observational results of large-scale structure. Here we will give some tools that might be necessary for the study of the early formation of structure.

In section 2 we study the cosmological evolution of the SF with a field and a fluid approach. We obtain a numerical solution for the scalar density as well as analytical expressions for the kinetic and potential energies of the SF. Also, we study the evolution for the density contrast of scalar fluctuations in the linear regime. In section 5 we study the rotation curves and the cusp/core discrepancy with SFDM haloes to compare with the predictions of CDM. Finally we give our conclusions in section 6.

2. The Scalar Field Dark Matter Cosmology

The framework of the standard cosmological model is of a homogeneous and isotropic Universe whose evolution is best described by Friedmann's equations that come from general relativity and whose main ingredients can be described by fluids whose characteristics are very similar to those we see in our Universe. Of course, the Universe is not exactly homogeneous and isotropic but this standard model does give us a framework within which we can study the evolution of the expansion rate of the Universe as well as the evolution of small fluctuations that give rise to gravitational structures in the Universe as galaxies and clusters of galaxies.

In this section, we study the cosmological dynamics of the SFDM/BEC model and the evolution of their scalar perturbations with two different frameworks: the field and fluid approach.

2.1. Background Universe: The Field Approach

We use the Friedmann-Lemaître-Robertson-Walker (FLRW) metric with scale factor $a(t)$. Our background Universe is composed by SFDM ($\Phi_0(t)$) endowed with a scalar potential $V \equiv V(\Phi_0)$, radiation (z), neutrinos (ν), baryons (b), and a cosmological constant (Λ) as dark energy. We begin by recalling the basic background equations. From the energy-momentum tensor $\mathbf{T}$ for a scalar field, the scalar energy density T^0_0 and the scalar pressure T^i_j are given by

$$T^0_0 = -\rho_{\Phi_0} = -\left(\frac{1}{2}\dot{\Phi}_0^2 + V\right), \tag{1}$$

$$T^i_j = P_{\Phi_0} = \left(\frac{1}{2}\dot{\Phi}_0^2 - V\right)\delta^i_j, \tag{2}$$

where the dots stand for the derivative with respect to the cosmological time and δ^i_j is the Kronecker delta. Thus, the Equation of State (EoS) for the scalar field is $P_{\Phi_0} = \omega_{\Phi_0}\rho_{\Phi_0}$ with

$$\omega_{\Phi_0} = \frac{\frac{1}{2}\dot{\Phi}_0^2 - V}{\frac{1}{2}\dot{\Phi}_0^2 + V}. \tag{3}$$

The radiation fields, the baryonic component and the cosmological constant are represented by perfect fluids with baryotropic equation of state $P_\gamma = (\gamma - 1)\rho_\gamma$, where γ is a constant, $0 \le \gamma \le 2$. For example, $\gamma_z = \gamma_\nu = 4/3$ for radiation and neutrinos, $\gamma_b = 1$ for baryons, and for a cosmological constant $\gamma_\Lambda = 0$.

The Einstein-Klein-Gordon equations that describe this Universe are (in units $c = \hbar = 1$).

$$\dot{H} = -\frac{\kappa^2}{2}\left(\dot{\Phi}_0^2 + \frac{4}{3}\rho_z + \frac{4}{3}\rho_\nu + \rho_b\right), \quad \ddot{\Phi}_0 + 3H\dot{\Phi}_0 + V_{,\Phi_0} = 0, \tag{4}$$

$$\dot{\rho}_z + 4H\rho_z = 0, \quad \dot{\rho}_\nu + 4H\rho_\nu = 0, \quad \dot{\rho}_b + 3H\rho_b = 0, \tag{5}$$

with the Friedmann constraint

$$H^2 = \frac{\kappa^2}{3}\left(\rho_{\Phi_0} + \rho_z + \rho_\nu + \rho_b + \rho_\Lambda\right), \tag{6}$$

being $\kappa^2 \equiv 8\pi G$, $H \equiv \dot{a}/a$ the Hubble parameter and the commas stand for the derivative with respect to scalar field. Notice that background scalar quantities at zero order have the subscript 0. In order to solve the system of equations (2.1), we define the following dimensionless variables

$$x \equiv \frac{\kappa}{\sqrt{6}}\frac{\dot{\Phi}_0}{H}, \quad u \equiv \frac{\kappa}{\sqrt{3}}\frac{\sqrt{V}}{H}, \quad z \equiv \frac{\kappa}{\sqrt{3}}\frac{\sqrt{\rho_z}}{H}, \tag{7}$$

$$\nu \equiv \frac{\kappa}{\sqrt{3}}\frac{\sqrt{\rho_\nu}}{H}, \quad b \equiv \frac{\kappa}{\sqrt{3}}\frac{\sqrt{\rho_b}}{H}, \quad l \equiv \frac{\kappa}{\sqrt{3}}\frac{\sqrt{\rho_\Lambda}}{H}. \tag{8}$$

Here we take the quadratic scalar potential $V = m^2\Phi_0^2/2$ with $m \sim 1 \times 10^{-22}$eV. Using these variables, the equations (2.1) for the evolution of the background Universe are transformed into

$$x' = -3x - su + \frac{3}{2}\Pi x, \quad u' = sx + \frac{3}{2}\Pi u, \quad z' = \frac{3}{2}\left(\Pi - \frac{4}{3}\right) z, \tag{9}$$

$$\nu' = \frac{3}{2}\left(\Pi - \frac{4}{3}\right)\nu, \quad b' = \frac{3}{2}(\Pi - 1)\, b, \quad l' = \frac{3}{2}\Pi l, \quad s' = s_0 s^{-k}, \tag{10}$$

where the prime denotes a derivative with respect to the e-folding number $N = \ln a$, and Π is defined as

$$-\frac{\dot{H}}{H^2} = \frac{3}{2}(2x^2 + b^2 + \frac{4}{3}z^2 + \frac{4}{3}\nu^2) \equiv \frac{3}{2}\Pi. \tag{11}$$

We have introduced the variable $s \equiv C_0/H$, where C_0 is a constant, as a control parameter for the dynamics of H. In Eq. (10), s_0 is a constant and the exponent k is ≤ 0 (see [94] for more details of the control parameter). With these variables, the density parameters Ω_i for each component i can be written as

$$\Omega_{\Phi_0} = x^2 + u^2, \quad \Omega_z = z^2 \quad \Omega_\nu = \nu^2, \quad \Omega_b = b^2, \quad \Omega_\Lambda = l^2, \tag{12}$$

subject to the Friedmann constraint

$$x^2 + u^2 + z^2 + \nu^2 + b^2 + l^2 = 1. \tag{13}$$

In addition, we may write the EoS of the scalar field as

$$\omega_{\Phi_0} = \frac{x^2 - u^2}{\Omega_{\Phi_0}}. \tag{14}$$

Since ω_{Φ_0} is a function of time, if its temporal average tends to zero, this would imply that Φ^2-dark matter can be able to mimic the EoS for CDM.

We solve the system of equations (2.1) for the background Universe numerically with a four order Adams-Bashforth-Moulton (ABM) method. We take as initial conditions the best estimates from 5-years [49] and 7-years WMAP [50] values to $\Omega_\Lambda^{(0)} = 0.73$, $\Omega_{DM}^{(0)} = 0.22994$, $\Omega_b^{(0)} = 0.04$, $\Omega_z^{(0)} = 0.00004$, $\Omega_\nu^{(0)} = 0.00002$ and the exponent $k = 0$. In Fig.1 we see the evolution of the scalar field Φ_0. This figure shows how the scalar field oscillates very stark about the minimum of the scalar potential V, on the bottom of this figure we show a zoom of these oscillations. In Fig.2 we show the evolutions of the kinetic ($\dot{\Phi}^2/2$) and potential energy ($m^2\Phi^2/2$) of the scalar field, as expected, the SF oscillations are translated into very stark oscillations for the kinetic and the potential energies of the SF. However, observe the evolution of the dark matter density of the scalar field, that means $\rho_\Phi = (\dot{\Phi}^2 + m^2\Phi^2)/2$ in Fig.3. Note the following crucial point, although the kinetic and potential energies show very stark oscillations, the sum of both energies does not oscillates at all, this sum is the density parameter Ω_{Φ_0}, which does not display any oscillation. The important point we have to take from now on into account is that the oscillations are not physical observables at all, they are a feature of the SF, what we observe in fact is the density of the SF which does not oscillates.

Fig.3 shows the numerical evolution of the density parameters in our model[2]. At early times, radiation dominates the evolution of the Universe. Later on, the Universe has an epoch where the energy density radiation is equal to the dark matter density, at z_{eq}, then dark matter begins to dominate the evolution. The recombination era in SFDM model occurs at $z \sim 1000$. At later times, the cosmological constant dominates the dynamics of the Universe at $z_\Lambda \sim 0.5$. The behavior is exactly the same as in the ΛCDM model. Furthermore, in the recombination era of the SFDM model, the neutrinos made up ~ 12 per cent of the Universe completely in agreement with the measurements of WMAP. Fig.4 shows the evolution of the EoS for the SF. Although the EoS varies with time (oscillations), the temporal average, $\langle\omega_{\Phi_0}\rangle$, drops to zero. Therefore, SFDM is like a pressureless fluid and behaves as CDM at cosmological scales ([93], [67], [68], [69]).

2.2. The Background: The Fluid Approach

In this section we perform a transformation in order to solve the Friedmann equations analytically with the approximation $H << m$.

Here we take the scalar potential as $V = m^2\Phi^2/2\hbar^2 + \lambda\Phi^4/4$, (we remain that we use units where $c = \hbar = 1$ unless it is stipulated). Then for the ultra-light boson particle we have that $m \sim 10^{-22}$ eV.

Now we express the SF, Φ_0, in terms of the new variables S and $\hat{\rho}_0$, where S is constant in the background and $\hat{\rho}_0$ will be the energy density of the fluid also in the background. So, our background field is proposed as

$$\Phi_0 = (\psi_0 \, e^{-imt/\hbar} + \psi_0^* \, e^{imt/\hbar}) \tag{15}$$

where,

$$\psi_0(t) = \sqrt{\hat{\rho}_0(t)} \, e^{iS/\hbar} \tag{16}$$

[2] The scale factor is such that $a = 1$ today, so that it relates to the redshift z by $a = (1+z)^{-1}$.

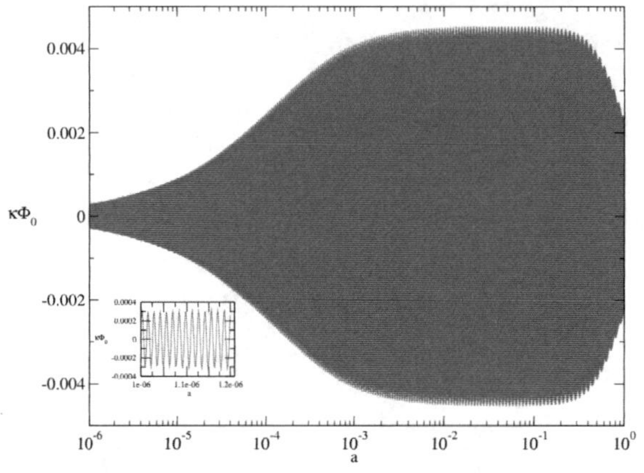

Figure 1. Evolution of the scalar field Φ_0 for the background Universe. The inset shows the fast oscillations of the scalar field.

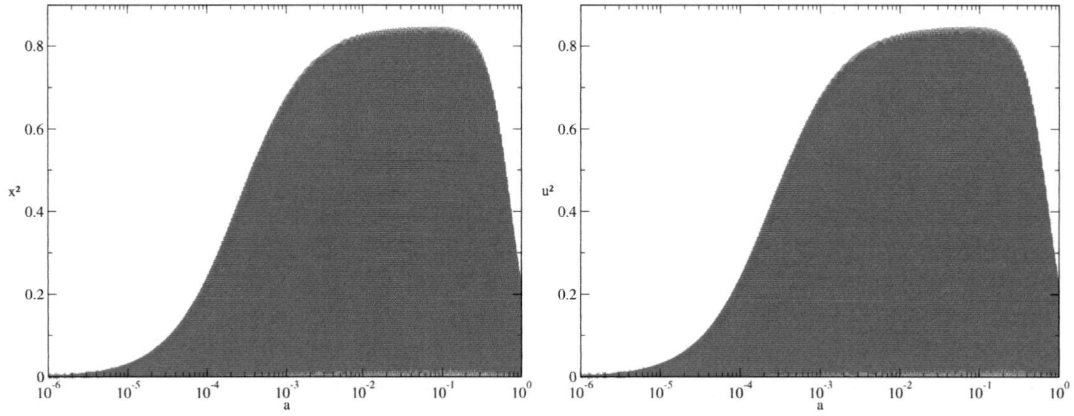

Figure 2. Evolution of the kinetic (top panel) and potential (bottom panel) energy of the scalar field Φ_0.

and with this our SF in the background can be finally expressed as,

$$\Phi_0 = 2\sqrt{\hat{\rho}_0}\cos(S - mt/\hbar), \tag{17}$$

with this we obtain

$$\begin{aligned} \dot{\Phi}_0^2 &= \hat{\rho}_0\left[\frac{\dot{\hat{\rho}}_0}{\hat{\rho}_0}\cos(S - mt/\hbar)\right. \\ &- \left. 2(\dot{S} - m/\hbar)\sin(S - mt/\hbar)\right]^2. \end{aligned} \tag{18}$$

To simplify, observe that the uncertanty relation implies that $m\Delta t \sim \hbar$, and for the background in the non-relativistic case the relation $\dot{S}/m \sim 0$ is satisfied. Notice also that for the background we have that the density goes as $(\ln\hat{\rho}_0)^{\cdot} = -3H$, but we also have that $H \sim 10^{-33}$ eV $<< m \sim 10^{-22}$ eV, so with these considerations at hand for the background, in (18) we have

$$\dot{\Phi}_0^2 = 4\frac{m^2}{\hbar^2}\hat{\rho}_0\sin^2(S - mt/\hbar) \tag{19}$$

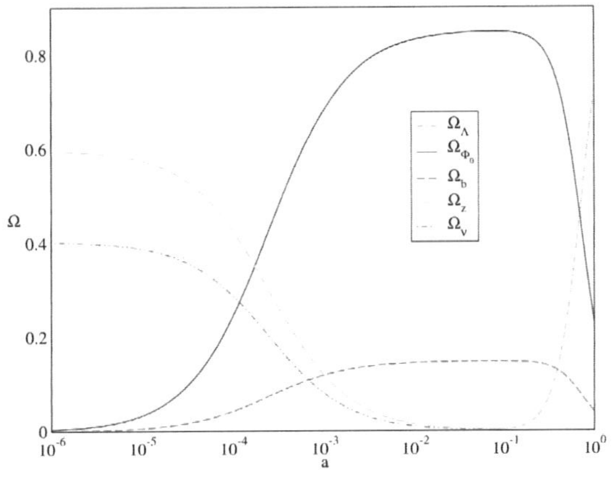

Figure 3. Evolution of the density parameters Ω_i for the background Universe. SFDM model mimics the standard ΛCDM behavior.

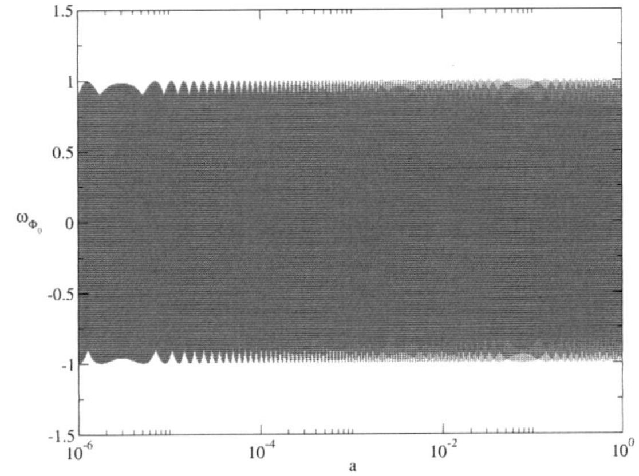

Figure 4. Evolution of the SFDM equation of state for the background Universe.

Finally, substituting this last equation and equation (17) into (1) when taking $\lambda = 0$, we obtain

$$\rho_{\Phi_0} = 2\frac{m^2}{\hbar^2}\hat{\rho}_0[\sin^2(S - mt/\hbar) + \cos^2(S - mt/\hbar)] = 2\frac{m^2}{\hbar^2}\hat{\rho}_0. \tag{20}$$

Comparing this result with (12) we have that the identity $\Omega_{\Phi_0} = 2m^2\hat{\rho}_0/\hbar^2$ holds for the background, so comparing with (20),

$$x = \sqrt{2\hat{\rho}_0}\frac{m}{\hbar}\sin(S - mt/\hbar) \tag{21}$$

$$u = \sqrt{2\hat{\rho}_0}\frac{m}{\hbar}\cos(S - mt/\hbar). \tag{22}$$

In terms of the two analytic results, we show the evolution of the kinetic and potential energies (21) and (22) in Fig. 5, where for the evolution we used the e-folding number N and the fact that $a \sim t^n \to t \sim e^{N/n}$.

Observe the excelent accordance with the numerical results in [68] for the kinetic and potential energies of the background respectively. Therefore, we expect that if we sum the analytical

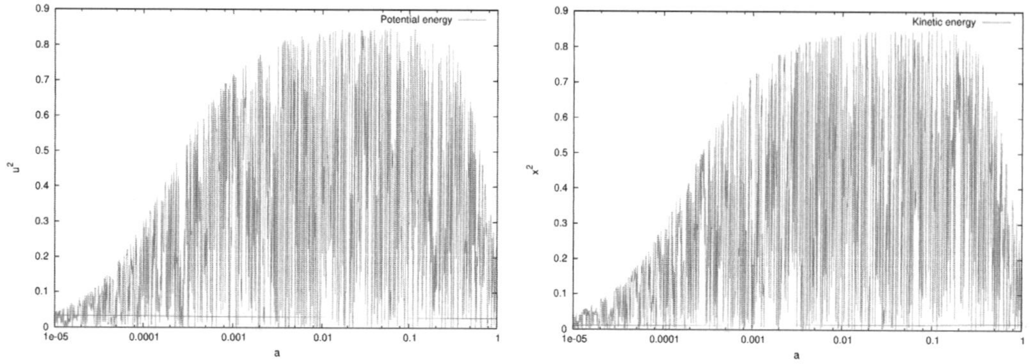

Figure 5. Analytical evolution of the potential (top panel) and kinetic (bottom panel) energies of the scalar field dark matter.

kinetic and the potential energies we will obtain the same behavior for the SFDM density shown in Fig. 3. In what follows we will use the analytic expressions (21) and (22) instead of the numerical ones.

3. Scalar Field Dark Matter Fluctuations

Nowadays it is known that our Universe is not exactly isotropic and spatially homogeneous like the FLRW metric describes. There exist small deviations of this model that were the cosmological seeds to lead the large-scale structure formation in the Universe. In this section we compute the growth of the SFDM overdensities $\delta\rho_\Phi$ at the stage when the density contrast $\delta \equiv \delta\rho_\Phi/\rho_{\Phi_0}$ is much smaller than unity. It is believed that the Universe was almost uniform after inflation, with very small density contrast. As the Universe expanded, the small over-densities grew until they began to collapse, leading to the formation of structure in the Universe. Here we consider small enough deviations in the FLRW model, so that they can be treated by linear perturbation theory.

3.1. The Linear Regime of SFDM perturbations: The Field approach

In the SFDM paradigm, the dark haloes of the galaxies are BECs composed of ultra light scalar particles with $m \sim 10^{-22}$eV then it would be about $10^{68} m$ GeV^{-1} scalar particles to follow in a single galaxy and the occupation numbers in galactic haloes are so big that the DM behaves as a coherent classical scalar field Φ that obeys the Klein-Gordon equation $(\Box^2 + m^2/\hbar^2)\Phi = 0$, where $\Box$ is the D'Alambertian. Therefore, to study the structure formation in the Universe we describe a non-interacting SFDM model such that: i) it can describe it more as a field than as particles and ii) we find a function that only depends on the three spatial coordinates and time.

First, we introduce the perturbed metric tensor in the FLRW background, we only consider scalar perturbations. We then give the equation of energy-momentum conservation and the Einstein field equations for the perturbed metric.

By definition, a perturbation done in any quantity, is the difference between its value in some event in real space-time, and its corresponding value in the background. So, for example for the SF we have

$$\Phi = \Phi_0(t) + \delta\Phi(\boldsymbol{x}, t), \tag{23}$$

where the background is only time dependent, while the perturbations also depend on the space coordinates. Similar cases apply for the metric;

$$g_{00} = -a^2(1 + 2\phi),$$

$$\begin{aligned} g_{0i} &= a^2 B_{,i}\,, \\ g_{ij} &= a^2[(1-2\psi)\delta_{ij} + 2E_{,ij}]. \end{aligned} \tag{24}$$

Here ψ is a perturbation associated to the curvature and E is asociated to the expansion. We will work under the Newtonian gauge, which is defined when $B = E = 0$. An advantage of using this gauge is that here the metric tensor $g_{\mu\nu}$ is diagonal, and so the calculations become much easier. We will only work with scalar perturbations, vector and tensor perturbations are eliminated from the beginning, so that only scalar perturbations are taken into account. Another advantage in using this gauge is that ϕ will play the role of the gravitational potential, this will help us to have a simpler physical interpretation, i.e., both potentials ϕ and ψ are then related. This metric has already been used in other works (see for example [9], [62] and [63]).

The perturbed energy-momentum tensor for the SF can be written as the background value $\mathbf{T}_0 \equiv \mathbf{T}_0(t)$ plus a perturbation $\delta\mathbf{T} \equiv \delta\mathbf{T}(x^\mu)$ where $x^\mu = [t, x^i]$, i. e.

$$\mathbf{T} = \mathbf{T}_0 + \delta\mathbf{T}. \tag{25}$$

We now derive the perturbed evolution equations for the different quantities mentioned above; the scalar perturbation $\delta\Phi$ and the scalar potential ψ. For the perturbed energy-momentum tensor, we get

$$\delta T^0_0 = -\delta\rho_\Phi = -(\dot\Phi_0\delta\dot\Phi - \dot\Phi_0^2\psi + V_{,\Phi_0}\delta\Phi), \tag{26}$$

$$\delta T^0_i = -\frac{1}{a}(\dot\Phi_0\delta\Phi_{,i}), \tag{27}$$

$$\delta T^i_j = \delta p_\Phi = (\dot\Phi_0\delta\dot\Phi - \dot\Phi_0^2\psi - V_{,\Phi_0}\delta\Phi)\delta^i_j. \tag{28}$$

Where in equations (3.1) the dot denotes differentiating with respect to cosmological time t, which is related to conformal time by the simple relation

$$\frac{d}{d\eta} = a\frac{d}{dt}. \tag{29}$$

In the Newtonian gauge, the metric tensor $g_{\mu\nu}$ becomes diagonal and from this, in the trace of the Einstein's equations, the scalar potentials ψ and ϕ are identical and, therefore, ψ plays the role of the gravitational potential

$$\psi - \phi = 0. \tag{30}$$

Usually this equation contains a term of anisotropic stress, which vanishes in the case of a SF. Altogether, the perturbed Einstein's equations $\delta G^i_j = \kappa^2\delta T^i_j$ to first order for a SF in the Newtonian gauge are

$$\begin{aligned} -8\pi G\delta\rho_\Phi &= 6H(\dot\phi + H\phi) - \frac{2}{a^2}\nabla^2\phi, \\ 8\pi G\dot\Phi_0\delta\Phi_{,i} &= 2(\dot\phi + H\phi)_{,i}\,, \\ 8\pi G\delta p_\Phi &= 2[\ddot\phi + 4H\dot\phi + (2\dot H + 3H^2)\phi]. \end{aligned} \tag{31}$$

These equations describe the evolution of the scalar perturbations. For the evolution of the perturbations in the SF we use the perturbed Klein-Gordon equation

$$\delta\ddot\Phi + 3H\delta\dot\Phi - \frac{1}{a^2}\nabla^2\delta\Phi + V_{,\Phi\Phi}\,\delta\Phi + 2V_{,\Phi}\,\phi - 4\dot\Phi_0\dot\phi = 0. \tag{32}$$

In order to solve the equations (31) and (32), we will turn to Fourier's space. The beauty of this expansion relaying on the fact that each Fourier mode will propagate independently. To first order, the derivation of Fourier's components is straightforward. The perturbation $\delta\Phi$ relates to its Fourier component $\delta\Phi_k$ by

$$\delta\Phi(t,x^i) = \int d^3k\delta\Phi(t,k^i)\exp(ik_ix^i) = \int d^3k\delta\Phi_k\exp(ik_ix^i), \tag{33}$$

where k is the wave number. Here the wave number is defined as $k = 2\pi/L$, being L the length scale of the perturbation.

The perturbed equations (31) altogether with the SF transformation read

$$\begin{aligned} 8\pi G(3H\dot{\Phi}_0\delta\Phi_k) + \frac{2k^2}{a^2}\phi &= -8\pi G(\dot{\Phi}_0\delta\dot{\Phi}_k - \phi\dot{\Phi}_0^2 + V_{,\Phi}\,\delta\Phi_k), \\ 2(H\phi + \dot{\phi}) &= 8\pi G\dot{\Phi}_0\delta\Phi_k, \\ 2[\ddot{\phi} + 4H\dot{\phi} + (2\dot{H} + 3H^2)\phi] &= 8\pi G(\dot{\Phi}_0\delta\dot{\Phi}_k - \phi\dot{\Phi}_0^2 - V_{\Phi}\delta\Phi_k), \end{aligned} \tag{34}$$

and the Klein-Gordon equation (32) transforms into

$$\delta\ddot{\Phi}_k + 3H\delta\dot{\Phi}_k + \frac{k^2}{a^2}\delta\Phi_k + V_{,\Phi\Phi}\,\delta\Phi_k - 4\dot{\phi}\dot{\Phi}_0 + 2\phi V_{,\Phi} = 0. \tag{35}$$

These set of equations describe the evolution of the perturbations. Eq. (34) makes reference to the evolution of the energy density, Eq. (34) to the evolution of the gravitational potential and finally, Eq. (35) refers to the perturbations over the SF.

Now, from the time derivative of (26), we have

$$\dot{\delta\rho_\Phi} = (\ddot{\Phi}_0 + V_\Phi)\delta\dot{\Phi}_k + (\delta\ddot{\Phi}_k + V_{,\Phi\Phi}\,\delta\Phi_k - \dot{\Phi}_0\dot{\phi})\dot{\Phi}_0 - 2\phi\dot{\Phi}_0\ddot{\Phi}_0. \tag{36}$$

Performing a Fourier transformation of the above equation and combining it with the unperturbed and perturbed Klein-Gordon equations (4) and (35) respectively, and with the use of equation (3.1) we arrive at

$$\dot{\delta\rho_\Phi} = -6H\dot{\Phi}_0\delta\dot{\Phi}_k + 6\phi\dot{\Phi}_0^2H - \frac{2k^2}{a^2\kappa^2}(H\phi + \dot{\phi}) + 3\dot{\phi}\dot{\Phi}_0^2. \tag{37}$$

And on the other hand we have,

$$\delta p_\Phi + \delta\rho_\Phi = 2\dot{\Phi}_0\delta\dot{\Phi}_k - 2\dot{\Phi}_0^2\phi, \tag{38}$$

Thus,

$$\dot{\delta\rho_\Phi} = -3H(\delta p_\Phi + \delta\rho_\Phi) - \frac{2k^2}{a^2\kappa^2}(H\phi + \dot{\phi}) + 3\dot{\phi}\dot{\Phi}_0^2. \tag{39}$$

This last equation can be expressed in terms of the density contrast δ_Φ making use of the equations from the background, we have

$$\dot{\delta}_\Phi + 3H(\frac{\delta p_\Phi}{\delta\rho_\Phi} - \omega_\Phi)\delta_\Phi = 3\dot{\phi}(1 + \omega_\Phi) - G_\phi, \tag{40}$$

where we have defined the function G_ϕ as

$$G_\phi = \frac{2k^2}{a^2\kappa^2}\frac{\dot{\phi} + H\phi}{\rho_{\Phi_0}}. \tag{41}$$

It is convenient to define the function F_ϕ as well,

$$F_\phi = 1 + \omega_\Phi. \tag{42}$$

Taking the time average of equation (40) we obtain

$$\dot{\delta}_\Phi + 3H(\frac{<\delta p_\Phi>}{<\delta\rho_\Phi>} - <\omega_\Phi>)\delta_\Phi = 3\dot{\phi} <F_\Phi> - <G_\phi>. \tag{43}$$

In equation (43) for the radiation and matter dominated eras the first term in the parenthesis goes as $<\delta p_\Phi> / <\delta\rho_\Phi> \approx 0$, see for example [67], also we can see from Fig.4 that $\langle\omega_{\Phi_0}\rangle \to 0$. It is easy to see the temporal average of the terms F_Φ and G_ϕ of the equation (43). From Fig.6 we observe that that $\langle F_\Phi\rangle$ tends to one. On the other hand, since we are using post-newtonian approximation $\langle G_\phi\rangle$ drops to zero, meaning that the second term on the right-hand side of Eq. (43) dissapears. So, due to the scalar oscillations around the minimum of the scalar potential, the unperturbed SF behaves very similar to the CDM model, but equation (43) tell us that their perturbations too, all the growing behavior for the k modes are recovered and preserved so far. Finally, we conclude that Eq. (43) resembles the equation for the density contrast as in the standard CDM model [17]. This means that the SFDM perturbations in the linear regime grow exactly as CDM perturbations.

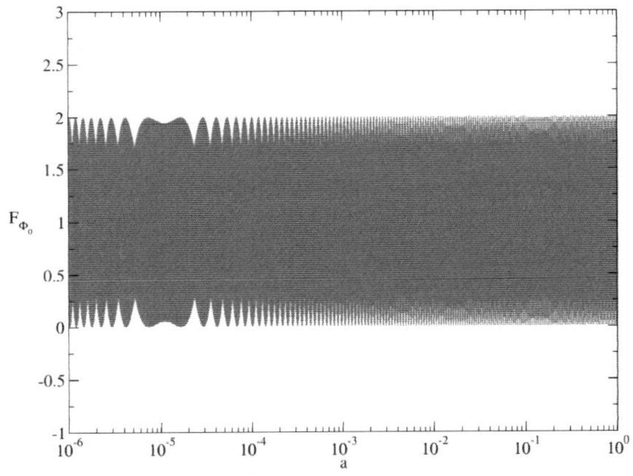

Figure 6. Evolution of $\langle F_\Phi\rangle$ term involved on the right-hand side of equation (43).

3.2. The rescaled equations for SFDM perturbations

We now study the cosmological evolution of our model through a dynamical system. In order to solve the system of equations for the perturbations given in Eqs. (3.1) and (35), we define the following dimensionless variables given by

$$U \equiv -\frac{\kappa}{\sqrt{6}}\frac{V_{,\Phi_0}}{H^2}, \quad l_1 \equiv \phi, \quad x_2 \equiv \frac{\dot{\phi}}{H}, \quad z_1 \equiv \frac{\kappa}{\sqrt{6}}\delta\Phi_k, \quad z_2 \equiv \frac{\kappa}{\sqrt{6}}\frac{\delta\dot{\Phi}_k}{H}. \tag{44}$$

Using the variables defined in Eq. (44), the equations can be transformed into an autonomus system with respect to the e-folding number N, then the density contrast can be written in terms of these variables

$$\delta = \frac{2[x(z_2 - xl_1) - Uz_1]}{\Omega_{\Phi_0}}. \tag{45}$$

Therefore the transition between the linear and non-linear regime of the SFDM perturbations can be studied through a numerical code.

3.3. The Natural Cut of the mass power spectrum

We rewrite equation (35) in the following form

$$\delta\ddot{\Phi}_k + (\frac{k^2}{a^2} + V_{,\Phi\Phi})\delta\Phi_k = -3H\delta\dot{\Phi}_k + 4\dot{\phi}\dot{\Phi}_0 - 2\phi V_{,\Phi}\,. \tag{46}$$

Equation (46) is the one of a harmonic oscillator with a damphing term $3H\delta\dot{\Phi}$ and an extra force. This equation contains oscillating, growing and decaying solutions, depending on the factor $\frac{k^2}{a^2} + V_{,\Phi\Phi}$. If this factor is positive, we expect that the fluctuations oscillate, only if the factor is negative we will have growing up fluctuations. This factor contains the scale k^2/a^2, which is always positive. Thus, if we expect to have growing up fluctuation of the scalar field, this happen only when the potential $V_{,\Phi\Phi}$ is negative. Then equation (46) will have growing up solutions for certain values of the scale k/a, this is, only for enough small values of the scale, which means for enough big fluctuations. This is the way we can fix the mass of the scalar field, the potential depends on the value of the scalar field mass. Observing the smallest structure we see in the Universe, we are able to fix the mass of the scalar field. Using this observations one can avoid the problem of substructure of the ΛCDM. In order to do this, in [67] the mass of the scalar field was fixed in 10^{-22}eV.

3.4. The Linear Regime of SFDM perturbations: The Fluid Approach

The SF Φ has very stark oscillations from the beginning, these oscillations are transmitted to the fluctuations which apparently seems to grow very fast and are too big. Nevertheless, this behavior is not physical, because we only see the oscillations of the fields, but we can not see clearly the evolution of its density, [69]. In order to drop out these oscillations, in what follows we perform two transformations. The first one changes the perturbed Klein-Gordon equation into a kind of 'Schrödinger' equation and the second transforms this last equation into a hydrodynamical system, where we can interpret the physical quantities easier and the observable quantities become much clear.

First, we consider the perturbed Klein-Gordon 32 where we have set $\dot{\phi} = 0$

$$\delta\ddot{\Phi} + 3H\delta\dot{\Phi} - \frac{1}{a^2}\hat{\nabla}^2\delta\Phi + V_{,\Phi\Phi}\,\delta\Phi + 2V_{,\Phi}\,\phi = 0 \tag{47}$$

Now we express the perturbed SF $\delta\Phi$ in terms of the field Ψ,

$$\delta\Phi = \Psi\, e^{-\,imt/\hbar} + \Psi^*\, e^{\,imt/\hbar}, \tag{48}$$

term which oscillates with a frequency proportional to m and $\Psi = \Psi(\boldsymbol{x}, t)$ which would be proportional to a wave function of an ensamble of particles in the condensate. With this equation and the expresion for the potential of the scalar field, (47) transforms into

$$-\,i\hbar(\dot{\Psi} + \frac{3}{2}H\Psi) + \frac{\hbar^2}{2m}(\Box\Psi + 9\lambda|\Psi|^2\Psi) + m\phi\Psi = 0, \tag{49}$$

where we have defined

$$\Box = \frac{d^2}{dt^2} + 3H\frac{d}{dt} - \frac{1}{a^2}\hat{\nabla}^2. \tag{50}$$

Notice that this last equation could represent a kind of 'Gross-Pitaevskii' equation in an expanding Universe. The only modification of equation (49) in comparison to the Schrödinger or the Gross-Pitaevskii equation is the scale factor a^{-1} associated to the co-moving spatial gradient and that the Laplacian $\hat{\nabla}^2 = \partial^2_{\rm X}$ transforms into the D' Alambertian □.

To explore the hydrodynamical nature of bosonic dark matter, we will use a modified fluid approach. Then, to make the connection between the theory of the field and the condensates waves function, the field is proposed as (Madelung transformation [79])

$$\Psi = \sqrt{\hat{\rho}}\, e^{iS}, \tag{51}$$

where Ψ will be the condensates wave function with $\hat{\rho} = \hat{\rho}(\boldsymbol{x}, t)$ and $S = S(\boldsymbol{x}, t)$, [34]. Here we have separated Ψ into a real phase S and a real amplitude $\sqrt{\hat{\rho}}$ and the condition $|\Psi|^2 = \Psi\Psi^* = \hat{\rho}$ is satisfied. From (51) we have

$$\begin{aligned} \dot{\hat{\rho}} &+ 3H\hat{\rho} - \frac{\hbar}{m}\hat{\rho}\Box S + \frac{\hbar}{a^2 m}\hat{\nabla}S\hat{\nabla}\hat{\rho} - \frac{\hbar}{m}\dot{\hat{\rho}}\dot{S} = 0 \\ \hbar\dot{S}/m &+ \omega\hat{\rho} + \phi + \frac{\hbar^2}{2m^2}\left(\frac{\Box\sqrt{\hat{\rho}}}{\sqrt{\hat{\rho}}}\right) + \frac{\hbar^2}{2a^2}[\hat{\nabla}(S/m)]^2 - \frac{\hbar^2}{2}(\dot{S}/m)^2 = 0 \end{aligned} \tag{52}$$

Now, taking the gradient of (52) then dividing by a and using the definition

$$\boldsymbol{v} \equiv \frac{\hbar}{ma}\hat{\nabla}S \tag{53}$$

we have,

$$\begin{aligned} \dot{\hat{\rho}} &+ 3H\hat{\rho} - \frac{\hbar}{m}\hat{\rho}\Box S + \frac{1}{a}\boldsymbol{v}\nabla\hat{\rho} - \frac{\hbar}{m}\dot{\hat{\rho}}\dot{S} = 0 \\ \dot{\boldsymbol{v}} &+ H\boldsymbol{v} + \frac{1}{2a\hat{\rho}}\nabla p + \frac{1}{a}\nabla\phi + \frac{\hbar^2}{2m^2 a}\nabla\left(\frac{\Box\sqrt{\hat{\rho}}}{\sqrt{\hat{\rho}}}\right) + \frac{1}{a}(\boldsymbol{v}\cdot\nabla)\boldsymbol{v} - \hbar(\dot{\boldsymbol{v}} + H\boldsymbol{v})(\dot{S}/m) = 0. \end{aligned} \tag{54}$$

where in (52) $\omega = 9\hbar^2\lambda/2m^2$ and in (54) we have defined $p = \omega\hat{\rho}^2$.

It is worth noting that, to this moment this last set of equations do not involve any approximations with respect to equation (49) and can be used in linear and non-linear regimes.

Now, neglecting squared terms, second order time derivatives and products of time derivatives in this last set of equations we get,

$$\frac{\partial\hat{\rho}}{\partial t} + \nabla\cdot(\hat{\rho}\boldsymbol{v}) + 3H\hat{\rho} = 0, \tag{55}$$

$$\frac{\partial\boldsymbol{v}}{\partial t} + H\boldsymbol{v} + (\boldsymbol{v}\cdot\nabla)\boldsymbol{v} - \frac{\hbar^2}{2m^2}\nabla\left(\frac{1}{2\hat{\rho}}\nabla^2\hat{\rho}\right) + \omega\nabla\hat{\rho} + \nabla\phi = 0, \tag{56}$$

$$\nabla^2\phi = 4\pi G\hat{\rho}, \tag{57}$$

where the equation for the gravitational field is given by the Poisson's equation (57). In these equations we have introduced $\boldsymbol{r} = a(t)\boldsymbol{x}$, such that $1/a\hat{\nabla} = \nabla = \partial_{\mathbf{r}}$.

Equation (53) shows the proportionality between the gradient of the phase and the velocity of the fluid. Note that $\boldsymbol{v}$ can represent the velocity field for the fluid and $\hat{\rho}$ will be the particles density number within the fluid. Also there exists an extra term of third order for the partial derivatives in the waves amplitude which goes as the gradient of $\frac{\hbar^2}{2m^2}\frac{\Box\sqrt{\hat{\rho}}}{\sqrt{\hat{\rho}}}$, this term would result in a sort of 'quantum pressure' that would act against gravity. We remain that ϕ represents the gravitational field. These two sets of equations (55) and (56) would be analogous to the Euler's equations of classical 'fluids', with the main difference that there exists a 'quantum term', which we will call Q and will be given by $Q = \frac{\hbar^2}{2m^2}\frac{\Box\sqrt{\hat{\rho}}}{\sqrt{\hat{\rho}}}$ which can describe a force or a sort of negative quantum pressure.

For equation (55) we have that $\hat{\rho}$ will represent the mass density or the particles density number of the fluid, where all the particles would have the same mass. Finally, these equations describe the dynamics of a great number of non-interacting identical particles that manifest themselves in the form of a fluid, also equation (49) can describe a great number of non-interacting but self-interacting identical particles in the way of a Bose gas, when the probability density is interpreted as the density number.

Now, these hydrodynamical equations are a set of complicated non-linear differential equations. To solve them we will restrict ourselves to a vecinity of total equilibrium.

For this let $\hat{\rho}_0$ be the mass density of the fluid in equilibrium, the average velocity $\boldsymbol{v}_0$ will be taken as zero in equilibrium, so we will only have $\boldsymbol{v}(\boldsymbol{x}, t)$ out of equilibrium. Then, the matter in the Universe will be considered as a hydrodynamical fluid inside an Universe in expansion. This system will then evolve in this Universe and later on they will collapse because of their gravitational attraction.

Then from (55) for the mass density of the fluid in equilibrium we have,

$$\frac{\partial \hat{\rho}_0}{\partial t} + 3H\hat{\rho}_0 = 0, \tag{58}$$

with solutions of the form

$$\hat{\rho}_0 = \frac{\rho_{0i}}{a^3}, \tag{59}$$

where as we know, in general if we have an equation of state of the form $\hat{p} = \omega\hat{\rho}$ and consider CDM or dust as dark matter such that $\hat{p} = 0$ it holds that $\hat{\rho} \propto a^{-3}$. Then, when the scale factor was small, the densities were necessarily bigger. Now, the particles density number are inversely proportional to the volume, and must be proportional to a^{-3}, therefore the matter energy density will also be proportional to a^{-3}, result that is consistent with our expression (59). In addition, the cosmological behavior of $\hat{\rho}_0$ is in agreement with the numerical results show in Fig. 3.

Now for the system out of equilibrium we have

$$\begin{aligned}
\frac{\partial \delta\hat{\rho}}{\partial t} &+ 3H\delta\hat{\rho} + \hat{\rho}_0 \nabla\cdot\delta\boldsymbol{v} = 0 \\
\frac{\partial \delta\boldsymbol{v}}{\partial t} &+ H\delta\boldsymbol{v} - \frac{\hbar^2}{2m^2}\nabla\left(\frac{1}{2}\nabla^2\frac{\delta\hat{\rho}}{\hat{\rho}_0}\right) + \omega\nabla\delta\hat{\rho} + \nabla\delta\phi = 0 \\
\nabla^2\delta\phi &= 4\pi G\delta\hat{\rho}
\end{aligned} \tag{60}$$

equations that are valid in a Universe in expansion. In order to solve system (60) we look for solutions in the form of plane waves, for this the convenient ansatz goes as

$$\delta\hat{\rho} = \hat{\rho}_1(t)\exp(i\boldsymbol{k}\cdot\boldsymbol{x}/a), \quad \delta\boldsymbol{v} = \boldsymbol{v}_1(t)\exp(i\boldsymbol{k}\cdot\boldsymbol{x}/a), \quad \delta\phi = \phi_1(t)\exp(i\boldsymbol{k}\cdot\boldsymbol{x}/a).$$

where $\boldsymbol{x}$ is the position vector and $\boldsymbol{k}$ is a real wavevector which corresponds to a wavelength L. If we substitute these ansatz in the set of equations (60), we then have

$$\frac{d\hat{\rho}_1}{dt} + 3H\hat{\rho}_1 + i\frac{\hat{\rho}_0}{a}\boldsymbol{k}\cdot\boldsymbol{v}_1 = 0, \tag{61}$$

$$\frac{d\boldsymbol{v}_1}{dt} + H\boldsymbol{v}_1 + i\frac{\hat{\rho}_1}{a}\left(\frac{v_q^2}{\hat{\rho}_0} - 4\pi G\frac{a^2}{k^2} + \omega\right)\boldsymbol{k} = 0, \tag{62}$$

$$\phi_1 + 4\pi G\frac{a^2}{k^2}\hat{\rho}_1 = 0. \tag{63}$$

Journal of Physics: Conference Series **378** (2012) 012012 doi:10.1088/1742-6596/378/1/012012

where we have defined the velocity

$$v_q^2 = \frac{\hbar^2 k^2}{4a^2 m^2}. \tag{64}$$

To solve the system is convenient to rotate the coordinate system so that the propagation of the waves will be along the direction of one of the axes. For this we know that the velocity vector can be divided into longitudinal (parallel to $\boldsymbol{k}$) and transverse (perpendicular to $\boldsymbol{k}$) parts, such that $\boldsymbol{v}_1 = \nu\boldsymbol{k} + \boldsymbol{v}_2$, where $\boldsymbol{v}_2$ is the vector perpendicular to the wave propagation vector $\boldsymbol{k}\cdot\boldsymbol{v}_2 = 0$. In terms of $\boldsymbol{v}_2$ for equations (61)-(63) we have

$$\frac{d\hat{\rho}_1}{dt} + 3H\hat{\rho}_1 + i\frac{\hat{\rho}_0}{a}k^2\nu = 0 \tag{65}$$

$$\frac{d\nu}{dt} + H\nu + \frac{i}{a}\left(\frac{v_q^2}{\hat{\rho}_0} - 4\pi G\frac{a^2}{k^2} + \omega\right)\hat{\rho}_1 = 0, \tag{66}$$

in addition to an equation for $\boldsymbol{v}_2$, $d\boldsymbol{v}_2/dt + H\boldsymbol{v}_2 = 0$, with solutions $\boldsymbol{v}_2 = C/a$ with C a constant of integration, i.e., perpendicular modes to the wave vector are eliminated with the expansion of the Universe. Now, if we use the result (59), then equation (65) can be written as

$$\frac{d}{dt}\left(\frac{\hat{\rho}_1}{\hat{\rho}_0}\right) = -\frac{ik^2\nu}{a} \tag{67}$$

System (65)-(66) can be treated as in the case of a Universe with no expansion, so combining the two equations and with the aid of (67), we get

$$\frac{d^2\delta}{dt^2} + 2H\frac{d\delta}{dt} + \left[(v_q^2 + \omega\hat{\rho}_0)\frac{k^2}{a^2} - 4\pi G\rho_0\right]\delta = 0, \tag{68}$$

where $\delta = \hat{\rho}_1/\hat{\rho}_0 = \rho_1/\rho_0$ is defined as the density contrast. This will be a fundamental equation in the understanding of the evolution of the primordial fluctuations.

4. Results

First we will give a brief summary of the results for the ΛCDM model, this will enable us to make a direct comparison with our results.

For CDM the equation for the evolution of the density contrast is given by,

$$\frac{d^2\delta}{dt^2} + 2H\frac{d\delta}{dt} + \left(c_s^2\frac{k^2}{a^2} - 4\pi G\hat{\rho}_0\right)\delta = 0, \tag{69}$$

where c_s is defined as the sound velocity (which in our case it is not). Now lets analyze equation (69) at the beginning of the matter dominated era a time just after the epoch of equality, and just before recombination when the radiation has cooled down and the photons do not interact with the electrons anymore, for a relativistic treatment see [36]. In this era, $a \geq a_{eq}$, practically all the interesting fluctuation modes are well within the horizon, and the evolution of the perturbations can be well described within the newtonian analysis. At this time, matter behaves like dust with zero pressure. So we have $a \sim t^{2/3}$, $c_s^2k^2/a^2 \approx 0$ and $\hat{\rho}_0 \sim t^{-2}$ therefore $H = (2/3)1/t$. For equation (69) we have

$$\frac{d^2\delta}{dt^2} + \frac{4}{3}\frac{1}{t}\frac{d\delta}{dt} - \frac{2}{3}\frac{1}{t^2}\delta = 0. \tag{70}$$

The solutions to this equation are of the form

$$\delta(t) \to t^{2/3}C_1 + \frac{C_2}{t}, \tag{71}$$

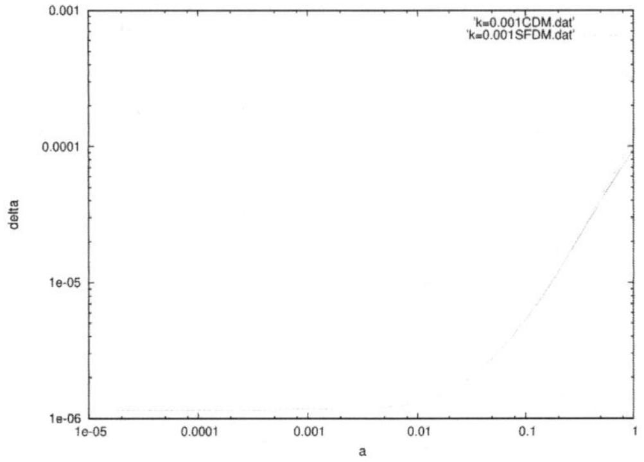

Figure 7. Evolution of the perturbations for the CDM model (dots) and SFDM model (lines) for $k = 1 * 10^{-3} hMpc^{-1}$. Notice how after the epoch of equality ($a_{eq} \sim 10^{-4}$) the evolution of both perturbations in nearly identical, $a = 1$ today. In this case we have taken $\lambda = 0$.

where C_1 and C_2 are integration constants, from this solution we can see that we have modes that will disappear as time goes by, and modes that grow proportionally to the expansion of the Universe. This is an important result, because then the density contrast will grow proportionally to the expansion of the Universe when this is dominated by matter. Then, these fluctuations can maybe grow and give life to the galaxies, clusters of galaxies and all the large-scale structure we see now a days.

Now lets see what happens to the SFDM at this epoch ($a \geq a_{eq}$). The evolution of the perturbations in this case will be given by equation (68).

In general we have that in equation (68) the term v_q is very small throughout the evolution of the pertubations ($v_q \leq 10^{-3} ms^{-1}$ for small k), so it really does not have a significant contribution on its evolution.

When the condition $\lambda = 0$ is taken we can have a BEC that might be or might not be stable, if there exists stability the results of SFDM are consistent with those obtained from CDM (in this case both equations (68) and (69) are almost equal), the condition of stability for the BEC in the SFDM case will come from the study of λ together with Q.

$$\frac{d^2\delta}{dt^2} + 2H\frac{d\delta}{dt} + \left(v_q^2\frac{k^2}{a^2} - 4\pi G\rho_0\right)\delta = 0, \tag{72}$$

As we can see in Fig. 7 the perturbations used for the ΛCDM model grow in a similar way for the SFDM model, when $\lambda = 0$, in this case both perturbations can give birth to structures quite similar in size, and this will happen with all the fluctuations as long as k is kept small.

When $\lambda \neq 0$ the results are quite different, so when discussing the evolution of the density perturbations, there are two different cases: i) In the case of $\lambda > 0$ the amplitude of the density contrast tends to decrease as λ grows bigger and bigger away from zero until the amplitude of the density takes negative values (around $\lambda \sim 10^8$), telling us that this kind of fluctuations can not grow in time, and hence do not form a BEC. ii) On the other hand if $\lambda < 0$ the fluctuations for the density contrast alway grow despite their size, this results means either than the fluctuations grow and form a stable BEC or than the density grows because it is collapsing into a single point and our BEC might be unstable, the study of the stability of this fluctuations needs then to be studied with non-linear perturbation theory. These results are shown in Fig. 8, in both figures 7 and 8 the initial condition for δ goes as $\delta \sim 1 * 10^{-5}$ in accordance with the data obtained from WMAP.

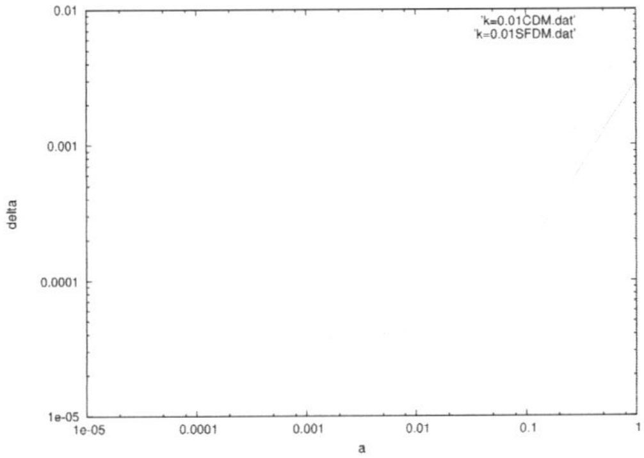

Figure 8. Evolution of the perturbations for the CDM model (dots) and SFDM model (lines) for $k = 1 * 10^{-2} hMpc^{-1}$ and $\lambda \neq 0$ and negative. Notice how after the epoch of equality ($a_{eq} \sim 10^{-4}$) the evolution of both perturbations is now different from the one in Fig. 7, $a = 1$ today. In this case we can clearly see that the SFDM fluctuations grow quicker than those for the CDM model.

If these fluctuations result stable and because they are big in size, this means that they can only give birth to large structures. These fluctuations can then help for the formation of large clusters or other large-scale structure in the Universe at its early stages (around $a \geq a_{eq}$). Then, as these kind of SFDM can only interact with radiation in a gravitational form it is not limited by its interaction with radiation, and the dark matter halos can then create potential wells that will collapse early in time giving enough time for the structures to form. Then if DM is some kind of SFDM, the luminous matter will follow the DM potentials giving birth to large-scale structure.

Finally, a further analysis is required to obtain better results of the growth of scalar perturbations with self-interacting SFDM as well as to study the transition between linear and no linear regime of scalar fluctuations. Nevertheless, we conclude that the SFDM model at cosmological scale is a plausible alternative to DM nature because it preserves all the success of the ΛCDM model.

5. Scalar Field Dark Matter in galactic dynamics

In this section we study the SFDM model at galactic scales, we will focus on the cusp/core discrepancy between the numerical predictions of the standard model and the astronomical observations.

As already said before, recent observations in far and nearby galaxies have shown that the ΛCDM model faces serious conflicts when trying to explain the galaxy formation at small scales [82, 31]. For instance, in the ΛCDM simulations the halos present rising densities towards the central region behaving as $\rho \sim r^{-1}$ well within 1 kpc [75]. On the other hand, several observations suggest that the dynamics (rotation curves) of dwarf and LSB galaxies are more consistent with a constant central density [52, 26, 86], this is most commonly known as the cusp/core problem [26].

Studying a wide range of galaxies of different morphologies and with magnitudes in the interval $-22 \leq M_B \leq -8$ [29] fit their rotation curves (RC) using a Burkert profile for the DM [16] and found that

$$\log(\mu_0/M_\odot pc^{-2}) = 2.15 \pm 0.2 \qquad (73)$$

remains approximately constant, where

$$\mu_0 = \rho_0 r_0 \tag{74}$$

with ρ_0 the central DM density and r_0 the core radius. Similar result where found in [51, 89]. Exploring further the constant value of μ_0 for the DM, [33] found that within r_0 the DM central surface density in terms of the mass inside it ($M_{<r_0}$) is $< \Sigma >_{0,DM}= M_{<r_0}/\pi r_0^2 \approx 72^{+42}_{-27}\mathrm{M}_\odot$ pc^{-2},the gravitational acceleration due to DM felt by a test particle at the radius r_0 was found to be

$$g_{DM}(r_0) = G\pi < \Sigma >_{0,DM}= 3.2^{+1.8}_{-1.2} \times 10^{-9}\mathrm{cm\ s}^{-2}, \tag{75}$$

additionally they reported the acceleration due to the luminous matter at r_0 to be $g_{bar}(r_0) = 5.7^{+3.8}_{-2.8} \times 10^{-10}\mathrm{cm\ s}^{-2}$.

In the ΛCDM model the galaxies have evolve through numerous mergers and grew in different environments, the star formation and basic properties of the galaxies are not expected to be a common factor among them. Therefore giving both the constancy of μ_0 and the core in the central regions of galaxies seems very unlikely in this model.

Any model trying to become a serious alternative to ΛCDM has to succeed in reproducing observations in which the standard model fails but also has to keep the solid description at large scale. It is always necesary to test the SFDM/BEC model with the two observations mentioned above, the cusp/core problem and the constant DM central surface density. In order to do this, in [82] it was used the Thomas-Fermi approximation and a static BEC DM halo to fit rotation curves of a set of galaxies. However, so far there was no comparison between the density profile and the data, in that work this was done by fitting rotation curves of 13 high resolution low surface brightness (LSB) galaxies and additionally compare the fits to two characteristic density profiles 1) the cuspy Navarro-Frenk-White (NFW) profile that results from N-body simulations using ΛCDM and 2) the Pseudo Isothermal (PI) core profile. The comparison allow to show that the model is in general agreement with the data and with a core in the central region. For the second goal the meaning of a core is somewhat ambiguous.

Large scale N-body simulations using collissionless CDM the internal region of DM halos show a density distribution described by a power law $\rho \sim r^\alpha$ with $\alpha \approx -1$, such behaviour is what is now called a cusp. On the other hand, observations mainly in dwarf and LSB galaxies seem to prefer a central density going as $\rho \sim r^0$. This discrepancy between observation and the CDM model receives the name of cusp/core problem. Among the empirical profiles most frequently used to describe the constant density behavior in this galaxies are the PI [10], the isothermal [8] and the Burkert profile [16]. Even though their behavior is similar in the central region and is specified by the central density fitting parameter, their second parameter called the core radius does not represent the same idea. For instance, in the PI profile (eq.(83)) we see that the core radius will be the distance in which the density is half the central density. For the Burkert profile the core radius R_c^{burk} will be when $\rho^{burk}(R_c^{burk}) = \rho_0^{burk}/4$ and for an isothermal profile (I) [89] $\rho^I(R_c^I) = \rho_0^I/2^{3/2}$. Hence, we see an ambiguity in the meaning of the core radius, they get the same name but the interpretation depends on the profile. If we want to compare the central density of LSB galaxies with that of NFW, it usually suffice to have a qualitative comparison, so far this is what we have been doing by fitting empirical profiles. However, high resolution rotation curves demand a more quantitative comparison. Indeed, if we want to test models by fitting RCs we have to know the specific meaning and size of the core, then we will be able to tell if a model is consistent with a cusp or not by making a direct comparison with the data.

In [82] it was proposed a new definition for the core and core radius that allow us to decide when a density profile is cusp or core. Using this definition in the SFDM/BEC model discussed above it is possible to find that the SFDM/BEC model can reproduced the constant value of

μ_0 and as a crosscheck with the PI profile, in [82] was found that the results are in very good agreement with observations. This argues in favor of the model and the core definition.

5.1. DM density profiles

In the following we follow [82] and provide the dark matter density profiles that will be used for the analysis.

The case in which the dark matter is in the form of a static BEC and the number of DM particles in the ground state is very large was considered in [14]. Following this paper and assuming the Thomas-Fermi approximation [25, 79] which neglects the anisotropic pressure terms that are relevant only in the boundary of the condensate, the system of equations describing the static BEC in a gravitational potential V is given by

$$\nabla p\left(\frac{\rho}{m}\right) = -\rho\nabla V, \quad \nabla^2 V = 4\pi G\rho, \tag{76}$$

with the following EoS

$$p(\rho) = U_0\rho^2, \tag{77}$$

where $U_0 = \frac{2\pi\hbar^2 a}{m^3}$, ρ is the mass density of the static BEC configuration and p is the pressure, as we are considering zero temperature p is not a thermal pressure but instead it is produced by the strong repulsive interaction between the ground state bosons. Assuming spherical symmetry and denoting R as the radius at which the pressure and density are zero, the density profile takes the form [14]

$$\rho_B(r) = \rho_0^B \frac{\sin(kr)}{kr} \tag{78}$$

where $k = \sqrt{Gm^3/\hbar^2 a} = \pi/R$ and $\rho_0^B = \rho_B(0)$ is the BEC central density, m is the mass of the DM particle and a is the scattering lenght. The mass at the radius r is given by

$$m(r) = \frac{4\pi\rho_0^B}{k^2} r\left(\frac{\sin(kr)}{kr} - \cos(kr)\right), \tag{79}$$

from here the tangential velocity V of a test particle at a distance r, is

$$V^2(r) = \frac{4\pi G\rho_0^B}{k^2}\left(\frac{\sin(kr)}{kr} - \cos(kr)\right). \tag{80}$$

The logaritmic slope of a density profile is defined as $\alpha = \frac{d(\log\ \rho)}{d(\log\ r)}$ using (7) in (10) we obtain

$$\alpha(r) = -\left[1 - \frac{\pi r}{R}\cot\left(\frac{\pi r}{R}\right)\right]. \tag{81}$$

We study now the Pseudo Isothermal (PI) profile. The empirical core profiles that exist in the literature fit two parameters, a scale radius and a scale density. A characteristic profile of this type is

$$\rho_{PI} = \frac{\rho_0^{PI}}{1 + (r/R_c)^2}, \tag{82}$$

this is the PI profile [10]. Here R_c is the scale radius and ρ_0^{PI} is the central density. The rotation curve is

$$V(r)_{PI} = \sqrt{4\pi G\rho_0^{PI} R_c^2\left(1 - \frac{R_c}{r}\arctan\left(\frac{r}{R_c}\right)\right)}. \tag{83}$$

The NFW profile emerges from numerical simulations that use only CDM and are based on the ΛCDM model [30, 73, 74]. In addition to this, we have chosen this profile because it is representative of what is called the cuspy behavior ($\alpha \approx -1$) in the center of galaxies due to DM. The NFW density profile[74] and the rotation curve are given respectively by

$$\rho_{NFW}(r) = \frac{\rho_i}{(r/Rs)(1\ +\ r/R_s)^2} \tag{84}$$

$$V_{NFW}(r) = \sqrt{4\pi G\rho_i R_s^3}\sqrt{\frac{1}{r}\left[ln\left(1\ +\ \frac{r}{R_s}\right) - \frac{r/R_s}{1\ +\ r/R_s}\right]}, \tag{85}$$

ρ_i is related with the density of the universe at the moment the halo collapsed and R_s^2 is a characteristic radius.

To solve the core radius ambiguity and to unify the concept for future comparison, in [82] it was found that a good definition for the *core* is a region where the density profile presents logarithmic slopes $\alpha \geq -1$ and the core radius will be the radius at which the core begins, that is to say, for radius smaller than the core radius we will have $\alpha \geq -1$, this means that its value r' is determined by the equation

$$\alpha(R) = -1. \tag{86}$$

The advantages of this definition are that the interpretation is independent of the profile chosen (also notice that it applies to the total density profile and is not restricted to that of DM) and in virtue of the same definition we can directly tell if a DM model profile is cored or cuspy. With this new definition the specific distance at which the core radius occurs still depends on the profile chosen but now the physical interpretation is only one. In the following when we refer to both the core and core radius we adopt the previous interpretation.

Applying the definition to (78) we get the core radius for the SFDM/BEC profile R_B and for comparison we use (83) because in turns out that the parameter R_c corresponds to the core radius as defined above. Finally, fitting the NFW profile provides a direct comparison between a cusp and core and hence to the cusp/core problem.

5.2. SFDM vs. PI and NFW

We see from (78) that the SFDM/BEC model satisfies $\rho \sim r^0$ near the origin, but a priori this does not imply consistency with observed RCs. Therefore we reproduce the fit of the profiles to thirteen high resolution observed RCs of a sample of LSB galaxies given in [82]. The RCs were taken from a subsample of [27], we chose galaxies that have at least 3 values within ~ 1 kpc, not presenting bulbs and the quality in the RC in Hα is good as defined in [70]. The RCs in this work omit galaxies presenting high asymmetries and included in the error bars are experimental errors in the velocity measurement, inclination and small asymmetries. Because the DM is the dominant mass component for these galaxies we adopt the minimum disk hypothesis (neglects baryon contribution to the observed RC).

As the difference between a core and a cusp is not overlapped only for data values inside 1kpc and given that in the interval ~ 1 to 10kpc the slopes of core and cusp profiles are very similar, which can lead to the wrong conclusion that cuspy halos are consistent with observations, we determined the logarithmic slope and the uncertainty following [27] with the difference that we fit only the data within 1 kpc and that there is no need of an uncertain "break radius".

In Fig. 9, we show the fits to the RC data and the density profiles, also shown are the core radius in the BEC (magenta) and in PI (blue) profiles. The gray arrow is the fit that determines α, the length denotes the fitted region and is bounded by R_1 that denotes the nearest radius to 1kpc where a data point is given. We show α by fitting values inside R_1 and we also show the core radius for the SFDM/BEC profile R_B in order to compare it with R_c (see [82] for more details of the fitting parameters).

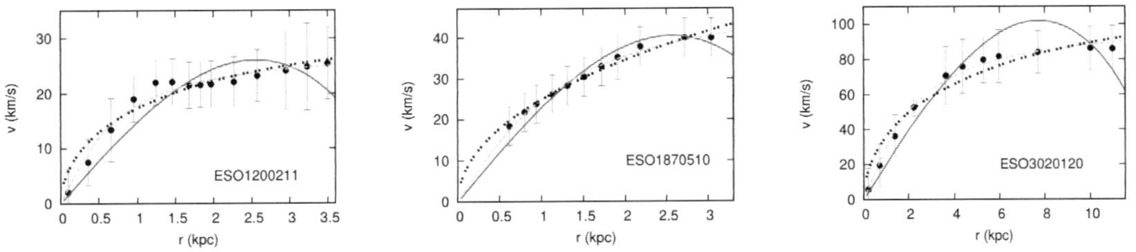

Figure 9. Observed LSB galaxy rotation curves and density profiles with the best halo fits. Below each RC is its density profile along with the fits. Shown are the PI (green dashed-line); the BEC (red solid line) and NFW (black dobble-dotted line) DM halo profiles, the observational data is drawn with error bars. The gray arrow denotes the best fit to the data within R_1 and the vertical arrows denote the PI (blue) and BEC (magenta) core radius.

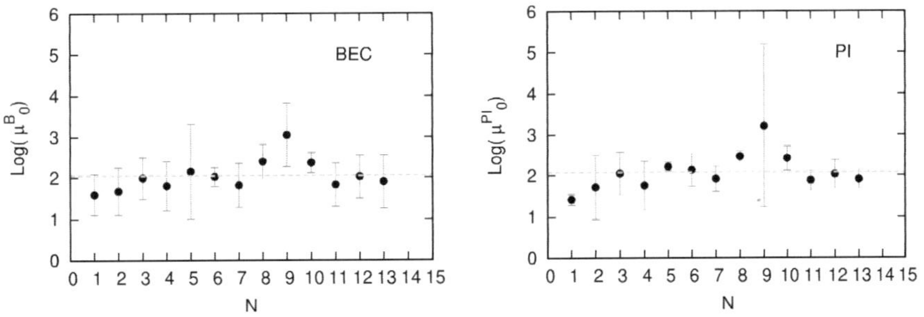

Figure 10. Plot of log $(\mu_0^B/M_\odot pc^{-2})$ and log $(\mu_0^{PI}/M_\odot pc^{-2})$ for each galaxy. N is an identifier for each galaxy. Here we observe that these values remain approximately constant in both profiles, this also serves as a crosscheck for our definition of R_B in the SFDM/BEC profile. The green dashed-line represents the mean values given in (88) and (89).

5.3. Discussion

The fits of the RCs in Fig 9 prove that the solid-body like behavior characterized by a linear increase in the velocity in the central region is more consistent with the core PI and SFDM/BEC profiles than the cuspy NFW. The fits within R_1 give an average value of α=−0.27 ± 0.18 consistent with those obtained in [27] α=−0.2 ± 0.2 and with α=−0.29 ± 0.07 reported by [76] analyzing 7 THINGS dwarf galaxies. The case of ESO1870510 might be considered to be consistent with NFW profile, however it is the innermost value that considerably decreases α, being an irregular galaxy more central data near the innermost region is required to discard the possibility of any violent event that might have caused such a slope value.

The density profiles corresponding to the RCs fits are also shown in Fig 9 for each galaxy. We see that the SFDM/BEC fit slightly deviates from the farthest data points as a result of having a finite radius R that is fixed by the same data. This discrepancy is due to the fact that the halo might be more extended than the value R. As a matter of a fact, the more the extended the "flat" outer region in the RCs the more conspicuous the discrepancy. [45] recently found finite temperature corrections to (83), this suffice to alleviate the latter problem in LSB galaxies and dwarfs but not for bigger galaxies. Some solutions have been proposed including vortex lattices [81, 100] and adding more nodes [88, 47] in the solution of system (76) and (76)), but so far no

Journal of Physics: Conference Series **378** (2012) 012012 doi:10.1088/1742-6596/378/1/012012

final conclusion has been reached.

When comparing the SFDM/BEC and PI core radius we find a general difference of $\sim$ 2kpc, the core size in the PI is approximately 2 kpc smaller than the BEC core size, but the PI central density is larger. In U4115, U11557 and U11583 both profiles are very similar which results in a similar core and central density values, this can also be taken as a consistency check for our core definition.

Comparing the values of R_B we did not find a tendency to a common value. Assuming that the core radius determines the transition where the DM distribution changes from the outer region to the inner constant central density, the lack of the unique value means that there is not a common radius at which this transition takes place.

For the second test [82] uses R_B to calculate (74). We have already seen that R_c and R_B are generally different and R_B is not a fit parameter. Hence *a priori* R_B is not expected to correlate with ρ_0^B. With the fitting parameters it was obtained

$$\log(\mu_0^B/M_\odot pc^{-2}) = \log \rho_0^B R_B = 2.05 \pm 0.56, \tag{87}$$

$$\log(\mu_0^{PI}/M_\odot pc^{-2}) = \log \rho_0^{PI} R_c = 2.08 \pm 0.46, \tag{88}$$

for the average values in the SFDM/BEC and PI profiles respectively. We see the excellent agreement of (88) with (89) that was used as a crosscheck and with (73) in which a much bigger sample was used. The agreement has shown that the SFDM/BEC model is capable of reproducing the constancy of the value μ_0, something that because of the cuspy nature is not possible in the NFW profile.

In Fig. 10 we plot the above values for each galaxy. If we define the DM central surface density mentioned in the introduction, we have for the BEC profile

$$< \Sigma >_{0,DM}^{B} = M_{<R_B}/\pi r_B^2, \tag{89}$$

where $M_{<R_B}$ is obtained from (79) evaluated at R_B. We show that for U11748 the value $log\mu_0^B$ is considerably above the rest and with the largest uncertainty. For this reason, in the following analysis we omit both, this value and the smallest one that corresponds to ESO1200211. Doing this we get $< \Sigma >_{0,DM}^{B} \approx$ 191.35 $M_\odot/pc^2$, and for the acceleration felt by a test particle located in R_B due to DM only we have $g_{DM}(R_B) \approx 5.2 \times 10^{-9}$ cms^{-2} broadly consistent with (75).

The fact that all galaxies present approximately the same order of magnitude in $g_{DM}(R_B)$ might suggest that R_B represents more than a transition towards a constant density, it can give us information about the close relation between DM and the baryons. Moreover, in view of the lack of a unique core radius, we can interpret the transition in DM distribution as an effect of crossing a certain acceleration scale instead of a radial length scale. Such interpretation reminds us that given in MOND but with the big difference that the acceleration scale found is for DM and is not a postulate of the model.

To determine which interpretation causes the transition whether an acceleration scale or a length scale we will need to study the properties of larger samples of galaxies with the new telescopes.

6. Conclusions

In this work, we revisit an alternative paradigm to dark matter nature known as scalar field dark matter or Bose-Einstein condensate dark matter model. In this model a fundamental scalar field plays the role of dark matter in the Universe. The hypothesis is that this scalar field undergoes a phase transition very early in the Universe leading to the formation of Bose-Einstein condensate drops. Therefore, the dark haloes of the galaxies in the Universe are huge drops of scalar field.

To explore if this hypothesis is a viable alternative to the standard model we developed both cosmological and galactic studies.

In the cosmological regime, we have shown, with a field approach, that the SFDM/BEC model with an ultralight mass of 10^{-22} eV mimics the behavior of the cosmological expansion rate predicted with the ΛCDM model. Moreover, although in general a scalar field is not a fluid, it can be treated as if it behaved like one. Thus, we performed a fluid approach for the cosmological evolution of the scalar field and we found that the analytical expressions for the kinetic and potential energies of the scalar field is in excellent agreement with our previous numerical results.

The interesting cosmological behavior of the scalar field indicates that their scalar fluctuations can be the appropriate for the purpose of structure formation, because overdense regions of SFDM/BEC can support the formation of galactic structure. Thus, we have revisited the growth of SFDM/BEC fluctuations in the linear regime. Within the linear theory of scalar perturbations we obtain an equation for the evolution of the density contrast. This equation differs from the density contrast equation for CDM, however the extra terms tend to the values of the standard equation. Therefore, as a goal of this model, we obtained that the scalar perturbations grow up exactly as the CDM paradigm. In addition, we study the liner growth of the scalar fluctuations with a fluid approach. In the case of the SFDM/BEC with $\lambda = 0$ we have shown that for the matter dominated era and for big structure this model simulates the behavior of CDM because in general in a matter dominated Universe for low-k, v_q tends to be a very small quantity tending to zero, so from (68) we can see that on this era we will have the CDM profile given by (69), i.e., the SFDM density contrast profile is very similar to that of the ΛCDM model (see Fig. 7). On the contrary for $\lambda \neq 0$ both models have different behavior as we can see from Fig. 8. The numerical results suggest that linear fluctuations on the SFDM can grow faster than those for CDM around $a \sim 10^{-2}$. Here an important point is that although CDM can grow it does so in a hierarchical way, while from Fig. 8 we can see that SFDM can have bigger fluctuations just before the ΛCDM model does, i.e., it might be that no hierarchical model of structure formation is needed for SFDM and is expected that for the non-linear fluctuations the behaviour will be quite the same as soon as the scalar field condensates, in a very early epoch when the energy of the Universe was about $\sim$ TeV. Thus, the standard and the SFDM/BEC model can be contrasted in their predictions concerning the formation of the first galaxies. If in the future we see more and more well formed and massive galaxies at high redshifts, this could be also a new indication in favour of the SFDM/BEC paradigm.

On the other hand, we also studied the implications of a SFDM/BEC model at galactic scales. We find that the SFDM/BEC model gives a constant density profile that is consistent with RCs of dark matter dominated galaxies. The profile is as good as one of the most frequently used empirical core profiles but with the advantage of coming from a solid theoretical frame. We fit data within 1kpc and found a logarithmic slope $\alpha = -\ 0.27 \pm\ 0.18$ in perfect agreement with a core. It is important to notice that the cusp in the central regions is not a prediction that comes from first principles in the CDM model, it is a property that is derived by fitting simulations that use only DM. In addition, we review a new definition for the core and core radius that allows a definite distinction when a density profile is core or cusp. Using this definition we find the core radius in the SFDM/BEC profile to be in most cases over 2kpc bigger than the core radius in the PI profile. As a second result of the core definition, we show the constant value of μ_0 which is proportional to the central surface density. This result is one of several conflicts that jeopardize the current standard cosmological model.

Finally, with all these intriguing results SFDM/BEC model could be a serious alternative to the dark matter problem in the Universe. Although the observational evidence seems to be in favor of some kind of cold dark matter, if we continue to observe even more galaxies at higher redshifts and if higher resolution observations of nearby galaxies exhibit a core density profile, this model can be a good alternative to ΛCDM. We expect that the close future observations in galaxies surveys can decide the nature of the dark matter.

Acknowledgments
We thank Victor Robles and Abril Suarez for many helpful discussions. This work was partially supported by CONACyT México under grants CB-2009-01, no. 132400, no. 166212, and I0101/131/07 C-234/07 of the Instituto Avanzado de Cosmologia (IAC) collaboration (http://www.iac.edu.mx/).

References

[1] Alcubierre M., Guzmán F. S., Matos T., Núñez D., Ureña L. A., 2002 Wiederhold P., Class Quant Grav, 19, 5017
[2] Arbey, A., Lesgourgues, J. and Salati, P., 2001, Phys. Rev. D **64** 123528 [astro-ph/0105564].
[3] Arbey, A., Lesgourgues, J. and Salati, P., 2002, Phys. Rev. D **65** 083514 [astro-ph/0112324].
[4] Arbey, A., Lesgourgues, J. and Salati, P., 2003, Phys. Rev. D **68** 023511 [astro-ph/0301533].
[5] Arbey, A., 2006 EAS Publ. Ser. **20** 257 [astro-ph/0509592].
[6] Arbey, A., 2006b Phys. Rev. D **74** 043516 [astro-ph/0601274].
[7] Arbey, A., 2009 EAS Publ. Ser. **36** 161 [arXiv:0812.3611 [astro-ph]].
[8] Athanassoula E., Bosma A., Papaioannou S., 1987, A&A, 179, 2340
[9] Bardeen J.M., 1980, Phys Rev D, 22, 1882
[10] Begeman K. G., Broeils A. H., Sanders R. H., 1991, MNRAS, 249, 523
[11] Bernal, A., & Guzmán, F. S.,2006, Phys. Rev. D 74, 063504
[12] Bernal, A., & Guzmán, F. S., 2006, Phys. Rev. D 74, 103002
[13] Bernal A., Matos T., Núñez D., 2008, Rev. Mex. A.A., 44, 149
[14] Böehmer C. G., Harko T., 2007, JCAP, 0706, 025
[15] Briscese F., 2011, Phys Lett B, 696, 315
[16] Burkert A., 1995, ApJ, 447, L25
[17] Ma C.P., Bertschinger E., 1995, ApJ, 455, 7.
[18] Clowe D. et al., 2006, ApJ, 648, 2
[19] Colín, P.; Avila-Reese, V.; Valenzuela, O., 2000, ApJ, 542, 622-630
[20] Colpi M., Shapiro S. L. and Wasserman I., 1986, Phys. Rev. Lett., 57, 2485
[21] Copeland E. S., Liddle A. R., Wands D., 1998, Phys. Rev. D, 57, 4686.
[22] Chavanis P. H., 2011, Phys. Rev. D, 84, 043531
[23] Chavanis P. H., 2011, Phys. Rev. D 84, 043532
[24] Chavanis P. H., 2012, A & A, 537, A127
[25] Dalfovo F., Giorgino S., Pitaevskii L. P., Stringari S., 1999, Rev. Mod. Phys. 71, 463
[26] de Blok W.J.G., 2010, Adv. Astron., 2010, Article ID 789293, 14 pages.
[27] de Blok W. J. G. et al., 2001, ApJ, 552, L23
[28] Dehnen H., Rose B., 1993, Astrophys. Sp Sci., 207, 133-144
[29] Donato et al., 2009, MNRAS, 397, 1169
[30] Dubinski J., Carlberg R. G., 1991, ApJ, 378, 496
[31] Friedmann D. E., 2011, arXiv:0912.1668v4
[32] García-Aspeitia, M.A., Magaña, J., Matos, T., 2011, GeRG Online, arXiv:1102.0825
[33] Gentile G. et al., 2009, Naure, 461, 627
[34] Ginzburg V. L., Landau L. D., 1950, Zh Eksp Teor Fiz, 20, 1064
[35] Gleiser M., 1988, Phys. Rev. D, 38, 2376
[36] Gorini V., Kamenshchik A. Y., Moschella U., Piattella O. F., Starobinsky A. A., 2008, JCAP, 0802, 016
[37] Guzmán, F. Siddhartha and Matos, Tonatiuh. Astron. Nachr. 320, 97.
[38] Guzman F. S., Matos T., 2000, Class. Quantum Grav., 17, L9-L16.
[39] Guzmán F.S., Ureña-López L. A., 2003, Phys. Rev. **D 68**, 024023.
[40] Guzmán F. S., Ureña-López L.A., 2004, Phys. Rev. **D 69**, 124033.
[41] Guzmán F. S., Ureña-López L. A., 2006, ApJ, 645, 814.
[42] Harko T., 2011, MNRAS, 413, 3095
[43] Harko T., 2011, JCAP, 05, 022
[44] Harko T., 2011, Phys. Rev. D, 83, 123515
[45] Harko T., Madarassy E. J. M., 2012, JCAP, 01, 020.
[46] Hu W., Barkana R., Gruzinov A., 2000, Phys. Rev. Lett., 85, 1158

[47] Ji S. U., Sin S. J., 1994, Phys. Rev. D, 50, 3650
[48] Klypin A., Kravtsov A. V., Valenzuela O., Prada F., 1999, ApJ, 522, 82
[49] E. Komatsu et al. 2009, ApJS **180**, 330-376.
[50] E. Komatsu et al. 2011, ApJS **192**, 18.
[51] Kormendy J., Freeman K. C., 2004, in Ryder S. D., Pisano D. J., Walker M. A., Freeman K. C., eds, IAU Symp. . 220, Dark Matter in Galaxies. Cambridge Univ. Press, Cambridge, p. 377
[52] Kuzio de Naray R., McGaugh S.S., & de Blok W.J.G., 2008, ApJ, 676, 920
[53] Lee J., Koh I., 1996, Phys Rev D, 53, 2236
[54] Lee J.W., Lim S., Choi D., 2008, arXiv:astro-ph/0805.3827.
[55] Lee, Jae-Weon, 2009, J. Korean Phys.Soc., 54, 2622
[56] Jae-Weon Lee, & Sooil Lim, 2010, JCAP, 1001, 007.
[57] Lee, Komatzu, 2010, ApJ, 718, 60.
[58] Lesgourgues, J., Arbey, A. and Salati, P., 2002, New Astron. Rev. **46** 791.
[59] Loeb, A.; Weiner, N.; 2011, Phys. Rev. Let., 106, 171302
[60] Lora V., Magaña J., Bernal, A., Sánchez-Salcedo F.J., Grebel, E.K., 2011, arXiv:1110.2684, to be published in JCAP
[61] Lundgren A. P., Bondarescu M., Bondarescu R., Balakrishna J., 2010, ApJ, 715, L35
[62] Ma C.-P., E. Bertschinger, 1995, ApJ, 455, 7
[63] Malik K. A., 2009, Phys Rept, 475, 1
[64] Matos, T., Guzmán, F. S., 2000, Class. Quant. Grav. 17, L9. arXiv:gr-qc /9810028.
[65] Matos, T., Ureña L. A., 2000, Class Quant Grav, 17, L75
[66] Matos, T., Guzmán F.S., 2001, Class. Quantum Grav., 18, 5055
[67] Matos, T., Ureña L. A., 2001, Phys Rev D, 63, 063506
[68] Matos T., Vázquez-González A.,Magaña J., 2009, MNRAS, 389, 13957
[69] Matos T., Magaña J., Suárez A., 2010, The Open Astron Journal, 3, 94
[70] McGaugh S. S. et al., 2001, ApJ, 122, 2381
[71] Milgrom M., 2010, AIP Conf.Proc. 1241:139 arXiv:0912.2678v2
[72] Moore B., Ghigna S., Governato F., Lake G., Quinn T., Stadel J., Tozzi P., 1999, ApJ, 524, L19
[73] Navarro J. F., Frenk C. S., White S. D. M., 1996, ApJ., 462, 563
[74] Navarro J. F., Frenk C. S., White S. D. M., 1997, ApJ, 490, 493
[75] Navarro J.F. et al., 2010, MNRAS, 402, 21
[76] Oh Se-Heon, Brook Chris et al., 2011, arXiv:1011.2777v2.
[77] Peebles P. J. E., Nusser A., 2010, Nat, 465, 565
[78] Penny S. J., Conselice C. J., De Rijcke S., Held E. V., 2009, MNRAS, 393, 1054
[79] Pitaevskii L., Stringari S., 2003, Bose-Einstein Condensation, Oxford University Press
[80] Ratra B., Peebles P. J. E., 1988, Phys. Rev. D, 37, 3406.
[81] Rindler-Daller T., Shapiro P. R., 2011, arXiv:1106.1256v2
[82] Robles V.H., Matos T., 2012, MNRAS, 392, (2012) in press. arXiv:1201.3032
[83] Rodríguez-Montoya, I., Magaña, J., Matos, T., & Pérez-Lorenzana, A., 2010, ApJ, 721, 1509
[84] Rodríguez-Montoya, I.; Pérez-Lorenzana A.; De La Cruz-Burelo, E.; Giraud-Héraud, Y.; Matos, T.; 2011, arXiv:1110.2751
[85] Sahni V., Wang L., 2000, Phys Rev D, 62, 103527
[86] Sánchez-Salcedo, F. J., Reyes-Iturbide, J. & Hernandez, X., 2006, MNRAS, 370, 1829.
[87] Sanders R. H., 2009, Adv. Astron., 2009, Article ID 752439, 9 pages.
[88] Sin S. J., 1994, Phys. Rev. D, 50, 3650
[89] Spano M. et al., 2008, MNRAS, 383, 297
[90] Spergel D., Steinhardt P., 2000, Phys. Rev. Let., 84, 3760-3763
[91] Suárez A., Matos T., 2011, MNRAS, 416, 87
[92] Thomas, S.A., Abdalla, F., Lahav, O., 2011, Phys. Rev. Let., 106, 241301
[93] Turner M.S., 1983, Phys. Rev. **D 28**, 1243.
[94] Ureña-López L.A., Reyes-Ibarra M. J., 2009, IJMPD, 18, 621.
[95] Ureña-López L.A., 2009, JCAP **01**, 014.
[96] Ureña L. A., 2010, AIP Conf Proc, 1318, 82
[97] Villaescusa-Navarro, F.; Dalal, N.; 2011, JCAP, 03, 024
[98] Wetterich C., 1988, Nucl. Phys. B, 302, 668.
[99] Woo T. P., Chiueh T., 2009, ApJ, 697, 850.
[100] Zinner N. T., 2011, arXiv:1108.4290v1

Journal of Physics: Conference Series **378** (2012) 012013 doi:10.1088/1742-6596/378/1/012013

Lepton electric and magnetic dipole moments induced by a vector unparticle

A. Moyotl[1], A. Rosado[1], G. Tavares-Velasco[2]
[1]Instituto de Física, Benemérita Universidad Autónoma de Puebla, Apartado Postal 72570 Puebla, México
[2]Facultad de Ciencias Físico Matemáticas, Benemérita Universidad Autónoma de Puebla, Apartado Postal 1152, Puebla, Pue., México

E-mail: amoyotl@sirio.ifuap.buap.mx

Abstract. We calculate the magnetic dipole moment (MDM) and the electric dipole moment (EDM) of a lepton induced by a vector unparticle with both vector and axial-vector couplings to leptons. We consider the most general scenario in which the unparticle induces lepton flavor violation (LFV) and CP violation. Some specific scenarios are examined to obtain constraints on the LFV unparticle couplings from the current limits on the muon MDM and the decay $\tau \to 3\mu$. While the experimental limit on the muon MDM favor the scenario in which there is dominance of the unparticle vector couplings over the axial-vector couplings, the experimental limit on the $\tau \to 3\mu$ decay strongly constraints the unparticle LFV couplings. We use these constraints to estimate the EDMs of the electron and the muon, which are negligible and far from the current experimental limits.

1. Introduction

Motivated by the Banks and Zaks model [1], Georgi conjectured recently the existence of a hidden scale invariant sector in the high energy theory that could interact with the standard model (SM) [2, 3]. The effects of such a hidden sector could show up at an energy scale of the order of a few TeVs. Since physical particles cannot exist in such a scale invariant sector, it would interact with the SM fields through scale invariant stuff known as unparticles. The details of such a theoretical framework are unknown, but it is possible to study its potential low energy effects via an effective field Lagrangian. The appropriate theoretical framework to describe unparticle physics is the one introduced in Ref. [1]. The hidden sector is a Banks and Zaks ($\mathcal{BZ}$) sector and the associated fields are described by renormalizable $\mathcal{O}_{\mathcal{BZ}}$ operators. It is assumed that these fields interact with the SM fields through the exchange of heavy particles at a very high energy $\mathcal{M}_{\mathcal{U}}$. Below this energy scale there emerge nonrenormalizable couplings between the fields of the $\mathcal{BZ}$ sector and the SM fields, which generically can be written as $\mathcal{O}_{SM}\mathcal{O}_{\mathcal{BZ}}/\mathcal{M}_{\mathcal{U}}^{d_{SM}+d_{\mathcal{BZ}}-4}$. Dimensional transmutation occurs through the renormalizable couplings of the $\mathcal{BZ}$ sector at an energy scale $\Lambda_{\mathcal{U}}$ as scale invariance emerges. At low energies, an effective theory can be used to describe the interactions between the SM fields and the $\mathcal{BZ}$ fields, which are associated with unparticles. The effective Lagrangian can be written as [2, 3]:

$$\mathcal{L}_{\mathcal{U}} = C_{\mathcal{O}_{\mathcal{U}}} \frac{\Lambda_{\mathcal{U}}^{d_{\mathcal{BZ}}-d_{\mathcal{U}}}}{\mathcal{M}_{\mathcal{U}}^{d_{SM}+d_{\mathcal{BZ}}-4}} \mathcal{O}_{SM}\mathcal{O}_{\mathcal{U}}, \tag{1}$$

where $C_{\mathcal{O}_\mathcal{U}}$ stands for the coupling constant and the operator dimension $d_\mathcal{U}$ is fractionary. Unparticle operators $\mathcal{O}_\mathcal{U}$ can be constructed from the nature of the primary operator $\mathcal{O}_{\mathcal{BZ}}$ and its transmutation. The Lorentz structure of these operators can be of scalar, vector, spinor or tensor type.

In this work we are interested in the spin-1 unparticle contribution to the magnetic dipole moment (MDM) and the electric dipole moment (EDM) of leptons in the most general case when there is lepton flavor violating (LFV) interactions.

The effective interactions of a spin-1 unparticle with a fermion-antifermion pair can be written as [2, 1]:

$$\mathcal{L}_{\mathcal{U}_V} = \frac{\lambda_V^{ij}}{\Lambda_\mathcal{U}^{d_\mathcal{U}-1}}\bar{f}_i\gamma_\mu f_j\mathcal{O}_\mathcal{U}^\mu + \frac{\lambda_A^{ij}}{\Lambda_\mathcal{U}^{d_\mathcal{U}-1}}\bar{f}_i\gamma_\mu\gamma^5 f_j\mathcal{O}_\mathcal{U}^\mu. \tag{2}$$

where i and j stand for the family index and $\lambda_{V,A}^{ij}$ stands for the associated coupling constant. Due to the invariant scale nature of unparticles, their propagators can be constructed by means of unitary cuts and the spectral decomposition formula. It can be shown that the propagator of a spin-1 unparticle is

$$\Delta_F^{\mu\nu}(p) = \Delta_F(p^2)\left(-g^{\mu\nu} + a\frac{p^\mu p^\nu}{p^2}\right), \tag{3}$$

with

$$\Delta_F(p^2) = \frac{A_{d_\mathcal{U}}}{2\sin(d_\mathcal{U}\pi)}(-p^2 - i\epsilon)^{d_\mathcal{U}-2}. \tag{4}$$

The $A_{d_\mathcal{U}}$ function, which is meant to normalize the spectral density, is given as follows:

$$A_{d_\mathcal{U}} = \frac{16\pi^2\sqrt{\pi}}{(2\pi)^{2d_\mathcal{U}}}\frac{\Gamma(d_\mathcal{U}+\frac{1}{2})}{\Gamma(d_\mathcal{U}-1)\Gamma(2d_\mathcal{U})} \tag{5}$$

If the spin-1 unparticle is taken as transverse, $p_\mu\pi^{\mu\nu}(p) = 0$, which translates into the condition $a = 1$. However, in the context of a specific conformal invariant theory, $a = 2(d_\mathcal{U}-2)/(d_\mathcal{U}-1)$ [4].

2. Unparticle contribution to electric and magnetic dipole moments of leptons

The spin-1 unparticle contribution to the electromagnetic vertex arises through the Feynman diagram shown in Fig. 1. The lepton MDM is given by [5]:

$$a_j^\mathcal{U} = \sum_{J=V,A}\sum_{i=e,\mu,\tau}|\lambda_J^{ij}|^2 F_J(m_i, d_\mathcal{U}), \tag{6}$$

where j is the flavor index and the F_J functions can be written as

$$F_J(m_i, d_\mathcal{U}) = \frac{A_{d_\mathcal{U}}}{16\pi^2\sin(\pi d_\mathcal{U})}\left(\frac{m_i^2}{\Lambda_\mathcal{U}^2}\right)^{d_\mathcal{U}-1} f_J\left(\sqrt{r_i}, d_\mathcal{U}\right), \tag{7}$$

with $\sqrt{r_i} = m_j/m_i$ and

$$\begin{aligned} f_V(z, d_\mathcal{U}) &= \frac{-z}{2-d_\mathcal{U}}\int_0^1 dx\,(1-x)^{d_\mathcal{U}-1}x^{2-d_\mathcal{U}}(1-z^2x)^{d_\mathcal{U}-3}\Big((3(3-x)-4d_\mathcal{U}) \\ &+ z\,(3(d_\mathcal{U}-1)x + d_\mathcal{U} - 3) + z^2\left((1+x)^2 d_\mathcal{U} + (x-5)x - 2\right) \\ &- z^3\,(3(d_\mathcal{U}-1)x + d_\mathcal{U} - 3)\,x\Big), \end{aligned} \tag{8}$$

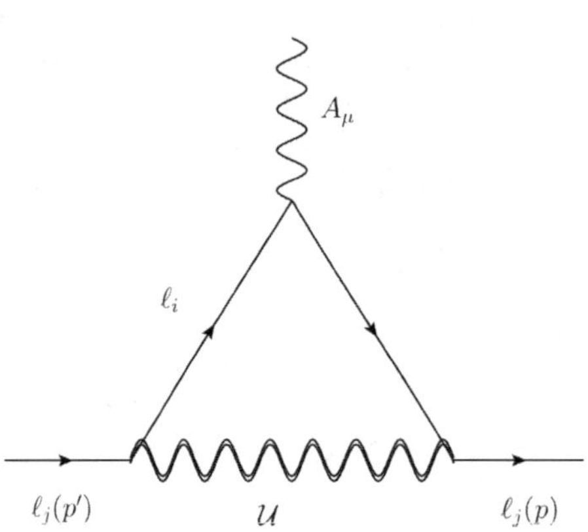

Figure 1. Feynman diagram for the lepton electromagnetic vertex induced by unparticle stuff.

also $f_A(z, d_\mathcal{U}) = f_V(-z, d_\mathcal{U})$.

As for the lepton EDM, which can only arise if both λ_V^{ij} and λ_A^{ij} are nonzero and have an imaginary phase, it is given by [5]:

$$d_j^{\mathcal{U}} = \sum_{i=e,\mu,\tau} \mathrm{Im}\left(\lambda_V^{ij}\lambda_A^{*ij}\right) G(m_i, d_\mathcal{U}), \tag{9}$$

with

$$G(m_i, d_\mathcal{U}) = \frac{e\,A_{d_\mathcal{U}}}{32\pi^2 \sin(\pi d_\mathcal{U})\, m_i}\left(\frac{m_i^2}{\Lambda_\mathcal{U}^2}\right)^{d_\mathcal{U}-1} g\left(\sqrt{r_i}, d_\mathcal{U}\right) \tag{10}$$

and

$$\begin{aligned} g(z, d_\mathcal{U}) &= \frac{1}{2-d_\mathcal{U}}\int_0^1 dx\,(1-x)^{d_\mathcal{U}-1}x^{2-d_\mathcal{U}}(1-z^2x)^{d_\mathcal{U}-3}\Big(4(1-z^2x)(2-d_\mathcal{U}) + (1-3x) \\ &+ z^2\left((x^2-1)d_\mathcal{U} + (x-1)x + 2\right)\Big). \end{aligned} \tag{11}$$

Below we will analyze the constraints on the LFV unparticle couplings from the muon EDM and LFV decays.

3. Numerical analysis and discussion

3.1. Muon anomalous magnetic moment

The muon MDM, a_μ, has been measured with an impressive accuracy of 0.54 ppm. The current world average, which is dominated by the measurements of the E281 collaboration at BNL, is given by [6]

$$a_\mu^{\mathrm{Exp.}} = 116592089\ (63) \times 10^{-11} \quad (0.54\mathrm{ppm}), \tag{12}$$

where the statistical and systematic errors, 0.46 ppm and 0.28 ppm, have been added in quadrature. On the other hand, the SM theoretical prediction, which is still under scrutiny as a large uncertainty arises from the hadronic contribution, is given by [7]

$$a_\mu^{\mathrm{SM}} = 116591834\ (48) \times 10^{-11}. \tag{13}$$

where the $\sigma(e^-e^+ \to \text{hadrons})$ data have been used to calculate the leading-order hadronic vacuum polarization contribution [8]. There is thus a discrepancy between the experimental and theoretical predictions larger than 3.6 standard deviations. It may be however that such a discrepancy will reduce to an acceptable level once the hadronic contribution is determined with best accuracy. As long as this disagreement is attributed to new physics, the allowed region for this class of contributions, with 95% C.L., is $\Delta a_\mu = (255 \mp 1.96 \times 80) \times 10^{-11}$.

We will consider some specific scenarios to obtain constraints on the LFV unparticle couplings. We will assume that there is a hierarchy in the LFV unparticle couplings as it is assumed in other models of LFV, i.e., $|\lambda^{e\mu}| < |\lambda^{e\tau}| < |\lambda^{\mu\tau}| \ll \lambda^{ii}$ for all the unparticle couplings. Also, we will consider that there are no large cancellations between distinct new physics contributions. Since experimental data strongly constrains LFV between the muon and the electron, we will concentrate on the scenario in which LFV transitions occur dominantly between the muon and tau leptons.

We first evaluate the contribution of the loops with internal muon and tau leptons to $a_\mu^{\mathcal{U}}$. The results are shown in Fig. 2 as functions of the dimension $d_{\mathcal{U}}$ and for $\Lambda_{\mathcal{U}} = 1$ TeV. While the contributions from vector couplings are positive, the axial-vector contributions are negative, with the contributions from axial-vector couplings being slightly larger in magnitude than the vector contributions. It means that, as long the vector and axial-vector unparticle couplings are of similar order of magnitude, the total contribution to the muon MDM could be negative. This is a scenario disfavored by the current experimental data, which requires a positive contribution to $a_\mu^{\mathcal{U}}$ from new physics. The situation is similar to that observed in the case of the scalar and pseudoscalar contributions [5, 9], although in this case the scalar contributions are slightly larger in magnitude than the pseudoscalar contributions. We thus conclude that the most favored scenario is that in which the strength of the axial-vector couplings is smaller than that of the vector couplings. In this way, we obtain a positive contribution to $a_\mu^{\mathcal{U}}$.

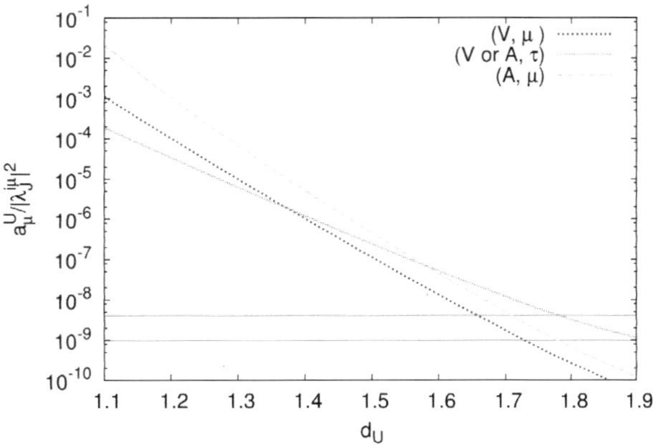

Figure 2. Contribution from each internal lepton loop to the muon MDM due to vector (V) or axial-vector (A) unparticle couplings as a function of the $d_{\mathcal{U}}$ dimension and for $\Lambda_{\mathcal{U}} = 1$ TeV. The absolute values of the (negative) (A) contributions are shown. The allowed region with 95% C.L. for new physics lies between the horizontal lines. The contribution from vector and axial-vector tau loops are indistinguishable.

If the current discrepancy between the theoretical SM prediction and the experimental value of the muon MDM is assumed to be due entirely to a spin-1 unparticle, we obtain the allowed area on the $|\lambda_V^{\tau\mu}|$ vs $|\lambda_V^{\mu\mu}|$ plane shown in Fig. 3. For $\Lambda_{\mathcal{U}}$ and $d_{\mathcal{U}}$ we used values that are consistent with the bounds obtained on the scale $\Lambda_{\mathcal{U}}$ from mono-photon production at LEP [10]. As inferred from Fig. 3, the experimental data allow a magnitude of the vector couplings as large as unity for $d_{\mathcal{U}} \simeq 2$ and $\Lambda_{\mathcal{U}} = 1$ TeV, whereas the most strong constraints are obtained for $d_{\mathcal{U}}$ close to unity. In general the limits on $|\lambda_V^{\mu\tau}|$ are slightly stronger than the limits on $|\lambda_V^{\mu\mu}|$, which is in agreement with our assumption. We also note that the bounds on the vector and axial-vector unparticle couplings are of similar order of magnitude than the bounds on the scalar and pseudoscalar unparticle couplings [5, 9].

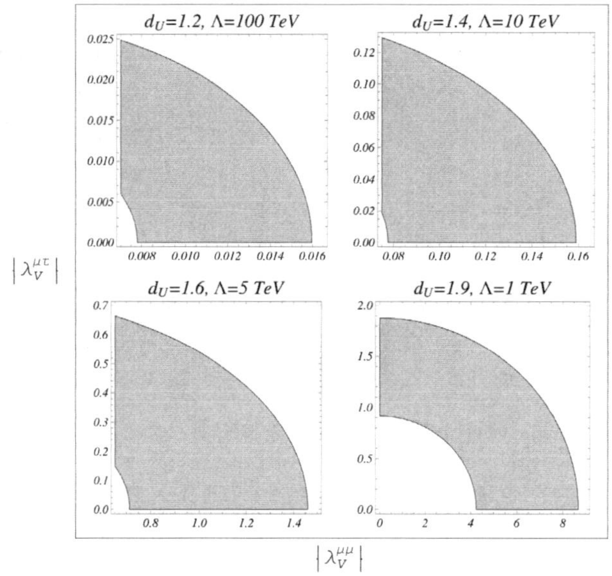

Figure 3. Allowed area on the $|\lambda_V^{\mu\tau}|$ vs $|\lambda_V^{\mu\mu}|$ plane consistent with the experimental limit on the muon MDM with 95% C.L. We used the indicated values of $\Lambda_{\mathcal{U}}$ and $d_{\mathcal{U}}$.

3.2. Decay $\tau \to 3\mu$

Apart from the muon MDM, constraints on LFV couplings can also be obtained from LFV processes involving the muon, which are strongly constrained by the experiment. The current constraints on LFV muon decays are $\mathrm{BR}(\mu \to e\gamma) < 2.4 \times 10^{-12}$ [11] and $\mathrm{BR}(\mu \to 3e) < 1.0 \times 10^{-12}$ [12]. On the other hand, less stringent constraints exist for LFV τ transitions: $\mathrm{BR}(\tau \to \mu\gamma) < 4.4 \times 10^{-8}$ [13], $\mathrm{BR}(\tau \to 3e) < 3.6 \times 10^{-8}$ [14], etc. We will now show that the $\tau \to 3\mu$ decay places strong constraints on the $\lambda_V^{\tau\mu}$ and $\lambda_V^{\mu\mu}$ couplings. If we neglect the axial-vector contributions, the $\tau \to 3\mu$ branching ratio can be written as follows [5]:

$$\mathrm{BR}(\tau \to 3\mu) = \frac{m_\tau \tau_\tau}{2^8 \pi^3} \left| \frac{A_{d_{\mathcal{U}}}}{\sin(d\pi)} \right|^2 \left(\frac{m_\tau}{\Lambda_{\mathcal{U}}} \right)^{4(d_{\mathcal{U}}-1)} |\lambda_V^{\mu\mu}|^2 |\lambda_V^{\mu\tau}|^2 \eta_1 \left(\frac{m_\mu}{m_\tau}, d_{\mathcal{U}} \right), \tag{14}$$

with τ_τ the tau mean life, while the η_1 function is

$$\begin{aligned} \eta_1(z, d_{\mathcal{U}}) &= -\frac{1}{2} \int dx_1 \int dx_2 \Big(2q_1^{2(d_{\mathcal{U}}-2)} \big((z+1)(3z-1)x_1 + x_1^2 + 2\,(x_1 + x_2 - 1)\, x_2 \\ &+ (6z-1)z^2 \big) + (q_1 q_2)^{d_{\mathcal{U}}-2} \,(x_1 + x_2 + z - 1)\,(x_1 + x_2 + (3z+1)z) \Big) \\ &+ (x_1 \leftrightarrow x_2), \end{aligned} \tag{15}$$

where $q_i = x_i + z^2$. The integration area on the x_1 vs. x_2 plane is

$$\begin{aligned} 3s_\tau^2 &\leq x_1 \leq 1 - 2s_\tau, \\ x_2 &\gtreqless \frac{1}{2}\left(1 - x_1 \mp \sqrt{\frac{((1-x_1)^2 - 4s_\tau^2)(x_1 - 3s_\tau^2)}{x_1 + s_\tau^2}}\right), \end{aligned} \tag{16}$$

with $s_\tau = m_\tau/m_\mu$.

The $\tau \to 3\mu$ branching ratio is constrained to be lower than about 10^{-8}. When we combine this bound with the one on the muon MDM, we find strong constraints on the LFV unparticle coupling constants. We show in Fig. 4 the allowed area on the $\lambda_V^{\tau\mu}$ vs $\lambda_V^{\mu\mu}$ plane for several values of $\Lambda_\mathcal{U}$ and $d_\mathcal{U}$. We observe that the $\lambda_V^{\mu\tau}$ parameter is tightly constrained for $d_\mathcal{U}$ close to unity, where $\lambda_V^{\mu\mu}$ is allowed to be considerably larger. There is also an area in which the opposite is true ($|\lambda_V^{\mu\tau}| \gg |\lambda_V^{\mu\mu}|$), but we will not consider it as conflicts with our previous assumption. Again, our bounds on the vector unparticle couplings are of similar order of magnitude than the ones obtained on the scalar unparticle couplings [5, 9].

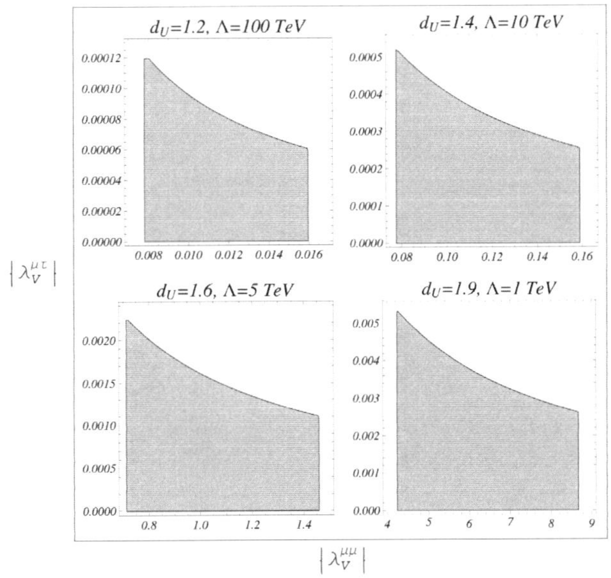

Figure 4. Allowed area on the $|\lambda_V^{\mu\tau}|$ vs $|\lambda_V^{\mu\mu}|$ plane consistent with the experimental constraints on the muon MDM and the $\tau \to 3\mu$ decay when the dominant contribution to these observables arise from the vector unparticle couplings.

3.3. Electric dipole moment

In the SM, the electron EDM is predicted to be negligibly small as it arises up to the three-loop level via the Cabbibo-Kobayashi-Maskawa (CKM) phase. A large fermion EDM, d_f, would thus be a clear indication of new sources of CP-violation [15]. The current experimental limit on the electron EDM with 90% C.L. is [16]:

$$|d_e| \leq 1.6 \times 10^{-27}\ \mathrm{e\,cm}, \tag{17}$$

whereas the experimental limits on the positive and negative muon EDMs with 95% C.L. are: [17]:

$$|d_\mu^+| \leq 2.1 \times 10^{-19} \,\mathrm{e\,cm}, \quad (18)$$
$$|d_\mu^-| \leq 1.5 \times 10^{-19} \,\mathrm{e\,cm}. \quad (19)$$

We will get an estimate of the order of magnitude for the electron and muon EDMs induced by a spin-1 unparticle. Apart from the magnitude of the $|\lambda_{V,A}^{ij}|$ parameters, the EDM depends on the associated CP-violating phases, so its analysis is even more complicated. Depending on the relative sign of the CP violating phases, the partial contributions can add coherently or destructively. We will content ourselves with considering a somewhat restrictive scenario, which will allows us to get an estimate of the order of magnitude of the EDM. First of all, as was the case for the muon MDM, we will assume that there is no large cancellation between different contributions to the EDM, so we can analyze each contribution separately. Also, to be consistent with the previous discussion, we will assume the following hierarchy $|\lambda_V^{\mu\tau}| \gg |\lambda_V^{\mu e}|$ and $|\lambda_V^{e\tau}| \gg |\lambda_V^{e\mu}|$. It means that the EDM would be entirely driven by the tau loop contributions. In view of this scenario, we numerically evaluate the tau loop contributions to Eq. (9). The results are shown in Fig. 5. We observe that the contribution from the tau loop to the electron and muon EDMs are indistinguishable, which means that the EDM is entirely controlled by the magnitude of the couplings constants and the imaginary phases.

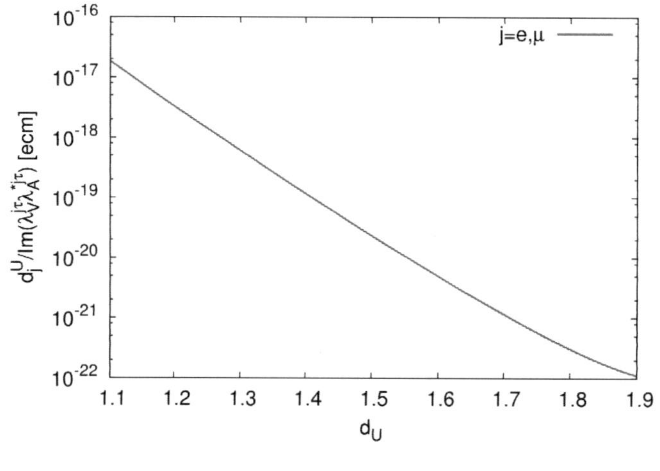

Figure 5. Contribution from the tau lepton loop to the electron EDM due to vector-axial unparticle couplings as a function of the $d_\mathcal{U}$ dimension and for $\Lambda_\mathcal{U} = 1$ TeV. This contribution is negative and thus its absolute value is shown. There is no distinction from the tau contribution to the muon EDM.

We now consider the constraints on the coupling constants obtained above to estimate the muon EDM in our working scenario. The spin-1 unparticle contribution is of the order of 10^{-17} e-cm for $d_\mathcal{U} \simeq 1.1$ and 10^{-21} e-cm for $d_\mathcal{U} \simeq 1.6$. The allowed magnitude of the $|\lambda_V^{\mu\tau}|$ parameter is of the order of 10^{-5} for $d_\mathcal{U} = 1.1$ and 10^{-3} for $d_\mathcal{U} = 1.6$. A good assumption for the $|\lambda_A^{\mu\tau}|$ coupling is that its magnitude is about one or two orders of magnitude below $|\lambda_V^{\mu\tau}|$, which is consistent with our previous discussion. Therefore the muon EDM is about $|d_\mu^\mathcal{U}| \simeq |\sin\delta_{(V,A)}^{\mu\tau}| \times 10^{-30}$ e-cm for $d_\mathcal{U} \simeq 1.1$, whereas $|d_\mu^\mathcal{U}| \simeq |\sin\delta_{(V,A)}^{\mu\tau}| \times 10^{-29}$ e-cm for $d_\mathcal{U} \simeq 1.6$. Here $\delta_{(V,A)}^{\mu\tau}$ is the relative CP-violating phase between the vector and axial-vector couplings. These results are several

Journal of Physics: Conference Series **378** (2012) 012013 doi:10.1088/1742-6596/378/1/012013

orders of magnitude smaller than the experimental limit on the muon EDM. The magnitude of the coupling constants for LFV transitions involving the electron are expected to be more suppressed as long as the hierarchy discussed above is respected. Then we can estimate that the electron EDM induced by vector unparticles is well below the 10^{-30} e-cm level and far from the experimental measurement.

4. Concluding remarks

The lepton magnetic and electric dipole moments were calculated in the framework of unparticle physics assuming interactions mainly driven by a spin-1 unparticle with both vector and axial-vector couplings to leptons. Analytical expressions were found that agree with previous calculations for the muon MDM in the limit of flavor conserving couplings.

The muon MDM was numerically evaluated to obtain bounds on the coupling constants from the experimental measurements assuming that LFV is mainly dominated by LFV couplings involving the muon and tau leptons. It was found that the latest experimental data for the muon MDM favor a scenario in which the spin-1 unparticle contribution is dominated by the vector couplings since the contribution from axial-vector couplings is negative. It is also found that the decay $\tau \to 3\mu$ can constrain considerably the allowed magnitude of the $\lambda_V^{\mu\mu}$ and $\lambda_V^{\mu\tau}$ couplings.

As far as the EDM is concerned, since it depends on several free parameters, we content ourselves with estimating its order of magnitude. It is found that the unparticle contributions are well below the experimental limits of the electron and muon EDMs.

Acknowledgments

We acknowledge support from Conacyt and SNI (México). Support from VIEP-BUAP is also acknowledged.

References

[1] T. Banks and A. Zaks, Nucl. Phys. B **196**, 189 (1982).
[2] H. Georgi, Phys. Rev. Lett. **98**, 221601 (2007) [arXiv:hep-ph/0703260].
[3] H. Georgi, Phys. Lett. B **650**, 275 (2007) [arXiv:0704.2457 [hep-ph]].
[4] B. Grinstein, K. A. Intriligator and I. Z. Rothstein, Phys. Lett. B **662**, 367 (2008) [arXiv:0801.1140 [hep-ph]].
[5] A. Moyotl, A. Rosado and G. Tavares-Velasco, Phys. Rev. D **84**, 073010 (2011) [arXiv:1109.4890 [hep-ph]].
[6] G. W. Bennett *et al.* [Muon G-2 Collaboration], Phys. Rev. D **73**, 072003 (2006) [arXiv:hep-ex/0602035].
[7] K. Nakamura *et al.* [Particle Data Group], J. Phys. G **37**, 075021 (2010).
[8] M. Davier, A. Hoecker, B. Malaescu, C. Z. Yuan and Z. Zhang, Eur. Phys. J. C **66**, 1 (2010) [arXiv:0908.4300 [hep-ph]].
[9] A. Hektor, Y. Kajiyama and K. Kannike, Phys. Rev. D **78**, 053008 (2008) [arXiv:0802.4015 [hep-ph]].
[10] K. Cheung, W. Y. Keung and T. C. Yuan, Phys. Rev. D **76**, 055003 (2007) [arXiv:0706.3155 [hep-ph]].
[11] J. Adam *et al.* [MEG collaboration], arXiv:1107.5547 [hep-ex].
[12] U. Bellgardt *et al.* [SINDRUM Collaboration], Nucl. Phys. B **299**, 1 (1988).
[13] B. Aubert *et al.* [BABAR Collaboration], Phys. Rev. Lett. **104**, 021802 (2010) [arXiv:0908.2381 [hep-ex]].
[14] B. Aubert *et al.* [BABAR Collaboration], Phys. Rev. Lett. **99**, 251803 (2007) [arXiv:0708.3650 [hep-ex]].
[15] W. Bernreuther and M. Suzuki, Rev. Mod. Phys. **63**, 313 (1991) [Erratum-ibid. **64**, 633 (1992)].
[16] B. C. Regan, E. D. Commins, C. J. Schmidt, D. DeMille, Phys. Rev. Lett. **88**, 071805 (2002).
[17] G. W. Bennett *et al.* [Muon (g-2) Collaboration], Phys. Rev. D **80**, 052008 (2009) [arXiv:0811.1207 [hep-ex]].

Journal of Physics: Conference Series **378** (2012) 012014 doi:10.1088/1742-6596/378/1/012014

The flavour symmetry S_3 and the neutrino mass matrix with two texture zeroes

F. González Canales* **and A. Mondragón****

Instituto de Física, Universidad Nacional Autónoma de México, 04510, México D.F., México

E-mail: *ffelix@fisica.unam.mx, **mondra@fisica.unam.mx

Abstract. In this work, discuss neutrino masses and mixings in the framework of a minimal S_3 symmetric extension of the Standard Model. In this model, the mass matrices of all fermions take the same generic form with two texture zeroes. The mass matrices of the neutrinos and charged leptons are re-parameterized in terms of their eigenvalues, then the neutrino mixing matrix, V_{PMNS}, is computed and exact, explicit analytical expressions for the neutrino mixing angles as functions of the masses of neutrinos and charged leptons are obtained in excellent agreement with the latest experimental data. We also compute the branching ratios of some selected flavour-changing neutral current (FCNC) processes, as well as the contribution of the exchange of neutral flavour-changing scalars to the anomaly of the magnetic moment of the muon, as functions of the masses of charged leptons and the neutral Higgs bosons. We find that the $S_3 \times Z_2$ flavour symmetry and the strong mass hierarchy of the charged leptons strongly suppress the FCNC processes in the leptonic sector, well below the present experimental bounds by many orders of magnitude. The contribution of FCNC's to the anomaly of the muon's magnetic moment is small, but not negligible.

1. Introduction

The observation of flavour oscillations of solar, atmospheric, reactor, and accelerator neutrinos established that they have non-vanishing masses and mix among themselves, much like the quarks do [1, 2, 3, 4, 5, 6, 7, 8, 9, 10, 11, 12, 13, 14, 15, 16, 17, 18, 19, 20]. This discovery brought out very forcibly the need of extending the Standard Model (SM) in order to accomodate in the theory the new data on neutrino physics in a consistent way that would allow for a unified and systematic treatment of the observed hierarchy of masses and mixings of all fermions. At the same time, the number of free parameters in the extended form of the SM had to be drastically reduced in order to give predictive power to the theory. These two seemingly contradictory demands are met by a flavour symmetry under which the families transfom in a non-trivial fashion. The observed pattern of neutrino mixing and, in particular, the non vanishing and sizeble value of the reactor mixings angle strongly suggest a flavour permutational symmetry.

The result of a combined analysis of all available neutrino oscillation data, including the recent results from long-baseline $\nu_\mu \longrightarrow \nu_e$ searches at the Tokai to Kamioka (T2K) and Main Injector Neutrino Oscillation Search (MINOS) experiments, give the following values the difference of the squared neutrino masses and the mixing angles in the lepton mixing matrix, U_{PMNS}, at 1σ

Journal of Physics: Conference Series **378** (2012) 012014 doi:10.1088/1742-6596/378/1/012014

confidence level [21]:

$$\Delta m_{21}^2 = 7.59^{+0.20}_{-0.18} \times 10^{-5}\ \mathrm{eV}^2,$$

$$\Delta m_{31}^2 = \begin{cases} -2.40^{+0.08}_{-0.09} \times 10^{-3}\ \mathrm{eV}^2, & (m_{\nu_2} > m_{\nu_1} > m_{\nu_3}). \\ +2.50^{+0.09}_{-0.16} \times 10^{-3}\ \mathrm{eV}^2, & (m_{\nu_3} > m_{\nu_2} > m_{\nu_1}). \end{cases} \tag{1}$$

$$\sin^2\theta_{12}^l = 0.312^{+0.017}_{-0.015}, \quad \sin^2\theta_{23}^l = \begin{cases} 0.52 \pm 0.06, \\ 0.52^{+0.06}_{-0.07}, \end{cases}, \quad \sin^2\theta_{13}^l = \begin{cases} 0.016^{+0.008}_{-0.006}, \\ 0.013^{+0.007}_{-0.005}, \end{cases}, \tag{2}$$

the upper (lower) row corresponds to inverted (normal) neutrino mass hierarchy.

In the last ten years, important theoretical advances have been made in the understanding of the mechanisms for the mass fermion generation and flavour mixing. A phenomenologically and theoretically meaningful approach for reducing the number of free parameters in the Standard Model is the imposition of texture zeroes [22, 23] and/or flavour symmetries. For a recent review of flavour symmetry models see [24]. Also, certain texture zeroes may be obtained from a flavour symmetry [25, 26]. In the case of the Minimal S_3-Invariant Extension of the Standard Model [27, 28, 29, 30, 31, 32, 33], the concept of flavour and generations is extended to the Higgs sector in such a way that all the matter fields – Higgs, quarks, and lepton fields, including the right-handed neutrino fields– have three species and therefore transform under the flavour symmetry group as the three dimensional representation $1 \oplus 2$ of the permutational group S_3. A model with more than one Higgs $SU(2)$ doublet has tree level flavour changing neutral currents whose exchange may give rise to lepton flavour violating processes and may also contribute to the anomalous magnetic moment of the muon. An effective test of the phenomenological success of the model is obtained by verifying that all flavour changing neutral current processes and the magnetic anomaly of the muon, computed in the S_3–Invariant extended form of the Standard Model, agree with the experimental values.

2. The Minimal S_3-invariant Extension of the Standard Model

In the Standard Model analogous fermions in different generations have identical couplings to all gauge bosons of the strong, weak and electromagnetic interactions. Prior to the introduction of the Higgs boson and mass terms, the Lagrangian is chiral and invariant with respect to permutations of the left and right fermionic fields.

The six possible permutations of three objects (f_1, f_2, f_3) are elements of the permutational group S_3. This is the discrete, non-Abelian group with the smallest number of elements. The three-dimensional real representation is not an irreducible representation of S_3. It can be decomposed into the direct sum of a doublet f_D and a singlet f_s, where

$$f_s = \frac{1}{\sqrt{3}}(f_1 + f_2 + f_3), \quad f_D^T = \left(\frac{1}{\sqrt{2}}(f_1 - f_2), \frac{1}{\sqrt{6}}(f_1 + f_2 - 2f_3)\right). \tag{3}$$

The direct product of two doublets $\mathbf{p_D}^T = (p_{D1}, p_{D2})$ and $\mathbf{q_D}^T = (q_{D1}, q_{D2})$ may be decomposed into the direct sum of two singlets $\mathbf{r_s}$ and $\mathbf{r_{s'}}$, and one doublet $\mathbf{r_D}^T$ where

$$\mathbf{r_s} = p_{D1}q_{D1} + p_{D2}q_{D2}, \quad \mathbf{r_{s'}} = p_{D1}q_{D2} - p_{D2}q_{D1}, \tag{4}$$

$$\mathbf{r_D}^T = (r_{D1}, r_{D2}) = (p_{D1}q_{D2} + p_{D2}q_{D1}, p_{D1}q_{D1} - p_{D2}q_{D2}). \tag{5}$$

The antisymmetric singlet $\mathbf{r_{s'}}$ is not invariant under S_3.

Since the Standard Model has only one Higgs $SU(2)_L$ doublet, which can only be an S_3 singlet, it can only give mass to the quark or charged lepton in the S_3 singlet representation, one in each family, without breaking the S_3 symmetry.

Hence, in order to impose S_3 as a fundamental symmetry, unbroken at the Fermi scale, we are led to extend the Higgs sector of the theory. The quark, lepton and Higgs fields are

$$Q^T = (u_L, d_L) \ , \ u_R \ , \ d_R \ , \ L^T = (\nu_L, e_L) \ , \ e_R \ , \ \nu_R \ \text{and} \ H, \tag{6}$$

in an obvious notation. All of these fields have three species, and we assume that each one forms a reducible representation $\mathbf{1}_S \oplus \mathbf{2}$. The doublets carry capital indices I and J, which run from 1 to 2, and the singlets are denoted by Q_3, u_{3R}, d_{3R}, L_3, e_{3R}, ν_{3R} and H_S. Note that the subscript 3 denotes the singlet representation and not the third generation. The most general renormalizable Yukawa interactions of this model are given by

$$\mathcal{L}_Y = \mathcal{L}_{Y_D} + \mathcal{L}_{Y_U} + \mathcal{L}_{Y_E} + \mathcal{L}_{Y_\nu}, \tag{7}$$

where

$$\begin{aligned} \mathcal{L}_{Y_D} = & -Y_1^d \overline{Q}_I H_S d_{IR} - Y_3^d \overline{Q}_3 H_S d_{3R} - Y_2^d [\, \overline{Q}_I \kappa_{IJ} H_1 d_{JR} + \overline{Q}_I \eta_{IJ} H_2 d_{JR} \,] \\ & -Y_4^d \overline{Q}_3 H_I d_{IR} - Y_5^d \overline{Q}_I H_I d_{3R} + \text{ h.c.}, \end{aligned} \tag{8}$$

$$\begin{aligned} \mathcal{L}_{Y_U} = & -Y_1^u \overline{Q}_I (i\sigma_2) H_S^* u_{IR} - Y_3^u \overline{Q}_3 (i\sigma_2) H_S^* u_{3R} - Y_4^u \overline{Q}_3 (i\sigma_2) H_I^* u_{IR} \\ & -Y_2^u [\, \overline{Q}_I \kappa_{IJ} (i\sigma_2) H_1^* u_{JR} + \eta \overline{Q}_I \eta_{IJ} (i\sigma_2) H_2^* u_{JR} \,] - Y_5^u \overline{Q}_I (i\sigma_2) H_I^* u_{3R} + \text{ h.c.}, \end{aligned} \tag{9}$$

$$\begin{aligned} \mathcal{L}_{Y_E} = & -Y_1^e \overline{L}_I H_S e_{IR} - Y_3^e \overline{L}_3 H_S e_{3R} - Y_2^e [\, \overline{L}_I \kappa_{IJ} H_1 e_{JR} + \overline{L}_I \eta_{IJ} H_2 e_{JR} \,] \\ & -Y_4^e \overline{L}_3 H_I e_{IR} - Y_5^e \overline{L}_I H_I e_{3R} + \text{ h.c.}, \end{aligned} \tag{10}$$

$$\begin{aligned} \mathcal{L}_{Y_\nu} = & -Y_1^\nu \overline{L}_I (i\sigma_2) H_S^* \nu_{IR} - Y_3^\nu \overline{L}_3 (i\sigma_2) H_S^* \nu_{3R} - Y_4^\nu \overline{L}_3 (i\sigma_2) H_I^* \nu_{IR} \\ & -Y_2^\nu [\, \overline{L}_I \kappa_{IJ} (i\sigma_2) H_1^* \nu_{JR} + \overline{L}_I \eta_{IJ} (i\sigma_2) H_2^* \nu_{JR} \,] - Y_5^\nu \overline{L}_I (i\sigma_2) H_I^* \nu_{3R} + \text{ h.c.}, \end{aligned} \tag{11}$$

and

$$\kappa = \begin{pmatrix} 0 & 1 \\ 1 & 0 \end{pmatrix} \quad \text{and} \quad \eta = \begin{pmatrix} 1 & 0 \\ 0 & -1 \end{pmatrix}. \tag{12}$$

Furthermore, we add to the Lagrangian the Majorana mass terms for the right-handed neutrinos

$$\mathcal{L}_M = -M_1 \nu_{IR}^T C \nu_{IR} - M_2 \nu_{3R}^T C \nu_{3R} - M_3 \nu_{3R}^T C \nu_{3R}. \tag{13}$$

Due to the presence of three Higgs fields, the Higgs potential $V_H(H_S, H_D)$ is more complicated than that of the Standard Model. This potential was analyzed by Pakvasa and Sugawara [34] who found that in addition to the S_3 symmetry, it has a permutational symmetry S_2': $H_1 \leftrightarrow H_2$, which is not a subgroup of the flavour group S_3. In this communication, we will assume that the vacuum respects the accidental S_2' symmetry of the Higgs potential and that $\langle H_1 \rangle = \langle H_2 \rangle$. With these assumptions, the Yukawa interactions, eqs. (8)-(11) yield mass matrices, for all fermions in the theory, of the general form [27]

$$\mathbf{M} = \begin{pmatrix} \mu_1 + \mu_2 & \mu_2 & \mu_5 \\ \mu_2 & \mu_1 - \mu_2 & \mu_5 \\ \mu_4 & \mu_4 & \mu_3 \end{pmatrix}. \tag{14}$$

The left-handed Majorana neutrinos ν_L naturally acquire their small masses through the see-saw mechanism type I of the form

$$\mathbf{M}_\nu = \mathbf{M}_{\nu_\mathbf{D}} \tilde{\mathbf{M}}_R^{-1} (\mathbf{M}_{\nu_\mathbf{D}})^T, \tag{15}$$

$-$	$+$
$H_S,\ \nu_{3R}$	$H_I,\ L_3,\ L_I,\ e_{3R},\ e_{IR},\ \nu_{IR}$

Table 1. Z_2 assignment in the leptonic sector.

where $\mathbf{M}_{\nu_\mathrm{D}}$ and $\tilde{\mathbf{M}}_R$ denote the Dirac and right handed Majorana neutrino mass matrices, respectively.

In principle, all entries in the mass matrices can be complex since there is no restriction coming from the flavour symmetry S_3. The mass matrices are diagonalized by bi-unitary transformations as

$$\begin{aligned} U^\dagger_{d(u,e)L}\mathbf{M}_{d(u,e)}U_{d(u,e)R} &= \mathrm{diag}(m_{d(u,e)}, m_{s(c,\mu)}, m_{b(t,\tau)}), \\ U^T_\nu \mathbf{M}_\nu U_\nu &= \mathrm{diag}(m_{\nu_1}, m_{\nu_2}, m_{\nu_3}). \end{aligned} \tag{16}$$

The entries in the diagonal matrices may be complex, so the physical masses are their absolute values.

The quark and lepton flavor mixing matrices, V_{PMNS} and V_{CKM}, arise from the mismatch between diagonalization of the mass matrices of u and d type quarks and the diagonalization of the mass matrices of charged leptons and left-handed neutrinos, respectively,

$$V_{CKM} = U^\dagger_{uL}U_{dL}, \quad V_{PMNS} = U^\dagger_{eL}U_\nu K, \tag{17}$$

where K is the diagonal matrix of the Majorana phase factors. Therefore, in order to obtain the unitary matrices appearing in eq. (17) and get predictions for the flavor mixing angles and CP violating phases, we should specify the mass matrices.

3. The mass matrices in the leptonic sector and Z_2 symmetry

A further reduction of the number of parameters in the leptonic sector may be achieved by means of an Abelian Z_2 symmetry. A possible set of charge assignments of Z_2, compatible with the experimental data on masses and mixings in the leptonic sector is given in Table 1.

These Z_2 assignments forbid the following Yukawa couplings

$$Y^e_1 = Y^e_3 = Y^\nu_1 = Y^\nu_5 = 0. \tag{18}$$

Therefore, the corresponding entries in the mass matrices vanish, *i.e.*, $\mu^e_1 = \mu^e_3 = 0$ and $\mu^\nu_1 = \mu^\nu_5 = 0$.

3.1. The mass matrix of the charged leptons

The mass matrix of the charged leptons takes the form

$$M_e = m_\tau \begin{pmatrix} \tilde{\mu}_2 & \tilde{\mu}_2 & \tilde{\mu}_5 \\ \tilde{\mu}_2 & -\tilde{\mu}_2 & \tilde{\mu}_5 \\ \tilde{\mu}_4 & \tilde{\mu}_4 & 0 \end{pmatrix}. \tag{19}$$

The unitary matrix U_{eL} that enters in the definition of the mixing matrix, V_{PMNS}, is calculated from

$$U^\dagger_{eL} M_e M^\dagger_e U_{eL} = \mathrm{diag}\left(m^2_e, m^2_\mu, m^2_\tau\right), \tag{20}$$

Journal of Physics: Conference Series **378** (2012) 012014 doi:10.1088/1742-6596/378/1/012014

where m_e, m_μ and m_τ are the masses of the charged leptons, and

$$\frac{M_e M_e^\dagger}{m_\tau^2} = \begin{pmatrix} 2|\tilde{\mu}_2|^2 + |\tilde{\mu}_5|^2 & |\tilde{\mu}_5|^2 & 2|\tilde{\mu}_2||\tilde{\mu}_4|e^{-i\delta_e} \\ |\tilde{\mu}_5|^2 & 2|\tilde{\mu}_2|^2 + |\tilde{\mu}_5|^2 & 0 \\ 2|\tilde{\mu}_2||\tilde{\mu}_4|e^{i\delta_e} & 0 & 2\,|\tilde{\mu}_4|^2 \end{pmatrix}. \tag{21}$$

Notice that this matrix has only one non-ignorable phase factor [29].

Once $M_e M_e^\dagger$ has been reparametrized in terms of the charged lepton masses, it is straightforward to compute M_e and U_{eL} also as functions of the charged lepton masses [29]. The resulting expression for M_e, written to order $\left(m_\mu m_e/m_\tau^2\right)^2$ and $x^4 = (m_e/m_\mu)^4$ is

$$M_e \approx m_\tau \begin{pmatrix} \frac{1}{\sqrt{2}}\frac{\tilde{m}_\mu}{\sqrt{1+x^2}} & \frac{1}{\sqrt{2}}\frac{\tilde{m}_\mu}{\sqrt{1+x^2}} & \frac{1}{\sqrt{2}}\sqrt{\frac{1+x^2-\tilde{m}_\mu^2}{1+x^2}} \\ \frac{1}{\sqrt{2}}\frac{\tilde{m}_\mu}{\sqrt{1+x^2}} & -\frac{1}{\sqrt{2}}\frac{\tilde{m}_\mu}{\sqrt{1+x^2}} & \frac{1}{\sqrt{2}}\sqrt{\frac{1+x^2-\tilde{m}_\mu^2}{1+x^2}} \\ \frac{\tilde{m}_e(1+x^2)}{\sqrt{1+x^2-\tilde{m}_\mu^2}}e^{i\delta_e} & \frac{\tilde{m}_e(1+x^2)}{\sqrt{1+x^2-\tilde{m}_\mu^2}}e^{i\delta_e} & 0 \end{pmatrix}. \tag{22}$$

This approximation is numerically exact up to order 10^{-9} in units of the τ mass. Notice that this matrix has no free parameters other than the Dirac phase δ_e.

The unitary matrix U_{eL} that diagonalizes $M_e M_e^\dagger$ and enters in the definition of the neutrino mixing matrix V_{PMNS} may be written as

$$U_{eL} = \begin{pmatrix} 1 & 0 & 0 \\ 0 & 1 & 0 \\ 0 & 0 & e^{i\delta_e} \end{pmatrix} \begin{pmatrix} O_{11} & -O_{12} & O_{13} \\ -O_{21} & O_{22} & O_{23} \\ -O_{31} & -O_{32} & O_{33} \end{pmatrix}, \tag{23}$$

where the orthogonal matrix $\mathbf{O}_{eL}$ in the right hand side of eq. (23), written to the same order of magnitude as M_e, is

$$\mathbf{O}_{eL} \approx \begin{pmatrix} \frac{1}{\sqrt{2}}x\frac{(1+2\tilde{m}_\mu^2+4x^2+\tilde{m}_\mu^4+2\tilde{m}_e^2)}{\sqrt{1+\tilde{m}_\mu^2+5x^2-\tilde{m}_\mu^4-\tilde{m}_\mu^6+\tilde{m}_e^2+12x^4}} & -\frac{1}{\sqrt{2}}\frac{(1-2\tilde{m}_\mu^2+\tilde{m}_\mu^4-2\tilde{m}_e^2)}{\sqrt{1-4\tilde{m}_\mu^2+x^2+6\tilde{m}_\mu^4-4\tilde{m}_\mu^6-5\tilde{m}_e^2}} & \frac{1}{\sqrt{2}} \\ -\frac{1}{\sqrt{2}}x\frac{(1+4x^2-\tilde{m}_\mu^4-2\tilde{m}_e^2)}{\sqrt{1+\tilde{m}_\mu^2+5x^2-\tilde{m}_\mu^4-\tilde{m}_\mu^6+\tilde{m}_e^2+12x^4}} & \frac{1}{\sqrt{2}}\frac{(1-2\tilde{m}_\mu^2+\tilde{m}_\mu^4)}{\sqrt{1-4\tilde{m}_\mu^2+x^2+6\tilde{m}_\mu^4-4\tilde{m}_\mu^6-5\tilde{m}_e^2}} & \frac{1}{\sqrt{2}} \\ -\frac{\sqrt{1+2x^2-\tilde{m}_\mu^2-\tilde{m}_e^2}(1+\tilde{m}_\mu^2+x^2-2\tilde{m}_e^2)}{\sqrt{1+\tilde{m}_\mu^2+5x^2-\tilde{m}_\mu^4-\tilde{m}_\mu^6+\tilde{m}_e^2+12x^4}} & -x\frac{(1+x^2-\tilde{m}_\mu^2-2\tilde{m}_e^2)\sqrt{1+2x^2-\tilde{m}_\mu^2-\tilde{m}_e^2}}{\sqrt{1-4\tilde{m}_\mu^2+x^2+6\tilde{m}_\mu^4-4\tilde{m}_\mu^6-5\tilde{m}_e^2}} & \frac{\sqrt{1+x^2}\tilde{m}_e\tilde{m}_\mu}{\sqrt{1+x^2-\tilde{m}_\mu^2}} \end{pmatrix}, \tag{24}$$

where, as before, $\tilde{m}_\mu = m_\mu/m_\tau$, $\tilde{m}_e = m_e/m_\tau$ and $x = m_e/m_\mu$.

The the mass values of the charged lepton masses [35]:

$$m_e = 0.51099891 \text{ MeV}, \quad m_\mu = 105.658367 \text{ MeV}, \text{ and } m_\tau = 1776.82 \text{ MeV}, \tag{25}$$

we obtain

$$\mathbf{O}_{eL} \approx \begin{pmatrix} \frac{1}{\sqrt{2}}\frac{\tilde{m}_e}{\tilde{m}_\mu} & -\frac{1}{\sqrt{2}} & \frac{1}{\sqrt{2}} \\ -\frac{1}{\sqrt{2}}\frac{\tilde{m}_e}{\tilde{m}_\mu} & \frac{1}{\sqrt{2}} & \frac{1}{\sqrt{2}} \\ -1 & -\frac{\tilde{m}_e}{\tilde{m}_\mu} & 0 \end{pmatrix} + \mathcal{O}\left(10^{-5}\right), \tag{26}$$

Journal of Physics: Conference Series **378** (2012) 012014 doi:10.1088/1742-6596/378/1/012014

which can be written in following form:

$$\mathbf{O}_{eL} \approx \begin{pmatrix} \frac{1}{\sqrt{2}} & 0 & \frac{1}{\sqrt{2}} \\ -\frac{1}{\sqrt{2}} & 0 & \frac{1}{\sqrt{2}} \\ 0 & 1 & 0 \end{pmatrix} \begin{pmatrix} \frac{\tilde{m}_e}{\tilde{m}_\mu} & -1 & 0 \\ 1 & -\frac{\tilde{m}_e}{\tilde{m}_\mu} & 0 \\ 0 & 0 & 1 \end{pmatrix}. \tag{27}$$

The second matrix on the right side is a numerically exact up to order $\mathcal{O}\left(10^{-7}\right)$ in units of the τ mass.

3.2. The mass matrix of the neutrinos

In the minimal S_3-invariant extension of Standard Model, the Yukawa interactions and the $S_3 \times Z_2$ flavour symmetry yield a mass matrix for the Dirac neutrinos of the form

$$\mathbf{M}_{\nu_\mathbf{D}} = \begin{pmatrix} \mu_2^\nu & \mu_2^\nu & 0 \\ \mu_2^\nu & -\mu_2^\nu & 0 \\ \mu_4^\nu & \mu_4^\nu & \mu_3^\nu \end{pmatrix}. \tag{28}$$

In principle, all entries in the mass matrix $\mathbf{M}_{\nu_\mathbf{D}}$ can be complex, since there is no restriction coming from the $S_3 \times Z_2$ flavour symmetry.

The mass of the left-handed Majorana neutrinos, $\mathbf{M}_\nu$, are generated by the see-saw mechanism type I,

$$\mathbf{M}_\nu = \mathbf{M}_{\nu_\mathbf{D}} \tilde{\mathbf{M}}_R^{-1} (\mathbf{M}_{\nu_\mathbf{D}})^T, \tag{29}$$

where $\tilde{\mathbf{M}}_R$ is the mass matrix of the right-handed neutrinos, which we take to be real and diagonal but non-degenerate

$$\tilde{\mathbf{M}}_R = \mathrm{diag}(M_1, M_2, M_3). \tag{30}$$

Then, the mass matrix $\mathbf{M}_\nu$ takes the form

$$\mathbf{M}_\nu = \begin{pmatrix} \frac{2\left(\mu_2^\nu\right)^2}{\overline{M}} & \frac{2\lambda\left(\mu_2^\nu\right)^2}{\overline{M}} & \frac{2\mu_2^\nu\mu_4^\nu}{\overline{M}} \\ \frac{2\lambda\left(\mu_2^\nu\right)^2}{\overline{M}} & \frac{2\left(\mu_2^\nu\right)^2}{\overline{M}} & \frac{2\mu_2^\nu\mu_4^\nu\lambda}{\overline{M}} \\ \frac{2\mu_2^\nu\mu_4^\nu}{\overline{M}} & \frac{2\mu_2^\nu\mu_4^\nu\lambda}{\overline{M}} & \frac{2\left(\mu_4^\nu\right)^2}{\overline{M}} + \frac{\left(\mu_3^\nu\right)^2}{M_3} \end{pmatrix} \tag{31}$$

where

$$\lambda = \frac{1}{2}\left(\frac{M_2 - M_1}{M_1 + M_2}\right) \quad \text{and} \quad \overline{M} = 2\frac{M_1 M_2}{M_2 + M_1} \tag{32}$$

When the first two right-handed neutrino masses are equal, the parameter λ vanishes and we recover the expresion for $\mathbf{M}_\nu$ given in Kubo *et al* [27].

As we have assumed the the right-handed neutrino mass matrix $\tilde{\mathbf{M}}_R$ to be real, the complex symetric neutrino mass matrix $\mathbf{M}_\nu$ has only three independent phase factor that come from the parameters μ_2, μ_3 and μ_4. Here, to simplify the analysis we will consider the case when $\arg\{\mu_2^\nu\} = \arg\{\mu_3^\nu\}$ o $2\arg\{\mu_4^\nu\} = \arg\left\{2\frac{\left(\mu_4^\nu\right)^2}{\overline{M}} + \frac{\left(\mu_3^\nu\right)^2}{M_3}\right\}$. The general case, with three independent phase factors, will be considered in detail elsewhere.

In the case considered here, the phase factors may be taken out of $\mathbf{M}_\nu$ as

$$\mathbf{M}_\nu = Q\overline{\mathbf{M}}_\nu Q \tag{33}$$

where

$$Q = \begin{pmatrix} e^{i\phi_2} & 0 & 0 \\ 0 & e^{i\phi_2} & 0 \\ 0 & 0 & e^{i\phi_4} \end{pmatrix} \quad \text{and} \quad \overline{\mathbf{M}}_\nu = \begin{pmatrix} a & d & b \\ d & a & e \\ b & e & c \end{pmatrix} \tag{34}$$

with $\phi_2 = \arg\{\mu_2^\nu\}$, $\phi_4 = \arg\{\mu_4^\nu\}$, $a = \frac{2|\mu_2^\nu|^2}{|M|}$, $b = \frac{2|\mu_4^\nu||\mu_2^\nu|}{|M|}$, $c = \frac{2|\mu_4^\nu|^2}{M} + \frac{|\mu_3^\nu|^2}{M_3}$, $d = \frac{2|\lambda||\mu_2^\nu|^2}{|M|}$ and $e = \frac{2|\mu_2^\nu||\mu_4^\nu||\lambda|}{|M|}$.

The real symetric matrix $\overline{\mathbf{M}}_\nu$ may be brought to diagonal form by means of a similarity transformation with an orthogonal matrix $\mathbf{O}_\nu$ as

$$\overline{\mathbf{M}}_\nu = \mathbf{O}_\nu \text{diag}\left\{m_{\nu_1}, m_{\nu_2}, m_{\nu_3}\right\} \mathbf{O}_\nu^T, \tag{35}$$

the columns in $\mathbf{O}_\nu$ are the normalized eigenvectors of $\overline{\mathbf{M}}_\nu$.

In order to compute $\mathbf{O}_\nu$ we notice that the diagonalization of $\overline{\mathbf{M}}_\nu$ is equivalent to the diagonalization of a mass matrix $\widehat{\mathbf{M}}$ with two texture zeroes.

First define a new mass matrix $\overline{\mathbf{M}}_\nu{}'$ obtained from $\overline{\mathbf{M}}_\nu$ by a $\frac{\pi}{4}$ rotation

$$\mathcal{U}_{\frac{\pi}{4}} = \begin{pmatrix} \frac{1}{\sqrt{2}} & 0 & \frac{1}{\sqrt{2}} \\ -\frac{1}{\sqrt{2}} & 0 & \frac{1}{\sqrt{2}} \\ 0 & 1 & 0 \end{pmatrix}, \tag{36}$$

through the similarity transformation $\overline{\mathbf{M}}_\nu{}' = \mathcal{U}_{\frac{\pi}{4}}^\dagger \overline{\mathbf{M}}_\nu \mathcal{U}_{\frac{\pi}{4}}$.Then, the matrix $\overline{\mathbf{M}}_\nu{}'$ can be written in the following form:

$$\overline{\mathbf{M}}_\nu{}' = \mu_0 \mathbb{I}_{3\times 3} + \widehat{\mathbf{M}} \tag{37}$$

where $\mathbb{I}_{3\times 3}$ is the identity matrix,

$$\mu_0 = a - d = \frac{2\left|\mu_2^\nu\right|^2}{|M|}\left(1 - |\lambda|\right) \qquad \text{and} \qquad \widehat{\mathbf{M}} = \begin{pmatrix} 0 & A & 0 \\ A & B & C \\ 0 & C & D \end{pmatrix} \tag{38}$$

with $A = \frac{b-e}{\sqrt{2}}$, $B = c + d - a$, $C = \frac{b+e}{\sqrt{2}}$ and $D = 2d$. As mentioned before the diagonalization of $\overline{\mathbf{M}}_\nu$ is reduced to diagonalization of the real symmetric matrix $\widehat{\mathbf{M}}$, which is a matrix with two texture zeroes of class I [25].

In fact, we can obtain a more general result by means of the similarity transformation, $\mathbf{M}' = \mathcal{R}_{\frac{\pi}{4}}^\dagger \mathbf{M} \mathcal{R}_{\frac{\pi}{4}}$, to applied to the general form of the mass matrices given in eq. (14), *i. e.*,

$$\mathbf{M}' = \mathcal{R}_{\frac{\pi}{4}}^\dagger \mathbf{M} \mathcal{R}_{\frac{\pi}{4}} = \widetilde{\mu}\mathbb{I}_{3\times 3} + \widetilde{M} \tag{39}$$

where $\widetilde{\mu} = \mu_2 - \mu_1$,

$$\mathcal{R}_{\frac{\pi}{4}} = \begin{pmatrix} \frac{1}{\sqrt{2}} & \frac{1}{\sqrt{2}} & 0 \\ -\frac{1}{\sqrt{2}} & \frac{1}{\sqrt{2}} & 0 \\ 0 & 0 & 1 \end{pmatrix} \qquad \text{and} \qquad \widetilde{M} = \begin{pmatrix} 0 & \mu_2 & 0 \\ \mu_2 & 2\mu_1 & \sqrt{2}\mu_5 \\ 0 & \sqrt{2}\mu_4 & \mu_3 + \mu_1 - \mu_2 \end{pmatrix}. \tag{40}$$

The basis in which the neutrino mass matrix takes the form (40) is a basis in the space of fermion flavors in which the mass matrices have the same form with two texture zeros for all fermions in the theory. However, this property does not imply special relations among mass eigenvalues and flavor mixing parameters. [36].

It is well know [37, 38] that in the hadronic sector the masses of quarks may be obtained from a matrix with two texture zeroes which succesfully reproduces the strong mass hierarchy of up and down type quarks. Also, the numerical values of the quark mixing angles determined in this framework are in good agreement with the experimental data [26]. Additionally, in a unified treatment in which the mass matrices of all fermions have a similar form with two texture zeroes class I and a normal hierarchy, the numerical values obtained for masses and mixing of the neutrinos are in very good agreement with all available experimental data [26, 23]. Therefore, we it is to be expected that the mixing matrix $PMNS$ that will be obtained from the mass matrices $\mathbf{M}_\nu$ and M_e could reproduce the current experimental data of the masses and mixings in the leptonic sector of the theory.

As in the case of the charged leptons, the matrices $\mathbf{M}_\nu$ and U_ν can be reparametrized in terms of the neutrino masses. For this we use the information that we already have about the diagonalization of a matrix with two texture zeroes of class I [25, 26, 37, 38]. Then, the mass matrix $\mathbf{M}_\nu$ for a Normal [Inverted] hierarchy in the mass spectrum takes the form

$$\mathbf{M}_\nu^{\mathbf{N[I]}} = \begin{pmatrix} \mu_0 + d & d & \frac{1}{\sqrt{2}}\left(C^{N[I]} + A^{N[I]}\right) \\ d & \mu_0 + d & \frac{1}{\sqrt{2}}\left(C^{N[I]} - A^{N[I]}\right) \\ \frac{1}{\sqrt{2}}\left(C^{N[I]} + A^{N[I]}\right) & \frac{1}{\sqrt{2}}\left(C^{N[I]} - A^{N[I]}\right) & m_{\nu_1} + m_{\nu_2} + m_{\nu_3} - 2(\mu_0 + d) \end{pmatrix} \tag{41}$$

where

$$A^{N[I]} = \sqrt{\frac{(m_{\nu_2} - \mu_0)\left(m_{\nu_{3[1]}} - \mu_0\right)\left(\mu_0 - m_{\nu_{1[3]}}\right)}{2d}}, \tag{42}$$

$$C^{N[I]} = \sqrt{\frac{(2d + \mu_0 - m_{\nu_1})\left(2d + \mu_0 - m_{\nu_{2[3]}}\right)\left(m_{\nu_{3[2]}} - \mu_0 - 2d\right)}{2d}}. \tag{43}$$

Also, the values allowed for the parameters μ_0 and $2d + \mu_0$ are in the following ranges

$$m_{\nu_{2[1]}} > \mu_0 > m_{\nu_{1[3]}} \quad \text{and} \quad m_{\nu_{3[2]}} > 2d + \mu_0 > m_{\nu_{2[1]}}. \tag{44}$$

Now, the unitary matrix U_ν takes the following form:

$$U_\nu = \begin{pmatrix} 1 & 0 & 0 \\ 0 & 1 & 0 \\ 0 & 0 & e^{i\delta_\nu} \end{pmatrix} \begin{pmatrix} \frac{1}{\sqrt{2}} & 0 & \frac{1}{\sqrt{2}} \\ -\frac{1}{\sqrt{2}} & 0 & \frac{1}{\sqrt{2}} \\ 0 & 1 & 0 \end{pmatrix} \begin{pmatrix} O_{11}^{N[I]} & O_{12}^{N[I]} & O_{13}^{N[I]} \\ O_{21}^{N[I]} & O_{22}^{N[I]} & O_{23}^{N[I]} \\ O_{31}^{N[I]} & O_{32}^{N[I]} & O_{33}^{N[I]} \end{pmatrix} \tag{45}$$

where $\delta_\nu = \phi_2 - \phi_4$ and $\mathbb{O}^{N[I]}$ is the matriz that diagonalizes $\widehat{\mathbf{M}}$. Explicit expresions for the elements of the orthogonal matrix $\mathbb{O}^{N[I]}$ reparametrized in terms of the neutrino masses are given in table 2, but in order to avoid a confused notation the dependence of $O_{ij}^{N[I]}$ on neutrinos masses is given in terms of the expresions:

$$\mathcal{D}_1^{N[I]} = 2d\left(m_{\nu_2} - m_{\nu_1}\right)\left(m_{\nu_{3[1]}} - m_{\nu_{1[3]}}\right), \qquad \mathcal{D}_2^{N[I]} = 2d\left(m_{\nu_2} - m_{\nu_1}\right)\left(m_{\nu_{3[2]}} - m_{\nu_{2[3]}}\right),$$
$$\mathcal{D}_3^{N[I]} = 2d\left(m_{\nu_{3[1]}} - m_{\nu_{1[3]}}\right)\left(m_{\nu_{3[2]}} - m_{\nu_{2[3]}}\right), \quad f_1 = \left(2d + \mu_0 - m_{\nu_1}\right),$$
$$f_2^{N[I]} = [-1]\left(2d + \mu_0 - m_{\nu_2}\right), \qquad f_3^{N[I]} = [-1]\left(m_{\nu_3} - \mu_0 - 2d\right). \tag{46}$$

see table 2. The superscripts N and I denote the normal and inverted hierarchy respectively.

4. The neutrino mixing matrix

The neutrino mixing matrix V_{PMNS}, is the product $U_{eL}^{\dagger}U_\nu K$, where K is the diagonal matrix of the Majorana phase factors, defined by $K = diag(1, e^{i\alpha}, e^{i\beta})$. Now, with the help of eqs. (23), (27) and (45), we obtain the theoretical expression of the elements of the lepton mixing matrix, V_{PMNS}^{th}. This expression has the following form:

$$V_{PMNS}^{th} = \begin{pmatrix} V_{e1}^{th} & V_{e2}^{th}e^{i\alpha} & V_{e3}^{th}e^{i\beta} \\ V_{\mu 1}^{th} & V_{\mu 2}^{th}e^{i\alpha} & V_{\mu 3}^{th}e^{i\beta} \\ V_{\tau 1}^{th} & V_{\tau 2}^{th}e^{i\alpha} & V_{\tau 3}^{th}e^{i\beta} \end{pmatrix} \tag{47}$$

where

$$V_{e1}^{th} = \tfrac{\tilde{m}_e}{\tilde{m}_\mu}O_{11}^{N[I]} - O_{21}^{N[I]}e^{i\delta_l}, \quad V_{e2}^{th} = \tfrac{\tilde{m}_e}{\tilde{m}_\mu}O_{12}^{N[I]} - O_{22}^{N[I]}e^{i\delta_l},$$
$$V_{e3}^{th} = \tfrac{\tilde{m}_e}{\tilde{m}_\mu}O_{13}^{N[I]} - O_{23}^{N[I]}e^{i\delta_l}, \quad V_{\mu 1}^{th} = -O_{11}^{N[I]} - \tfrac{\tilde{m}_e}{\tilde{m}_\mu}O_{21}^{N[I]}e^{i\delta_l},$$
$$V_{\mu 2}^{th} = -O_{12}^{N[I]} - \tfrac{\tilde{m}_e}{\tilde{m}_\mu}O_{22}^{N[I]}e^{i\delta_l}, \quad V_{\mu 3}^{th} = -O_{13}^{N[I]} - \tfrac{\tilde{m}_e}{\tilde{m}_\mu}O_{23}^{N[I]}e^{i\delta_l},$$
$$V_{\tau 1}^{th} = O_{31}^{N[I]}, \quad V_{\tau 2}^{th} = O_{32}^{N[I]}, \quad V_{\tau 3}^{th} = O_{33}^{N[I]} \tag{48}$$

with $\delta_l = \delta_\nu - \delta_e$, the elements O^e and $O^{N[I]}$ in the eqs. (48) are given in the eq. (24) and the table 2, respectively.

4.1. The Mixing Angles

In the standard PDG parametrization, the entries in the lepton mixing matrix are parametrized in terms of the mixing angles and phases. Thus, the mixing angles are related to the observable moduli of lepton V_{PMNS} through the relations:

$$\sin^2\theta_{12}^l = \tfrac{|V_{e2}|^2}{1-|V_{e3}|^2}, \quad \sin^2\theta_{23}^l = \tfrac{|V_{\mu 3}|^2}{1-|V_{e3}|^2}, \quad \sin^2\theta_{13}^l = |V_{e3}|^2. \tag{49}$$

Then, theoretical expression for the lepton mixing angles as functions of the lepton mass ratios are readily obtained when the theoretical expressions for the modulii of the entries in the $PMNS$ mixing matrix, given in eqs. (48), are substituted for $|V_{ij}|$ in the right hand side of eqs. (49).

$$\sin^2\theta_{12}^l = \frac{\left(\frac{\tilde{m}_e}{\tilde{m}_\mu}\right)^2\left(O_{12}^{N[I]}\right)^2 + \left(O_{22}^{N[I]}\right)^2 - 2\frac{\tilde{m}_e}{\tilde{m}_\mu}O_{12}^{N[I]}O_{22}^{N[I]}\cos\delta_l}{1 - \left(\frac{\tilde{m}_e}{\tilde{m}_\mu}\right)^2\left(O_{13}^{N[I]}\right)^2 - \left(O_{23}^{N[I]}\right)^2 + 2\frac{\tilde{m}_e}{\tilde{m}_\mu}O_{13}^{N[I]}O_{23}^{N[I]}\cos\delta_l} \tag{50}$$

Set	Elements O_{ij}
I	$O_{11}^{N[I]} = -\sqrt{\frac{[-1]\left(m_{\nu_3}-\mu_0\right)\left(m_{\nu_2}-\mu_0\right)f_1}{\mathcal{D}_1^{N[I]}}}$ $\quad O_{12}^{N[I]} = \sqrt{\frac{\left(m_{\nu_{3[1]}}-\mu_0\right)\left(\mu_0-m_{\nu_{1[3]}}\right)f_2^{N[I]}}{\mathcal{D}_2^{N[I]}}}$ $O_{13}^{N[I]} = \sqrt{\frac{[-1]\left(\mu_0-m_{\nu_1}\right)\left(m_{\nu_2}-\mu_0\right)f_3^{N[I]}}{\mathcal{D}_3^{N[I]}}}$ $\quad O_{21}^{N[I]} = \sqrt{\frac{[-1]2d\left(\mu_0-m_{\nu_1}\right)f_1}{\mathcal{D}_1^{N[I]}}}$ $O_{22}^{N[I]} = \sqrt{\frac{2d\left(m_{\nu_2}-\mu_0\right)f_2^{N[I]}}{\mathcal{D}_2^{N[I]}}}$ $\quad O_{23}^{N[I]} = \sqrt{\frac{[-1]2d\left(m_{\nu_3}-\mu_0\right)f_3^{N[I]}}{\mathcal{D}_3^{N[I]}}}$ $O_{31}^{N[I]} = -\sqrt{\frac{[-1]\left(\mu_0-m_{\nu_1}\right)f_2^{N[I]}f_3^{N[I]}}{\mathcal{D}_1^{N[I]}}}$ $\quad O_{32}^{N[I]} = -1[+1]\sqrt{\frac{\left(m_{\nu_2}-\mu_0\right)f_1 f_3^{N[I]}}{\mathcal{D}_1^{N[I]}}}$ $O_{33}^{N[I]} = [-1]\sqrt{\frac{\left(m_{\nu_3}-\mu_0\right)f_1 f_2^{N[I]}}{\mathcal{D}_1^{N[I]}}}$
II	$O_{11}^{N[I]} = \sqrt{\frac{[-1]\left(m_{\nu_3}-\mu_0\right)\left(m_{\nu_2}-\mu_0\right)f_1}{\mathcal{D}_1^{N[I]}}}$ $\quad O_{12}^{N[I]} = -\sqrt{\frac{\left(m_{\nu_{3[1]}}-\mu_0\right)\left(\mu_0-m_{\nu_{1[3]}}\right)f_2^{N[I]}}{\mathcal{D}_2^{N[I]}}}$ $O_{13}^{N[I]} = \sqrt{\frac{[-1]\left(\mu_0-m_{\nu_1}\right)\left(m_{\nu_2}-\mu_0\right)f_3^{N[I]}}{\mathcal{D}_3^{N[I]}}}$ $\quad O_{21}^{N[I]} = \sqrt{\frac{[-1]2d\left(\mu_0-m_{\nu_1}\right)f_1}{\mathcal{D}_1^{N[I]}}}$ $O_{22}^{N[I]} = \sqrt{\frac{2d\left(m_{\nu_2}-\mu_0\right)f_2^{N[I]}}{\mathcal{D}_2^{N[I]}}}$ $\quad O_{23}^{N[I]} = \sqrt{\frac{[-1]2d\left(m_{\nu_3}-\mu_0\right)f_3^{N[I]}}{\mathcal{D}_3^{N[I]}}}$ $O_{31}^{N[I]} = -1[+1]\sqrt{\frac{[-1]\left(\mu_0-m_{\nu_1}\right)f_2^{N[I]}f_3^{N[I]}}{\mathcal{D}_1^{N[I]}}}$ $\quad O_{32}^{N[I]} = -\sqrt{\frac{\left(m_{\nu_2}-\mu_0\right)f_1 f_3^{N[I]}}{\mathcal{D}_1^{N[I]}}}$ $O_{33}^{N[I]} = [-1]\sqrt{\frac{\left(m_{\nu_3}-\mu_0\right)f_1 f_2^{N[I]}}{\mathcal{D}_1^{N[I]}}}$
III	$O_{11}^{N[I]} = \sqrt{\frac{[-1]\left(m_{\nu_3}-\mu_0\right)\left(m_{\nu_2}-\mu_0\right)f_1}{\mathcal{D}_1^{N[I]}}}$ $\quad O_{12}^{N[I]} = \sqrt{\frac{\left(m_{\nu_{3[1]}}-\mu_0\right)\left(\mu_0-m_{\nu_{1[3]}}\right)f_2^{N[I]}}{\mathcal{D}_2^{N[I]}}}$ $O_{13}^{N[I]} = -\sqrt{\frac{[-1]\left(\mu_0-m_{\nu_1}\right)\left(m_{\nu_2}-\mu_0\right)f_3^{N[I]}}{\mathcal{D}_3^{N[I]}}}$ $\quad O_{21}^{N[I]} = \sqrt{\frac{[-1]2d\left(\mu_0-m_{\nu_1}\right)f_1}{\mathcal{D}_1^{N[I]}}}$ $O_{22}^{N[I]} = \sqrt{\frac{2d\left(m_{\nu_2}-\mu_0\right)f_2^{N[I]}}{\mathcal{D}_2^{N[I]}}}$ $\quad O_{23}^{N[I]} = \sqrt{\frac{[-1]2d\left(m_{\nu_3}-\mu_0\right)f_3^{N[I]}}{\mathcal{D}_3^{N[I]}}}$ $O_{31}^{N[I]} = -\sqrt{\frac{[-1]\left(\mu_0-m_{\nu_1}\right)f_2^{N[I]}f_3^{N[I]}}{\mathcal{D}_1^{N[I]}}}$ $\quad O_{32}^{N[I]} = \sqrt{\frac{\left(m_{\nu_2}-\mu_0\right)f_1 f_3^{N[I]}}{\mathcal{D}_1^{N[I]}}}$ $O_{33}^{N[I]} = -\sqrt{\frac{\left(m_{\nu_3}-\mu_0\right)f_1 f_2^{N[I]}}{\mathcal{D}_1^{N[I]}}}$

Table 2. Sets of $O_{ij}^{N[I]}$. The superscripts N and I denote the normal and inverted hierarchy respectively.

$$\sin^2\theta_{23}^l = \frac{\left(O_{13}^{N[I]}\right)^2 + \left(\frac{\tilde{m}_e}{\tilde{m}_\mu}\right)^2\left(O_{23}^{N[I]}\right)^2 + 2\frac{\tilde{m}_e}{\tilde{m}_\mu}O_{13}^{N[I]}O_{23}^{N[I]}\cos\delta_l}{1 - \left(\frac{\tilde{m}_e}{\tilde{m}_\mu}\right)^2\left(O_{13}^{N[I]}\right)^2 - \left(O_{23}^{N[I]}\right)^2 + 2\frac{\tilde{m}_e}{\tilde{m}_\mu}O_{13}^{N[I]}O_{23}^{N[I]}\cos\delta_l} \tag{51}$$

$$\sin^2\theta_{13}^l = \left(\frac{\tilde{m}_e}{\tilde{m}_\mu}\right)^2\left(O_{13}^{N[I]}\right)^2 + \left(O_{23}^{N[I]}\right)^2 - 2\frac{\tilde{m}_e}{\tilde{m}_\mu}O_{13}^{N[I]}O_{23}^{N[I]}\cos\delta_l \tag{52}$$

where, as before, $\tilde{m}_\mu = m_\mu/m_\tau$ and $\tilde{m}_e = m_e/m_\tau$.

Journal of Physics: Conference Series **378** (2012) 012014 doi:10.1088/1742-6596/378/1/012014

In a first, preliminary analysis for the mixing angle θ^l_{13} and for an inverted neutrino mass hierarchy ($m_{\nu_2} > m_{\nu_1} > m_{\nu_3}$) the eq. (52) takes the form:

$$\sin^2\theta^l_{13} \approx \frac{(\mu_0 + 2d - m_{\nu 3})(\mu_0 - m_{\nu_3})}{(m_{\nu_1} - m_{\nu_3})(m_{\nu_2} - m_{\nu_3})}. \tag{53}$$

Now, with the following values for neutrino masses $m_{\nu_2} = 0.056eV$, $m_{\nu_1} = 0.053eV$ and $m_{\nu_3} = 0.048eV$, and the parameters values $\delta_l = \pi/2$, $\mu_0 = 0.049$ and $d = 8\times 10^{-5}$ we get $\sin^2\theta^l_{13} \approx 0.029 \longrightarrow \theta^l_{13} \approx 9.8°$. In a more complete analysis, where will be make a χ^2 fit of the exact theoretical expressions for the modulii of the entries of the lepton mixing matrix $|(V^{th}_{PMNS})_{ij}|$ to the experimental values (for example the values given by Gonzalez-Garcia [14]) will be considered in detail elsewhere.

5. Flavour Changing Neutral Currents (FCNC)

Models with more than one Higgs $SU(2)$ doublet have tree level flavour changing neutral currents. In the Minimal S_3-invariant Extension of the Standard Model considered here, there is one Higgs $SU(2)$ doublet per generation coupling to all fermions. The flavour changing Yukawa couplings may be written in a flavour labelled, symmetry adapted weak basis as

$$\begin{aligned}\mathcal{L}_Y^{\rm FCNC} &= \left(\overline{E}_{aL}Y^{ES}_{ab}E_{bR} + \overline{U}_{aL}Y^{US}_{ab}U_{bR} + \overline{D}_{aL}Y^{DS}_{ab}D_{bR}\right)H^0_S \\ &+ \left(\overline{E}_{aL}Y^{E1}_{ab}E_{bR} + \overline{U}_{aL}Y^{U1}_{ab}U_{bR} + \overline{D}_{aL}Y^{D1}_{ab}D_{bR}\right)H^0_1 + \\ &\left(\overline{E}_{aL}Y^{E2}_{ab}E_{bR} + \overline{U}_{aL}Y^{U2}_{ab}U_{bR} + \overline{D}_{aL}Y^{D2}_{ab}D_{bR}\right)H^0_2 + \text{h.c.}\end{aligned} \tag{54}$$

The Yukawa couplings of immediate physical interest in the computation of the flavour changing neutral currents are those defined in the mass basis, according to $\tilde{Y}^{EI}_m = U^\dagger_{eL}Y^{EI}_w U_{eR}$, where U_{eL} and U_{eR} are the matrices that diagonalize the charged lepton mass matrix defined in eqs. (16). We obtain [30]

$$\tilde{Y}^{E1}_m \approx \frac{m_\tau}{v_1}\begin{pmatrix} 2\tilde{m}_e & -\frac{1}{2}\tilde{m}_e & \frac{1}{2}x \\ -\tilde{m}_\mu & \frac{1}{2}\tilde{m}_\mu & -\frac{1}{2} \\ \frac{1}{2}\tilde{m}_\mu x^2 & -\frac{1}{2}\tilde{m}_\mu & \frac{1}{2}\end{pmatrix}_m \quad\text{and}\quad \tilde{Y}^{E2}_m \approx \frac{m_\tau}{v_2}\begin{pmatrix} -\tilde{m}_e & \frac{1}{2}\tilde{m}_e & -\frac{1}{2}x \\ \tilde{m}_\mu & \frac{1}{2}\tilde{m}_\mu & \frac{1}{2} \\ -\frac{1}{2}\tilde{m}_\mu x^2 & \frac{1}{2}\tilde{m}_\mu & \frac{1}{2}\end{pmatrix}_m, \tag{55}$$

where $\tilde{m}_\mu = 5.94\times 10^{-2}$, $\tilde{m}_e = 2.876\times 10^{-4}$ and $x = m_e/m_\mu = 4.84\times 10^{-3}$. All the non-diagonal elements are responsible for tree-level FCNC processes. If the S'_2 symmetry in the Higgs sector is preserved [34], $\langle H^0_1\rangle = \langle H^0_2\rangle = v$.

The amplitude of the flavour violating process $\mu \to 3e$, is proportional to $\tilde{Y}^E_{\mu e}\tilde{Y}^E_{ee}$ [39]. Then, the leptonic branching ratio,

$$Br(\mu\to 3e) = \frac{\Gamma(\mu\to 3e)}{\Gamma(\mu\to e\nu\bar{\nu})} \quad\text{and}\quad \Gamma(\mu\to 3e) \approx \frac{m^5_\mu}{3\times 2^{10}\pi^3}\frac{\left(Y^{1,2}_{\mu e}Y^{1,2}_{ee}\right)^2}{M^4_{H_{1,2}}}, \tag{56}$$

which is the dominant term, and the well known expression for $\Gamma(\mu\to e\nu\bar{\nu})$ [40], give

$$Br(\mu\to 3e) \approx 2(2+\tan^2\beta)^2\left(\frac{m_e m_\mu}{m^2_\tau}\right)^2\left(\frac{m_\tau}{M_H}\right)^4, \tag{57}$$

Table 3. Leptonic FCNC processes, calculated with $M_{H_{1,2}} \sim 120\ GeV$.

FCNC processes	Theoretical BR	Experimental upper bound BR	References
$\tau \to 3\mu$	8.43×10^{-14}	2×10^{-7}	B. Aubert *et. al.* [42]
$\tau \to \mu e^+ e^-$	3.15×10^{-17}	2.7×10^{-7}	B. Aubert *et. al.* [42]
$\tau \to \mu\gamma$	9.24×10^{-15}	6.8×10^{-8}	B. Aubert *et. al.* [43]
$\tau \to e\gamma$	5.22×10^{-16}	1.1×10^{-11}	B. Aubert *et. al.* [44]
$\mu \to 3e$	2.53×10^{-16}	1×10^{-12}	U. Bellgardt *et al.* [41]
$\mu \to e\gamma$	2.42×10^{-20}	1.2×10^{-11}	M. L. Brooks *et al.* [45]

taking for $M_H \approx 120\ GeV$ and $\tan\beta = 1$ we obtain $Br(\mu \to 3e) = 2.53 \times 10^{-16}$, well below the experimental upper bound for this process, which is 1×10^{-12} [41].

Similar computations give the numerical estimates of the branching ratios for some others flavour violating processes in the leptonic sector. These results, and the corresponding experimental upper bounds are shown in Table 3. In all cases considered, the theoretical estimations made in the framework of the minimal S_3-invariant extension of the SM are well below the experimental upper bounds [30].

6. Muon anomalous magnetic moment

In the minimal S_3-invariant extension of the Standard Model we are considering here, we have three Higgs $SU(2)$ doublets, one in the singlet and the other two in the doublet representations of the S_3 flavour group. The Z_2 symmetry decouples the charged leptons from the Higgs boson in the S_3 singlet representation. Therefore, in the leading order of perturbation theory there are two neutral scalars and two neutral pseudoscalars whose exchange will contribute to the anomalous magnetic moment of the muon. Since the heavier generations have larger flavour-changing couplings, the largest contribution comes from the heaviest charged leptons coupled to the lightest of the neutral Higgs bosons.

A straightforward computation gives

$$\delta a_\mu^{(H)} = \frac{Y_{\mu\tau} Y_{\tau\mu}}{16\pi^2} \frac{m_\mu m_\tau}{M_H^2} \left(log \left(\frac{M_H^2}{m_\tau^2} \right) - \frac{3}{2} \right). \tag{58}$$

With the help of (55) we may write $\delta a_\mu^{(H)}$ as

$$\delta a_\mu^{(H)} = \frac{m_\tau^2}{(246\ GeV)^2} \frac{(2 + \tan^2\beta)}{32\pi^2} \frac{m_\mu^2}{M_H^2} \left(log \left(\frac{M_H^2}{m_\tau^2} \right) - \frac{3}{2} \right), \tag{59}$$

Taking again $M_H = 120\ GeV$ and the upper bound for $\tan\beta = 14$ gives an estimate of the largest possible contribution of the FCNC to the anomaly of the muon's magnetic moment $\delta a_\mu^{(H)} \approx 1.7 \times 10^{-10}$. This number has to be compared with the difference between the experimental value and the Standard Model prediction for the anomaly of the muon's magnetic moment [46]

$$\Delta a_\mu = a_\mu^{exp} - a_\mu^{SM} = (28.7 \pm 9.1) \times 10^{-10}, \tag{60}$$

which means

$$\frac{\delta a_\mu^{(H)}}{\Delta a_\mu} \approx 0.06. \tag{61}$$

Hence, the contribution of the flavour changing neutral currents to the anomaly of the magnetic moment of the muon is smaller than or of the order of 6% of the discrepancy between the experimental value and the Standard Model prediction.

7. Conclusions

A well defined structure of the Yukawa couplings is obtained, which permits the calculations of mass and mixings matrices for quark and leptons in a unified way. A further reduction of redundant parameters is achieved in the leptonic sector by introducing a Z_2 symmetry. The flavour symmetry group $S_3 \times Z_2$ relates the nuetrino mass spectrum and mixings. This allowed us to compute the neutrino mixing matrix, V_{PMNS}, explicitly in terms of the masses of the charged leptons and neutrinos and one phase δ_l. In this model, we obtained a general result for mass matrices of all femions in the theory. This result is that through of a similarity transformation we can written all mass matrices of fermions in the same generic form with two texture zeros. We also found that V_{PMNS} has three CP-violating phases, namely, one Dirac phase $\delta_l = \delta_\nu - \delta_e$ and two Majorana phases, α and β, which are functions of the neutrino and lepton masses. The numerical values of the reactor, θ_{13}, mixing angle is determined by the masses of the neutirnos which have an inverted hierarchy, with the values $|m_{\nu_2}| = 0.056\ eV$, $|m_{\nu_1}| = 0.053\ eV$ and $|m_{\nu_3}| = 0.048\ eV$, obtained $\theta^l_{13} \approx 9.8^\circ$ in agreement with the latest analysis of the experimental data on neutrino oscillations and mixings. We also obtained explicit expressions for the matrices of the Yukawa couplings of the lepton sector parametrized in terms of the charged lepton masses and the VEV's of the neutral Higgs bosons in the S_3-doublet representation. These Yukawa matrices are closely related to the fermion mass matrices and have a structure of small and very small entries reflecting the observed charged lepton mass hierarchy. With the help of the Yukawa matrices, we computed the branching ratios of a number of FCNC processes and found that the branching ratios of all FCNC processes considered here are strongly suppressed by powers of the small mass ratios m_e/m_τ and m_μ/m_τ, and by the ratio $(m_\tau/M_{H_{1,2}})^4$, where $M_{H_{1,2}}$ is the mass of the neutral Higgs bosons in the S_3-doublet. Taking for $M_{H_{1,2}}$ a very conservative value ($M_{H_{1,2}} \approx 120\ GeV$), we found that the numerical values of the branching ratios of the FCNC in the leptonic sector are well below the corresponding experimental upper bounds by many orders of magnitude. It has already been argued that small FCNC processes mediating non-standard quark-neutrino interactions could be important in the theoretical description of the gravitational core collapse and shock generation in the explosion stage of a supernova [50, 51, 52]. Finally, the contribution of the flavour changing neutral currents to the anomalous magnetic moment of the muon is small but non-negligible, and it is compatible with the best, state of the art measurements and theoretical computations.

8. Acknowledgements

We thank Prof. Jens Erler and Dr. Genaro Toledo-Sánchez for helpful discussions about $g-2$. This work was partially supported by DGAPA-UNAM under contract PAPIIT-IN113712-3.

References

[1] Jung C K, McGrew C, Kajita T and Mann T 2001 *Ann. Rev. Nucl. Part. Sci.* **51** 451–488
[2] Mohapatra R N and Smirnov A Y 2006 *Ann. Rev. Nucl. Part. Sci.* **56** 569–628 (*Preprint* hep-ph/0603118)
[3] Altmann M *et al.* (GNO) 2005 *Phys. Lett.* **B616** 174–190 (*Preprint* hep-ex/0504037)
[4] Smy M B *et al.* (Super-Kamiokande) 2004 *Phys. Rev.* **D69** 011104 (*Preprint* hep-ex/0309011)
[5] Ahmad Q R *et al.* (SNO) 2002 *Phys. Rev. Lett.* **89** 011302 (*Preprint* nucl-ex/0204009)
[6] Ahmad Q R *et al.* (SNO) 2002 *Phys. Rev. Lett.* **89** 011301 (*Preprint* nucl-ex/0204008)
[7] Aharmim B *et al.* (SNO) 2005 *Phys. Rev.* **C72** 055502 (*Preprint* nucl-ex/0502021)
[8] Fukuda S *et al.* (Super-Kamiokande) 2002 *Phys. Lett.* **B539** 179–187 (*Preprint* hep-ex/0205075)
[9] Ashie Y *et al.* (Super-Kamiokande) 2005 *Phys. Rev.* **D71** 112005 (*Preprint* hep-ex/0501064)
[10] Bemporad C, Gratta G and Vogel P 2002 *Rev. Mod. Phys.* **74** 297 (*Preprint* hep-ph/0107277)

[11] Araki T *et al.* (KamLAND) 2005 *Phys. Rev. Lett.* **94** 081801 (*Preprint* hep-ex/0406035)
[12] Maltoni M, Schwetz T, Tortola M A and Valle J W F 2004 *New J. Phys.* **6** 122 (*Preprint* hep-ph/0405172)
[13] Schwetz T 2005 *Acta Phys. Polon.* **B36** 3203–3214 (*Preprint* hep-ph/0510331)
[14] Gonzalez-Garcia M C and Maltoni M 2008 *Phys. Rept.* **460** 1–129 (*Preprint* 0704.1800)
[15] Apollonio M *et al.* (CHOOZ) 2003 *Eur. Phys. J.* **C27** 331–374 (*Preprint* hep-ex/0301017)
[16] K and Eitel 2005 *Nuclear Physics B - Proceedings Supplements* **143** 197 – 204 ISSN 0920-5632 URL http://www.sciencedirect.com/science/article/pii/S0920563205001076
[17] Elliott S R and Engel J 2004 *J. Phys.* **G30** R183 (*Preprint* hep-ph/0405078)
[18] Seljak U, Slosar A and McDonald P 2006 *JCAP* **0610** 014 (*Preprint* astro-ph/0604335)
[19] Elgaroy O and Lahav O 2005 *New J. Phys.* **7** 61 (*Preprint* hep-ph/0412075)
[20] Lesgourgues J and Pastor S 2006 *Phys. Rept.* **429** 307–379 (*Preprint* astro-ph/0603494)
[21] Schwetz T, Tortola M and Valle J W F 2011 *New J. Phys.* **13** 109401 (*Preprint* 1108.1376)
[22] Fritzsch H and Xing Z z 2000 *Prog. Part. Nucl. Phys.* **45** 1–81 (*Preprint* hep-ph/9912358)
[23] Xing Z z and Zhang H 2003 *Phys. Lett.* **B569** 30–40 (*Preprint* hep-ph/0304234)
[24] Ishimori H *et al.* 2010 *Prog. Theor. Phys. Suppl.* **183** 1–163 (*Preprint* 1003.3552)
[25] Gonzalez Canales F and Mondragon A 2011 *J. Phys. Conf. Ser.* **287** 012015 (*Preprint* 1101.3807)
[26] Barranco J, Gonzalez Canales F and Mondragon A 2010 *Phys. Rev.* **D82** 073010 (*Preprint* 1004.3781)
[27] Kubo J, Mondragon A, Mondragon M and Rodriguez-Jauregui E 2003 *Prog. Theor. Phys.* **109** 795–807 (*Preprint* hep-ph/0302196)
[28] Kubo J *et al.* 2005 *J. Phys. Conf. Ser.* **18** 380–384
[29] Felix O, Mondragon A, Mondragon M and Peinado E 2007 *AIP Conf. Proc.* **917** 383–389 (*Preprint* hep-ph/0610061)
[30] Mondragon A, Mondragon M and Peinado E 2007 *Phys. Rev.* **D76** 076003 (*Preprint* 0706.0354)
[31] Mondragon A, Mondragon M and Peinado E 2008 *J. Phys.* **A41** 304035 (*Preprint* 0712.1799)
[32] Mondragon A, Mondragon M and Peinado E 2008 *AIP Conf. Proc.* **1026** 164–169 (*Preprint* 0712.2488)
[33] Mondragon A, Mondragon M and Peinado E 2008 *Rev. Mex. Fis.* **S54N3** 81–91 (*Preprint* 0805.3507)
[34] Pakvasa S and Sugawara H 1978 *Phys. Lett.* **B73** 61
[35] Nakamura K *et al.* (Particle Data Group) 2010 *J. Phys.* **G37** 075021
[36] Fritzsch H and Xing Z Z 1997 *Phys. Lett.* **B413** 396–404 (*Preprint* hep-ph/9707215)
[37] Mondragon A and Rodriguez-Jauregui E 1999 *Phys. Rev.* **D59** 093009 (*Preprint* hep-ph/9807214)
[38] Mondragon A and Rodriguez-Jauregui E 2000 *Phys. Rev.* **D61** 113002 (*Preprint* hep-ph/9906429)
[39] Sher M and Yuan Y 1991 *Phys. Rev.* **D44** 1461–1472
[40] Yao W M *et al.* (Particle Data Group) 2006 *J. Phys.* **G33** 1–1232
[41] Bellgardt U *et al.* (SINDRUM) 1988 *Nucl. Phys.* **B299** 1
[42] Aubert B *et al.* (BABAR) 2004 *Phys. Rev. Lett.* **92** 121801 (*Preprint* hep-ex/0312027)
[43] Aubert B *et al.* (BABAR) 2005 *Phys. Rev. Lett.* **95** 041802 (*Preprint* hep-ex/0502032)
[44] Aubert B *et al.* (BABAR) 2006 *Phys. Rev. Lett.* **96** 041801 (*Preprint* hep-ex/0508012)
[45] Brooks M L *et al.* (MEGA) 1999 *Phys. Rev. Lett.* **83** 1521–1524 (*Preprint* hep-ex/9905013)
[46] Jegerlehner F 2007 *Acta Phys. Polon.* **B38** 3021 (*Preprint* hep-ph/0703125)
[47] Fogli G L, Lisi E, Marrone A, Palazzo A and Rotunno A M 2009 (*Preprint* 0905.3549)
[48] Roa J E, Latimer D C and Ernst D J 2010 *Phys. Rev.* **C81** 015501 (*Preprint* 0904.3930)
[49] Ge H L, Giunti C and Liu Q Y 2009 *Phys. Rev.* **D80** 053009 (*Preprint* 0810.5443)
[50] Raffelt G G 2007 (*Preprint* astro-ph/0701677)
[51] Amanik P S, Fuller G M and Grinstein B 2005 *Astropart. Phys.* **24** 160–182 (*Preprint* hep-ph/0407130)
[52] Esteban-Pretel A, Tomas R and Valle J W F 2007 *Phys. Rev.* **D76** 053001 (*Preprint* 0704.0032)

Journal of Physics: Conference Series **378** (2012) 012015 doi:10.1088/1742-6596/378/1/012015

The process $D^{\pm} \to K^{\mp}\pi^{\pm}\pi^{\pm}$

D. Delepine

Departamento de Física, DCI, Campus León, Universidad de Guanajuato, C.P. 37150, León, Guanajuato, México

E-mail: delepine@fisica.ugto.mx

Gaber Faisel

Department of Physics and Center for Mathematics and Theoretical Physics, National Central University, Chung-li, TAIWAN 32054.
and Egyptian Center for Theoretical Physics, Modern University for Information and Technology, Cairo, Egypt

E-mail: gfaisel@ncu.edu.tw

Carlos A. Ramirez

Escuela de Fisica, Universidad Industrial de Santander, Bucaramanga, Colombia.

E-mail: jpjdramirez@yahoo.com

Abstract. The main ingredients and the importance of the process $D^{\pm} \to K^{\mp}\pi^{\pm}\pi^{\pm}$ are described: its contribution to reveal the true nature of the scalar K_0^* particle, what we know about the parametrization of its dynamics, its importance for CPV in the charm physics and possible new physics contributions.

1. Introduction

The decay $D^{\pm} \to K^{\mp}\pi^{\pm}\pi^{\pm}$ is a Cabibbo Favored (CF [1]) process with large branching ratio (9.51(34) [2, 3]). It is dominated by the $K-\pi$ S-wave (7.3(2) % of the total), with two resonances: the $\kappa = K_0^*(800)$ and the $K_0^*(1430)$ (1.21(6) %). There is also a P-wave contribution by the $K^*(892)$ (1.1(1) %) and smaller components like the $\pi^+\pi^-$ ($I = 2$, 1.4(3) %), etc. That makes this processes much simpler than for example $D^{\pm} \to K_s^0\pi^{\pm}\pi^0$ where one has the additional $\pi^{\pm} - \pi^0$ resonant interaction with scalar ($\sigma = f_0(600)$, etc.) and vector (ρ, etc.) contributions. Given that the dominant contribution is the $\kappa = K_0^*(800)$, this is a good place (maybe the best available) to study it. The same can be said about the strangeness scalar form factor [4]. Additionally one may be interested to see if the Watson's theorem is violated [4, 5] and how, for the s-wave phase of the $K - \pi$ interaction [6]. Possible CP violations are interesting too, given that CPV has not been observed in any of the up quark sector.

The momenta involved in this process are defined as $p_D = p_K + p_1 + p_2 \equiv p_{K1} + p_2 \equiv p_{K2} + p_1 = p_K + p_{12}$. The Mandelstam variables and the partial decay rate are

$$
\begin{aligned}
s = p_{K1}^2 &= (p_K + p_1)^2,\ t = p_{K2}^2 = (p_K + p_2)^2,\ u = p_{12}^2 = (p_1 + p_2)^2 \\
& s + t + u = m_D^2 + m_K^2 + 2m_\pi^2 \\
\mathrm{d}\Gamma &= = \frac{|\mathcal{M}|^2}{256\pi^3 m_D^3}\mathrm{d}s\mathrm{d}t = \frac{|\mathcal{M}|^2\sqrt{s}}{64\pi^3 m_D^3}|\mathbf{p_K}'||\mathbf{p_2}'|\mathrm{d}\sqrt{s}\ \mathrm{d}\cos\theta' \qquad (1)
\end{aligned}
$$

In order to obtain the amplitude $\mathcal{M}$ one has to use the effective hamiltonian [1]

$$
\mathcal{H} = \frac{G_F}{\sqrt{2}}V_{cs}^* V_{ud}\left(c_1 \bar{s}c_L \bar{u}d_L + c_2 \bar{u}c_L \bar{s}d_L\right) + \mathrm{H.C.} = \frac{G_F}{\sqrt{2}}V_{cs}^* V_{ud}\left(c_1\mathcal{O}_1 + c_2\mathcal{O}_2\right) + \mathrm{H.C.} \qquad (2)
$$

with $a_1 = 1.2(1)$ and $a_2 = -0.5(1)$ [1]. In this work we are going to neglect three body [7] and pion-pion final state interactions [6, 8] given that they are small. Thus the amplitudes can be reduced to form factors and decay constants as follows [9]

$$
\begin{aligned}
\mathcal{M} &= \frac{G_F}{\sqrt{2}}V_{cs}V_{ud}^*\left[c_1 < K^-\pi^+\pi^+|\bar{s}c\bar{u}d|D^+ > + c_2 < K^-\pi^+\pi^+|\bar{u}c\bar{s}d|D^+ >\right] \\
&= \frac{G_F}{\sqrt{2}}V_{cs}V_{ud}^*\left[a_1 < K^-\pi_1^+|\bar{s}c|D^+ >< \pi_2^+|\bar{u}d|0 > + a_2 < K^-\pi_1^+|\bar{s}d|0 >< \pi_2^+|\bar{u}c||D^+ >\right] + (\pi_1 \leftrightarrow \pi_2) \\
&\equiv \frac{G_F}{\sqrt{2}}V_{cs}V_{ud}^*\left[a_1 A_1(s,\ t) + a_2 A_2(s,\ t)\right] + (s \leftrightarrow t) \qquad (3)
\end{aligned}
$$

with

$$
\begin{aligned}
A_2 &= \left[t - u - \frac{\Delta_{D\pi}^2\Delta_{K\pi}^2}{m_{K1}^2}\right]F_+^{D\pi}(s)F_+^{K\pi}(s) + \frac{\Delta_{D\pi}^2\Delta_{K\pi}^2}{m_{K1}^2}F_0^{D\pi}(s)F_0^{K\pi}(s) \\
&= -4F_+^{D\pi}(s)F_+^{K\pi}(s)|\mathbf{p_K}'|\mathbf{p_2}'||\cos\theta' + \frac{\Delta_{D\pi}^2\Delta_{K\pi}^2}{s}F_0^{D\pi}(s)F_0^{K\pi}(s) \qquad (4)
\end{aligned}
$$

The heavy quark FF can be taken as those of the dipole approximation:

$$
F_{+,\ 0}^{D\pi}(s) = \frac{F_+^{D\pi}(0)}{1 - s/m_{+,\ 0}^2},\ F_+^{D\pi}(0) = F_0^{D\pi}(0) \simeq 0.624 \qquad (5)
$$

with the masses $m_+ = m_{D^{*0}} = 2007$ MeV, $m_0 = m_0^{*0} = 2352$ MeV [9, 2]. The A_1 is the dominant contribution and can be reduced to the form factors

$$
\begin{aligned}
A_1 &= < K^-\pi_1^+|\bar{s}c|D^+ >< \pi_2^+|\bar{u}d|0 > \\
&= \frac{-if_\pi p_2^\mu}{m_D}\left[f(p_K + p_1)_\mu + g(p_K - p_1)_\mu + rp_{2\mu} + \frac{ih}{m_D^2}\epsilon_{\mu\nu\alpha\beta}p_2^\nu(p_K + p_1)^\alpha(p_K - p_1)^\beta\right] \\
&= \frac{if_\pi}{2m_D}\left[(s_{K1} - \Delta_{D\pi}^2)f + (u - t + \Delta_{K\pi}^2)g - 2m_\pi^2 r\right] \qquad (6)
\end{aligned}
$$

2. Form Factors

In the previous section the amplitude was reduced to several Form factors (FF). There are two kinds of them: those depending on the heavy physics (charm or D), like $F^{D\pi}_{+,\,0}(x)$ and those depending on the light quarks physics. The former can be parametrized by a dipole dominance (a single resonance) as in eq. (5 and so on). The remaining light quark form factors: F_+, F_0, f and g (r is suppressed by the small pion mass) are more complicate. However one step further can be obtained by reducing f and g to the F_+, F_0 form factors. In general f and g are functions of s, p_2^2 and θ'. The p_2 dependence is of course trivial given the on-shell condition $p_2^2 = m_\pi^2$. The angular dependence is simple and can be taken into account by a partial wave expansion (PWE), with only the s and p waves, given that tensor contributions are less than 1 % [9]. Thus the only remaining dependence is on s.

Let us consider fist the intermediate scalar contribution (see fig. 1), or s wave:

$$\begin{aligned}
J_\mu &\simeq\ < p_K,\ p_1|J_{L,\ \mu}|p_D >= \sum_{K_0^*} < p_K,\ p_1|K_0^* > BR_{K_0^*}(s) < K_0^*|J_{L,\ \mu}|p_D > \\
&= c_d BR_{K_0^*}(-i)\left[F_+^{DK_0^*}(p_2^2)(p_D+p_{K1})_\mu + f_-^{DK_0^*}(p_2^2)p_{2\mu}\right] \\
&= -ic_d BR_{K_0^*}\left(2F_+^{DK_0^*}p_{K1\mu} + [F_+^{DK_0^*} + f_-^{DK_0^*}]p_{2\mu}\right) \\
&= \frac{1}{m_D}\left[f p_{K1\mu} + g(p_K-p_1)_\mu + r p_{2\mu} + \frac{ih}{m_D^2}\epsilon_{\mu\nu\alpha\beta}q^\nu p_{K1}^\alpha(p_K-p_1)^\beta\right] \qquad (7)
\end{aligned}$$

and

$$\begin{aligned}
\frac{f}{m_D} &= -2ic_d BW_{K_0^*}F_+^{DK_0^*},\ \frac{r}{m_D} = -ic_d BW_{K_0^*}[F_+^{DK_0^*} + f_-^{DK_0^*}],\ g = h = 0 \\
A_{1S} &= f_\pi c_d BW_{K_0^*}\left[(s-m_D^2)F_+^{DK_0^*}(m_\pi^2) + m_\pi^2 f_-^{DK_0^*}(m_\pi^2)\right] \\
&\simeq f_\pi\frac{\Delta^2_{K\pi}}{f_{K_0^*}s}(s-m_D^2)F_+^{DK_0^*}(0)F_0^{K\pi}(s) \qquad (8)
\end{aligned}$$

with $p_2 \equiv p_D - p_K - p_1 \equiv p_D - p_{K1}$, $s = p_{K1}^2$, $p_2^2 = m_\pi^2$, $BW_m(s) = [m^2 - s - i\sqrt{s}\,\Gamma(s)]^{-1}$ and given that $c_d BW_{K_0^*}(s) = \Delta^2_{K\pi}F_0^{K\pi}(s)/sf_{K_0^*}$.

This approximation (one resonance) seems too drastic but one has to remember that according to the Watson theorem the $\pi - K$ scattering phase is the same in all cases, as long as one is in the elastic range ($\sqrt{s} \leq 1.3$ GeV). Given analiticiy and unitarity one can show that the form factors are determinated up to multiplicative constant. This is what has been obtained in the last eq. (8) and it is realized in the Omnes representation for the FF given below [4, 5]:

$$F_0(s) = F_0(0)\exp\left[\frac{s}{\pi}\int_{s_{K\pi}}^{\infty}\frac{\phi_0 \mathrm{d}s'}{s'(s'-s-i\epsilon)}\right],\ \sigma t_0 = \sin\phi_0\ \mathrm{e}^{i\phi_0} \qquad (9)$$

where the elastic phase are measured in [6], and can be fitted as

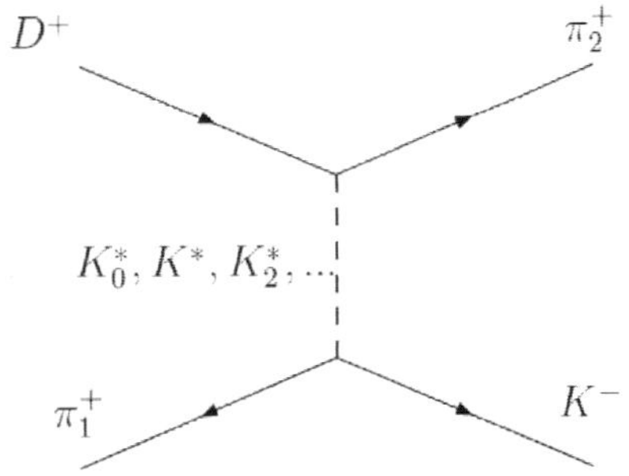

Figure 1. Resonance (scalar, vector, etc.) approximation for the Form Factors

$$
\begin{aligned}
<\pi^+K^-|\ T\ \ |\pi^+K^- >= & \frac{1}{3}\left(T^{3/2}+2T^{1/2}\right)=\frac{2}{3}a_0 e^{i\phi_0}\\
a_0 e^{i\phi_0} \quad & = \frac{1}{2i}\left[\left(\eta_0^{1/2}e^{2i\delta_0^{1/2}}-1\right)+\frac{1}{2}\left(\eta_0^{3/2}e^{2i\delta_0^{3/2}}-1\right)\right] \simeq \frac{1}{2i}\left[e^{2i\delta_0^{1/2}}+\frac{1}{2}e^{2i\delta_0^{3/2}}-\frac{3}{2}\right]\\
\delta_0^{1/2} \quad & = \tan^{-1}\left[\frac{\Gamma_{K_0^*}m^2_{K_0^*}p/\sqrt{s}\ p_{K_0^*}}{m^2_{K_0^*}-s}\right]+\tan^{-1}\left[\frac{ap}{1+bp^2+cp^4}\right]\\
\delta_0^{3/2} \quad & = -\tan^{-1}\left[\frac{a^{3/2}p}{1+b^{3/2}p^2+c^{3/2}p^4}\right] \qquad (10)
\end{aligned}
$$

where $p = p(s)$ is the CM momenta and $p_{K_0^*} = p(s = m^2_{K_0^*})$. The parameters obtained in the fit are (all in the corresponding units of GeV): $m_{K_0^*} = 1.44$ GeV, $\Gamma_{K_0^*} = 0.32$ GeV, $a = 1.91$, $b = 1.5$, $c = 0.82$, $a^{3/2} = 1.01$, $b^{3/2} = 0.5$, $c^{3/2} = 0.9$. This expression is valid in the elastic range, $4m_\pi^2 < s \leq (m_K + m_{\eta'})^2 \simeq (1.3\ \text{GeV})^2$, before the $\eta' - K$ channel is open. The inelasticity contribution from the $\eta - K$ channel seems to be negligible and it is not going to be considered in this work [4].

The vector contribution or p wave, with $p_2 = p_D - p_K - p_1 \equiv p_D - p_V$, $s = p_V^2$

$$
\begin{aligned}
J_\mu \quad \equiv \quad & < p_K,\ p_1|J_{L,\ \mu}|p_D > \simeq \sum_{K^*} < p_K,\ p_1|K^* > BR_{K^*}(s) < K^*|J_{L,\ \mu}|p_D >\\
= \quad & \sum_{\lambda^*} g^*(p_K - p_1)\cdot\epsilon^* BR_{K^*}\left[F_{1A}\epsilon^*_\mu + F_{2A}p_D\cdot\epsilon^* p_{D\mu} + F_{3A}p_D\cdot\epsilon^* p_2^\mu + iF_V\epsilon_{\mu\rho\lambda\phi}\epsilon^{*\rho}p_D^\lambda p_V^\phi\right]\\
= \quad & -g^* BR_{K^*}\left[F_{1A}(p_K - p_1)_\mu + F_{2A}p_D\cdot(p_K - p_1)p_{D\mu} + F_{3A}p_D\cdot(P_K - p_1)q_\mu\right.\\
& \left. + iF_V\epsilon_{\mu\rho\lambda\phi}(p_K - p_1)^\rho p_D^\lambda p_V^\phi - \frac{\Delta^2_{K\pi}}{m_*^2}\left(F_{1A}p_{V\mu} + F_{2A}p_D\cdot p_V p_{D\mu} + F_{3A}p_D\cdot p_V q_\mu\right)\right]\\
= \quad & \frac{1}{m_D}\left[f p_{V\mu} + g(p_K - p_1)_\mu + r p_{2\mu} + \frac{ih}{m_D^2}\epsilon_{\mu\nu\alpha\beta}q^\nu p_V^\alpha (p_K - p_1)^\beta\right] \qquad (11)
\end{aligned}
$$

to obtain

$$\frac{f}{m_D g^* BR_{K^*}(s)} = \frac{\Delta^2_{K\pi}}{m_*^2} F_{1A} - \left[p_D\cdot(p_K-p_1) - \frac{\Delta^2_{K\pi}}{m_*^2} p_D\cdot p_V\right] F_{2A}, \quad \frac{g}{m_D g^* BR_{K^*}} = -F_{1A}$$
$$\frac{r}{m_D g^* BR_{K^*}} = -\left[p_D\cdot(p_K-p_1) - \frac{\Delta^2_{K\pi}}{m_*^2} p_D\cdot p_V\right](F_{2A}+F_{3A}), \quad \frac{h}{m_D^3 g^* BR_{K^*}} = -F_V \quad (12)$$

where the FF F_{iA} and F_V depend on the heavy quark physics and are evaluated at $p_2^2 = m_\pi^2$. Using the identities

$$p_D\cdot(p_K-p_1) = \frac{1}{2}(t-u+\Delta^2_{K\pi}), \; p_D\cdot p_V = \frac{1}{2}(m_V^2+\Delta^2_{D\pi})$$
$$p_D\cdot(p_K-p_1) - \frac{\Delta^2_{K\pi}}{m_*^2} p_D\cdot p_V = \frac{1}{2}\left(t-u-\frac{\Delta^2_{K\pi}\Delta^2_{D\pi}}{m_V^2}\right) = -2|\mathbf{p_K}'|\mathbf{p_2}'||\cos\theta' \quad (13)$$

one can obtain the vector contribution

$$\begin{aligned} A_{1V}(s) &= \langle K^-\pi_1^+|\bar{s}c|D^+\rangle\langle \pi_2^+|\bar{u}d|0\rangle = \frac{if_\pi}{2m_D}\left[(s-\Delta^2_{D\pi})f + (u-t+\Delta^2_{K\pi})g - 2m_\pi^2 r\right] \\ &= -if_\pi g^* BW^*(s)\left[2F_{1A} - (s-\Delta^2_{D\pi})F_{2A} + 2m_\pi^2(F_{2A}+F_{3A})\right]|\mathbf{p_K}'|\mathbf{p_2}'||\cos\theta' \\ &= -4f_\pi g^* BW^* A_0(s)\sqrt{s}\,|\mathbf{p_K}'|\mathbf{p_2}'|\cos\theta' = -4f_\pi\frac{F_+(s)}{f_* m_*}A_0(s)\sqrt{s}\,|\mathbf{p_K}'|\mathbf{p_2}'|\cos\theta' \quad (14)\end{aligned}$$

and the FF are parametrized again in the dipole approximation

$$V(s) = \frac{V(0)}{1-s/m_V^2}, \; A_i(s) = \frac{A_i(0)}{1-s/m_A^2} \quad (15)$$

with $m_A = 2.63(10)(13) \simeq m_{D_s1} = 2.5$ GeV, $A_1(0) = 0.6226(56)(65)(74)$ and $m_V = m_{D_s^*} = 2.1$ GeV, $r_V = V(0)/A_1(0) = 1.463(17)(31)$, $r_2 = A_2(0)/A_1(0) = 0.801(20)(20)$ [11].
The Vector FF have been taken as ($q = P_P - P_V$)

$$\begin{aligned} \langle V(p_V,\epsilon)|J^L_\mu|P(p_P)\rangle &= -i\left[(m_V+m_P)\left(\epsilon^* - \frac{\epsilon^*\cdot q}{q^2}q\right)_\mu A_1(q^2) - \left(p_P+p_V-\frac{\Delta^2_{PV}}{q^2}q\right)_\mu\right. \\ &\left.\cdot\frac{\epsilon^*\cdot q A_2(q^2)}{m_V+m_P} + 2m_V q_\mu\frac{\epsilon^*\cdot q}{q^2}A_0(q^2)\right] - \frac{2V(q^2)}{m_P+m_V}\epsilon_{\mu\nu\alpha\beta}\epsilon^{*\nu}p_P^\alpha p_V^\beta \\ &= F_{1A}\epsilon^*_\mu + F_{2A}p_P\cdot\epsilon^* p_{P\mu} + F_{3A}p_P\cdot\epsilon^* q^\mu + iF_V\epsilon_{\mu\nu\alpha\beta}\epsilon^{*\nu}p_P^\alpha p_V^\beta \quad (16)\end{aligned}$$

and the two representation are related as

$$\begin{aligned} F_{1A} &= -i(m_V+m_P)A_1, \; F_{2A} = \frac{2iA_2}{m_V+m_P}, \; F_V = \frac{2iV}{m_V+m_P} \\ F_{3A} &= i\frac{m_V+m_P}{q^2}A_1 - \frac{2im_V}{q^2}A_0 - i\frac{1+\Delta^2_{PV}/q^2}{m_V+m_P}A_2 \quad (17)\end{aligned}$$

and to cancel the poles at $q^2 = 0$ one has that $2m_V A_0(0) = (m_P+m_V)A_1(0) - (m_P - m_V)A_2(0)$, $A_3(0) = A_0(0) = 0.71$ [12] (1991)), in agreement with Babar measurement [11]!.

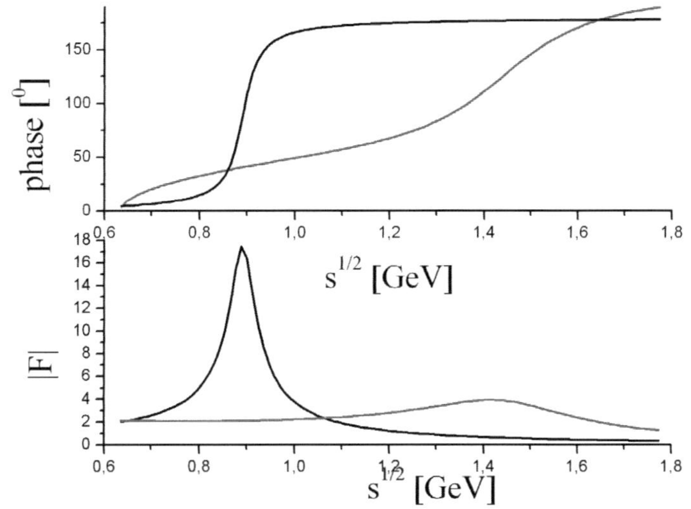

Figure 2. Form Factors with their phases. The black (red) line is the Vector (scalar)

Again one resonance has been taken into account (a much better approximation than in the scalar case), but the same considerations mentioned for the scalar case are valid here too.

The vector form factor F_+ is dominated by the $K^*(890)$ resonance, with $m_* \simeq g_* f_* = 891.7$ MeV and $\Gamma_* = 51$ MeV [2]

$$\begin{aligned} F_+(s) &= \frac{m_*^2}{m_*^2 - s - i\sqrt{s}\,\Gamma_*(s)}, \; \Gamma_*(s) = \Gamma_* \frac{m_*^2}{s}\left(\frac{p}{p_*}\right)^3 \\ p &= \frac{1}{2\sqrt{s}}\lambda^{1/2}\left(s,\, m_K^2,\, m_\pi^2\right),\; p_* = \frac{1}{2m_*}\lambda^{1/2}\left(m_*^2,\, m_K^2,\, m_\pi^2\right) \end{aligned} \tag{18}$$

Escribanos's paper [9] obtains

$$A_1 \simeq f_\pi \chi_S^{\text{eff.}}(m_D^2 - s)F_0^{K\pi}(s) - 4 f_\pi \chi_V^{\text{eff.}} N(s)^2 F_+^{K\pi}(s)|\mathbf{p_K}'|\mathbf{p_2}'|| \cos\theta' \tag{19}$$

with [9]

$$\chi_S^{\text{eff.}} \simeq \frac{g_{K_0^* K\pi} F_0^{DK_0^*}(m_\pi^2)}{\Gamma_{K_0^*}(m_{K_0^*}^2)|F_0^{K\pi}(m_{K_0^*}^2)|} = 4.4(28)\ \text{GeV}^{-1},\ \chi_V^{\text{eff.}} \simeq f_{K^*}^{-1} = 4.9(2)\ \text{GeV}^{-1} \tag{20}$$

and $F_0^{DK_0^*}(m_\pi^2) = 1.24(7)$ and $F_+^{DK^*}(m_\pi^2) = 0.76(7)$ [9]

3. Experimental data and fitting

At the moment the best experimental results are those from Cleo [8], but earlier results and several analysis are available like those of the Focus collab. [10] and so on. The extraction of

Journal of Physics: Conference Series **378** (2012) 012015 doi:10.1088/1742-6596/378/1/012015

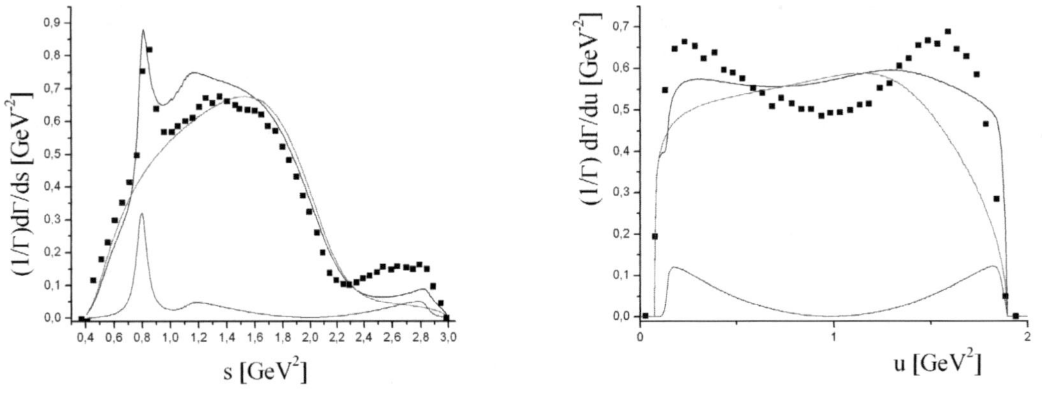

Figure 3. The spectra: from Escribano, Total one (data are from Cleo), low and high energy. The black lines are the total, the red ones the dominant scalar contributions and the blue lines are the vector contributions. The BR=7.0 %, $\chi_S = 4.2$ and $\chi_P = 4.8$

Mode	BR[%]	$A_{\rm CP}$ [%]
$D^0 \to K^-\pi^+$	3.95(5)	-
$D^0 \to \pi^-\pi^+$	0.143(3)	0.4(5)(1)
$D^0 \to K^-K^+$	0.394(7)	-0.04(3)(1)
$D^+ \to K^0_S\pi^+$	1.47(7)	-0.71(19)(20)
$D^\pm \to \pi^+\pi^-\pi^\pm$	0.327(22)	1.7(42)
$D^\pm \to K^\mp\pi^\pm\pi^\pm$	9.51(34)	-0.5(4)(9)
$D^\pm \to K^0_s\pi^\pm\pi^0$	6.90(32)	0.3(9)(3)
$D^\pm \to K^+K^-\pi^\pm$	0.98(4)	0.39(61)

Table 1. Direct CP in D nonleptonic decays, from table 156, p. 218 in HAFG [2]

the s wave phase (to compare with Watson theorem) is not clear as the parametrization of the dominant s-wave is cumbersome, given that it is dominated by the broad κ resonance with a large overlapping with the $K_0^*(1430)$ resonance. This is the case in the Isobar parametrization. For this reason another parametrization (the so called MIPWA) was proposed that allows to extract the s-wave phase and amplitude in cleaner way. Unfortunately it seems that the Watson theorem is violated (there is no agreement with the phases obtained by the Lass coll. [6]), and additional contributions are present like $\pi-\pi$ interactions [8], three body Final state interactions [7] and so on. From the theoretical side, the paper of Escribano, that we follow here seems to do a good job. The obtained fit is shown in Fig. 3. Unfortunately the $\pi - \pi$ (this contribution is not included) spectra is not well fitted and there is no explanation for the violation of the Watson theorem (even if it is by an overall phase). In order to obtain a better fit it is necessary to include these contributions in a proper manner. New results from LHCB [13, 14, 15] are expected and certainly will contribute to clarify the situation.

4. CP and New Physics

CP has been predicted in the Standard Model (SM) to be of order 10^{-3} [16, 1] in several processes. This is of course interesting given that it may be at the reach of future experiments, as it is shown in table 1. However for this process no calculation has been done and we are exploring the SM and SUSY contributions in the present work. A more detailed calculation will be given in a future work [19]

In the case of CF processes, the contributions of the SM tree diagram have no weak CP violating phase required for non vanishing CP asymmetry. One has to look for loop diagrams

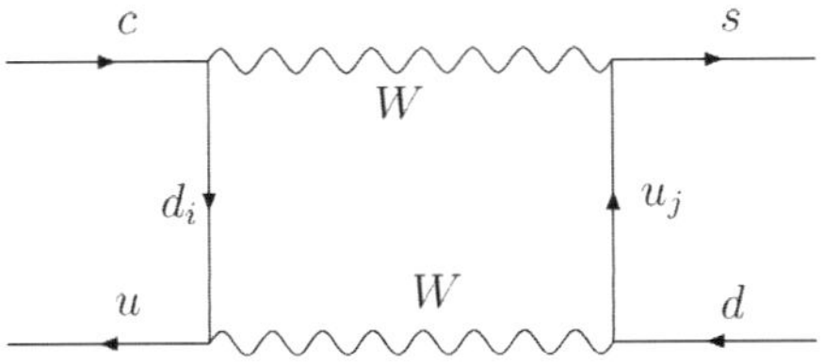

Figure 4. Box diagram contribution to CPV in the Standard Model

that can lead to the required weak CP violating phase. At one loop level, the only contribution can by obtained by considering the box diagram in Fig. 4 [18]

$$\Delta\mathcal{H} = \frac{G_F^2 m_W^2}{2\pi^2}\left[V_{cd}V_{du}\left(V_{su}V_{ud}f_{ud}+V_{sc}V_{cd}f_{cd}+V_{st}V_{td}f_{td}\right)+V_{cs}V_{su}\left(V_{su}V_{ud}f_{us}\right.\right.$$
$$\left.\left.+V_{sc}V_{cd}f_{cs}+V_{st}V_{td}f_{ts}\right)+V_{cb}V_{bu}\left(V_{su}V_{ud}f_{ub}+V_{sc}V_{cd}f_{cb}+V_{st}V_{td}f_{tb}\right)\right]\bar{u}c_L\bar{s}d_L \quad (21)$$

using the unitarity relations for the CKM matrix one gets the box contribution for the Wilson coefficient

$$\begin{aligned}\Delta c_2 &= \frac{G_F m_W^2}{\sqrt{2}\,\pi^2 V_{cs}^* V_{ud}}\{V_{cd}V_{du}\left[V_{sc}V_{cd}\left(f_{cd}-f_{ud}\right)+V_{st}V_{td}\left(f_{td}-f_{ud}\right)\right]+V_{cs}V_{su}\left[V_{sc}V_{cd}\right.\\ &\quad\left.\cdot\left(f_{cs}-f_{us}\right)+V_{st}V_{td}\left(f_{ts}-f_{us}\right)\right]+V_{cb}V_{bu}\left[V_{sc}V_{cd}\left(f_{cb}-f_{ub}\right)+V_{st}V_{td}\left(f_{tb}-f_{ub}\right)\right]\}\\ &= \frac{G_F m_W^2}{\sqrt{2}\,\pi^2 V_{cs}^* V_{ud}}\{V_{cs}V_{su}\left[V_{sc}V_{cd}\left(f_{cs}-f_{cd}-f_{us}+f_{ud}\right)+V_{st}V_{td}\left(f_{ts}-f_{td}-f_{us}+f_{ud}\right)\right]\\ &\quad+V_{cb}V_{bu}\left[V_{sc}V_{cd}\left(f_{cb}-f_{cd}-f_{ub}+f_{ud}\right)+V_{st}V_{td}\left(f_{tb}-f_{td}-f_{ub}+f_{ud}\right)\right]\}\\ &\simeq \frac{G_F m_W^2}{\sqrt{2}\,\pi^2 V_{cs}^* V_{ud}}\cdot 3.8\cdot 10^{-7}\mathrm{e}^{0.05\cdot i}\simeq 10^{-8}\mathrm{e}^{0.05\cdot i}\end{aligned} \quad (22)$$

with

$$f_{xy} = \frac{7xy-4}{4(1-x)(1-y)}+\frac{1}{x-y}\left[\frac{y^2\log y}{(1-y)^2}\left(1-2x+\frac{xy}{4}\right)-\frac{x^2\log x}{(1-x)^2}\left(1-2y+\frac{xy}{4}\right)\right] \quad (23)$$

It should be noted that penguin diagrams at one loop level do not contribute to the process under this study and can contribute only at two loop level (dipenguin diagrams). However, their weak phase is estimated to be one order of magnitude less than that generated by the box diagram with their absolute value of the same order as box diagram. Clearly from eq.22, the CP violating weak phase is so small and lead to a negligible CP asymmetry. Thus, it is interesting to look for new physics contributions that can enhance the weak CP phase. As an example of weak phases producing CPV, one can mention SUSY models. In the case of R-parity conservation SUSY models, the contributions to Wilson coefficients are of the same order of the SM and thus we do not expect them to enhance the CP asymmetry. The charged Higgs contributions also are suppressed by the light quark masses. A larger contribution can be obtained in SUSY models when R-parity is violated [19].

Acknowledgement
Gaber Faisel's work is supported by the National Science Council of R.O.C. under grants NSC 99-2112-M-008-003-MY3 and NSC 99-2811-M-008-085. D.D. has been supported by PROMEP and DINPO project from Guanajuato University. C.A.R. wish to thank to D.D. and to the Departamento de Física, DCI, Campus León, Universidad de Guanajuato, for its hospitality.

References

[1] M. Artuso, B. Meadows and A. A. Petrov, Ann. Rev. Nucl. Part. Sci. **58**, 249 (2008) [arXiv:0802.2934 [hep-ph]].
[2] K. Nakamura et al. (Particle Data Group), J. Phys. G 37, 075021 (2010).
[3] D. Asner *et al.* [Heavy Flavor Averaging Group HFAG], arXiv:1010.1589 [hep-ex].
[4] M. Jamin, J. A. Oller and A. Pich, Nucl. Phys. B **622**, 279 (2002) [arXiv:hep-ph/0110193].
[5] J. F. Donoghue, J. Gasser and H. Leutwyler, Nucl. Phys. B **343**, 341 (1990).
[6] D. Aston *et al.*, Nucl. Phys. B **296**, 493 (1988).
[7] B. Kubis, arXiv:1108.5866 [hep-ph].
[8] G. Bonvicini *et al.* [CLEO Coll.], Phys. Rev. D **78**, 052001 (2008) [arXiv:0802.4214 [hep-ex]].
[9] D. R. Boito and R. Escribano, Phys. Rev. D **80**, 054007 (2009) [arXiv:0907.0189 [hep-ph]].
[10] J. M. Link *et al.* [The FOCUS Collab.], Phys. Lett. B **681**, 14 (2009) [arXiv:0905.4846 [hep-ex]].
[11] P. del Amo Sanchez *et al.* [The BABAR Coll.], Phys. Rev. D **83**, 072001 (2011) [arXiv:1012.1810 [hep-ex]].
[12] J. G. Korner, K. Schilcher, M. Wirbel and Y. L. Wu, Z. Phys. C **48**, 663 (1990).
[13] M. Gersabeck [LHCb Collaboration], PoS B **EAUTY2011**, 020 (2011).
[14] B. O'Leary *et al.* [SuperB Collaboration], [arXiv:1008.1541 [hep-ex]]. Super-B
[15] T. Aushev *et al.*, arXiv:1002.5012 [hep-ex]. Super KEKB
[16] R. Aaij *et al.* [LHCb Collaboration], arXiv:1112.0938 [hep-ex].
[17] O. Gedalia and G. Perez, *TASI 2009 Lectures - Flavor Physics*, arXiv:1005.3106 [hep-ph].
[18] T. Inami and C. S. Lim, Prog. Theor. Phys. **65**, 297 (1981) [Erratum-ibid. **65**, 1772 (1981)].
[19] D. Delepine, G. Faisel and C. Ramirez, work in progress.

XIII Mexican Workshop on Particles and Fields IOP Publishing
Journal of Physics: Conference Series **378** (2012) 012016 doi:10.1088/1742-6596/378/1/012016

Physical constraints on Higgs masses and quartic couplings in the two Higgs doublet model

D Morales[1], S R Juárez[1] and P Kielanowski[2]

[1]Departamento de Física, Escuela Superior de Física y Matemáticas, Instituto Politécnico Nacional, U P "Adolfo López Mateos" C P 07738, México D F
[2]Departamento de Física, CINVESTAV IPN, Av. Instituto Politécnico Nacional 2508, C P 07360, México D F

E-mail: damoralesc@hotmail.con, rebecajw@gmail.com, kiel@fis.cinvestav.mx

Abstract. In the framework of the two Higgs doublet model we graphically illustrate the energy dependence of the Higgs masses and the quartic couplings. Our results are based on the theoretical constraints on the quartic Higgs couplings through the stability and triviality principles that are obtained, using the Lagrange multipliers method and the numerical solutions of the Renormalization Group Equations that give the energy dependence of the parameters of the Lagrangian. From these solutions we can also establish the region of validity of the model in terms of the initial conditions of the parameters.

1. Introduction

For over three decade and until now the standard model (SM) of particle physics [1 - 3] has provided us with an excellent theoretical framework consistent with the experiments [4]. However it has some limitations since it does not explain the hierarchy of the masses, the masses of the neutrinos and has 19 free parameters which are adjusted by the experiment and the detection of the Higgs particle [5] is an open problem for the experimental verification of the SM.

Although there are these limitations, the work of theorists has continued to find solution to these problems. A non-super symmetric extension of the SM that arises in a natural way is known in the literature as the two Higgs doublet model (2HDM), which incorporate an extra scalar field doublet.

In this model there are different ways of coupling the scalar field with the up-type quarks and down-type quarks [6 - 9], depending on the type of symmetry that is imposed on the scalar field. Here we consider the 2HDM type-II model, where the first scalar field is coupled only to the down-type quarks and second field is coupled to the up-type quarks.

2. The Extension of the Standard Model

The Higgs sector of the 2HDM consists of two scalar doublets Φ_1 y Φ_2, both having hypercharge $Y = 1$, which written in terms eight real scalars become:

$$\Phi_1 = \begin{pmatrix} \phi_1 + i\phi_2 \\ \phi_3 + i\phi_4 \end{pmatrix}, \quad \Phi_2 = \begin{pmatrix} \phi_5 + i\phi_6 \\ \phi_7 + i\phi_8 \end{pmatrix}. \tag{1}$$

Journal of Physics: Conference Series **378** (2012) 012016 doi:10.1088/1742-6596/378/1/012016

The most general Higgs potential which can be constructed with these two scalar doublets, that is renormalizable and compatible with the SM gauge symmetry contains 14 parameters. This number of parameters is reduced by imposing the Z_2 discrete symmetry [10 - 12]:

$$\Phi_1 \to \Phi_1, \Phi_2 \to -\Phi_2 \quad \text{or} \quad \Phi_1 \to -\Phi_1, \Phi_2 \to \Phi_2 \; . \tag{2}$$

This Z_2 symmetry choice automatically excludes the violation of CP symmetry, which simplifies the analysis.

The gauge invariant and renormalizable Higgs potential can be written as:

$$V(\Phi_1,\Phi_2) = \mu_1^2\Phi_1^+\Phi_1 + \mu_2^2\Phi_2^+\Phi_2 + \lambda_1(\Phi_1^+\Phi_1)^2 + \lambda_2(\Phi_2^+\Phi_2)^2 + \lambda_3(\Phi_1^+\Phi_1)(\Phi_2^+\Phi_2)$$
$$+ \lambda_4(\Phi_1^+\Phi_2)(\Phi_2^+\Phi_1) + \frac{1}{2}\lambda_5\left[(\Phi_1^+\Phi_2)^2 + (\Phi_2^+\Phi_1)^2\right]. \tag{3}$$

The neutral scalar fields of both doublets acquire non-zero vacuum expectation values (VEV) v_i, just as in the standard model, and $v_1^2+v_2^2 = v^2 = (246 \text{ GeV})^2$. After the electroweak symmetry breaking, in the potential, the resulting physical particles in the Higgs sector are: two CP-even neutral Higgs scalar (H^0, h^0), one CP-odd neutral Higgs scalar (A^0), two charged Higgs bosons ($H^{\pm}$), and three Goldstone bosons ($G^{\pm}$, G^0) that contribute to the mass generation of the gauge vector bosons $W^{\pm}$ and Z^0 [13].

2.1. Scalars masses

The masses of the massive Higgs scalars, in terms of the quartic couplings and VEV of the potential, satisfy the following relations:

$$M^2_{H^0,h^0} = \lambda_1 v_1^2 + \lambda_2 v_2^2 \pm \left[(\lambda_1 v_1^2 - \lambda_2 v_2^2)^2 + (2\Lambda v_1 v_2)^2\right]^{\frac{1}{2}} , \tag{4}$$

$$M^2_{A^0} = -\frac{4M_W^2}{g^2}\lambda_5 , \tag{5}$$

$$M^2_{H^\pm} = -\frac{2M_W^2}{g^2}(\lambda_4 + \lambda_5) , \tag{6}$$

where $\Lambda = \frac{1}{2}(\lambda_3 + \lambda_4 + \lambda_5)$.

2.1.1. *Bounds for the λ parameters.* The minimum of the potential must be stable, and it is required to have positive masses and furthermore the potential must be bounded from below. We obtain the following boundary values for the quartic couplings from the positive values of the above expressions for the masses:

$$\lambda_1 > 0, \quad \lambda_2 > 0, \quad \lambda_4 + \lambda_5 < 0, \quad \lambda_5 < 0, \quad \lambda_4 < |\lambda_5|, \quad \lambda_1\lambda_2 > \frac{1}{4}(\lambda_3 + \lambda_4 + \lambda_5)^2 \; . \tag{7}$$

3. Lagrange Multipliers Method

A rigorous way to obtain the quartic couplings restrictions is through the use of the Lagrange multipliers and the implementation of the vacuum stability conditions. The problem we consider involves constraints with equality and inequality relations, to be solved through the Karush-Kuhn-Tucker method.

Equation (3) can be written in terms of four gauge invariant combinations x_i of the scalar doublets and the parameters b_i, defined as:

$$x_1 = \Phi_1^+\Phi_1, \quad x_2 = \Phi_2^+\Phi_2, \quad x_3 = \text{Re}(\Phi_1^+\Phi_2), \quad x_4 = \text{Im}(\Phi_1^+\Phi_2) \; , \tag{8}$$

$$b_{11} = \lambda_1, \quad b_{22} = \lambda_2, \quad b_{12} = \lambda_3, \quad b_{33} = (\lambda_4 + \lambda_5), \quad b_{44} = (\lambda_4 - \lambda_5) . \tag{9}$$

Then the Higgs potential becomes:

$$V = \mu_1^2 x_1 + \mu_2^2 x_2 + b_{11} x_1^2 + b_{22} x_2^2 + b_{12} x_1 x_2 + b_{33} x_3^2 + b_{44} x_4^2 . \tag{10}$$

As one can see the Higgs potential consists of a quadratic part and a quartic part. We will focus our attention on the quartic part of the potential and construct a Lagrangian function to be analyzed.

3.1. Lagrange function

The Lagrange function of our problem has the form:

$$L(x) = F(x) + \sum_{i=1}^{n} \Lambda_i g_i(x) + \sum_{j=1}^{m} \Gamma_j h_j(x) . \tag{11}$$

Where $F(x)$ is the function to be minimized (quartic potential), Λ_i and Γ_j are the Lagrange multipliers, $g_i(x)$ are equality constraints and h_j are inequality constraints. We build the constraints by using the Schwarz inequality and the vacuum expectation values restrictions, as follows:

$$h(x) = x_1 x_2 - x_3^2 - x_4^2 \geq 0, \qquad g(x) = x_1 + x_2 - v^2 = 0 . \tag{12}$$

Then the Lagrange function can be written as:

$$L(x) = b_{11} x_1^2 + b_{22} x_2^2 + b_{33} x_3^2 + b_{44} x_4^2 + b_{12} x_1 x_2 + \Lambda_1 (x_1 + x_2 - v^2) + \Lambda_2 (x_1 x_2 - x_3^2 - x_4^4) . \tag{13}$$

The vacuum stability condition implies that:

$$L(x_{\min}) > 0 . \tag{14}$$

3.1.1. *Conditions of Karush-Kuhn-Tucker.*

The Karush-Kuhn-Tucker conditions arise by imposing the following conditions:

$$\frac{\partial L}{\partial x_1} = 2b_{11} x_1 + b_{12} x_2 + \Lambda_2 x_2 + \Lambda_1 = 0 , \tag{15}$$

$$\frac{\partial L}{\partial x_2} = 2b_{22} x_2 + b_{12} x_1 + \Lambda_2 x_1 + \Lambda_1 = 0 , \tag{16}$$

$$\frac{\partial L}{\partial x_3} = 2b_{33} x_3 - 2\Lambda_2 x_3 = 0 , \tag{17}$$

$$\frac{\partial L}{\partial x_4} = 2b_{44} x_4 - 2\Lambda_2 x_4 = 0 , \tag{18}$$

$$\frac{\partial L}{\partial \Lambda_1} = x_1 + x_2 - v^2 = 0 , \tag{19}$$

$$\frac{\partial L}{\partial \Lambda_2} = x_1 x_2 - x_3^2 - x_4^2 = 0 . \tag{20}$$

Now we explore the various solutions. There are four possible cases:

$$x_3 \neq 0, x_4 = 0 \Rightarrow \Lambda_2 = b_{33} , \tag{21}$$

$$x_4 \neq 0, x_3 = 0 \Rightarrow \Lambda_2 = b_{44} , \tag{22}$$

$$x_3 \neq 0, x_4 = 0 \Rightarrow \Lambda_2 = 0, \qquad x_3 = 0, x_4 \neq 0 \Rightarrow \Lambda_2 = 0 , \tag{23}$$

$$x_3 \neq 0, x_4 \neq 0 \Rightarrow \Lambda_2 = b_{33} = b_{44} . \tag{24}$$

Solving the system of equations (15) – (20) for the variables x_1, x_2, and Λ_1, we obtain:

$$x_1 = \frac{(b_{12}+b_{33})-2b_{22}}{4b_{11}b_{22}-(b_{12}+b_{33})^2}\Lambda_1 > 0 \,, \tag{25}$$

$$x_2 = \frac{(b_{12}+b_{33})-2b_{11}}{4b_{11}b_{22}-(b_{12}+b_{33})^2}\Lambda_1 > 0 \,, \tag{26}$$

$$\Lambda_1 = \frac{4b_{11}b_{22}-(b_{12}+b_{33})^2}{2(b_{12}+b_{33}-b_{11}-b_{22})}v^2 \,, \tag{27}$$

Equations (25) – (27) imply that:

$$b_{11}+b_{22}-b_{33}-b_{12} > 0 \,, \quad 4b_{11}b_{22}-(b_{12}+b_{33})^2 > 0 \,, \tag{28}$$

$$b_{12}+b_{33}-2b_{11} < 0, \quad b_{12}+b_{33}-2b_{22} < 0 \,. \tag{29}$$

Then the constraints that satisfy the conditions of positivity of the masses and stability for quartic couplings are:

$$\lambda_1 > 0, \quad \lambda_2 > 0 \,, \tag{30}$$

$$\lambda_5 < 0, \quad \lambda_4 + \lambda_5 < 0 \,, \tag{31}$$

$$\lambda_3 + \lambda_4 \pm |\lambda_5| < \lambda_1 + \lambda_2 \,, \tag{32}$$

$$-2\sqrt{\lambda_1\lambda_2} < \lambda_3 + \lambda_4 \pm |\lambda_5| \,. \tag{33}$$

3.1.2. *Extreme case.* In this case the Higgs quartic potential has its lowest possible value, i. e., $V = 0$, then:

$$\lambda_1 x_1^2 + \lambda_2 x_2^2 + (\lambda_1+\lambda_2+\lambda_3)x_1x_2 = \left(\sqrt{\lambda_1}x_1 - \sqrt{\lambda_2}x_2\right)^2 + \left((\lambda_1+\lambda_2+\lambda_3)+2\sqrt{\lambda_1\lambda_2}\right)x_1x_2 \geq 0 \,. \tag{34}$$

And we get:

$$\sqrt{\frac{\lambda_1}{\lambda_2}} = \frac{x_2}{x_1} = \frac{v_2^2}{v_1^2}, \quad \lambda_3+\lambda_4+\lambda_5 = -2\sqrt{\lambda_1\lambda_2}\,, \quad \rightarrow \quad \frac{\lambda_1}{\lambda_2} = \frac{v_2^4}{v_1^4} = (\tan\beta)^4 \tag{35}$$

The Higgs masses become:

$$M_{H^0}\begin{cases}(4\lambda_1\lambda_2)^{\frac{1}{4}}v, & \lambda_1 \neq \lambda_2, \quad v_1 \neq v_2, \\ (2\lambda)^{\frac{1}{2}}v, & \lambda_1 = \lambda_2 = \lambda, \quad v_1 = v_2 = \frac{v}{\sqrt{2}},\end{cases} \tag{36}$$

$$M_{h^0} = 0, \quad M_{H^\pm} = \left(\frac{1}{2}|\lambda_4+\lambda_5|\right)^{\frac{1}{2}}v, \quad M_{A^0} = (|\lambda_5|)^{\frac{1}{2}}v \,. \tag{37}$$

4. Triviality constraints

We explore the asymptotic behavior of the parameters in the model, and their relations, through the Renormalization Group Equations (RGE) [8]. The RGE are important in studying the physics of the standard model and its extensions. The RGE determine the dependence of the coupling constants and other parameters of the Lagrangian on t, defined as $t = \ln(E/m_t)$, where E is the renormalization point energy. To numerically evaluate the energy dependence of the λ_i quartic couplings, we consider the RGE of the parameters, i.e., the gauge couplings, and the Yukawa couplings of the top and the down quark sectors.

$$\frac{dg_k}{dt} = \frac{1}{(4\pi)^2} b_k g_k^3, (k = 1,2,3), \tag{38}$$

$$\frac{dg_t}{dt} = \frac{1}{(4\pi)^2} [\frac{9}{2} g_t^2 + \frac{1}{2} g_b^2 - (\frac{17}{20} g_1^2 + \frac{9}{4} g_2^2 + 8g_3^2)]g_t, \tag{39}$$

$$\frac{dg_b}{dt} = \frac{1}{(4\pi)^2} [\frac{9}{2} g_b^2 + \frac{1}{2} g_t^2 - (\frac{1}{4} g_1^2 + \frac{9}{4} g_2^2 + 8g_3^2)]g_b, \tag{40}$$

$$\frac{d\lambda_1}{dt} = \frac{1}{(4\pi)^2} [24\lambda_1^2 - 3\lambda_1(3g^2 + g'^2 - 4g_t^2) + 2\lambda_3^2 + \lambda_4^2 + \lambda_5^2 - 2\lambda_3\lambda_4 + \frac{3}{8}(g^2 + g'^2)^2 + \frac{3}{4} g^4 - 6g_t^4], \tag{41}$$

$$\frac{d\lambda_2}{dt} = \frac{1}{(4\pi)^2} [24\lambda_1^2 - 3\lambda_2(3g^2 + g'^2 - 4g_b^2) + 2\lambda_3^2 + \lambda_4^2 + \lambda_5^2 - 2\lambda_3\lambda_4 + \frac{3}{8}(g^2 + g'^2)^2 + \frac{3}{4} g^4 - 6g_b^4], \tag{42}$$

$$\frac{d\lambda_3}{dt} = \frac{1}{(4\pi)^2} [4\lambda_3^2 + 4(3\lambda_3 + \lambda_4)(\lambda_1 - \lambda_2) - 3\lambda_3(3g^2 + g'^2 - 2(g_t^2 + g_b^2)) + 2\lambda_4^2 + 2\lambda_5^2$$
$$+ \frac{3}{4}(g^2 - g'^2)^2 + \frac{3}{2} g^4 - 12g_t^2 g_b^2], \tag{43}$$

$$\frac{d\lambda_4}{dt} = \frac{1}{(4\pi)^2} [4\lambda_4^2 + 4\lambda_4(\lambda_1 + \lambda_2 + 3\lambda_3) - 3\lambda_4(3g^2 + g'^2 - 2(g_t^2 + g_b^2)) + 8\lambda_5^2 + 3g^2 g'^2 - 12g_t^2 g_b^2], \tag{44}$$

$$\frac{d\lambda_5}{dt} = \frac{1}{(4\pi)^2} \lambda_5[4(\lambda_1 + \lambda_2 + 2\lambda_3 + 3\lambda_4) - 3(3g^2 + g'^2 - 2(g_t^2 + g_b^2))]. \tag{45}$$

The range of values, for the energy in GeV and the variable t are ($E_0 = M_t$, E_u=1.234×10^{13}), and (t_0 = 0, t_u = 25), where M_t is the mass of the quark top and E_u corresponds to the electroweak unification energy, where $g_1(E_t) = g_2(E_t)$. Now we explore the energy bounds for the 2HDM, through the running of the quartic couplings which determine the mass values of the Higgs. For our analysis, we set the masses $M_H^{\pm}$ = 609 GeV and M_A = 621.9 GeV given in [14].

The figure 1 and figure 2 show that the range of validity of the 2HDM is short, here $M_t < E <$ 292 GeV, i.e., $0 < t < 0.52$.

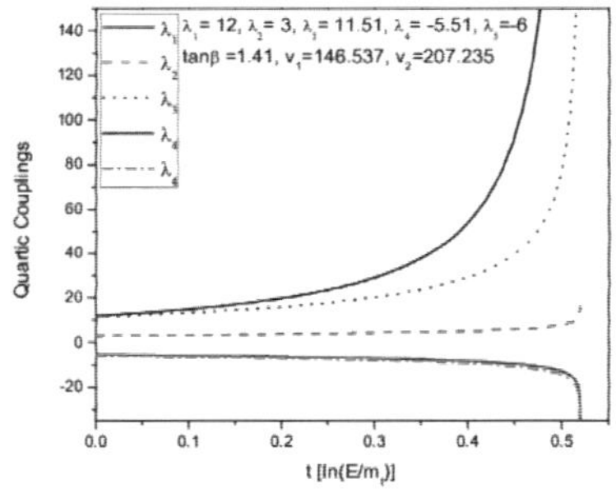

Figure 1. The energy dependence of the quartic couplings.

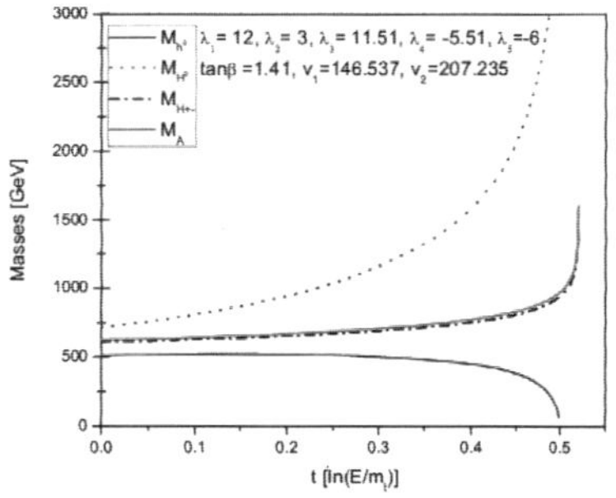

Figure 2. The energy dependence of Higgs masses.

Journal of Physics: Conference Series **378** (2012) 012016 doi:10.1088/1742-6596/378/1/012016

In the case of figure 3 and figure 4 we see that the 2HDM is valid in the whole range of energies then $M_t < E < E_u$.

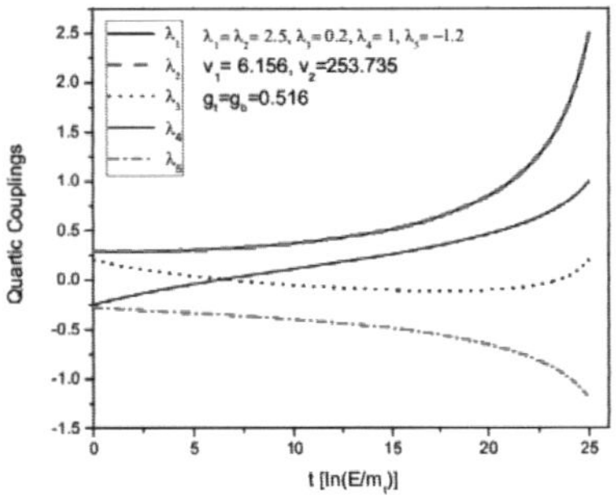

Figure 3. The energy dependence of the quartic couplings.

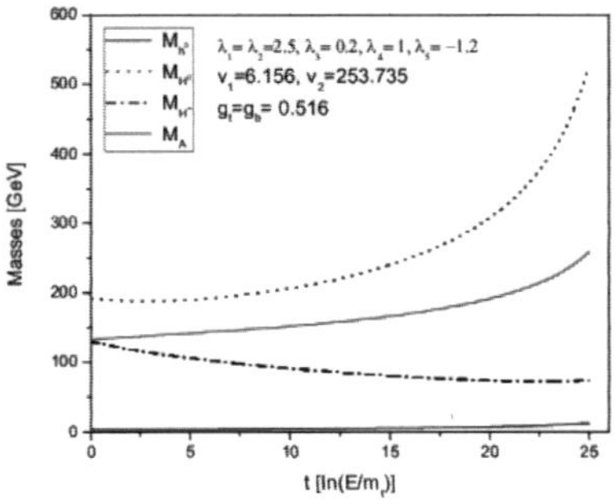

Figure 4. The energy dependence of Higgs masses.

5. Conclusions

We examined the mass content of the 2HDM extension of the SM, considering real vacuum expectation values of both Higgs fields in the case with CP symmetry. We have obtained through the mass formulas and through the Lagrange multipliers method, a set of constraints to be satisfied by the quartic couplings in the Higgs sector of the model. With the numerical solutions to the RGE, as shown in the figures, we conclude that the range of validity of the 2HDM depends on the initial values of the quartic parameters.

For small Higgs masses and small quartic couplings, the range of validity of the model goes up to the electro-weak unification scale and further.

References

[1] Glashow S 1961*Nucl. Phys.* **22** 579
[2] Weinberg S 1967 *Phys. Rev. Lett.* **19** 1264
[3] Salam A 1968 *Elementary Particle Theory* ed Svarthholm (Almquist and Wiksell Stockholm) *Proc. Of the Nobel Symposium held in 1968 at Lerum Sweden Stockholm* 8 p367
[4] Nakamura K *et al* 2010 *J. Phys.* G **37** 075021
[5] Higgs P 1964 *Phys. Lett.* **12** 132
[6] Davidson S and Haber H E 2005 *Phys. Rev.*D **72** 035004
[7] Krawczyk M 1998 *Acta Phys. Polon.* B **29** 3543-68
[8] Inoue K, Kakuto A and Nakano Y 1980 *Prog. Theor. Phys.* **63** 234-45
[9] Kominis D and Sekhar R 1993 *Phys. Lett.*B **304** 152-8
[10] Deshpande N G and Ma E 1978 *Phys. Rev.* D **18** 2574-76
[11] Babu K S and Ma E 1986 *Phys. Rev* D **31** 2861-65
[12] Ferreira P M and Jones D R 2009 Bounds on scalar masses in two Higgs doublet model *J. Higgs Energy Phys.* JHEP08(2009) 069
[13] Juárez S R and Morales D 2008 *PARTICLES AND FIELDS*: *XI Mexican Workshop on Particles and Fields* ed H C Valdez *et al* (Melville New York) AIP *Conf. Proc.* 1026 pp 294-6
[14] Mahmoudi F and Stäl O 2010 *Phys. Rev.*D **81** 035016

Journal of Physics: Conference Series **378** (2012) 012017 doi:10.1088/1742-6596/378/1/012017

Neutrino-electron scattering and tensor couplings

Estela A. Garcés[1], J. Barranco[2], A. Bolaños [3],O. G. Miranda[1], and T. I. Rashba[4]

[1]Departamento de Física, Centro de Investigación y de Estudios Avanzados del IPN, Apdo. Postal 14-740 07000 México, D F, Mexico
[2]Division de Ciencias e Ingenierías, Universidad de Guanajuato, Campus Leon, C.P. 37150, León, Guanajuato, México.
[3]Facultad de Ciencias Físico Matemáticas,
Benemérita Universidad Autónoma de Puebla,
Apartado Postal 1152, Puebla, Puebla, México.
[4]IZMIRAN, Institute of Terrestrial Magnetism, Ionosphere and Radio Wave Propagation of the Russian Academy of Sciences, 142190, Troitsk, Moscow region, Russia

E-mail: egarces@fis.cinvestav.mx

Abstract. In this work we study the possibility of new tensorial interactions in neutrino electron scattering. Two new physics schemes are considered, the non standard interactions formalism (NSI) as well as the Unparticle stuff case. Constraints in tensorial couplings are obtained for both cases focusing on the TEXONO collaboration recent results. For the case of tensorial unparticles we find that previous limits are improved, for a given region of parameters, with the analysis of TEXONO data, and for the NSI tensorial parameters, the limit obtained is more restrictive than previous bounds reported in the literature.

1. Introduction

Neutrino physics has been a very exciting topic of research in the last decades. Thanks to a huge amount of experimental evidence of neutrino flavor conversions we now know that neutrinos have mass and their flavors mix [1], therefore the Standard Model must be extended to incorporate this fact.

This challenge has motivated a wide range of phenomenological and theoretical research in order to incorporate the massive nature of neutrinos within structure that contains the Standard Model as a low energy theory, since it has been a extremely successful model with a high power of prediction [2, 3]. In this work we study two new physics schemes from a phenomenological point of view: on the one hand, we have non standard interactions, that are very well stablished in neutrino phenomenology and their parameters can be probed in the experiments designed to do precision measurements of the oscillation parameters, and, on the other hand, the unparticle physics scheme, that is a totally different approach inspired in scale invariant theories and has been more recently proposed.

The formalism of non-standard four-fermion interactions provides an appropriate way of parameterizing different kinds of new physics scenarios, being suitable for different massive neutrino models and having also the advantage of being model independent.

In previous works, non standard interactions have been studied in the neutrino sector, but mostly those operators with $(V - A)$ Lorentz structure have been of interest [4, 5, 6, 7, 8], while less attention has been given to non standard interactions with tensor structure. In this work we study this kind of effective interactions in the case of neutrino electron scattering.

Unparticle physics is based in the possible existence of a new kind of fields in nature that unlike those in the Standard Model are scale invariant and do not manifest themselves as particles. This kind of fields were studied first in the Bank and Zaks model [9], and later in 2007 Georgi proposed a scenario where the Bank-Zaks Fields can interact with the Standard Model fields in an ultra violet sector [10, 11].

This proposal has motivated many new phenomenological analyses that involve the presence of unparticle operators, the plethora of analysis in unparticle physics cover a wide range of energy regimes, signals of unparticles can be searched in collider experiments and could be found in the LHC.

Another interesting place to look for evidence of unparticle physics signals is in the modification to the cross sections at low energy. Here we analyze the case of neutrino electron scattering and how possible interactions mediated by unparticles could modify the SM cross section and therefore the neutrino spectrum.

Previous analysis have mostly considered unparticle operators with scalar and vector structure, only a few cases have studied the possibility of tensor unparticle operators. Here we explore the inclusion of an antisymmetric rank 2 unparticle operator and derive the neutrino electron cross section due to the new interaction mediated by unparticles.

The content of this conference proceedings is organized as follows, en section 2 we introduce the statistical tools that are used for the analysis, and also the relevant details of the data analyzed, after that in section 3 and 4 we derive the contribution to the neutrino-electron cross section due to the tensorial NSI and tensorial unparticle respectively. Finally in section 5 we present and discuss the results of both analyses, we also compare our results with previous works in the literature.

2. Statistical analysis and the TEXONO experiment

Since the discovery of electroweak interactions neutrino electron scattering is a topic of high interest in elementary particle physics, given the fact that it is a clean process free from hadronic difficulties in its theoretical calculation, it can be used as a clean probe of the Standard Model and also is very suitable to study many of its extensions. Neutrino electron scattering has been recently measured with improved accuracy by the TEXONO collaboration [12], and since the detector is very close to the reactor neutrino source and therefore neutrino oscillation effects can be avoided, this experimental setup is very appropriate to the study of new physics [13, 14].

The cross section for neutrino electron scattering in the Standard Model is given by

$$\frac{d\sigma(\bar{\nu}_e)}{dT} = \frac{2G_F^2 m_e}{\pi}\left[g_R^2 + g_L^2(1-\frac{T}{E_\nu})^2 - g_R g_L \frac{m_e T}{E_\nu^2}\right], \tag{1}$$

where G_F is the fermi constant, m_e is the electron mass, and the coupling constants are given by $g_L = \frac{1}{2} + \sin^2\theta_W$, $g_R = \sin^2\theta_W$.

The Kuo Sheng reactor is used as an antineutrino source, it has a power of 2.9 GW and provides an average antineutrino flux of $6.4 \times 10^{12} cm^{-2}s^{-1}$ [15].

In this section we introduce the necessary details for the statistical analysis of the neutrino data collected by the CsI(Tl) scintillator detector.

The TEXONO collaboration also reported recent results in the framework of NSI and in the unparticle physics scheme, where operators of scalar and vectorial structure are analyzed [15], an analysis of the tensor operators is missing, and in this work we introduce the results of such an analysis. We will see that it is possible to obtain strong constraints to the parameters in both physics approaches.

In order to estimate the expected number of antineutrino events detected in an energy bin, we compute the following integral

$$N_i = K \int_{T_i}^{T_{i+1}} \int_{E_\nu} \frac{d\sigma}{dT} \frac{d\phi(\bar{\nu}_e)}{dE_\nu} dE_\nu dT, \tag{2}$$

where the cross section is the one that corresponds to the new physics scheme under study, NSI or unparticle, plus the Standard Model contribution given in equation 1. The factor K accounts for the time exposure and the number of electron targets; $d\phi(\bar{\nu}_e)/dE_\nu$ is the neutrino spectrum which can be parametrized as the exponential of a polynomial of order five as has been recently discussed in the literature [16].

We havealso considered the relative abundances of each radioactive isotope in the nuclear reactor, ^{235}U(98%), ^{238}U(1.5%), and ^{239}Pu(0.4%). The electron recoil energy is divided into ten bins, T_i, running from 3 to 8 MeV.

Once we have computed the theoretically expected events per bin we can compute the χ^2 function as

$$\chi^2 = \sum_{i=1} \left[\frac{N_{expt}(i) - [N_{NSI,U}(i)]}{\Delta_{stat}(i)} \right]^2, \tag{3}$$

where $N_{NSI,U}(i)$, is the calculated event rate in the ith energy data bin for the Tensorial NSI or the unparticle cases, $N_{expt}(i)$ is the observed event rate for the corresponding energy bin given by the experiment, and $\Delta_{stat}(i)$ is the statistical uncertainty of the associated measurement.

3. Limits to Unparticle interactions in neutrino electron scattering

At energies above certain Λ, a hidden sector operator $\mathcal{O}_{UV}$ of dimension d_{UV} could couple to the SM operators $\mathcal{O}_{SM}$ of dimension d_{SM} via the exchange of heavy particles of mass M, in the ultra violet energy regime the interaction Lagrangian is

$$\mathcal{L}_{UV} = \frac{\mathcal{O}_{UV}\mathcal{O}_{SM}}{M^{d_{UV}+d_{SM}-4}}. \tag{4}$$

The hidden sector becomes scale invariant at Λ and then the interactions become of the following form

$$\mathcal{L}_{\mathcal{U}} = C_{\mathcal{O}_{\mathcal{U}}} \frac{\Lambda^{d_{UV}-d}}{M^{d_{UV}+d_{SM}-4}} \mathcal{O}_{\mathcal{U}}\, \mathcal{O}_{SM}, \tag{5}$$

where $\mathcal{O}_{\mathcal{U}}$ is the unparticle operator of scaling dimension d in the low energy limit and $C_{\mathcal{O}_{\mathcal{U}}}$ is a dimensionless coupling constant.

The unparticle sector could appear at low energies in the form of new massless fields coupled very weakly to the SM particles, having important manifestations in neutrino electron scattering.

Scalar and vectorial unparticle interactions have been studied previously in the context of neutrino electron scattering [17, 18, 15]. In what follows we will derive the neutrino electron cross section for the case of tensor unparticle operators.

The effective interactions for the tensor unparticle operators have the form in the low enegy regime [19, 20]

$$\frac{-i}{4}\frac{\kappa_T}{\Lambda_{\mathcal{U}}^{d_{\mathcal{U}}}} \bar{f}\,(\gamma_\mu \overleftrightarrow{D}_\nu + \gamma_\nu \overleftrightarrow{D}_\mu)\,\psi_f\, \mathcal{O}_{\mathcal{U}}^{\mu\nu} \text{ and } \frac{\kappa_T}{\Lambda_{\mathcal{U}}^{d_{\mathcal{U}}}} \mathcal{F}_{\mu\alpha}\mathcal{F}_\nu^\alpha\, \mathcal{O}_{\mathcal{U}}{}^{\mu\nu}, \tag{6}$$

where $\mathcal{F}_{\mu\alpha}$ denotes the gauge field strength. The unparticle propagator for the tensorial field is given by

$$[\mathcal{A}_{\mathcal{F}}(P^2)]_{\mu\nu,\rho\sigma} = \frac{\mathcal{A}_d}{2\sin(d\pi)}(-P^2)^{d-2} T_{\mu\nu,\rho\sigma}(P), \tag{7}$$

where

$$\mathcal{A}_d = \frac{16\pi^{5/2}}{(2\pi)^{2d}} \frac{\Gamma(d+1/2)}{\Gamma(d-1)\Gamma(2d)} \tag{8}$$

and the tensor operator $T_{\mu\nu\rho\sigma}$ is given by

$$T_{\mu\nu\rho\sigma} = \frac{1}{2}\left\{\pi_{\mu\rho}\pi_{\nu\sigma} + \pi_{\mu\sigma}\pi_{\nu\rho} - \frac{2}{3}\pi_{\mu\nu}\pi_{\rho\sigma}\right\}, \tag{9}$$

with

$$\pi_{\mu\nu} = -g_{\mu\nu} + \frac{P_\mu P_\nu}{P^2}. \tag{10}$$

If we consider the case of neutrino electron scattering we will arrive to the amplitude

$$M_T = \frac{f(d)}{\Lambda_U^{2d-2}}\left\{\bar{\nu}(k')\sigma^{\mu\nu}\nu(k)\right\}\left(-(P)^2 - i\varepsilon\right)^{d_u-2} T_{\mu\nu\rho\sigma}\left\{e(p')\sigma^{\rho\sigma}e(p)\right\}, \tag{11}$$

where

$$f(d) = \frac{\lambda_{i\nu}^{\alpha\beta}\lambda_{if}}{2\sin(d\pi)}\mathcal{A}_d. \tag{12}$$

In what follows we will use the definition for the couplings $\lambda = \sqrt{\lambda_{i\nu}^{\alpha\beta}\lambda_{if}}$ and we will leave the scale $\Lambda = 1$ TeV fixed. From this amplitude, it is possible to obtain the differential cross section for neutrino electron scattering for the tensorial unparticle interactions. The electric part which takes the form:

$$\frac{d\sigma_T}{dT} = \frac{f(d)^2}{\pi\Lambda_U^{4d-4}} 2^{2d-3} m_e^{2d-3} T^{2d-4}\left[\left(1 - \frac{T}{2E_\nu}\right)^2 - \frac{mT}{2E_\nu^2}\right] \tag{13}$$

and the magnetic contribution to the cross section is given by

$$\frac{d\sigma_T}{dT} = \frac{f(d)^2}{\pi\Lambda_U^{4d-4}} 2^{2d-2} m_e^{2d-3} T^{2d-4}\left(1 - \frac{T}{2E_\nu}\right)^2. \tag{14}$$

For the total cross section we add this two contributions plus the one calculated from the Standard Model.

For not only scale but also conformally invariant theories, there is a bound from unitarity constraints in the dimension d [21], in this phenomenological approach we leave free the value of the dimension parameter for a broader range where the case of a theory scale invariant but not conformally invariant is possible.

After performing a χ^2 analysis of the TEXONO data, we have found that, the unparticle tensor interaction, our constraints are more restrictive than previous limits reported for the case when $d > 2.03$. This result is very encouraging and the details of the analysis will be published elsewhere.

4. Tensorial neutrino electron NSI

It is a common characteristic of many extensions of the Standard Model to introduce new interactions that can be parametrized with the help of an effective Lagrangian. The most studied case regards with $V - A$ extensions of the Standard Model. It is well known that in this case the effective four fermion Lagrangian takes the form: [22, 23, 24, 25, 26]

$$-\mathcal{L}_{\mathrm{V-A}}^{eff} = \varepsilon_{\alpha\beta}^{fP} 2\sqrt{2}G_F(\bar{\nu}_\alpha\gamma_\rho L\nu_\beta)(\bar{f}\gamma^\rho P f), \tag{15}$$

where f is a first generation SM fermion: e, u or d, and $P = L$ or R are the chiral projectors.

In particular for the neutrino electron scattering the neutrino electron cross section for the $(V-A)$ NSI case is given by

$$\begin{aligned}\frac{d\sigma(E_\nu,T)}{dT} &= \frac{2G_F^2 m_e}{\pi}[(\tilde{g}_L^2 + \sum_{\alpha\neq e}|\epsilon_{\alpha e}^{eL}|^2) \\ &+ (\tilde{g}_R^2 + \sum_{\alpha\neq e}|\epsilon_{\alpha e}^{eR}|^2)\left(1-\frac{T}{E_\nu}\right)^2 - (\tilde{g}_L\tilde{g}_R + \sum_{\alpha\neq e}|\epsilon_{\alpha e}^{eL}||\epsilon_{\alpha e}^{eR}|)m_e\frac{T}{E_\nu^2}],\end{aligned} \tag{16}$$

with the new effective couplings left and right given by $\tilde{g}_L = g_L + \epsilon_{\alpha\alpha}^{eL}$ and $\tilde{g}_R = g_R + \epsilon_{\alpha\alpha}^{eR}$.

Previous study to this interaction in neutrino electron scattering has been given in different articles [27] and there are some constraints reported in the literature [28].

The Lagrangian describing this type of interaction is given by

$$-\mathcal{L}_{\mathrm{T}}^{eff} = \varepsilon_{\alpha\beta}^{fT} 2\sqrt{2}G_F(\bar{\nu}_\alpha\sigma^{\mu\nu}\nu_\beta)(\bar{f}\sigma_{\mu\nu}f), \tag{17}$$

with $\sigma_{\mu\nu} = \gamma_\mu\gamma_\nu + \gamma_\nu\gamma_\mu$. In this case, for the tensorial contribution the amplitude for the process is given by

$$|M|^2 = \varepsilon_{\alpha\beta}^{fT\,2}\frac{G_F^2}{2}128m_e^2(4E_\nu^2 + T^2 - (4E_\nu + m_e)T) \tag{18}$$

and therefore the differential NSI neutrino electron cross section takes the form:

$$\frac{d\sigma_T^{NSI}}{dT} = \frac{|M|^2}{64\pi m_e E_\nu^2} = \varepsilon_{\alpha\beta}^{fT\,2}\frac{4G_F^2 m_e}{\pi}\left[\left(1-\frac{T}{2E_\nu}\right)^2 - \frac{m_e T}{4E_\nu^2}\right]. \tag{19}$$

In this case, $\varepsilon_{\alpha\beta}^{fT}$ parametrizes the strength of the tensorial NSI coupling.

The limit obtained for the tensorial non standard parameter on the tensorial coupling gives the bound, at 90 % CL, $g_T \leq 0.20$ which is much better than the previously reported constraint by the LAMPF collaboration [28]. We obtained the value of $\chi^2_{min} = 5.43$ for 9 d.o.f. We have verified explicitly that previous reactor experiments give weaker constraints than those obtained here using the TEXONO data, when we perform the same analysis for the MUNU experiment we find that $g_T \leq 0.25$ at 90 % CL, in an energy range of 0.2 to 2.0 MeV [14].

5. Discussion and summary

The recent experimental results put neutrino physics in a time of precision measurements, where important guidance for physics beyond the Standard Model could be found.

For the unparticle tensor interaction, when $d > 2.03$ we find that our constraints are more restrictive than previous limits reported, for the tensor NSI previous constraints are also improved by this analysis.

As can be seen the results are encouraging and future neutrino electron scattering experiments could give even stronger constraints.

Another possible place to search for this type of interaction in the future could be the coherent neutrino nuclei scattering that is also part of the TEXONO low energy neutrino physics program [29] and other proposals [30, 31, 32].

Acknowledgments

This work has been supported by CONACyT. J.B. is partially supported by UNAM-DGAPA PAPIIT IN113211. A. B. acknowledges RED-FAE CONACYT for Postdoctoral Grant. This work is an updated version of our previous preprint [33]

References

[1] T. Schwetz, M. Tortola and J. W. F. Valle 2011 *New J. Phys.* **13** 063004
[2] H. Nunokawa, S. J. Parke and J. W. F. Valle, 2008 *Prog. Part. Nucl. Phys.* **60** 338
[3] R. N. Mohapatra and A. Y. Smirnov 2006 *Ann. Rev. Nucl. Part. Sci.* **56** 569
[4] J. Barranco, O. G. Miranda, C. A. Moura and J. W. F. Valle 2006 *Phys. Rev.* D **73** 113001
[5] J. Barranco, O. G. Miranda, C. A. Moura and J. W. F. Valle 2008 *Phys. Rev.* D **77** 093014
[6] A. Bolanos, O. G. Miranda, A. Palazzo, M. A. Tortola and J. W. F. Valle, 2009 *Phys. Rev.* D **79** 113012
[7] J. Kopp, P. A. N. Machado and S. J. Parke 2010 *Phys. Rev.* D **82** 113002
[8] N. C. Ribeiro, H. Nunokawa, T. Kajita, S. Nakayama, P. Ko and H. Minakata 2008 *Phys. Rev.* D **77** 073007
[9] T. Banks and A. Zaks, Nucl. Phys. B **196**, 189 (1982).
[10] H. Georgi 2007 *Phys. Rev. Lett.* **98** 221601
[11] H. Georgi, 2007 *Phys. Lett.* B **650** 275
[12] M. Deniz *et al.* [TEXONO Collaboration] 2007 *Phys. Rev.* D **81** 072001 [arXiv:0911.1597 [hep-ex]].
[13] Z. Daraktchieva *et al.* [MUNU Collaboration] 2005 *Phys. Lett.* B **615** 153
[14] H. T. Wong *et al.* [TEXONO Collaboration] 2007 *Phys. Rev.* D **75** 012001
[15] M. Deniz *et al.* [TEXONO Collaboration] 2010 *Phys. Rev.* D **82** 033004 [arXiv:1006.1947 [hep-ph]].
[16] T. .A. Mueller, D. Lhuillier, M. Fallot, A. Letourneau, S. Cormon, M. Fechner, L. Giot and T. Lasserre *et al.*, Phys. Rev. C **83**, 054615 (2011) [arXiv:1101.2663 [hep-ex]].
[17] A. B. Balantekin and K. O. Ozansoy, Phys. Rev. D **76**, 095014 (2007) [arXiv:0710.0028 [hep-ph]].
[18] J. Barranco, A. Bolanos, O. G. Miranda, C. A. Moura and T. I. Rashba, Phys. Rev. D **79**, 073011 (2009) [arXiv:0901.2099 [hep-ph]].
[19] K. Cheung, W. Y. Keung and T. C. Yuan, Phys. Rev. D **76**, 055003 (2007) [arXiv:0706.3155 [hep-ph]].
[20] T. -i. Hur, P. Ko and X. -H. Wu, Phys. Rev. D **76**, 096008 (2007) [arXiv:0709.0629 [hep-ph]].
[21] B. Grinstein, K. A. Intriligator and I. Z. Rothstein, Phys. Lett. B **662**, 367 (2008) [arXiv:0801.1140 [hep-ph]].
[22] V. D. Barger, R. J. N. Phillips and K. Whisnant 1991 *Phys. Rev.* D **44** 1629
[23] S. Bergmann, M. M. Guzzo, P. C. de Holanda, P. I. Krastev and H. Nunokawa 2000 *Phys. Rev.* D **62** 073001 [arXiv:hep-ph/0004049].
[24] P. Huber and J. W. F. Valle 2001 *Phys. Lett.* B **523** 151
[25] Z. Berezhiani and A. Rossi 2002 *Phys. Lett.* B **535** 207; Z. Berezhiani, R.S. Raghavan and A. Rossi 2002 *Nucl. Phys.* B **638** 62
[26] S.Davidson, C. Pena-Garay, N. Rius and A. Santamaria 2003 *J. High Energy Phys.* JHEP03(2003)011
[27] B. Kayser, E. Fischbach, S. P. Rosen and H. Spivack, Phys. Rev. D **20**, 87 (1979).
[28] R. C. Allen *et al.*, Phys. Rev. D **47**, 11 (1993).
[29] H. T. Wong, H. B. Li, J. Li, Q. Yue and Z. Y. Zhou 2006 *J. Phys. Conf. Ser.* **39** 266
[30] J. I. Collar [CoGeNT Collaboration] 2008 *J. Phys. Conf. Ser.* **136** 022009
[31] A. J. Anderson, J. M. Conrad, E. Figueroa-Feliciano, K. Scholberg and J. Spitz Coherent Neutrino Scattering in Dark Matter Detectors *Preprint* arXiv:1103.4894 [hep-ph]
[32] K. Scholberg 2006 *Phys. Rev.* D **73** 033005
[33] J. Barranco, A. Bolanos, E. A. Garces, O. G. Miranda and T. I. Rashba, arXiv:1108.1220 [hep-ph].

Journal of Physics: Conference Series **378** (2012) 012018 doi:10.1088/1742-6596/378/1/012018

Single Top quark production at lepton colliders

F. Peñuñuri[1], **F. Larios**[2] **and Antonio O. Bouzas**[2]
[1] Facultad de Ingeniería, Universidad Autónoma de Yucatán, A.P. 150, Cordemex, Mérida, Yucatan, México
[2] Departamento de Física Aplicada, CINVESTAV-Mérida, A.P. 73, 97310 Mérida, Yucatán, México

E-mail: larios@mda.cinvestav.mx

Abstract. Single top quark production at lepton (or non-hadronic) colliders like the e^+e^- International Linear Collider (ILC) can be used to obtain high precision measurements of the V_{tb} CKM matrix element as well as the effective tbW coupling. For the ILC we have calculated the QCD correction for the cross section in the context of an effective vector boson approximation. Our results show a $\sim 10\%$ increase due to the strong interaction.

1. Single Top production at e^+e^-, e^-e^-, γe and $\gamma\gamma$ colliders

The top quark is likely to provide us with the first clues of physics beyond the Standard Model [1]. The cross section for $t\bar{t}$ production being much greater than for single top has made this mode not only the one of initial discovery but also the one for further studies. In fact, new physics effects are probably already manifest in the recent $t\bar{t}$ forward-backward asymmetry observed at the Tevatron [2, 3]. On the other hand, single top production originates from the weak interaction and is directly proportional to the left handed tbW coupling. Just as it was at the Tevatron, in the LHC this process will be extensively studied [4].

The planned International Linear Collider (ILC) will collide electron and positron beams at an initial energy of 500 GeV and higher. It will provide a clean environment for the study of precision measurements. There are however other three possible modes of operation depending on the initial beams to be used like e^-e^-, γe and $\gamma\gamma$.

Table 1. Summary of results for the four types of lepton colliders studied in Ref. [5].

beams	Number of diagrams	$t\bar{t}$ background	σ(fb) ($\sqrt{s} = 0.5$ TeV)	σ(fb) ($\sqrt{s} = 1.0$ TeV)
$e^+ e^-$	20	yes	3.1	6.7
$\gamma\gamma$	21	yes	9.2	18.8
$e\ e$	20	no	1.7	9.1
γe	4	no	30.3	67.6

The single top production processes at lepton and photon (e^+e^-, e^-e^-, γe and $\gamma\gamma$) colliders have been extensively studied at tree level in [5]. We can summarize their results in Table 1

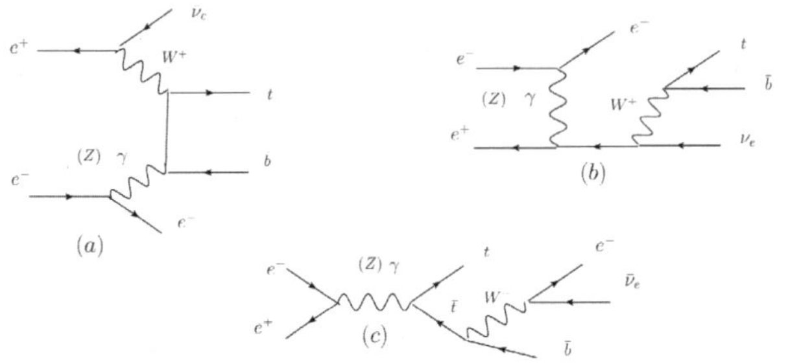

Figure 1. The three types of diagrams for the process $e^+e^- \to t\bar{b}e^-\bar{\nu}_e$: (a) vector boson fusion, (b) vector boson exchange and (c) e^+e^- annihilation.

where the four possible colliders are presented. The reaction $\gamma e^- \to \bar{t}b\nu_e$, is particularly suitable for precision studies, as it does not have the $t\bar{t}$ background. Compared to the ILC $e^+e^- \to t\bar{b}e^-\bar{\nu}_e$ process the γe^- reaction can yield a larger production rate and is directly proportional to the V_{tb} term. Further studies have thus been done for this reaction. In particular, the QCD corrections have been studied in [6]. Their conclusion is that the QCD correction is not very large ($\sim 5\%$) so that this mode remains very well suited for a precise measurement of V_{tb}. The approach of [6] is to use the effective vector boson approximation, or *effective W approximation* [7] (EWA) and to compute the QCD loop corrections for the $W^+\gamma \to t\bar{b}$ fusion process. Then, the convolution with the $f_{W^+/e^+}(x)$ distribution function is applied to obtain the correction to the actual $e^+\gamma$ process. We would like to point out that the authors of [6] have made a very clear and thorough presentation of the calculation. In this work we use their analysis of the $W^+\gamma \to t\bar{b}$ process to estimate the QCD correction for the $e^+e^- \to t\bar{b}e^-\bar{\nu}_e$ process of the ILC. Here, in addition to the convolution with the W^+ boson distribution function we will use the effective photon (as well as the effective Z boson) approximation to obtain the QCD correction. We will use the same input values as in [6] for masses and coupling constants, except for the masses of top and bottom quarks we take $m_t = 173$ GeV and $m_b = 4.2$ GeV.

2. Vector boson contributions at tree level

At tree level there are 20 diagrams for the process $e^+e^- \to t\bar{b}e^-\bar{\nu}_e$ [5]. We can list them in three different types: (a) vector boson fusion, (b) vector boson exchange and (c) e^+e^- annihilation (see Figure1). For the energy range we consider one of the diagrams actually corresponds to $t\bar{t}$ production, where one of the tops decays leptonically. In order to exclude $t\bar{t}$ production from the single top process we discard all events where the invariant mass of the decay products (e^-,$\bar{\nu}_e$,$\bar{b}$) falls inside an interval around the top mass $m_t - \Delta M \leq M_{e\nu b} \leq m_t + \Delta M$. We take the value $\Delta M = 20$GeV as in [5].

The effective W approximation relies on the fact that the vector boson fusion diagrams become dominant when heavy particles are produced in very high energy collisions [7]. In general, three conditions should be met for the EWA to work well: (1) The mass of the vector boson (M_W or M_Z) should be much smaller than its energy, and this can be met if we require $M_V \ll \sqrt{s}/2$, (2) For $q\bar{q}$ production $m_q \gg M_V$, this is true for the top quark but not for the bottom quark, and (3) One polarization mode should be dominant so that interference effects can be neglected. Fortunately, in our case the mode $W\gamma \to t\bar{b}$ dominates for longitudinal W, and the modes with the Z boson $WZ \to t\bar{b}$ give even lower contributions.

As expected, this method works very well for $t\bar{t}$ production at high $\sqrt{s}$ and to a lesser degree for single top, which in our case can be seen as $t\bar{b}$ production. In [6] the QCD correction to the

Journal of Physics: Conference Series **378** (2012) 012018 doi:10.1088/1742-6596/378/1/012018

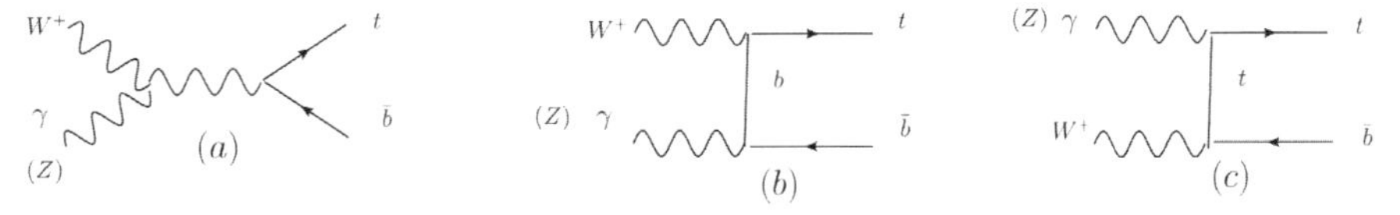

Figure 2. Diagrams for the $W^+\gamma(Z) \to t\bar{b}$ process.

process $e^+\gamma \to t\bar{b}\bar{\nu}_e$ was calculated by doing first the QCD correction to the $W^+\gamma$ fusion into $t\bar{b}$ and then by taking the convolution with an effective W^+ coming from the initial positron (see Figure 2). We follow the same approach by doing the one loop QCD correction to $W^+\gamma \to t\bar{b}$ as well as $W^+Z \to t\bar{b}$ and then convoluting with the effective distribution functions for W^+, γ and Z:

$$\begin{aligned} \sigma(e^+e^- \to t\bar{b}\bar{\nu}_e e^-) \quad &= \quad (1) \\ &\sum_{W_L,W_T} \int_{x_W^{\min}}^1 dx_W f_{W^+/e^+}(x_W) \int_0^1 dx_\gamma f_{\gamma/e^-}(x_\gamma)\, \sigma(W^+\gamma \to t\bar{b})(\hat{s}) \\ + \quad &\sum_{W_{L,T},Z_{L,T}} \int_{x_W^{\min}}^1 dx_W f_{W^+/e^+}(x_W) \int_{x_Z^{\min}}^1 dx_Z f_{Z/e^-}(x_Z)\, \sigma(W^+Z \to t\bar{b})(\hat{s}) \end{aligned}$$

Where, $x_V^{\min} = 2M_V/\sqrt{s}$, $\hat{s} = x_W x_\gamma s$ or $x_W x_Z s$, and the structure functions can be found in [7]. The tree-level cross section for single top production at the ILC is shown in Fig. 3. The exact Born-level calculation for the $e^+e^- \to t\bar{b}e^-\bar{\nu}_e$ process is obtained with CalcHEP [8] and is shown by the solid line. We can see that the prediction of the EWA (dot-dashed curve) is in very good agreement with the exact result for center of mass energies above 1.5 TeV. However, for the energy range of the ILC the EWA values can be significantly lower. In particular, for $\sqrt{s} = 1000$GeV there is a 15% difference and for $\sqrt{s} = 500$GeV the EWA result is about one half of the exact value.

Some kinematical aspects of our calculation are worth discussing in more detail. Because in the dominant diagrams for the complete process $e^+e^- \to W^{+*}Z^* \to t\bar{b}e^-\bar{\nu}_e$ the virtual vector bosons get space-like momenta $k_V^2 \leq 0$ ($V = W$, Z) and are always far from their mass shell, the EWA is known to work better when the vector-boson squared momenta are set to $k_V^2 = 0$ in the external legs, as discussed in [10]. Nevertheless, when dealing with a process like $t\bar{t}$ production one may set $k_V^2 = M_V^2$, as this introduces only a small error of order $M_V/\sqrt{s}$. For this reason it is customary [7, 6] to set the external massive W^+ and Z on-shell for convenience. We stress here, however, that the EWA requires [7, 10] the limits of integration to be taken as defined in (1) for all polarization states of the massive vector bosons, regardless of whether one chooses to set $k_V^2 = 0$ or M_V^2 [1]. In our study we keep the EWA condition $k_Z^2 = 0$ for the Z boson, but choose $k_W^2 = M_W^2$ for the initial-state W^+ for calculational convenience. We have numerically checked that indeed by setting $k_W^2 = 0$ we don't find a significant change in the result.

Because of the kinematics of the $W^+Z \to t\bar{b}$ process, its tree-level scattering amplitude has a pole singularity within its physical region [9], leading to divergent behavior as we integrate it over the Mandelstam variable t (or the polar angle of the outgoing quark). This is due to the fact that, for energy values from slightly above the threshold $m_t + m_b$ and up to ~ 3 TeV, there exists a certain value of t such that the massive Z boson can actually decay into $b\bar{b}$, causing the bottom quark propagator to hit its pole at that t. By strictly adhering to the EWA requirement $k_Z^2 = 0$,

[1] Notice that in [6] the lower limit is set to $x_W^{\min} = 0$ for the longitudinal W^+ distribution, a procedure that is only justified a posteriori by their results.

Journal of Physics: Conference Series **378** (2012) 012018 doi:10.1088/1742-6596/378/1/012018

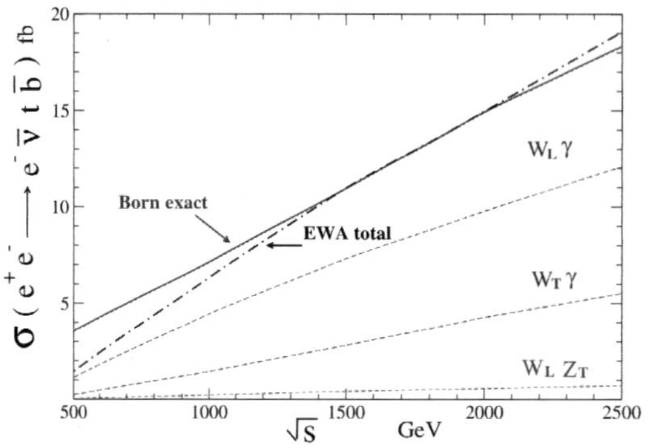

Figure 3. The Born cross section for single top production at the ILC.

the singularity is pushed outside the physical region and no divergent integration appears in the computation. Notice that a similar situation does not occur with the W^+ boson, as it cannot decay into $t\bar{b}$.

Below, we will describe the QCD corrections to the $W^+\gamma$ and W^+Z processes, including the Dipole Subtraction Method of infrared divergences. We have followed closely the analysis of the $W^+\gamma$ mode done by Kuhn et.al. in [6].

3. QCD correction to the $W^+\gamma(Z) \to t\bar{b}$ process.

The QCD loop correction to the $W^+\gamma(Z) \to t\bar{b}$ process is given by 9 Feynman diagrams (see Fig.2 of [6]). The renormalization procedure involves only the quark's wave function and mass parameter. Specific formulas can be found in [6]. Concerning the renormalization scale dependence we have also set α_s at the scale $\mu = \sqrt{s}$ for our numerical calculation (it becomes $\sqrt{\hat{s}}$ under the convolution). The extraction of IR singularities is done with the subtraction method of the dipole formalism [11]. This method consists of adding and subtracting a so-called dipole term:

$$\sigma^{NLO}(W^+\gamma \to t\bar{b}) = \int_{tbg} \left[(d\sigma^R)_{\epsilon=0} - (d\sigma^B \otimes dV_{dipole})_{\epsilon=0}\right] + \int_{tb} [d\sigma^V + d\sigma^B \otimes \mathbf{I}]_{\epsilon=0} \quad (2)$$

Where $d\sigma^R$ comes from the real emission $W^+\gamma(Z) \to t\bar{b}g$ process and $d\sigma^B \otimes dV_{dipole}$ is the subtracting dipole term that matches pointwise the singularities associated with the soft and/or collinear gluon. Both terms are calculated in $d = 4$ dimensions. In the second integral the same dipole term has been partially integrated in the gluon phase space and then added to the virtual correction $d\sigma^V$. This sum is performed in $d = 4 - 2\epsilon$ dimensions (consistent with dimensional regularization).

The general formula for the dipole term is found in Eq. (5.16) of [11]. The specific expression in our case is:

$$d\sigma^B \otimes dV_{dipole} = \frac{\langle V_{gt,b}\rangle}{2k_g \cdot k_t} |\mathcal{M}_0(\tilde{k}_{gt}, \tilde{k}_b)|^2 + \{t \leftrightarrow b\}, \quad (3)$$

where

$$\langle V_{gt,b}\rangle = 8\pi\alpha_s C_F\{\frac{2}{1 - \tilde{z}_t(1 - y_{gt,b})} - \frac{\tilde{v}_{gt,b}}{v_{gt,b}}[1 + \tilde{z}_t + \frac{m_t^2}{k_g \cdot k_t}]\},$$

$$\begin{aligned}
\tilde{z}_t &= \frac{k_t \cdot k_b}{(k_t + k_g) \cdot k_b}\,, \qquad y_{gt,b} = 2\frac{k_g \cdot k_t}{s x_{tb}}\,, \qquad \tilde{v}_{gt,b} = \frac{\lambda_{tb}}{x_{tb}}\,, \\
v_{gt,b} &= \sqrt{(1+a_{gt,b})^2 - a_{gt,b}^2/z_b}\,, \qquad a_{gt,b} = \frac{2z_b}{x_{tb}(1-y_{gt,b})}\,, \\
\tilde{k}_b &= \frac{x_b}{2}P + \frac{\lambda_{tb}}{\lambda_{gt}}(k_b - \frac{P \cdot k_b}{s}P)\,, \qquad \tilde{k}_{gt} = P - \tilde{k}_b\,, \qquad P = k_W + k_\gamma\,,
\end{aligned}$$

and $\mathcal{M}_0(\tilde{k}_{gt}, \tilde{k}_b)$ is the Born-level $W^+\gamma \to t\bar{b}$ amplitude with one modification: the final state momenta k_t and k_b have been replaced by $\tilde{k}_{gt}$ and $\tilde{k}_b$ respectively. The other variables are defined as in [6]: $\mu_q = m_q/\sqrt{s}$, $z_q = \mu_q^2$, $x_t = 1 + z_t - z_b$, $x_b = 1 + z_b - z_t$, $x_{tb} = 1 - z_t - z_b$, $\lambda_{tb} = \lambda(1, z_t, z_b)$, $\lambda_{gt} = \lambda(1, (k_g + k_t)^2/s, z_b)$, and $\lambda(x, y, z) = \sqrt{x^2 + y^2 + z^2 - 2xy - 2xz - 2yz}$.

For the real emission correction we have prepared a Fortran program that integrates the cross section for the $W^+\gamma \to t\bar{b}g$ process along with dipole subtraction. As it turns out, the subtraction term defined by the dipole formalism in the first integral of Eq. (2) is actually a very good approximation to the real emission cross section in an important part of the tbg phase space, so that the numerical results we obtained were very small: about two orders of magnitude below the values obtained for the virtual correction.

The expression for the dipole term in the virtual correction is:

$$d\sigma^B \otimes \mathbf{I} = |\mathcal{M}_d(W^+\gamma \to t\bar{b})|^2 \frac{\alpha_s}{2\pi} \frac{1}{\Gamma(1-\epsilon)} \left(\frac{4\pi\mu^2}{s}\right)^\epsilon (\mathbf{I}_{gt,b} + \mathbf{I}_{gb,t})\,, \tag{4}$$

where $\mathcal{M}_d(W^+\gamma \to t\bar{b})$ is the Born-level amplitude in $d = 4 - 2\epsilon$ dimensions (the flux term of the $t\bar{b}$ phase space integration is understood). The dipole function is given by $\mathbf{I}_{gt,b} = C_F[2I^{eik} + I^{coll}_{gt,b}]$ (also $\mathbf{I}_{gb,t} = \mathbf{I}_{gt,b}\{t \leftrightarrow b\}$), where I^{eik} and $I^{coll}_{gt,b}$ are given by Eqs. (5.34) and (5.35) in [11]:

$$\begin{aligned}
I^{eik} &= \frac{x_{tb}}{\lambda_{tb}}\{\frac{\ln\rho}{2\epsilon} + \frac{\pi^2}{6} - \ln\rho\ln[1 - (\mu_t + \mu_b)^2] - \frac{1}{2}ln^2\rho_t - \frac{1}{2}ln^2\rho_b \\
&+2Li_2(-\rho) - 2Li_2(1-\rho) - \frac{1}{2}Li_2(1-\rho_t^2) - \frac{1}{2}Li_2(1-\rho_b^2)\} \\
I^{coll}_{gt,b} &= \frac{1}{\epsilon} + 3 + \ln\mu_t + \ln(1-\mu_b) - 2\ln[(1-\mu_b)^2 - z_t] - \frac{\mu_b}{1-\mu_b} \\
&- \frac{2}{x_{tb}}\left[\mu_b(1-2\mu_b) + z_t\ln\frac{\mu_t}{1-\mu_b}\right]
\end{aligned} \tag{5}$$

where $\rho^2 = (x_{tb} - \lambda_{tb})/(x_{tb} + \lambda_{tb})$, $\rho_t = (x_{tb} - \lambda_{tb} + 2z_t)/(x_{tb} + \lambda_{tb} + 2z_t)$, and $\rho_b = \rho_t\{t \leftrightarrow b\}$. These formulas also appear in [6], except that in their Eq. (4.14) in $I^{coll}_{gt,b}$ the constant term should not be 5 but 3.

Concerning the calculation of $d\sigma^V$, the details can be found in [6]. We actually worked out this same computation before doing the case for the Z boson. As expected from the results shown in Fig. 3 the contribution from W^+Z fusion is much smaller than the one from $W^+\gamma$. In fact, we only considered the correction for the polarizations W^+ longitudinal and Z transversal as the other possibilities yield negligible contributions.

Our results are shown in Figs. 4 and 5. The solid line in Fig. 4 is the same exact Born-level result shown in Fig. 3. The dashed line is obtained by adding to the solid line the QCD correction computed within the EWA method. Fig. 5 shows the ratio of the QCD correction to the exact Born cross section (solid line). As seen there, for $\sqrt{s} > 1$ TeV the QCD correction remains roughtly stable about 11%. Its slow decrease above ~ 2 TeV is due to the running of $\alpha_s(\mu = \sqrt{\hat{s}})$. On the other hand, below 1 TeV the correction drops to 7% with decreasing energy. This is an effect of the Born-level EWA predicted cross section, that drops to $\sim 50\%$ of the exact Born-level value as shown in Fig. 3. If instead of using the exact Born value as denominator we use the EWA Born prediction (see dot-dashed line in Fig. 3) our results (dashed line) show a 17% increase in the Born level cross section at $\sqrt{s} = 500$ GeV.

It will be interesting to compare this result based on the effective W approximation with a future more robust calculation based on the complete $e^+e^- \to t\bar{b}e^-\bar{\nu}_e$ process.

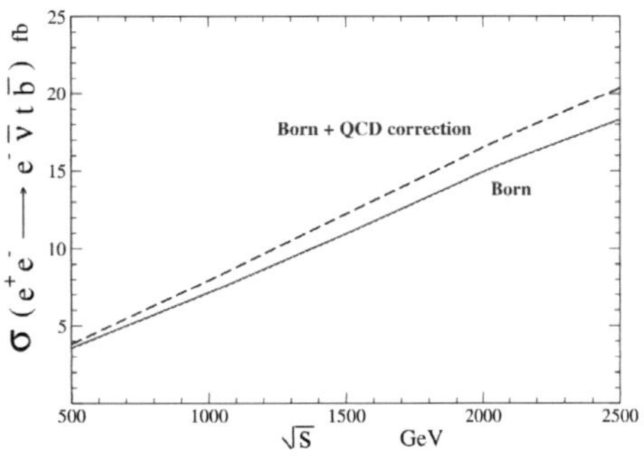

Figure 4. The QCD correction from $W^+\gamma$ and W^+Z fusion to the $e^+e^- \to t\bar{b}e^-\bar{\nu}_e$ process. The exact Born-level calculation (solid line) and Born plus QCD correction (dashed line).

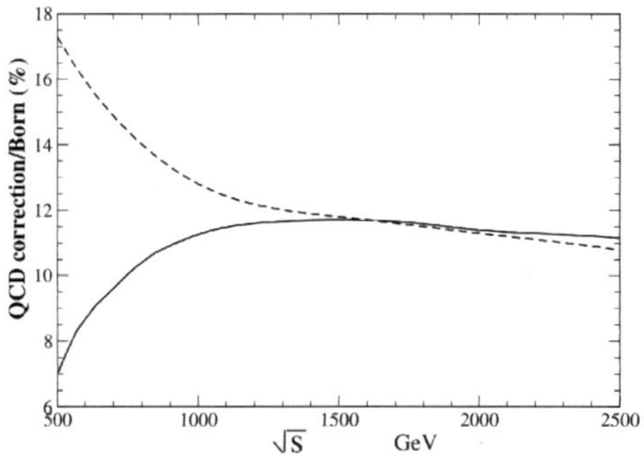

Figure 5. Ratio of QCD correction to exact Born cross section (solid line) and ratio of QCD correction to EWA Born cross section (dashed line).

Acknowledgments

We thank RedFAE, Conacyt and SNI for support. F.L. thanks C.-P. Yuan for useful discussions.

References

[1] Kumar K, Tait TMP and Vega-Morales R 2009, *JHEP* **0905**, 022; Berger EL, Cao QH and Low I 2009, *Phys. Rev.* D**80**, 074020; Bernreuther W, Gonzalez P and Wiebusch M 2009, *Eur. Phys. J.* C**60**, 197; del Aguila F, Aguilar-Saavedra J, Moretti M, Piccinini F, Pittau R and Treccani M 2010, *Phys. Lett.* B**685**, 302; Liu GL 2010, *Phys. Rev.* D**82**, 115032; Cakir IT, Cakir O and Sultansoy S 2010, *Phys. Lett.* B**685**, 170; Zhang C and Willenbrock S 2011, *Phys. Rev.* D**83**, 034006; Groote S, Korner JG, Melic B and Prelovsek S 2011, *Phys. Rev.* D**83** 054018; Martinez R, Perez MA and Poveda N 2008, *Eur. Phys. J.* C**53**, 221-230; Cordero-Cid A, Perez MA, Tavares-Velasco G *et al.* 2004, *Phys. Rev.* D**70**, 074003.

[2] Abazov VM *et al.* [D0 Collaboration] 2008, *Phys. Rev. Lett.* **100**, 142002; Aaltonen T *et al.* [CDF Collaboration] 2008, *Phys. Rev. Lett.* **101**, 202001; *ibid* 2011 *Phys. Rev.* D**83** 112003.

[3] Chivukula RS, Simmons EH and Yuan CP 2010, *Phys. Rev.* D**82**, 094009; Shu J, Tait TMP and Wang K 2010, *Phys. Rev.* D**81**, 034012; Berger EL, Cao QL, Chen CR, Li CS and Zhang H 2011, *Phys. Rev. Lett.* **106** 201801; Cao J, Wang L, Wu L and Yang JM 2011, *Phys. Rev.* D**84** 074001.

[4] Heim S, Cao Q. -H., Schwienhorst R and Yuan C.-P. 2010 *Phys. Rev.* D**81**, 034005; Cao Q. -H., Schwienhorst R, Benitez J. A. and Yuan C.-P. 2005, *Phys. Rev.* D**72**, 094027; Cao Q. -H., Schwienhorst R and Yuan C. -P. 2005, *Phys. Rev.* D**71**, 054023; Campbell J. M., Frederix R, Maltoni F and Tramontano F 2009, *Phys. Rev. Lett.* **102**, 182003; Kidonakis N 2010, *Phys. Rev.* D**81**, 054028; Falgari P, Mellor P and Signer A 2010, *Phys. Rev.* D**82**, 054028; Wang J, Li CS, Zhu HX and Zhang JJ 2010, *Preprint* hep-ph/1010.4509; Barger V, McCaskey M and Shaughnessy G 2010, *Phys. Rev.* D**81**, 034020; Bordes G and van Eijk B 1995, *Nucl. Phys.* B**435** 23; Smith MC and Willenbrock S 1996, *Phys. Rev.* D**54** 6696; Stelzer T, Sullivan Z and Willenbrock S 1997, *Phys. Rev.* D**56** 5919; B.W. Harris et al 2002, *Phys. Rev.* D**66** 054024.

[5] Boos E et al. 2001, *Eur. Phys. J.* C**21**, 81.

[6] Kuhn JH, Sturm C and Uwer P 2003, *Eur. Phys. J.* C**30**, 169.

[7] Dawson S 1985, *Nucl. Phys.* B249, 42.

[8] A.Pukhov et al 1999, *Preprint* hep-ph/9908288; A.Pukhov 2004 *Preprint* hep-ph/0412191.

[9] Coleman S and Norton RE 1965, *Nuov. Cim.* **38** 438.

[10] Kauffman R 1989, Ph.D. thesis, SLAC-0348.

[11] Catani S, Dittmaier S, Seymour MH and Trocsanyi Z 2002, *Nucl. Phys.* B**627**, 189.

XIII Mexican Workshop on Particles and Fields IOP Publishing
Journal of Physics: Conference Series **378** (2012) 012019 doi:10.1088/1742-6596/378/1/012019

New Physics contribution to $B \to \rho K, \pi K^*$ decays

D. Delepine

Departamento de Física, DCI, Campus León, Universidad de Guanajuato, C.P. 37150, León, Guanajuato, México

E-mail: delepine@fisica.ugto.mx

Gaber Faisel

Department of Physics and Center for Mathematics and Theoretical Physics, National Central University, Chung-li, TAIWAN 32054.
and Egyptian Center for Theoretical Physics, Modern University for Information and Technology, Cairo, Egypt

E-mail: gfaisel@ncu.edu.tw

M. Shalaby

Ain Shams University, Faculty of Science, Cairo 11566, Egypt.

E-mail: mstfshalaby@yahoo.com

Abstract. In this paper we discuss supersymmetric contributions to the direct CP asymmetries of $B \to \pi K^*$ and $B \to \rho K$ decays. We use Soft Collinear Effective Theory as a framework for our study. We show that within the standard model and including the next leading order QCD corrections, the predicted CP asymmetries can not accommodate the experimental measurements. We show that non-minimal flavor SUSY contributions mediated by gluino exchange can enhance the CP asymmetries significantly to accommodate the experimental measurements.

1. Introduction

The decay modes $B \to \pi K^*$ and $B \to \rho K$ are generated at the quark level in the same way as $B \to K\pi$ decay. As a consequence, studying these decays can shed light of possible new physics that can solve the $K\pi$ puzzle. These decay modes are studied within SM in framework of QCDF [1], PQCD [2, 3, 4, 5] and Soft Collinear Effective Theory (SCET) [6, 7]. In SCET, the direct CP asymmetries of $B^- \to \pi^- \bar{K}^{*0}$ and $B^- \to \rho^- \bar{K}^0$ are zero while the CP asymmetries in other channels are small. Recently, in Ref.([8]) fits to $B \to \pi K^*$ and $B \to \rho K$ decays are performed where data can be accommodated within the standard model due principally to the large experimental uncertainties, particularly in the CP-violating asymmetries.

Supersymmetry (SUSY) is one of the most interesting candidates for physics beyond the standard model as it naturally solves the hierarchy problem. In addition, SUSY has new sources for CP violation which can account for the baryon number asymmetry and affect other CP violating observables in the B and K decays. The effects of these phases on the CP asymmetries in semi-leptonic τ decays has been studied in Refs.([9, 10, 11]).

Journal of Physics: Conference Series **378** (2012) 012019 doi:10.1088/1742-6596/378/1/012019

Decay channel	Exp.	SM prediction
$\pi^0 K^{(*)+}$	6.9 ± 2.3	$7.2^{+0.3+1.1}_{-0.2-0.9}$
$\pi^- K^{(*)+}$	8.6 ± 0.9	$7.8^{+0.2+1.1}_{-0.2-1.0}$
$\pi^0 \bar{K}^{(*)0}$	2.4 ± 0.7	$7.8^{+0.5+1.2}_{-0.5-1.0}$
$\pi^+ \bar{K}^{(*)0}$	$9.9^{+0.8}_{-0.9}$	$10.3^{+0.7+1.7}_{-0.7-1.4}$
$\rho^0 K^+$	$3.81^{+0.48}_{-0.46}$	$4.8^{+0.6+0.8}_{-0.6-0.7}$
$\rho^+ \bar{K}^0$	$8.0^{+1.5}_{-1.4}$	$10.9^{+0.6+1.7}_{-0.6-1.5}$
$\rho^0 \bar{K}^0$	4.7 ± 0.7	$10.2^{+0.6+1.6}_{-0.6-1.4}$
$\rho^- K^+$	$8.6^{+0.9}_{-1.1}$	$2.6^{+0.5+0.4}_{-0.4-0.4}$

Table 1. Branching ratios in units 10^{-6} of $B \to \pi K^*$ and $B \to \rho K$ decays. The first uncertainty in the predictions is due to the uncertainties in SCET parameters while the second uncertainty is due to the uncertainties in the CKM matrix elements.

In this paper, we discuss SUSY contributions to the CP asymmetries of the $B \to \pi K^*$ and $B \to \rho K$ decays in the framework of SCET[12, 13, 14, 15]. SCET is an effective field theory describing the dynamics of highly energetic particles moving close to the light-cone interacting with a background field of soft quanta[16]. It provides a systematic and rigorous way to deal with the decays of the heavy hadrons that involve different energy scales. The scaling of fields and momenta in SCET depends on a small parameter λ. Generally λ is defined as the ratio of the smallest and the largest energy scales in the given process. Then, the SCET Lagrangian and effective Hamiltonian are expanded in terms of λ that help to reduce the complexity of the calculations. In addition, the factorization formula provided by SCET is perturbative to all powers in α_s expansion. For details about SCET formalism we refer to Refs.[12, 13, 14, 15]

2. $B \to \pi K^*$ and $B \to \rho K$ decays within Standard Model

Within the framework of the standard model, the branching ratios of $B \to \pi K^*$ and $B \to \rho K$ are in agreements with their corresponding experimental values in most of the decay modes as given in Table 1. In that table, the first uncertainty in the predictions is due to the uncertainties in SCET parameters while the second uncertainty is due to the uncertainties in the CKM matrix elements. On the other hand, the SM predictions for the CP asymmetries of $B \to \pi K^*$ and $B \to \rho K$ are presented in Table 2. Clearly from that Table, the CP asymmetry of $B^+ \to \pi^0 K^{*+}$ has different sign in comparison with the experimental measurement. Moreover, the predicted CP asymmetries in many of the decay modes are in agreement with the experimental measurements due to the large errors in these measurements. We see also from the Table that, the predicted CP asymmetry of $\bar{B} \to \pi^0 \bar{K}^{*0}$ and $B^+ \to \rho^0 K^+$ disagree with the experimental results within 1σ. error of the experimental data. This can be attributed to the lack of the weak CP violating phases as SM Wilson coefficients are real and the only source of the weak phase is the phase of the CKM matrix.

3. $B \to \rho K$ and $B \to \pi K^*$ CP asymmetries including Supersymmetry

In this section we analyze the SUSY contributions to the CP asymmetries of $B^- \to \pi^- \bar{K}^{(*)0}, B^- \to \rho^- \bar{K}^0, \bar{B}^0 \to \rho^+ K^-$ and $B^- \to \rho^0 K^-$ as their SM predictions are very small as have been shown in the previous section. In SUSY, Flavor Changing Neutral Current(FCNC) and CP quantities are sensitive to particular entries in the mass matrices of the scalar fermions. Thus it is useful to adopt a model independent- parametrization, the so-called Mass Insertion Approximation (MIA) where all the couplings of fermions and sfermions to neutral gauginos are flavor diagonal [17].

Decay channel	Exp.	SM prediction
$\pi^0 K^{*+}$	0.04 ± 0.29	$-0.08^{+0.03+0.002}_{-0.03-0.002}$
$\pi^- K^{*+}$	-0.18 ± 0.07	$-0.12^{+0.04+0.01}_{-0.03-0.001}$
$\pi^0 \bar{K}^{*0}$	-0.15 ± 0.12	$-0.01^{+0.002+0.0003}_{-0.003-0.003}$
$\pi^+ \bar{K}^{*0}$	-0.038 ± 0.042	$-0.004^{+0.001+0.001}_{-0.001-0.0003}$
$\rho^0 K^+$	0.37 ± 0.11	$0.06^{+0.07+0.002}_{-0.08-0.002}$
$\rho^+ \bar{K}^0$	-0.12 ± 0.17	$-0.005^{+0.001+0.0004}_{-0.001-0.0001}$
$\rho^0 \bar{K}^0$	$-0.02 \pm 0.27 \pm 0.08 \pm 0.06$	$-0.02^{+0.01+0.002}_{-0.01-0.001}$
$\rho^- K^+$	0.15 ± 0.06	$0.14^{+0.11+0.004}_{-0.11-0.01}$

Table 2. Direct CP asymmetries of $B \to \pi K^*$ and $B \to \rho K$ decays. The first uncertainty in the predictions is due to the uncertainties in SCET parameters while the second uncertainty is due to the uncertainties in the CKM matrix elements.

At next leading order in α_s expansion, the dominant SUSY contributions to our decay modes are originated from diagrams mediated by the exchange of gluino and chargino. The complete expressions for the gluino and chargino contributions to the Wilson coefficients can be found in Refs. [18, 19, 20, 21].

After including SUSY contributions to the mentioned decays and keeping the dominant terms we find [7]

$$\begin{aligned}
A(B^- \to \pi^- \bar{K}^{(*)0}) \times 10^7 &\simeq -0.0178(\delta^d_{LL})_{23} - 6.6914(\delta^d_{LR})_{23} - 1.5857(\delta^d_{RL})_{23} - (0.0052 + 0.0003i)(\delta^u_{LR})_{32} \\
&- (0.0046 - 0.0003i)(\delta^u_{RL})_{32} + (0.3319 - 0.0612i), \\
A(B^- \to \pi^0 \bar{K}^{(*)-}) \times 10^7 &\simeq 0.0125(\delta^d_{LL})_{23} + 4.7315(\delta^d_{LR})_{23} + 1.1212(\delta^d_{RL})_{23} + (0.0056 - 0.0001i)(\delta^u_{LR})_{32} \\
&- (0.0223 - 0.0001i)(\delta^u_{RL})_{32} + (0.2508 - 0.1259i), \\
A(B^0 \to \pi^0 \bar{K}^{(*)0}) \times 10^7 &\simeq -0.0127(\delta^d_{LL})_{23} - 4.7315(\delta^d_{LR})_{23} - 1.1212(\delta^d_{RL})_{23} + (0.0094 + 0.0001i)(\delta^u_{LR})_{32} \\
&- (0.0185 + 0.0001i)(\delta^u_{RL})_{32} + (0.2949 - 0.0707i), \\
A(B^0 \to \pi^+ \bar{K}^{(*)-}) \times 10^7 &\simeq 0.0178(\delta^d_{LL})_{23} + 6.6914(\delta^d_{LR})_{23} + 1.5857(\delta^d_{RL})_{23} - (0.0106 + 0.0005i)(\delta^u_{LR})_{32} \\
&- (0.0099 - 0.0005i)(\delta^u_{RL})_{32} + (0.2695 - 0.1392i), \\
A(\bar{B}^- \to \rho^- K^0) \times 10^7 &\simeq 0.0043(\delta^d_{LL})_{23} + 1.6190(\delta^d_{LR})_{23} - 1.0851(\delta^d_{RL})_{23} - (0.0001 + 0.0005i)(\delta^u_{LR})_{32} \\
&- (0.0021 - 0.0005i)(\delta^u_{RL})_{32} - (0.3473 + 0.0111i), \\
A(B^- \to \rho^0 K^-) \times 10^7 &\simeq -0.0031(\delta^d_{LL})_{23} - 1.1448(\delta^d_{LR})_{23} + 0.7673(\delta^d_{RL})_{23} - (0.0037 + 0.0006i)(\delta^u_{LR})_{32} \\
&- (0.0120 - 0.0006i)(\delta^u_{RL})_{32} - (0.2232 + 0.0501i), \\
A(\bar{B}^0 \to \rho^0 K^0) \times 10^7 &\simeq 0.0030(\delta^d_{LL})_{23} + 1.1448(\delta^d_{LR})_{23} - 0.7673(\delta^d_{RL})_{23} - (0.0032 + 0.0003i)(\delta^u_{LR})_{32} \\
&- (0.0108. - 0.0003i)(\delta^u_{RL})_{32} - (0.3470 + 0.0307i), \\
A(B^- \to \rho^+ K^-) \times 10^7 &\simeq -0.0043(\delta^d_{LL})_{23} - 1.6190(\delta^d_{LR})_{23} + 1.0851(\delta^d_{RL})_{23} \\
&- (0.0008 + 0.0010i)(\delta^u_{LR})_{32} - (0.0037 - 0.0010i)(\delta^u_{RL})_{32} \\
&- (0.1723 + 0.0386i), \qquad (1)
\end{aligned}$$

The mass insertions $(\delta^u_{RL})_{32}$ and $(\delta^u_{LR})_{32}$ are not constrained by $b \to s\gamma$ and so we can set them as $(\delta^u_{RL})_{32} = (\delta^u_{LR})_{32} = e^{i\delta_u}$ where δ_u is the phase that can vary from $-\pi$ to π. It should be noted that in order to have a well defined Mass Insertion Approximation scheme, it is necessary to have

$|(\delta^f_{AB})_{ab}| < 1$ but here in order to maximize the SUSY CP-violating contributions we take it of order one. Applying $b \to s\gamma$ constraints leads to the following parametrization [22]

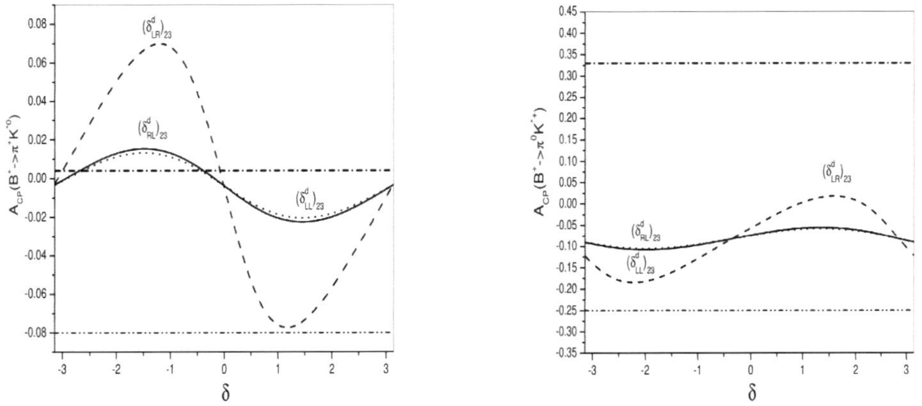

Figure 1. CP asymmetries versus the phase of the $(\delta^d_{AB})_{23}$ where A and B denote the chirality i.e. L, R. for 3 different mass insertions. The left diagram corresponds to $A_{CP}(B^+ \to \pi^+ \bar{K}^{*0})$ while the right diagram corresponds to $A_{CP}(B^+ \to \pi^0 K^{*+})$. In both diagrams we take only one mass insertion per time and vary the phase of from $-\pi$ to π. The horizontal lines in both diagrams represent the experimental measurement to 1σ.

$$(\delta^d_{LL})_{23} = e^{i\delta_d} \qquad (\delta^d_{LR})_{23} = (\delta^d_{RL})_{23} = 0.01 e^{i\delta} \tag{2}$$

In our analysis we consider two scenarios, the first one with a single mass insertion where we keep only one mass insertion per time and take the other mass insertions to be zero and the second scenario with two mass insertions will be considered only in the cases when one single mass insertion is not sufficient to accommodate the experimental measurement. After setting the different mass insertions as mentioned above, we see from Eq.(1) that, the terms that contain the mass insertions $(\delta^u_{RL})_{32}$ and $(\delta^u_{LR})_{32}$ will be small in comparison with the other terms and thus we expect that their contributions to the asymmetries will be small. These terms are obtained from diagrams mediated by the chargino exchange and thus we see that gluino contributions give the dominant contributions as known in the literature.

We start our analysis of the direct CP asymmetries by considering the first scenario in which we take only one mass insertion corresponding to the gluino mediation and set the others to be zero.

After substituting the mass insertions given in eq.(2) in eq.(1) we find that the first and third terms in the amplitudes $B^+ \to \pi^+ \bar{K}^{*0}$ and $B^+ \to \pi^0 K^{*+}$ will be approximately equal and both of them will be smaller than the second term. As a consequence, one predicts that the asymmetries generated by the mass insertions $(\delta^d_{LL})_{23}$ and $(\delta^d_{RL})_{23}$ will be equal and in the same time these asymmetries will be smaller than the case of using $(\delta^d_{LR})_{23}$ which can be seen from Fig.1. In that Figure,we plot the CP asymmetries, $A_{CP}(B^+ \to \pi^+ \bar{K}^{*0})$ and $A_{CP}(B^+ \to \pi^0 K^{*+})$ versus the phase of the $(\delta^d_{AB})_{32}$ where A and B denote the chirality i.e. L and R. for 3 different mass insertions. The horizontal lines in both diagrams represent the experimental measurements to 1σ. As can be seen from Figure1 left, for all gluino mass insertions, the value of the CP asymmetry $A_{CP}(B^+ \to \pi^+ \bar{K}^{*0})$ is enhanced to accommodate the experimental measurement of the asymmetry within 1σ for many values of the phase of the mass insertions. On the other hand,Figure 1right shows that the CP asymmetry $A_{CP}(B^+ \to \pi^0 K^{*+})$

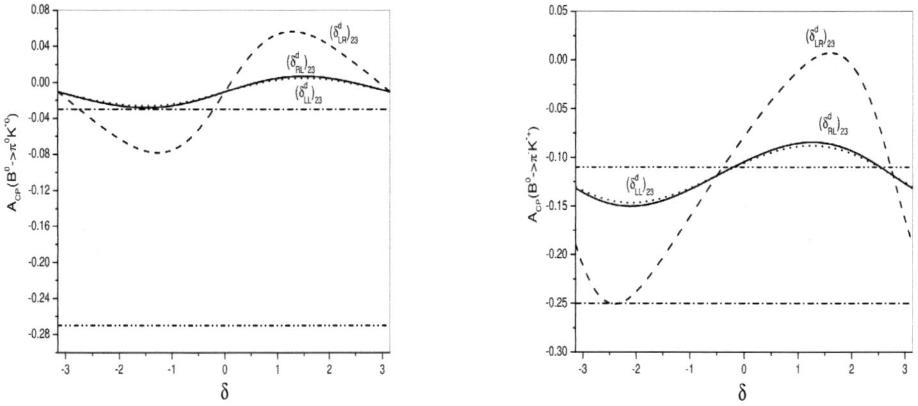

Figure 2. CP asymmetries versus the phase of the $(\delta^d_{AB})_{23}$ where A and B denote the chirality i.e. L, R. for 3 different mass insertions. The left diagram corresponds to $A_{CP}(B^0 \to \pi^0 \bar{K}^{*0})$ while the right diagram corresponds to $A_{CP}(B^0 \to \pi^- K^{*+})$. In both diagrams we take only one mass insertion per time and vary the phase of from $-\pi$ to π. The horizontal lines in both diagrams represent the experimental measurement to 1σ.

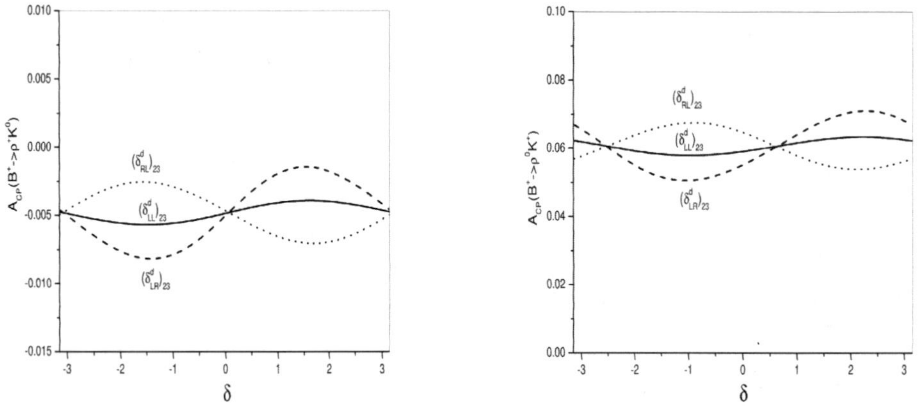

Figure 3. CP asymmetries versus the phase of the $(\delta^d_{AB})_{23}$ where A and B denote the chirality i.e. L, R. for 3 different mass insertions. The left diagram corresponds to $A_{CP}(B^+ \to \rho^+ K^0)$ while the right diagram corresponds to $A_{CP}(B^+ \to \rho^0 K^+)$. In both diagrams we take only one mass insertion per time and vary the phase of from $-\pi$ to π.

is enhanced to accommodate the experimental measurement within 1σ for all values of the phase of the mass insertions. The point we stress here is that SUSY Wilson coefficients provide source of large weak phases, which are needed for accommodation of CP asymmetries.

In Fig.2 we plot the two asymmetries, $A_{CP}(B^0 \to \pi^0 \bar{K}^{*0})$ and $A_{CP}(B^0 \to \pi^- K^{*+})$ versus the phase of the $(\delta^d_{AB})_{32}$ as before. As can be seen from Fig.2 left, $A_{CP}(B^0 \to \pi^0 \bar{K}^{*0})$ lies within 1σ range of its experimental value for many values of the phase of the mass insertion $(\delta^d_{LR})_{23}$ only. The reason for that is as before,(see eq.(1)) the two mass insertions $(\delta^d_{LL})_{23}$ and $(\delta^d_{RL})_{23}$ will give equal contributions to the CP asymmetries which will be smaller than the case of using $(\delta^d_{LR})_{23}$. On the other hand, Fig.2 right, we see that $A_{CP}(B^0 \to \pi^- K^{*+})$ can be accommodated within 1σ for many values of the phase of the three gluino mass insertions.

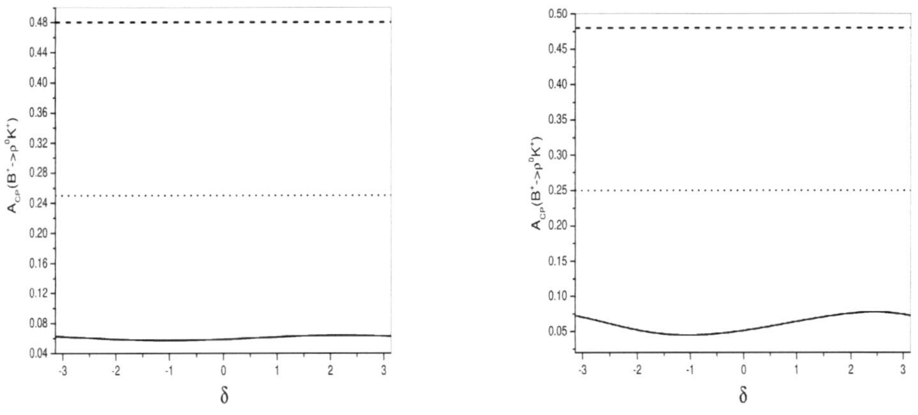

Figure 4. CP asymmetry of $A_{CP}(B^+ \to \rho^0 K^+)$ versus the phase of the mass insertion for 2 different mass insertions. The left diagram correspond to gluino contributions where we keep the two mass insertions $(\delta^d_{LR})_{23}$ and $(\delta^d_{RL})_{23}$ and set the other mass insertions to zero. The right diagram correspond to both gluino and chargino contributions where we keep the two mass insertions $(\delta^d_{LR})_{23}$ and $(\delta^u_{RL})_{32}$ and set the other mass insertions to zero. In both diagrams we assume that the two mass insertion have equal phases and we vary the phase from $-\pi$ to π. The horizontal lines in both diagrams represent the experimental measurements to 1σ.

Finally we discuss the CP asymmetries of the decay modes $B^+ \to \rho^+ K^0$ and $B^+ \to \rho^0 K^+$. After substituting the mass insertions given in eq.(2) in eq.(1), we find that the first and third terms in the amplitudes $B^+ \to \rho^+ K^0$ and $B^+ \to \rho^0 K^+$ will be no longer equal as previous cases and thus we expect their contributions to the asymmetries will be different which can be seen from Fig.(3) where, as before, we plot $A_{CP}(B^+ \to \rho^+ K^0)$ and $A_{CP}(B^+ \to \rho^0 K^+)$ versus the phase of the $(\delta^d_{AB})_{23}$. In Fig.(3) we do not show the horizontal lines representing the 1σ range of the experimental measurement as the three curves of the $A_{CP}(B^+ \to \rho^+ K^0)$ corresponding to the three gluino mass insertions totally lie in this 1σ range for all values of the phase of the mass insertions. On the other hand, Fig.3 right, we see that $A_{CP}(B^+ \to \rho^0 K^+)$ can not be accommodated within 1σ for any value of the phase of all gluino mass insertions. This motivates us to consider the second scenario with two mass insertions.

In Fig.4, we plot the CP asymmetry, $A_{CP}(B^+ \to \rho^0 K^+)$ versus the phase of the mass insertion for 2 different mass insertions. The left diagram correspond to gluino contributions where we keep the two mass insertions $(\delta^d_{LR})_{23}$ and $(\delta^d_{RL})_{23}$ and set the other mass insertions to zero.

The right diagram correspond to both gluino and chargino contributions where we keep the two mass insertions $(\delta^d_{LR})_{23}$ and $(\delta^u_{RL})_{32}$ and set the other mass insertions to zero. In both diagrams we assume that the two mass insertion have equal phases and we vary the phase from $-\pi$ to π. As before, the horizontal lines in both diagrams represent the experimental measurement to 1σ. As can be seen from Fig.4 left, two gluino mass insertions can not accommodate the experimental measurement for any value of the phase of the mass insertion. On the other hand from Fig.4 right, two mass insertions one corresponding to chargino contribution and the other corresponding to gluino contribution can not accommodate the experimental measurements. We find that in order to accommodate the CP symmetry in this case the Wilson coefficient $C_9^{\tilde{g}}$ should be increased at least by a factor $-6\pi/\alpha$ without violating any constraints on the SUSY parameter space. We show the corresponding diagram in Fig.5.

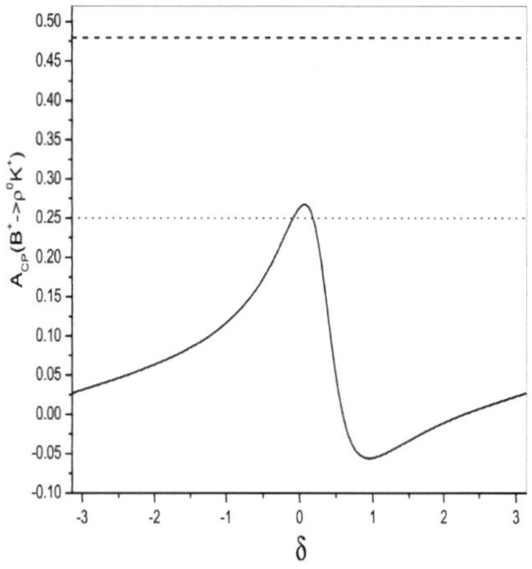

Figure 5. CP asymmetry of $A_{CP}(B^+ \to \rho^0 K^+)$ versus the phase of the mass insertion for 2 different mass insertions correspond to gluino contributions where we keep the two mass insertions $(\delta^d_{LR})_{23}$ and $(\delta^d_{LL})_{23}$ and set the other mass insertions to zero. We assume that the two mass insertion have equal phases and we vary the phase from $-\pi$ to π. The horizontal lines in the diagram represent the experimental measurements to 1σ.

4. Conclusion

In this paper we studied the decay modes $B \to \pi K^*$ and $B \to \rho K$ within the framework of Soft Collinear Effective Theory. We have shown that within the standard model, including the next leading order QCD corrections, their direct CP asymmetries can not accommodate the experimental measurements. We have discussed also SUSY contributions to the direct CP asymmetries using the Mass Insertion Approximation. In contrast to SM, we found that direct CP asymmetries can be significantly enhanced by SUSY contributions mediated by gluino exchange and accommodate the experimental measurements.

Acknowledgement

Gaber Faisel's work is supported by the National Science Council of R.O.C. under grants NSC 99-2112-M-008-003-MY3 and NSC 100-2811-M-008-036.

[1] M. Beneke and M. Neubert, Nucl. Phys. B **675**, 333 (2003) [arXiv:hep-ph/0308039].
[2] C. D. Lu and M. Z. Yang, Eur. Phys. J. C **23**, 275 (2002) [arXiv:hep-ph/0011238].

[3] X. Liu, H. s. Wang, Z. j. Xiao, L. Guo and C. D. Lu, Phys. Rev. D **73**, 074002 (2006) [arXiv:hep-ph/0509362].
[4] L. Guo, Q. g. Xu and Z. j. Xiao, Phys. Rev. D **75**, 014019 (2007) [arXiv:hep-ph/0609005].
[5] D. Q. Guo, X. F. Chen and Z. J. Xiao, Phys. Rev. D **75**, 054033 (2007) [arXiv:hep-ph/0702110].
[6] W. Wang, Y. M. Wang, D. S. Yang and C. D. Lu, Phys. Rev. D **78**, 034011 (2008) [arXiv:0801.3123 [hep-ph]].
[7] G. Faisel, D. Delepine and M. Shalaby, Phys. Lett. B **705**, 361 (2011) [arXiv:1101.2710 [hep-ph]].
[8] C. W. Chiang and D. London, Mod. Phys. Lett. A **24**, 1983 (2009) [arXiv:0904.2235 [hep-ph]].
[9] D. Delepine, G. Faisl, S. Khalil and G. L. Castro, Phys. Rev. D **74**, 056004 (2006) [arXiv:hep-ph/0608008].
[10] D. Delepine, G. Faisel and S. Khalil, Phys. Rev. D **77**, 016003 (2008) [arXiv:0710.1441 [hep-ph]].
[11] D. Delepine, G. Faisel, S. Khalil and M. Shalaby, Int. J. Mod. Phys. A **22**, 6011 (2007).
[12] C. W. Bauer, S. Fleming and M. E. Luke, Phys. Rev. D **63**, 014006 (2000) [arXiv:hep-ph/0005275].
[13] C. W. Bauer, S. Fleming, D. Pirjol and I. W. Stewart, Phys. Rev. D **63**, 114020 (2001) [arXiv:hep-ph/0011336].
[14] J. Chay and C. Kim, Phys. Rev. D **68**, 071502 (2003) [arXiv:hep-ph/0301055].
[15]
[15] J. Chay and C. Kim, Nucl. Phys. B **680**, 302 (2004) [arXiv:hep-ph/0301262].
[16] S. Fleming, PoS E **FT09**, 002 (2009) [arXiv:0907.3897 [hep-ph]].
[17] L. J. Hall, V. A. Kostelecky and S. Raby, Nucl. Phys. B **267**, 415 (1986).
[18] S. Bertolini, F. Borzumati, A. Masiero and G. Ridolfi, Nucl. Phys. B **353**, 591 (1991).
[19] E. Gabrielli and G. F. Giudice, Nucl. Phys. B **433**, 3 (1995) [Erratum-ibid. B **507**, 549 (1997)] [arXiv:hep-lat/9407029].
[20] F. Gabbiani, E. Gabrielli, A. Masiero and L. Silvestrini, Nucl. Phys. B **477**, 321 (1996) [arXiv:hep-ph/9604387].
[21] A. J. Buras, P. Gambino, M. Gorbahn, S. Jager and L. Silvestrini, Nucl. Phys. B **592**, 55 (2001) [arXiv:hep-ph/0007313].
[22] K. Huitu and S. Khalil, Phys. Rev. D **81**, 095008 (2010) [arXiv:0911.1868 [hep-ph]].

Journal of Physics: Conference Series **378** (2012) 012020 doi:10.1088/1742-6596/378/1/012020

Flavor changing neutral current in rare top quark decay $t \to c\bar{l}l$ in generic Z' and left-right symmetric models

R. Gaitán-Lozano

Departamento de Física, Facultad de Estudios Superiores Cuautitlán, UNAM, C.P. 54770, Estado de México, México

E-mail: rgaitan@servidor.unam.mx

J.H. Montes de Oca Y.

Departamento de Física, Facultad de Estudios Superiores Cuautitlán, UNAM, C.P. 54770, Estado de México, México

E-mail: josehalim@comunidad.unam.mx

L. Díaz-Cruz

Cuerpo Académico de Partículas, Campos y Relatividad Facultad de Ciencias Físico-Matemáticas, BUAP. Apdo. Postal 1364, C.P. 72000 Puebla, Pue., México

E-mail: jldiaz@fcfm.buap.mx

Abstract. We study the rare top decay $t \to c\bar{l}l$, which involves flavor violation, as a possible probe of new physics. This decay is analyzed with the simplest Standard Model (SM) extensions with additional gauge symmetry $U(1)$ or $SU(2)$, known as generic Z' model or left-right symmetric model, respectively. The considered models allow to obtain this decay at tree level through Flavor Changing Neutral Currents (FCNC). We have considered diagonal non-universal couplings for up-type quark sector and found that BR($t \to c\bar{l}l$) for 1 TeV$< M_{Z'} <$ 3 TeV can be of order $\sim 10^{-11}$ in generic Z' model or $\sim 10^{-12}$ in left-right symmetric model.

1. Introduction

Processes with Flavor Changing Neutral Current (FCNC) give an important signal for physics beyond Standard Model (SM). One of them is the top quark goes to charm quark and a lepton-antilepton pair, $t \to c\bar{l}l$. Rare top decay is important not only as a test of FCNC but also as a means to identify an extension of the SM. FCNC are absent in the SM at the tree-level due to the Glashow-Iliopoulos-Maiani (GIM) mechanism. However, new FCNC states can appear in top decays if there is physics beyond the SM. In this context, rare top quark decays are interesting because they might be a source of possible new physics effects [1]. In some particular models beyond the SM, rare top decays may be significantly enhanced to reach detectable levels [2]. In the Large Hadron Collider (LHC) or future International Linear Collider (ILC) an eventual signal of FCNC in the top quark decay will have to be ascribed to new physics.

One of the puzzle of the SM has to do with the understanding of the origin of parity violation in low-energy weak interaction processes. Within the framework of left-right symmetric models, based on the gauge group $SU(2)_L \otimes SU(2)_R \otimes U(1)_{B-L}$, this problem finds a natural answer [3].

No study is complete without including the generic Z' model based on the gauge group $SU(2)_L \otimes U(1)_Y \otimes U(1)$ which is defined to have identical couplings as for usual Z SM boson [4]. In this paper we consider both extensions and the non universal family couplings in order to induce the FCNC for the rare top decay [5], [6].

This paper is organized as follows. In Sec. 2 we introduce the family non-universal couplings and define some notation. In Sec 3 we analyze the rare top decay and in Sec. 4 we discuss and summarize our result.

2. FCNC and family non-universal couplings

The generic Z' and left-right symmetric models introduce two neutral gauge bosons, one of them comes from SM gauge group and the other one comes from extra gauge group. If there is no mixing between the neutral gauge bosons Z_1 and Z_2, then the neutral current terms can be written as

$$\mathcal{L}_{NC} = -eJ^{\mu}_{EM}A_{\mu} - gJ^{\mu}_{1}Z_{1\mu} - gJ^{\mu}_{2}Z_{2\mu}, \tag{1}$$

where $Z_{1\mu}$ is the usual electroweak neutral gauge boson, $Z_{2\mu}$ is the neutral gauge boson associated with extra gauge symmetry and $g_{1,2}$ are the respective gauge couplings. The current associated with the additional gauge symmetry is

$$J_{2\mu} = \sum_{i,j} \overline{\psi}^0_i \gamma_\mu \left[\epsilon_{\psi L ij} P_L + \epsilon_{\psi R ij} P_R\right] \psi^0_j, \tag{2}$$

where i, j running over all quarks and leptons, $P_{L,R} = \frac{1}{2}(1 \mp \gamma_5)$ and $\epsilon_{\psi L,R}$ are chiral couplings between fermions and neutral gauge boson $Z_{2\mu}$. The superscript zero denotes the interaction basis.

We shall assume that couplings of the Z_2 with leptons and quarks are diagonal family non-universal. The simplest case is obtained when only the left-handed top quark coupling is diagonal family non-universal [5]. Under last assumption the up quark sector chiral couplings are written as $\epsilon_{uRij} = Q^u_R \delta_{ij}$ and

$$\epsilon_{uL} = Q^u_L \begin{pmatrix} 1 & 0 & 0 \\ 0 & 1 & 0 \\ 0 & 0 & x \end{pmatrix}. \tag{3}$$

The Q^u_L and Q^u_R are chiral couplings of the up-type quarks in the additional gauge symmetry and x is a parameter which can be of order 1 but not equal to 1. In generic Z' model the chiral couplings are the same as the SM Z boson. The chiral couplings in LRSM are obtained when neutral gauge bosons are written in mass basis through the orthogonal transformation

$$\begin{pmatrix} W_{L3} \\ W_{R3} \\ B \end{pmatrix} = \begin{pmatrix} -s_W & c_W & 0 \\ -s_W & \frac{-s_W^2}{c_W} & \sqrt{c_W^2 - s_W^2}/c_W \\ \sqrt{c_W^2 - s_W^2} & \frac{s_W}{c_W}\sqrt{c_W^2 - s_W^2} & \frac{s_W}{c_W} \end{pmatrix} \begin{pmatrix} A \\ Z_1 \\ Z_2 \end{pmatrix}, \tag{4}$$

where $s_W = \sin\theta_W$, $c_W = \cos\theta_W$ and θ_W is the Weinberg angle. Therefore, the chiral couplings in LRSM are given by

$$Q^i_L = -\sqrt{\frac{3}{5}}\left(\frac{1}{2\alpha}\right)(B-L)_i \tag{5}$$

and

$$Q^i_R = \sqrt{\frac{3}{5}}\left(\alpha T^i_{3R} - \frac{1}{2\alpha}(B-L)_i\right), \tag{6}$$

Journal of Physics: Conference Series **378** (2012) 012020 doi:10.1088/1742-6596/378/1/012020

Table 1. Numerical values of the chiral couplings.

Chiral coupling	$U(1)$	$SU(2)_R$
Q^u_L	0.3456	-0.0849
Q^u_R	-1.544	0.5038
Q^d_L	-0.4228	-0.0849
Q^d_R	0.0772	-0.6736
Q^e_L	-0.2684	0.2548
Q^e_R	0.2316	-0.3339
Q^ν_L	0.5	0.2548

where B and L denote the baryon and lepton numbers of fermion i, respectively. T_{3R} is the third component of its right-handed isospin in the $SU(2)_R$ group and the parameter $\alpha = \sqrt{(1-2s_W^2)/s_W^2}$. Numerical values of the chiral couplings for generic Z' and left-right symmetric models are shown in table 1. Therefore the neutral current associated with extra neutral gauge boson and up-type quarks in the mass basis is

$$\mathcal{L}_{NC}^{Z_2} = -g_2 Z'^{\mu} \left(\bar{u}, \quad \bar{c}, \quad \bar{t} \right) \gamma_\mu \left[B_L^u P_L + B_R^u P_R \right] \begin{pmatrix} u \\ c \\ t \end{pmatrix}, \tag{7}$$

where $B_L^u = V_L \epsilon_{uL} V_L^\dagger$ and $B_R^u = \epsilon_{uR}$. The assumption of the down quark sector has no mixing and the unitary conditions of CKM matrix give

$$B_L^u \approx Q_L^u \begin{pmatrix} 1 & (1-x) V_{ub} V_{cb}^* & (x-1) V_{ub} V_{tb}^* \\ (1-x) V_{cb} V_{ub}^* & 1 & (x-1) V_{cb} V_{tb}^* \\ (x-1) V_{tb} V_{ub}^* & (x-1) V_{tb} V_{cb}^* & x \end{pmatrix} \tag{8}$$

If the Z_2 couplings are diagonal but family non-universal, FCNC are induced by fermion mixing.

3. Rare top decay

We now proceed to calculate the branching ratio for the rare top decay $t \to c\bar{l}l$ using the formalism introduced in the previous section. The diagram for this decay are shown in figure 1.

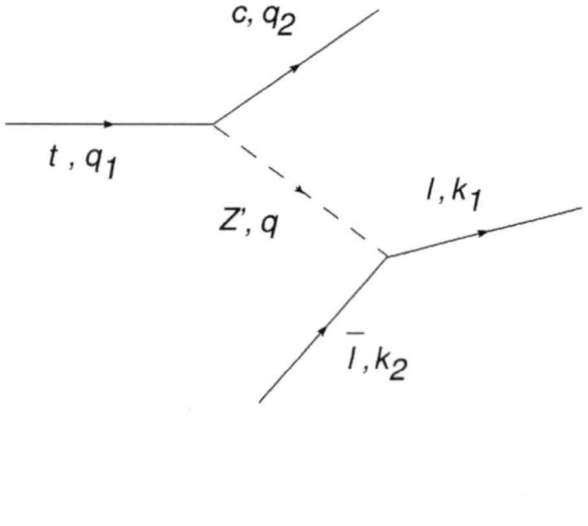

Figure 1. Feynman diagram for rare top decay.

Journal of Physics: Conference Series **378** (2012) 012020 doi:10.1088/1742-6596/378/1/012020

Table 2. Numerical values of the integrals for each charged lepton.

Lepton	I_1	I_2	I_3	I_4	I_5
Tau	0.083207	0.08327	0.33240	0.16647	0.49901
Muon	0.083314	0.083314	0.33332	0.16663	0.49999
Electron	0.083314	0.083314	0.33333	0.16663	0.5

If we consider that $q^2 << M_{Z'}^2$, then the average of amplitude and width decay can be written as

$$\begin{aligned}\left|\overline{\mathcal{M}}_{t\to c\bar{l}l}\right|^2 &= \frac{4g^2}{M_{Z'}^4}[x_1(q_1\cdot k_1)(q_2\cdot k_2)+x_2(q_1\cdot k_2)(q_2\cdot k_1)+x_3 m_l^2(q_1\cdot q_2)\\ &\quad -x_4 m_t m_c(k_1\cdot k_2)-m_t m_c m_l^2 x_5],\end{aligned} \tag{9}$$

and

$$\Gamma_{t\to c\bar{l}l}=\frac{g^2 m_t^5}{256\pi^3 M_{Z'}^4}\left(x_1I_1+x_2I_2+\frac{2m_l^2}{m_t^2}x_3I_3-\frac{2m_c}{m_t}x_4I_4-\frac{8m_c m_l^2}{m_t^3}x_5I_5\right), \tag{10}$$

respectively, where I_i, $i=1,...,5$, are integrals obtained in the kinematic development and their numerical values are shown in table 2. The parameters $x_{1,2,3,4,5}$ are defined as

$$x_1 = 2\left(|(B_L^u)_{23}|^2\left|\left(B_L^l\right)_{33}\right|^2+|(B_R^u)_{23}|^2\left|\left(B_R^l\right)_{33}\right|^2\right), \tag{11}$$

$$x_2 = 2\left(|(B_L^u)_{23}|^2\left|\left(B_R^l\right)_{33}\right|^2+|(B_R^u)_{23}|^2\left|\left(B_L^l\right)_{33}\right|^2\right), \tag{12}$$

$$x_3 = \mathrm{Re}\left[\left(B_L^l\right)_{33}(B_R^{u*})_{23}\right]\left(|(B_L^u)_{23}|^2+|(B_R^u)_{23}|^2\right), \tag{13}$$

$$x_4 = \mathrm{Re}\left[(B_L^u)_{23}\left(B_R^{l*}\right)_{33}\right]\left(\left|\left(B_L^l\right)_{33}\right|^2+\left|\left(B_R^l\right)_{33}\right|^2\right), \tag{14}$$

$$x_5 = \mathrm{Re}\left[(B_L^u)_{23}\left(B_L^l\right)_{33}(B_R^{u*})_{23}\left(B_R^{l*}\right)_{33}\right]+\mathrm{Re}\left[(B_L^u)_{23}\left(B_L^{l*}\right)_{33}(B_R^u)_{23}\left(B_R^{l*}\right)_{33}\right]. \tag{15}$$

Therefore, the branching ratio as function of the model parameters and Z' mass is

$$\begin{aligned}BR(t\to cl^+l^-) &= \frac{g^2 m_t^5}{128\pi^3 M_{Z'}^4\Gamma_{top}}[Q_L^{l2}Q_L^{u2}|V_{tb}|^2|V_{cb}|^2 I_1(x-1)^2\\ &\quad +\frac{m_c}{m_t}Re\left(V_{tb}V_{cb}\right)I_4 Q_R^l Q_L^u(Q_L^{l2}+Q_R^{l2})(x-1)].\end{aligned} \tag{16}$$

The SM parameters used in this calculations are $m_t = 172.9$ GeV, $m_c = 1.29$ GeV, $V_{tb} = 0.99$, $V_{cb} = 0.041$, $\Gamma_{top} = 1.2$ GeV and $\sin\theta_W = 0.2316$ [7]. The figures 2, 3, 4 and 5 show the behavior of the branching ratio of the rare top decay $t \to c l\bar{l}$ in both models.

4. Conclusion

The family non-universal couplings is important in order to induce the FCNC from gauge bosons in extra gauge groups models. The branching ratio of the rare top decay $t \to c l\bar{l}$ can be of the order of 10^{-12} for 1 TeV $\leq m_{Z'} \leq$ 3 TeV and $0.1 \leq x \leq 0.9$ in the framework of Z' models that we have discussed. Note that the branching ratio in LRSM is approximately 10 times smaller than the branching ratio in generic Z' model. Is is found that rare top decay is extremely suppressed due to the Z' boson mass and the family non-universal parameter x.

Journal of Physics: Conference Series **378** (2012) 012020 doi:10.1088/1742-6596/378/1/012020

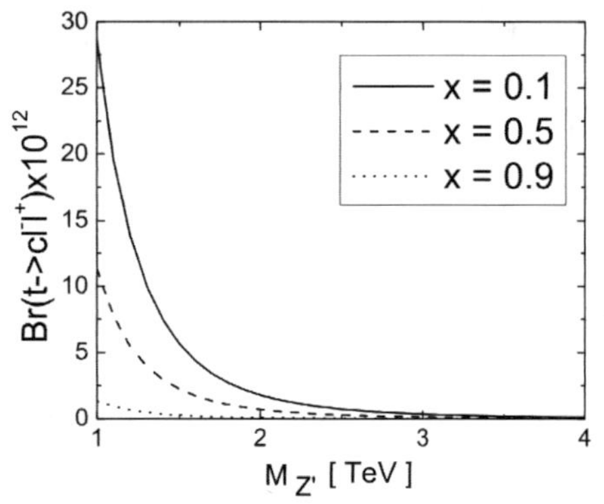

Figure 2. Branching ratio as function of the Z' mass in generic Z' model.

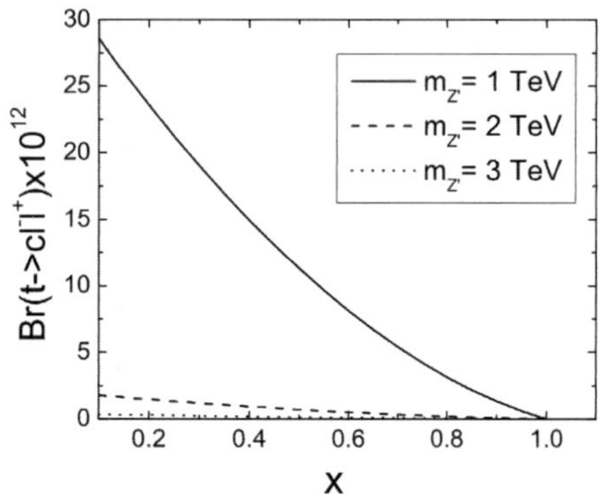

Figure 3. Branching ratio as function of the family non-universal parameter x in generic Z' model.

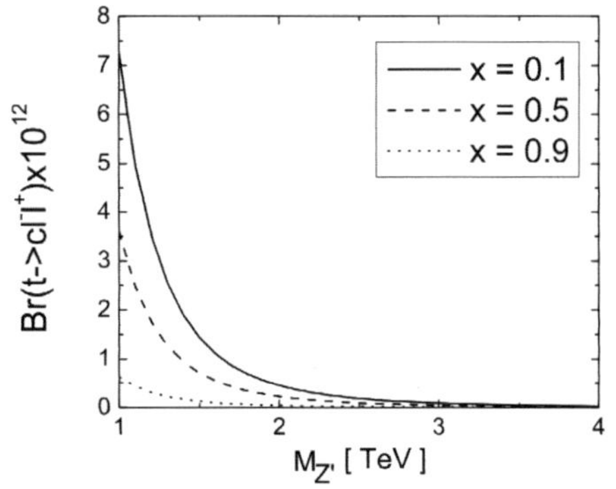

Figure 4. Branching ratio as function of the Z' mass in LRSM.

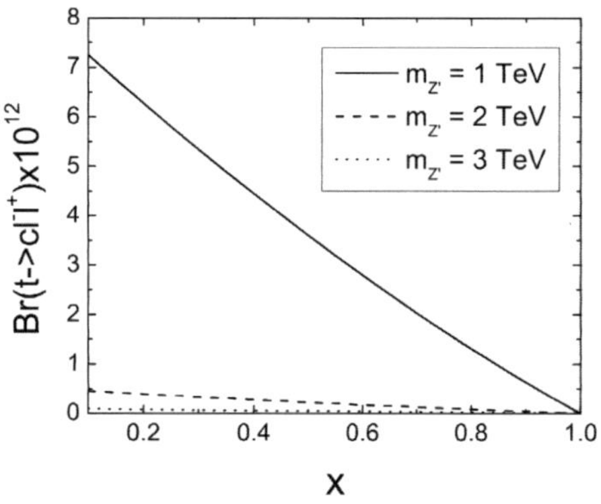

Figure 5. Branching ratio as function of the family non-universal parameter x in LRSM.

Acknowledgments

This work is supported in part by PAPIIT project IN117611, Sistema Nacional de Investigadores (SNI) in México. J.H.Montes de Oca Y. is thankful for support from the postdoctoral DGAPA-UNAM grant.

References

[1] Accomando E, Belyaev A, Fedeli L, King S F and Shepherd-Themistocleous C, 2011 *Phys. Rev.* D **83**, 075012; Drobnak J, Faijer S and Kamenik J F, 2009 *JHEP* **03**, 077.

[2] Béjar S., Guasch J. and Solá J. 2001 *Nucl. Phys.* B **600**, 21. Gaitán R., Miranda O. G. and Cabral-Rosetti L.G. 2006 *AIP Conf.Proc.* **857** 179-185.

[3] Mohapatra R N, Pati J C, 1975 *Phys. Rev.* D **11**, 566; Mohapatra R N, Pati J C, 1975 *Phys. Rev.* D **11**, 2558; Senjanovic G, Mohapatra R N, 1975 *Phys. Rev.* D **12**, 1502; Senjanovic G, Mohapatra R N, 1981

Phys. Rev. D **23**, 165; For a review R N Mohapatra, in: Branco G. C., Lavoura L. and Silva J. P., 1999 *CP Violation*, World Scientific.

[4] Langacker P, 2009 *Rev. Mod. Phys.* **81**, 1199; Langacker P and Plumacher M, 2000 *Phys. Rev.* D **62**, 013006;Langacker P and Luo M, 1992 *Phys. Rev.* D **45**, 278.

[5] Arhrib A, Cheung K, Chiang C W, Yuan T C, 2006 *Phys. Rev.* D **73**, 075015.

[6] Çakir O, Çakir L T, Senol A, Tasci A T, 2010 *Eur. Phys. J. C.* **70**, 293.

[7] Nakamura K *et al.* (Particle Data Group), 2010 *J. Phys. G* **37**, 075021 and 2011 partial update for the 2012 edition.

Journal of Physics: Conference Series **378** (2012) 012021 doi:10.1088/1742-6596/378/1/012021

Decays $Z_H \to Z\gamma$ and $Z_H \to ZZ$ in the littlest Higgs model[1]

I. Cortés-Maldonado, A. Fernández-Tellez, G. Tavares-Velasco

Facultad de Ciencias Físico Matemáticas, Benemérita Universidad Autónoma de Puebla, Apartado Postal 1152, Puebla, Pue., México

E-mail: icortesm@cern.ch

Abstract. The study of the phenomenology of an extra neutral gauge boson, Z_H, can help us to unravel the underlying theory from which this particle arises. We study the decay of such a particle into two neutral gauge bosons, $Z_H \to Z\gamma$ and $Z_H \to ZZ$, in the littlest Higgs model. These decays are induced at the one-loop level by a fermion triangle and are interesting as they are strongly dependent on the mechanism of anomaly cancellation of the model. Other relevant tree-level two- and three-body Z_H decays are also calculated. It is found that the branching ratios for the $Z_H \to \gamma Z$ decays can be as large as that of a tree-level three-body decay but the $Z_H \to ZZ$ decay is very suppressed. We also discuss the experimental prospects for detecting these decays at the LHC.

1. Introduction

Little Higgs models [1, 2, 3, 4, 5, 6, 7] have emerged recently as an interesting alternative to solve the little hierarchy problem without the need of fine-tuning. This class of theories are based on the old idea that the Higgs boson is a pseudo-Goldstone boson arising from a global symmetry spontaneously broken at a scale of the order of a few TeVs. In these models there is a set of new particles that play the role of partners of the standard model (SM) gauge bosons and the top quark. These new particles cancel the quadratic divergences to the Higgs boson mass, m_H, arising at one-loop from the exchange of SM particles, thereby allowing a naturally light Higgs boson. The most popular realization of this idea is the littlest Higgs model (LHM). Apart from reproducing the SM at the electroweak scale, the LHM predicts heavy partners for the top quark and the SM gauge bosons, which are necessary to cancel the quadratic divergences of the Higgs boson mass at the one-loop level.

An extra neutral gauge boson arises in models in which the SM group is extended with an extra gauge group or if it is embedded into a larger gauge group. The study of the phenomenology of such a particle may be helpful to identify the model from which it arises. In the LHM there are two extra neutral gauge bosons: the Z gauge boson partner, Z_H, which is associated with an additional $SU(2)$ gauge group, and the photon partner, A_H, which is associated with an extra $U(1)$ gauge group. Although the latter is the lightest new particle and has a great potential to show up at a particle collider, it would not offer a robust signal of the model due to the arbitrariness of the charge assignments of the SM fermions under the extra $U(1)$ gauge group.

[1] Presented at the XIII Mexican Workshop on particles and Fields, Mexico.

Journal of Physics: Conference Series **378** (2012) 012021 doi:10.1088/1742-6596/378/1/012021

In this work we will present the calculation of the the decays $Z_H \to V_i Z$ ($V_i = \gamma, Z$) in the framework of the LHM. These decays, which are interesting as their rate is dictated by the mechanism of anomaly cancellation, have been already studied in the context of a superstring-inspired E_6 model [8], the minimal 331 model [9], and 5D warped-space models [10].

2. The littlest Higgs model

The LHM is a nonlinear sigma model with a global symmetry under the $SU(5)$ group and a gauged subgroup $[SU(2) \otimes U(1)]^2$. The Goldstone bosons are parametrized by the following Σ field

$$\Sigma = e^{i\Pi/f}\, \Sigma_0\, e^{i\Pi^T/f} \tag{1}$$

where Π is the pion matrix. The Σ field transforms under the gauge group as $\Sigma \to \Sigma' = U\,\Sigma\,U^T$, with $U = L_1 Y_1 L_2 Y_2$ an element of the gauge group.

The pattern of symmetry breaking is as follows. The $SU(5)$ global symmetry is broken down to $SO(5)$ by the sigma field VEV, Σ_0, which is of the order of the scale of the symmetry breaking. After the global symmetry is broken, 14 Goldstone bosons arise accommodated in multiplets of the electroweak gauge group: a real singlet, a real triplet, a complex triplet and a complex doublet. The latter will be identified with the SM Higgs doublet. At the same scale, the gauge symmetry is broken down to its diagonal subgroup, $SU(2) \times U(1)$. The real singlet and the real triplet are absorbed by the gauge bosons associated with the broken gauge symmetry, which acquire their masses. A Coleman-Weinberg potential induces a VEV for the complex doublet and masses for the components of the complex triplet. Finally, electroweak symmetry breaking (EWSB) proceeds in the usual way at the Fermi scale.

The LHM effective Lagrangian is composed of the kinetic energy Lagrangian of the Σ field, $\mathcal{L}_{\rm K}$, the Yukawa Lagrangian, $\mathcal{L}_{\rm Y}$, and the kinetic terms of the gauge and fermion sectors. The sigma field kinetic Lagrangian is given by

$$\mathcal{L}_{\rm K} = \frac{f^2}{8}{\rm Tr}|D_\mu \Sigma|^2, \tag{2}$$

with the $[SU(2) \times U(1)]^2$ covariant derivative defined by

$$D_\mu \Sigma = \partial_\mu \Sigma - i \sum_{j=1}^{2} \left[g_j W^a_{j\,\mu}(Q^a_j \Sigma + \Sigma Q^{a\,T}_j) + g'_j B_{j\,\mu}(Y_j \Sigma + \Sigma Y^T_j) \right]. \tag{3}$$

The heavy $SU(2)$ and $U(1)$ gauge bosons are $W^\mu_j = \sum_{a=1}^3 W^{\mu\,a}_j Q^a_j$ and $B^\mu_j = B^\mu_j Y_j$, with Q^a_j and Y_j the gauge generators, while g_i and g'_i are the respective gauge couplings. The VEV Σ_0 generates masses for the gauge bosons and mixing between them. The heavy gauge boson mass eigenstates are given

$$W'^a = -cW^a_1 + sW^a_2, \tag{4}$$
$$B' = -c'B_1 + s'B_2, \tag{5}$$

with masses $m_{W'} = \frac{f}{2}\sqrt{g_1^2 + g_2^2}$ and $m_{B'} = \frac{f}{\sqrt{20}}\sqrt{g_1'^2 + g_2'^2}$.

The orthogonal combinations of gauge bosons are identified with the SM gauge bosons:

$$W^a = sW^a_1 + cW^a_2, \tag{6}$$
$$B = s'B_1 + c'B_2, \tag{7}$$

which remain massless at this stage.

The gauge and Yukawa interactions that break the global $SO(5)$ symmetry induce radiatively a Coleman-Weinberg potential for the complex doublet, h, and triplet, ϕ, whose explicit form can be obtained after expanding the Σ field:

$$V_{\rm CW} = \lambda_{\phi^2} f^2 {\rm Tr}|\phi|^2 + i\lambda_{h\phi h} f \left(h\phi^\dagger h^T - h^*\phi h^\dagger\right) - \mu^2|h|^2 + \lambda_{h^4}|h|^4, \quad (8)$$

where λ_{ϕ^2}, $\lambda_{h\phi h}$, and λ_{h^4} depend on the fundamental parameters of the model, whereas μ^2 is treated as a free parameter of the order of $f^2/16\pi^2$. The Coleman-Weinberg potential induces a mass term for the complex triplet, whose components acquire a mass of the order of f. The neutral component of the complex doublet develops a VEV, v, of the order of the electroweak scale, which is responsible for EWSB.

At the electroweak scale, EWSB proceeds as usual, yielding the final mass eigenstates: the three SM gauge bosons are accompanied by three heavy gauge bosons which are their counterpart, A_H, W_H and Z_H. In this stage the masses of the heavy gauge bosons get corrected by terms of the order of $(v/f)^2$ and so are the masses of the weak gauge bosons. The heavy gauge boson masses are given by

$$m^2_{Z_H} \simeq m^2_{W_H} = m^2_W\left(\frac{f^2}{s^2c^2v^2} - 1\right), \qquad m^2_{A_H} \simeq m^2_Z s^2_W\left(\frac{f^2}{5s'^2c'^2v^2} - 1\right), \quad (9)$$

with $t_W = s_W/c_W$, being s_W and c_W the sine and cosine of the Weinberg angle θ_W.

The fermion sector of the LHM is identical to the SM one except in the top sector, which requires a new vector-like top quark T, which is known as the top partner. The T loops cancel the quadratically divergent contribution to the Higgs mass arising from the top quark loops. This fixes the Yukawa interactions, given by [2]

$$\mathcal{L}_{\rm Y} = \frac{1}{2}\lambda_1 f \epsilon_{ijk}\epsilon_{xy}\chi_i\Sigma_{jx}\Sigma_{ky}u'^c_3 + \lambda_2 f\tilde{t}\tilde{t}^c + {\rm H.c.}, \quad (10)$$

where ϵ_{ijk} and ϵ_{xy} are antisymmetric tensors. The subscripts i, j (x, y) are summed over 1..3 (4..5). In addition, t_3 is the SM top quark, u'_3 is the SM right-handed top quark, $(\tilde{t}, \tilde{t}'^c)$ is a new vector-like top quark and $\chi = (b_3, t_3, \tilde{t})$. The first term of $\mathcal{L}_{\rm Y}$ induces the couplings of the Higgs boson to the fermions such that the quadratic divergences from the top quark loop are canceled by the top partner loop. The expansion of the Σ field leads to the physical states, t and T, after diagonalizing the mass matrix. At the leading order in v/f, the masses of the SM top quark and the new top quark T are given by [11]

$$m_t = \frac{\lambda_1\lambda_2}{\sqrt{\lambda_1^2+\lambda_2^2}}v, \quad m_T = f\sqrt{\lambda_1^2+\lambda_2^2}. \quad (11)$$

There is no need to introduce extra vector-like quarks for the first two quark generations as the quadratic divergences arising from light fermions are not important below the cutoff scale $\Lambda_S = 4\pi f$.

The couplings of the heavy neutral gauge bosons to the fermions depend on the isospin and hypercharge of the fermions, which is dictated by the gauge invariance of the scalar couplings to the fermions under $U(1)_1 \times U(1)_2$. These couplings can be written as

$$\mathcal{L} = \frac{g'}{s'c'}\left(-c'^2 J^\mu_{B_1} + s'^2 J^\mu_{B_2}\right) A_{H\mu} + \frac{gc}{s} J^\mu_{W^3} Z_{H\mu} + {\rm H.c.}, \quad (12)$$

with $J^\mu_{W^3} = \bar{Q}_L\gamma^\mu(T^3)Q_L$ and $J^\mu_{B_{1,2}}\bar{f}\gamma^\mu Y_{1,2}f$, while $Y_{1,2}$ represent the $U(1)_{i,j}$ quantum number assignments of the Σ field. The fermion hypercharges are given in terms of two free parameters,

y_u and y_e, which can be fixed to $y_u = 2/5$ and $y_e = 3/5$ by requiring anomaly cancellation under both $U(1)$ groups.

The remaining terms of the LHM Lagrangian and all the Feynman rules for the new interactions can be found in [11, 12]. In particular, the Feynman rules for all the couplings necessary for the calculation of the decays of the Z_H gauge boson are summarized in Ref. [13].

3. Decays of the extra neutral Z_H gauge boson

We will now present the decay widths of the most relevant tree-level decay channels of the Z_H gauge boson together with those of the one-loop decays $Z_H \to V_i Z$ ($V_i = \gamma, Z$). For more details, the interested reader is referred to [13].

3.1. Tree-level decays

The dominant decays of the extra neutral gauge boson Z_H are the tree-level induced two-body decays $Z_H \to \bar{f}f$, $Z_H \to W^+W^-$, $Z_H \to ZH$, and $Z_H \to A_H H$. The latter is the only kinematically allowed tree-level two-body decay involving a new particle as a final state.

The decay width into the fermion pair $\bar{f}f$ can be written as:

$$\Gamma(Z_H \to f\bar{f}) = \frac{g^2 m_{Z_H} N_c^f}{24\pi c_W^2}\sqrt{1-4y_f}\,(1-y_f)\,{g'}_L^{2}, \tag{13}$$

where we introduced the notation $y_a = (m_a/m_{Z_H})^2$, ${g'}^f_L = \frac{c_W\, c}{s} T^3$, with $T_f^3 = 1\ (-1)$ for up (down) type fermions, and N_c^f is the fermion color number.

On the other hand, the $Z_H \to W^+W^-$ and $Z_H \to ZH$ decay widths are given by

$$\Gamma(Z_H \to WW) = \frac{g^2_{Z_H WW} m_{Z_H}}{192\pi y_W^2}\,(1-4y_W)^{3/2}\left(1+20y_W+12y_W^2\right), \tag{14}$$

and

$$\begin{aligned}\Gamma(Z_H \to ZH) &= \frac{g^2_{Z_H ZH}}{192\pi m_{Z_H} y_Z}\sqrt{(1-(\sqrt{y_H}-\sqrt{y_Z})^2)(1-(\sqrt{y_H}+\sqrt{y_Z})^2)} \\ &\times \left(1+(y_H-y_Z)^2+y_Z^2-2(y_H-5y_Z)\right),\end{aligned} \tag{15}$$

where $g_{Z_H WW} = \frac{g c\, s\,(c^2-s^2)\, v^2}{2f^2}$ and $g_{Z_H ZH} = -\frac{g^2 v}{2c_W}\frac{(c^2-s^2)}{2\, c\, s}$. The $Z_H \to A_H H$ decay width can be obtained from $\Gamma(Z_H \to ZH)$ after the replacements $y_Z \to y_{A_H}$ and $g_{Z_H ZH} \to g_{Z_H A_H H} = -\frac{g g' v (s^2 {c'}^2 + c^2 {s'}^2)}{4\, c\, s\, c'\, s'}$ are done.

For completeness we will also present numerical results for the following tree-level three-body decays into SM particles: $Z_H \to \bar{f}f\gamma$, $Z_H \to \bar{f}fZ$, $Z_H \to \bar{t}tH$, $Z_H \to ZHH$, $Z_H \to ZW^-W^+$, $Z_H \to \gamma W^-W^+$, and $Z_H \to ZZZ$. Other kinematically allowed three-body decays involve a heavy photon: $Z_H \to A_H HH$, $Z_H \to A_H WW$, $Z_H \to A_H ZZ$, and $Z_H \to A_H A_H A_H$. To obtain the decay widths, we squared the decay amplitude for each set of Feynman diagrams with the aid of the FeynCalc package [14], and the integration over the three-body phase space was performed numerically. We would like to note that the decays widths for $Z_H \to \bar{f}f\gamma$ and $Z_H \to \gamma W^-W^+$ were obtained with the assumption of $E_\gamma \geq 10$ GeV to avoid infrared divergences.

3.2. One-loop decays

We now turn to the one-loop level two-body decays $Z_H \to \gamma Z$ and $Z_H \to ZZ$, which are induced by the fermion triangle shown in Fig. 1. It is worth mentioning that the Z_H decay into a photon pair is forbidden by the Landau-Yang theorem. The $Z_H \to \gamma Z$ decay amplitude can be written as

Journal of Physics: Conference Series **378** (2012) 012021 doi:10.1088/1742-6596/378/1/012021

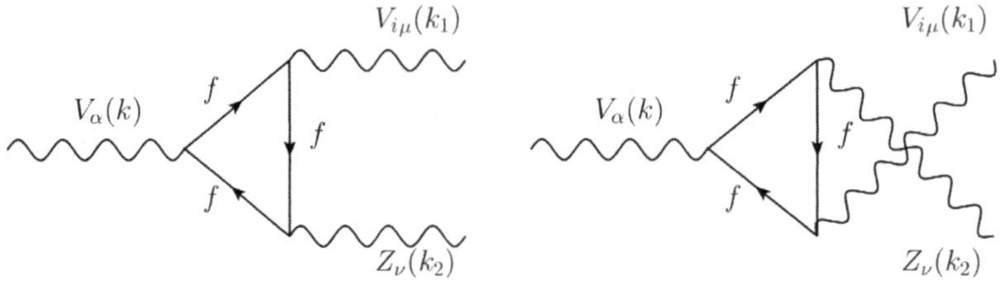

Figure 1. One-loop Feynman diagrams contributing to the extra neutral gauge boson decay $Z_H \to V_i Z$, with $V_i = \gamma, Z$ in the LHM.

$$\begin{aligned}\mathcal{M}(Z_H \to \gamma Z) &= \frac{i}{m_{Z_H}^2}\left(A_1^{\gamma Z}\left(k_1^\nu \epsilon^{\alpha\mu\lambda\rho} + k_1{}^\alpha \epsilon^{\mu\nu\lambda\rho}\right) k_{1\lambda} k_{2\rho} + A_2^{\gamma Z} k_1 \cdot k_2\, \epsilon^{\alpha\mu\nu\lambda} k_{1\lambda}\right)\\ &\times\ \epsilon_\alpha(k)\epsilon_\mu(k_1)\epsilon_\nu(k_2), \end{aligned} \tag{16}$$

where the four-momenta $k_{1\mu}$ and $k_{2\nu}$ correspond to the outgoing γ and Z gauge bosons, respectively. This amplitude displays explicitly electromagnetic gauge invariance. The $A_i^{\gamma Z}$ coefficients can be written in terms of Passarino-Veltman scalar functions as follows

$$\begin{aligned}A_1^{\gamma Z} &= \left(\frac{g}{8\pi c_W}\right)^2 \frac{2}{(1-y_Z)^3}\sum_f \xi_{\gamma Z}^f \Big(y_Z^2\left(2 + B_c - B_a + 2(2y_f+1)C_a\right) + B_a - B_c\\ &- 2y_Z\left(1 + 3(B_b - B_c) + (2y_f+1)C_a\right)\Big), \end{aligned} \tag{17}$$

$$\begin{aligned}A_2^{\gamma Z} &= \left(\frac{g}{8\pi c_W}\right)^2 \frac{2}{(1-y_Z)^3}\sum_f \Bigg(\xi_{\gamma Z}^f \Big(y_Z^2\left(B_a - B_c - 2(1 + C_a)\right) + B_c - B_a - 4y_f C_a\\ &+\ 2y_Z\left(1 + B_b - B_c + (2y_f+1)C_a\right)\Big) + 4\lambda_{\gamma Z}^f (1-y_Z)^2 y_f C_a\Bigg), \end{aligned} \tag{18}$$

where $\xi_{\gamma Z}^f = N_c^f Q^f g_L'^{\,f} g_L{}^f$ and $\lambda_{\gamma Z}^f = N_c^f Q^f g_L'^{\,f} g_R{}^f$, with Q_f the fermion electric charge, and $g_{L,R}^f$ the couplings of the Z gauge boson to the fermions. The sum is over all the charged fermions. Anomaly cancellation requires that $\sum_f \xi_{\gamma Z}^f = 0$. B_i stands for the following two-point scalar functions $B_a = B_0(0, m_f^2, m_f^2)$, $B_b = B_0(m_{Z_H}^2, m_f^2, m_f^2)$, and $B_c = B_0(m_Z^2, m_f^2, m_f^2)$, while $C_a = m_{Z_H}^2 C_0(0, m_Z^2, m_{Z_H}^2, m_f^2, m_f^2, m_f^2)$ is a three-point scalar function scaled by the m_{Z_H} mass. It is evident that the $A_i^{\gamma Z}$ coefficients are free of ultraviolet divergences.

The $Z_H \to \gamma Z$ decay width follows easily and it is given by

$$\Gamma(Z_H \to \gamma Z) = \frac{1}{3}\frac{(1-y_Z)^5 (1+y_Z)\, m_{Z_H}}{2^5 \pi y_Z}|A_1^{\gamma Z} - A_2^{\gamma Z}|^2. \tag{19}$$

We now concentrate on the $Z_H \to ZZ$ decay. The respective amplitude must obey Bose symmetry and can be written as

$$\mathcal{M}(Z_H \to ZZ) = \frac{i A^{ZZ}}{m_{Z_H}^2}\left(k_1^\nu \epsilon^{\alpha\mu\lambda\rho} + k_2^\mu \epsilon^{\alpha\nu\lambda\rho}\right) k_{1\lambda} k_{2\rho} \epsilon_\alpha(k)\epsilon_\mu(k_1)\epsilon_\nu(k_2), \tag{20}$$

where $k_{1\mu}$ and $k_{2\nu}$ are the four-momenta of the outgoing Z gauge bosons. The coefficient A^{ZZ} is

$$\begin{aligned} A^{ZZ} &= \frac{g^3}{8\pi^2 c_W^3}\frac{1}{(1-4y_Z)^2}\sum_f \Bigg((1+4y_Z^2\,(B_b - B_c + (1+2y_f)C_b + 2) \\ &- 2y_Z\,(3-(1-4y_f)(B_b-B_c)+5y_f C_b) + 2y_f\,((B_b - Bc) + C_b) - 4y_Z^3 C_b)\xi^f_{ZZ} \\ &+ 2y_f\,(1-4y_Z)\,(B_b - B_c + y_Z C_b)\,\lambda^f_{ZZ} \\ &- 2y_f\,(1-4y_Z)\,(2(B_b-B_c)+(1-2y_Z)\,C_b)\,\rho^f_{ZZ}\Bigg), \end{aligned} \tag{21}$$

with $\xi^f_{ZZ} = N_c^f g_L'^f g_L^{f\,2}$, $\lambda^f_{ZZ} = N_c^f g_L'^f g_R^{f\,2}$, and $\rho^f_{ZZ} = N_c^f g_L^f g_R^f g_L'^f$. The two-point scalar functions B_b and B_c were given above, while the scaled three-point scalar function is $C_b = m^2_{Z_H} C_0(m_Z^2, m_Z^2, m^2_{Z_H}, m_f^2, m_f^2, m_f^2)$. The sum is now over all fermions. Anomaly cancellation requires that $\sum_f \xi^f_{ZZ} = 0$.

The $Z_H \to ZZ$ decay width is given by

$$\Gamma(Z_H \to ZZ) = \frac{1}{3}\frac{m_{Z_H}}{2^7 \pi y_Z}(1-4y_Z)^{5/2}|A^{ZZ}|^2. \tag{22}$$

We will examine below the behavior of the branching ratios of all the above decays as functions of the symmetry breaking scale f and the mixing angle c.

4. Numerical Results

In Fig. 2 we show the corresponding results for the Z_H branching ratios discussed above as a function of the mixing angle c and for $f = 4$ TeV, which corresponds to the strongest constraint on the scale f [15]. For simplicity we used $\tan\theta' = c'/s' = 1$, although there is little dependence on this parameter. Also, the value 120 GeV is used for the mass of the Higgs boson. We observe that around $c = 1/\sqrt{2}$, several couplings of the Z_H gauge boson vanish and so it decays mainly into fermions. Apart from the color factor, the branching ratios for the light fermions are almost identical since the fermion mass is negligible. The subdominant decays are $Z_H \to \bar{t}tH$, $Z_H \to \bar{f}f\gamma$, $\bar{f}fZ$, $Z_H \to A_H H$, $Z_H \to A_H WW$, $Z_H \to A_H ZZ$, and $Z_H \to A_H HH$. The branching ratios for the last two decays are of similar magnitude and the latter was not included in the plot. Other tree-level decays such as $Z_H \to WW$, $Z_H \to ZH$, $Z_H \to ZHH$, $Z_H \to ZWW$, and $Z_H \to \gamma WW$ exactly vanish when $c = 1/\sqrt{2}$. On the other hand, when c is not close to $1/\sqrt{2}$, the decay $Z_H \to WW$ width can be as dominant as the fermion decays. As far as the one-loop decays are concerned, the $Z_H \to \gamma Z$ branching ratio can be as high as the tree-level decays $Z_H \to A_H H$ or $Z_H \to \bar{l}lZ$, but the $Z_H \to ZZ$ decay has a very small branching ratio. The former has a branching ratio of the order of 10^{-3} while the latter has a rate of about 10^{-5}.

We now show the Z_H branching ratios as functions of the scale f and for $c = 1/\sqrt{2}$ in Fig. 3. In this case, several decay channels vanish due to the vanishing of the Z_H couplings. It is interesting to note that the $Z_H \to \gamma Z$ branching ratio is even larger than the ones for the tree-level decays involving a heavy photon. The latter are suppressed by phase space. However, the decay $Z_H \to ZZ$ has a negligible branching ratio and it would hardly have the chance of being detected.

5. Experimental perspectives

At the LHC, the production of an extra neutral gauge boson would proceed mainly via the Drell-Yan process [12, 11]. We have calculated the number of $Z_H \to V_i Z$ ($V_i = \gamma, Z$) events

Journal of Physics: Conference Series **378** (2012) 012021 doi:10.1088/1742-6596/378/1/012021

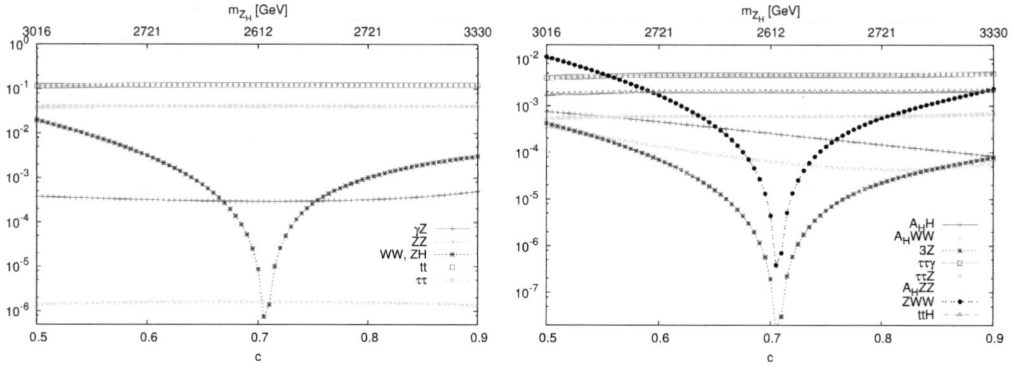

Figure 2. Branching ratios for the one-loop decays $Z_H \to \gamma Z$ and $Z_H \to ZZ$ in the LHM as a function of the mixing angle c. We also include the main tree-level two- and three-body decays. We used the value $m_H = 120$ GeV for the Higgs boson mass. The branching ratios for the decays $Z_H \to ZHH$, $Z_H \to \gamma WW$, and $Z_H \to A_H HH$ are not shown in the plot but $Br(Z_H \to ZHH) \sim Br(Z_H \to 3Z)$, $Br(Z_H \to \gamma WW) \sim Br(Z_H \to ZWW)$ and $Br(Z_H \to A_H HH) \sim Br(Z_H \to A_H ZZ)$. For the one-loop decays we used the package LoopTools [16, 17] to numerically evaluate the Passarino-Veltman scalar functions.

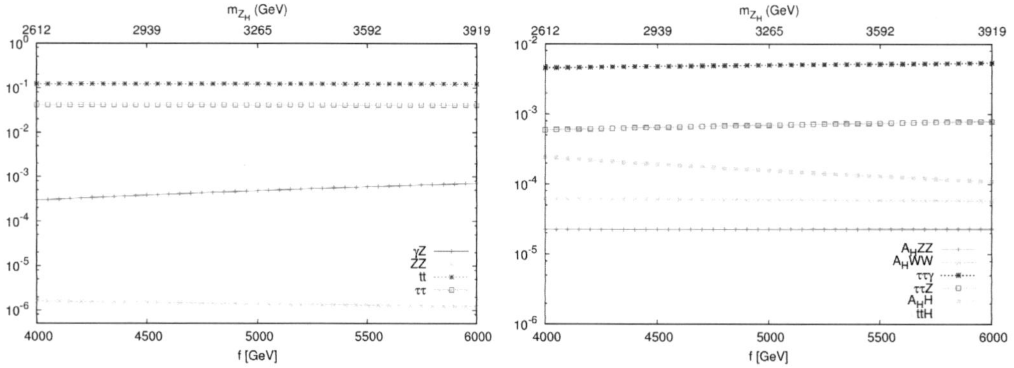

Figure 3. Branching ratios for the one-loop decays $Z_H \to \gamma Z$ and $Z_H \to ZZ$ in the LHM as a function of the scale of symmetry breaking f and for $c = 1/\sqrt{2}$. We also include the main tree-level two- and three-body decays. We used the value $m_H = 120$ GeV for the Higgs boson mass. The branching ratios not shown vanish exactly for this value of c.

at the LHC and the results are presented in [13]. For comparison purpose, we also calculated the number of $Z_H \to \ell\ell$ events. There are promising expectations for the discovery of an extra neutral gauge boson decaying into a lepton pair. As soon as the LHC reaches the nominal $\sqrt{s}=$ 14 TeV energy, it will take a few years to collect 100 fb^{-1} of integrated luminosity. For $f = 4$ TeV and $c = 1/\sqrt{2}$, which corresponds to the value $m_{Z_H} = 2.6$ GeV, there would be a few hundreds of $Z_H \to \ell\ell$ events. These results are similar to those obtained in other models [18, 19, 20], though they are one or two orders of magnitude smaller than those reported by ATLAS and CMS, using the SSM model. As far as the potential observation of the γZ and ZZ decay channels is concerned, LHC experiments have studied the sensitivity to the production of SM diboson events, including the γZ and ZZ signals with the Z gauge boson decaying into a highly energetic lepton pair (ee and $\mu\mu$) accompanied by an isolated high p_T photon [21, 22, 23, 24].

Journal of Physics: Conference Series **378** (2012) 012021 doi:10.1088/1742-6596/378/1/012021

The s-channel γZ production is of special interest due to its sensitivity to the $Z\gamma V_i$ ($V_i = Z, \gamma$) vertex [25], which is forbidden at tree-level in the SM. In Ref. [13] it was shown that there would be about about one dozen of $Z_H \to Z\gamma$ events at the LHC. For a heavier Z_H, an increase of about one order of magnitude in the integrated luminosity would be necessary to look for this decay process. As for the $Z_H \to ZZ$ decay channel, where the final state includes four leptons, it is clear that an experimental signature is not favorable. However, if we consider that the total LHC integrated luminosity will be of the order of 3000 fb^{-1}, there is still some chance to observe the $Z_H \to ZZ$ decay channel.

6. Conclusions

We have calculated the one-loop decays $Z_H \to V_i Z$ ($V_i = \gamma, Z$) in the framework of the LHM. While the branching ratio for the $Z_H \to \gamma Z$ decay can be as large as 10^{-3}, the $Z_H \to ZZ$ decay has a branching ratio of the order of 10^{-5}. We have also discussed the prospects for the experimental observation of these decays at the LHC. We conclude that the detection of the $Z_H \to V_i Z$ decays would not be favorable. In fact, it would be necessary to collect more than 1000 fb^{-1} of integrated luminosity in order to have a few handful of $Z_H \to \gamma Z$ candidate events. The situation of the $Z_H \to ZZ$ decay is less favorable.

Acknowledgment

We acknowledge support from Conacyt and SNI (México). Support from VIEP-BUAP is also acknowledged.

References

[1] N. Arkani-Hamed, A. G. Cohen and H. Georgi, Phys. Lett. B **513**, 232 (2001) [arXiv:hep-ph/0105239].
[2] N. Arkani-Hamed, A. G. Cohen, E. Katz and A. E. Nelson, JHEP **0207**, 034 (2002) [arXiv:hep-ph/0206021].
[3] N. Arkani-Hamed, A. G. Cohen, T. Gregoire and J. G. Wacker, JHEP **0208**, 020 (2002) [arXiv:hep-ph/0202089].
[4] N. Arkani-Hamed, A. G. Cohen, E. Katz, A. E. Nelson, T. Gregoire and J. G. Wacker, JHEP **0208**, 021 (2002) [arXiv:hep-ph/0206020].
[5] I. Low, JHEP **0410**, 067 (2004) [arXiv:hep-ph/0409025].
[6] D. E. Kaplan and M. Schmaltz, JHEP **0310**, 039 (2003) [arXiv:hep-ph/0302049].
[7] M. Schmaltz, JHEP **0408**, 056 (2004) [arXiv:hep-ph/0407143].
[8] D. Chang, W. Y. Keung and S. C. Lee, Phys. Rev. D **38**, 850 (1988).
[9] M. A. Perez, G. Tavares-Velasco and J. J. Toscano, Phys. Rev. D **69**, 115004 (2004) [arXiv:hep-ph/0402156].
[10] M. Perelstein and Y. H. Qi, Phys. Rev. D **82**, 015004 (2010) [arXiv:1003.5725 [hep-ph]].
[11] T. Han, H. E. Logan, B. McElrath and L. T. Wang, Phys. Rev. D **67**, 095004 (2003) [arXiv:hep-ph/0301040].
[12] G. Burdman, M. Perelstein and A. Pierce, Phys. Rev. Lett. **90**, 241802 (2003) [Erratum-ibid. **92**, 049903 (2004)] [arXiv:hep-ph/0212228].
[13] I. Cortes-Maldonado, A. Fernandez-Tellez and G. Tavares-Velasco, J. Phys. G **39**, 015003 (2012) [arXiv:1109.4390 [hep-ph]].
[14] R. Mertig, M. Bohm and A. Denner, Comput. Phys. Commun. **64**, 345 (1991).
[15] C. Csaki, J. Hubisz, G. D. Kribs, P. Meade and J. Terning, Phys. Rev. D **67**, 115002 (2003) [arXiv:hep-ph/0211124].
[16] G. J. van Oldenborgh and J. A. M. Vermaseren, Z. Phys. C **46**, 425 (1990).
[17] T. Hahn, Nucl. Phys. Proc. Suppl. **89**, 231 (2000) [arXiv:hep-ph/0005029].
[18] P. Nath *et al.*, Nucl. Phys. Proc. Suppl. **200-202**, 185 (2010) [arXiv:1001.2693 [hep-ph]].
[19] P. Langacker, arXiv:0911.4294 [hep-ph].
[20] T. G. Rizzo, arXiv:hep-ph/0610104.
[21] S. Chatrchyan *et al.* [CMS Collaboration], arXiv:1105.2758 [hep-ex].
[22] G. Aad *et al.* [ATLAS Collaboration], arXiv:1106.1592 [hep-ex].
[23] V. I. M. Outschoorn [On behalf of the ATLAS Collaboration], arXiv:1011.0164 [hep-ex].
[24] T. Aaltonen *et al.* [CDF Collaboration], Phys. Rev. D **82**, 031103 (2010) [arXiv:1004.1140 [hep-ex]].
[25] F. Larios, M. A. Perez, G. Tavares-Velasco and J. J. Toscano, Phys. Rev. D **63**, 113014 (2001) [arXiv:hep-ph/0012180].

Journal of Physics: Conference Series **378** (2012) 012022 doi:10.1088/1742-6596/378/1/012022

Higgs mediated flavor-violating transitions of the top quark

J. I. Aranda[1], A. Cordero-Cid[2], F. Ramírez-Zavaleta[1], E. S. Tututi[1] and J. J. Toscano[3]

[1]Facultad de Ciencias Fśico-Matemáticas, Universidad Michoacana de San Nicolás de Hidalgo, Avenida Francisco J. Mújica S/N, 58060, Morelia, Michoacán, México.

[2]Facultad de Ciencias de la Electrónica, Benemérita Universidad Autónoma de Puebla, Blvd. 18 Sur y Av. San Claudio, 72590, Puebla, Puebla, México.

[3]Facultad de Ciencias Físico-Matemáticas,Benemérita Universidad Autónoma de Puebla, Apartado Postal 1152, Puebla, Puebla, México.

E-mail: jarandas@umich.mx

Abstract. Flavor violating tu_iH vertex induces top quark couplings $tc\gamma$ and $tu_i\gamma\gamma$ ($u_i = u, c$) at the one-loop level, in the effective Yukawa sector through dimension six $SU_L(2) \times U_Y(1)$-invariant operators. The $t \to cH$ and $t \to c\gamma\gamma$ decays can reach branching ratios about 5×10^{-3} and 10^{-4}, respectively. As for tc production, it is found that, due to the presence of a resonant effect in the convoluted cross section $\sigma(e^+e^- \to \gamma\gamma \to t\bar{c} + \bar{t}c)$, about $(0.5 - 2.7) \times 10^3$ tc events may be produced at the ILC for $m_H \sim m_t$.

1. Introduction

The fact that the top quark is the heaviest elementary particle known suggests that it would be quite sensitive to new physics effects. Although very suppressed in the SM, the flavor violating $t \to u_iV$ decays can have sizable branching ratios in many well motivated of its extensions. On the other hand, it is expected that new physics effects on top quark physics can be observed in both LHC and ILC colliders. In this work it is intended to study the rare top quark decays $t \to u_iH$, $t \to u_i\gamma$, and $t \to u_i\gamma\gamma$, as well as to investigate the $\gamma\gamma \to \bar{t}u_i + t\bar{u}_i$ reaction in the context of the ILC operating in the $\gamma\gamma$ mode, which has shown to be the most efficient mechanism in producing tc events, as the corresponding cross section can be up to 2 orders of magnitude and 1 order of magnitude greater than those associated with the production mechanism in the e^+e^- and $e\gamma$ collision modes, respectively [1].

2. Study of the coupling f_if_jH

We incorporate into the classical action the virtual effects of the heavy degrees by introducing $SU_L(2) \times U_Y(1)$-invariant operators of dimension higher than four. To extend the Yukawa sector with dimension-six operators to induce the most general coupling of the Higgs boson to quarks and leptons. A Yukawa sector with these features has the following structure

$$\mathcal{L}^Y_{eff} = -Y^q_{ij}(\bar{Q}_i\Phi_q q_j) - \frac{\alpha^q_{ij}}{\Lambda^2}(\Phi^\dagger\Phi)(\bar{Q}_i\Phi_q q_j) + H.c., \quad (1)$$

where Y_{ij}, Q_i, Φ_q ($\Phi_q = \Phi, \tilde{\Phi}$ for $q = d, u$ respectively and a sum over q is implied), d_i, and u_i stand for the usual components of the Yukawa matrix, the left-handed quark doublet, the Higgs doublet, and the right-handed quark singlets of down and up type, respectively. The α_{ij} numbers are the components of a 3×3 general matrix, which parametrize the details of the underlying physics, whereas Λ is the typical scale of these new physics effects. This effective Yukawa sector induces a flavor and CP violating coupling $f_i f_j H$ of renormalizable type given by

$$\Gamma_{f_i f_j H} = -i(\omega_R^{ij} P_R + \omega_L^{ij} P_L), \tag{2}$$

where $\omega_R^{ij} = \frac{g m_i}{2 m_W}\delta_{ij} + \Omega_{ij}$ and $\omega_L^{ij} = \frac{g m_i}{2 m_W}\delta_{ij} + \Omega_{ij}^*$, with $P_{R,L} = (1 \pm \gamma_5)/2$. In these expressions, $\Omega^{(u,d)} = \frac{1}{\sqrt{2}}\left(\frac{v}{\Lambda}\right)^2 V_L^{(u,d)} \alpha^{(u,d)} V_R^{(u,d)\dagger}$, being $V_L^{(u,d)}$ and $V_R^{(u,d)}$ the usual unitary matrices, which correlate gauge states to mass eigenstates [1].

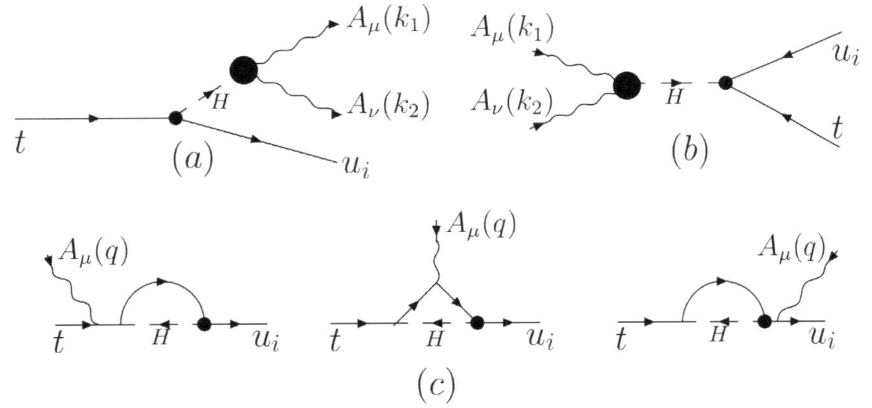

Figure 1. Feynman diagrams contributing to the $t \to c\gamma\gamma$ (a) and $t \to c\gamma$ (c) decays and to the scattering process $\gamma\gamma \to tc$ (b).

The $f_i f_j H$ vertex induces the flavor violating vertices $f_i f_j \gamma$ and $f_i f_j \gamma\gamma$ at the one-loop level. The contribution to $f_i f_j \gamma\gamma$ occurs through two set of Feynman diagrams, each of them giving a finite and gauge invariant contribution. The contribution to $f_i f_j \gamma\gamma$ occurs through two set of Feynman diagrams, each of them giving a finite and gauge invariant contribution. One of these sets, whose contribution is marginal, include box diagrams, reducible diagrams characterized by the one-loop $f_i f_j \gamma$ coupling, and reducible diagrams composed by the one-loop $f_i - f_j$ bilinear coupling. The other set of diagrams, which gives the dominant contribution, is characterized by the SM one-loop $\gamma\gamma H^*$ coupling, with $*$ stands for a virtual Higgs. We will study the decays $t \to u_i \gamma$ and $t \to u_i \gamma\gamma$, as well as the scattering process $\gamma\gamma \to t u_i$ (see Fig. 1).

We are also interested in studying the $t \to u_i H$ decay, whose branching ratio can be written as

$$Br(t \to u_i H) = \left(\frac{\Omega_{tu_i}^2}{32\pi}\right)\left(\frac{m_t}{\Gamma_t}\right)\left(1 - \left(\frac{m_H}{m_t}\right)^2\right)^2. \tag{3}$$

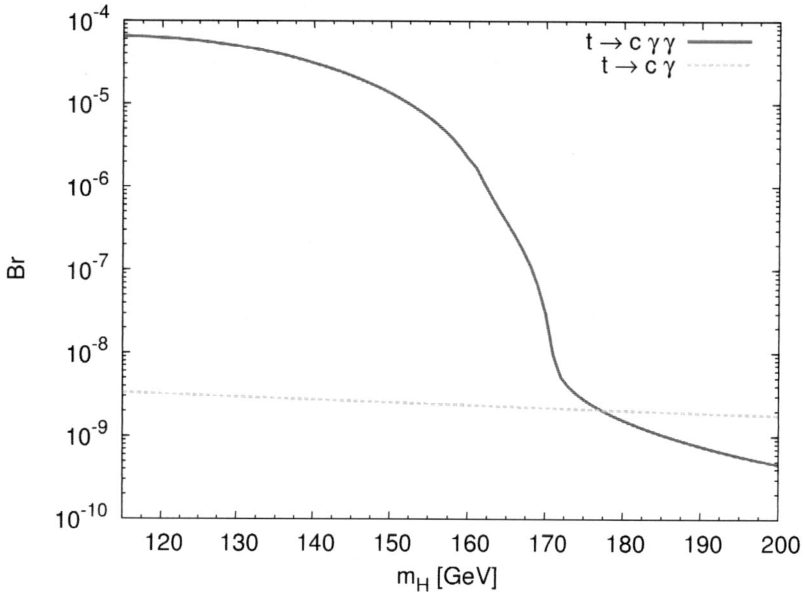

Figure 2. The branching ratios for the $t \to c\gamma$, and $t \to c\gamma\gamma$ decays as a function of the Higgs mass. The thin lines correspond to a Higgs mass in the range 115 GeV $< m_H <$ 171.3 GeV.

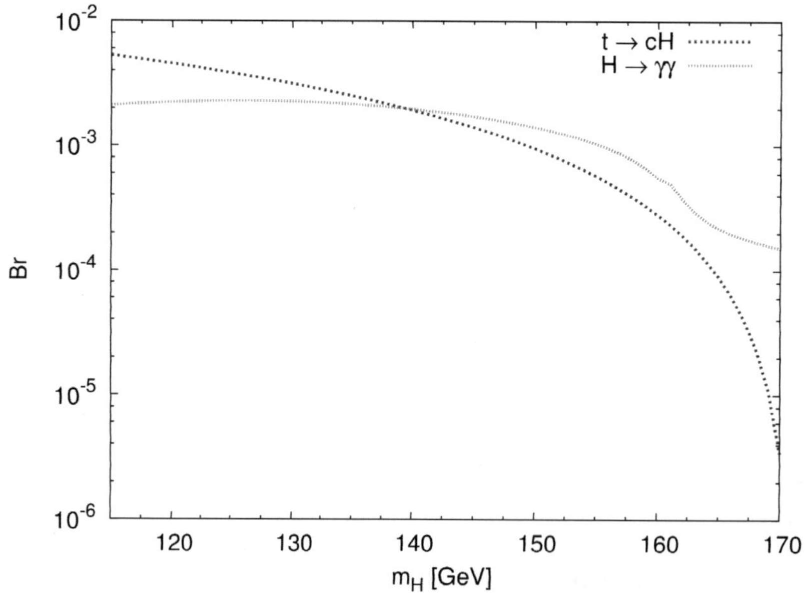

Figure 3. The branching ratios for the $t \to cH$, and $H \to \gamma\gamma$ decays as a function of the Higgs mass. The thin lines correspond to a Higgs mass in the range 115 GeV $< m_H <$ 171.3 GeV.

We will consider only the contribution arising from the diagrams characterized by the one-loop $\gamma\gamma H^*$ coupling, as it dominates. In order to make predictions, some value for the Ω_{tu_i} parameter must be assumed. In order to make predictions, some value for the Ω_{tu_i} parameter must be assumed. By assuming $\lambda_{tc} \sim 1$, branching ratios for the $t \to cH$ and $t \to c\gamma$ transitions of the order of 10^{-4} and 10^{-8}, respectively, were found. On the other hand, the branching ratio for the $t \to c\gamma\gamma$ decay was calculated within the context of the THDM-III, which has the simplest extended Higgs sector that naturally incorporates Higgs mediated flavor changing neutral current at the tree level [1]. It was found that this decays can reach a branching as high as 10^{-4}.

We have found that the best constraint arises from the recently observed $D^0 - \overline{D^0}$ mixing [1]. Since this observable receives contributions simultaneously from both the tuH and tcH couplings, it is only possible to obtain $|\Omega_{tc}\Omega_{tu}| < 10^{-3}$. We will assume that $\Omega_{tu} = 10^{-1}\Omega_{tc}$, which implies the bounds $\Omega_{tc}^2 < 10^{-2}$ and $\Omega_{tu}^2 < 10^{-4}$.

We will present results only for those processes involving the quark c, which can be translated to the corresponding transitions involving the quark u simply multiplying by a factor of 10^{-2}. In Fig. 3, the branching ratio for the $t \to cH$ decay and for the $H \to \gamma\gamma$ decay are shown as a function of the Higgs mass. We can see that $Br(t \to cH)$ ranges from 5×10^{-3} to 5×10^{-4} for a Higgs mass in the range 115 GeV $< m_H < 2m_W$. On the other hand, from Fig.2 we can see that the branching ratio for the $t \to c\gamma$ decay is suppressed, as it has a value essentially constant of about of 5×10^{-8}, which however is about 6 orders of magnitude larger than the SM prediction. In contrast, the $t \to c\gamma\gamma$ decay has a branching ratio which can be as high as $10^{-4} - 10^{-5}$. It should be noticed that $Br(t \to c\gamma\gamma) \sim Br(t \to cH)Br(H \to \gamma\gamma)$, which is evident from graphs in Fig. 2 [1]. From the amplitude for the $f_i f_j \gamma\gamma$ coupling, one can construct the unpolarized

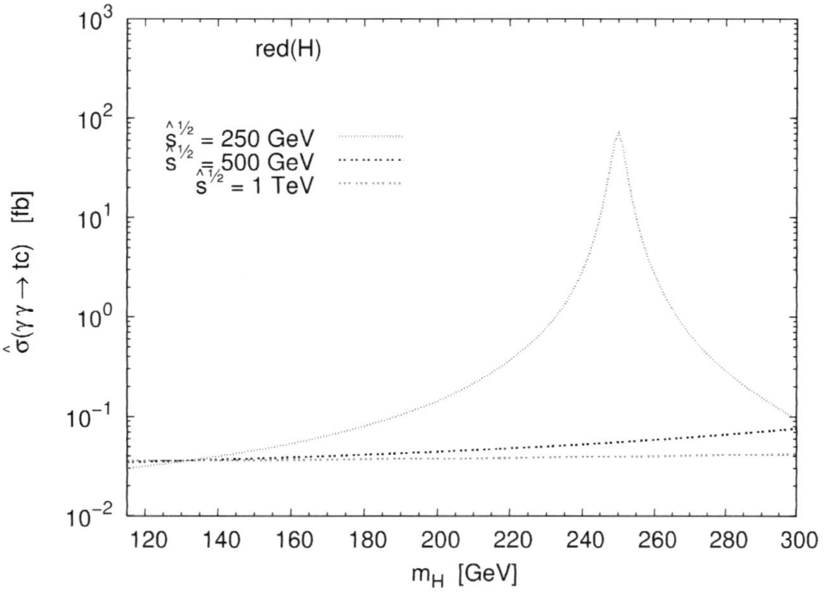

Figure 4. The cross section for the $\gamma\gamma \to t\bar{c} + \bar{t}c$ processes as a function of the Higgs mass for some values of the c.m.s energy..

cross section of the $\gamma\gamma \to \bar{t}c + t\bar{c}$ collision, which in turn defines the cross section of the complete

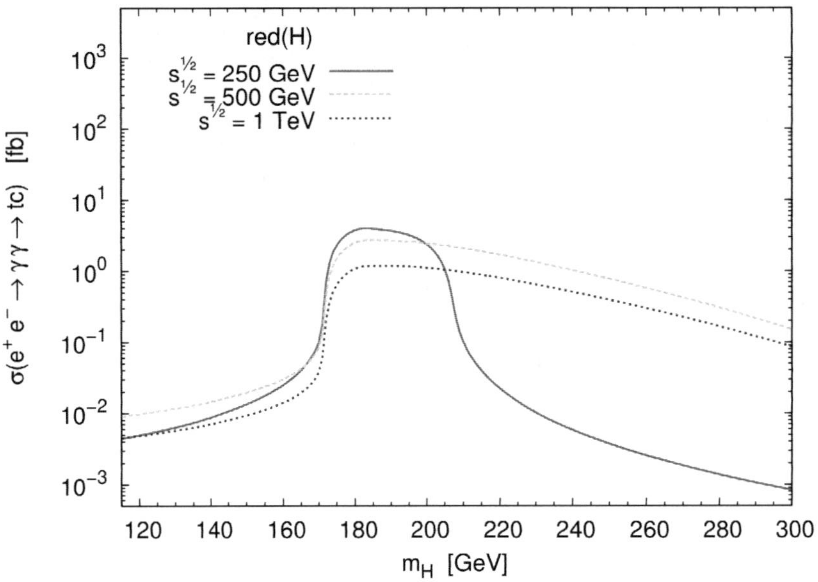

Figure 5. The cross section for the $e^+e^- \to \gamma\gamma \to t\bar{c} + \bar{t}c$ processes as a function of the Higgs mass for some values of the c.m.s energy

process $e^+e^- \to \gamma\gamma \to \bar{t}c + t\bar{c}$ through the relation [1]

$$\sigma(s) = \int_{(m_c+m_t)/\sqrt{s}}^{y_{max}} dz \, \frac{d\mathcal{L}_{\gamma\gamma}}{dz} \, \hat{\sigma}(\gamma\gamma \to \bar{t}\, c + t\, \bar{c}), \tag{4}$$

where $z = \sqrt{\hat{s}}/\sqrt{s}$, being $\sqrt{s}$ and $\sqrt{\hat{s}}$ the c.m.s. energies of the e^+e^- and $\gamma\gamma$ collisions, respectively. In the above expression $\frac{d\mathcal{L}_{\gamma\gamma}}{dz}$ represents the photon luminosity [1]. The optimum value for the upper limit in the above integral is $y_{max} \approx 0.83$. As in the case of the $t \to c\gamma\gamma$ decay, we are considering only the contribution given by the diagrams characterized by the SM $\gamma\gamma H^*$ vertex. Initially, the ILC will operate from a $\sqrt{s}$ of 200-500 GeV, with a luminosity of 500 fb^{-1} within the first years of operation and 1000 fb^{-1} during the second phase of operation at 500 GeV.

We will present results for energies in the range 250 GeV $< \sqrt{s} <$ 1000 GeV. In Fig. 5 is shown the pure cross section as a function of the Higgs mass for some values of $\sqrt{\hat{s}}$ and $\sqrt{s}$. As it is shown Fig. 4, the convoluted cross section appears as a function of the Higgs mass for some values of $\sqrt{\hat{s}}$ and $\sqrt{s}$. It can be appreciated a significant enhancement of the convoluted cross section for values of the Higgs mass near the value of the top mass. This peculiar effect, which, as it can be appreciated from the graphic in Fig. 5, is not present in the pure $\hat{\sigma}(\gamma\gamma \to t\bar{c} + \bar{t}c)$ cross section (see Fig. 4), has its origin in a resonance that occurs for a Higgs mass near the top mass value [1].

The maximum value of the cross section corresponds to m_H = 184 GeV. The respective number of events is 460, 2720, 2460, 2090, and 1700 if a luminosity of 1000 fb^{-1} is assumed. Our results are 1 order of magnitude lower than those derived within the context of top-color-assisted technicolor models and several orders of magnitude lower than those obtained in the littlest Higgs model [1].

3. Conclusions.

We have investigated some phenomenological implications of a flavor violating tcH vertex induced by an extended Yukawa sector that incorporates $SU_L(2) \times U_Y(1)$-invariant operators of up to dimension six. By bounding the tcH vertex from the recently observed $D^0 - \overline{D^0}$ mixing, the $t \to cH$, $t \to c\gamma$, $t \to c\gamma\gamma$, and $\gamma\gamma \to t\bar{c} + \bar{t}c$ processes were predicted. Sizable branching ratios for the $t \to cH$ and $t \to c\gamma\gamma$ decays of the order of 10^{-3} and 10^{-4}, respectively, were found. A branching ratio of about 5×10^{-8} for the $t \to c\gamma$ decays was found, which though it is about 6 orders of magnitude larger than the SM prediction, it is not large enough to be detected. It was found that about of $(0.5 - 2.7) \times 10^3$ tc events may be produced for a Higgs boson with $m_H \sim m_t$.

3.1. Acknowledgments

Authors wishing to acknowledge financial support from CIC (UMSNH).

References

[1] J. I. Aranda *et al.*, Phys. Rev. **D81**, 077701 (2010) and references there in.

Journal of Physics: Conference Series **378** (2012) 012023 doi:10.1088/1742-6596/378/1/012023

Flavour parametrization of couplings with fermions in the 2HDM: the case with the 4-zero texture mass matrix

L. López-Lozano

Departamento de Física, DCI, Campus León, Universidad de Guanajuato, C.P. 37150, León, Guanajuato, México.

Abstract. The 2HDM has been one of the most studied extensions of the SM, due to its simple form and wealthy new phenomenology. Most versions are characterized by the mechanism through which the FCNC are controlled. In order to give to the 2HDM the status of an intermediate step in the rank of energies (that is, a low energy limit between the electroweak scale and the characteristic scale of the general theory), a mechanism to relate several versions of the 2HDM is given. This mechanism exploits the flavour symmetry of the Yukawa Lagrangian and the fact that the mass generation comes from a yet unknown property of the vacuum but related with the CKM entries in the case of quarks.

1. Introduction

In the last years, models that extend the Standard Model (SM) have concentrated much attention. One of motivations for this is the fact that in the next years crucial experimental results, coming from LHC, could demonstrate the existence of physics beyond SM [1, 2]. There are signs that the last SM particle to be discovered, the Higgs boson, could have a mass around 125 GeV [3, 2] consistent with many theoretical predictions of such model [4], thus unless a big surprise arises, it is expected that the signs of new physics phenomena will be manifest in small deviations from the SM at such scales; in such a case, the models that describe slight deviations are good benchmarks to find theoretical or phenomenological hints that help us understand the whole description of particle physics beyond the electroweak scale. New experimental results will shed light on the nature of the scalar sector, that could be the starting point to understand new physics in terms of a non-SM Higgs sector. There are two strong constraints that bound the model that extend SM Higgs sector: the electroweak ρ parameter (that is $\rho_{\text{exp}} \sim 1$) and the highly suppressed flavour changing neutral current (FCNC) [5, 6].

The 2HDM [7] is one of the simplest extensions of SM that includes phenomenology beyond SM. The introduction of an additional doublet generates new physics in the interaction of fermions with scalars, and deviations from the SM phenomenology of gauge bosons due to the presence of five physical scalar fields (h^0, H^0, A^0 and $H^{\pm}$). Nevertheless, some mechanism must be used to explain the experimental constraints that are highly accurate at the electroweak scale. Different versions of the 2HDM have been proposed, and they can be classified by the mechanism through which they suppress the FCNC. A very interesting review of all these versions can be seen in [8] and references therein.

Journal of Physics: Conference Series **378** (2012) 012023 doi:10.1088/1742-6596/378/1/012023

Version	Y_1^u	Y_2^u	Y_1^d	Y_2^d	Y_1^ℓ	Y_2^ℓ
I	$0_{3\times3}$	-	$0_{3\times3}$	-	$0_{3\times3}$	-
II	$0_{3\times3}$	-	-	$0_{3\times3}$	-	$0_{3\times3}$
L.S.	$0_{3\times3}$	-	$0_{3\times3}$	-	-	$0_{3\times3}$
Flipped	$0_{3\times3}$	-	-	$0_{3\times3}$	$0_{3\times3}$	-
III	-	-	-	-	-	-

Table 1. Conditions on the Yukawa matrices, in order to reproduce versions of 2HDM. Empty spaces mean that the Yukawa matrices can be different from zero. By convention, the second Higgs doublet couples always with the up-type quarks[8]

There are versions of 2HDM with Natural Flavour Conservation (NFC)[9, 10, 11], that use a Z_2 symmetry imposed on the Higgs doublets in order to eliminate FCNC at tree level, before diagonalizing trough a bi-unitary transformation. On the other hand some models introduce alignment in order to diagonalized simultaneously both Yukawa matrices with the same effect [12] with some withdraws [13]. In the version III [14, 15, 16] no Yukawa matrix can be in general zero, because both doublets contribute to generate the quark masses.

A more general approach can be constructed assuming one mechanism that make the Yukawa matrix elements equal to zero, the suppression of FCNC is reached trough a suitable parametrization of the Yukawa matrix entrances without any additional assumptions on the symmetry of Higgs doublets. This approach is not equivalent to NFC, as it has consequences on the Higgs potential, where the terms that violate CP symmetry are present because, in this case, there is not any symmetry that suppress them. In the table 1 it is shown, under the convention that the second Higgs doublet couples with the up-type quark, how different versions of the 2HDM can be generated by choosing combinations of zero Yukawa matrices. An early interesting and thoroughly analysis about a general parametrization of the 2HDM is given in [17] where a suitable selection of parameters reduce the general parametrization into specific versions.

At this work we introduce a parametrization, that relates through flavour transformations on one Yukawa matrix, one version into another that have the advantage of show how different versions of 2HDM could correspond to a specific wake-basis for the fermions and suggest that all the version could corresponds to the same physical process, the flavour transformation respect the vacuum, which particular version is determined by nature. This parametrization enables us to separate the problem of the mass hierarchy from the flavour violation that comes from the scalar interactions.

2. Flavour parametrization in the 2HDM

The most general Yukawa lagrangian in the 2HDM for quarks and leptons is given by

$$\begin{aligned} -\mathcal{L}_Y &= \bar{Q}_L^0 \left[\left(Y_1^d \Phi_1 + Y_2^d \Phi_2 \right) d_R^0 + \left(Y_1^u \tilde{\Phi}_1 + Y_2^u \tilde{\Phi}_2 \right) u_R^0 \right] \\ & \quad + \bar{L}_L^0 \left(Y_1^\ell \Phi_1 + Y_2^\ell \Phi_2 \right) \ell_R^0 + \text{H.c} \end{aligned} \quad (1)$$

It could be possible that the Yukawa matrix elements are directly related with the mass spectrum of fermions because masses define the scale of the interactions. There exist many efforts to express the mixing of fermions in terms of their masses even in the SM (see [18, 19, 20] and references therein), thus we can expect parametrizations in the 2HDM of scalar interactions to be related with the mass matrix too. In the 2HDM the mass matrix has two contributions, weighted by the Vacuum Expectation Value (VEV) of the Higgs doublets, *i.e.*

$$M^f = \frac{1}{\sqrt{2}}\left(v_1 Y_1^f + v_2 e^{i\theta} Y_2^f\right), \tag{2}$$

where θ is the CP violating phase that come from the complex VEV's of the Higgs doublets and $f = u, d, \ell$ is flavour index. The equation (2) can be interpreted as a relation that imposes conditions on the Yukawa matrices, once the general form of the mass matrix is known, in such a way that only one Yukawa matrix is free. The table 1 suggests that it is possible to choose a parametrization that transforms one version into another through an independent flavour transformation of Yukawa matrices. The flavour freedom of the 2HDM Yukawa lagrangian is manifest when the two Yukawa matrices that conform the mass matrix are simultaneously transformed. In the Higgs basis, where only the first doublet has a non-zero VEV, it is clear that, once the mass matrix is known, some assumptions have to be made on, at least, one of the Yukawa matrices to that end. In order to generate the scalar interactions that inherits the hierarchy of the mass matrices we can express in general the Yukawa matrices as a bilinear flavour transformation of the mass matrix, *i.e.*

$$Y_a^f = A_{aL}^{f\dagger} M^f A_{aR}^f, \tag{3}$$

where $a = 1, 2$ and A_{aL}^f and A_{aR}^f are unitary matrices that mix the flavour of fermions. As have been point out in several places [21, 22] the mass matrix elements are determined by the CKM matrix entrances. Thus here we have translated the problem of finding the Yukawa matrix elements into that of finding the parameters of a bilinear-transformation. In this way we can explore several models that generate scalar interactions with fermions using the structure of flavour transformations. If M^f represents a SM mass matrix, we can generate from this, Yukawa matrices that represents couplings with non-standard Higgs doublets. It is possible to show how a particular group of transformations can generate several versions of the 2HDM. Particular cases of (3) as unitary or orthogonal transformations are very attractive because reduces the number of free parameters. So far we analyse the consequences on the lagrangian the election of particular form for A_{aL}^f and A_{aR}^f in the mentioned particular cases.

2.1. Unitary parametrization

For three generations of fermions we can restrict ourselves to unitary transformations and use the $SU(3)$ group to write

$$A_{aL}^f = A_{aR}^f = \sum_{j=0}^{8} C_{aj}^f \lambda_j, \tag{4}$$

where λ_j are the Gell-Mann matrices and $j = 0, \ldots, 8$ is a group index. So far we omit the subscript L and R. As said before the expression (2) reduce the number of free parameters reducing the problem of find only one Yukawa matrix, thus without loss of generality, it possible to choose the flavour basis where $A_1^f = A_1^f = \frac{\sqrt{\cos\gamma_f}}{v}\lambda_0$ and $A_2^f = A_2^f$ in such way that the equation for the rest of free parameters in A_2^f, is given by

$$\sqrt{2}M^f = \frac{v_1}{v}\cos\gamma_f M^f + \frac{v_2}{v}\sin\gamma_f A_2^{f\dagger} M^f A_2^f, \tag{5}$$

where the Yukawa matrix has been written as

$$Y_2^f = \frac{\sin\gamma_f}{v} A_2^{f\dagger} M^f A_2^f. \tag{6}$$

As we can see if $\gamma_f = n\pi$, where n integer, can reproduce the zeros of the table 1, that corresponds to different versions of 2HDM with NFC. In principle M^f is determined by the CKM matrix thus this former equation is enough to find the product of the parameter $C^f_{2ij} \equiv C^{f*}_{2i} C^f_{2j}$, nevertheless it is expected that future experiments gives information on the Yukawa couplings directly, thus multiplying last equation by Gell-Mann matrices and taking the trace, we find that the relation between the phenomenological information with the free parameters can be cast in

$$\mathrm{Tr}(\lambda_i Y_2^f \lambda_j) = \frac{\sin\gamma_f}{v} \sum_{k,l} C^f_{2kl} \mathrm{Tr.}(\lambda_i \lambda_k M^f \lambda_l \lambda_j) \tag{7}$$

This general expression gets simplified when we introduce the assumption of the preservation of textures coming from the mass matrix, in this case only the diagonal Gell-Mann matrices contributes, that is, λ_3 and λ_8, reducing drastically the free parameters required.

2.2. Orthogonal transformations with a 4-zero mass matrix

An interesting particular case is given by an infinitesimal orthogonal transformations with a 4-zero texture mass matrix. The 4-zero texture mass matrix is given by

$$M^f = \begin{pmatrix} 0 & C_f & 0 \\ C_f^* & D_f & B_f \\ 0 & B_f^* & A_f \end{pmatrix}, \tag{8}$$

where we take into account that this parameters are known. To simplify it is possible to extract the phases of non-diagonal elements through $M^f = P^f \bar{M}^f P$ where $P = \mathrm{diag}(1, \exp(i\theta_C), \exp[i(\theta_B + \theta_C)])$. Here $\bar{M}$ is a symmetric matrix with real and positive non-diagonal elements. If we introduce a infinitesimal orthogonal transformation in order to generate small deviations from the SM scalar interactions, that is

$$\bar{Y}_2^f = \mathcal{O}^T(\theta_1, \theta_2, \theta_3) \bar{M}^f \mathcal{O}(\theta_1, \theta_2, \theta_3), \tag{9}$$

where θ_1, θ_2 and θ_3 are small Euler angles of the infinitesimal rotation. In this case the Yukawa matrix without phases at leading order in the angles is

$$\bar{Y}_2^f = \begin{pmatrix} -2(\theta_1+\theta_3)|C_f| & |C_f|\left(1-(\theta_1+\theta_3)\frac{D_f}{|C_f|}\right) & \theta_2|C_f| - (\theta_1+\theta_3)|B_f| \\ |C_f|\left(1-(\theta_1+\theta_3)\frac{D_f}{|C_f|}\right) & |D_f| + 2(\theta_1+\theta_3)|C_f| - 2\theta_2|B_f| & |B_f| + \theta_2(D_f - A_f) \\ \theta_2|C_f| - (\theta_1+\theta_3)|B_f| & |B_f| + \theta_2(D_f - A_f) & A_f + 2\theta_2|B_f| \end{pmatrix}. \tag{10}$$

We observe that the scalar couplings have only three free parameters and the Yukawa matrix inherits the hierarchy of the mass matrix.

3. Conclusions

We have shown the flavour parametrizations of the 2HDM where the Yukawa matrices are constructed by a flavour transformations of the mass matrix of the SM. In this formalism the couplings of scalars with fermions are given in terms of the parameters of a bilinear transformation. This parametrization have essentially two advantage: On the one hand generates in a systematic way the non-SM scalar couplings through perturbations of the properties of vacuum that are expressed in the hierarchy of the mass matrix. On the other hand explains the different versions of the 2HDM with only one principle, for instance the alignment between Yukawa matrices and versions similar to those with NFC are described by one parameter per sector γ_f. Because different versions with NFC have different phenomenology, experiments can determined easily the value of this parameter. Likewise this parametrization is enough general in order to describe more generic versions as the version III where both Yukawa matrices contributes to the generation of mass. Reasonable assumptions can be used in order to reduce the number of free parameters, for instance the use of unitary transformations that are suitable even for non-hermitian Yukawa matrices. Also the infinitesimal orthogonal matrices are a good benchmark to describe small deviations from the SM. One additional relative *advantage* is that is preserved the source of CP in the scalar potential that was absent when a discrete symmetry Z_2 were introduced on the Higgs doublets.

References

[1] Dittmaier S *et al.* (LHC Higgs Cross Section Working Group) 2011 (*Preprint* 1101.0593)
[2] Chatrchyan S *et al.* (CMS Collaboration) 2012 (*Preprint* 1202.1488)
[3] Aad G *et al.* (ATLAS Collaboration) 2012 (*Preprint* 1202.1408)
[4] Erler J 2010 *Phys.Rev.* **D81** 051301 (*Preprint* 1002.1320)
[5] Glashow S, Iliopoulos J and Maiani L 1970 *Phys.Rev.* **D2** 1285–1292
[6] Gunion J F, Haber H E, Kane G L and Dawson S 2000 *Front. Phys.* **80** 1–448
[7] Lee T 1973 *Phys.Rev.* **D8** 1226–1239
[8] Branco G, Ferreira P, Lavoura L, Rebelo M, Sher M *et al.* 2011 (*Preprint* 1106.0034)
[9] Glashow S L and Weinberg S 1977 *Phys.Rev.* **D15** 1958
[10] Paschos E 1977 *Phys.Rev.* **D15** 1966
[11] Aoki M, Kanemura S, Tsumura K and Yagyu K 2009 *Phys.Rev.* **D80** 015017 (*Preprint* 0902.4665)
[12] Pich A and Tuzon P 2009 *Phys. Rev.* **D80** 091702 (*Preprint* 0908.1554)
[13] Ferreira P, Lavoura L and Silva J P 2010 *Phys.Lett.* **B688** 341–344 (*Preprint* 1001.2561)
[14] Cheng T and Sher M 1987 *Phys.Rev.* **D35** 3484
[15] Diaz-Cruz J L, Noriega-Papaqui R and Rosado A 2005 *Phys. Rev.* **D71** 015014
[16] Diaz-Cruz J L, Noriega-Papaqui R and Rosado A 2004 *Phys. Rev.* **D69** 095002
[17] Diaz-Cruz J, Diaz-Furlong A and Montes de Oca J 2010 (*Preprint* 1010.0950)
[18] Fritzsch H and Xing Z z 2003 *Phys.Lett.* **B555** 63–70 (*Preprint* hep-ph/0212195)
[19] Fritzsch H 1977 *Phys. Lett.* **B70** 436
[20] Fritzsch H and Xing Z Z 1997 *Phys.Lett.* **B413** 396–404 (*Preprint* hep-ph/9707215)
[21] Fritzsch H and Xing Z z 1999 *Nucl.Phys.* **B556** 49–75 (*Preprint* hep-ph/9904286)
[22] Fritzsch H and Xing Z z 2000 *Prog.Part.Nucl.Phys.* **45** 1–81 (*Preprint* hep-ph/9912358)

Journal of Physics: Conference Series **378** (2012) 012024 doi:10.1088/1742-6596/378/1/012024

Effects of heavy Majorana neutrinos in semileptonic heavy quark decays

D Delepine[1], **G López Castro**[2] **and N Quintero**[2]

[1] Division de Ciencias e Ingenierías, Universidad de Guanajuato, Campus Leon, C.P. 37150, León, Guanajuato, México

[2] Departamento de Física, Centro de Investigación y de Estudios Avanzados, Apartado Postal 14-740, 07000 México D.F., México

E-mail: nquintero@fis.cinvestav.mx

Abstract. The experimental observation of lepton number violating (LNV) processes, where the total lepton number is violated by two units ($\Delta L = 2$), represents the most appropriate way to address the question of the nature of the neutrinos as Majorana or Dirac particles. LNV processes mediated by the exchange of heavy Majorana neutrinos, such as three-body decays of τ leptons and charged pseudoscalar mesons, have been widely studied. In this work we study the contribution of heavy Majorana neutrinos in LNV four-body semileptonic decays of neutral B mesons and top quarks. We focus in a scenario where a single heavy neutrino can enhance the decay rates of these processes via the resonant mechanism. Using current bounds on heavy neutrino mixings, we find that the branching ratios of these processes can be at the level of 10^{-6} to 10^{-7}. These decay modes seem to be at the reach of the current and forthcoming experiment, and their experimental search can provide complementary constraints on masses and mixings of heavy Majorana neutrinos.

1. Introduction

The neutrino oscillation experiments have established that the observed light neutrinos are massive. A massive neutrino is called a Dirac neutrino if its antiparticle is different from it ($\nu \neq \bar{\nu}$), while it is called a Majorana neutrino if it coincides with its antiparticle ($\nu = \bar{\nu}$). Establishing if neutrinos are Dirac or Majorana particles is one of the most important puzzles in contemporary particle physics. The standard model (SM) itself contains only three massless neutrinos. The simplest way to generate non-zero but small neutrino masses is to extend the SM by introducing right-handed (RH) singlet neutrinos N_R. The new neutrinos are imposed to obey the Majorana condition, in which the charged conjugated field $N_R^c \equiv \mathcal{C}\bar{N}_R^T = N_R$, with $\mathcal{C} = -i\gamma^2\gamma^0$, are referred to as Majorana neutrinos. This is the well-known type-I seesaw mechanism [1, 2] in which, besides the Dirac mass term $\bar{\nu}_L M_D N_R$ + h.c. from the Yukawa interactions, there is also a heavy Majorana mass term $\bar{N}_R^c M_R N_R/2$ + h.c.. The interesting fact that a Majorana mass term violates lepton number, has clean phenomenological signatures that can be used to distinguish Majorana from Dirac neutrinos.

In terms of the mass eigenstates, the gauge interaction Lagrangian of the charged currents can be written as [3]

Journal of Physics: Conference Series **378** (2012) 012024 doi:10.1088/1742-6596/378/1/012024

$$\mathcal{L}_W = -\frac{g}{2} W_\mu^+ \sum_{\ell=e,\mu,\tau} \Big[\sum_{j=1}^{3} U_{\ell j} (\bar{\ell} \gamma^\mu P_L \nu_j) + \sum_{k=1}^{n} V_{\ell k} (\bar{\ell} \gamma^\mu P_L N_k) \Big] + h.c., \quad (1)$$

with $P_L = (1 - \gamma_5)/2$ the left-handed chirality projector and g the $SU(2)_L$ (weak) coupling constant. In Eq. (1), $U_{\ell j}$ is the neutrino mixing matrix responsible for neutrino oscillations, while $V_{\ell k}$ describes the strength of charged-current interactions between leptons ($\ell = e, \mu, \tau$) and heavy neutrinos $N_1, ..., N_n$. In the phenomenological applications of the present work, we will assume that only one heavy neutrino with mass m_N and charged current couplings $V_{\ell N}$ to leptons, dominates the decay amplitudes via the resonant enhancement mechanism.

The most extensively studied of these lepton number violating (LNV) processes is the nuclear neutrinoless double-beta ($0\nu\beta\beta$) decay. Observation of this kind of processes will prove that the total lepton number is not conserved and will establish the Majorana nature of light neutrinos (Schechter-Valle theorem) [2]. Up to now, nuclear $0\nu\beta\beta$ decays have not been observed experimentally and just upper limits have been imposed on their decay rates [4, 5].

Alternatively, other LNV ($\Delta L = 2$) processes have been studied in the literature. They include: three-body τ lepton and charged pseudoscalar mesons ($K^\pm, D^\pm, D_s^\pm, B^\pm$, and $B_c^\pm$) decays [3, 6, 7, 8, 9, 10, 11, 12, 13], hyperon decays $\Sigma^- \to \Sigma^+ e^- e^-$, $\Xi^- \to p\mu^-\mu^-$ [14], lepton nuclear conversion processes $e^- \to \mu^+$ [15], $\mu^- \to e^+$ [7, 16] and $\mu^- \to \mu^+$ [7, 17]. These processes can be mediated by the exchange of (light or heavy) virtual massive Majorana neutrinos. In this work, we will study alternative $\Delta L = 2$ processes in four-body decays of the neutral B meson and top quark.

2. $\Delta L = 2$ processes and the resonant mechanism

The Majorana nature of neutrinos can be experimentally verified via LNV processes involving the production of same-sign dileptons in the final state. The underlying subprocess of LNV is the dilepton production from two virtual W bosons, $W^-W^- \to \ell_i^- \ell_j^-$ [3] (as is shown in Figure 1). The structure of this subprocess changes the lepton number from $L = 0$ to $L = 2$ ($\Delta L = 2$) and the observation of LNV processes will establish that neutrinos are Majorana particles.

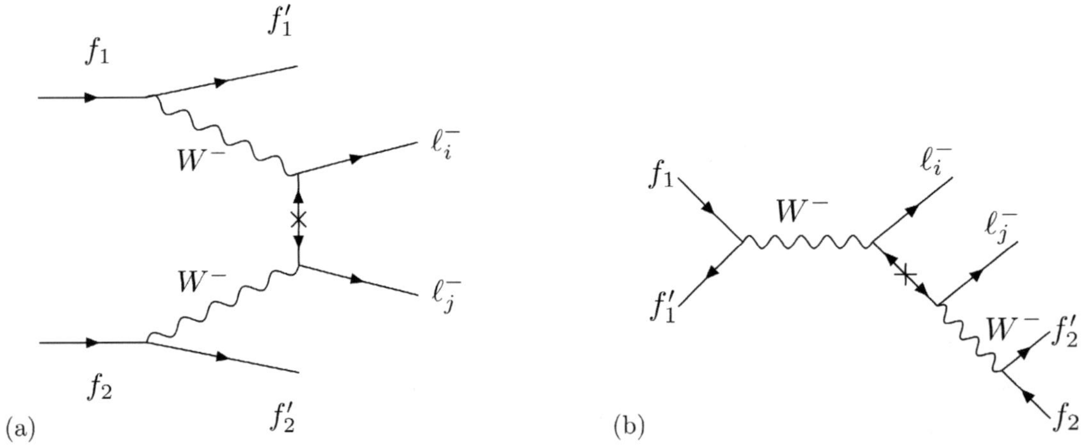

Figure 1. Feynman diagram representing the production of same-sign dileptons in the final state (a) through t-channel and (b) through s-channel, in which $\times$ denote either the light ν_i (for $i = 1, 2, 3$) or heavy N_k (for $k = 1, ..., n$) Majorana neutrinos.

The underlying dynamics due to the exchange of the virtual Majorana neutrino is codified in the leptonic tensor current [3]

$$
\begin{aligned}
L^{\mu\nu} &= \frac{g^2}{2}\Bigg\{\sum_{j=1}^{3} U_{\ell_1 j}U_{\ell_2 j}\, m_{\nu_j}\, \bar{u}_{\ell_1}\left[\frac{\gamma^\mu\gamma^\nu}{q^2 - m_{\nu_j}^2 + i\Gamma_{\nu_j} m_{\nu_j}} + \frac{\gamma^\nu\gamma^\mu}{\tilde{q}^2 - m_{\nu_j}^2 + i\Gamma_{\nu_j} m_{\nu_j}}\right] P_R u_{\ell_2}^c \\
&+ \sum_{k=1}^{n} V_{\ell_1 k}V_{\ell_2 k}\, m_{N_k}\, \bar{u}_{\ell_1}\left[\frac{\gamma^\mu\gamma^\nu}{q^2 - m_{N_k}^2 + i\Gamma_{N_k} m_{N_k}} + \frac{\gamma^\nu\gamma^\mu}{\tilde{q}^2 - m_{N_k}^2 + i\Gamma_{N_k} m_{N_k}}\right] P_R u_{\ell_2}^c\Bigg\}, \quad (2)
\end{aligned}
$$

which receives contributions from light ν_i (for $i = 1, 2, 3$) and heavy N_k (for $k = 1, ..., n$) Majorana neutrinos. Here, q is the four-momentum carried by the Majorana neutrino, $\Gamma_\nu(\Gamma_N)$ and $m_\nu(m_N)$ are the decay width and the mass of the light (heavy) Majorana neutrino, respectively, and P_R is the right-handed projector operator. Given that we have identical leptons in the final state, $\tilde{q}$ represents the four-momentum of the Majorana neutrino which is obtained from q in the antisymmetrized amplitude.

From Eq. (2) we can see that the dynamics for $\Delta L = 2$ processes will be driven by the properties of the exchanged neutrinos. For a light Majorana neutrino compared to the energy scale in the process ($q^2,\ \tilde{q}^2 \gg m_{\nu_j}^2$), we can neglect the mass and width of the neutrino in the denominator of the propagators and the decay amplitude for LNV processes becomes proportional to

$$\langle m_{\ell_1\ell_2}\rangle \equiv \sum_{j=1}^{3} U_{\ell_1 j}U_{\ell_2 j}\, m_{\nu_j}, \quad (3)$$

where $\langle m_{\ell_1\ell_2}\rangle$ is the the so-called *effective mass of Majorana neutrino*. If neutrinos are very heavy compared to the momentum transfer ($m_{N_k}^2 \gg q^2,\ \tilde{q}^2$), the contribution of the second term in Eq. (2) is dominat and it scales as $\sum_{k=1}^{n} V_{\ell_1 k}V_{\ell_2 k}/m_{N_k}$ where $V_{\ell k}$ is the mixing matrix between the charged leptons and heavy neutrinos. Unfortunately, in both cases the transition rate of this $\Delta L = 2$ process encounters a severe suppression either due to the small neutrino mass, or due to the heavy neutrino mass. However, when the heavy neutrino mass is kinematically accessible ($q^2, \tilde{q}^2 \simeq m_{N_k}^2$), the process may undergo a resonant production of the heavy neutrino which substantially enhances the transition rate. Therefore, the decay amplitude for the process under consideration can be enhanced through the resonant production of a heavy Majorana neutrino. This enhancement of the amplitude is known as the *resonant mechanism* [3].

The resonant contributions of heavy Majorana neutrinos to $\Delta L = 2$ processes have been widely considered in three-body decays of τ lepton and charged pseudoscalar mesons ($K^\pm, D^\pm, D_s^\pm, B^\pm$, and $B_c^\pm$) [3, 8, 9, 10, 11, 12]. These works have shown that the associated decay rates receive a huge resonant enhancement. In this work, we extend the study of the heavy Majorana neutrinos to LNV four-body semileptonic decays of neutral B mesons and top quarks [18].

Searches for LNV charged meson decays have been performed by several experiments. Table 1 summarizes the current experimental limits on branching ratios given in Refs. [4, 19]. These limits, along with theoretical predictions, allow to derive stringent limits on the heavy neutrino mass m_N and its mixing $V_{\ell N}$ with charged leptons [3, 8]. Thus, Ref. [3] has reported the strong constraints $|V_{\ell N}|^2 \leq 10^{-9}$ ($\ell = e, \mu$) for the mass region ($m_\pi + m_\ell \leq m_N \leq m_K - m_\ell$) from $K \to \ell\ell\pi$ decays. Less restrictive bounds on the mixing angles ($|V_{\ell N}| \leq 10^{-3} \sim 10^{-1}$) [3], although in a wider range of neutrino masses, have been reported also from similar decays of charged heavy mesons.

Journal of Physics: Conference Series **378** (2012) 012024 doi:10.1088/1742-6596/378/1/012024

Table 1. Experimental bounds on branching fractions of $\Delta L = 2$ rare meson decays [4, 19].

Decay mode	$\mathcal{B}_{exp}$
$K^+ \to \pi^- e^+ e^+$	6.4×10^{-10}
$K^+ \to \pi^- \mu^+ \mu^+$	3.0×10^{-9}
$K^+ \to \pi^- e^+ \mu^+$	5.0×10^{-10}
$D^+ \to \pi^- e^+ e^+$	9.6×10^{-5}
$D^+ \to \pi^- \mu^+ \mu^+$	4.8×10^{-6}
$D^+ \to \pi^- e^+ \mu^+$	5.0×10^{-5}
$D^+ \to K^- e^+ e^+$	1.2×10^{-4}
$D^+ \to K^- \mu^+ \mu^+$	1.3×10^{-5}
$D^+ \to K^- e^+ \mu^+$	1.3×10^{-4}
$D_s^+ \to \pi^- e^+ e^+$	6.9×10^{-4}
$D_s^+ \to \pi^- \mu^+ \mu^+$	2.9×10^{-5}
$D_s^+ \to \pi^- e^+ \mu^+$	7.3×10^{-4}
$D_s^+ \to K^- e^+ e^+$	6.3×10^{-4}
$D_s^+ \to K^- \mu^+ \mu^+$	1.3×10^{-5}
$D_s^+ \to K^- e^+ \mu^+$	6.8×10^{-4}
$B^+ \to \pi^- e^+ e^+$	1.6×10^{-6}
$B^+ \to \pi^- \mu^+ \mu^+$	5.8×10^{-8}
$B^+ \to \pi^- e^+ \mu^+$	1.3×10^{-6}
$B^+ \to K^- e^+ e^+$	1.0×10^{-6}
$B^+ \to K^- \mu^+ \mu^+$	5.4×10^{-8}
$B^+ \to K^- e^+ \mu^+$	2.0×10^{-6}

3. Semileptonic four-body decays

In this section, we study the contribution of heavy Majorana neutrinos to LNV four-body semileptonic decays of neutral B mesons and top quarks. As it was pointed out in the previous section, we focus in a scenario where a single heavy neutrino can enhance the decay rates of these processes via the resonant mechanism. As far as we know this kind of $\Delta L = 2$ processes in neutral B meson decays have not been considered previously neither from a theoretical nor from an experimental point of view. We will also re-calculate the $\Delta L = 2$ top quark decays. These processes were studied previously in Ref. [20]; here we use a different method to integrate the decay rate and compare both results.

3.1. LNV $\bar{B}^0 \to D^+ \ell^- \ell^- \pi^+$ decays

Let us first consider the LNV four-body $\bar{B}^0(p) \to D^+(p_1)\ell^-(p_2)\ell^-(p_3)\pi^+(p_4)$ decays, where p denotes the four-momentum of the decaying particle and p_i $(i = 1, 2, 3, 4)$ denote the four-momenta of final state particles. Note that a final state containing a pair ofτ leptons is not kinematically allowed in this decay. In Figure 2 we show the Feynman diagrams that contribute to this process at the lowest order. Given that the CKM quark mixing factor $V_{ub}V_{cd}$ related to diagram 2(b) is suppressed with respect to the factor $V_{cb}V_{ud}$ related to diagram 2(a) ($|V_{ub}V_{cd}|/|V_{cb}V_{ud}| \sim 0.02$), we can neglect its contribution.

According to Ref. [13], in the range of neutrino masses m_N where the resonance effects dominate the decay amplitude, the diagrams of Figure 2(c) and 2(d) will give very small

Journal of Physics: Conference Series **378** (2012) 012024 doi:10.1088/1742-6596/378/1/012024

contributions. In addition to that, they receive a loop suppression. Thus, we expect that amplitude 2(a) will be the dominant contribution and we do not include the other diagrams in our calculations. Let us notice that in this work, we will only consider channels with same-sign dileptons in the final state and we will disregard lepton flavor changing process.

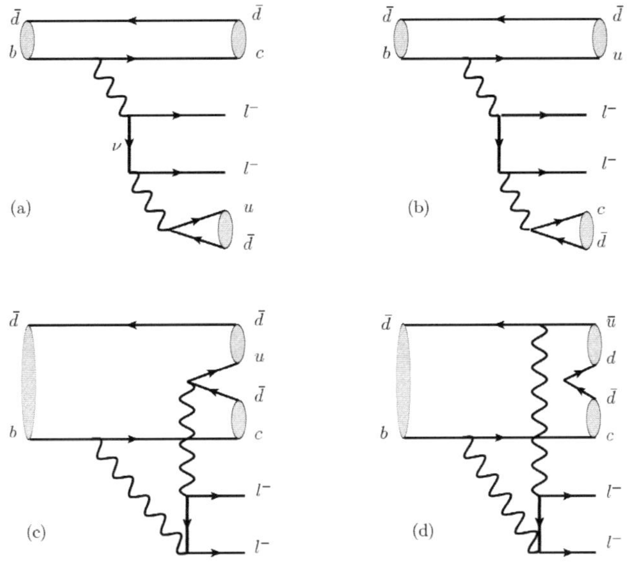

Figure 2. Feynman diagrams for the LNV four-body neutral B meson decay.

The Lorentz-invariant decay amplitude corresponding to diagram 1(a) is given by

$$\mathcal{M} = G_F^2 V_{cb}^* V_{ud}\, H_\mu^1 L^{\mu\nu} H_\nu^2, \tag{4}$$

where $|V_{cb}| = 40.6 \times 10^{-3}$, $|V_{ud}| = 0.97425$ are the elements of the CKM matrix [4], G_F is the Fermi constant and the leptonic tensor current $L^{\mu\nu}$ is defined in Eq. (2). The hadronic current H_μ^1, related to the $b \to c$ underlying quark transition, is expressed as

$$\mathrm{H}_\mu^1 = \langle D(p_1)|\bar{c}\gamma_\mu b|B(p)\rangle = \left[(p+p_1)_\mu - \frac{(m_B^2-m_D^2)}{t}Q_\mu\right]F_1(t) + \left[\frac{(m_B^2-m_D^2)}{t}\right]Q_\mu\, F_0(t),$$

where F_0 and F_1 are the form factors for the $B \to D$ transition, with $t = (p-p_1)^2 \equiv Q^2$, and m_B (m_D) is the mass of the B (D) meson, respectively.

In Eq. (4), H_ν^2 is the hadronic current describing the pion production from the vacuum

$$H_\nu^2 = \langle \pi(p_4)|\bar{d}\gamma_\nu\gamma_5 u|0\rangle = i f_\pi (p_4)_\nu, \tag{5}$$

which is expressed in terms of the decay constant $f_\pi = 130.4$ MeV [4] and its four-momentum p_4. For the purposes of a numerical evaluation, we will use two common parametrizations of the form factors[1] $F_{1(0)}(t)$, namely, the one provided by the Wirbel-Stech-Bauer (WSB) model [21] and, just for comparison, we will use also the parametrization provided by the covariant light front (CLF) model [22].

[1] A model independent evaluation can be done by using recent measurements of the form factors reported by the BaBar Collaboration [23]. These measurements seem to be in good agreement with calculations based on lattice calculations, and to depart from the predictions based on Refs. [21, 22]. We are grateful to A. Kronfeld for pointing this out to us.

The mixings of the heavy neutrino with the three charged leptons will be taken from the currently most restrictive bounds as reported in Ref. [24]

$$\mathrm{SetI}: \quad |U_{eN}|^2 < 3 \times 10^{-3}, \quad |U_{\mu N}|^2 < 3 \times 10^{-3}, \quad |U_{\tau N}|^2 < 6 \times 10^{-3} \,. \tag{6}$$

According to the resonant mechanism, the heavy Majorana neutrino can produce a resonant enhancement of the LNV four-body neutral B meson decays for neutrino masses in the range $(m_\pi + m_\ell) \leq m_N \leq (m_B - m_D - m_\ell)$. The decay width of the intermediate neutrino state is obtained by adding up the contributions of all the neutrino decay channels that can be opened at the neutrino mass value m_N [3]:

$$\Gamma_N = \sum_f \Gamma(N \to f)\theta(m_N - \sum_i m_{f_i}) \,, \tag{7}$$

where m_{f_i} in the argument of the step function are the masses of the final state particles in the neutrino decay channel f. The dominant decay modes of the neutrino in the range of masses relevant for resonant B meson decays are the following: $\ell^\mp P^\pm$, $\nu_\ell P^0$, $\ell^\mp V^\pm$, $\nu_\ell V^0$, $\ell_1^\mp \ell_2^\pm \nu_{\ell_2}$, $\nu_{\ell_1} \ell_2^- \ell_2^+$, and $\nu_{\ell_1} \nu \bar{\nu}$, where ℓ, ℓ_1, $\ell_2 = e$, μ, τ, and P (V) denotes a pseudoscalar (vector) meson state. The expressions for the partial decay rates of these channels can be found in Appendix C of Ref. [3]. We have re-evaluated the decay width of the neutrino which is plotted in Figure 3 for neutrino masses m_N in the range where it can produce a resonant enhancement of the B meson decay amplitude. The decay width of the heavy Majorana neutrino is very small compared to its mass ($\Gamma_N \ll m_N$), thus we can apply the narrow-width approximation (NWA) method [3], which is particularly convenient in order to perform the numerical integration over the phase space.

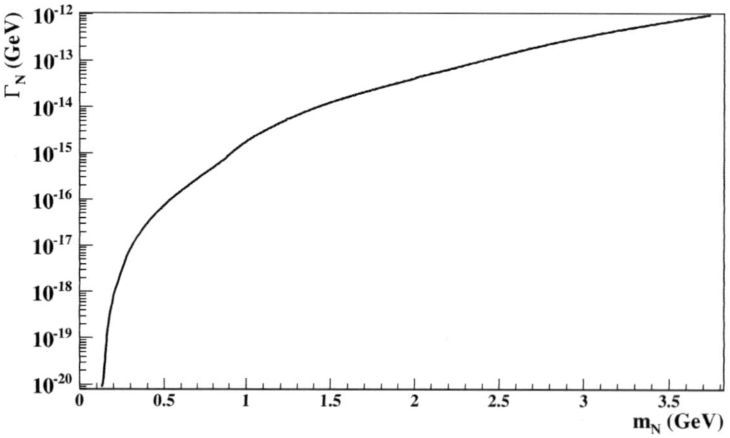

Figure 3. Neutrino decay width as a function of neutrino masses relevant to produce resonant enhancement in LNV four-body decays of the neutral B meson.

Using the WSB model predictions for the form factors and using the updated set of neutrino couplings shown in Eq. (6), in Figure 4 we plot the numerical results for the branching ratios of $\bar{B}^0 \to D^+ \ell^- \ell^- \pi^+$ decays for the ee (dashed-line) and $\mu\mu$ (solid-line) same-sign dilepton channels as a function of the neutrino mass m_N [18]. From Figure 4, we can see that the branching ratios in both channels have a maximum value for neutrino masses close to $m_N \simeq m_\ell + m_\pi$ $(\ell = e, \mu)$.

In addition, we note that as long as the value of neutrino mass m_N increases the branching ratios decreases up to several orders of magnitude along the range of neutrino masses under consideration.

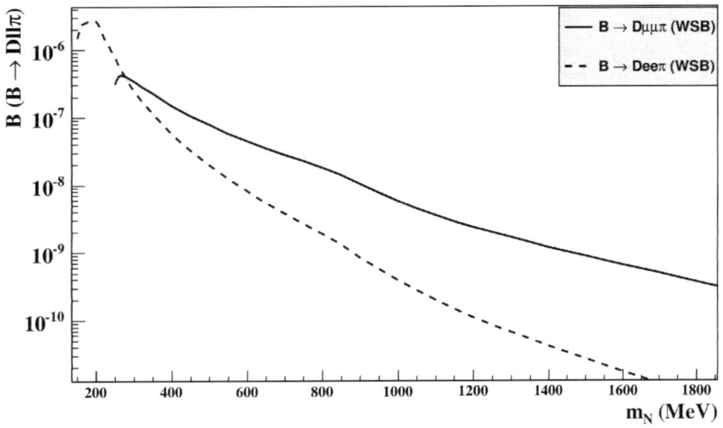

Figure 4. Branching ratios for $\bar{B}^0 \to D^+e^-e^-\pi^+$ (dashed line) and $\bar{B}^0 \to D^+\mu^-\mu^-\pi^+$ (solid line) decays as function of m_N, using the set of neutrino coupling, Eq. (6). The WSB model is used for evaluating $B \to D$ form factors.

In Table 2 we display the largest possible values of the branching ratios of $\bar{B}^0 \to D^+\ell^-\ell^-\pi^+$ ($\ell = e, \mu$) decays, which correspond to the lowest values of resonant Majorana neutrino masses [18]. For comparison, we have used the WSB and CLF models for the evaluation of the form factors. In general, the results obtained for the same-sign dielectron and dimuon channels in both models are in agreement for almost all values of neutrino masses under consideration in Table 2.

Table 2. Branching ratios for $\bar{B}^0 \to D^+\ell^-\ell^-\pi^+$ decays using the set of the heavy neutrino mixings given in Eq. (6). The WSB [21] and CLF [22] models are used for evaluating $B \to D$ form factors.

$\bar{B}^0 \to D^+e^-e^-\pi^+$			$\bar{B}^0 \to D^+\mu^-\mu^-\pi^+$		
m_N (MeV)	WSB	CLF	m_N (MeV)	WSB	CLF
170	2.6×10^{-6}	3.4×10^{-6}	250	3.0×10^{-7}	3.9×10^{-7}
190	2.8×10^{-6}	3.6×10^{-6}	270	4.1×10^{-7}	5.4×10^{-7}
200	2.6×10^{-6}	3.4×10^{-6}	300	3.4×10^{-7}	4.3×10^{-7}
220	1.5×10^{-6}	2.0×10^{-6}	400	1.4×10^{-7}	1.9×10^{-7}
250	7.3×10^{-7}	9.7×10^{-7}	500	7.0×10^{-8}	1.0×10^{-7}
300	2.5×10^{-7}	3.3×10^{-7}	600	4.0×10^{-8}	6.0×10^{-8}

We note that the values of the branching ratios shown in Table 2 are at the level of the experimental upper bounds of LNV three-body charged B meson decays reported in Table 1. Therefore, we can expect that searches at current (LHCb) and future (SuperB) experiments can reach a sensitivity that is able to improve current bounds on the mixing angles $V_{\ell N}$ ($\ell = e, \mu$).

Journal of Physics: Conference Series **378** (2012) 012024 doi:10.1088/1742-6596/378/1/012024

3.2. LNV $t \to b\ell^+\ell^+W^-$ decays

Now we will examine the LNV processes $t(p) \to b(p_1)\ell^+(p_2)\ell^+(p_3)W^-(p_4)$. In Figure 5 we show the corresponding Feynman diagram. A similar diagram obtained from the exchange of leptons in the final state must be added to this. We will only consider decay channels with same-sign dileptons in the final state.

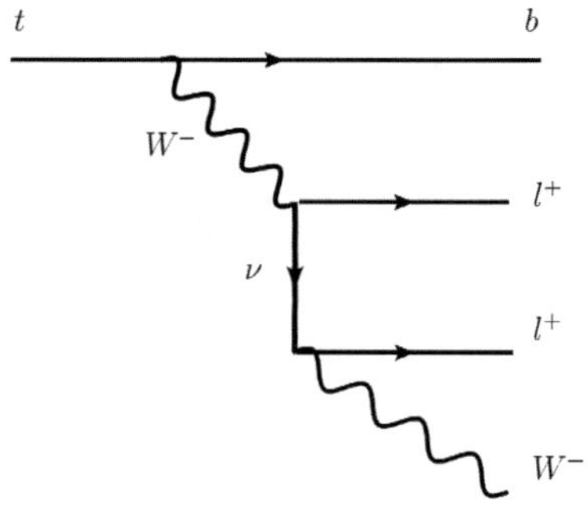

Figure 5. Feynman diagram for the LNV four-body top quark decay.

The Lorentz-invariant decay amplitude for this process is given by

$$\mathcal{M} = G_F m_W^2 \sqrt{2}\Big(\frac{g}{\sqrt{2}}\Big) V_{tb}\; H_\mu L^{\mu\nu} \varepsilon_\nu^*, \tag{8}$$

where g is the SU(2) weak coupling, $\varepsilon_\nu^* = \varepsilon_\nu^*(p_4)$ is the real W boson polarization 4-vector, $L^{\mu\nu}$ is given by expression (2) and $V_{tb} \simeq 1$ is the CKM matrix related to the $t \to b$ weak transition, which can be expressed as

$$H_\mu = \bar{u}_t(p_1)\gamma^\sigma(1-\gamma_5)u_b(p)\Pi^W_{\sigma\mu}, \tag{9}$$

with $Q = p - p_1$ and $\Pi^W_{\sigma\mu}$ is the W boson propagator defined as

$$\Pi^W_{\sigma\mu} = \Big[-g_{\sigma\mu} + \frac{Q_\sigma Q_\mu}{m_W^2}\Big]\frac{i}{(Q^2 - m_W^2) + i\Gamma_W m_W}, \tag{10}$$

where m_W and Γ_W are the mass and the decay width of the W boson, respectively.

If the heavy Majorana neutrino mass lies in the range $(m_W + m_\ell) \leq m_N \leq (m_t - m_b - m_\ell)$, the LNV four-body top quark decays can be resonantly enhanced, according to the resonant mechanism. These $\Delta L = 2$ top quark decays have been considered previously in the work of Bar-Shalom *et al.* [20], and similar top quark decays that also includes the final W gauge boson decaying into two jets were studied in Ref. [25].

For a heavy neutrino with mass larger than the W boson mass ($m_N > m_W$), which is relevant for top quark decays, the decay width of the Majorana neutrino is not very small compared to its mass [3]. Thus, for all the channels related to LNV top quark decays, the decay widths can be computed numerically by integrating over the five-dimensional phase space without using the narrow width approximation. Using the set of neutrino mixings given in Eq. (6), in Table 3 we display the largest values for the branching ratios of $t \to b\ell^+\ell^+W^-$ ($\ell = e, \mu, \tau$) decays, for a specific range of neutrino masses which allows a comparison with the results of Ref. [20]. We observe an overall good agreement for the ee and $\mu\mu$ channels as expected from the new values of the mixing angles (Eq. (6)) and a small difference for the $\tau\tau$ channel.

Table 3. Branching ratios (units of 10^{-6}) for $t \to b\ell^+\ell^+W^-$ decays [18].

m_N (GeV)	ee	$\mu\mu$	$\tau\tau$
90	0.29	0.29	1.12
100	0.12	0.12	0.47
110	0.05	0.05	0.19

4. Concluding remarks

In this work we have studied the resonant contributions of heavy Majorana neutrinos in semileptonic LNV four-body decays of neutral the B meson and the top quark. Using current bounds on heavy neutrino mixings we have calculated numerically the enhanced branching ratios as a function of the mass of Majorana neutrino. Both, in neutral B meson and top quark decay,s we have found branching ratios of order $10^{-6} - 10^{-7}$. Thus, their experimental search can provide complementary constraints on masses and mixings of heavy Majorana neutrinos to the ones derived from LNV three-body decays of τ leptons and charged pseudoscalar mesons.

Let us close this section with an estimate of the expected sensitivities to the signals of our $\Delta L = 2$ decays. Using as a (conservative) reference the $\sim 450 \times 10^6$ $B\bar{B}$ pairs accumulated by the BABAR collaboration at the $\Upsilon(4S)$, we can provide an estimate of the sensitivity to the branching ratio of $\bar{B}^0 \to D^+\ell^-\ell^-\pi^+$ decay at the B factories by using the $K^-\pi^+\pi^+$ mode to reconstruct the charged D meson. By assuming a 70% efficiency for the identification and reconstruction of each of the six charged tracks in the final state one can reach a sensitivity of $\sim 2.0 \times 10^{-7}$ which can test some range of our upper limits for the dileptonic channels. Slightly better sensitivities can be reached using Belle data which is about B meson pairs produced by BABAR. Improved sensitivities can be obtained at the LHCb, Super-KEKB, Super-B factories which are expected to accumulate a larger data set with respect to B factories.

Acknowledgements
The authors are grateful to Conacyt (México) and DAIP project (Guanajuato University) for financial support. They also very grateful to Jose Benítez and Carlos Chávez for useful discussions.

References

[1] P. Minkowski, Phys. Lett. B **67**, 421 (1977); M. Gell-Mann, P. Ramond, R. Slansky, Proceedings of the Supergravity Stony Brook Workshop, New York, 1979; T. Yanagida, Proceedings of the Workshop on the Baryon Number of the Universe and Unified Theories, Tsukuba, Japan, 1314 February 1979; R. N. Mohapatra, G. Senjanovic, Phys. Rev. Lett. **44**, 912 (1980).
[2] J. Schechter and J. W. F. Valle, Phys. Rev. D **22**, 2227 (1980).
[3] A. Atre, T. Han, S. Pascoli, and B. Zhang, JHEP **0905**, 030 (2009).
[4] K. Nakamura, *et al.*, J. Phys. G. **37**, 075021 (2010).
[5] For a recent reviews on neutrinoless double beta decay, see e.g. : W. Rodejohann, Int. J. Mod. Phys. E **20**, 1833 (2011); S. M. Bilenky, Phys. Part. Nucl. **41**, 690-715 (2010);
[6] A. Ilakovac, B. A. Kniehl and A. Pilaftsis, Phys. Rev. D **52**, 3993 (1995); A. Ilakovac and A. Pilaftsis, Nucl. Phys. **B437**, 491 (1995); A. Ilakovac, Phys. Rev. D **54**, 5653 (1996); V. Gribanov, S. Kovalenko and I. Schmidt, Nucl. Phys. **B607**, 355 (2001).
[7] A. Atre, V. Barger, and T. Han, Phys. Rev. D **71**, 113014 (2005).
[8] J. C. Helo, S. Kovalenko, and I. Schmidt, Nucl. Phys, **B853**, 80 (2011).
[9] C. Dib, V. Gribanov, S. Kovalenko, and I. Schmidt, Phys. Lett. B **493**, 82 (2000).
[10] A. Ali, A. V. Borisov, and N. B. Zamorin, Eur. Phys. J. C **21**, 123 (2001).
[11] G. Cvetic, C. Dib, S. K. Kang, and C. S. Kim, Phys. Rev. D **82**, 053010 (2010).

[12] J. M. Zhang and G. L. Wang, Eur. Phys. J. C **71**, 1715 (2011).
[13] M. A. Ivanov, S. G. Kovalenko, Phys. Rev. D **71**, 053004 (2005).
[14] L. S. Littenberg and R. E. Shrock, Phys. Rev. D **46**, R892 (1992); C. Barbero, G. López Castro, and A. Mariano, Phys. Lett. B **566**, 98 (2003).
[15] C. S. Lim, E. Takasugi, and M. Yoshimura, Prog. Theor. Phys. **113**, 1367 (2005).
[16] P. Domin, S. Kovalenko, A. Faessler and F. Simkovic, Phys. Rev. C **70**, 065501 (2004).
[17] F. Simkovic, A. Faessler, S. Kovalenko and I. Schmidt, Phys. Rev. D **66**, 033005 (2002); E. Takasugi, Nucl. Instrum. Meth. A **503**, 252 (2003).
[18] D. Delepine, G. López Castro, and N. Quintero, Phys. Rev. D **84**, 096011 (2011).
[19] R. Aaij *et al.*, LHCb Collaboration, arXiv:1110.0730.
[20] S. Bar-Shalon, N. G. Deshpande, G. Eilam, J. Jiang, and A. Soni, Phys. Lett. B **643**, 342 (2006).
[21] M. Wirbel, B. Stech and M. Bauer, Z. Phys. C **29**, 637 (1985); M. Bauer, B. Stech and M. Wirbel, Z. Phys. C **34**, 103 (1987).
[22] H. Y. Cheng, C. K. Chua and C. W. Hwang, Phys. Rev. D **69**, 074025 (2004).
[23] B. Aubert *et al.*, BABAR Collaboration, Phys. Rev. Lett. **104**, 011802 (2010).
[24] F. del Aguila, J. de Blas and M. Perez-Victoria, Phys. Rev. D **78**, 013010 (2008).
[25] Z. Si and K. Wang, Phys. Rev. D **79**, 014034 (2009).

Journal of Physics: Conference Series **378** (2012) 012025 doi:10.1088/1742-6596/378/1/012025

Bounding the $Z'\mu\tau$ coupling from lepton flavor violating transitions

J. I. Aranda[(a)], **J. Montaño**[(b)], **F. Ramírez-Zavaleta**[(a)], **J. J. Toscano**[(c)] **and E. S. Tututi**[(a)]

[(a)]Facultad de Ciencias Físico Matemáticas, Universidad Michoacana de San Nicolás de Hidalgo, Avenida Francisco J. Mújica S/N, 58060, Morelia, Michoacán, México.
[(b)]Departamento de Física, CINVESTAV, Apartado Postal 14-740, 07000, México, D. F., México.
[(c)]Facultad de Ciencias Físico Matemáticas, Benemérita Universidad Autónoma de Puebla, Apartado Postal 1152, Puebla, Puebla, México.

E-mail: feramirez@umich.mx

Abstract. We present the calculation for the strength of the lepton flavor violating $Z'\mu\tau$ coupling, in which the experimental result of the muon magnetic dipole moment was used to bound this vertex, by means of a model-independent approach. The $Z'\mu\tau$ vertex is represented by the $\Omega_{\mu\tau}$ parameter, which is of the order of 10^{-2} for a Z' boson mass of 1 TeV.

1. Introduction

The presence of a new neutral massive gauge boson (Z') is predicted in the context of numerous extensions of the Standard Model (SM). The $SU_C(3) \times SU_L(2) \times U_Y(1) \times U'(1)$ extended electroweak gauge group is the simplest model capable to predict the existence of the Z' boson [1].

Although absent in the SM, the phenomena of lepton flavor violation can arise in many of its well-motived extensions. One interesting feature of most models beyond the SM is the presence of generalized Yukawa sectors, which favor nondiagonal transitions mediated by neutral massive spin-1 particles [1, 2].

Currently, the experimental data have proven the existence of neutrino oscillations, which in simple words, tell us that lepton flavor conservation is violated in nature. However, the only signal of lepton flavor violation comes from transitions between neutral leptons. Lepton flavor violating transitions between charged leptons constitute an interesting object of study, since if they occur in nature are further evidence of lepton flavor violation.

We are interested in study the impact of lepton flavor violation on the $\tau \to \mu\gamma$ decay mediated by a Z' gauge boson. To this end, we estimate the strength of the $Z'\mu\tau$ coupling in the context of the simplest extension of the SM.

2. The extended model

We consider the neutral current lagrangian contained in the simplest model based on the $SU_C(3) \times SU_L(2) \times U_Y(1) \times U'(1)$ electroweak gauge group [2, 3]

$$\mathcal{L}_{NC} = -eJ^{\alpha}_{EM}A_{\alpha} - g_1 J^{\alpha}_1 Z_{\alpha,1} - g_2 J^{\alpha}_2 Z_{\alpha,2}, \quad (1)$$

where e is the electromagnetic coupling, J^{α}_{EM} is the electromagnetic neutral current, g_1 is the gauge coupling of the SM, J^{α}_1 is the weak neutral current of the SM, g_2 is the gauge coupling of the $U'(1)$ group and J^{α}_2 is the new weak neutral current given as

$$J^{\alpha}_2 = \sum_{i,j} \overline{\psi'_i}\, \gamma^{\alpha}(\epsilon^{\psi}_{Lij}\, P_L + \epsilon^{\psi}_{Rij}\, P_R)\, \psi'_j, \tag{2}$$

where $\epsilon^{\psi}_{L,Rij}$ are the chiral couplings of Z_2 with i,j running over all quarks and leptons, ψ'_i represents a fermion in the gauge interaction basis and $P_{L,R} = \frac{1}{2}(1 \pm \gamma_5)$ are the chiral projectors. The interaction between Z_1 and Z_2 is so weak that it can be neglected. Thus, the mass eigenstates are Z^0 and Z', respectively [1].

The $\epsilon^{l}_{L,Rij}$ matrix elements for the charged leptons are represented in the gauge eigenstates basis. This matrix is transformed into the mass eigenstates one by diagonalizing the mass matrix in the Yukawa sector [2]. Therefore, the couplings in the mass eigenstates basis (Ω_L, Ω_R) can be written as [4]

$$\Omega_{Lij} = g_2\,(V_L\, \epsilon^{l}_L\, V^{\dagger}_L)_{ij},$$
$$\Omega_{Rij} = g_2\,(V_R\, \epsilon^{l}_R\, V^{\dagger}_R)_{ij},$$

where $V_{L,R}$ are the unitary matrices that diagonalize the mass matrix in the Yukawa sector of the SM. The lepton current is transformed into the mass eigenstates basis as

$$J^{\alpha} = \sum_{i,j} \bar{l}_i\, \gamma^{\alpha}(\Omega_{Lij}\, P_L + \Omega_{Rij}\, P_R)\, l_j,$$

where $l_i = e, \mu, \tau$.

This current allow us to analyze the $Z'\mu\tau$ coupling, which is given by the strength of the $\Omega_{L,R\tau\mu}$ matrix elements. Therefore, the lagrangian containing the relevant information is

$$\begin{aligned}\mathcal{L}^{Z'l_il_j}_{NC} = &- \left[\bar{l}_i \gamma^{\alpha}(\Omega_{Ll_il_j}\, P_L + \Omega_{Rl_il_j}\, P_R)\, l_j \right. \\ &\left. + \bar{l}_j\, \gamma^{\alpha}(\Omega_{Ll_jl_i}\, P_L + \Omega_{Rl_jl_i}\, P_R)\, l_i \right] Z'_{\alpha}.\end{aligned} \tag{3}$$

Since we are interested in bounding the strength of the $Z'l_il_j$ coupling, ignoring possible CP-violating effects, we assume for simplicity that $\Omega_{L,Rl_il_j} = \Omega_{L,Rl_jl_i}$ and $\Omega_{R\,l_il_j} = \Omega^{*}_{L\,l_il_j}$. We denote $\Omega_{L\,l_il_j}$ simply as $\Omega_{l_il_j}$. After some algebra, the last equation becomes

$$\begin{aligned}\mathcal{L}^{Z'l_il_j}_{NC} = &- \left[\bar{l}_i\, \gamma^{\alpha}(Re(\Omega_{l_il_j}) + iIm(\Omega_{l_il_j})\, \gamma_5)\, l_j \right. \\ &\left. + \bar{l}_j\, \gamma^{\alpha}(Re(\Omega_{l_il_j}) - iIm(\Omega_{l_il_j})\, \gamma_5)\, l_i \right] Z'_{\alpha}.\end{aligned} \tag{4}$$

The $Z'\tau\mu$ vertex that emerges from the $\mathcal{L}_{NC}$ induces the lepton flavor violating $Z' \to \tau\mu$ decay at the one-loop level. Thus, the estimation of the strength of the $\Omega_{\mu\tau}$ parameter is a crucial information to compute the branching ratio for the $Z' \to \tau\mu$ process. We present two different methods in order to obtain the $\Omega_{\mu\tau}$ strength. The first one consists in bounding from the experimental result for the muon anomalous magnetic dipole moment in a model-independent way. The second one relies on bounding from the experimental constraint for the $\tau \to \mu\gamma$ decay, by using several extended models.

3. Calculation of the $\Omega_{\mu\tau}$ parameter

The contribution of the $Z'\mu\tau$ vertex to the anomalous magnetic dipole moment of the muon is calculated using the Feynman diagram shown in Figure 1(a), which corresponds to the $\mu\mu\gamma$ electromagnetic vertex at the one-loop level. To define the magnetic dipole moment contribution $i(a_\mu/2m_\mu)\sigma_{\alpha\beta}q^\beta$, we must calculate the one-loop amplitude for the on-shell $\mu\mu\gamma$ vertex. The contribution to the magnetic dipole moment is free of ultraviolet divergences. We find that contribution to the a_μ form factor can be written as

$$a_\mu = -\frac{\sqrt{\alpha}\, x_\mu\, x_\tau}{\pi^{3/2}} \mathrm{Re}(\Omega_{\mu\tau}^2)\, g(x_\mu, x_\tau), \tag{5}$$

where α is the fine structure constant and

$$g(x_\mu, x_\tau) = \int_0^1 dx \int_0^{1-x} dy \frac{1-2x-2y}{(x+y)(1-x_\tau^2 - x_\mu^2\,(x+y-1)) - 1}. \tag{6}$$

Here, we introduced the dimensionless variables $x_i = m_i/m_{Z'}, i = \mu, \tau$.

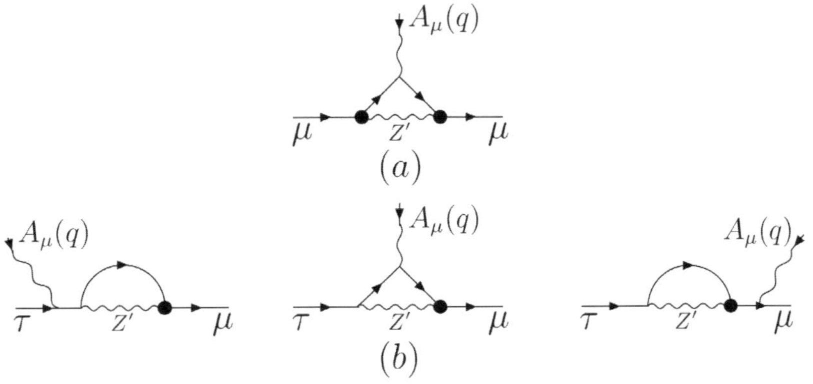

Figure 1. Feynman diagrams contributing to the muon anomalous magnetic dipole moment (a) and lepton flavor violating $\tau \to \mu\gamma$ decay (b).

Nowadays, the anomalous magnetic dipole moment of the muon is one of the physical observables best measured. Therefore, we employ the experimental uncertainty $|\Delta a^{\mathrm{Exp}}|$ [5] to get a bound on the $\mathrm{Re}(\Omega_{\mu\tau}^2)$ parameter, where

$$|\Delta a_\mu^{\mathrm{Exp}}| < 6 \times 10^{-10}. \tag{7}$$

In Figure 2 is shown with a solid line, the behavior of the maximum of $\mathrm{Re}(\Omega_{\mu\tau}^2)$ as a function of the Z' boson mass. It is important to notice that the growth of this parameter starts at 10^{-3} and reaches values of 10^{-2} for a Z' mass interval of 500 GeV $< m_{Z'} <$ 1.5 TeV.

In this paragraph we proceed to compute the contribution of the lepton flavor violating $Z'\mu\tau$ vertex to the $\tau \to \mu\gamma$ decay. The $\tau \to \mu\gamma$ process corresponds to the Feynman diagrams shown in Figure 1(b). Notice that this transition is of dipolar type, finite and model dependent since the $Z'\tau\mu$ vertex is found in terms of the g_2 coupling constant and the Q_R (Q_L) chiral couplings [2, 3, 6]. After having calculated the amplitude of the $\tau \to \mu\gamma$ process and then apply the Passarino-Veltman reduction scheme, we arrive to the associated branching ratio

$$\mathrm{Br}(\tau \to \mu\gamma) = \frac{\alpha g_2^2\, \Omega_{\mu\tau}^2}{2^{11}\pi^4} \left[(Q_L + Q_R)^2\, |F_V|^2 + (Q_L - Q_R)^2\, |F_A|^2\right] \frac{m_\tau}{\Gamma_\tau}, \tag{8}$$

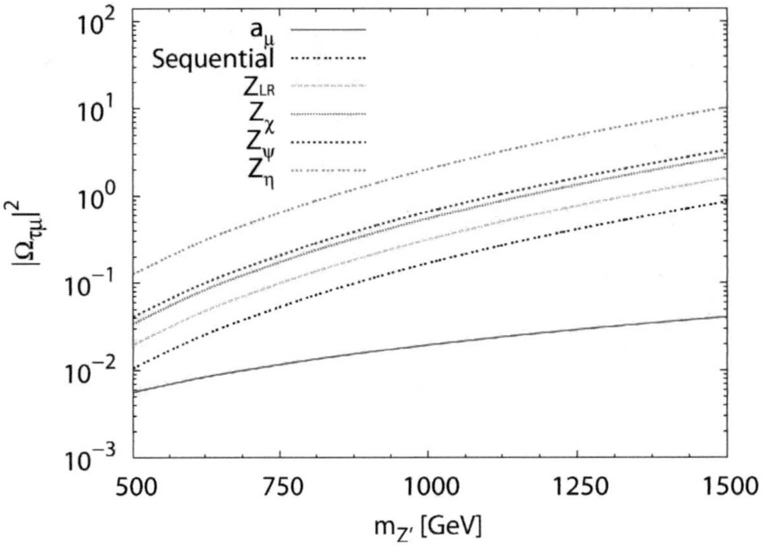

Figure 2. Behavior of $\text{Re}(\Omega^2_{\mu\tau})$, $|\Omega_{\mu\tau}|^2$ as a function of the Z' boson mass. Notice that $\text{Re}(\Omega^2_{\mu\tau})$ is shown with continuous line, while the dotted lines corresponds to the different models.

where Γ_τ is the total decay width of the tau lepton and

$$F_V = 1 + 2\,m_\tau^2\,C_0(1) + 2\frac{m_{Z'}^2}{m_\tau^2}(B_0(1) - B_0(2)), \tag{9}$$

$$F_A = 1 + \frac{m_\tau^2}{m_{Z'}^2} + 2\Big(m_\tau^2\,C_0(1) + \Big(\frac{m_{Z'}^2}{m_\tau^2} - 3\Big)\Big(B_0(1) - B_0(2)\Big)\Big), \tag{10}$$

where $B_0(1) \equiv B_0(0, m_\tau^2, m_{Z'}^2)$, $B_0(2) \equiv B_0(m_\tau^2, m_\tau^2, m_{Z'}^2)$ and $C_0(1) \equiv C_0(m_\tau^2, 0, 0, m_\tau^2, m_{Z'}^2, m_\tau^2)$. Here, the muon mass has been neglected.

The branching ratio in Eq. (8) must be less than the corresponding experimental constraint $\text{Br}_{\text{Exp}}(\tau \to \mu\gamma) < 4.4 \times 10^{-8}$ [5], so this restriction allows us to bound the $|\Omega_{\mu\tau}|^2$ parameter as a function of the Z' boson mass. The behavior of $|\Omega_{\mu\tau}|^2$ for different models can be appreciated in Figure 2. It can be seen from this figure that the sequential model offers the most restrictive constraint and its strength varies between 10^{-1} to 10^0 for a Z' mass in the range of 500 GeV $< m_{Z'} <$ 1.5 TeV.

4. Conclusions

In this work we resort to the simplest extension of the SM to estimate the strength of the $Z'\mu\tau$ coupling, represented by the $\Omega_{\tau\mu}$ parameter. For this purpose, we used two different methods: by using the experimental result for the anomalous magnetic dipole moment of the muon and by means of the experimental constraint for the $\tau \to \mu\gamma$ transition. By using the experimental uncertainty of the muon anomalous magnetic moment, which is measured with amazing precision, we obtain the maximum allowed value for $\text{Re}(\Omega^2_{\mu\tau})$ as a function of Z' boson mass, which ranges from 10^{-3} to 10^{-2}. The results indicate that the bound for the $\text{Re}(\Omega^2_{\mu\tau})$ parameter is more restrictive than that obtained by the $\tau \to \mu\gamma$ dipolar transition calculation.

Acknowledgments
This work has been partially supported by CIC-UMSNH and SNI (México).

References
[1] P. Langacker, Rev. Mod. Phys. **81**, 1199 (2008).
[2] L. S. Durkin and P. Langacker, Phys. Lett. B **166**, 436 (1986); M. Cvetic and P. Langacker, Proceedings of Ottawa 1992: Beyond the standard model 3, 454-458, (1992).
[3] P. Langacker and M. Luo, Phys. Rev. D **45**, 278 (1992).
[4] J. I. Aranda, *et. al.*, J. Phys. G: Nucl. Part. Phys. **38**, 045006 (2011).
[5] Nakamura K. *et. al.*, J. Phys. G: Nucl. Part. Phys. **37** 075021, (2010).
[6] A. Arhrib, *et al.*, Phys. Rev. D **73**, 075015 (2006).

Journal of Physics: Conference Series **378** (2012) 012026 doi:10.1088/1742-6596/378/1/012026

Signal of FCNC and CP violation through $h \to c\bar{b}W^-$ in two Higgs doublet model

L. Díaz-Cruz

Cuerpo Académico de Partículas, Campos y Relatividad Facultad de Ciencias Físico-Matemáticas, BUAP. Apdo. Postal 1364, C.P. 72000 Puebla, Pue., México

E-mail: jldiaz@fcfm.buap.mx

R. Gaitán-Lozano

Departamento de Física, Facultad de Estudios Superiores Cuautitlán, UNAM, C.P. 54770, Estado de México, México

E-mail: rgaitan@servidor.unam.mx

J.H. Montes de Oca Y.

Departamento de Física, Facultad de Estudios Superiores Cuautitlán, UNAM, C.P. 54770, Estado de México, México

E-mail: josehalim@comunidad.unam.mx

Abstract. We discuss the formulation of the general two-Higgs doublet model type III, which incorporates flavor changing neutral scalar interactions (FCNSI) and CP violation from several sources. CP violation can arise either from Yukawa terms or from the Higgs potential, be it explicit or spontaneous. We discuss the limit that includes CP violation and Yukawa four textures to control FCNSI and evaluate the CP asymmetry for the decay $h \to bcW$, which may allow to test the patterns of FCNSI and CP violation, that arises in these models.

1. Introduction

Despite the success of the Standard Model (SM) in the gauge and fermion sectors, the Higgs sector remains the least tested aspect of the model [1], which leaves the puzzles associated with the mechanism of electroweak symmetry breaking (EWSB) still unsolved. On one hand, the analysis of radiative corrections within the SM [2, 3, 4, 5], points towards the existence of a Higgs boson, with a mass of the order of the EW scale, which in turn could be detected at the LHC [6, 7]. Nowdays, the Higgs boson mass has a experimental bound of the order of 125 GeV [CMS , ATLAS]. On the other hand, the SM is often considered as an effective theory, valid up to an energy scale of $O(TeV)$, that eventually will be replaced by a more fundamental theory [8], which will explain, among other things, the physics behind EWSB and perhaps even the origin of flavor. Many examples of candidate theories, which range from supersymmetry [9, 10, 11] to strongly interacting models [12, 13] as well as some extra dimensional scenarios [14, 15, 16], include a multi-scalar Higgs sector. In particular, models with two scalar doublets have been

Journal of Physics: Conference Series **378** (2012) 012026 doi:10.1088/1742-6596/378/1/012026

studied extensively [17, 18, 19], as they include a rich structure with interesting phenomenology [20, 21, 22].

Several versions of the 2HDM have been studied in the literature [23]. Some models (known as 2HDM-I and 2HDM-II) involve natural flavor conservation [24], while other models (known as 2HDM-III) [23], allow for the presence of flavor changing scalar interactions (FCNSI) at a level consistent with low-energy constraints [25]. There are also some variants (known as top, lepton, neutrino), where one Higgs doublet couples predominantly to one type of fermion [27], while in other models it is even possible to identify a candidate for dark matter [26]. The definition of all these models, depends on the Yukawa structure and symmetries of the Higgs sector [28, 29, 30, 31, 32], whose origin is still not known. The possible appearance of new sources of CP violation is another characteristic of these models [33].

In this paper we aim to discuss FCNC at general version of the Two-Higgs doublet model (2HDM-III), which incorporates flavor or CP violation from all possible sources [34, 35, 36].

Within model I (2HDM-I) where only one Higgs doublet generates all gauge and fermion masses [6], while the second doublet only knows about this through mixing, and thus the Higgs phenomenology will share some similarities with the SM, although the SM Higgs couplings will now be shared among the neutral scalar spectrum. The presence of a charged Higgs boson is clearly the signal beyond the SM. Within 2DHM-II one also has natural flavor conservation [24], and its phenomenology will be similar to the 2HDM-I, although in this case the SM couplings are shared not only because of mixing, but also because of the Yukawa structure. On the other hand, the distinctive characteristic of 2HDM-III is the presence of FCNSI, which require a certain mechanism in order to suppress them, for instance one can imposes a certain texture for the Yukawa couplings [37], which will then predict a pattern of FCNSI Higgs couplings [38]. Within all those models (2HDM I,II,III) [39, 40, 41], the Higgs doublets couple, in principle, with all fermion families, with a strength proportional to the fermion masses, modulo other parameters.

There are also other models where the Higgs doublets couple non-universally to the fermion families, which have also been discussed in the literature [27, 42, 43]. In principle, the general model includes CPV, which could arise from the same CPV phase that appears in the CKM matrix, as in the SM, from some other extra phase coming from the Yukawa sector or from the Higgs potential [44]. However, in order to discuss which type of CP violation can appear in each case besides containing a generic pattern of FCNSI, moduled by certain texture, will include new sources of CPV as well.

2. General two Higgs doublet model with textures in Yukawa matrices

The scalar field content of the model is two doublets under $SU(2)_L$ with hypercharges $Y_1 = Y_2 = 1/2$. The model is classified by choice of the Higgs potential and the scalar-fermion couplings. The simpler versions are considered when the potential is invariant under Z_2 discrete symmetry and none of the doublets couples simultaneously to the up-type and down-type fermions. These are known as 2HDM type I and II. In type I, only one of the doublets couples to fermions. In type II, one doublet couples to up-type fermions, the other to down-type fermions to prevent tree-level FCNCs. Type III is more general and realistic model which allows all possible Higgs-fermion couplings. This work is based on 2HDM type III with a general potential which contains explicit and spontaneous CP violation.

2.1. *General Higgs potential and Higgs mass-eigenstates*

We follow closely the formalism developed and notation introduced by [45]. The most general gauge invariant renormalizable Higgs scalar potential in a covariant form with respect to global $U(2)$ transformation is given by

$$V = Y_{a,\bar{b}}\Phi^{\dagger}_{\bar{a}}\Phi_b + \frac{1}{2}Z_{a\bar{b}c\bar{d}}\left(\Phi^{\dagger}_{\bar{a}}\Phi_b\right)\left(\Phi^{\dagger}_{\bar{c}}\Phi_d\right), \tag{1}$$

r	q_{r1}	q_{r2}
1	$\cos\theta_{12}\cos\theta_{13}$	$-\sin\theta_{12} - i\cos\theta_{12}\sin\theta_{13}$
2	$\sin\theta_{12}\cos\theta_{13}$	$\cos\theta_{12} - i\sin\theta_{12}\sin\theta_{13}$
3	$\sin\theta_{13}$	$i\cos\theta_{13}$

Table 1. Mixing angles for Higgs bosons which consider spontaneous and explicit CPV [45].

where $\Phi_a = (\phi_a^+, \phi_a^0)^T$ and a, b, c, d are labels with respect to two dimensional Higgs flavor space. The index conventions means that replacing an unbarred index with a barred index is equivalent to complex conjugation and barred.unbarred index pair denotes a sum. The most general $U(1)_{EM}$-conserving vacuum expectation values are

$$\langle \Phi_a \rangle = \frac{v}{\sqrt{2}} \begin{pmatrix} 0 \\ \hat{v}_a \end{pmatrix}, \tag{2}$$

where $\begin{pmatrix} \hat{v}_1, & \hat{v}_2 \end{pmatrix} = e^{i\eta} \begin{pmatrix} \cos\beta, & \sin\beta e^{i\xi} \end{pmatrix}$ and $v = 246$ GeV. The covariant form for the scalar potential minimum conditions is

$$v\hat{v}_{\bar{a}}^* \left[Y_{a\bar{b}} + \frac{1}{2} v^2 Z_{a\bar{b}c\bar{d}} \hat{v}_{\bar{c}}^* \hat{v}_d \right] = 0. \tag{3}$$

The Higgs mass-eigenstates of the neutral Higgs bosons are explicitly derived in [45]. The expressions for Higgs bosons in terms of the generic basis is

$$\begin{aligned} h_k &= \frac{1}{\sqrt{2}} \overline{\Phi}_{\bar{a}}^{0\dagger} (q_{k1}\hat{v}_a + q_{k2}\hat{w}_a e^{-i\theta_{23}}) \\ & \frac{1}{\sqrt{2}} (q_{k1}^* \hat{v}_{\bar{a}}^* + q_{k2} * \hat{w}_{\bar{a}}^* e^{i\theta_{23}}) \overline{\Phi}_{\bar{a}}^0 \end{aligned} \tag{4}$$

for $k = 1, ..., 4$, where $h_{1,2,3}$ are the neutral physical Higgs bosons and $h_4 = G^0$ is the goldstone boson. The parameters $q_{k1,2}$ are functions of the neutral Higgs mixing angles and the explicit values are shown in table 1. It is possible to invert the expression (4) and the result is given by

$$\Phi_a = \begin{pmatrix} G^+\hat{v}_a + H^+\hat{w}_a \\ \frac{v}{\sqrt{2}}\hat{v}_a + \frac{1}{\sqrt{2}} \sum_{k=1}^4 \left(q_{k1}\hat{v}_a + q_{k2} e^{-\theta_{23}} \hat{w}_a \right) h_k \end{pmatrix}, \tag{5}$$

where $\hat{w}_a^T = e^{-i\eta} \begin{pmatrix} -\sin\beta e^{-i\xi}, & \cos\beta \end{pmatrix}$.

2.2. Higgs-fermions couplings

We focus on quarks fields with analogous treatment for leptons. The most general structure of the Yukawa lagrangian for the quark fields, can be written as follows:

$$\mathcal{L}_Y^{quarks} = \overline{q}_L^0 Y_1^D \phi_1 d_R^0 + \overline{q}_L^0 Y_2^D \phi_2 d_R^0 + \overline{q}_L^0 Y_1^U \widetilde{\phi}_1 u_R^0 + \overline{q}_L^0 Y_2^U \widetilde{\phi}_2 u_R^0 + h.c., \tag{6}$$

where $Y_{1,2}^{U,D}$ are the 3×3 Yukawa matrices, q_L denotes the left handed quarks doublets and u_R, d_R correspond to the right handed singlets. Here $\widetilde{\phi}_{1,2} = i\sigma_2\phi_{1,2}^*$. The superscript zero means that the quarks are weak eigenstates. After getting a correct SSB [60, 61, 62, 63], the Higgs doublets are decomposed as shown in (5) and the neutral scalar and pseudoscalar couplings within up-type quarks in mass eigenstate basis are

$$\mathcal{L}_{up}^{neutral} = \overline{u}_i \left(S_{ijr}^u + \gamma^5 P_{ijr}^u \right) u_j H_r + \overline{u}_i M_{ij}^U u_j, \tag{7}$$

where we have denoted the scalar and pseudoscalar couplings as

$$\begin{aligned} S_{ijr}^u &= \frac{1}{v} M_{ij}^U \left(q_{r1} - \tan\beta \mathrm{Re}\left[q_{r2} e^{-i(\theta_{23}+\xi)} \right] \right) \\ &+ \frac{1}{2\sqrt{2}\cos\beta} \left(q_{r2}^* e^{i\theta_{23}} \widetilde{Y}_{2ij}^U + q_{r2} e^{-i\theta_{23}} \widetilde{Y}_{2ij}^{U\dagger} \right) \end{aligned} \tag{8}$$

and

$$\begin{aligned} P_{ijr}^u &= \frac{1}{v} M_{ij}^U \tan\beta \mathrm{Im}\left[q_{r2} e^{-i(\theta_{23}+\xi)} \right] \\ &+ \frac{1}{2\sqrt{2}\cos\beta} \left(q_{r2}^* e^{i\theta_{23}} \widetilde{Y}_{2ij}^U - q_{r2} e^{-i\theta_{23}} \widetilde{Y}_{2ij}^{U\dagger} \right), \end{aligned} \tag{9}$$

respectively. The mass matrices are given as follows:

$$M^U = \frac{v_1}{\sqrt{2}} \widetilde{Y}_1^U + e^{-i\xi} \frac{v_2}{\sqrt{2}} \widetilde{Y}_2^U \tag{10}$$

and

$$M^D = \frac{v_1}{\sqrt{2}} \widetilde{Y}_1^D + e^{i\xi} \frac{v_2}{\sqrt{2}} \widetilde{Y}_2^D, \tag{11}$$

where $\widetilde{Y}_{1,2}^U = U_L Y_{1,2}^U U_R^\dagger$ and $\widetilde{Y}_{1,2}^D = D_L Y_{1,2}^D D_R^\dagger$ with $u_{L,R} = U_{L,R} u_{L,R}^0$ and $d_{L,R} = D_{L,R} d_{L,R}^0$. The v.e.v. v_1 and v_2 are real and positive, while the phase ξ introduces spontaneous CP violation.

Analogously, the down-type quarks are

$$\mathcal{L}_{down}^{neutral} = \overline{d}_i \left(S_{ijr}^d + \gamma^5 P_{ijr}^d \right) d_j H_r + \overline{d}_i M_{ij}^D d_j, \tag{12}$$

with

$$\begin{aligned} S_{ijr}^d &= \frac{1}{v} M_{ij}^D \left(q_{r1} - \tan\beta \mathrm{Re}\left[q_{k2} e^{-i(\theta_{23}+\xi)} \right] \right) \\ &+ \frac{1}{2\sqrt{2}\cos\beta} \left(q_{r2} e^{-i\theta_{23}} Y_2^D + q_{r2}^* e^{i\theta_{23}} \widetilde{Y}_2^{D\dagger} \right) \end{aligned} \tag{13}$$

and

$$\begin{aligned} P_{ijr}^d &= -\frac{1}{v} M_{ij}^D \tan\beta \mathrm{Im}\left[q_{k2} e^{-i(\theta_{23}+\xi)} \right] \\ &+ \frac{1}{2\sqrt{2}\cos\beta} \left(q_{r2} e^{-i\theta_{23}} \widetilde{Y}_2^D - q_{r2}^* e^{i\theta_{23}} \widetilde{Y}_2^{D\dagger} \right). \end{aligned} \tag{14}$$

For completeness we write the Yukawa couplings for charged Higgs bosons,

$$\begin{aligned} \mathcal{L}_Y^{H^+} &= \overline{u} \left[H^+ e^{-i\xi} M^U V \frac{I-\gamma^5}{\sqrt{2}} - H^+ e^{-i\xi} V M^D \frac{I+\gamma^5}{\sqrt{2}} \right. \\ &\left. + \frac{1}{\cos\beta} H^+ \left(V \widetilde{Y}_2^D \frac{I+\gamma^5}{2} - \widetilde{Y}_2^{U\dagger} V \frac{I-\gamma^5}{2} \right) \right] d \\ &+ h.c. \end{aligned} \tag{15}$$

where V denotes the CKM matrix and physical eigenstates for the charged Higgs boson ($H^{\pm}$) can be obtain through (5).

Journal of Physics: Conference Series **378** (2012) 012026 doi:10.1088/1742-6596/378/1/012026

2.3. Universal Yukawa textures

Suppression for FCNC can be achieved when a certain form of the Yukawa matrices that reproduce the observed fermion masses and mixing angles is implemented in the model. This could be done either by implementing the Frogart-Nielsen mechanism to generate the fermion mass hierarchies [52], or by studying a certain ansatz for the fermion mass matrices [37]. The first proposal for the Higgs boson couplings [38], the so called Cheng-Sher ansazt, was based on the Fritzsch six-texture form of the mass matrices, namely:

$$M_l = \begin{pmatrix} 0 & C_q & 0 \\ C_q^* & 0 & B_q \\ 0 & B_q^* & A_q \end{pmatrix}. \tag{16}$$

Then, by assuming that each Yukawa matrix $Y_{1,2}^q$ has the same hierarchy, one finds: $A_q \simeq m_{q3}$, $B_q \simeq \sqrt{m_{q2}m_{q3}}$ and $C_q \simeq \sqrt{m_{q1}m_{q2}}$. Then, the fermion-fermion′-Higgs boson (ff'^0) couplings obey the following pattern: $Hf_if_j \sim \sqrt{m_{f_i}m_{f_j}}/m_W$, which is also known as the Cheng-Sher ansatz. This brings under control the FCNC problem, and it has been extensively studied in the literature to search for flavor-violating signals in the Higgs sector

In our previous work we considered in detail the case of universal four-texture Yukawa matrices [25], and derived the scalar-fermion interactions, showing that it was possible to satisfy current constraints from LFV and FCNC [53, 54]. Predictions for Higgs phenomenology at the LHC was also studied in ref. [55, 56]. We can consider this a universal model, in the sense that it was assumed that each Yukawa matrix $Y_{1,2}^q$ has the same hierarchy.

3. FCNC and CP violation sources

The Higgs-fermions interactions and potential of the general 2HDM contain several sources of CPV and FCNC. In order to explore these sources we consider some limiting cases. As it is discussed in previous sections, the assumption of universal 4-textures for the Yukawa matrices, allows to express one Yukawa matrix in terms of the quark masses, and parametrize the FCNSI in terms of the unknown coefficients χ_{ij}, namely $\widetilde{Y}_{2ij}^U = \chi_{ij}\frac{\sqrt{m_i m_j}}{v}$, where the hermiticity condition reads $\chi_{ij} = \chi_{ij}^\dagger$. These parameters can be constrained by considering all types of low energy FCNC transitions. Although these constraints are quite strong for transitions involving the first and second families, as well as for the b-quark, it turns out that they are rather mild for the top quark.

3.1. FCNC and explicit CP violation

In this case we assume the hermiticity condition for the Yukawa matrices, but the Higgs sector could be only explicit CP violating. Then, from (8) and (9), one obtains the following expressions for the couplings of the neutral Higgs bosons with up-type quarks, namely:

$$S_{ijr}^u = \frac{1}{v}M_{ij}^U\left(q_{r1} - \tan\beta \mathrm{Re}\left[q_{r2}\right]\right) + \frac{\sqrt{m_i m_j}}{\sqrt{2}v\cos\beta}\chi_{ij}\left(\mathrm{Re}\left[q_{r2}\right]\right) \tag{17}$$

and

$$P_{ijr}^u = \frac{1}{v}M_{ij}^U\left[\tan\beta \mathrm{Im}\left[q_{r2}\right]\right] - \frac{\sqrt{m_i m_j}}{\sqrt{2}v\cos\beta}\chi_{ij}\mathrm{Im}\left[q_{r2}\right], \tag{18}$$

similar expressions can be obtained for the down-type quarks.

Journal of Physics: Conference Series **378** (2012) 012026 doi:10.1088/1742-6596/378/1/012026

3.2. FCNC and CP conserving

In this case we shall consider that the Higgs sector is CP conserving, while the Yukawa matrices could be non-hermitian. Then, without loss of generality, we can assume that h_3 is CP odd, while h_1 and h_2 are CP even. Then: $\cos\theta_{12} = \sin(\beta - \alpha)$, $\sin\theta_{12} = \cos(\beta - \alpha)$, $\sin\theta_{13} = 0$, and $e^{-i\theta_{13}} = 1$. The mixing angles α and β that appear in the neutral Higgs mixing, corresponds to the standard notation. Additionally, when one assumes a 4-texture for the Yukawa matrices, the Higgs-fermion couplings further simplify as $\widetilde{Y}^U_{2ij} = \chi_{ij}\frac{\sqrt{m_i m_j}}{v}$. Then, the corresponding coefficient expressions for up sector and h^0 $(r = 1)$ are

$$S^u_{ij1} = \frac{1}{v} M^U_{ij} \left[\sin(\beta-\alpha) - \tan\beta\cos(\beta-\alpha)\right] + \frac{\sqrt{m_i m_j}}{2\sqrt{2}v}\frac{\cos(\beta-\alpha)}{\cos\beta}\left(\chi_{ij} + \chi^\dagger_{ij}\right) \tag{19}$$

$$P^u_{ij1} = \frac{\sqrt{m_i m_j}}{2\sqrt{2}v}\frac{\cos(\beta-\alpha)}{\cos\beta}\left(\chi_{ij} - \chi^\dagger_{ij}\right) \tag{20}$$

For H^0 $(r = 2)$ one finds:

$$S^u_{ij2} = \frac{1}{v} M^U_{ij} \left[\cos(\beta-\alpha) - \tan\beta\sin(\beta-\alpha)\right] + \frac{\sqrt{m_i m_j}}{2\sqrt{2}v}\frac{\sin(\beta-\alpha)}{\cos\beta}\left(\chi_{ij} + \chi^\dagger_{ij}\right), \tag{21}$$

$$P^u_{ij2} = \frac{\sqrt{m_i m_j}}{2\sqrt{2}v}\frac{\sin(\beta-\alpha)}{\cos\beta}\left(\chi_{ij} - \chi^\dagger_{ij}\right) \tag{22}$$

Finally, for A^0 $(r = 3)$ one obtains:

$$S^u_{ij3} = i\frac{\sqrt{m_i m_j}}{2\sqrt{2}v\cos\beta}\left(\chi_{ij} - \chi^\dagger_{ij}\right), \tag{23}$$

$$P^u_{ij3} = \frac{i}{2v} M^U_{ij}\tan\beta - i\frac{\sqrt{m_i m_j}}{2\sqrt{2}v\cos\beta}\left(\chi_{ij} + \chi^\dagger_{ij}\right) \tag{24}$$

Note that under hermiticity on Yukawa matrices the P_{ijr} couplings for the h^0 and H^0 and S_{ijr} for the A^0 are vanished.

4. Higgs decay $h \to c\bar{b}W^-$

In this section we shall evaluate the asymmetry coefficient for the decay $h \to c\bar{b}W$ in order to analyze presence of both FCNSI and CPV within the 2HDM. In the SM the FCNC are suppressed, but in the 2HDM extensions these processes are found even at tree level. We consider the neutral Higgs boson decay $h \longrightarrow W\bar{b}c$ at tree level. Two diagrams contribute to this decay, the first one is through the FCNC coupling $h \longrightarrow \bar{t}^* c \longrightarrow W^-\bar{b}c$, its Feynman diagram is shown on left figure 1. The other one is through $h \longrightarrow W^{+*}W^- \longrightarrow W^-\bar{b}c$, also shown in figure 2.

The couplings of the neutral Higgs with the quarks and the charged boson W with the neutral Higgs are written as $i\left(S^u_{231} + \gamma^5 P^u_{231}\right)$ and $igM_W q_{11} g^{\mu\nu}$, respectively. The other vertices are the usual SM contribution. The average amplitude for these diagrams is thus

$$\overline{|\mathcal{M}|}^2 = \overline{|\mathcal{M}_1|}^2 + \overline{|\mathcal{M}_2|}^2 + \overline{\mathcal{M}_1^\dagger\mathcal{M}_2} + \overline{\mathcal{M}_2^\dagger\mathcal{M}_1} \tag{25}$$

We can obtain an approximation when the terms proportional to the charm and bottom masses

Journal of Physics: Conference Series **378** (2012) 012026 doi:10.1088/1742-6596/378/1/012026

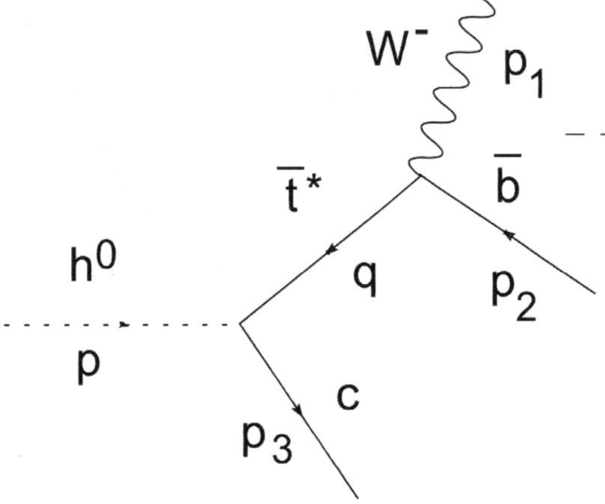

Figure 1. Tree level Feynman diagrams for $h \longrightarrow W^-\bar{b}c$.

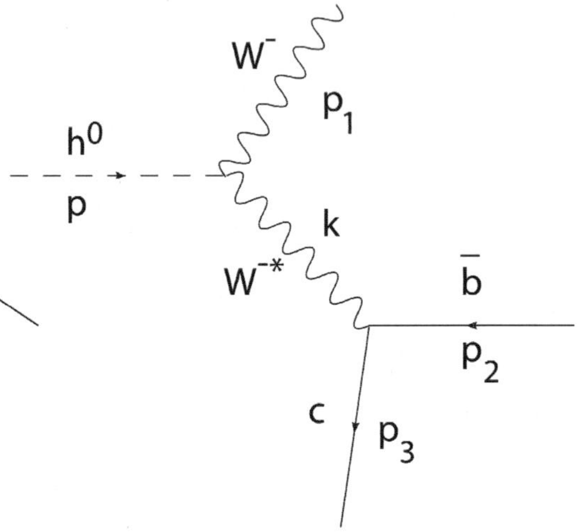

Figure 2. Tree level Feynman diagrams for $h \longrightarrow W^-\bar{b}c$.

are neglected. Then, the explicit expressions are

$$\begin{aligned} \overline{|M_1|}^2 &= \frac{g^4}{4M_W^2}\,|P_t|^2 \Big\{ |S_{231}^u + P_{231}^u|^2 \left[4\,(p_1\cdot p_2)\,(p_1\cdot q)\,(p_3\cdot q)\right. \\ &\quad \left. +2M_W^2\,(p_2\cdot q)\,(p_3\cdot q) - M_W^2 q^2\,(p_2\cdot p_3) - 2q^2\,(p_1\cdot p_2)\,(p_1\cdot p_3)\right] \\ &\quad +m_t^2\,|S_{231}^u - P_{231}^u|^2 \left[M_W^2\,(p_2\cdot p_3) + 2\,(p_1\cdot p_2)\,(p_1\cdot p_3)\right]\Big\}, \end{aligned} \tag{26}$$

$$\overline{|\mathcal{M}_2|}^2 = g^4\,(q_{11})^2\,|V_{cb}|^2\,|P_W(k)|^2 \left(M_W^2 p_2\cdot p_3 + 2p_1\cdot p_2 p_1\cdot p_3\right), \tag{27}$$

$$\begin{aligned} \overline{\mathcal{M}_1^\dagger\mathcal{M}_2} + \overline{\mathcal{M}_2^\dagger\mathcal{M}_1} &= \frac{g^4 m_t}{M_W} q_{11} V_{cb} P_W(k) \left(M_W^2 p_2\cdot p_3 + 2p_1\cdot p_2 p_1\cdot p_3\right) \\ &\quad \left[\mathrm{Re}\,(S_{231}^u + P_{231}^u)\,\mathrm{Re}P_t - \mathrm{Im}\,(S_{231}^u + P_{231}^u)\,\mathrm{Im}P_t\right], \end{aligned} \tag{28}$$

where $P_W(k) = \left(k^2 - M_W^2 + iM_W\Gamma_W\right)^{-1}$ and $P_t\,(q) = \left(q^2 - m_t^2 + im_t\Gamma_t\right)^{-1}$. Then, the width for the decay is

$$\begin{aligned} \Gamma_{h\longrightarrow W\bar{b}c} &= \frac{g^4 m_h}{256\pi^3}\big[|S_{231}^u - P_{231}^u|^2 I_1 + |S_{231}^u + P_{231}^u|^2 I_2 + q_{11}^2\,|V_{cb}|^2 I_3 \\ &\quad +\mathrm{Re}(S_{231}^u + P_{231}^u) q_{11} V_{cb} I_4 - \mathrm{Im}(S_{231}^u + P_{231}^u) q_{11} V_{cb} I_5\big] \end{aligned} \tag{29}$$

where the the analytical expressions for the dimensionless integrals $I_{1,\ldots,5}$ are given in appendix and numerical values as function of littlest neutral Higgs bosons are shown in table. In case of CP explicit violation or CP conserving the width decay is written as

$$\begin{aligned} \Gamma_{CPV} &= \frac{g^4 m_h}{256\pi^3}\Big\{(I_1 + I_2)\,|\chi_{23}|^2 \frac{m_c m_t}{2v^2\cos^2\beta}\left(\sin^2\theta_{12} + \cos^2\theta_{12}\sin^2\theta_{13}\right) \\ &\quad -\frac{\sqrt{m_c m_t}}{2\sqrt{v}\cos\beta} V_{cb}\cos\theta_{12}\cos\theta_{13}\left[\mathrm{Re}\,(\chi_{23})\,(I_4\sin\theta_{12} + I_5\cos\theta_{12}\sin\theta_{13})\right. \\ &\quad \left.-\mathrm{Im}\,(\chi_{23})\,(I_4\cos\theta_{12}\sin\theta_{13} + I_5\sin\theta_{12})\right] + I_3\,|V_{cb}|^2\cos^2\theta_{12}\cos^2\theta_{13}\Big\} \end{aligned} \tag{30}$$

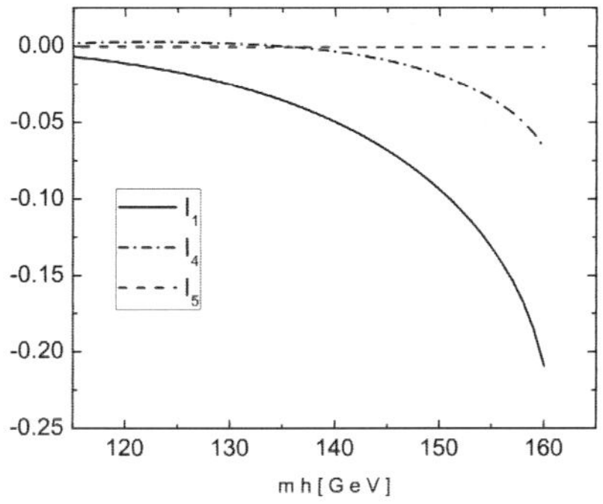

Figure 3. Numerical integrals $I_{1,4,5}$.

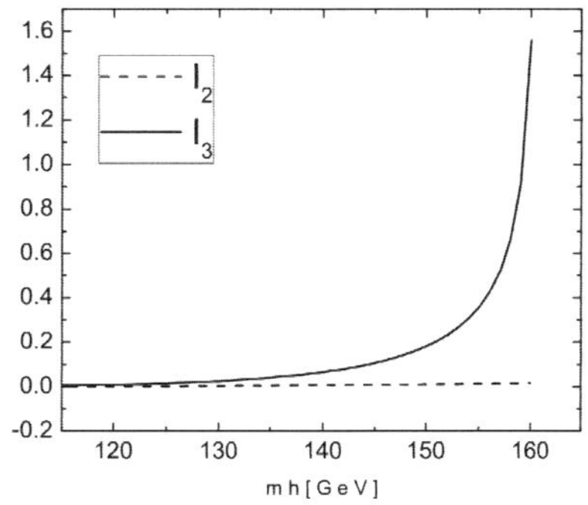

Figure 4. Numerical integrals $I_{2,3}$.

or

$$\begin{aligned}\Gamma_{CPC} &= \frac{g^4 m_h}{256\pi^3}\left[(I_1+I_2)\left|\chi_{23}\right|^2 \frac{m_c m_t \cos^2(\beta-\alpha)}{2v^2\cos^2\beta} + I_3\left|V_{cb}\right|^2 \sin^2(\beta-\alpha)\right.\\ &\left. -\frac{\sqrt{m_c m_t}V_{cb}}{2\sqrt{v}}\frac{\sin(\beta-\alpha)\cos(\beta-\alpha)}{\cos\beta}\left(I_4\mathrm{Re}\left(\chi_{23}\right) - I_5\mathrm{Im}\left(\chi_{23}\right)\right)\right], \end{aligned} \tag{31}$$

respectively.

4.1. CP asymmetry coefficient

In order to find the CP asymmetry coefficient we also need to calculate the conjugate decay. We denote the average amplitude of the conjugate decay as

$$\overline{\left|\widetilde{\mathcal{M}}\right|}^2 = \overline{\left|\widetilde{\mathcal{M}}_1\right|}^2 + \overline{\left|\widetilde{\mathcal{M}}_2\right|}^2 + \overline{\widetilde{\mathcal{M}}_1{}^\dagger\widetilde{\mathcal{M}}_2} + \overline{\widetilde{\mathcal{M}}_2{}^\dagger\widetilde{\mathcal{M}}_1}. \tag{32}$$

The square terms are the same as the above, $\overline{\left|\widetilde{\mathcal{M}}_{1,2}\right|}^2 = \overline{\left|\mathcal{M}_{1,2}\right|}^2$, while for the interference terms we have

$$\begin{aligned}\overline{\widetilde{\mathcal{M}}_1^\dagger\widetilde{\mathcal{M}}_2} + \overline{\widetilde{\mathcal{M}}_2^\dagger\widetilde{\mathcal{M}}_1} &= \frac{g^4 m_t}{M_W} q_{11} V_{cb} P_W(k)\left(M_W^2 p_2\cdot p_3 + 2p_1\cdot p_2 p_1\cdot p_3\right)\\ &\left[\mathrm{Re}\left(S_{231}^u + P_{231}^u\right)\mathrm{Re}P_t + \mathrm{Im}\left(S_{231}^u + P_{231}^u\right)\mathrm{Im}P_t\right]. \end{aligned} \tag{33}$$

The definition for the CP asymmetry coefficient is

$$A_{CP} = \frac{\Gamma_{h\to W^+\bar{b}c} - \Gamma_{h\to W^- b\bar{c}}}{\Gamma_{h\to W^+\bar{b}c} + \Gamma_{h\to W^- b\bar{c}}}. \tag{34}$$

Therefore, the CP asymmetry coefficient of the decay in general 2HDM is given by:

$$A_{CP} = \frac{V_{cb} q_{11} I_5 \mathrm{Im}\left(S_{231}^u + P_{231}^u\right)}{f\left(S_{231}^u, P_{231}^u, q_{11}, m_h\right)}, \tag{35}$$

where

$$f\left(S^u_{231}, P^u_{231}, q_{11}, m_h\right) = \left|S^u_{231} - P^u_{231}\right|^2 I_1 + \left|S^u_{231} + P^u_{231}\right|^2 I_2 + \left|V_{cb}\right|^2 q^2_{11} I_3 + q_{11} V_{cb} \mathrm{Re}\left(S^u_{231} + P^u_{231}\right) I_4. \quad (36)$$

In case of CP explicit violation or CP conserving the asymmetry coefficient is written as

$$A_{CPV} = \frac{\sqrt{m_c m_t} I_5 \cos\theta_{12}\cos\theta_{13}}{\sqrt{2} v \cos\beta} \frac{\mathrm{Re}\left(\chi_{23}\right)\cos\theta_{12}\sin\theta_{13} - \mathrm{Im}\left(\chi_{23}\right)\sin\theta_{12}}{f_{CPV}\left(\theta_{12}, \theta_{13}, \beta, \chi_{23}\right)} \quad (37)$$

or

$$A_{CPC} = \frac{\sqrt{m_c m_t} I_5 \sin\left(\beta - \alpha\right)}{\sqrt{2} v \cos\beta} \frac{\cos\left(\beta - \alpha\right)\mathrm{Im}\left(\chi_{23}\right)}{f_{CPC}\left(\alpha, \beta, \chi_{23}, m_h\right)}, \quad (38)$$

respectively. The f_{CPV} and f_{CPC} are defined as

$$f_{CPV} = \left(I_1 + I_2\right)\left|\chi_{23}\right|^2 \frac{m_c m_t}{2v^2\cos^2\beta}\left(\sin^2\theta_{12} + \cos^2\theta_{12}\sin^2\theta_{13}\right) - \frac{\sqrt{m_c m_t} V_{cb}}{2\sqrt{v}\cos\beta}\cos\theta_{12}\cos\theta_{13}\mathrm{Re}\left(\chi_{23}\right)\left(I_4\sin\theta_{12} + I_5\cos\theta_{12}\sin\theta_{13}\right) + I_3\left|V_{cb}\right|^2\cos^2\theta_{12}\cos^2\theta_{13} \quad (39)$$

and

$$f_{CPC} = \left(I_1 + I_2\right)\left|\chi_{23}\right|^2 \frac{m_c m_t \cos^2\left(\beta - \alpha\right)}{2v^2\cos^2\beta} + I_3\left|V_{cb}\right|^2\sin^2\left(\beta - \alpha\right) - \frac{\sqrt{m_c m_t} V_{cb}}{2\sqrt{v}} \frac{\sin\left(\beta - \alpha\right)\cos\left(\beta - \alpha\right)}{\cos\beta} I_4 \mathrm{Re}\left(\chi_{23}\right) \quad (40)$$

4.2. Numerical results

We shall discuss in detail the result for FCNC and CP conserving case. The asymmetry depends of the five free parameters. One of them is the Higgs boson mass which appears in the J integrals, the other ones are the mixing angles α, β and the complex parameter χ_{23}. The mixing angle β is taken within the values $1 < \tan\beta < 50$ [64]. For the mixing angle α we study three scenarios, $\alpha < \beta$, $\alpha \approx \beta$ and $\alpha > \beta$. For each scenario we take two possible values for $|\chi_{23}|$. Therefore, the scenarios studied here are:

i) $\alpha < \beta$ for $|\chi_{23}| = 0.9$ and $|\chi_{23}| = 0.1$, figures 5 and 6.
ii) $\alpha \approx \beta$ for $|\chi_{23}| = 0.9$ and $|\chi_{23}| = 0.1$, figure 7 and 8.
iii) $\alpha > \beta$ for $|\chi_{23}| = 0.9$ and $|\chi_{23}| = 0.1$, figure 9 and 10.

We use the reported values $m_t = 171.2$ GeV, $m_b = 4.2$ GeV, $m_c = 1.27$ GeV, $M_W = 80.39$ GeV and $\sin\theta_W = 0.231$ [64].

The figures 5, , 6, 7, 8, 9 and 10 show that the asymmetry has values of the order 10^{-3} to 10^{-2} (10^{-4} to 10^{-3} and 10^{-3}) for scenario i) (ii) and iii)).

5. Conclusion

The FCNC in Yukawa sector has been considered in a basis independent formalism and general mixing amount neutral scalar bosons has been used in such a way to introduce CP violation. The CP violation sources can then be related to scalar mixing or textures of Yukawa matrices. In the case of scalar mixing are the possibilities to have explicit and spontaneous CP violation, here we just considered the first one possibility.

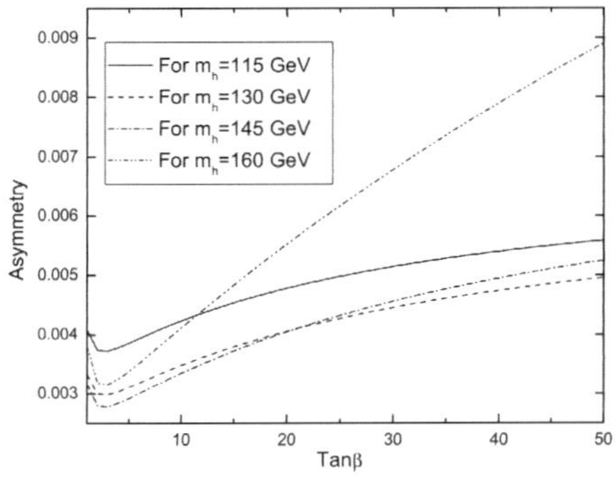

Figure 5. The asymmetry as function of $\tan\beta$ for scenario 1 for $|\chi_{23}| = 0.1$

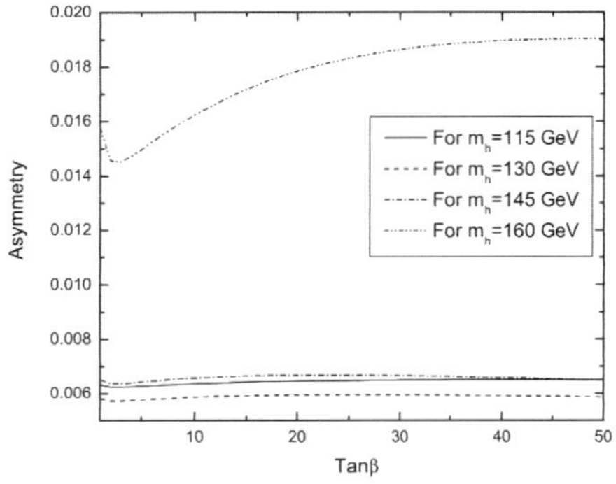

Figure 6. The asymmetry as function of $\tan\beta$ for scenario 1 for $|\chi_{23}| = 0.9$

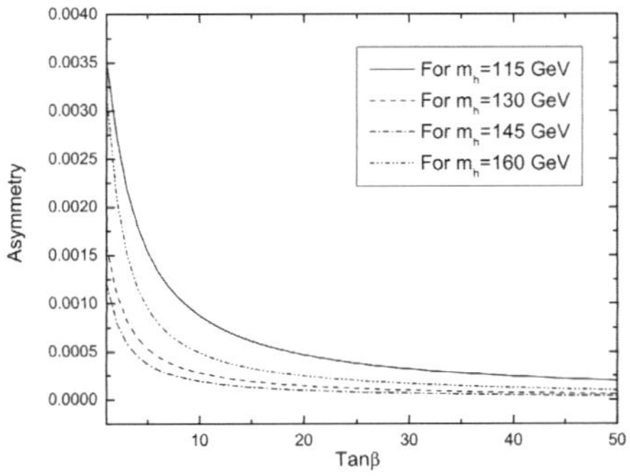

Figure 7. The asymmetry as function of $\tan\beta$ for scenario 2 for $|\chi_{23}| = 0.1$

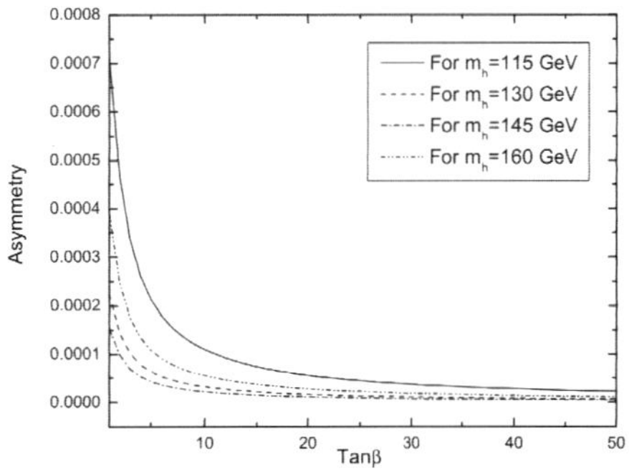

Figure 8. The asymmetry as function of $\tan\beta$ for scenario 2 for $|\chi_{23}| = 0.9$

We have also evaluated the CPV asymmetry for the decay $h \to c\bar{b}W$, which allows to test the presence of both FCNC and CPV. We found that for certain optimal range of parameters the decay asymmetry could be of $O(10^{-2})$ to $O(10^{-4})$ in the FCNC and CP conserving case. The asymmetry behavior has a dependency proportional to the mixing complex parameter χ_{23} too. The mixing angles α and β control the shape of the graphs. The asymmetry keeps same shape for the Higgs boson mass range between 115 GeV and 160 GeV.

In order to detect this asymmetry we could have to resort to a linear collider, since the final state seems difficult to reconstruct at a hadron collider. Although a final conclusion would require a detailed study, which we plan to address in a future publication.

Journal of Physics: Conference Series **378** (2012) 012026 doi:10.1088/1742-6596/378/1/012026

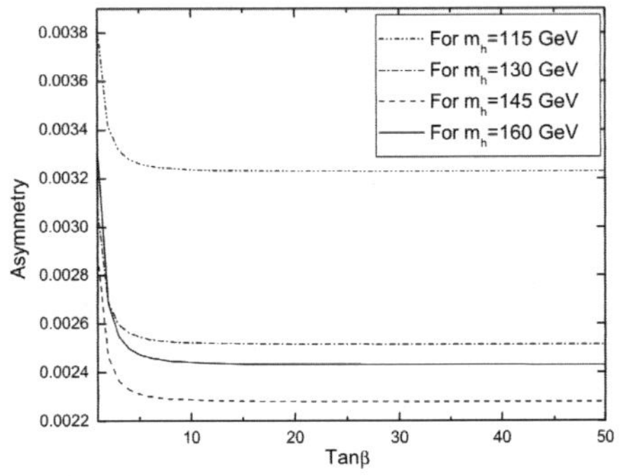

Figure 9. The asymmetry as function of $\tan\beta$ for scenario 3 for $|\chi_{23}| = 0.1$

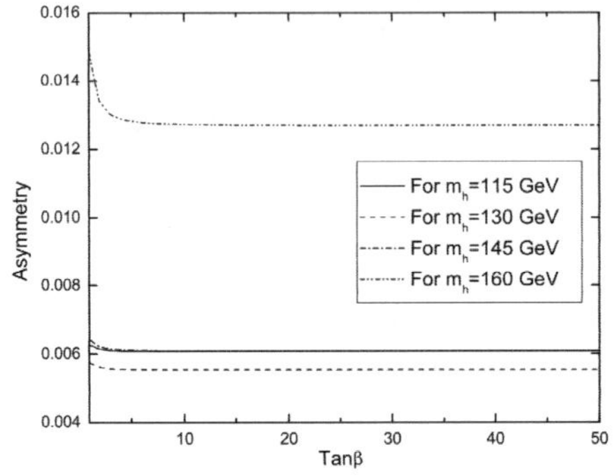

Figure 10. The asymmetry as function of $\tan\beta$ for scenario 3 for $|\chi_{23}| = 0.9$

Acknowledgments

This work is supported in part by PAPIIT project IN117611, Sistema Nacional de Investigadores (SNI) in México. J.H.Montes de Oca Y. is thankful for support from the postdoctoral DGAPA-UNAM grant.

References

[1] J. F. Gunion, H. E. Haber, G. L. Kane and S. Dawson
[2] J. Erler, Phys. Rev. D **81**, 051301 (2010) arXiv:1002.1320 [hep-ph].
[3] H. Flacher, M. Goebel, J. Haller, A. Hocker, K. Moenig and J. Stelzer, Eur. Phys. J. C **60**, 543 (2009) [arXiv:0811.0009 [hep-ph]].
[4] J. L. Diaz-Cruz and D. A. Lopez-Falcon, Phys. Lett. B **568**, 245 (2003) [arXiv:hep-ph/0304212].
[5] U. Baur,eConfC010630:P1WG1,2001. hep-ph/0202001
[6] M. S. Carena and H. E. Haber, Prog. Part. Nucl. Phys. **50**, 63 (2003) [arXiv:hep-ph/0208209].
[7] P. Nath *et al.*, Nucl. Phys. Proc. Suppl. **200-202**, 185 (2010) [arXiv:1001.2693 [hep-ph]].
[8] M. Bustamante, L. Cieri and J. Ellis, arXiv:0911.4409 [hep-ph].
[9] J. Ellis, Int. J. Mod. Phys. A **25**, 2409 (2010) [arXiv:1004.0648 [hep-ph]].
[10] See, for instance, recent reviews in "Perspectives on Supersymmetry", ed. G. L. Kane, World Scientific Publishing Co., 1998; H. E. Haber, Nucl. Phys. Proc. Suppl. **101**, 217 (2001), [hep-ph/0103095.]
[11] H. E. Haber, Nucl. Phys. Proc. Suppl. **101** 217 (2001), hep-ph/0103095.
[12] N. Arkani-Hamed, A. Cohen and H. Georgi, Phys. Lett. B**513** (2001) 232 [arXiv: hep-h/0105239].
[13] A. Aranda, J. L. Diaz-Cruz, J. Hernandez-Sanchez and R. Noriega-Papaqui, Phys. Lett. B **658**, 57 (2007) [arXiv:0708.3821 [hep-ph]].
[14] A. Aranda, C. Balazs and J. L. Diaz-Cruz, Nucl. Phys. B **670**, 90 (2003) [arXiv:hep-ph/0212133].
[15] W. F. Chang, J. N. Ng and A. P. Spray, arXiv:1004.2953 [hep-ph].
[16] E. O. Iltan, Eur. Phys. J. C **51**, 689 (2007) [arXiv:hep-ph/0511241].
[17] H. E. Haber, G. L. Kane and T. Sterling, Nucl. Phys. B **161**, 493 (1979).
[18] J. Liu and L. Wolfenstein, Nucl. Phys. B **289**, 1 (1987).
[19] Y. L. Wu and L. Wolfenstein, Phys. Rev. Lett. **73**, 1762 (1994) [arXiv:hep-ph/9409421].
[20] M. Carena and others, Report of the Tevatron Higgs working group (2000), hep-ph/0010338.
[21] I. F. Ginzburg, M. Krawczyk and P. Osland, arXiv:hep-ph/0211371.
[22] I. F. Ginzburg and M. Krawczyk, Phys. Rev. D **72**, 115013 (2005) [arXiv:hep-ph/0408011].
[23] E. Accomando *et al.*, arXiv:hep-ph/0608079.
[24] S. L. Glashow and S. Weinberg, Phys. Rev. D **15**, 1958 (1977).
[25] J. L. Diaz-Cruz, R. Noriega-Papaqui and A. Rosado, Phys. Rev. D **69**, 095002 (2004) [arXiv:hep-ph/0401194].
[26] J. L. Diaz-Cruz, Phys. Rev. Lett. **100**, 221802 (2008) [arXiv:0711.0488 [hep-ph]].

[27] D. Atwood, S. Bar-Shalom and A. Soni, Phys. Lett. B **635**, 112 (2006) [arXiv:hep-ph/0502234].
[28] J. L. Diaz-Cruz, Mod. Phys. Lett. A **20**, 2397 (2005) [arXiv:hep-ph/0409216].
[29] A. Aranda, J. L. Diaz-Cruz and A. Rosado, Int. J. Mod. Phys. A **22**, 1417 (2007) [arXiv:hep-ph/0507230].
[30] J. L. Diaz-Cruz and A. Rosado, Rev. Mex. Fis. **53**, 396 (2007) [arXiv:hep-ph/0610167].
[31] R. Barbieri and L. J. Hall, arXiv:hep-ph/0510243.
[32] C. D. Froggatt, R. Nevzorov, H. B. Nielsen and D. Thompson, Phys. Lett. B **657**, 95 (2007) [arXiv:0708.2903 [hep-ph]].
[33] I. F. Ginzburg, Acta Phys. Polon. B **37**, 1161 (2006) [arXiv:hep-ph/0512102].
[34] M. Maniatis, A. von Manteuffel and O. Nachtmann, Eur. Phys. J. C **57**, 719 (2008) [arXiv:0707.3344 [hep-ph]].
[35] J. M. Gerard and M. Herquet, Phys. Rev. Lett. **98**, 251802 (2007) [arXiv:hep-ph/0703051].
[36] A. W. El Kaffas, P. Osland and O. M. Ogreid, Nonlin. Phenom. Complex Syst. **10**, 347 (2007) [arXiv:hep-ph/0702097].
[37] H. Fritzsch, Phys. Lett. **B**70 (1977) 436.
[38] T. P. Cheng and M. Sher, Phys. Rev. D **35**, 3484 (1987).
[39] A. E. Carcamo Hernandez, R. Martinez and J. A. Rodriguez, Eur. Phys. J. C **50**, 935 (2007) [arXiv:hep-ph/0606190].
[40] Y. F. Zhou, J. Phys. G **30**, 783 (2004) [arXiv:hep-ph/0307240].
[41] M. Aoki, S. Kanemura, K. Tsumura and K. Yagyu, Phys. Rev. D **80**, 015017 (2009) [arXiv:0902.4665 [hep-ph]].
[42] H. E. Logan and D. MacLennan, Phys. Rev. D **81**, 075016 (2010) [arXiv:1002.4916 [hep-ph]].
[43] H. E. Logan and D. MacLennan, Phys. Rev. D **79**, 115022 (2009) [arXiv:0903.2246 [hep-ph]].
[44] J. F. Gunion and H. E. Haber, Phys. Rev. D **72**, 095002 (2005) [arXiv:hep-ph/0506227].
[45] H. E. Haber and D. O'Neil, Phys. Rev. D **74**, 015018 (2006) [arXiv:hep-ph/0602242].
[46] J. L. Diaz-Cruz and A. Mendez, Nucl. Phys. B **380**, 39 (1992).
[47] T.D. Lee, Phys. Rev. D8, 1226 (1973).
[48] L. J. Hall and M. B. Wise, Nucl. Phys. **B**187, 397, (1981).
[49] J.F. Donoghue and L. F. Li, Phys. Rev. **D**19, 945 (1979).
[50] P. Osland, P. N. Pandita and L. Selbuz, Phys. Rev. D **78**, 015003 (2008) [arXiv:0802.0060 [hep-ph]].
[51] J. L. Diaz-Cruz, JHEP **0305**, 036 (2003) [arXiv:hep-ph/0207030].
[52] C. D. Frogatt and H. B. Nielsen, Nucl. Phys. B**147**, 277 (1979).
[53] J. L. Diaz-Cruz and G. Lopez Castro, Phys. Lett. B **301**, 405 (1993);
J. L. Diaz-Cruz, R. Noriega-Papaqui and A. Rosado, Phys. Rev. D **71**, 015014 (2005) [arXiv:hep-ph/0410391];
A. Aranda, J. Lorenzo Diaz-Cruz, J. Hernandez-Sanchez and E. Ma, Phys. Rev. D **81**, 075010 (2010) [arXiv:1001.4057 [hep-ph]];
[54] J. L. Diaz-Cruz and J. J. Toscano, Phys. Rev. D **62**, 116005 (2000) [arXiv:hep-ph/9910233];
U. Cotti, J. L. Diaz-Cruz, R. Gaitan, H. Gonzales and A. Hernandez-Galeana, Phys. Rev. D **66**, 015004 (2002) [arXiv:hep-ph/0205170];
J. L. Diaz-Cruz, D. K. Ghosh and S. Moretti, Phys. Lett. B **679**, 376 (2009) [arXiv:0809.5158 [hep-ph]];
[55] J. L. Diaz-Cruz, J. Hernandez–Sanchez, S. Moretti, R. Noriega-Papaqui and A. Rosado, Phys. Rev. D **79**, 095025 (2009) [arXiv:0902.4490 [hep-ph]];
C. Balazs, J. L. Diaz-Cruz, H. J. He, T. M. P. Tait and C. P. Yuan, Phys. Rev. D **59**, 055016 (1999) [arXiv:hep-ph/9807349];
J. L. Diaz-Cruz and O. A. Sampayo, Int. J. Mod. Phys. A **8**, 4339 (1993);
J. L. Diaz-Cruz and O. A. Sampayo, Phys. Rev. D **50**, 6820 (1994).
[56] A. Arhrib, M. Capdequi Peyranere, W. Hollik and S. Penaranda, arXiv:hep-ph/0307391;
L. Randall, JHEP **0802**, 084 (2008) [arXiv:0711.4360 [hep-ph]];
S. Bejar, J. Guasch and J. Sola, Nucl. Phys. B **675**, 270 (2003) [arXiv:hep-ph/0307144];
A. Arhrib, Phys. Rev. D **72**, 075016 (2005) [arXiv:hep-ph/0510107].
[57] A. Pich and P. Tuzon, Phys. Rev. **D**80 091702 (2009), hep-ph/09081554.
[58] , M. Jung, A. Pich and P. Tuzon, hep-ph/10060470
[59] Braeuninger, Carolin B. and Ibarra, Alejandro and Simonetto, Cristoforo, Phys. Lett. **B**692 189 (2010), hep-ph/1005.5706
[60] N. Barros e Sa, A. Barroso, P. Ferreira and R. Santos, PoS C **HARGED2008**, 014 (2008) [arXiv:0906.5453 [hep-ph]];
A. Barroso, P. M. Ferreira and R. Santos, Phys. Lett. B **652**, 181 (2007) [arXiv:hep-ph/0702098];
A. Barroso, P. M. Ferreira and R. Santos, Afr. J. Math. Phys. **3**, 103 (2006) [arXiv:hep-ph/0507329];
A. Barroso, P. M. Ferreira and R. Santos, Phys. Lett. B **632**, 684 (2006) [arXiv:hep-ph/0507224].

[61] I. P. Ivanov, Phys. Rev. D **75**, 035001 (2007) [Erratum-ibid. D **76**, 039902 (2007)] [arXiv:hep-ph/0609018].
[62] M. Maniatis, A. von Manteuffel, O. Nachtmann and F. Nagel, Eur. Phys. J. C **48**, 805 (2006) [arXiv:hep-ph/0605184].
[63] E. Ma and M. Maniatis, arXiv:1005.3305 [hep-ph].
[64] K. Nakamura et al. (Particle Data Group), J. Phys. G 37, 075021 (2010)

XIII Mexican Workshop on Particles and Fields IOP Publishing
Journal of Physics: Conference Series **378** (2012) 012027 doi:10.1088/1742-6596/378/1/012027

On the Higgs potential in the MS_3IESM

A Mondragón, M Mondragón and U J Saldaña Salazar

Instituto de Física, Universidad Nacional Autónoma de México, Apdo. Postal 20-364, 01000, México D.F., México.

E-mail: ulisesjesus@fisica.unam.mx

Abstract. We will briefly present a new derivation of the most general form of the Higgs potential with the highest degree of S_3 flavour symmetry. The discussion will be made in the framework of the Minimal S_3-Invariant Extension of the Standard Model (MS_3IESM). In this form of the theory, the Higgs sector has three flavoured Higgs fields that are $SU(2)_L$ doublets and belong to a reducible $\mathbf{1} \oplus \mathbf{2}$ representation of the S_3 flavour symmetry. Requiring the highest degree of S_3 flavour symmetry, the Higgs potential, in its most general form, has only six self-couplings.

1. Introduction

The Standard Model (SM), which successfully describes the fundamental interactions between elementary particles, is being experimentally tested in its still unknown scalar sector. It has become of major importance to experimentally proof the existence of the SM single massive scalar field. The explanation, to whether the massive nature of fundamental particles is being originated from their Yukawa interactions with a scalar field, depends upon the latter experimental observation. Apart from the unknown scalar sector, there are yet a few other open questions that emerge from the fact that there are three generations of matter. It is desirable, that any flavour extension of the SM, should have a unified treatment for both the scalar sector and the flavoured Yukawa interactions.

This paper is organized as follows. In section 2, we briefly comment on what the flavour puzzle is and its relation to the Higgs mechanism. In section 3, we discuss and sketch the construction of the Minimal S_3-Invariant Extension of the Standard Model (MS_3IESM). In section 4, the derivation of the most general S_3-invariant Higgs potential is presented. In section 5, we summarize and conclude with some remarks.

2. The flavour puzzle and its relation to the Higgs mechanism

In 1973, Kobayashi and Maskawa first predicted that the SM needed to have at least three generations of matter, in order to theoretically accommodate Charge-Parity (CP) violation without having any further complications [1]. Nevertheless, a sign of having only three families was not found until later on, when the experiments operating at the electron-positron colliders Stanford Linear Collider (SLC) and Large Electron-Positron collider (LEP) by looking at Z decays showed that there are only three flavours of active neutrinos [2]. The latter fact seems to be pointing out that matter appears in three generations. This came to be known as the flavour puzzle of high energy physics. Why are there three generations of matter? Do we have

Journal of Physics: Conference Series **378** (2012) 012027 doi:10.1088/1742-6596/378/1/012027

any other information regarding this problem? In order to start moving efficiently along the direction of the real explanation, the right questions have to be asked.

In the following subsections the features of the flavour puzzle and how they are related with the Higgs mechanism are briefly commented.

2.1. Some features of the flavour puzzle

There are some phenomenological aspects, that clearly show interrelations between the different families. It is through the study of this interrelations, that we can come closer to find, a solution to the flavour puzzle.

- **Flavour mixing:**
  The experimentally observed flavour oscillations, for both the quark and lepton sector, imply, that weak eigenstates are not mass eigenstates. Matter generations may then be considered as independent entities that interrelate only through their dynamics. The proportion of this interrelationship or *mixing* is expressed by an angle. The reported mixing angles in the Particle Data Group (PDG) parametrization, for both the quark and lepton sector, are:
  - quark weak mixing angles at 2σ [3]:
    * $\theta^q_{12} = 13.021^o \pm 0.039^o$,
    * $\theta^q_{23} = 2.350^o \pm 0.052^o$,
    * $\theta^q_{13} = 0.199^o \pm 0.008^o$;
  - lepton weak mixing angles with normal neutrino mass hierarchy at 1σ [4]:
    * $\theta^l_{12} = 33.95^{o}{}^{+1.05^o}_{-0.93^o}$,
    * $\theta^l_{23} = 46.14^{o}{}^{+3.46^o}_{-3.44^o}$,
    * $\theta^l_{13} = 6.54^{o}{}^{+1.59^o}_{-1.41^o}$.

  In 1967, Sakharov proposed a set of physical conditions on the early universe evolution that could give an explanation to the nowadays observed baryon asymmetry [5]. One of these physical conditions is the presence of CP violation. The SM with massive neutrinos has two possible sources of CP violation: the quark and lepton mixing matrices. As Kobayashi and Maskawa first noticed [1], a 3×3 complex unitary matrix has one non-removable phase besides the three mixing angles. It is through this phase that one is able to introduce CP violation into the SM weak processes. So this means, that apart from the mixing angles, flavour oscillations between three mass states also bring about a desired source for CP violation.
- **Mass spectrum:**
  Fundamental fermions in the SM without the Higgs mechanism are massless. Once the Yukawa interactions are introduced, through the spontaneous electroweak symmetry breaking, which implies a non zero vacuum expectation value for the Higgs boson, fermions become massive.
  At present time, all known fundamental fermions are of massive nature. A semi-log plot of the most recent charged fermions mass values [3] divided over the top quark mass, see Figure 1, clearly shows a strong hierarchical structure. Neutrino masses can not be seen in Figure 1 due to their smallness; their presence in the semi-log plot would only strengthen the strong hierarchical structure. *In which way the mass distribution could help us to discern the underlying flavour structure?* This feature of the flavour puzzle requires to be approached in a very careful manner. It is here where flavour theories depart from each other.

Therefore, what may be seen from these aspects of the flavour puzzle is that all the known reciprocal relations between matter generations are observed in the weak sector. This last statement is what motivates the next subsection.

Journal of Physics: Conference Series **378** (2012) 012027 doi:10.1088/1742-6596/378/1/012027

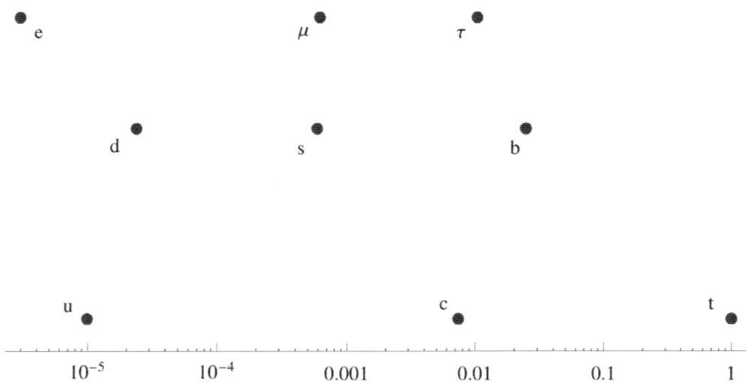

Figure 1. Semi-log plot of the most recent charged fermions mass values divided over the top quark mass.

2.2. The relation with the Higgs mechanism

The SM gauge group is given by $G_{SM} = SU(3)_C \otimes SU(2)_L \otimes U(1)_Y$. The transition, by spontaneous symmetry breaking, from the SM gauge group, G_{SM}, to $SU(3)_C \otimes U(1)_{EM}$ is produced by the Higgs mechanism. Through this mechanism, the massive vector bosons and the fundamental fermions acquire their mass. Besides this, a massive scalar boson is left. The experimentally elusive remnant field is called the Higgs boson.

In the Higgs mechanism the local symmetry group that gets spontaneously broken is $SU(2)_L$. It is from this symmetry group from which the characteristics of the flavour puzzle emerge. In fact, the strong relation of these features with the Higgs mechanism, as already seen by F. Wilczek in 1977 [6], can be clearly seen in the limit of massless fundamental fermions, because the weak mixing angles lose all meaning, and thereby, without the Higgs mechanism the aspects of the flavour puzzle vanish. Thus, a complete study of the flavour puzzle should require a careful examination of its features along with the Higgs mechanism.

3. The Minimal S_3-Invariant Extension of the Standard Model (MS_3IESM)

A vast literature exists on trying to solve the flavour puzzle by using a flavour symmetry, whether continuous or discrete, but given its size, we only refer to some recent reviews on non-Abelian discrete flavour symmetries [7–9]. It has been shown in [10] that by demanding simplicity to a flavour extension of the SM the resulting flavour framework is that of the MS_3IESM. The same result can be obtained by a different approach; somewhat similar to the one shown in [11]. Through the following subsections the derivation of the essential ingredients of the MS_3IESM will be presented, after that, the construction of the flavour framework is briefly sketched.

3.1. Derivation of S_3 as the flavour symmetry

In section 2, it was highlighted the importance of asking the right questions in order to efficiently work towards the solution of the flavour puzzle. If a flavour symmetry really exists, then, the aspects there mentioned could help to point out what the flavour symmetry group is. It is a well known fact, that when particles are described by a symmetry group, their mass distributes according to the irreducible representation (irrep) they belong. Any strong evidence could then come from the mass spectrum.

Matter generations are found to be replicated over each individual sector of fundamental particles: u type quarks, d type quarks, charged leptons and neutrinos. Therefore, the flavour

Journal of Physics: Conference Series **378** (2012) 012027 doi:10.1088/1742-6596/378/1/012027

symmetry could be acting independently on each fermion sector.

How is mass being distributed along each independent fermion sector? According to this question, the mass spectrum, see Figure 1, transforms into what we call the flavour spectrum, see Figure 2, which was already shown in [10]. Neutrino masses were not considered in the plot due to the uncertainty in their individual mass values, although we believe their masses also distribute as the charged fermions masses.

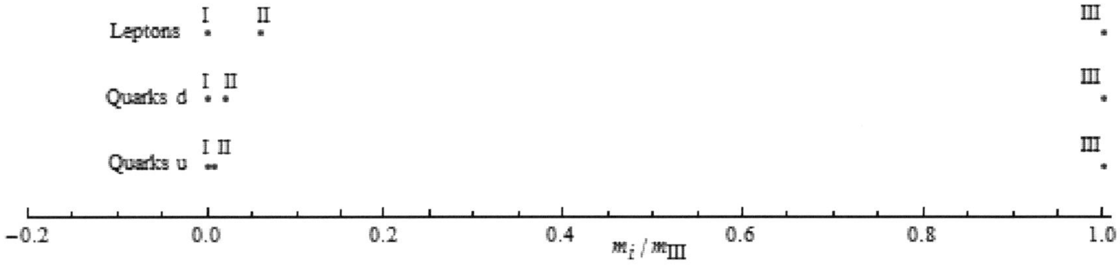

Figure 2. The flavour structure unveiled from the charged fermions mass spectrum. The indices I, II and III refer to the family index. Plot and legend taken from Figure 27.1 in [10].

From this picture, a clear pattern of group symmetry representations of doublets and singlets, $\mathbf{2} \oplus \mathbf{1}$, emerges; which suggests to accommodate the particles in these representations. For the massive neutrinos, depending on whether they have normal or inverted hierarchy, we believe that two kinds of pattern should exist: $\mathbf{2} \oplus \mathbf{1}$ or $\mathbf{1} \oplus \mathbf{2}$, respectively.

In section 2, it was stated that a complete study of the flavour puzzle should require a careful examination of its aspects along with the Higgs mechanism. This means, that it is necessary to survey the mass spectrum in relation to the Higgs mechanism: how things are before and after its introduction. Before the introduction of the Yukawa interactions families are undistinguishable. A flavour symmetry naturally arises from the SM matter Lagrangian as an invariance under permutations of the three families. Seen in this way, this fact points naturally to a symmetry group which represents the permutations of three objects. S_3, the smallest non-Abelian discrete symmetry group, naturally describes this scenario. This particular characteristic of S_3 of expressing the similarity of three different flavoured fermions, with regard to their interactions with gauge bosons, was already stressed out in [12]. After the Higgs mechanism breaks the gauge symmetry, families become distinguishable because their experimental observed masses are different. The invariance under permutations of the three families is now hidden. Could these two seemingly unrelated facts be connected? Let us explore this question.

A group theory analysis reveals that S_3 has three distinct irreps: two singlets, $\mathbf{1_S}$ and $\mathbf{1_A}$, symmetric and antisymmetric, respectively, and one doublet, $\mathbf{2}$. A space of three families implies the use of the three dimensional real representation of S_3, which has as invariant subspaces: a one dimensional space corresponding to the $\mathbf{1_S}$ and a two dimensional space, orthogonal to the singlet's space, corresponding to the $\mathbf{2}$. Hence, a three dimensional representation of S_3, 3, reduces to a direct sum of S_3 irreps: $\mathbf{2} \oplus \mathbf{1_S}$. This means, that, after the introduction of the S_3-invariant Yukawa interactions, the mass spectra would then have the flavour structure $\mathbf{2} \oplus \mathbf{1}$. The latter fact is in very good agreement with the experimentally determined flavour spectrum, see Figure 2. As a consequence, S_3 then becomes the symmetry which naturally reconciles, without any further assumptions, two seemingly unrelated facts: three undistinguishable families before the introduction of the Yukawa interactions with three discernible massive families after the Higgs mechanism.

Journal of Physics: Conference Series **378** (2012) 012027 doi:10.1088/1742-6596/378/1/012027

3.2. Construction of the flavour framework

An addition of a flavour symmetry to the SM is done through two theoretical elements: the Yukawa Lagrangian and the Higgs potential. When adding a flavour symmetry to the SM the first step corresponds to the assignment of the particles to the irreps of the flavour symmetry group. From the flavour spectrum, see Figure 2, it is evident how the assignment should be done: the first two families go into the doublet representation and the third family to the symmetric singlet representation. The second step is to assign the Higgs weak doublet field to a flavour irrep. The SM has a single weak doublet Higgs field, therefore, in order to construct S_3 invariant terms, in both the Yukawa Lagrangian and the Higgs potential, the weak doublet Higgs field must be assigned to the symmetric singlet irrep of S_3. Subsequent to the breaking of the gauge symmetry by the Higgs mechanism, the generic mass matrix for Dirac fermions that results:

$$\mathbf{M}_f = \begin{pmatrix} \mu_1 & 0 & 0 \\ 0 & \mu_1 & 0 \\ 0 & 0 & \mu_3 \end{pmatrix},$$

shows a degenerate spectrum in the first two families, which is in clear disagreement with the experimentally observed mass spectrum. The latter result was already observed and proof as a theorem in [12]. The mass differences between the first two families can be explained by breaking the flavour symmetry [13]. But, if S_3 is to be *preserved*, how to incorporate the mass differences without breaking the flavour symmetry? If two more Higgs weak doublets are added, in a S_3 doublet irrep, in addition to the one from the SM, it can be shown that the generic mass matrix for Dirac fermions transforms into [14]:

$$\mathbf{M}_f = \begin{pmatrix} \mu_1 + \mu_2 & \mu_4 & \mu_5 \\ \mu_4 & \mu_1 - \mu_2 & \mu_6 \\ \mu_7 & \mu_8 & \mu_3 \end{pmatrix},$$

which implies not only a non-degenerate mass spectrum but also a mixing between the families. Furthermore, as an important consequence, *the concepts of flavour and generations are now extended to a more fundamental level*, because now, the Higgs sector also has flavour as well as all fermions.

When finding a correct parametrization of the mass matrices in terms of their mass eigenvalues, the mixing angles may now be expressed in terms of mass ratios. This new possibility is usually considered the primary objective of constructing flavour frameworks [15–24]. Accidentally, this intention of preserving the flavour symmetry, which meant the introduction of two more SM Higgs fields, satisfies the condition of needing at least three Higgs weak doublets, in order to be able to express the fermion mixing angles in terms of mass ratios [21].

4. Derivation of the most general S_3-invariant Higgs potential

Within the MS_3IESM, *which is the most general S_3-invariant Higgs potential?, and, in which sense is it being referred as the most general?* Work on this topic, the most general S_3-invariant Higgs potential, has been done before [12, 14, 21, 25–33]. A direct comparison between the expressions of the distinct authors of the aforecited literature shows a confusing scenario: different numbers of invariant terms and self-couplings. Any comparisons between theory and experiment depend upon the form of the Higgs potential, therefore, in order to avoid misleading conclusions a search for an answer to both questions is needed.

What are the prominent attributes of the most general expression? In which sense the term general is to be understood? We propose the following convention:

- *that it has the highest degree of flavour symmetry, and, within this*
- *the highest possible degree of arbitrariness.*

We give now a systematic procedure, following the previous convention, to construct the Higgs potential:

1. In order to guarantee the highest degree of flavour symmetry, all the linearly independent S_3-invariant terms with two and four Higgs weak doublet fields ought to be found,
   $n = 2$:
   - $\mathbf{1_S} \otimes \mathbf{1_S}$
   - $[\mathbf{2} \otimes \mathbf{2}]_\mathbf{S}$

   $n = 4$:
   - $\mathbf{1_S} \otimes \mathbf{1_S} \otimes \mathbf{1_S} \otimes \mathbf{1_S}$
   - $[(\mathbf{1_S} \otimes \mathbf{2}) \otimes (\mathbf{1_S} \otimes \mathbf{2})]_\mathbf{S}$
   - $[(\mathbf{1_S} \otimes \mathbf{2}) \otimes (\mathbf{2} \otimes \mathbf{2})_\mathbf{2}]_\mathbf{S}$
   - $(\mathbf{2} \otimes \mathbf{2})_\mathbf{S} \otimes (\mathbf{2} \otimes \mathbf{2})_\mathbf{S}$
   - $(\mathbf{2} \otimes \mathbf{2})_\mathbf{A} \otimes (\mathbf{2} \otimes \mathbf{2})_\mathbf{A}$
   - $[(\mathbf{2} \otimes \mathbf{2})_\mathbf{2} \otimes (\mathbf{2} \otimes \mathbf{2})_\mathbf{2}]_\mathbf{S}$,

   where the braces $[\ \ ]_x$ refer to the action of just taking the x part of the tensor product.
2. Make an explicit convention of where to place the symmetric and antisymmetric S_3 doublet components.
$$H_D = \begin{pmatrix} H_{DA} \\ H_{DS} \end{pmatrix},$$
   this matters, because, the mass matrix for Dirac fermions depends on this convention.
3. For each S_3-invariant term, make all the different contractions of independent weak indexes, $X_\mathbf{w} X_\mathbf{w}$, for example:
   - $(\mathbf{2} \otimes \mathbf{2})_\mathbf{S} \otimes (\mathbf{2} \otimes \mathbf{2})_\mathbf{S} = \mathbf{1_{S_3}}$
     * $(\mathbf{2_w} \otimes \mathbf{2_w})_\mathbf{S} \otimes (\mathbf{2_{w'}} \otimes \mathbf{2_{w'}})_\mathbf{S} = \mathbf{1_{S_3 \otimes G_{SM}}}$
     * $(\mathbf{2_w} \otimes \mathbf{2_{w'}})_\mathbf{S} \otimes (\mathbf{2_w} \otimes \mathbf{2_{w'}})_\mathbf{S} = \mathbf{1_{S_3 \otimes G_{SM}}}$
     * $(\mathbf{2_w} \otimes \mathbf{2_{w'}})_\mathbf{S} \otimes (\mathbf{2_{w'}} \otimes \mathbf{2_w})_\mathbf{S} = \mathbf{1_{S_3 \otimes G_{SM}}}$,

   the next expressions clearly show how different structures may emerge from the same S_3-invariant term in terms of the weak doublet Higgs fields:
     * $\frac{1}{2}(H_{1w}^\dagger H_{1w} + H_{2w}^\dagger H_{2w})^2$
     * $\frac{1}{2}[(H_{1w}^\dagger H_{1w})^2 + (H_{2w}^\dagger H_{2w})^2 + (H_{1w}^\dagger H_{2w})^2 + (H_{2w}^\dagger H_{1w})^2]$
     * $\frac{1}{2}[(H_{1w}^\dagger H_{1w})^2 + (H_{2w}^\dagger H_{2w})^2 + (H_{1w}^\dagger H_{2w})^2 + (H_{2w}^\dagger H_{1w})^2]$

   where the convention $H_{1w} = H_{DAw}$, $H_{2w} = H_{DSw}$ was taken in order to simplify the notation.
4. Assign the same self-coupling to each different weak indexes contraction of the same S_3-invariant term.

The necessary elements to carry out all the aforementioned steps are: compute the direct products of S_3 irreps and carefully make explicit the weak, $SU(2)_L$, indexes in all terms.

The expression for the most general S_3-invariant Higgs potential, already shown in [10], has only *six independent self-couplings* and it is given by:

$$V_H = \mu_S^2(H_S^\dagger H_S) + \mu_D^2(H_1^\dagger H_1 + H_2^\dagger H_2) + a(H_S^\dagger H_S)^2 + b f_{ijk}[(H_S^\dagger H_i)(H_j^\dagger H_k) + h.c.] + c[(H_S^\dagger H_1)(H_1^\dagger H_S) + (H_S^\dagger H_2)(H_2^\dagger H_S) + (H_S^\dagger H_1)^2 + (H_S^\dagger H_2)^2 + (H_1^\dagger H_S)^2 + (H_2^\dagger H_S)^2 + (H_S^\dagger H_S)(H_1^\dagger H_1 + H_2^\dagger H_2)] + d[2(H_1^\dagger H_1 + H_2^\dagger H_2)^2 + (H_1^\dagger H_1 - H_2^\dagger H_2)^2 + (H_1^\dagger H_2 + H_2^\dagger H_1)^2 + (H_1^\dagger H_2 - H_2^\dagger H_1)^2] + e(H_1^\dagger H_2 - H_2^\dagger H_1)^2 + f[(H_1^\dagger H_1 - H_2^\dagger H_2)^2 + 2(H_1^\dagger H_1 + H_2^\dagger H_2)^2 - 2(H_1^\dagger H_2 - H_2^\dagger H_1)^2],$$

Journal of Physics: Conference Series **378** (2012) 012027 doi:10.1088/1742-6596/378/1/012027

where $a, b, c, ..., f$ are constants; μ_S^2, μ_D^2 are negative real numbers; 1, 2, and S are the flavour indexes for the components of the doublet and the symmetric singlet irrep of S_3, respectively; $f_{112} = f_{121} = f_{211} = -f_{222} = 1$, all other tensor components vanish.

5. Conclusions

The flavour puzzle still remains an open problem of high energy physics. However, the addition of a flavour symmetry to the SM is the simplest approach known in order to try to give an explanation to the flavour puzzle. We have emphasized the importance of looking for the right questions in order to start moving efficiently along the direction of the real explanation.

Some of the aspects of the flavour puzzle, from our point of view, were discussed in Section 2. It was highlighted the strong relation with the Higgs mechanism that exists, such that any complete study of the flavour puzzle should require a careful examination of its aspects along with it.

A different approach to derive the essential ingredients of the MS_3IESM was presented. The latter was made by taking into consideration some of the features of the flavour puzzle along with the Higgs mechanism, in particular, the mass spectrum, see Figure 1, which is the strongest suggestion of flavour structure we have nowadays. The mass spectrum was related with a flavour structure through the question: how is mass being distributed along each independent fermion sector?, see Figure 2. A $\mathbf{2} \oplus \mathbf{1}$ structure can be readily seen from it. When studying the latter along with the Higgs mechanism, it was necessary to reconcile two seemingly unrelated facts: three undistinguishable families before the introduction of the Yukawa interactions with three discernible massive families after the breaking of the gauge symmetry by the Higgs mechanism. From this quest, we found that: S_3 is the only symmetry that naturally reconciles two seemingly unrelated facts without any further assumptions. The scalar sector was extended, by introducing two more Higgs weak doublets into the SM in a flavour doublet irrep of S_3, and this led to a non-degenerate mass spectra with flavour mixing; furthermore, the concepts of flavour and generations were extended to a more fundamental level, because the Higgs sector was flavoured as well as the fermion sector.

Finally, the steps in the computation of the most general S_3-invariant Higgs potential were presented and each one of those steps was illustrated with an example. This calculation is important for a meaningful and significative comparison of theory and experiment, in this respect, any conclusions on the Higgs phenomenology depend strongly on having the form of the Higgs potential which is in agreement with the flavour symmetry proposed.

Acknowledgements

U. J. S. S. thanks the XIII-MWPF local organizing committee for the giving hospitality in León, Guanajuato. Also he thanks the Instituto de Física-UNAM for supporting his attendance to this interesting and stimulating workshop.

We acknowledge support from the Mexican grant PAPIIT IN113712.

References

[1] Kobayashi M and Maskawa T 1973 *Prog. Theor. Phys.* **49** 652–657
[2] 2006 *Phys. Rept.* **427** 257–454 (*Preprint* hep-ex/0509008)
[3] Nakamura K *et al.* (Particle Data Group) 2010 *J. Phys. G* **G37** 075021
[4] Schwetz T, Tortola M and Valle J 2011 *New J. Phys.* **13** 109401 (*Preprint* 1108.1376)
[5] Sakharov A 1967 *Pisma Zh. Eksp. Teor. Fiz.* **5** 32–35 reprinted in *Kolb, E.W. (ed.), Turner, M.S. (ed.): The early universe* 371-373, and in *Lindley, D. (ed.) et al.: Cosmology and particle physics* 106-109, and in Sov. Phys. Usp. 34 (1991) 392-393 [Usp. Fiz. Nauk 161 (1991) No. 5 61-64]
[6] Wilczek F and Zee A 1977 *Phys. Rev.* **D15** 3701
[7] Ma E 2007 (*Preprint* 0705.0327)
[8] Altarelli G and Feruglio F 2010 *Rev. Mod. Phys.* **82** 2701–2729 (*Preprint* 1002.0211)

[9] Ishimori H, Kobayashi T, Ohki H, Shimizu Y, Okada H *et al.* 2010 *Prog.Theor.Phys.Suppl.* **183** 1–163 (*Preprint* 1003.3552)
[10] Saldaña Salazar U J To be published in the Proceedings of FLASY 2011.
[11] Hall L J and Murayama H 1995 *Phys.Rev.Lett.* **75** 3985–3988 (*Preprint* hep-ph/9508296)
[12] Derman E 1979 *Phys.Rev.* **D19** 317
[13] Mondragon A and Rodriguez-Jauregui E 1999 *Phys.Rev.* **D59** 093009 (*Preprint* hep-ph/9807214)
[14] Yahalom R 1984 *Phys.Rev.* **D29** 536
[15] Gatto R, Sartori G and Tonin M 1968 *Phys.Lett.* **B28** 128–130
[16] Cabibbo N and Maiani L 1968 *Phys.Lett.* **B28** 131–135
[17] Pagels H 1975 *Phys.Rev.* **D11** 1213
[18] Weinberg S 1977 *Trans.New York Acad.Sci.* **38** 185–201
[19] Wilczek F and Zee A 1977 *Phys.Lett.* **B70** 418
[20] Fritzsch H 1977 *Phys.Lett.* **B70** 436
[21] Wyler D 1979 *Phys.Rev.* **D19** 330
[22] Ebrahim A 1978 *Phys.Lett.* **B73** 181–184
[23] Mohapatra R N and Senjanovic G 1978 *Phys.Lett.* **B73** 176
[24] Froggatt C and Nielsen H B 1979 *Nucl.Phys.* **B147** 277
[25] Pakvasa S and Sugawara H 1978 *Phys.Lett.* **B73** 61
[26] Koide Y 1999 *Phys.Rev.* **D60** 077301 revised version (*Preprint* hep-ph/9905416)
[27] Kubo J, Okada H and Sakamaki F 2004 *Phys.Rev.* **D70** 036007 (*Preprint* hep-ph/0402089)
[28] Chen S L, Frigerio M and Ma E 2004 *Phys.Rev.* **D70** 073008 (*Preprint* hep-ph/0404084)
[29] Kimura T 2005 *Prog.Theor.Phys.* **114** 329–358
[30] Koide Y 2006 *Phys.Rev.* **D73** 057901 (*Preprint* hep-ph/0509214)
[31] Beltran O, Mondragon M and Rodriguez-Jauregui E 2009 *J.Phys.Conf.Ser.* **171** 012028
[32] Meloni D, Morisi S and Peinado E 2011 *J.Phys.G* **G38** 015003 (*Preprint* 1005.3482)
[33] Bhattacharyya G, Leser P and Pas H 2011 *Phys.Rev.* **D83** 011701 (*Preprint* 1006.5597)

Journal of Physics: Conference Series **378** (2012) 012028 doi:10.1088/1742-6596/378/1/012028

Status of the SuperB project

G. Wormser

LAL, Univ Paris-Sud, CNRS/IN2P3, Orsay, France

E-mail: wormser@lal.in2p3.fr

Abstract.
The SuperB project is a very ambitious program whose goal is to build, in the immediate vicinity of the Frascati National Laboratory, a e^+e^-collider operating in the $\Upsilon(4S)$ region with a luminosity in excess of 10^{36} Hz/cm^{-2}, surpassing by two orders of magnitude the present generation.Such a progress has ben made possible by the new Crab Waist colliding scheme together with the design of very low emittance rings. The physics goal of this machine is to determine the structure of the new physics (NP) at the Terascale that is likely to show up at the LHC. This will be possible through a very detailed scrutiny of all NP induced indirect effects in rare (or even forbidden in the Standard Model)decays and precision measurements in the quark and lepton sectors. The project, an official element of the European HEP Strategy, has been recently approved and fully funded by the Italian government. The site has been selected and the detector collaboration is currently being formed. The first beams are expected in 2016, with a yearly integrated luminosity of 15 ab^{-1}.

1. Introduction

All major discoveries concerning the flavour sector have, as strange as it may seem, been first made by indirect observations. The existence and properties of the 4th, 5th and 6th quarks have been demonstrated a few years before their direct observations through the non observation of neutral currents reactions (GIM mechanism), CP violation in the K sector (KM hypothesis), rate of the B mixing. These indirect measurements not only provided information on the existence and mass of these quarks but also on the organisation of the quark sector in the Standard Model. The SuperB project goal is to repeat this brilliant history but now for the New Physics Beyond the Standard Model (BSM or NP). There is good hope that this approach will be very successful since we already know that strong constraints already exist when building any BSM model to avoid producing effects in the flavour sector in contradiction with current observations. The detailed understanding of the unknown physical source of these constraints and the identification of the BSM model that Nature has chosen to implement is the ultimate goal of SuperB. This task will be made easier when LHC will have discovered some new particles in the mass range between 200 GeV and 1 TeV, because this will precisely set the mass scale for the indirect effects induced by this new particle at SuperB. The requirement which was used to specify the integrated luminsoty needed for the SuperB physics case is that a 3 σ effect should be detected at SuperB in one of the accessible channels for a 1 TeV particle given the fact that its couplings phase should be at minimum the SM phase (its couplings can not be lower as is explained in the so called Minimum Flavour Violation scheme). The minimum integrated luminosity to reach this goal is 75 ab^{-1}, which can be recorded in 5 years of data taking at SuperB.

	Observable/mode	H^+ high $\tan\beta$	MFV	non-MFV	NP Z penguins	Right-handed currents	LTH	SUSY AC	SUSY RVV2	SUSY AKM	SUSY δLL	SUSY FBMSSM
✓	$\tau \to \mu\gamma$							⋆⋆⋆	⋆⋆⋆	⋆	⋆⋆⋆	⋆⋆⋆
✓	$\tau \to \ell\ell\ell$						⋆⋆⋆					
✓	$B \to \tau\nu, \mu\nu$	⋆⋆⋆(CKM)										
✓	$B \to K^{(*)+}\nu\bar{\nu}$			⋆	⋆⋆⋆			⋆	⋆	⋆	⋆	⋆
✓	S in $B \to K_S^0\pi^0\gamma$					⋆⋆⋆						
✓	S in other penguin modes			⋆⋆⋆(CKM)		⋆⋆⋆		⋆⋆⋆	⋆⋆	⋆	⋆⋆⋆	⋆⋆⋆
✓	$A_{CP}(B \to X_s\gamma)$			⋆⋆⋆		⋆⋆		⋆	⋆	⋆	⋆⋆⋆	⋆⋆⋆
✓	$BR(B \to X_s\gamma)$		⋆⋆⋆	⋆		⋆						
✓	$BR(B \to X_s\ell\ell)$			⋆	⋆	⋆						
✓	$B \to K^{(*)}\ell\ell$ (FB Asym)							⋆	⋆	⋆	⋆⋆⋆	⋆⋆⋆
	$B_s \to \mu\mu$							⋆⋆⋆	⋆⋆⋆	⋆⋆⋆	⋆⋆⋆	⋆⋆⋆
	β_s from $B_s \to J/\psi\phi$							⋆⋆⋆	⋆⋆⋆	⋆⋆⋆	⋆	⋆
✓	a_{sl}						⋆⋆⋆					
✓	Charm mixing							⋆⋆⋆	⋆	⋆	⋆	⋆
✓	CPV in Charm	⋆⋆									⋆⋆⋆	

Figure 1. List of golden mode channels for SuperB and their relevance for various New Physics Models. The modes with a red tick are accessible to SuperB, the ones without are only available with hadron machines.

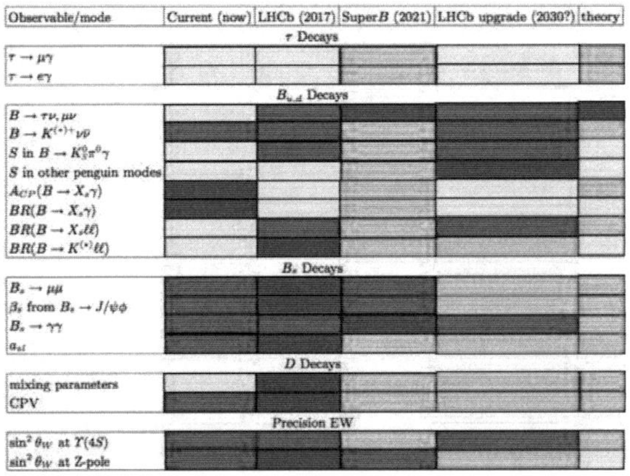

Figure 2. Sensivity of SuperB and LHCb to various physics modes. The colour code is as follows : Red : No measurement, Yellow : Not precise, Blue : Precise, Green : Very precise

2. SuperB Physics Goals

The general goal of the SuperB physics program is to understand in depth the underlying structure of the new physics (NP) beyond the Standard Model (BSM). It has been demonstrated in great detail([1]) that by measuring deviations from SM expectations in a variety of different channels, one could gather very precious information about the structure of the NP. This is because all new particles with masses below 1 TeV will generate deviations from SM even if their couplings to normal quarks and leptons is minimum. The table 1 shows the relative sensitivity of the key superB observables with respect to various theoretical frameworks. Several lessons can be learned from this table : firstly, the golden channels lists must be quite comprehensive to be able to disentangle all the various scenarios; secondly that this golden channel list contains many modes that are only accessible to a SuperB factory. Some modes will be well measured by the LHCb experiment which will have completed its first data taking phase when SuperB will start and integrated around 5 fb^{-1}, as can be seen from the table 2 but many others just can't be measured at all or measured with the needed precision without SuperB. It has to be noted that the physics program of SuperB is not restricted to the study of B decays at the $\Upsilon(4S)$ resonance. The search for lepton flavour violation is one of the major physics goals and the present limits on the decay $\tau \to \mu\gamma$ and τ to three charged leptons can be improved by a factor 10, reaching BR level from 10^{-9} to a few 10^{-10} becoming quite competitive and complementary with the related search $\mu \to e\gamma$. A key asset is this search is the fact that SuperB will benefit of a 80 % polarized electron beam. This will allow in the search for $\tau \to \mu\gamma$ to benefit from a very significant background rejection. This extra factor will be quite important since many NP models predict BR for such very rare decays in the 10^{-9} range. The beam polarization also offers the possibility to perform search for CP violation in τ decays,and to try to perform the first measurement of τ magnetic moment. In addition to this key role in the lepton sector, the beam polarization will allow to measure the $\mu^+\mu^-$ forward-backward asymmetry at 10 GeV, adding a measurement of comparable precision to th eone performed at LEP. This will again allow a good sensitivity to NP. Another feature of the SuperB physics program is to run at charm thresold with asymmetric beam energies, giving access for the first time to time-dependant CP asymmetries in the charm sector. The very high luminosity (10^{35}) foreseen at 4 GeV will allow unprecedented precision in the charm sector.

3. The SuperB Accelerator

The SuperB accelerator concept stems from a series of very innovative ideas developped in the accelerator community in the last ten years. Previous experience with PEP-II and KEKB rings showed it would be extremely difficult to increase in a significant manner the beam currents. Therefore the only solution left to reach 10^{36} luminosity is to collide nanometer size beam. This requires the production of extremely low emittance beams in the electron and positron sources and in the linac, the low emittance conservation in the rings and the design a final focus system capable to generating and putting in stable collisions 50 nm vertical size beams.This has been made possible in particular thanks to the development of the all the accelerator R&D performed in the framework of the International Linear Collider project where the issues are exactly the same. One must also point out that SuperB rings have also benefitted from the development of very high brilliance 3rd generation light sources. But the collision of such dense beams creates, if no counter measures are taken, very large beam-beam resonances which will immediately blow up the beams. The Crab waist scheme, which consists in the addition of two sextupoles very near the Interaction point, has been precisely invented by Pantaleo Raimondi,[2]in order to suppress these beam-beam effects by predistorting the beams before the collision in order to minimize the beam-beam effects. The effect of the crab waist sextupoles of the beam profiles are displayed in Fig. 3. The tune space, which was previously heavily populated by beam killing resonances, offers now very large resonance-free zones. The crab waist scheme was very successfully demonstrated on the DAFNE storage ring in 2008-2009 [3]. DAFNE luminosity has been increased by a factor 3 compared to the absence of sextupoles, as predicted by the simulations. It is often asked how to compare this sizable but modest gain with the factor 100 expected between SuperB and present colliders. As mentioned above, the very large luminosity increase will be due to the very small size of the beams, going from micrometer to nanometer vertical size. The Crab waist technique is the key enabling factor, making possible the collision of such dense beams without detrimental beam-beam effects.

The detailed description of the accelerator design has been recently updated and is described in [4]. The preliminary layout of the accelerator on the SuperB site (see below) is indicated on the figure 4. It consists of a polarized electron source, a high yield positon source coupled to a positron damping ring,, a 7 GeV Linac and of two 1250 m rings. As a result of the partnership with the Italian Institute of Technology, SuperB rings will also be used to produce top class synchrotron beam lines. 6 such beam lines are tentatively indicated on the figure 4, located on the High Energy Ring, since 7 GeV light sources are much more difficult to find than 4 GeV

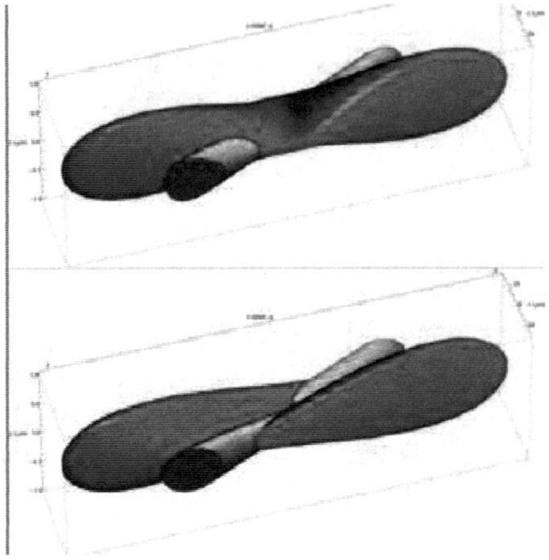

Figure 3. Profiles of the two colliding beams witout (top) and with (bottom) crab waist scheme. The predistortion of the beams allow the minimum waist of one beam to be aligned with the other.

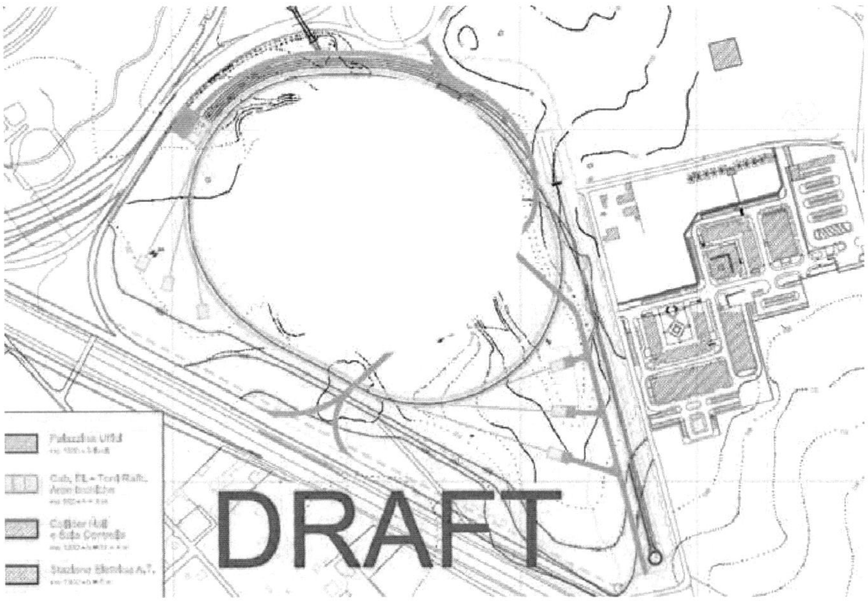

Figure 4. Layout of the SuperB accelerator on the Tor Vergata (preliminary)

ones. The brilliance of such beamlines will be higher than any other presently running machine, given the very low emittance and relatively high current (2 Amps) of the Super B ring.

4. The SuperB Site

Given the revolutionnary character of the SuperB machine, no present ring or tunnel in Europe is capable of hosting the SuperB project. It has been therefore necessary to explore green field scenarios. the SuperB project has issued a Site Specifications document, which has been reviewed by an international Site Committee chaired by John Osborne,from CERN. The specifications

Journal of Physics: Conference Series **378** (2012) 012028 doi:10.1088/1742-6596/378/1/012028

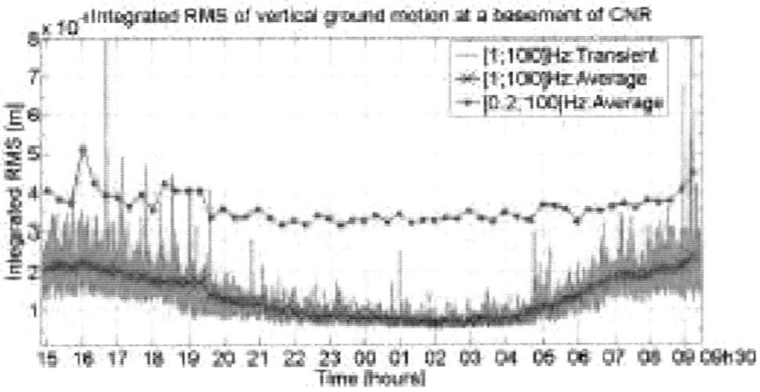

Figure 5. Long term measurement of the RMS vertical displacement at a point on Tor Vergata site, close to the future SuperB Interaction Region. The RMS varies between 10 and 20 nm, when the noise is integrated between 1 and 100 Hz.

adress the following issues : site size given the machine circumference (1250 meters), the space needed for synchrotron beam lines, and office and utility spaces, geology, vibration level, constructibility, ease of access for people and components, availability of energy and cooling, proximity of a nearby INFN structure, industrial expertise level in the vicinity. A clear desire has also been expressed towards a site allowing a shallow tunnel since the presence of several synchrotron light beamlines make this possibility extremely cost effective. Although several site candidates have been identified throughout Italy, the INFN management indicated its marked preference for a site in the immediate vicinity of the Frascati National Laboratory where is located the accelerator design team. Two sites have been finally considered, the LNF site itself wheer the machine has to be deeply underground given the strong slopes on this site and the nearby Tor-Vergata University site (Fig.4). The geological structure of the Tor Vergata is very favorable because of its flatness, uniformity and the good vibration damping properties of the pyroclastic material. A intensive vibration measurement campaign has been performed by the LAPP Annecy team and the results show that the RMS verticla motion along the ring or at the IP is between 20 and 40 nm, much below the 300 nm required, in spite of the presence of the Rome-Naples highway 100 m away!(Fig.5). Given all these nice properties (the site has also been checked against archeological remains), the Tor Vergata site ahs been formally selected and officially proposed by the Tor Vergata University to INFN for SuperB use. INFN has endorsed this proposal and therefore, the Tor vergata site has been definively selected as SuperB site.

5. The SuperB Detector

Since the BABAR and SuperB physics requirtements are very similar, the SuperB detector can and will reuse very significant parts of the BABAR detector in order to save costs and time, and because its very good properties : magnet, iron yoke, CsI(Tl) based crystal calorimeter, and the quartz bars of the Particle IDentification system (PID). The overall structure of the SuperB detector shown in Fig. 3 is therefore very similar to BABAR. However, sevral main differences exist. The much higher luminosity in SuperB will force the SuperB vertex detector to be much more radiation resistant both in terms of integrated dose and instantaneous occupancy thhan the BABAr one. In addition, the smaller boost (0.238 instead of 0.56) in SuperB forces the first SVT layer to be closer to the beam pipe, in order to retain a comparable or better proper time resolution. The SuperB physics program calls for a better hermeticty of the detector because of the renewed importance of the mode involving one or two neutrinos in the final state. SuperB will therefore be equipped, budget permitting, of a forward PID device and of a backward

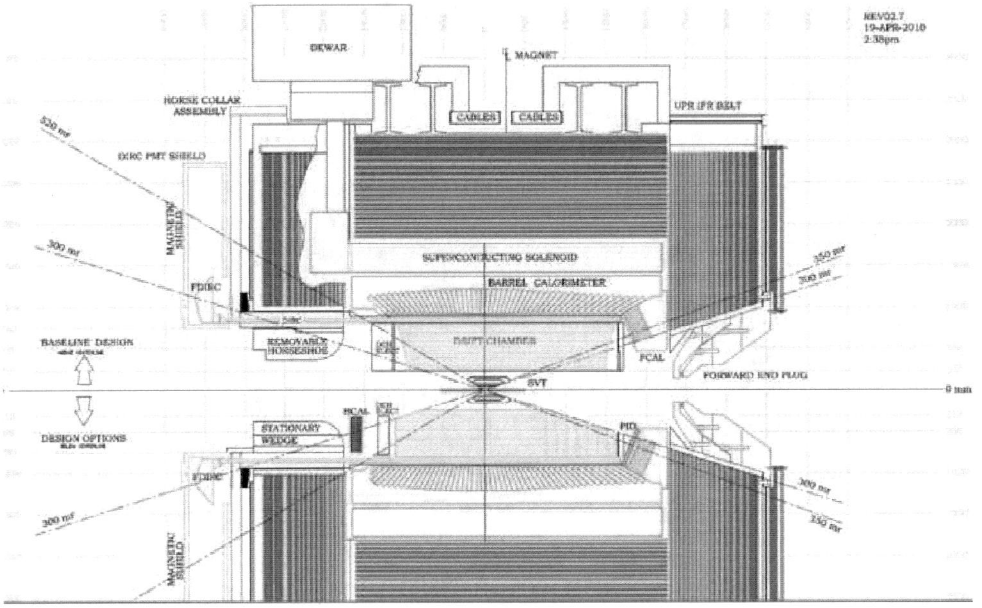

Figure 6. Cross section of the SuperB detector. the bootom part includes teh two options considered to improve SuperB hermeticity: a backward calorimeter and a forward PID device.

calorimeter. These two additions will increase in total by 10% superB efficiency to the very rare modes described above. The detailed description of the SuperB detector can be found in [5],[6]. It is displayed in Fig.6.

Some modifications will be implemented on the parts coming from BABAR: the flux-return will be augmented with additional absorber to increase the number of interaction lengths for muons to roughly 7λ; the DIRC camera will be replaced by a twelve-fold modular camera using multi-channel plate (MCP) photon detectors in a focusing configuration using fused silica optics to reduce the impact of beam related backgrounds and improve performance; the forward EMC will feature cerium-doped LYSO (lutetium yttrium orthosilicate) crystals, which have a much shorter scintillation time constant, a lower Molière radius and better radiation hardness than the current CsI(Tl) crystals, again for reduced sensitivity to beam backgrounds and better position resolution.

The tracking detectors for SuperB will be new. To maintain sufficient proper-time difference (Δt) resolution for time-dependent *CP* violation measurements with the SuperB boost of $\beta\gamma = 0.24$, the vertex resolution will be improved by reducing the radius of the beam pipe, placing the innermost layer of the SVT at a radius of roughly 1.2 cm. This innermost layer of the SVT will be constructed of either silicon striplets or Monolithic Active Pixel Sensors (MAPS) or other pixelated sensors, depending on the estimated occupancy from beam-related backgrounds. Likewise, the design of the cell size and geometry of the DCH will be driven by occupancy considerations. The hermeticity of the SuperB detector, and, thus, its performance for certain physics channels will be improved by including a backwards veto-quality EMC detector comprising a lead-scintillator stack.

Journal of Physics: Conference Series **378** (2012) 012028 doi:10.1088/1742-6596/378/1/012028

6. The SuperB Project Status

Between December 2010 and May 2011, the SuperB project has succesfully cleared all the govermental milestones to become a fully funded project. SuperB has been ranked first among a list of 14 Flagship Projects forming the core of the 2011-2013 National Research Plan in Italy. This Plan has been approved by the government and voted at both Houses of the Italian parliament in December 2010, together with a generic funding mechanism. This resulted in a immediate release of 19 M towards the construction of SuperB. The details of the SuperB funding, and especially its pluriannual investment profile have been recently endorsed by the interministerial CIPE committee. A budget of 50 M in 2011 has been allocated to the project. according to the INFN triennal plan indicated below. The scope of teh SuprB project consisting of a HEP accelerator and a synchrotron light source, a consortium will be made within the following months between INFN, IIT (The Italian Institute of Technology, in charge of the light source sector), Tor Vergata University (mainly in charge of the civil construction on the site) and the Italian Ministry for Research. The decision has been taken to name this consortium Nicola Cabibbo laboratory. The intention of the Italian Governement is to make this structure evolve into an ERIC (European Research Infrastructre Consortium) in a few years. The Nicola Cabibbo Laboratory wil be very active to seek international partnerships for the construction of the SuperB accelerator.

7. Conclusion

The launch of the SuperB project is a very rare and important event in European particle physics. Only once every 25 years or so, a large scale project is created in Europe besides the CERN projects, the last exemple of such an initiative being HERA in Hamburg. It is of course very clear that the concentration of European investments at CERN is essential, but the added diversity by such national or regional programs is also important and was recognized as sucvh in the European Strategy document adopted in Lisbon in July 2006. The exceptional scientific and technological merits of the SuperB project make this new adventure particularly exciting. Recent decisive progress have been accomplished with the final approval and funding of the machine by the Italian Governnement, the final site selection on the campus of the Tor Vergata University, the decision to create the Nicola Cabbibbo Laboratory consortium, the formal launch of the detector collaboration formation. In parallel to these major milestones, the detailed technical design of the accelerator, synchrotron light beam lines and detecor is taking place, with the publication of Technical Design Reports documents expected early 2012. Given all this activity, the first beams can be expected in 2016, in a time frame comparable to the similar program, SUPERKEKB and BELLE-II, being pursued in Japan. The competition with this project will be accompanied with a very substantive collaboration, as was already the case in the PEP-II/BABAR-KEKB/BELLE lifetime and will certainly turn out to be very fruitful. Major scientific breakthroughs, in terms of a deep understanding of the organisation of the physics beyond the Standard Model, in partnership with the LHC results, can therefore be confidently expected around 2020, when SuperB will have collected its nominal integrated luminosity of 75 ab^{-1}.

8. References

[1] SuperB Progress Reports: Physics, SuperB Collaboration, INFN/AE-10/2,LAL 10-110, arXiv:1008.1541[hep-ex]
[2] Suppression of beam-beam resonances in crab waist collisions, P. Raimondi et al., EPAC08-WEPP045, July 2008.
[3] Tests of crab-waist collisions at DAFNE Phi Factory, M. Zobov et al., PhysReV Lett.104:174801, 2010.
[4] SuperB Progress Reports: Accelerator, SuperB Collaboration,, arXiv:1009.6178(physics.acc-ph)
[5] M. Bona *et al.*, *SUPERB : A High-Luminosity Heavy Flavour Factory. Conceptual Design Re-*

port, arXiv:0709.0451v2 [hep-ex], INFN/AE-07/2, SLAC-R-856, LAL 07-15, also available at http://www.pi.infn.it/SuperB/CDR .

[6] SuperB Progress Reports: Detector, SuperB Collaboration, E. Grauges et al., INFN/AE-10/4,LAL 2010-115, arXiv:1007.4241[physics.ins-det]

Journal of Physics: Conference Series **378** (2012) 012029 doi:10.1088/1742-6596/378/1/012029

Nuclear modification factor for heavy flavors: An energy loss effect or more baryons than mesons?

Alejandro Ayala[1], Eleazar Cuautle[1], J. Magnin[2], Luis Manuel Montaño[3] and G. Toledo Sánchez[4]

[1]Instituto de Ciencias Nucleares, Universidad Nacional Autónoma de México, Apartado Postal 70-543, México Distrito Federal 04510, Mexico.
[2]Centro Brasileiro de Pesquisas Físicas, CBPF, Rua Dr. Xavier Sigaud 150, 22290-180, Rio de Janeiro, Brazil.
[3] Centro de Investigación y de Estudios Avanzados del IPN, Apartado Postal 14-740, México Distrito Federal 07000, Mexico.
[4]Instituto de Física, Universidad Nacional Autónoma de México, Apartado Postal 20-364, México Distrito Federal 01000, Mexico.

E-mail: ecuautle@nucleares.unam.mx

Abstract. The properties of the nuclear modification factor for heavy flavors are usually attributed to the energy loss suffered by the heavy quark propagating in a QCD plasma. Nevertheless it is a bit surprising that the suppression of this factor is as strong as the one suffered by light flavors. In this work we show that when accounting for the momentum shift associated to the opening of the recombination channel to produce hadrons in the QCD plasma, it is not necessary to invoke such a strong energy loss. We show that when the heavy baryon to meson ratio is larger in nuclear than in proton collisions, data from RHIC and LHC for the nuclear modification factor of electrons coming from heavy flavor decays as well as for charmed masons can be accounted for.

1. Introduction

The medium created during relativistic heavy-ion collisions has been a subject of intense studies since the early experiments at the GSI-SIS and the CERN-SPS. The heavy-ion program has been successfully pursued reaching the so called ultrarelativistic domain at RHIC and LHC. The results indicate that a new state of the matter has been created, where the fundamental QCD degrees of freedom are at play, the so called quark-gluon plasma. A paradigm for the interpretation of these results is the energy loss suffered by fast moving partons. This energy loss is often times associated to the nuclear modification factor (R_{AA}), defined as the ratio of single hadron transverse momentum spectra in heavy-ion collisions to the same quantity obtained in proton proton collisions, normalized to the number of binary collisions. First results on R_{AA} indicated a suppression for increasing transverse momentum (p_t). An intriguing result is that heavy flavor hadrons are equally as suppressed as light hadrons [1, 2]. Such behavior was first obtained from the analysis of electrons from the decay of heavy flavors and later confirmed from the analysis of charm mesons [3, 4]. The last results are preliminary and, with more statistics, they could be modified, although the overall behavior is not expected to change significantly.

Journal of Physics: Conference Series **378** (2012) 012029 doi:10.1088/1742-6596/378/1/012029

When the suppression of heavy flavors is only attributed to energy loss in the QCD medium, the above result is surprising for if the main contribution comes from radiative processes, the *dead cone effect* [5] should prevent heavy quarks from losing as much energy as light ones. This motivated the reviewing of energy loss scenarios to incorporate contributions from collisional processes, diffusion, geometry, as well as dynamical properties of the medium [6]. However, even these refined scenarios do not yet provide a fully convincing explanation for the properties of the heavy flavor R_{AA}.

Much less attention has been paid to the fact that a shift of the hadron momentum in the nuclear medium can come not only from a loss of energy but also from a momentum redistribution when the quarks from the medium form either baryons or mesons. This is the central idea behind the recombination/coalescence scenario as a new channel for hadron production in a heavy-ion environment [7].

The recombination model [8] was put forward to explain the production of mesons at intermediate p_t. Recombination becomes less important at large p_t due to the opening of the fragmentation channel with increasing p_t. Recently, the model has been modified accordingly to incorporate the description of hadron production in heavy-ion collisions. Among the results of this model are the description of spectra of identified hadrons [9] at LHC energies, as well as the baryon to meson ratio. Although the model works well, it depends on unknown experimental parameters like two or three-quark distribution functions, to make mesons or baryons respectively, in addition to other not well controlled approximations.

In order to overcome some of these limitations, in Refs. [10] we used Monte Carlo techniques to estimate the probability to form clusters of two and three quarks. The physical picture is that, in average, the three quarks forming a baryon come from lower momentum bins than the two quarks making up a meson. Since there are more quarks with lower momenta there is a larger chance to form baryons than mesons. A direct consequence of this momentum redistribution is an increase of the baryon to meson ratio in ion-ion (AA) with respect to proton-proton (pp) collisions. This ratio has been measured for a large variety of light and strange hadrons in high-energy nuclear collisions [11]. The upshot is that for intermediate transverse momenta, the ratio is enhanced in nuclear with respect to the corresponding one in proton collisions. Although no measurements exist for the case of heavy flavors, there are model calculations that describe under different approaches this enhancement [12, 13, 14].

In the present work we use the Dynamical Quark Recombination Model (DQRM) [10] to compute the heavy baryon to meson ratio which in turn is used to compute R^e_{AA} and R^D_{AA}. We show that when this increase is accounted for, only a moderate energy loss is needed to reproduce the data.

2. Model for nuclear modification factor

Measurements of the nuclear modification factor have been made for charged light hadrons and more recently for heavy flavor hadrons. The present work develops a model to quantify this nuclear modification factor. It relies on counting the hadron species containing a heavy flavor, when these hadrons are produced in AA and comparing this number when produced in pp. For definitiveness, let us concentrate on describing the nuclear modification factor for a single heavy flavor, say charm (c) quarks. Considering that the number of hadrons containing c-quarks produced in AA or pp collisions in a given momentum bin can be obtained from counting the number of open charm mesons ($N^D_{AA\ /\ pp}$), charm baryons ($N^\Lambda_{AA\ /\ pp}$) and hidden charm mesons ($N^{c\bar{c}}_{AA\ /\ pp}$), the follow relation is valid,

$$N^{H^c}_{AA\ /\ pp} = (N^D_{AA\ /\ pp} + N^\Lambda_{AA\ /\ pp} + N^{c\bar{c}}_{AA\ /\ pp}). \tag{1}$$

In the case of pp, we normalize these numbers to the average number of binary collisions, $\langle n_b \rangle$. A simple possibility to account for the shift in energy is considered by means of introducing the energy loss parameter ε. Then the number of produced hadrons containing charm in both environments will be related by

$$(N_{AA}^{D} + N_{AA}^{\Lambda} + N_{AA}^{c\bar{c}}) = \varepsilon\langle n_b\rangle (N_{pp}^{D} + N_{pp}^{\Lambda} + N_{pp}^{c\bar{c}}). \tag{2}$$

From Eq. (2), we can build the nuclear modification factor for D-mesons in terms of the number of charm mesons, obtaining (see Ref [15] for details),

$$R_{AA}^{D} \simeq \varepsilon \left(1 + \frac{N_{pp}^{\Lambda}}{N_{pp}^{D}}\right) \Big/ \left(1 + \frac{N_{AA}^{\Lambda}}{N_{AA}^{D}}\right). \tag{3}$$

where we have neglected the term describing the ratio of hidden charm to D mesons in pp collisions, $N_{pp}^{c\bar{c}}/N_{pp}^{D}$, since this ratio is very small. From this equation, we can see that even in the absence of energy loss ($\varepsilon = 1$) the nuclear modification factor for D mesons is smaller than one, provided the ratio of charm baryons to open charm mesons is enhanced in AA with respect to pp collisions.

In the case of non-photonic single electrons, the nuclear modification factor R_{AA}^{e} can be expressed as [13, 16],

$$\begin{aligned} R_{AA}^{e} &= \frac{1}{\langle n_b\rangle}\frac{N_{AA}^{\Lambda}B^{\Lambda\to e} + N_{AA}^{D}B^{D\to e}}{N_{pp}^{\Lambda}B^{\Lambda\to e} + N_{pp}^{D}B^{D\to e}} \\ &= \frac{1}{\langle n_b\rangle}\left(\frac{N_{AA}^{D}}{N_{pp}^{D}}\right)\left(\frac{B^{D\to e} + \frac{N_{AA}^{\Lambda}}{N_{AA}^{D}}B^{\Lambda\to e}}{B^{D\to e} + \frac{N_{pp}^{\Lambda}}{N_{pp}^{D}}B^{\Lambda\to e}}\right), \end{aligned} \tag{4}$$

where $B^{D,\Lambda\to e}$ is the branching ratio for the decay of D mesons and charm baryons into electrons, respectively. Using Eq.(3), we can write Eq. (4) as

$$R_{AA}^{e} \equiv \varepsilon\, T_{AA}^{e}, \tag{5}$$

where again the energy loss factor ε has been introduced and the contribution from hidden charm mesons was ignored (see Ref [15] for details). The function T_{AA}^{e} is given by

$$T_{AA}^{e} \equiv \left[\left(1 + \frac{N_{pp}^{\Lambda}}{N_{pp}^{D}}\right) \Big/ \left(1 + \frac{N_{AA}^{\Lambda}}{N_{AA}^{D}}\right)\right] \times \left(\frac{1 + xN_{AA}^{\Lambda}/N_{AA}^{D}}{1 + xN_{pp}^{\Lambda}/N_{pp}^{D}}\right) \tag{6}$$

where $x = B^{\Lambda\to e}/B^{D\to e}$

The result in Eq. 6 indicates that when $x < 1$, T_{AA}^{e} is also smaller than one when the ratio of charm hadrons to mesons is enhanced in AA with respect to pp collisions [13]. Therefore Eq. (5) states that R_{AA}^{e} is less than one, even in the absence of energy loss, since the electrons are more copiously produced from open charm mesons that baryons ($x < 1$).

3. Dynamical quark recombination model and nuclear modification factor

In general terms, recombination can be understood as a means to explain hadron production whereby after a heavy-ion reaction, several liberated partons, sharing a phase space cell transform into hadrons. However, this scenario has no quantitative significance without some meaningful input on the parton distributions at earlier times. On this bases, the recombination model, can be given quantitative meaning in terms of different hadronization scenarios. One of such is the dynamical quark recombination model (DQRM). This is thermal model that is

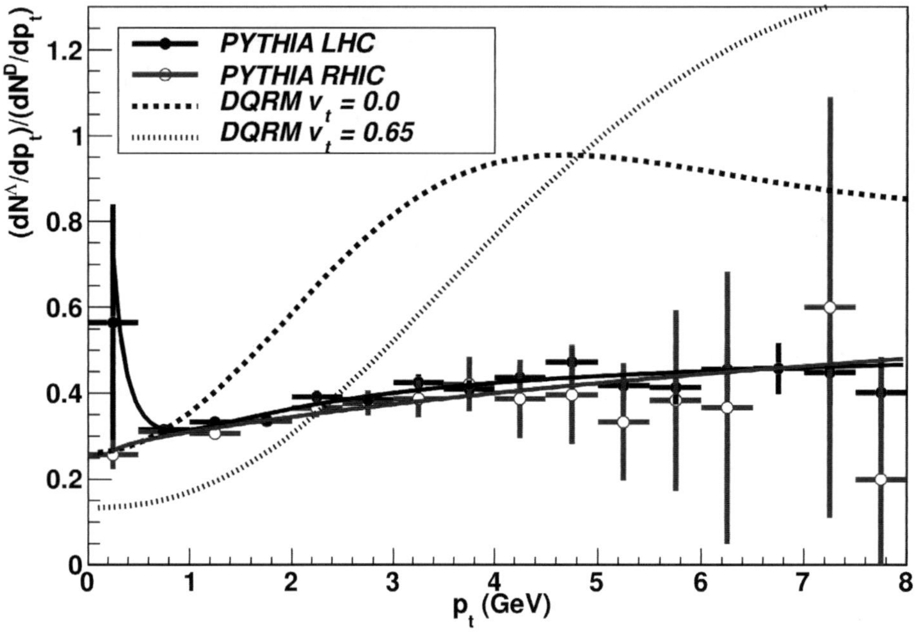

Figure 1. (Color on line) DQRM charm baryon to meson ratio in AA compared to the same ratio in pp collisions. The DQRM curves are computed for two transverse expansion velocities $v_t = 0$ and 0.65. For the ratio in pp we use PYTHIA simulations at $\sqrt{s_{NN}} = 200$ GeV and 2.7 TeV with 35×10^6 and 15×10^6 events, respectively. Shown are also fits to the simulations.

able to give hadron abundances. The model uses a Monte Carlo simulation to compute the probability $\mathcal{P}$ for quarks to coalesce and form cluster of two quarks (mesons) and of three quarks (baryons). This probability in turn depends on density and temperature and thus on the proper time hadronization. Once these probabilities are computed, they are shown to be larger for baryons than for mesons at low energy densities [13]. The relative population of one or the other kind of cluster at low densities can be fixed by combinatorial arguments, describing the baryon and meson production and its ratio at low and intermediate p_t (see [13, 10] for details). In order to estimate the hadron transverse momentum distribution in central AA, we take into account the space time evolution of the collision. Assuming Bjorken scenario which incorporates the fact that initially the expansion is along the beam direction, and considering the situation of central collisions, and assuming that there is not dependence of the particle yield on the transverse polar coordinates, the transverse momentum distribution can be written as: (see details in Ref [13])

$$\frac{dN}{p_t dp_t dy} = g\frac{m_t \Delta y}{4\pi}\frac{\rho_{\rm nucl}}{\Delta\tau}\int_{\tau_0}^{\tau_f} \tau d\tau \mathcal{P}(\tau) \times I_0(p_t \sinh\eta_t/T)e^{-m_t\cosh\eta_t/T}, \quad (7)$$

where v_t is the transverse velocity expansion, m_t is the transverse mass, Δy the rapidity interval, $\rho_{\rm nucl}$ the nuclear radius, $\Delta\tau = \tau_f - \tau_0$ the proper time interval and T the proper time dependent

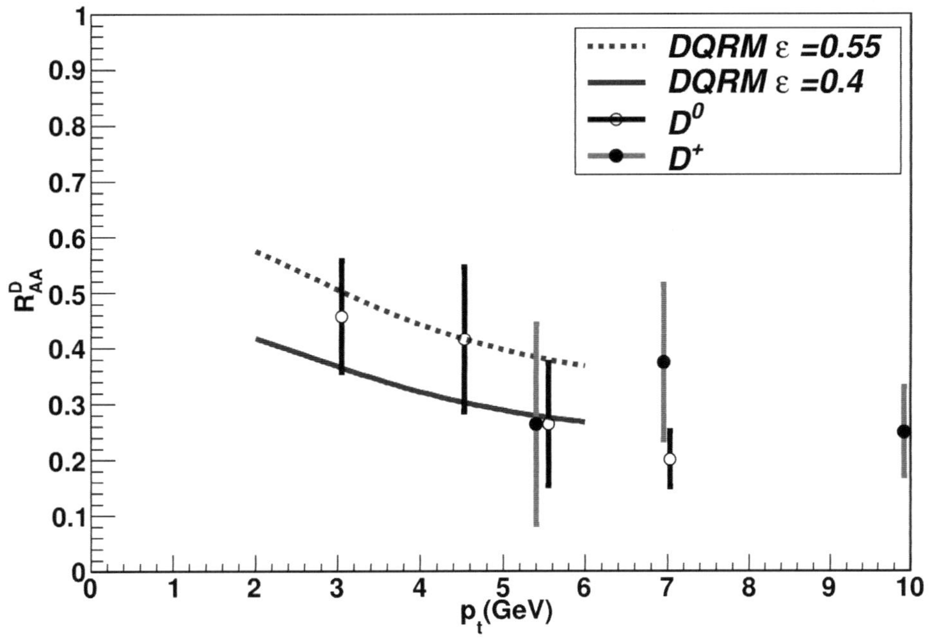

Figure 2. (Color on line) Nuclear modification factor for D mesons compared to ALICE data. The curves are computed using the DQRM with $v_t = 0.65$ as appropriate for LHC energies. The charm baryon to meson ratio in pp is taken from the PYTHIA simulation described in Fig. 1. For simplicity, the energy loss parameter is taken as two constant values, $\varepsilon = 0.55$ (upper curve) and $\varepsilon = 0.4$ (lower curve).

temperature

$$T = T_0 \left(\frac{\tau_0}{\tau}\right)^{v_s^2}, \tag{8}$$

with $v_s^2 = 1/3$. I_0 is a Bessel function I of order zero. v_t and η_t are related through $v_t = \tanh \eta_t$. g is the degeneracy factor that takes care of the spin degree of freedom.
This method to compute the transverse momentum distribution in AA collision is used to get the results of the next section.

4. Results

Figure 1 shows the DQRM charm baryon to meson ratio. We set the masses of the charm baryon and mesons to $m^\Lambda = 2.29$ GeV and $m^D = 1.87$ GeV. We take the initial hadronization proper time $\tau_0 = 1$ fm, at an initial temperature $T_0 = 175$ MeV and the final hadronization temperature $T_f = 100$ MeV that, according to Eq. (8), corresponds to $\tau_f = 8$ fm. Shown are the cases between $v_t = 0$ and $v_t = 0.65$. The values are in agreement with LHC results.
The baryon to meson ratio in pp at $\sqrt{s_{NN}} = 200$ GeV and 2.7 TeV, obtained from PYTHIA simulations with 35×10^6 and 15×10^6 events, respectively, is also shown. The fit of the simulation present not dependence on the energy, at RHIC and LHC seems to be almost the same, contrary

to different theoretical models which suggest that the maximum of the ratio is shifted to higher transverse momentum as the energy increase. The behaviors of the ratio simulated is that we expect, it present an enhancement in AA with respect to pp collisions.

Figure 2 shows R^D_{AA} compared to ALICE data [4]. The theoretical curves are computed using Eq. (3) with the heavy baryon to meson ratio obtained in AA from the DQRM with the same parameters as before and the particular value $v_t = 0.65$, which is a standard choice for the transverse expansion velocity at LHC energies. The heavy baryon to meson ratio in pp is obtained from the PYTHIA simulation shown in Fig. 1 for LHC energies. To see the effects of the energy loss parameter, for simplicity, we take two constant values, $\varepsilon = 0.55$ (upper curve) and $\varepsilon = 0.4$ (lower curve). We notice that even in this simple scenario, data are well described and the energy loss parameter does not need to be as small as in the case of light flavors, which in this language means $\varepsilon \simeq 0.22$, to account for the suppression in R^D_{AA}.

In fact, the experimental results show that the nuclear modification factor at low p_t is not suppressed. It may seem that our calculation does not properly describe this behavior. However, notice that the R^D_{AA} rise with decreasing p_t, is not an energy loss effect. This rise is known to be due to the Cronin effect, that is to soft multiple interactions in a nuclear environment with respect to pp collisions. There is also evidence that in this region, gluon shadowing also contributes to the rise of the yield.

Since our goal was to concentrate on effects at intermediate p_t, and not on a detailed model for R_{AA}, we did not attempt to address the low p_t region, given that there different effects from the one studied, are at work.

Figure 3 shows R^e_{AA} compared to data from STAR [2] and ALICE [4]. The theoretical curves are computed using Eq. (5) with the heavy baryon to meson ratio obtained in AA from the DQRM and in pp from the PYTHIA simulations of Fig. 1. To account for the finding that electrons from heavy flavor decays come almost in equal proportions from the decays of charm and beauty hadrons for $p_t \geq 5$ GeV [17], here we consider a single species of heavy baryons and mesons with effective masses. We take $m^D = 3.57$ GeV, the average between the masses of the D^0 and the B^0 mesons and $m^\Lambda = 3.95$ GeV, the average between the masses of the Λ_c and the Λ_b. Also, we consider that the possible charm and beauty mesons decaying into electrons or positrons are $D^\pm$ ($B^{D^\pm \to e^\pm} = 16\%$), D^0, $\bar{D}^0$ ($B^{D^0,\ \bar{D}^0 \to e^\pm} = 6.53\%$), $D^\pm_s$ ($B^{D^\pm_s \to e^\pm} = 8\%$) and $B^\pm$ ($B^{B^\pm \to e^\pm} = 10.8\%$), B^0, $\bar{B}^0$ ($B^{B^0,\ \bar{B}^0 \to e^\pm} = 10.1\%$). The possible charm and beauty baryons decaying into electrons or positrons are Λ_c, $\bar{\Lambda}_c$ ($B^{\Lambda_c,\ \bar{\Lambda}_c \to e^\pm} = 4.5\%$), Λ_b and $\bar{\Lambda}_b$ ($B^{\Lambda_b,\ \bar{\Lambda}_b \to e^\pm} = 5.35\%$) (the experimentally reported branching ratio corresponds to the semileptonic decay $\Lambda_b \to \Lambda_c\ l\ \bar{\nu}_l$. Here we consider that half of this comes from the decay into electrons). The other parameters used for the AA case are as before, with $v_t = 0.55$ for the RHIC case and $v_t = 0.65$ for the LHC case. For simplicity, the energy loss parameter is taken also as two constant values $\varepsilon = 0.55$ (upper curves) and $\varepsilon = 0.4$ (lower curves). Once again, even in this simple scenario, data are well described for $p_t \geq 2$ GeV and the energy loss parameter does not need to be as small as in the case of light flavors to account for the suppression in R^e_{AA}. For $p_t \leq 2$ GeV, the rise in the data is usually attributed to other effects like shadowing, which is not considered in our approach. We have checked that the model results are not affected if the effective masses used are slightly varied.

5. Conclusions

In summary, we have shown that considering a momentum redistribution of quarks in a medium, caused by their recombination to form mesons and baryons, the nuclear modification factor for heavy flavors can be described without the need of a large energy loss. This momentum redistribution is manifested through the heavy baryon to meson ratio. We have shown that an increase of this ratio in nuclear, with respect to proton collisions is directly related to the suppression of the nuclear modification factor. This result is not model depent, since the effect

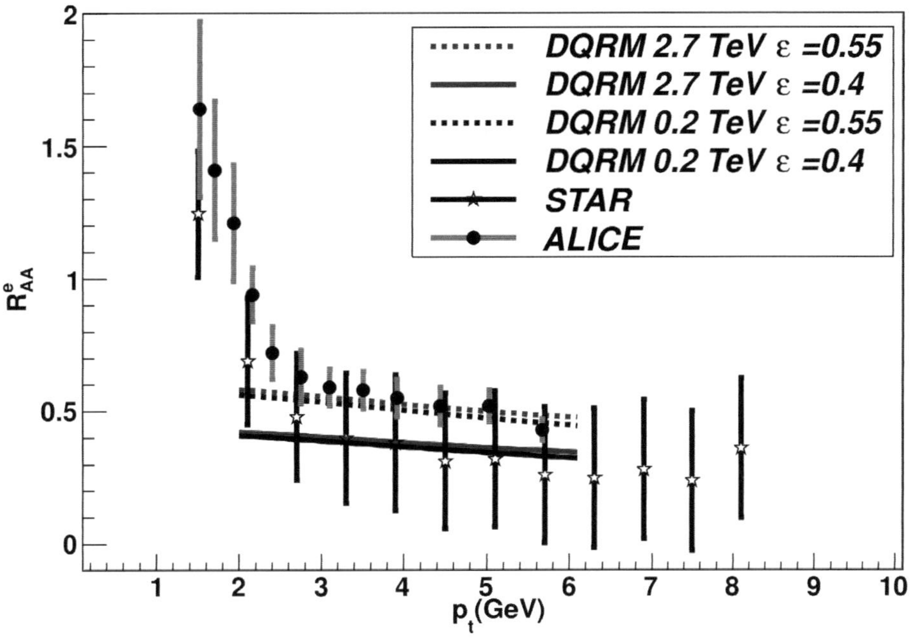

Figure 3. (Color on line) Nuclear modification factor for non-photonic single electrons compared to STAR and ALICE data. The curves are computed using the DQRM with $v_t = 0.65$ for the LHC case and $v_t = 0.55$ for the RHIC case. The charm baryon to meson ratio in pp is taken from PYTHIA simulations for $\sqrt{s_{NN}} = 200$ GeV and $\sqrt{s_{NN}} = 2.7$ TeV for the sum of charm and beauty hadrons. For simplicity the energy loss parameter is taken as two constant values, $\varepsilon = 0.55$ (upper curves) and $\varepsilon = 0.4$ (lower curves).

will be present whenever the baryon to meson ratio increases in nuclear with respect to proton collisions. Details on this increase are of course model dependent. Given that we are dealing with a hadronization process, which belongs to the realm of non-perturbative effects, we have resorted to using the dynamical quark recombination model to make explicit calculations. We expect that this model works at intermediate p_t. At low p_t the problem should take into account Cronin and shadowing effects, while at high momentum, fragmentation should be the dominant effect. The energy loss considered in the present work was introduced as a constant parameter, with the purpose to better illustrate the effect of a baryon to meson ratio enhancement. The baryon to meson ratio has a dependence on collective phenomena such as transverse flow, which is known to make the momentum distribution for heavier hadrons to fall less steeply than for lighter ones. Since we want to keep matters as simple as possible, we did not address all the above mentioned issues, concentrating only on the increase of the baryon to meson ratio and the probability to form clusters of two and three quarks within a thermal environment. Whether the observed trend of baryon to meson ratio increase is also present in data for heavy flavors will ultimately be determined by the upcoming data from the LHC.

Acknowledgments
Support for this work has been received in part by CONACyT (Mexico) under grant numbers 128534 and 101597 and PAPIIT-UNAM under grant numbers IN103811 and IN107911.

References

[1] S.S. Adler *et al.*, (PHENIX Collaboration), Phys. Rev. Lett. **96**, 032301 (2006); A. Adare *et al.* (PHENIX Collaboration), Phys. Rev. Lett. **98**, 172301 (2007); Y. Pachmayer (for the ALICE Collaboration) J. Phys. **G 38**, 124186 (2011).

[2] B.I. Abelev *et al.* (STAR Collaboration), Phys. Rev. Lett. **98**, 192301 (2007).

[3] A. Adare *et al.* (PHENIX Collaboration), Phys. Rev. Lett. **98**, 162301 (2007).

[4] A. Dainese (for the ALICE Collaboration), J. Phys. **G 38**, 124032 (2011).

[5] Y. L. Dokshitzer and D. E. Kharzeev, Phys. Lett. B **519**, 199 (2001); B.-W. Zhang, E. Wang, and X.-N. Wang, Phys. Rev. Lett. **93**, 072301 (2004).

[6] H. van Hees and R. Rapp, Phys. Rev. C **71**, 034907 (2005); H. van Hees, M. Mannarelli, V. Greco and R. Rapp, Phys. Rev. Lett **100**, 192301 (2008). M. Djordjevic, M. Gyulassy, S. Wicks, Eur. Phys. J. C **43**, 135 (2005); G. D. Moore and D. Teaney, Phys. Rev. C **71**, 064904 (2005); N. Armesto, M. Cacciari, A. Dainese, C. Salgado and U. Wiedemann, Phys. Lett. B **637**, 362 (2006); M. Djordjevic, Phys. Rev. C **74**, 064907 (2006); Nucl. Phys. A **783**, 197c (2007); Nucl. Phys. A **830**, 163c (2009); S. Wicks, W. Horowitz, M. Djordjevic and M. Gyulassy, Nucl. Phys. A **784**, 426 (2007); S. Wicks and M. Gyulassy, J. Phys. G **34**, (2007) S989; M. Djordjevic, and U. Heinz, Phys. Rev. Lett. **101**, 022302 (2008); A. Ayala, J. Magnin, L. M. Montaño, and E. Rojas, Phys. Rev. C **77**, 044904 (2008); B. Z. Kopeliovich, I. K. Potashnikova, and I. Schmidt, Phys. Rev. C **82**, 037901 (2010); J. Uphoff, O. Fochler, Z. Xu and C. Greiner, Phys. Rev. C **84**, 024908 (2011); W. M. Alberico, A. Beraudo, A. De Pace, A. Molinari, M. Montero, M. Nardi and F. Prino, Eur. Phys. J. C **71**, 1666 (2011).

[7] R. C. Hwa and C. B. Yang, Phys. Rev. C **67**, 034902 (2003); R. J. Fries, B. Muller, C. Nonaka, and S. A. Bass, Phys. Rev. Lett. **90**, 202303 (2003); V. Greco, C. M. Ko, and P. Levai, ibid. **90**, 202302 (2003); R.C. Hwa and C.B. Yang, Phys. Rev. C **70**, 024904 (2004).

[8] K.P. Das and Rudolph Hwa, Phys. Lett. B 68B, 459 (1977).

[9] Rudolph C. Hwa and Lilin Zhu, Phys. Rev.C 84, 064914 (2011).

[10] A. Ayala, M. Martinez, G. Paic, and G. T. Sánchez, Phys. Rev. C **77**, 044901 (2008).

[11] S.S. Adler *et al.* (PHENIX Collaboration), Phys. Rev. C **69**, 034909 (2004); J. Adams *et al.* (STAR Collaboration) *Measurements of identified particles at intermediate transverse momentum in the STAR experiment from Au+Au collisions at* $\sqrt{s_{NN}}$ = 200 *GeV*, nucl-ex/0601042; A. László and T. Schuster (NA49 Collaboration), Nucl. Phys. A **774**, 473 (2006); B.I. Abelev *et al.* (STAR Collaboration), Phys. Lett. B **655**, 104 (2007).

[12] Y. Oh, C. M. Ko, S. H. Lee, and S. Yasui, Phys. Rev. C **79**, 044905 (2009). However see also Y. Oh, C. M. Ko, Phys. Rev. C **79**, 067902 (2009).

[13] A. Ayala, J. Magnin, L. M. Montaño, and G. Toledo Sánchez, Phys. Rev. C **80**, 064905 (2009).

[14] I. Bautista and C. Pajares, Phys. Rev. C **82**, 034912 (2010).

[15] Alejandro Ayala, Eleazar Cuautle, J. Magnin, Luis Manuel Montano, G.Toledo Sanchez, *Heavy flavor nuclear modification factor: more baryons than mesons less energy loss* (arXiv:1110.4587 [nucl-th]).

[16] P. R. Sorensen and X. Dong, Phys. Rev. C **74**, 024902 (2006); G. Martinez-Garcia, S. Gadrat, and P. Crochet, arXiv:hep-ph/0702035; G. Martinez-Garcia, S. Gadrat, and P. Crochet, Phys. Lett. B **663**, 55 (2008); [Erratum-*ibid.* B **666**, 533 (2008)].

[17] M. M. Aggarwal *et al.* (STAR Collaboration), Phys. Rev. Lett. **105**, 202301 (2010).

Journal of Physics: Conference Series **378** (2012) 012030 doi:10.1088/1742-6596/378/1/012030

Recent results from the LHCb experiment

E. Polycarpo

LAPE/Instituto de Fsica, Universidade Federal do Rio de Janeiro, CP 68528, Rio de Janeiro, 21941-972, Brasil

E-mail: poly@if.ufrj.br

Abstract. The LHCb experiment was presented with emphasis on the performance during the 2010 and 2011 runs. In addition, a selection of the first few competitive physics results were presented.

1. Introduction

The LHCb experiment is dedicated to the study of heavy flavour physics at the Large Hadron Collider (LHC), being designed to perform precision measurements of charge-parity (CP) violation and rare decays involving transitions of the b and c quarks. The violation of the CP symmetry is one of the ingredients necessary for the generation of the baryon asymmetry observed in the universe [1]. Within the Standard Model (SM), CP is only violated in the flavor changing processes and all the CP violation observables can be written in terms of only 4 parameters, the parameters of the Cabibbo-Kobayashi-Maskawa (CKM) matrix, making the model highly predictive. However, the amount of CP violation provided by the CKM mechanism is much too tiny to explain the observed baryon asymmetry [2], therefore new sources of CP violation are needed and might show up in flavor changing processes. These indirect searches for new physics are complementary to the direct searches performed at the general purpose experiments and are sensitive to higher energy scales.

In order to identify these new sources of CP violation, an excellent control of the SM contributions is needed. In the last decade, huge progress has been achieved on this field by the B factories and the Tevatron experiments CDF and D0 and also by lattice QCD calculations. This progress can be illustrated by the evolution of the constraints imposed by measurements to the sides and angles of one of the CKM unitarity triangles, from 2001 to 2011, shown in Fig. 1 [4], which also show constraints due to observations in the kaon system. The agreement and consistency of the SM predictions with measurements is remarkable. Nevertheless, it is still possible that new physics (NP) phenomenae contribute to $b \to s$ transitions, predicted to be small in the SM, and experimentally not probed with the same precision as the $b \to d$ transitions. Thanks to the large b and c cross-sections in high energy proton collisions and a dedicated trigger, LHCb will make important contributions to this sector.

With the first data collected at a centre-of-mass energy of 7 TeV at the LHC, LHCb published many results on heavy flavor production, including some first observations. The focus of this presentation was on the detector performance and on a few of the first competitive results on CP violation and rare decays.

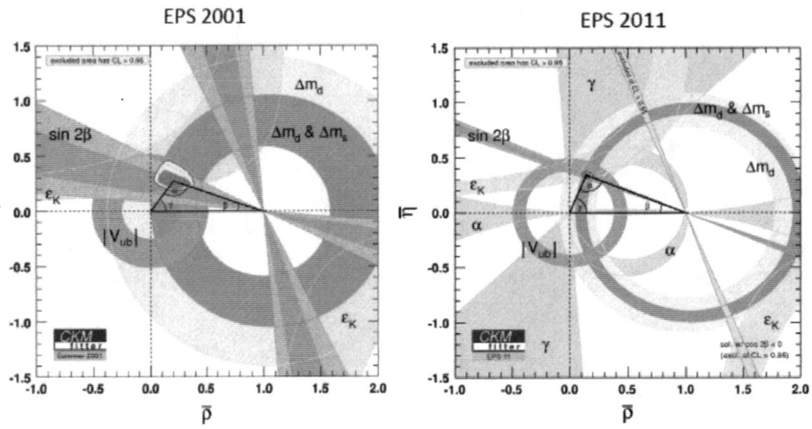

Figure 1. Evolution of the constraints to the unitarity triangle sides and angles from 2001 (left) to 2011 (right), from [4] (updated results and plots available at `http://ckmfitter.in2p3.fr/`).

2. LHCb detector and performance

The LHCb detector [3] is a single-arm forward spectrometer (see Fig. 2) covering the region $(1.9 \leq \eta \leq 4.9)$ where the $b\bar{b}$ (and $c\bar{c}$) production is peaked. It incorporates precision vertexing and tracking systems, particle identification over a wide momentum spectrum and relies on relatively soft transverse momentum triggers, efficient for both leptonic and purely hadronic decays of the B and D mesons.

In 2011, the LHC machine provided proton-proton collisions at a center-of-mass energy of 7 TeV and increasing luminosity. LHCb can work at constant luminosity independently from the other intersections, thus optimizing the number of interactions per bunch crossing for collecting clean events. In spite of running at an instantaneous luminosity of about 3.5×10^{32} cm^{2}s^{-1} and an average number of interactions per crossing of μ ~1.5, about twice the original planned instantaneous luminosity and four times the expected μ, the experiment performed very well during this period.

A very high precision measurement of the primary and secondary vertices is essential for the determination of the B or D meson decays and to resolve the fast B_s oscillations. A very good momentum resolution is also crucial to the obtention of good mass resolutions and background discrimination. These measurements are performed by the micro-strip silicon vertex detector (VELO) and the tracking system. The latter consists of one silicon micro- strip detector, named the Tracker Turicensis (TT), located in front of the spectrometer magnet and three tracking stations downstream of the magnet. The stations are composed of a silicon micro- strip detector in the inner parts (IT) and of straw-tubes in the less occupied outer parts (OT).

The VELO, the highest resolution vertex detector at the LHC, is a retractible device that surrounds the collision region and is closed at every beam injection. The VELO sensors achieve a single hit resolution of 4 μm in the direction transverse to the beam for the smallest pitch region and optimal angle. A primary vertex resolution of about 16 μm in X/Y and 76 μm in Z has been obtained for 25 track vertices, see plot of Fig. 3(a). The corresponding proper-time resolution of about 50 fs allows for an accurate determination of the B-lifetimes (of about 1500 fs). Another important performance figure for a B-physics experiment is the impact parameter resolution which is found to be 14.4 μm + 19.5/p_T μm (preliminary determination).

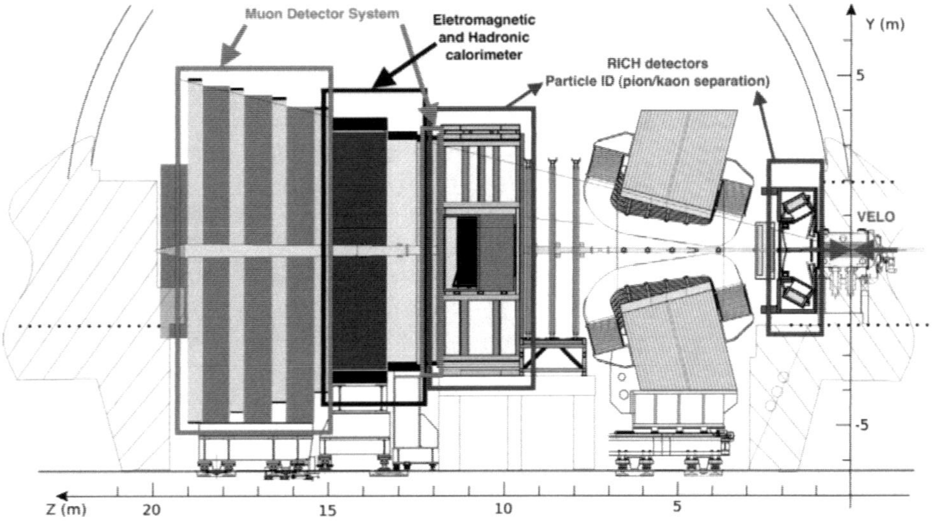

Figure 2. Schematic overview of the LHCb detector

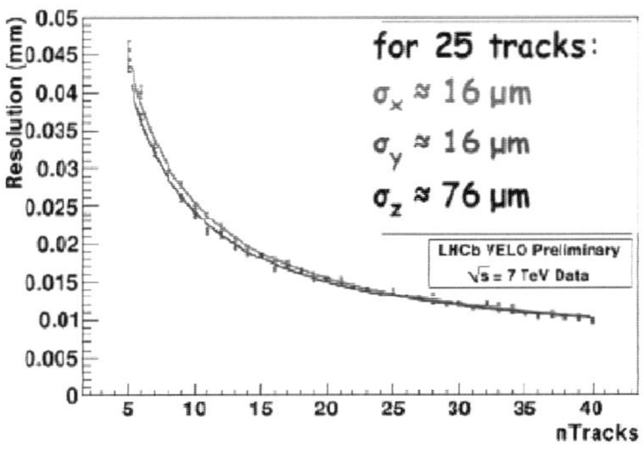

(a)

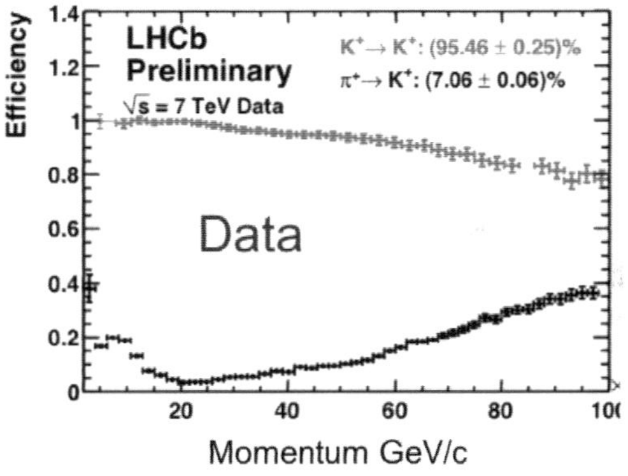

(b)

Figure 3. Primary vertex resolution as a function of the number of tracks (a) and efciency and purity of the Ring Imaging Cherenkov system (b).

The tracking system together with the 4 Tm magnet provide a momentum resolution of 0.3% to 0.5%. Continuous improvements in the alignment of the tracking stations during the 2010 data taking have resulted in significant improvements in the mass resolution, resulting in for example a J/ψ mass resolution of 13 MeV/c^2.

Another important requirement for LHCb analyses is the particle identification, which is provided by the Ring Imaging CHerenkov (RICH), the calorimeters and the muon sub-detectors. Hadron identification, provided by the RICH system, in particular the ability to distinguish kaons and pions, is crucial to many LHCb analyses, particularly where the final states of interest are purely hadronic. The calorimeters provide identification of electrons, photons and hadrons in addition to the measurement of their energies and positions. As well as being part of the

Journal of Physics: Conference Series **378** (2012) 012030 doi:10.1088/1742-6596/378/1/012030

LHCb trigger, the muon system provides identification of muons to a very high level of purity, fundamental for many CP-sensitive measurements that have J/Ψ in their final states and also for those involving rare leptonic or semileptonic decays.

In order to achieve π/K separation over the extensive momentum range of $\sim$ 1 to 100 GeV/c, two separate RICH detectors are employed, using three separate radiators: aerogel, C_4F_{10} and CF_4 . The aerogel radiator of RICH-1, located upstream of the LHCb dipole magnet, is composed of 5 cm thick tiles arranged around the LHC beam pipe. Located directly behind the aerogel is $\sim$1 m of C_4F_{10}. Together, the radiators of RICH-1 provide PID for tracks from 1 to approximately 60 GeV/c. RICH-2, located downstream of the magnet, contains the CF_4 gas radiator and provides PID over the momentum range of approximately 50 to 100 GeV/c. The arrangement of optics is similar in both sub-detectors; spherical focusing mirrors project the Cherenkov photons onto a series of flat mirrors which then reflect them onto a set of photon detector arrays, located outside the detector acceptance. The photon detector used is the Hybrid Photon Detector (HPD). Angular resolutions of 1.62 and 0.62 mrad have been achieved for the RICH1 and RICH2, respectively. These numbers are very close the values obtained from Monte Carlo simulations. The efficiency and purity of π/K identification over the entire momentum range is shown in Fig. 3(b), which are also in good agreement with simulation.

The LHCb Calorimetry system (CALO) takes the classical form of an electromagnetic calorimeter (ECAL) followed by a hadron calorimeter (HCAL) and is located downstream of RICH-2. To help in the distinction of $e^{\pm}$ from the overwhelming background of π^0 and $\pi^{\pm}$ mesons, longitudinal separation of the EM showers is needed. This is achieved by using two additional detectors in front of the ECAL: a Scintillator Pad (SPD) and Pre-Shower (PS) detector. The ECAL is a shashlik type sampling calorimeter of thickness 25 X_0, composed of 66 alternating layers of lead absorber and scintillator, achieving an energy resolution of $\frac{\sigma_E}{E} = \frac{9\%}{\sqrt{E}} \oplus 0.8\%$. The HCAL, also a sampling type detector, is composed of alternating layers of iron and scintillator with a resolution of $\frac{\sigma_E}{E} = \frac{69\%}{\sqrt{E}} \oplus 9\%$. Thanks to the good performance of the CALO system, LHCb has reconstructed 1599 $B \to {K^*}^0\gamma$ decays and 210 $B_s^0 \to \phi\gamma$ decays in the first 340 pb^{-1} of pp collisions recorded in 2011 (see Fig. 4), allowing the measurement of the ratio of the branching fractions, found to be $\frac{\mathcal{B}(B\to {K^*}^0\gamma)}{\mathcal{B}(B_s^0\to\phi\gamma)} = 1.52 \pm 0.14(stat) \pm 0.10(syst) \pm 0.12(fs/fd)$ [5]. This measurement is in good agreement with previous measurements and show the potential of the experiment to make important contributions to the study of these radiative $b \to s\gamma$ transitions, which may have its dynamics modified by new virtual heavy particles propagating within the loop.

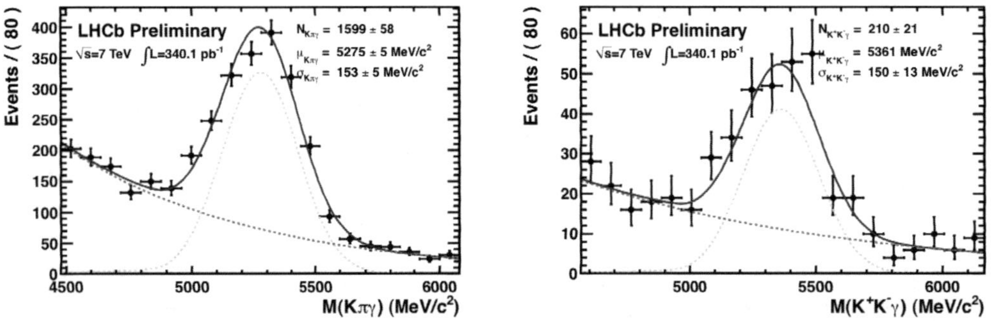

(a) (b)

Figure 4. Invariant mass of $B \to {K^*}^0\gamma$ (a) and $B_s^0 \to \phi\gamma$ (b) candidates.

The muon system is composed of five stations of Multi-Wire Proportional Chambers (MWPCs), labelled M1-M5, positioned around the beam axis. Stations M2 to M5 are located downstream of the calorimeters, with 80 cm thick iron plates interspersed between each. These iron plates act as absorbers to reduce any hadronic background that survives past the calorimeters. The first station, M1, sits immediately in front of the calorimeters. Due to the high particle fluxes experienced around the inner region of this upstream station, a technology with extended longevity is used: triple-GEM (Gas Electron Multiplier) detectors. Based on a sample of $J/\psi \rightarrow \mu^+\mu^-$ decays the efficiency has been determined at $97.3 \pm 1.2\%$ which is in close agreement with the MC expectations. Incorrect identification rates of kaons and pions as muons are at the level of 1% and below that for particles with momentum above 10 GeV/c. An illustration of the muon identification performance is given by the clear peaks corresponding to the different di-muon resonances J/ψ, $\psi(2S)$, $\Upsilon(1S)$, $\Upsilon(1S)$ and $\Upsilon(1S)$ shown in the mass spectra of Fig. 5.

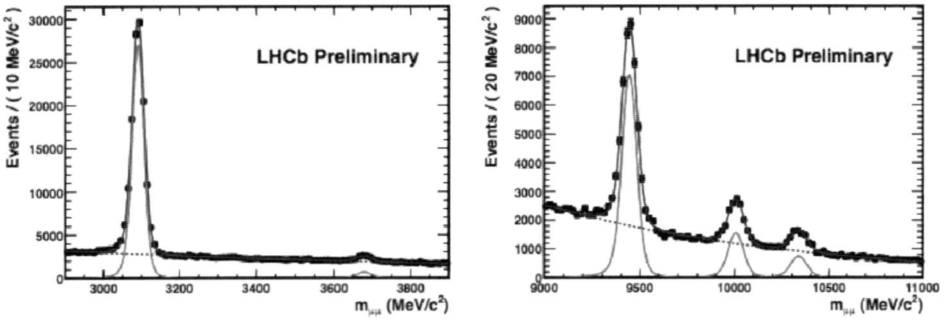

Figure 5. Di-muon invariant mass spectrum in the ranges (2.9 - 3.9) GeV/c^2 (left) and (9-11) GeV/c^2 (right).

3. Search for the rare decays $B^0_{d,s} \rightarrow \mu^+\mu^-$

The search for the rare decays $B^0_{s,d} \rightarrow \mu^+\mu^-$ is one of the most promising ways to test the SM. Within this model, the branching ratios of these flavour changing neutral current transitions, which occur only at loop level, are helicity suppressed and are computed with good precision: $\mathcal{B}(B^0_s \rightarrow \mu^+\mu^-) = (3.2 \pm 0.2) \times 10^{-9}$ and $\mathcal{B}(B^0_d \rightarrow \mu^+\mu^-) = (0.010 \pm 0.001) \times 10^{-9}$ [6]. Large enhancements are possible in many variants of SuperSymmetry and alternative new physics models.

Before the 2011 summer, the best published limits from the Tevatron at 95% CL were obtained using 6.1 fb^{-1} by the D0 collaboration [7], and using 2 fb^{-1} by the CDF collaboration[8]. The CDF collaboration had also presented a preliminary result[9] with 6.9 fb^{-1} in which an excess of $B^0_s \rightarrow \mu^+\mu^-$ candidates was reported, compatible with a $\mathcal{B}(B^0_s \rightarrow \mu^+\mu^-) = (1.8^{+1.1}_{-0.9}) \times 10^{-8}$. The LHCb Collaboration had obtained the limits $\mathcal{B}(B^0_s \rightarrow \mu^+\mu^-) < 5.4 \times 10^{-8}$ and $\mathcal{B}(B^0_d \rightarrow \mu^+\mu^-) < 1.5 \times 10^{-8}$ at 95% CL based on 37 pb^{-1} of luminosity collected in the 2010 run [10]. At the summer conferences, LHCb presented a measurement based on 300 pb^{-1} of integrated luminosity collected between March and June 2011 [11] (accepted for publication in Physics Letters B).

The LHCb analysis is done by classifying selected candidates in bins of a 2D parameter space made by the invariant mass and a multivariate classifier (Boosted Decision Tree - BDT) which condensates geometrical and kinematical information of the event. The signal and background expectation in each bin is calculated using data from control channels. The signal BDT shape is obtained from $B^0_{d,s} \to h^+h^-$ free from trigger biases, whilst the background expectation is calculated by interpolating from mass sidebands. The parameters describing the invariant mass line shape of the signal are obtained from data: the average mass values are extracted from $B^0_{d,(s)} \to K^+\pi^-(K^+K^-)$ samples and the resolutions are estimated by interpolating the ones obtained with the dimuon resonances J/ψ, $\psi(2S)$ and $\Upsilon(1S,2S,3S)$ (see Fig. 5). The number of expected signal events is then obtained by normalizing to channels of known branching fractions: $B^+ \to J/\psi K^+$, $B^0_s \to J/\psi\phi$ and $B^0 \to K^+\pi^-$, that are selected with a strategy as similar as possible to the signal. The peaking background yield is calculated using $K \to \mu$ and $\pi \to \mu$ misidentification rates obtained from data using decays such as $\Lambda \to p\pi^-$ and $D^0 \to K^+\pi^-$.

The signal and background expectations are compared with the distribution of observed events, and the limits are set using the CL_s method [12]. The number of observed events for the $B^0_s \to \mu^+\mu^-$ decay is in good agreement with the background expectation and the SM signal branching fraction. The upper limits obtained after combination with the results obtained in 2010 are [11]

$$\mathcal{B}(B^0_s \to \mu^+\mu^-) < 1.2(1.4) \times 10^{-8} \text{at } 90\%\ (95\%)\ \text{CL},$$
$$\mathcal{B}(B^0_d \to \mu^+\mu^-) < 2.6(3.2) \times 10^{-9} \text{at } 90\%\ (95\%)\ \text{CL}.$$

In Fig. 6 the luminosity needed for a 3σ evidence as a function of the $\mathcal{B}(B^0_s \to \mu^+\mu^-)$ is shown. Approximately 2 fb^{-1} are needed to reach the SM prediction [13]. As an example of how the upper limits on $\mathcal{B}(B^0_s \to \mu^+\mu^-)$ impose constraints on extensions of the SM, it is shown in Fig. 7 the constraints imposed by upper limits down to $2\times$ SM level in the region around the current NUHM1 best fit point [14]. In this figure, tanβ is the ratio of the vacuum expectation values of the two neutral CP-even Higgs fields and M_A is the pseudoscalar Higgs mass.

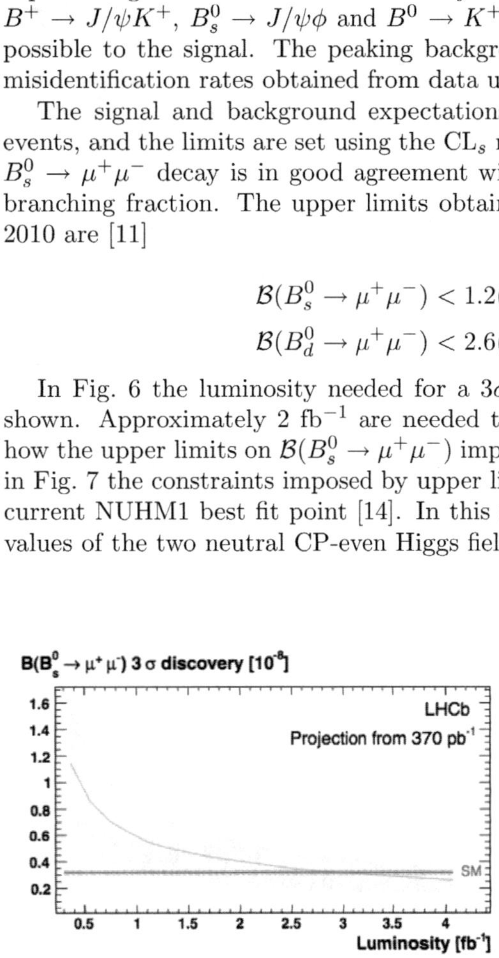

Figure 6. Luminosity needed in order to get a $B^0_s \to \mu^+\mu^-$ 3σ evidence.

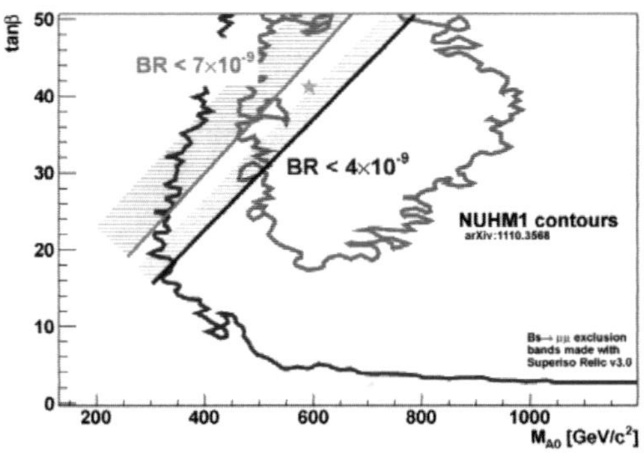

Figure 7. Constraints imposed by upper limits on $\mathcal{B}(B^0_s \to \mu^+\mu^-)$ at the 10^9 level on the region around the minimum of the NUMH1 fit (from [14]).

4. Measurement of the CP violation phase ϕ_s

The measurement of ϕ_s, the relative phase between the direct decay amplitude and the amplitude of decay via mixing in the process $B_s \to J/\psi\phi$, is one of the key measurements for LHCb. It gives rise to time-dependent CP violation, a difference in the proper decay time distribution of

Journal of Physics: Conference Series **378** (2012) 012030 doi:10.1088/1742-6596/378/1/012030

B and $\bar{B}$ decays. In the SM, it is predicted to be $\phi_s = -2\beta_s = arg(-V_{ts}V_{tb}^*/V_{cs}V_{cb}^*)$, neglecting penguin contributions. Global fits to experimental data give a small and precise value $2\beta_s = (0.0363 \pm 0.0017)$ rad [4]. New particles could contribute to the mixing box diagram modifying the SM prediction, by adding a new phase. Before the 2011 summer conferences, the results from the experiments at Tevatron were both above the SM prediction [15, **?**], although not with enough precision to rule out the SM value. The values were compatible with the 68% confidence intervals obtained by LHCb with the first 36 pb^{-1} of data collected in 2010 and, when combined with the semi-leptonic asymmetry measurement by D0 [17] could be considered as a possible hint for NP.

The latest results reported by LHCb [18] were obtained with $\sim$340 pb^{-1}, from which $\sim$ 8300 candidates are selected. The fit to the decay time distributions provide a simultaneous measurement of the phase ϕ_s, of the average lifetime Γ_s and of $\Delta\Gamma_s$, the difference between the decay widths of the mass eigenstates. The measurement requires several inputs. The oscillation frequency Δm_s is obtained from an unbinned likelihood fit to the mass and proper time distribution of about 9200 $B_s^0 \to D_s^- \pi^+$ candidates [19], $\Delta m_s = (17.725 \pm 0.041 \pm 0.026)$ ps^{-1}. The proper time resolution is measured by using J/ψ originated directly from the interaction point and the acceptance is determined from the relative efficiency of events passing a lifetime unbiased selection and a lifetime biased selection. Since the final state of the decay $B_s^0 \to J/\psi\phi$ of a pseudo-scalar to two vector mesons is not a CP eigenstate, an angular analysis is needed to disentangle the CP odd and even states. The angular acceptances for the three transversity angles ϕ, θ and ψ are determined from simulation and cross-checked in the decay $B_d^0 \to J/\psi K^*$. The flavour of the B_s^0 meson is determined by algorithms exploiting several properties of each event. In order to enhance the tagging power $\epsilon(1-2\omega)^2$, where ϵ is the tagging efficiency and ω is the probability of incorrect tagging (mistag), a per-event mistag probability is used in the fits. This mistag probability is calibrated on data, using $B^+ \to J/\psi K^+$ events and validated on $B_d^0 \to J/\psi K^*$. The results from the unbinned maximum likelihood fit of the mass, proper time and angular distributions are given in Fig. 8. The signal sample contains a 4% contribution from non-resonant s-wave decays, which is also taken into account in the fit.

The ϕ_s and $\Delta\Gamma_s$ values extracted from the fit are $\phi_s = (0.13 \pm 0.18 \pm \pm 0.07)$ rad, $\Gamma_s = (0.656 \pm 0.009 \pm 0.008)$ ps^{-1} and $\Delta\Gamma_s = (0.123 \pm 0.029 \pm 0.011)$ ps^{-1}. The dominant sources of systematic uncertainties are the background description and the angular acceptances. These are the most precise measurements of ϕ_s and γ_s and the first significant measurement of a non-zero $\Delta\Gamma_s$ value, all compatible with SM predictions, as can be seen from Fig. 9.

The decay $B_s^0 \to J/\psi f_0$, first observed by LHCb, is also sensitive to the phase ϕ_s. Since the final state is a pure CP odd state, no angular analysis is needed. The value of Γ_s is given as input to the unbinned maximum likelihood fit to the mass and proper time distributions, which provides $\phi_s = (-0.44 \pm 0.44 \pm 0.02)$ rad [21]. A first naive combination of this result with the ϕ_s obtained from the $B_s^0 \to J/\psi\phi$, assuming Gaussian errors, yields $\phi_s = (0.03 \pm 0.16 \pm 0.07)$ rad [22].

5. Search for direct CP violation in the decay $D^+ \to K^- K^+ \pi^+$

Within the SM, hadronic charm decays occur dominantly via tree level amplitudes. Since all quarks building up initial and final states belong to the first two generations, the transitions are governed by a 2×2 Cabibbo quark mixing matrix, which is real. CP violating penguin contributions are strongly suppressed and therefore, CP violation is expected to be very small [23]. For D^+ decays, CP asymmetries predictions are maximal for singly cabibbo suppressed modes and are of the order of 10^{-3} [24]. Any observation larger than this would be an unambiguous sign for new physics.

LHCb performed a model independent search for local CP asymmetries across the Dalitz Plot (DP), based on the work in [25], using a sample that corresponds to about 35 pb^{-1} collected in

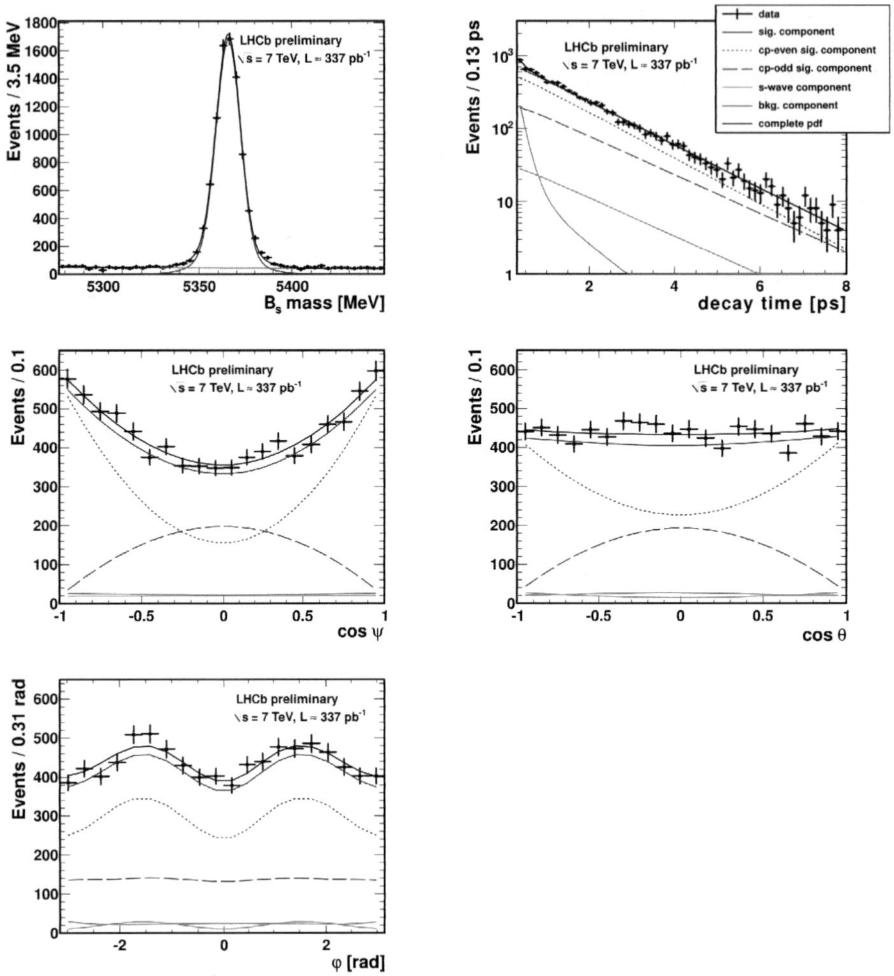

Figure 8. Fit projections to invariant mass (top left), proper time (top right) and angular distributions.

2010 [26]. Approximately 400,000 $D^{\pm} \to K^{\mp}K^{\pm}\pi^{\pm}$ decays have been selected for this analysis. The basic strategy is to divide the D^{+} and D^{-} Dalitz plots into bins and for each bin compute the anisotropy variable $S^{i}_{CP} = \frac{N^{i}_{+} - N^{i}_{-}}{\sqrt{N^{i}_{+} + N^{i}_{-}}}$, where N^{i}_{+} and N^{i}_{-} are the number of D^{+} and D^{-} in bin i. The DP are normalized to the same number of entries to eliminate global asymmetries. In the absence of local asymmetries, the distribution of the S^{i}_{CP} is a Gaussian centered at zero and with width $\sigma = 1$. Deviations from this distribution would be the evidence for CP violation. A χ^2 test is also used to evaluate the compatibility with the hypothesis of null local asymmetry, where $\chi^2 = \sum_i (S^{i}_{CP})^2$. The number of degrees of freedom is the number of bins minus one, due to the normalisation. The p-value calculated from this gives the probability of observing a χ^2 greater than or equal to that observed under the hypothesis of no CP-violation. Several arrangements of bins are tested, uniform and adaptative, determined from pseudo-experiments in order to optimize the sensitivity to different sources of CP violation. The main advantage

Journal of Physics: Conference Series **378** (2012) 012030 doi:10.1088/1742-6596/378/1/012030

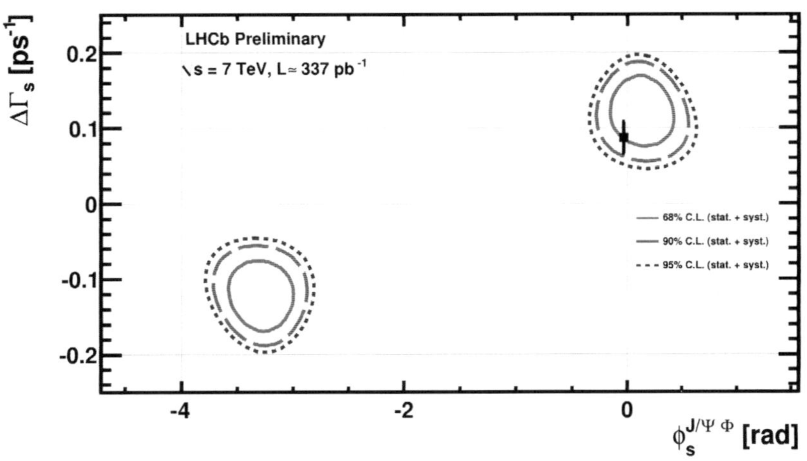

Figure 9. Constraints of LHCb's measurement on the ϕ_s–$\Delta\Gamma_s$ plane from $B_s^0 \to J/\psi\phi$.

of this method is to draw a map of the regions in the Dalitz plot in which CPV occurs. The method, however, is not capable to quantify the strenght of CPV. For that one needs to perform a full Dalitz plot analysis.

The technique relies on careful accounting for local asymmetries that could be induced by sources such as the different production mechanisms for D^+ and D^-, differences in the K-nucleon inelastic cross-section, differences in the reconstruction or trigger efficiencies, left-right detector asymmetries, etc. The existence of these local asymmetries are investigated using the Cabibbo favoured control channels $D_s \to K^-K^+\pi^+$ and $D^+ \to K^-\pi^+\pi^+$. The 3-hadron invariant mass distributions for both signal and control channel candidates is shown in Fig. 10. The method is also applied to the events in the side bands shown in Fig. 10 to investigate possible asymmetries that could arise due to the background contamination. All these tests are fully consistent with no asymmetry, thus the method is determined to be very robust against systematic effects. No evidence for CP violation was found in the decay $D^+ \to K^-K^+\pi^+$ and p-values ranging from 4% to 99% have been observed for the different binning schemes tested. In particular, for a uniform grid of 530 equaly sized bins and an adaptative configuration of 106 bins of sizes optimized to have about the same number of events, the p-values obtained are 10.6% and 65%. The S_{CP} values across the DP are shown in Fig. 11. The corresponding S_{CP} distributions are shown in Fig. 12, with a behavior perfectly compatible to normal Gaussians.

This analysis will be performed with the 1 fb^{-1} dataset collected in 2011, from which up to 10 M $D \to K^{\mp}K^{\pm}\pi^{\pm}$ are expected to be selected. Moreover, additional singly and doubly Cabibbo suppressed modes will be available which might be sensitive to different NP effects.

6. Conclusions

Thanks to the large b and c cross-sections at the LHC and to the excellent performance of the machine and the experiment itself, LHCb could already perform competitive measurements involving rare decays and CP violation. No deviation from the SM was observed, but there is still room for NP observations, since the precision is still dominated by statistical uncertainties. Moreover, the experiment has many other studies on the physics programme, several of them under preparation with the whole data collected in 2011.

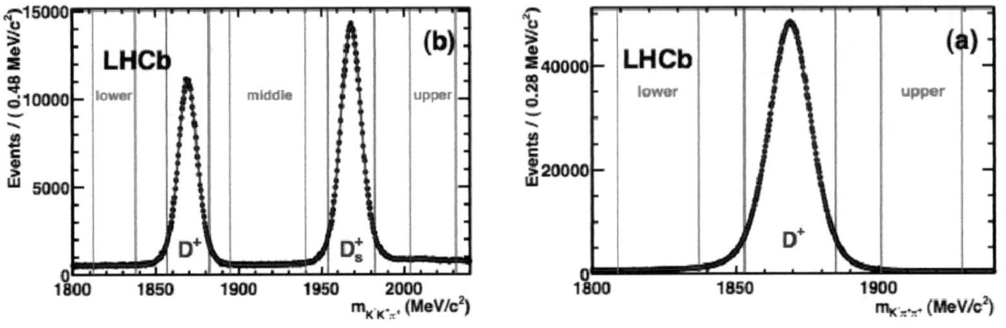

(a) (b)

Figure 10. Mass spectra of $K^{\mp}K^{\pm}\pi^{\pm}$ (a) and $K^{\mp}\pi^{\pm}\pi^{\pm}$ (b) candidates.

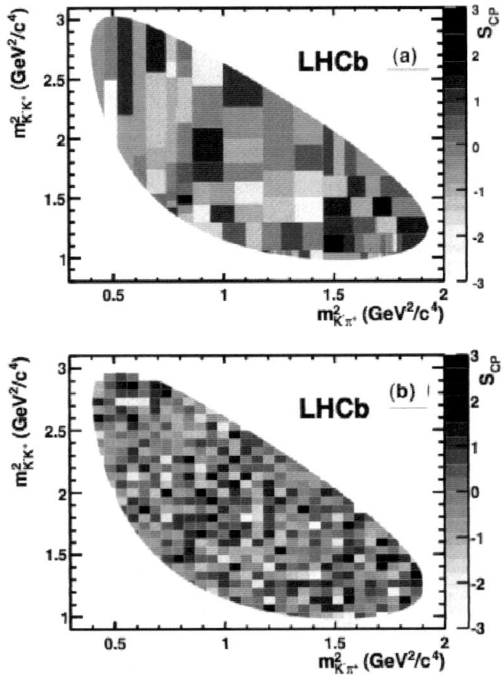

Figure 11. Distribution of S_{CP} in the Dalitz plot for (a) *adaptative* binning and (b) uniform grid of 530 bins.

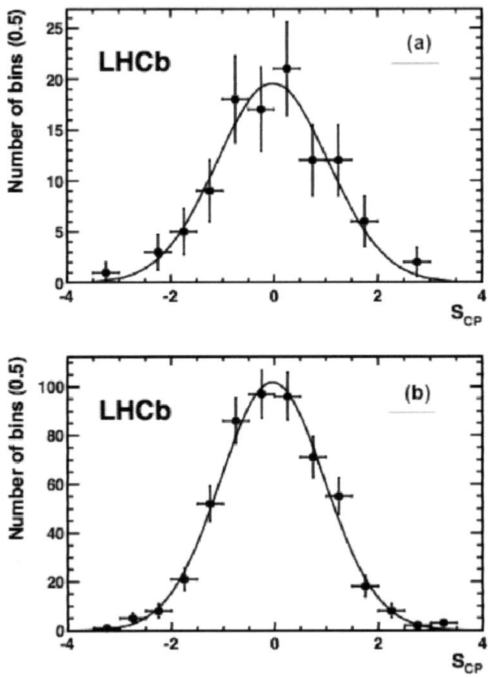

Figure 12. S_{CP} distributions fitted to Gaussian functions for (a) *adaptative* binning and (b) uniform grid of 530 bins.

7. Acknowledgments

I would like to thank the organizers for the invitation, in particular Dr. W. Bietenholz, and Dr. G. Contreras for the support to my participation on the workshop. I also want to thank the whole LHCb team for a warm and fructiferous collaboration for more than a decade now. My research activities at LHCb are supported by CAPES, CNPq, FAPERJ and FINEP.

References

[1] A. D. Sakharov, *JETP Lett.* **5** 24 (1967).
[2] M. B. Gavela *et al.*, *Nucl. Phys. B* **430** (1994) 382.
[3] A. Alves *et al.* [LHCb Collaboration], *JINST* **3** (2008) S08005.
[4] J. Charles et al. (CKMfitter group), *Eur.Phys.J.* **C**41, 1-131(2005), *Preprint* hep-ph/0406184.
[5] R. Aaiji *et al.* [LHCb Collaboration], LHCb-CONF-2011-055.
[6] A.J. Buras, G.Isidori and P.Paradisi, *arXiv:1007.5291i* [hep-ph] (2010).
[7] V. Abazov *et al.* [D0 Collaboration],*Phys. Lett.* B **693** (2010) 539.
[8] T. Aaltonen *et al.* [CDF Collaboration],*Phys. Rev. Lett.* B **100** (2008) 101802.
[9] T. Aaltonen *et al.* [CDF Collaboration],*arXiv:1107.2304v1* [hep-ex] (2011).
[10] R. Aaiji *et al.* [LHCb Collaboration],*Phys. Lett.* B **699** (2011) 330.
[11] R. Aaiji *et al.* [LHCb Collaboration],*Preprint* arXiv:1112.1600 [hep-ex] (2011).
[12] A. Read, *J. Phys. G* **28** (2002) 2693.
[13] D. Martinez Santos, *Preprint* arXiv:1201.5359 [hep-ex] (2011).
[14] O. Buchmueller *et al.*, *Preprint* arXiv:1110.3568 [hep-ph] (2011).
[15] A. Abulencia *et al.* [CDF Collaboration], CDF Public Note 10206.
[16] V. M. Abazov *et al.* [D0 Collaboration], D0 Conference note 6098-CONF (2010).
[17] V. M. Abazov *et al.* [D0 Collaboration], *Preprint* arXiv:1005.2757v1 [hep-ex] (2010).
[18] R. Aaiji *et al.* [LHCb Collaboration], LHCb-CONF-2011-049.
[19] R. Aaiji *et al.* [LHCb Collaboration], LHCb-CONF-2011-050.
[20] R. Aaiji *et al.* [LHCb Collaboration], *Phys. Lett. B***698** (2011) 115.
[21] R. Aaiji *et al.* [LHCb Collaboration], LHCb-CONF-2011-051.
[22] R. Aaiji *et al.* [LHCb Collaboration], LHCb-CONF-2011-056.
[23] A. A. Petrov, *Preprint* arXiv:1101.3822 [hep-ph] (2011).
[24] F. Bucella *eal.*, *Preprint* arXiv:9212253 [hep-ph] (1992).
[25] I. Bediaga *et al.*, *Phys. Rev. D* **80** (2009) 096006.
[26] R. Aaiji *et al.* [LHCb Collaboration], *Phys. Rev. D***84** (2011) 112008.

Journal of Physics: Conference Series **378** (2012) 012031 doi:10.1088/1742-6596/378/1/012031

Search for Z rare decays on CDF: $Z^0 \rightarrow J/\psi\gamma$ and $Z^0 \rightarrow \Upsilon\gamma$

Fernanda Psihas, S. Carillo, J. Morales, F. Vazquez
Universidad Iberoamericana, Mexico D.F., Mexico

P. Murat
Fermi National Accelerator Laboratory, Batavia, Illinois 60510

E-mail: psihas@fnal.gov

Abstract. We present status of a search for two rare of the Z boson, $Z^0 \rightarrow J/\psi\gamma$ and $Z^0 \rightarrow \Upsilon\gamma$ at CDF. The state of analysis is described at the time of this worksop.

1. Introduction

Fermilab (Fermi National Accelerator Laboratory) is a High Energy Physics Laboratory located in Batavia, Illinois, United States. The highest energy Fermilab accelerator, the Tevatron, collides protons and antiptotons at energies of up to 1.96 TeV. In this process, conditions similar to those of the very early universe are reproduced.

One of the interaction points of the particle beams is the location of the CDF (Collider Detector at Fermilab) Experiment. Its goal is to discover the identity and the properties of the elementary particles that form the universe and carry the fundamental forces. CDF is a general purpose detector that allows the identification of different particles, products of the interaction and some of their characteristics. The search for rare decays in this work is done using data from the CDF Run II detector (collected from 2001 to 2011).

1.1. Understanding the CDF Detector

In order to work with data from the detector, an understanding of it is required. Some of its characteristics establish observational limitations. Understanding its behavior helps us determine the expected distributions of important traits in the particles we want to observe.

The CDF detector is divided into layered components. Each layer detects distinct characteristics of the particles. Combining the data obtained from each layer allows us to properly identify particles.

Journal of Physics: Conference Series **378** (2012) 012031 doi:10.1088/1742-6596/378/1/012031

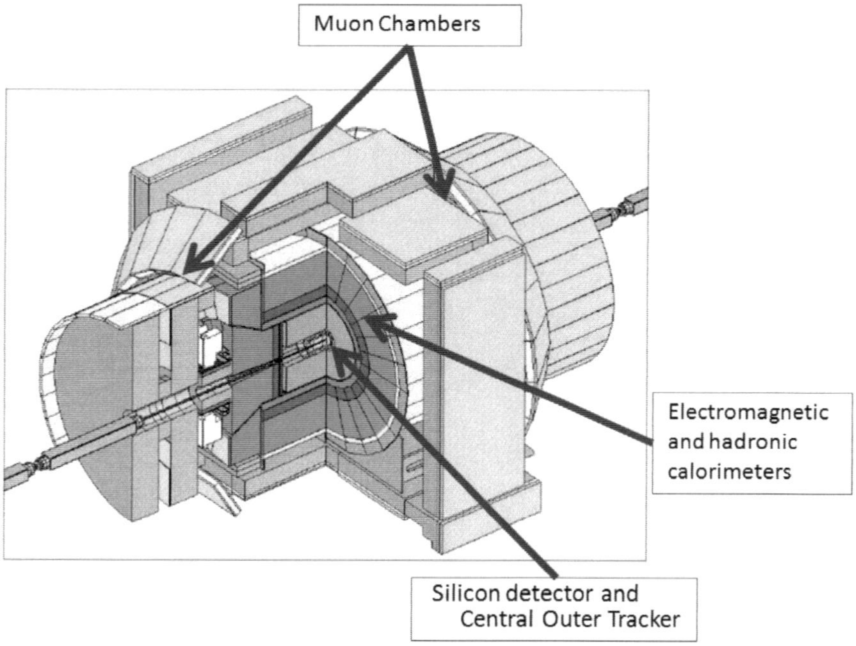

Figure 1. Diagram of the CDF detector.

Layers of the CDF detector

- *SVX (Silicon Vertex Tracker)*: Detects only charged particles as they ionize the silicon. It can also determine whether a track comes directly from the collision point.

- *COT (Central Outer Tracker)*: Detects the position of the charged particles. A magnetic field forces particles to deviate from their original trajectories, allowing this part of the detector to also determine their momentum.

- *Electromagnetic Calorimeter*: Consists of lead plates, alternated with scintillating material. The amount of light emitted is related to the energy the particle deposits in the material. This calorimeter detects this amount of energy deposited, which helps us identify different particles like electrons and photons.

- *Hadronic Calorimeter*: Consists of iron plates, alternated with scintillating material. It works under the same principle as the electromagnetic calorimeter. In this calorimeter we many observe the energy deposited by hadrons and other particles, given that all the other particles tend to deposit all their energy in the electromagnetic calorimeter.

- *Muon Chambers*: Most hadrons do not reach these chambers. In other words, any particle that reaches this part of the detector can be considered a good muon candidate, provided that it satisfies other requirements.

1.2. Particle decays

The decay channels that we will search for are the following rare decays of the Z boson:

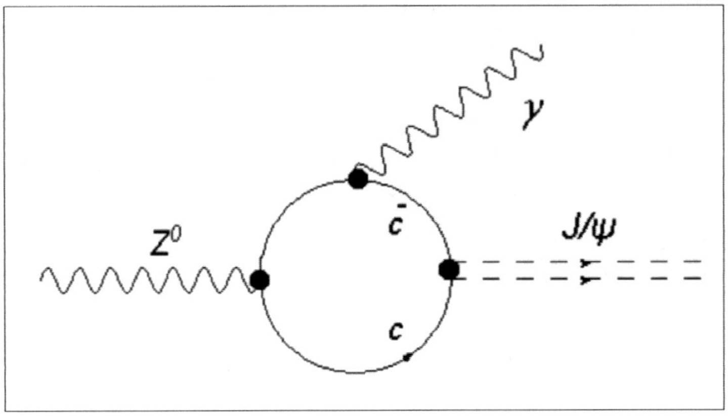

Figure 2. Feynman diagram for the $Z^0 \to J/\psi\gamma$ decay

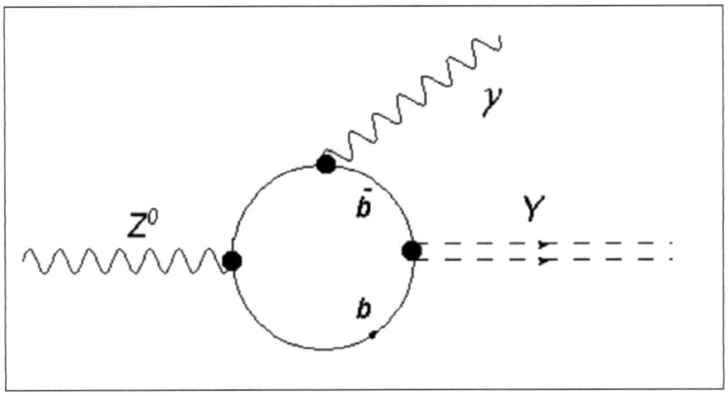

Figure 3. Feynman diagram for the $Z^0 \to \Upsilon\gamma$ decay

This search is done for an energetic photon and a pair of oppositely charged muons in the final state for both decays.
The branching ratios have been predicted theoretically [2] to be within the following ranges:

$$B(Z^0 \to J/\psi\gamma) = [1.1 \times 10^{-5}, 1.64 \times 10^{-7}]$$

$$B(Z^0 \to \Upsilon\gamma) = [1.28 \times 10^{-4}, 4.27 \times 10^{-7}]$$

Journal of Physics: Conference Series **378** (2012) 012031 doi:10.1088/1742-6596/378/1/012031

1.3. Physics Motivation

Number of the W and Z boson decays decays which have have been studied experimentally is suprizingly small [3]. Among yet unobserved decay modes, the radiative decays $V \to P\gamma$, where V is a Z or a W boson and P is a light pseudoscalar boson, could have branching ratios close to the limit of the experimental sensitivity of the Tevatron experiments. The most recent Tevatron searches for $W^{\pm} \to \pi^{\pm}\gamma$ decay [1] have significantly improved previously existing limits. However no experimental searches for the $Z^0 \to J/\psi\gamma$ or $Z^0 \to \Upsilon\gamma$ decays have been published. Together with the clean experimental signatures of these decays, this gives a strong motivation to search for them. It is also worth mentioning that the background and trigger considerations could favor searches for the rare radiative decays of W and Z bosons at the Tevatron, compared to LHC.

2. Analysis Strategy

The search for these rare decays must begin with a search for the particles in their final states. A photon and a pair of oppositely charged muons must be identified from the data. In order to do this, we must determine how they interact with the detector and how that behavior will manifest itself in the data that we analyze.

The first step towards identifying the particles is deciding which dataset should be used. The data, coming from the detector is stored in different datasets depending on particular characteristics of the data. When the characteristics of a particle observed in the detector coincide with the expected behavior of a specific particle, it is stored in a dataset associated with it. For example, if a particle in an event is identified to be a good photon candidate the information for that event will be stored in a set called *photon-triggered dataset1*. Because the high Pt photon trigger identification efficiency is high, we have chosen to use datasets with this trigger for the analysis. With this selection, only the muon identification is required.

In order to identify muons from the data, we must establish certain requirements or cuts that events must meet in order to be considered a good J/ψ or Υ candidate. The process to establish these cuts requires us to study the decay through simulations. Once cuts are established we can take a better look at the data and search for the expected decay. Finally, the branching ratio of the decay will be calculated if the decay is found, or an upper limit to it in case it is not found. The steps in the methodology have been split into different core activities. They will be described in the following sections.

2.1. Monte Carlo Event Generation

Monte Carlo (MC) simulation of the signal processes is important to reach a better understanding of the behavior we will expect from them. It helps us to take a closer look at the distribution of different variables, allowing us to get familiarized with the decay mode and specific traits of the outcome. We also use simulations to calculate the signal acceptance and study the efficiencies of different cuts.

The MC events are generated with PYTHIA event generator, customized to include the radiative decays of the Z boson into heavy quarkonia. The generated events are simulated in the CDF detector using the GEANT3-based simulation code. We use the MC samples to optimize the analysis selection cuts. Several examples are shown in the following figures.

Acceptance of the muon chambers in limited to $-1.5 < \eta < 1.5$, as shown in figure 4. Only particles in that area will be detected. Knowing this, we show the fraction of the generated J/ψ that will reach the detector in black. A distribution for their Pt (transversal momentum) is shown in figure 5.

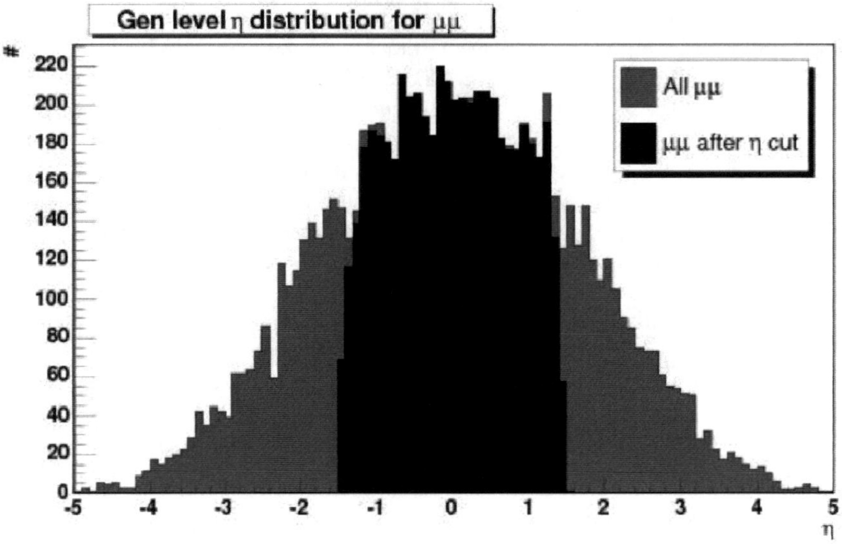

Figure 4. Distribution of the quantity η for every pair of muons in events generated by Pythia. In blue: The entire distribution. In black: The events for which $-1.5 < \eta < 1.5$ for both muons.

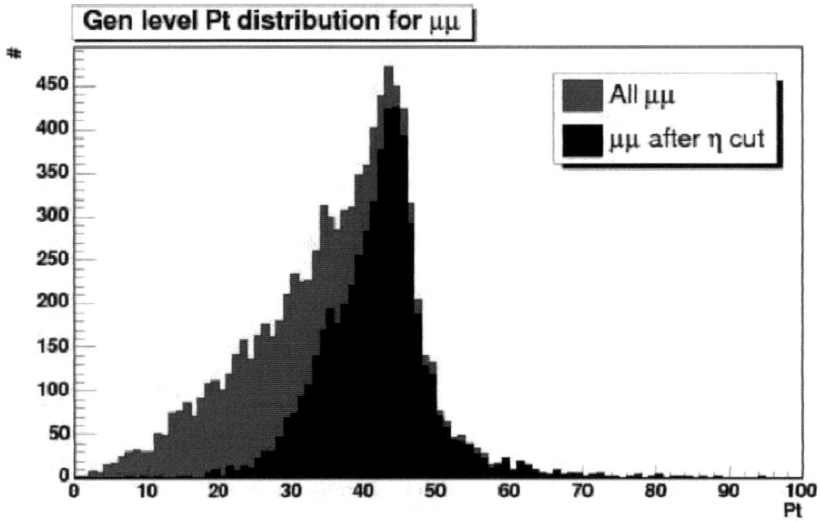

Figure 5. Distribution of transverse momentum for every pair of muons in events generated by Pythia. In blue: The entire distribution. In black: The events for which $-1.5 < \eta < 1.5$ for both muons.

From this we can observe that the efficiency of the Pt $>$ 15 GeV cut is close to 100%. As a consequence, we choose to reject or cut out of our data sample all muon candidate pairs which do not have this minimum Pt value

Journal of Physics: Conference Series **378** (2012) 012031 doi:10.1088/1742-6596/378/1/012031

2.2. Establishing cuts

Based on what we learned from the simulations, we established the necessary cuts that would be made to the data from the detector in order to have a clean signal from the particles of interest. It is important to note that for every pair of muon candidates in the data, different cuts are made to each of the two. We call the leading muon the one to which stricter, or tight cuts will be made. The following is a list of the cuts made.

- *Leading Muon Chamber Stubs:* This cut requires all leading muon candidates to have been detected by the CMX, CMUP or BMU muon chambers.

- *Subleading Muon Chamber Stubs:* This cut requires all leading muon candidates to have been detected by any of the muon chambers.

- *Pt* > 3*:* Transversal momentum is required for the second muon due to an observation made on the simulations. A peak is observed in 5 at Pt of around 2 GeV. We assume this peak to be background from hadrons so we want to reject it.

- *Energy deposition cut:* Requires all muon candidate to have the following minimum values of deposited energy in the electromagnetic and hadronic calorimeter:

$$E_{em} < 2 + 0.0115(p - 100)GeV \qquad E_{had} < 6 + 0.028(p - 100)GeV$$

  When a relativistic particle passes through matter, it ionizes the atoms or molecules it encounters, losing energy. The mean rate at which it loses its energy depends of the kind of particle it is and its momentum. If a particle loses a very small amount of energy through matter, it is called a minimum ionizing particle or MIP. Given that muons are MIPs, we expect their deposited energy to be low.

- *COT track quality cuts:* The COT is segmented into 8 super layers, alternating stereo and axial, with an angle difference of 2 degrees. Each super layer has 12 sense wires which detect passing particles. A minimum number of hits is required on each type of layer to ensure track quality.

$$NAxialHits > 20 \qquad NStereoHits > 20$$

- $\chi^2 < 4$ *per degree of freedom:* This is a cut related to track reconstruction quality. χ^2 is the average deviation of a detected track hits from its reconstructed track. It is normalized over the COT resolution so that for a perfectly modeled track, the distribution for χ^2 would peak at $\chi^2 = 1$.

- *Distance to interaction point* $DZ0 < 5cm$*:* Reconstructed tracks of the particles are required to be close to the interaction point to ensure that they come from the collision.

- *Dimuon* $Pt > 15$ *GeV:* This requirement comes from the distribution observed in figure5 from generator level data.

- *Dimuon Isolation* < 0.1*:* This is another cut related to data quality. It is required that the energy coming from the muon pair in a part of the detector has a more significant value than energy coming from any other particle. Energy deposited by other particles in a defined area around the dimuon must not be greater than 0.1 of the total energy in that area.

Figure 6. $\mu\mu\gamma$ invariant mass distribution in the data. The data in this histogram has passed all the previously mentioned selection cuts

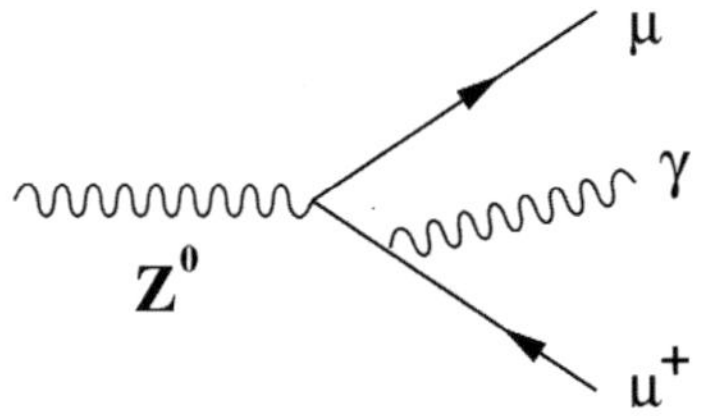

Figure 7. Example decay diagram.

The events in figure 6 have passed the selection cuts, however, there are still some that are shown in that histogram which are not of our interest. Figure 7 is an example of a decay that does not get filtered with any of the cuts applied to the data, so far. In order to filter out these background from the signal we want to look for, an additional selection cut was needed. With this purpose the 3 dimensional angle distribution for the dimuons was examined.
The histogram in figure 8 shows what is already implicit from the decay of the energetic Υ meson: The muons product of this decay are separated by a small angle. Based on this an extra requirement was decided upon:

- *3D angle between muons > 60 degrees:* This requirement is established due to the expected distribution for the angle of separation of the muons coming from the J/ψ or Υ decay.

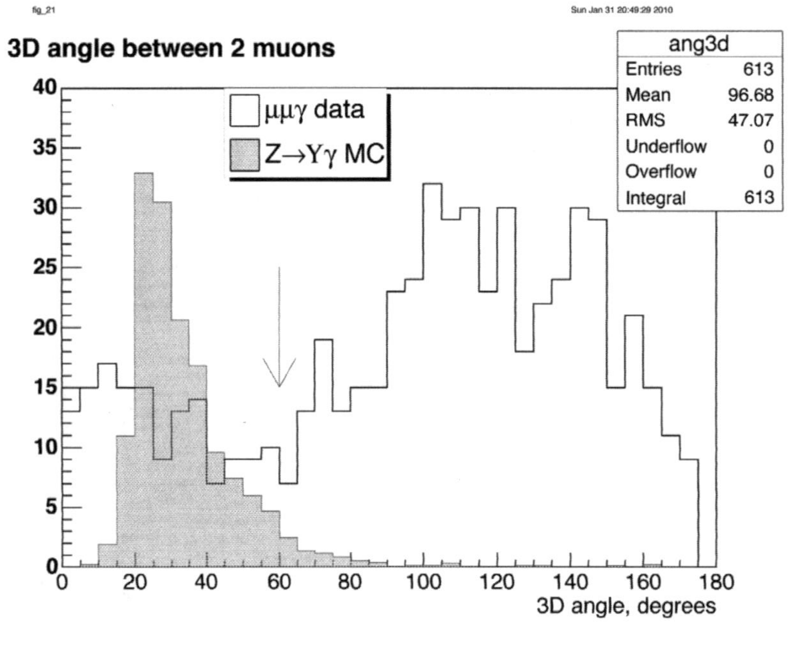

Figure 8. Angular distribution in 3D for the pairs of muons. Brown: $Z^0 \to \Upsilon\gamma$ MC. White: data.

2.3. *Full Simulation*

While the events generated by Pythia may describe the behavior of the particles, that alone does not give us enough information about the signal. With the geometry of the detector and the read out process

Journal of Physics: Conference Series **378** (2012) 012031 doi:10.1088/1742-6596/378/1/012031

involved it is hard to determine what percentage of the events generated in a collision will be lost with the cuts and what percentage will be lost in the detector. We use a full CDF detector simulation program to simulate and study the performance and efficiency of the detector. This helps us achieve a better understanding of our expected signal. The generated events by Pythia are run though a detector simulation. This introduces numerical boundary conditions of the detector to the data. It models the interaction of the particles with the detector. The output contains simulated digitized hits, as if that information was coming from the detector itself. This output is then run through the offline reconstruction code, as would be done with real detector data. The final result is a sample of simulated events with information given exactly as it would be for real events observed by the CDF detector.

The analysis code, which has in it the selection cuts, is then run in the MC (MonteCarlo) events. The outcome can be studied to observe the behavior of the expected signal in detail. After running this code in the MC for $Z^0 \to J/\psi\gamma$ and $Z^0 \to \Upsilon\gamma$ events, we looked into the reconstructed invariant mass distributions that is shown in figure 9:

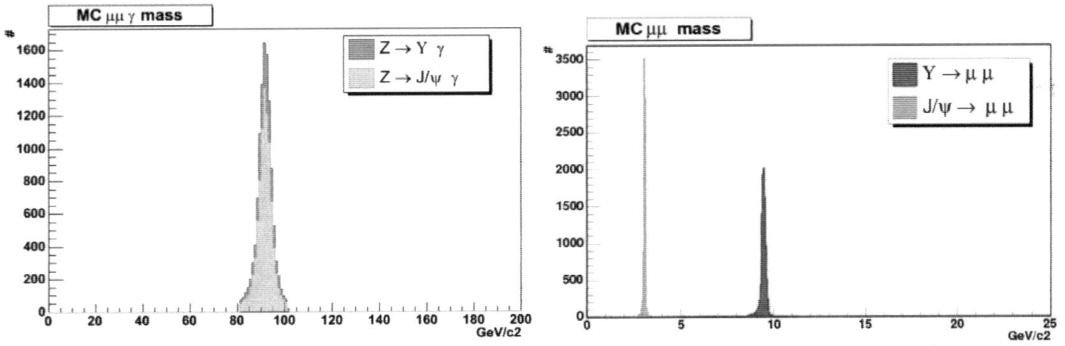

Figure 9. MC Events histograms. Left: $\mu\mu\gamma$ invariant mass distribution. Right: $\mu\mu$ invariant mass distribution. Green: $Z^0 \to J/\psi\gamma$ MC. Red: $Z^0 \to \Upsilon\gamma$ MC

The distributions in figure 9 show the mass peaks from the simulated data. These are the distributions that we must look for in the data; it is what our expected signal will look like if the decay is found.

2.4. Data

Once we had reached a better understanding of the expected signal and different distributions it was time to run the analysis code on real data. We analyzed the data coming from the detector in the same way we did for the MC simulations. We ran the analysis code on approximately 5 fb^{-1} of integrated luminosity. The results are shown on the following histograms.

The histogram in figure 10 shows how we found the Z mass peak in the data, reconstructed from the photon and the two muons. However, this is not enough to claim that this is indeed the final state of either the $Z^0 \to J/\psi\gamma$ or the $Z^0 \to \Upsilon\gamma$ decays. In order to do that we must confirm that the muon-antimuon pair are coming from either a J/ψ or an υ decay. The following histograms show the results for the expected distributions.

As we can see in figure 10, there is no strong evidence for the expected decays. We can also confirm, by the small amount of events in the region, that we are close to the experimental sensitivity for this decay. There is, however, one event which is found close to the J/ψ peak. Regardless of the analysis process

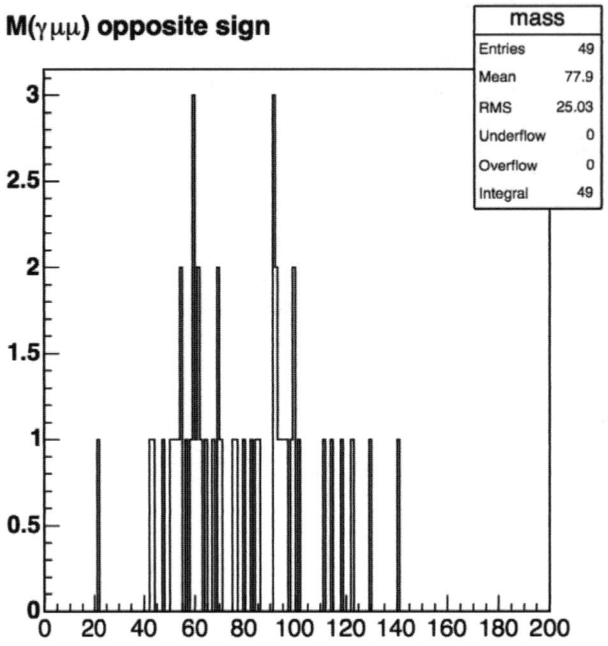

Figure 10. Final data $\mu\mu\gamma$ invariant mass distribution. This histogram shows the events that have met all requirements, including the 3D angle requirement.

that we have used in the data, more information is needed to establish whether or not this dimuon pair comes from a J/ψ decay.

2.4.1. Branching Ratio The following equation will be used for the branching ratio calculation:

$$N = L_{integ} \cdot \sigma_{p\bar{p}\rightarrow Z^0} \cdot B(Z^0 \rightarrow J/\psi + \gamma) \cdot B(J/\psi \rightarrow \mu^+\mu^-) \cdot A$$

$$\text{where } A = A_{MC} \cdot SF \quad \text{and} \quad SF = \frac{ID_{Eff\ Data}}{ID_{Eff_{MC}}}$$

N — Number of events with this decay
A — Acceptance
L_integ — Integrated luminosity
SF — Scale factor
$\sigma_{p\bar{p}\rightarrow Z^0}$ — $p - \bar{p}$ to Z cross section
ID_{Eff} — Identification efficiency

3. Future Work

We plan to complete the analysis using full statistics of approximately 10 fb^{-1} collected by the CDF experiment in Run II.

References

[1] T. et Aaltonen. Search for the rare radiative decay $w \to \pi\gamma$ in $p\bar{p}$ collisions at $\sqrt{s} = 1.96$ TeV. *Phys. Rev. D*, 85:032001, Feb 2012.

[2] N.N. Achasov. The $Z \to c\bar{c} \to \gamma\gamma^*$, $Z \to b\bar{b} \to \gamma\gamma^*$ triangle diagrams and the $Z \to \gamma\psi$, $Z \to \gamma\Upsilon$ decays. *Phys.Atom.Nucl.*, 74:437–445, 2011.

[3] K. Nakamura et al (Particle Data Group). The review of particle properties. *J. Phys.*, G 37, 2010.

Journal of Physics: Conference Series **378** (2012) 012032 doi:10.1088/1742-6596/378/1/012032

Noise and ultraviolet divergences in the dynamics of the chiral condensate in QCD

G. Krein

Instituto de Física Teórica, Universidade Estadual Paulista
Rua Dr. Bento Teobaldo Ferraz, 271 - Bloco II, 01140-070 São Paulo, SP, Brazil

E-mail: gkrein@ift.unesp.br

Abstract. The time evolution of the matter produced in high energy heavy-ion collisions seems to be well described by relativistic viscous hydrodynamics. In addition to the hydrodynamic degrees of freedom related to energy-momentum conservation, degrees of freedom associated with order parameters of broken continuous symmetries must be considered because they are all coupled to each other. Of particular interest is the coupling of degrees of freedom associated with the chiral symmetry of QCD. Quantum and thermal fluctuations of the chiral fields act as noise sources in the classical equations of motion, turning them into stochastic differential equations in the form of Ginzburg-Landau-Langevin (GLL) equations. Analytic solutions of GLL equations are attainable only in very special circumstances and extensive numerical simulations are necessary, usually by discretizing the equations on a spatial lattice. However, a not much appreciated issue in the numerical simulations of GLL equations is that ultraviolet divergences in the form of lattice-spacing dependence plague the solutions. The divergences are related to the well-known Rayleigh-Jeans catastrophe in classical field theory. In the present communication we present a systematic lattice renormalization method to control the catastrophe. We discuss the implementation of the method for a GLL equation derived in the context of a model for the QCD chiral phase transition and consider the nonequilibrium evolution of the chiral condensate during the hydrodynamic flow of the quark-gluon plasma.

1. Introduction

Experiments of high energy heavy-ion collisions conducted at the Relativistic Heavy Ion Collider (RHIC) have produced a state of matter that is strongly interacting and evolves as a low-viscosity fluid [1, 2, 3, 4, 5, 6] – the strongly interacting quark-gluon plasma (sQGP). The characterization of the produced matter as strongly interacting came as a great surprise, since earlier expectations based on the property of asymptotic freedom of quantum chromodynamics (QCD) were that RHIC would produce a gas-like system of weakly interacting quarks and gluons. The case for a strongly-interacting system was made on the observation that the space-time evolution of the produced sQGP is well described by relativistic viscous hydrodynamics. In particular, hydrodynamics accounts for the measured momentum anisotropies of the detected hadrons produced in the collision. The momentum anisotropies are the translation to momentum space of the initial spatial eccentricity of the collision; they are encoded in the Fourier moments v_2, v_3, v_4, $\cdots$ of the measured azimuthal distribution of particles [7]. A weakly–interacting, gas–like fluid would have no means to induce momentum anisotropies unless the particles of the fluid are strongly interacting.

Journal of Physics: Conference Series **378** (2012) 012032 doi:10.1088/1742-6596/378/1/012032

Crucial physics input to a hydrodynamic description of the sQGP are the equation of state (EOS) and transport coefficients, as shear (η) and bulk (ζ) viscosities; they reflect the properties of the flowing hadronic matter and should be calculable from QCD. Assuming validity of hydrodynamics for the sQGP, extraction of transport coefficients from measured momentum anisotropy patterns can pinpoint the location of the QCD phase transition or of a crossover from hadronic to quark-gluon matter, as suggested in Ref. [8]. However, it should be noted that in addition to the hydrodynamic degrees of freedom related to energy-momentum conservation, degrees of freedom associated with order parameters of broken continuous symmetries must be considered because they are all coupled to each other. Of particular interest is the coupling of degrees of freedom associated with chiral symmetry, which is an approximate symmetry of QCD that is responsible for the smallness of the pion mass and is supposed to be restored at high temperatures and baryon density. A first work in the direction of coupling chiral degrees of freedom to viscous hydrodynamics was recently conducted in Ref. [9]. That work investigated the influence of the coupling of chiral fields (described by the linear σ model) to viscous hydrodynamic equations on the charged-hadron elliptic flow v_2 and on the ratio $v_4/(v_2)^2$ considering the kinematical conditions of a Au + Au collision at $\sqrt{s} = 200$ GeV per nucleon. The conclusions of Ref. [9] were that v_2 is very sensitive to the temperature dependence of the (shear) viscosity to entropy ratio η/s, while v_2 is not very sensitive to the coupling of the chiral fields. In addition, it was found that the ratio $v_4/(v_2)^2$ is much more sensitive than v_2 to both the coupling of the chiral sources and the temperature dependence of η/s.

One important limitation of the study in Ref. [9] was the neglect of quantum and thermal fluctuations of the chiral fields. Temperature effects in that reference were taken into account via the one-loop effective potential for the chiral fields. Fluctuations, on the other hand, act as noise sources in the classical equations of motion, turning them stochastic differential equations in the form of Ginzburg-Landau-Langevin (GLL) equations. Such equations describe the time evolution of an order parameter under the influence of an environment which is encoded usually in the form of dissipation and noise terms - for a good text book and references on the subject, see Ref. [10]. Typically, when a system consisting of a two-phase mixture in a homogeneous phase is rapidly driven across the critical coexistence temperature into a nonequilibrium state, fluctuations around the initial homogeneous state will develop, the system will break into domains of different phases in space, and ultimately will reach a new equilibrium state.

In the present communication we concentrate on a difficulty that appears in the numerical simulation of GLL equations. Analytic solutions of GLL equations are attainable only in very special circumstances, like when nonlinear couplings are neglected. Linearization of the equations are usually valid in a restricted time interval only [11] and extensive numerical simulations, usually by discretizing the equations on a space lattice, are necessary for obtaining equilibrium solutions. One difficulty is related to the well-known Rayleigh-Jeans ultraviolet catastrophe of classical field theory [12], which manifest themselves via severe lattice-spacing dependence of the solutions of the GLL equations. This is an intrinsic feature of the problem, not related to a particular discretization method of the equations. In order to obtain sensible results in numerical simulations of GLL equations, a renormalization procedure has to be implemented. The implementation of a simple and efficient method [13, 14] to control the lattice-spacing dependence of the solutions of a GLL derived from the linear σ model will be one of the main topics of the present communication.

2. Ginzburg-Landau-Langevin equation - chiral symmetry

Let us start with a brief review of the usage of a GLL equation in statistical physics. A GLL equation for an order parameter field $\varphi(x,t)$ in presence of a noise field $\zeta(x,t)$ is typically given by [10]

$$\Gamma\,\frac{\partial\varphi(x,t)}{\partial t} = -\frac{\delta H[\varphi]}{\delta\varphi(x,t)} + \zeta(x,t), \tag{1}$$

where Γ is an Onsager (dissipation) coefficient and $H[\varphi]$ is the Ginzburg-Landau-Wilson (GLW) Hamiltonian, which is typically of the form

$$H[\varphi] = \int dx\,\left[\frac{1}{2}\kappa(\nabla\varphi)^2 + U_0(\varphi)\right], \tag{2}$$

with κ a positive constant (the range of the interaction in e.g. Ising-like models). A prototype $U(\varphi)$ is

$$U_0(\varphi) = \frac{1}{2}\,r_0\,\varphi^2 + \frac{1}{4}\,u_0\,\varphi^4, \tag{3}$$

with r_0 and $u_0 > 0$ constants that can depend on the temperature. For $r_0 < 0$, $U_0(\varphi)$ is of a double-well shape. For simplicity, let us consider a simple white-noise correlation function for the noise field:

$$\langle\zeta(x,t)\rangle_\zeta = 0, \qquad \langle\zeta(x,t)\zeta(x',t')\rangle_\zeta = 2\Gamma T\,\delta(t-t')\,\delta(x-x'), \tag{4}$$

where $\langle\cdots\rangle_\zeta$ means average over noise realizations and T is the external temperature. From the Fokker-Planck equation associated with the GLL equation of Eq. (1), it is not difficult to show that the equilibrium probability distribution of field configurations $\varphi(x)$ is the Boltzmann factor

$$P_{\rm eq}[\varphi] = e^{-H[\varphi]/T}. \tag{5}$$

The equilibrium partition function is then given by the functional integral

$$Z[\varphi] = \int \mathcal{D}\varphi\, e^{-H[\varphi]/T}. \tag{6}$$

The meaning of this is that a correlation function, e.g. a two-point correlation function $\langle\varphi(x_1)\varphi(x_2)\rangle$, defined as

$$\langle\varphi(x_1)\varphi(x_2)\rangle = \frac{1}{Z}\int \mathcal{D}\varphi\,\varphi(x_1)\varphi(x_2)\,e^{-H[\varphi]/T}, \tag{7}$$

can also be calculated via the long-time solutions of the GLL equation as

$$\langle\varphi(x_1)\varphi(x_2)\rangle = \lim_{t\to\infty}\frac{1}{N_R}\sum_{r=1}^{N_R}\varphi_r(x_1,t)\varphi_r(x_2,t), \tag{8}$$

where $\varphi_r(x,t)$ is a solution of the GLL equation of Eq. (1), with the index r indicating r-th noise realization, and N_r is the total number of realizations (N_r is supposed to be large). However, despite straightforward, the calculation of the averages in Eq. (8) face the problem that the solutions of the GLL equations suffer from sever lattice spacing dependence, as mentioned in the previous Section. Before discussing this, let us consider the derivation of GLL equations in the context of the quark-gluon plasma.

GLL equations can be derived from a microscopic Hamiltonian describing interparticle interactions by means of the Schwinger-Keldysh (closed-time-path) effective action formalism [15]. In general, the resulting equations can be rather involved, with the order parameter appearing nonlinearly in functionals nonlocal in space and time (memory). Simple forms like the ones shown in

Eqs. (1) - (4) are obtained only after a series of approximations and truncations. In the context of the strongly-interacting quark-gluon plasma (SQGP), Ref. [16] is a recent example where such a simple equation can be derived. Let us review the logic behind such approaches, starting with the hydrodynamic description of the sQGP used in Ref. [9]. Quarks and antiquarks are assumed to constitute a heat bath in local thermal equilibrium, propagated via fluid dynamic equations at the the level of densities, while the chiral fields are propagated via medium-corrected classical equations of motion. Specifically, the energy-momentum tensor $T^{\mu\nu}$ of the combined quark $q=(u,d)$ and chiral $\phi=(\phi^0,\phi^i)=(\sigma,\pi^i), i=1,2,3$ fields is split as

$$T^{\mu\nu}=T_q^{\mu\nu}+T_\phi^{\mu\nu}, \tag{9}$$

with

$$T_q^{\mu\nu}=(\epsilon+p)u^\mu u^\nu-pg^{\mu\nu}+\Pi^{\mu\nu}, \tag{10}$$

where the unknowns are the fluid four-velocity u_μ (normalized as $u_\mu u^\mu=1$), the viscous shear-tensor $\Pi^{\mu\nu}$, the energy density $\epsilon=\epsilon(\phi,T)$, and the pressure $p=p(\phi,T)$ - in the spirit of a hydrodynamics description, they are supposed in local equilibrium at a temperature T. The fluid equations of motion are obtained from the condition of energy-momentum conservation $D_\mu T^{\mu\nu}=0$, where D_μ is the geometric covariant derivative [9] – the equations of motion for the components of $\Pi^{\mu\nu}$ are those of the dissipative hydrodynamic formalism of Refs. [17, 18, 19]. To close the set of equations, one needs explicit expressions for the temperature dependence of ϵ and p, and the equations of motion for the chiral fields.

Here, like in Ref. [9], we use the linear σ model, so that the contribution from the chiral fields to the energy-momentum tensor in Eq. (9) is given by

$$T_\phi^{\mu\nu}=\sum_{a=0}^{4}\frac{\partial\langle\mathcal{L}_\phi\rangle}{\partial(\partial_\mu\phi_a)}\partial^\nu\phi_a-g^{\mu\nu}\langle\mathcal{L}_\phi\rangle, \tag{11}$$

where $\langle\cdots\rangle$ means local thermal average and

$$\mathcal{L}=\mathcal{L}_q+\mathcal{L}_\phi, \tag{12}$$

with

$$\mathcal{L}_q=\bar{q}\left[i\gamma^\mu\partial_\mu-g\left(\sigma+\gamma_5\vec{\tau}\cdot\vec{\pi}\right)\right]q, \tag{13}$$

and

$$\mathcal{L}_\phi=\frac{1}{2}\left(\partial_\mu\sigma\partial^\mu\sigma+\partial_\mu\vec{\pi}\cdot\partial^\mu\vec{\pi}\right)-U(\sigma,\vec{\pi}), \tag{14}$$

where

$$U(\sigma,\vec{\pi})=\frac{\lambda^2}{4}\left(\phi^2-v^2\right)^2-h_q\sigma-U_0\ , \tag{15}$$

is the potential which exhibits chiral symmetry breaking, and $\phi^2=\sigma^2+\vec{\pi}^2$. The parameters of the model are fixed as follows. The vacuum expectation values of the fields (condensates) are taken $\langle\sigma\rangle=f_\pi$ and $\langle\vec{\pi}\rangle=0$, with the pion decay constant $f_\pi=93$ MeV. Also, the partially conserved axial–vector current relation yields $h_q=f_\pi m_\pi^2$ with $m_\pi=138$ MeV. This leads to $v^2=f_\pi^2-m_\pi^2/\lambda^2$. Using $\lambda^2=20$, the value of the mass of the σ field is $m_\sigma=\sqrt{2\lambda^2 f_\pi^2+m_\pi^2}\sim 600$ MeV. The constant U_0 is chosen such that the potential energy vanishes in the ground state. Note that these are fitted to experimental values, because at the level of approximation worked in Ref. [9], they do not become corrected by quantum or thermal effects.

Journal of Physics: Conference Series **378** (2012) 012032 doi:10.1088/1742-6596/378/1/012032

The equations of motion of the chiral fields are given by

$$\Box\sigma + \frac{\delta V_{\rm eq}(\sigma,\vec{\pi})}{\delta\sigma} = -g\rho_s\,, \tag{16}$$

$$\Box\vec{\pi} + \frac{\delta V_{\rm eq}(\sigma,\vec{\pi})}{\delta\vec{\pi}} = -g\vec{\rho}_{ps}\,, \tag{17}$$

where $\Box$ is the D'Alambertian and ρ_s and $\vec{\rho}_{ps}$ are the equilibrium thermal averages of the scalar and pseudo-scalar chiral condensates :

$$\rho_s = \langle\bar{q}q\rangle = I_\sigma\,\sigma, \qquad \vec{\rho}_{ps} = \langle\bar{q}\gamma_5\vec{\tau}q\rangle = I_\pi\,\vec{\pi}, \tag{18}$$

where

$$I_\sigma = 2gd_q\int\frac{d^3p}{(2\pi)^3}\frac{1}{E}\left(1+e^{E/T}\right)^{-1}, \tag{19}$$

$$I_\pi = 2gd_q\int\frac{d^3p}{(2\pi)^3}\frac{1}{E}\left(1+e^{E/T}\right)^{-1}. \tag{20}$$

Here, $d_q = 12$ is the color–spin–isospin charge degeneracy of the quarks and $E = (p^2+m_q^2)^{1/2}$ is the single-quark energy, with the quark mass given by $m_q^2 = g^2(\sigma^2+\pi^2)$. We consider the case of zero net baryon number (zero baryon chemical potential). In addition, $V_{\rm eq}$ is the one-loop equilibrium effective potential

$$V_{\rm eq}(\sigma,\vec{\pi}) = U(\sigma,\vec{\pi}) - 2d_qT\int\frac{d^3p}{(2\pi)^3}\,\ln\left(1+e^{-E/T}\right). \tag{21}$$

Here we have changed the definition of d_q as compared to the d_q in Ref. [9] by a factor of two, incorporating this factor in the definition of ρ_s, $\vec{\rho}_{ps}$ and $V_{\rm eq}$.

In Eqs. (16) and (17), the influence of the quark fluid on the chiral fields is approximated by an equilibrium average over quantum and thermal fluctuations, encoded by the thermal averages of the scalar and pseudo-scalar chiral condensates ρ_s and $\vec{\rho}_{ps}$. A mean field treatment like this neglects physical effects like the decay of mesons into quark-antiquark pairs and the scattering of mesons by the quarks and antiquarks of the fluid. Such microscopic processes lead to dissipation and noise terms in the effective effective equations of motion for the coarse-grained chiral fields, which are of the form of GLL equations. Within the linear sigma model, a GLL equation can be derived using the Schwinger-Keldysh (closed-time-path) effective action formalism[15]. Written in the rest frame of the fluid, the GLL equation for the σ field derived in Ref. [16] is given by

$$\frac{\partial^2\sigma}{\partial t^2} - \nabla^2\sigma + \eta\frac{\partial\sigma}{\partial t} + \frac{\delta U}{\delta\sigma} + g\rho_s = \xi, \tag{22}$$

with the noise field satisfying

$$\langle\xi(x,t)\rangle_\xi = 0, \qquad \langle\xi(x,t)\xi(x',t')\rangle_\xi = m_\sigma\eta\coth\left(\frac{m_\sigma}{2T}\right)\delta(t-t')\,\delta^3(x-x'), \tag{23}$$

where the dissipation coefficient is given by

$$\eta = g^2\frac{d_q}{\pi}\frac{1}{m_\sigma^2}\left(\frac{m_\sigma^2}{4}-m_q^2\right)^{3/2}\left[1-2\left(1+e^{m_\sigma/2T}\right)^{-1}\right]. \tag{24}$$

Here the processes $\sigma \to \bar{q}q$ are included; but processes $\sigma \leftrightarrow 2\pi$ are not. The last ones are extremely important when $m_\sigma > 2m_\pi$, which occur at low temperatures. On the other hand, at low temperatures η as given in Eq. (24) is very small and σ decay and recombination processes must be included to take into account of the correct physics. Using the result of a study in Ref. [20] in the context of the σ model, processes $\sigma \leftrightarrow 2\pi$ can be taken into account with a noise correlation of the form

$$\langle \xi(x,t)\xi(x',t')\rangle_\xi = 2\eta_\sigma T\, \delta(t-t')\, \delta^3(x-x'), \tag{25}$$

with $\eta_\sigma = 2.2\ \text{fm}^{-1}$ for a temperature $T = 123$ MeV .

From Eq. (15), one has

$$\frac{\delta U}{\delta \sigma} = -\lambda v^2 \sigma + \lambda \sigma^3, \tag{26}$$

where we used for simplicity $h_q = 0$ in the equation of motion. The GLL equation of Eq. (22) can then be written as

$$\frac{\partial^2 \sigma}{\partial t^2} - \nabla^2 \sigma + \eta \frac{\partial \sigma}{\partial t} - \left(\lambda v^2 - I_\sigma\right)\sigma + \lambda \sigma^3 = \xi. \tag{27}$$

Note that the presence of the second order time derivative turns the GLL equation a hyperbolic partial differential equation, in contrast to the purely diffusive parabolic equation of Eq. (1). The presence of the second order time derivative is a consequence of causality imposed by relativity. In the context of statistical physics, hyperbolic GLL equations appear when memory effects are incorporated in the phase separation and phase ordering processes [21, 22]. One consequence of the hyperbolic nature of the equation is that time-dependent oscillations appear in the evolution of the order parameter toward equilibrium, as we shall see below. Regarding the equilibrium probability distribution of field configurations $\sigma(x)$, one can show that it is given by the Bolztmann factor (the proof is a little more elaborated than in the case of a parabolic equation, see e.g. Ref. [23]):

$$P_{\text{eq}}[\sigma] = e^{-H[\sigma]/T'}, \tag{28}$$

where $T' = (m_\sigma/2)\coth(m_\sigma/2T)$ and

$$H[\sigma] = \int d^3x \left[\frac{1}{2}(\nabla\sigma)^2 - \frac{1}{2}\left(\lambda v^2 - I_\sigma\right)\sigma^2 + \frac{1}{4}\lambda\sigma^4\right]. \tag{29}$$

As already mentioned analytic solutions of such nonlinear equations are attainable only in very special circumstances, like when nonlinear terms in $H[\sigma]$ are dropped. Such linear approximations are usually valid in a restricted time interval only [11] and extensive numerical simulations, usually by discretizing the equations on a space lattice, are necessary for obtaining solutions over a large time interval. In the next Section, we will obtain explicit numerical solutions of Eq. (27) and discuss a method do regularize the equation so that results independent of the lattice spacing are obtained.

3. Lattice simulation of the GLL equation

In this Section we discuss the physical situation inwhich the system initially at a high temperature, where the σ field is almost zero, is instantaneously quenched to a low temperature phase; e.g. below the critical (or crossover) temperature.

It is useful to work with dimensionless quantities. We rescale space-time coordinates (t, x) and the field σ: $x = x'/(\lambda v^2)^{1/2}$, $t = t'/(\lambda v^2)^{1/2}$, $\sigma' = \sigma/v$. The GLL equation in Eq. (27) can be rewritten in terms of these dimensionless variables as (henceforth we drop the primes)

$$\frac{\partial^2 \sigma}{\partial t^2} - \nabla^2 \sigma + \bar{\eta}\frac{\partial \sigma}{\partial t} - \epsilon\,\sigma + \sigma^3 = \zeta, \tag{30}$$

Journal of Physics: Conference Series **378** (2012) 012032 doi:10.1088/1742-6596/378/1/012032

where the dimensionless noise field $\zeta(x,t)$ has zero mean and correlation

$$\langle\zeta(x,t)\zeta(x',t')\rangle_\zeta = 2\gamma\,\delta(t-t')\,\delta^3(x-x'), \tag{31}$$

with

$$\bar{\eta} = \eta/(\lambda v^2)^{1/2}, \qquad \epsilon = -1 + I_\sigma/(\lambda v^2), \qquad \gamma = \eta T/v^2. \tag{32}$$

Here we used Eq. (25) for the noise correlation, appropriate for the physical situation we are considering. In the present work we use γ varying the range of $0.5 \leq \gamma \leq 2$, $\bar{\eta} = 1$, and $\epsilon = -1$ (corresponding to the neglect of $I_\sigma/(\lambda v^2)$, which should not be a bad approximation for low temperatures). Evidently when hydrodynamic equations are coupled to the chiral equations, one is not free to choose the values of $\bar{\eta}$, ϵ and γ, since they are all functions of the local temperature determined by the hydrodynamic equations.

We discretize Eq. (30) on a three-dimensional spatial lattice and use a leapfrog algorithm for the time evolution. We use the notation

$$\sigma(x,t) = \sigma(x,y,z,t) \rightarrow \sigma_{ijk}(n), \qquad t = n\Delta t, \quad n = 0,1,\cdots,N_t - 1, \tag{33}$$

$$(x,y,x) = (ia,ja,ka), \quad i,j,k = 0,1,\cdots,N-1,, \tag{34}$$

where a is the (dimensionless) lattice spacing. The leapfrog algorithm is defined by the following iteration scheme:

$$\frac{\partial\sigma(x,t)}{\partial t} \rightarrow \dot{\sigma}_{ijk}(n) = \frac{1}{2}\left[\dot{\sigma}_{ijk}(n+1/2) + \dot{\sigma}(n-1/2)\right], \tag{35}$$

$$\frac{\partial^2\sigma(x,t)}{\partial t^2} \rightarrow \ddot{\sigma}_{ijk}(n) = \frac{1}{\Delta t}\left[\dot{\sigma}_{ijk}(n+1/2) - \dot{\sigma}_{ijk}(n-1/2)\right], \tag{36}$$

$$\dot{\sigma}_{ijk}(n+1/2) = \frac{1}{\Delta t}\left[\sigma_{ijk}(n+1) - \sigma_{ijk}(n)\right], \tag{37}$$

Now we rewrite Eq. (30) in terms of the discrete field variables $\sigma_{ijk}(n)$:

$$\frac{1}{\Delta t}\left[\dot{\sigma}_{ijk}(n+1/2) - \dot{\sigma}_{ijk}(n-1/2)\right] = A(\sigma,\zeta)_{ijk}(n) - \frac{\bar{\eta}}{2}\left[\dot{\sigma}_{ijk}(n+1/2) + \dot{\sigma}_{ijk}(n-1/2)\right], \tag{38}$$

where

$$A(\sigma,\zeta) = (\nabla^2\sigma) - \epsilon\,\sigma + \sigma^3 + \zeta. \tag{39}$$

The equation that has to be solved iteratively is then

$$\dot{\sigma}_{ijk}(n+1/2) = \frac{1}{1+(\bar{\eta}/2)\Delta t}\left\{\left[1-(\bar{\eta}/2)\Delta t\right]\dot{\sigma}_{ijk}(n-1/2) + A(\sigma,\zeta)_{ijk}(n)\,\Delta t\right\}, \tag{40}$$

with the field updated as

$$\sigma_{ijk}(n+1) = \sigma_{ijk}(n) + \dot{\sigma}_{ijk}(n+1/2)\,\Delta t. \tag{41}$$

The noise field is generated at each time and lattice point as

$$\zeta_{ijk}(n) = \sqrt{\frac{2\gamma}{a^3\Delta t}}\,G_{ijk}(n), \tag{42}$$

where $G_{ijk}(n)$ is a zero-mean unit-variance Gaussian random number.

In Fig. 1 we present results for the spatially averaged value of the (noise-averaged) field as a function of time

$$\bar{\sigma}(t) = \frac{1}{N^3} \sum_{i,j,k=1}^{N} \langle \sigma_{ijk}(n) \rangle_\zeta, \tag{43}$$

where $\langle \sigma_{ijk}(n) \rangle_\zeta$ is the average over noise realizations. Results are shown for three values of lattice spacings, $a = 0.5, 0.8$ and 1.0 – in physical units, they correspond roughly to 0.35 fm, 0.57 fm and 0.7 fm, respectively. For the time spacing, we use $\Delta t = 0.001$ – corresponding to 7×10^{-4} fm/c. We show results for $\gamma = 1$ and $\gamma = 2$.

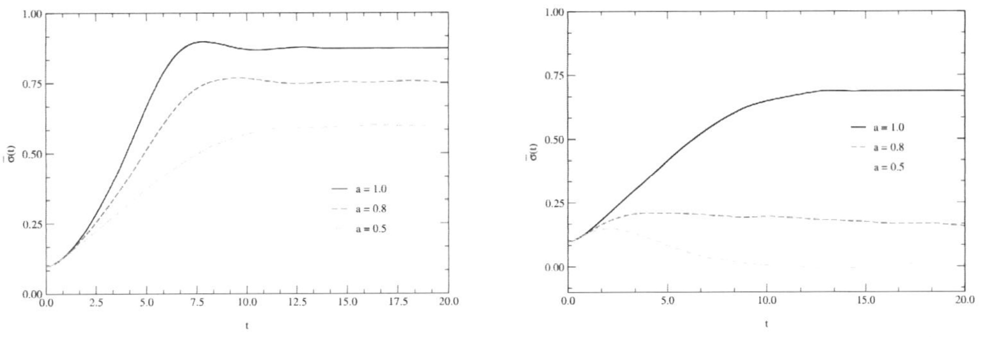

Figure 1. Volume average of the noise averaged σ field as a function of the time, Eq. (43). The left panel is for $\gamma = 1$ and the right panel is for $\gamma = 2$.

The results in Fig. 1 show very clearly the severe lattice spacing dependence of the solutions. The sensitivity is stronger for larger values of γ – note that γ is a measure of the strength of the quantum and thermal fluctuations. As said previously, the sensitivity of the solutions on a is due to the noise sources and is a manifestation of the ultraviolet divergences of classical field theory; when $\zeta = 0$, the sensitivity is absent. Although we have solved the GLL equation for fixed values of $\bar{\eta}$, ϵ and γ, it should be clear that in the coupled hydrochiral evolution, the lattice spacing sensitivity of the GLL equation can introduce large errors in the outcome of the simulation. In the next Section we discuss a method to control the lattice spacing sensitivity.

4. Renormalization of the GLL equation

The equilibrium probability distribution for the field configurations σ that are solutions of Eq. (22) is the Boltzmann factor given in Eq. (28). Equilibrium expectation values and correlation functions of σ with this partition function leads to ultraviolet divergences in spatial dimension larger than one – these divergences are completely unrelated to the usual divergences of the quantum theory. In the dynamics toward equilibrium, the divergences manifest themselves in the wild lattice spacing dependence of the solutions as $a \to 0$, as seen in the previous Section. The lattice dependence can be eliminated by renormalizing the GLL equation through the addition of appropriate counterterms. In three-dimensions, the theory is super-renormalizable, only the term proportional to σ^2 in Eq. (29) needs renormalization.

In the calculation of the macroscopic free energy (or the effective potential, $\Omega = -T \log Z$), there are only two divergent perturbative graphs in three spatial dimensions, the ones shown in Fig. 2: the tadpole diagram (left graph), which is linearly divergent, and the setting-sun diagram (right graph), which is logarithmically divergent. The divergent parts of these graphs can be calculated as a function of the lattice spacing – for an explicit and detailed derivation, see Ref. [13]. They are given in terms of the dimensionless quantities introduced above as [13, 24]:

$$\Omega_{div}(\sigma) = \frac{1}{2}\epsilon(\mu)\sigma^2, \tag{44}$$

with

$$\epsilon(\mu) = 3\gamma\frac{\Sigma}{4\pi a} - 6\gamma^2\left\{\frac{1}{16\pi^2}\left[\log\left(\frac{6}{\mu a}\right) + 0.09\right]\right\}, \tag{45}$$

where $\Sigma \simeq 3.1759$, and μ is a renormalization scale. Finite results are then obtained by subtracting this term from the original (rescaled) GLL equation; the parameter v^2 then becomes renormalized and μ dependent, it has to be fitted to a measurable quantity. Once it is fitted at some scale μ, its value at another scale μ' is given by renormalization group equations, as usual [10].

Figure 2. Divergent Feynman graphs that contribute to the effective potential in three spatial dimensions. Left graph is the tadpole and the right graph is the setting-sun.

In Fig. 3 we present the results with the above counter-term added to the GLL equation – we used the value $\mu = 1$, but the results are not much dependent on the value of μ, since the first term in Eq. (45) dominates the divergence. As seen, the subtraction leads to solutions that are independent of the lattice spacing within very good precision. Similar results are obtained for more intense noises, i.e. for larger values of γ.

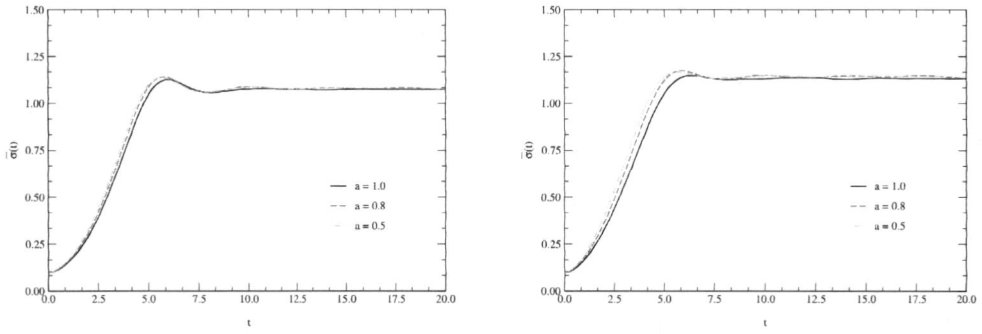

Figure 3. Volume average of the noise averaged σ field as a function of the time, Eq. (43), obtained with the counter-term added to the GLL equation. The left panel is for $\gamma = 1$ and the right panel is for $\gamma = 2$.

Journal of Physics: Conference Series **378** (2012) 012032 doi:10.1088/1742-6596/378/1/012032

5. Conclusions

In summary, we have discussed a simple and systematic method to renormalize GLL equations on a spatial lattice. Numerical results of simulations were presented for a GLL equation derived in the context of the linear σ model. The usage of the renormalization method when coupling the chiral equations to hydrodynamic fluid equations is straightforward. Work in this direction is in progress.

Acknowledgments

A good deal of the material presented here is the result of collaborative work with N.C. Cassol-Seewald, R.L.S. Farias, E.S. Fraga, R.S. Marques de Carvalho, J. Peralta and R.O. Ramos. Work partially financed by Conselho Nacional de Desenvolvimento Científico e Tecnológico - CNPq, and Fundação de Amparo à Pesquisa no Estado de São Paulo - FAPESP.

References

[1] Adler S S *et al* (PHENIX Collaboration) 2003 *Phys. Rev. Lett.* **91** 182301
[2] Adams J *et al* (STAR Collaboration) 2004 *Phys. Rev. Lett.* **92** 112301
[3] Adcox K *et al* (PHENIX Collaboration) 2004 *Phys. Rev.* C **69** 024904
[4] Adams J *et al* (STAR Collaboration) 2005 *Phys. Rev.* C **72** 014904
[5] Arsene I *et al* (BRAHMS Collaboration) 2005 *Phys. Rev.* C **72** 014908
[6] Back B B *et al* (PHOBOS Collaboration) 2005 *Phys. Rev.* C **72**, 051901(R)
[7] J. -Y. Ollitrault 1992 *Phys. Rev.* D **46** 229
[8] Csernai L P, Kapusta J I and McLerran L D 2006 *Phys. Rev. Lett.* **97** 152303
[9] J. Peralta-Ramos and G. Krein 2011 *Phys. Rev.* C **84** 044904
[10] Onuki A 2002 *Phase Transition Dynamics* (Cambridge: Cambridge University Press)
[11] Copetti M I M, Krein G, Machado J M and Marques de Carvalho R S 2005 *Phys. Lett.* A **338** 232
[12] Parisi G 1988 *Statistical Field Theory* (New York: Addison-Wesley)
[13] Cassol-Seewald N C, Farias R L S, Fraga E S, Krein G and Ramos R O 2012 *Physica* A (to appear) *Preprint* arXiv:0711.1866 [hep-ph]
[14] Farias R L S, Cassol-Seewald N C, Krein G and Ramos R O 2007 *Nucl. Phys.* A **782** 33
[15] Calzetta E and Hu B L 2008 *Nonequilibrium Quantum Field Theory* (Cambridge: Cambridge Monographs on Mathematical Physics)
[16] Nahrgang M, Leupold S, Herold C and Bleicher M 2011 *Phys. Rev.* C **84** 024912
[17] Baier R, Romatschke P, Son D T, Starinets A O and Stephanov M A 2008 *J. High Energy Phys.* **04** 100
[18] Bhattacharyya S, Ehubeny V E, Minwalla S and Rangamani M 2008 *J. High Energy Phys.* **02** 045
[19] Natsuume M and Okamura T 2008 *Phys. Rev.* D **77** 066014; **78** 089902(E).
[20] Biró T and Greiner C 1997 *Phys. Rev. Lett.* **79** 3138
[21] Koide T, Krein G and Ramos R O 2006 *Phys. Lett.* B **636** 96
[22] Cassol-Seewald N C, Copetti M I M and Krein G 2008 *Comp. Phys. Comm.* **179** 297
[23] Horowitz A M 1985 *Phys. Lett.* B **156** 89
[24] Fraga E S, Krein G and Mizher A J 2007 *Phys. Rev. D* **76** 034501

Journal of Physics: Conference Series **378** (2012) 012033 doi:10.1088/1742-6596/378/1/012033

Electromagnetic currents of elementary particles from a second order formalism

E G Delgado-Acosta[1], M Napsuciale[2] and S Rodriguez[3]

[1] Instituto de Física, Universidad Autónoma de San Luis Potosí, Av. Manuel Nava 6, San Luis Potosí, SLP 78290, México

[2] Departamento de Física, Universidad Guanajuato, Loma del Bosque 103, León, GTO 37150, México

[3] Facultad de Ciencias Fiísico Matemáticas, Universidad Autónoma de Coahuila, Edificio D, Unidad Camporredondo, Saltillo, COAH 25280, México

E-mail: [1]german@ifisica.uaslp.mx, [2]mauro@fisica.ugto.mx, [3]simonrodriguez@uadec.edu.mx

Abstract. We use a second order formalism based on Poincaré symmetry, and known as Covariant Projector Formalism to calculate the electromagnetic currents of spin 1/2, spin 1 and spin 3/2 particles. We restrict to parity conserving processes and obtain the currents in terms of the gyromagnetic factor g alone. This parameter is then fixed by calculating the forward Compton scattering cross section. In the end we reproduce the Dirac moments for the electron from a more general second order Lagrangian, we also reproduce the Standard Model results for the W boson. In the spin 3/2 case we show that the RS current is incomplete as it does not include the spin 1 piece of the vector-spinor representation. As a consequence, when we calculate the spin 3/2 EM moments, instead of the universal value $g = 2$, we find $g = 2/3$ as the gyromagnetic ratio for a Rarita-Schwinger particle.

1. Introduction

The present study is devoted to the electromagnetic (EM) interaction of particles with spin 1/2, spin 1, and spin 3/2. The interest in this matter is motivated by existing problems of the conventional description. For example, the Proca equation describes spin 1 massive particles [1]

$$[(-p^2 + m^2)g_{\mu\nu} + p_\mu p_\nu]A^\nu = 0, \tag{1}$$

but in the presence of an EM field it fails to predict the correct EM moments of a charged spin 1 massive particle such as the W boson. Spin 3/2 particles are usually described by the Rarita-Schwinger equation [2]

$$(\gamma \cdot p - m)\psi^\alpha = 0, \quad p^\alpha \psi_\alpha = 0, \tag{2}$$

which is plagued by the Velo-Zwanziger problem, allowing for a super-luminal propagation of the wave fronts when coupled to an EM field. We show that the above problems are avoided within a more elaborated formalism suggested in [3], based on covariant projectors. The advantages of the covariant projector formalism can be summarized as follows:

- It is based on minimal coupling and allows for the construction of an EM interaction written in terms of undetermined parameters that can be fixed by physical requirements, among them, well behaved forward Compton scattering cross sections.
- In the $(1/2, 0) \oplus (0, 1/2)$ Dirac representation of the homogeneous Lorentz group (HGL) it reproduces, upon an appropriate choice of parameters, the EM properties of a spin 1/2 fermion described by the Dirac Lagrangian.
- It adequately describes the EM interaction of massive spin 1 particles in the $(1/2, 1/2)$ representation of the HLG.
- In the spin 3/2 case it provides an equation of motion with causal propagation of the wave fronts by restricting the gyromagnetic factor to the universal value of $g = 2$.

In the next section we shall highlight the essentials of the covariant projector formalism.

2. Covariant projector formalism

The second order formalism used here consists in the construction of an equation of motion for a state $\psi^{(m,s)}$ transforming in a given multi-spin representation, $\psi^{(m,s_1,s_2,..s)}$, of the Poincaré group as a projection onto its spin s sector, and the mass, m. The projector is expressed in terms of the two Casimir operators of the group, the squared momentum operator P^2 and the squared Pauli-Lubanski operator W^2. The equation of motion for a mass m and spin s has the initial form

$$\mathcal{P}^{(m,s)}\psi^{(m,s)} = \psi^{(m,s)}, \tag{3}$$

and in general can be rewritten as

$$(\Gamma_{\mu\nu}p^\mu p^\nu - m^2)\psi^{(m,s)} = 0, \tag{4}$$

where the $\Gamma_{\mu\nu}$ tensor is expressed in terms of the P^2 and W^2 operators, its explicit form however has to be worked out in the particular representation of interest. It is important to note that the projection does not restrict the form of the antisymmetric part of the $\Gamma_{\mu\nu}$ tensor, to see this we write the equation as

$$(\Gamma_{\mu\nu}\partial^\mu\partial^\nu + m^2)\psi = 0. \tag{5}$$

It is obvious that the antisymmetric part of $\Gamma_{\mu\nu}$ is irrelevant in the free case since the derivatives ∂^μ and ∂^ν commute, however it becomes activated when introducing interactions via the gauge principle, leading to non-commuting covariant derivatives, $\partial \rightarrow D = \partial + ieA$,

$$[D^\mu, D^\nu] = ieF^{\mu\nu}. \tag{6}$$

In effect, in order to preserve generality of the coupled theory, one has to define the $\Gamma_{\mu\nu}$ tensor with the most general antisymmetric part allowed in a given representation. We will refer to this method either as covariant projector formalism, [3], or, NKR formalism.

2.1. Spin 1/2 in the $(1/2, 0) \oplus (0, 1/2)$ representation

Following the method described above, we can write an equation of motion for a state ψ in this representation as

$$(\Gamma^S_{\mu\nu}p^\mu p^\nu - m^2)\psi = 0. \tag{7}$$

In the present work we restrict to parity conserving processes so that the most general tensor to be considered is

$$\Gamma^S_{\mu\nu} = g_{\mu\nu} - ig_S M^S_{\mu\nu}, \tag{8}$$

Journal of Physics: Conference Series **378** (2012) 012033 doi:10.1088/1742-6596/378/1/012033

where g_S is an arbitrary parameter and $M^S_{\mu\nu}$ are the generators of the representation $(1/2,0)\oplus(0,1/2)$ of the HLG

$$M^S_{\mu\nu} = \frac{i}{4}(\gamma_\mu\gamma_\nu - \gamma_\nu\gamma_\mu) = \frac{i}{4}[\gamma_\mu,\gamma_\nu] = \frac{1}{2}\sigma_{\mu\nu}, \tag{9}$$

with γ^μ being the Dirac gamma matrices. The associated free Lagrangian is

$$\mathcal{L}^S_{\text{free}} = (\partial^\mu\overline{\psi})\Gamma_{\mu\nu}\partial^\nu\psi - m^2\overline{\psi}\psi. \tag{10}$$

The EM interaction is introduced by minimal coupling,

$$\partial_\mu \to D_\mu = \partial_\mu + ieA_\mu, \tag{11}$$

where e is the charge of the particle, the resulting gauged Lagrangian being

$$\mathcal{L}^S = \mathcal{L}^S_{\text{free}} - ej_\mu A^\mu + e^2\overline{\psi}(\Gamma^S_{\mu\nu} + \Gamma^S_{\nu\mu})\psi. \tag{12}$$

Then the EM current in momentum space is found to be

$$j^S_\mu(\mathbf{p}',\lambda';,\mathbf{p};\lambda) = e\overline{u}(\mathbf{p}',\lambda')(\Gamma^S_{\nu\mu}p'^\nu + \Gamma^S_{\mu\nu}p^\nu)u(\mathbf{p},\lambda), \tag{13}$$

where where the $u(\mathbf{p},\lambda)$ spinors transform in the $(1/2,0)\oplus(0,1/2)$ representation of the HLG. Insertion of the explicit form of the $\Gamma^S_{\mu\nu}$ tensor leads to the following current,

$$j^S_\mu(\mathbf{p}',\lambda';,\mathbf{p};\lambda) = e\overline{u}(\mathbf{p}',\lambda')\left[(p'+p)_\mu + ig_S M^S_{\mu\nu}(p'-p)^\nu\right]u(\mathbf{p},\lambda). \tag{14}$$

In effect, the method used directly provides a current that parallels the Gordon decomposition of the Dirac current.

2.2. Spin 1 in the (1/2, 1/2) representation

The equation of motion relevant to this case has the form

$$(\Gamma^V_{\alpha\beta\mu\nu}p^\mu p^\nu - m^2 g_{\alpha\beta})V^\beta = 0. \tag{15}$$

For particles in parity conserving processes we define the general $\Gamma^V_{\alpha\beta\mu\nu}$ tensor as

$$\Gamma^V_{\alpha\beta\mu\nu} = g_{\alpha\beta}g_{\mu\nu} - \frac{1}{2}(g_{\alpha\nu}g_{\beta\mu} + g_{\alpha\mu}g_{\beta\nu}) - i\left(g - \frac{1}{2}\right)[M^V_{\mu\nu}]_{\alpha\beta}, \tag{16}$$

where $M^V_{\mu\nu}$ are the generators of the representation $(1/2,1/2)$,

$$[M^V_{\mu\nu}]_{\alpha\beta} = i(g_{\alpha\mu}g_{\beta\nu} - g_{\alpha\nu}g_{\beta\mu}). \tag{17}$$

Following the same procedure as in the previous section, the EM current is derived as

$$j^V_\mu(\mathbf{p}',\lambda';\mathbf{p},\lambda) = -e\,\overline{\eta}^\alpha(\mathbf{p}',\lambda')\left[(p'+p)_\mu g_{\alpha\beta} + ig[M^V_{\mu\nu}]_{\alpha\beta}(p'-p)^\nu\right]\eta^\beta(\mathbf{p},\lambda), \tag{18}$$

where the $\eta^\alpha(\mathbf{p},\lambda)$ transform in the $(1/2,1/2)$ representation of the HLG.

Journal of Physics: Conference Series **378** (2012) 012033 doi:10.1088/1742-6596/378/1/012033

2.3. Spin 3/2 in the $(1/2,1/2)\otimes[(1/2,0)\oplus(0,1/2)]$ representation
In this case the equation of motion has the form:

$$(\Gamma_{\alpha\beta\mu\nu}p^\mu p^\nu - m^2 g_{\alpha\beta})\psi^\beta = 0. \tag{19}$$

There are various antisymmetric structures allowed in this representation, and for a parity conserving particle the $\Gamma_{\alpha\beta\mu\nu}$ tensor is written in terms of five undetermined parameters:

$$\begin{aligned}\Gamma_{\alpha\beta\mu\nu} = & -\frac{1}{3}g_{\alpha\nu}g_{\beta\mu} - \frac{1}{6}i\sigma_{\alpha\nu}g_{\beta\mu} - \frac{1}{3}g_{\alpha\mu}g_{\beta\nu} + \frac{2}{3}g_{\alpha\beta}g_{\mu\nu} + \frac{i}{3}g_{\mu\nu}\sigma_{\alpha\beta} \\ & - \frac{i}{6}g_{\beta\nu}\sigma_{\alpha\mu} + \frac{i}{6}g_{\alpha\nu}\sigma_{\beta\mu} + \frac{i}{6}g_{\alpha\mu}\sigma_{\beta\nu} \\ & - g_S\frac{i}{2}g_{\alpha\beta}\sigma_{\mu\nu} + g_V\left(g_{\alpha\mu}g_{\beta\nu} - g_{\beta\mu}g_{\alpha\nu}\right) + id\left(g_{\beta\nu}\sigma_{\alpha\mu} - g_{\beta\mu}\sigma_{\alpha\nu}\right) \\ & + ic\left(g_{\alpha\mu}\sigma_{\beta\nu} - g_{\alpha\nu}\sigma_{\beta\mu}\right) + if\gamma^5\epsilon_{\alpha\beta\mu\nu},\end{aligned} \tag{20}$$

The number of unknown coefficients gets reduced by eliminating the spin 1/2 sector coupling contained in the ψ^μ representation and identifying the gyromagnetic factor as $g = 2f + g_V$. In so doing, the EM current for a particle on mass-shell is found as

$$j_\mu(\mathbf{p}',\lambda';\mathbf{p},\lambda) = -e\,\overline{u}^\alpha(\mathbf{p}',\lambda')\left[(p'+p)_\mu g_{\alpha\beta} + ig[M^{3/2}_{\mu\nu}]_{\alpha\beta}(p'-p)^\nu\right]u^\beta(\mathbf{p},\lambda), \tag{21}$$

where

$$[M^{3/2}_{\mu\nu}]_{\alpha\beta} = M^S_{\mu\nu}g_{\alpha\beta} + [M^V_{\mu\nu}]_{\alpha\beta}, \tag{22}$$

are the generators of the $(1/2,1/2)\otimes[(1/2,0)\oplus(0,1/2)]$ representation of the HLG. This current also decomposes in a similar way to the cases seen before. This is on variance with the Rarita-Schwinger current which on-shell is written in terms of an A-parameter (see [4] for example) as

$$j^{RS}_\mu(\mathbf{p}',\lambda';\mathbf{p},\lambda) = -e\,\overline{u}^\alpha(\mathbf{p}',\lambda')\left[2m\gamma_\mu g_{\alpha\beta}\right]u^\beta(\mathbf{p},\lambda). \tag{23}$$

The decomposition of the latter current,

$$j^{RS}_\mu(\mathbf{p}',\lambda';\mathbf{p},\lambda) = -e\,\overline{u}^\alpha(\mathbf{p}',\lambda')\left[(p'+p)_\mu g_{\alpha\beta} + ig_S M^S_{\mu\nu}g_{\alpha\beta}(p'-p)^\nu\right]u^\beta(\mathbf{p},\lambda), \tag{24}$$

makes it visible that it does not include any term due to underlying vector representation. As a consequence, this current fails to describe a magnetic coupling proportional to $g_{3/2}\mathbf{S}\cdot\mathbf{B}$.

3. Multipole expansions

The electromagnetic moments of a particle are defined by means of a multipole expansion of a corresponding current density. We here design such a current density in transforming our electromagnetic currents from above to the Breit frame:

$$J^B_\mu(s,\lambda,\mathbf{q}) = \frac{1}{\omega}j^s_\mu(\mathbf{p}',\lambda;\mathbf{p},\lambda), \qquad p' = (\omega/2,\mathbf{q}/2), \quad p = (\omega/2,-\mathbf{q}/2). \tag{25}$$

The Cartesian electromagnetic moments for a particle of spin s and polarization λ are then defined as [5]:

$$Q^l_E(s,\lambda) = b^{l0}(-i\partial_\mathbf{q})\varrho_E(s,\lambda,\mathbf{q})\Big|_{\mathbf{q}=0}, \tag{26}$$

$$Q^l_M(s,\lambda) = \frac{1}{l+1}\,b^{l0}(-i\partial_\mathbf{q})\varrho_M(s,\lambda,\mathbf{q})\Big|_{\mathbf{q}=0}, \tag{27}$$

Journal of Physics: Conference Series **378** (2012) 012033 doi:10.1088/1742-6596/378/1/012033

where the electric density $\varrho_E(s,\lambda)$ and the so called magnetic density $\varrho_M(s,\lambda)$ read:

$$\varrho_E(s,\lambda,\mathbf{q}) = j^0_B(s,\lambda,\mathbf{q}), \qquad \varrho_M(s,\lambda,\mathbf{q}) = \partial_{\mathbf{q}}\cdot[\mathbf{j}_B(s,\lambda,\mathbf{q})\times\mathbf{q}], \tag{28}$$

the b^{l0} coefficients are obtained from the spherical harmonics as

$$b^{l0}(\mathbf{r}) = l!\sqrt{4\pi/(2l+1)}r^lY_{l0}(\Omega), \tag{29}$$

so that

$$b^{00}(-i\partial_{\mathbf{q}}) = 1, \tag{30}$$

$$b^{10}(-i\partial_{\mathbf{q}}) = -i\frac{\partial}{\partial q_3}, \tag{31}$$

$$b^{20}(-i\partial_{\mathbf{q}}) = \frac{\partial^2}{\partial q_1^2} + \frac{\partial^2}{q_2^2} - 2\frac{\partial^2}{\partial q_3^2}, \tag{32}$$

$$b^{30}(-i\partial_{\mathbf{q}}) = -3\,i\frac{\partial}{\partial q_3}\left(3\frac{\partial^2}{\partial q_1^2} + 3\frac{\partial^2}{\partial q2^2} - 2\frac{\partial^2}{\partial q_3^2}\right), \tag{33}$$

and so on.

3.1. Electromagnetic moments of spin 1/2 and spin 1 particles in the covariant projector formalism

We begin with the spin 1/2 case. In order to obtain explicit expressions for the moments, we need knowledge of the $u(\mathbf{p},\lambda)$, we can use the textbook basis:

$$u\left(\mathbf{p},+\frac{1}{2}\right) = \mathcal{N}_S\begin{pmatrix} m+p_0 \\ 0 \\ p_3 \\ p_1+ip_2 \end{pmatrix}, \tag{34}$$

$$u\left(\mathbf{p},-\frac{1}{2}\right) = \mathcal{N}_S\begin{pmatrix} 0 \\ m+p_0 \\ p_1-ip_2 \\ -p_3 \end{pmatrix}, \tag{35}$$

with $\mathcal{N}_S = [2m(m+p_0)]^{-1/2}$. Inserting this spinors into (14), we find the current needed to obtain the Breit densities (28) and the resulting set of moments as

$$Q^0_E\left(\frac{1}{2},\lambda\right) = e, \tag{36a}$$

$$Q^1_M\left(\frac{1}{2},\lambda\right) = \frac{eg_S}{2m}\left\langle\frac{1}{2},\lambda\right|S_z\left|\frac{1}{2},\lambda\right\rangle \equiv \frac{eg_S}{2m}\langle S_z\rangle. \tag{36b}$$

Here Q^0_E, Q^1_M stand for the charge and dipole magnetic moment of the particle respectively, higher moments are found to be zero. The Dirac results [6, 7] are recovered for $g = 2$, the Compton scattering process providing an additional argument in favor of this value. We now turn to the spin 1 case. In repeating the procedure for the current in (18), again the explicit form of $\eta^\alpha(\mathbf{p},\lambda)$ is required,

$$\eta(\mathbf{p},+1) := \frac{\mathcal{N}_V}{\sqrt{2}}\begin{pmatrix} -(m+p_0)(p_1+ip_2) \\ -m^2-p_0m-p_1^2-ip_1p_2 \\ -i\left(p_2^2-ip_1p_2+m(m+p_0)\right) \\ -(p_1+ip_2)p_3 \end{pmatrix}, \tag{37a}$$

$$\eta(\mathbf{p},0) := \mathcal{N}_V\begin{pmatrix} (m+p_0)p_3 \\ p_1p_3 \\ p_2p_3 \\ p_3^2+m(m+p_0) \end{pmatrix}, \tag{37b}$$

$$\eta(\mathbf{p},-1) := \frac{\mathcal{N}_V}{\sqrt{2}}\begin{pmatrix} (m+p_0)(p_1-ip_2) \\ m^2+p_0m+p_1^2-ip_1p_2 \\ -i\left(p_2^2+ip_1p_2+m(m+p_0)\right) \\ (p_1-ip_2)p_3 \end{pmatrix}, \tag{37c}$$

with $\mathcal{N}_V=[m(m+p_0)]^{-1}$. In terms of expectation values between vectors of polarization λ, for a spin 1 particle we obtain:

$$Q_E^0(1,\lambda) = e, \tag{38a}$$

$$Q_M^1(1,\lambda) = \frac{eg_V}{2m}\langle S_z\rangle, \tag{38b}$$

$$Q_E^2(1,\lambda) = \frac{e(1-g_V)}{m^2}\langle 3S_z^2-\mathbf{S}^2\rangle, \tag{38c}$$

where Q_E^2 denotes the electric quadrupole moment. These expressions match the known results for the electroweak W gauge boson for $g=2$ as described by the standard model Lagrangian [8, 9]. We conclude from this that the covariant projector formalism gives an adequate description of spin 1/2 and spin 1 particles. However the result is not a surprising one because the currents in (14) and (18) parallel their counterparts from Dirac and SM Lagrangians [7, 10], in the sense that they decompose into a spinless part proportional to $(p'+p)^\mu$ and a spin part associated with the term $igM_{\mu\nu}(p'-p)^\nu$.

3.2. Spin 3/2 particles in the Rarita-Schwinger formalism

To make calculations in the $(1/2,1/2)\otimes[(1/2,0)\oplus(0,1/2)]$ representation we need the spin 3/2 basis previously employed in [4],

$$u^\alpha(\mathbf{p},+3/2) = \eta^\alpha(\mathbf{p},+1)u(\mathbf{p},+1/2), \tag{39a}$$

$$u^\alpha(\mathbf{p},+1/2) = \frac{1}{\sqrt{3}}\eta^\alpha(\mathbf{p},+1)u(\mathbf{p},-1/2)+\sqrt{\frac{2}{3}}\eta^\alpha(\mathbf{p},0)u(\mathbf{p},+1/2), \tag{39b}$$

$$u^\alpha(\mathbf{p},-1/2) = \frac{1}{\sqrt{3}}\eta^\alpha(\mathbf{p},-1)u(\mathbf{p},+1/2)+\sqrt{\frac{2}{3}}\eta^\alpha(\mathbf{p},0)u(\mathbf{p},-1/2), \tag{39c}$$

$$u^\alpha(\mathbf{p},-3/2) = \eta^\alpha(\mathbf{p},-1)u(\mathbf{p},-1/2), \tag{39d}$$

where $u(\mathbf{p},\lambda)$ and $\eta^\alpha(\mathbf{p},\lambda)$ have been previously introduced in (35) and (37). The calculation of the multipole decompositions of the Rarita-Schwinger spin-3/2 current amounts to the following set of multipole moments,

$$\begin{aligned}
\text{RS:}\quad Q_E^0\left(\frac{3}{2},\lambda\right) &= e, &\text{(40a)}\\
Q_M^1\left(\frac{3}{2},\lambda\right) &= \frac{eg_S}{2m}\frac{1}{3}\langle S_z\rangle, &\text{(40b)}\\
Q_E^2\left(\frac{3}{2},\lambda\right) &= \frac{e}{m^2}\frac{1}{3}\langle \mathcal{A}\rangle, &\text{(40c)}\\
Q_M^3\left(\frac{3}{2},\lambda\right) &= \frac{eg_S}{2m^3}\langle \mathcal{B}\rangle, &\text{(40d)}
\end{aligned}$$

with $g_S = 2$. The latter equations show that only the gyromagnetic factor, g_S, of the Dirac-spinor piece of the four- vector–spinor contributes to the magnetic dipole and octupole moments, with the result that one has to ascribe $g = 2/3$ to the spin-3/2 field. The electric quadrupole moment turns out to be indifferent to both the g_S and g_V values. In a forthcoming paper, we shall argue that this shortcoming of the Rarita-Schwinger framework can be avoided within the covariant projector formalism, where the g_V and g_S factors will be shown to participate the expressions for the spin-3/2 multipoles on equal footings.

4. Compton scattering

4.1. Spin 1/2 particles in the $(1/2, 0) \oplus (0, 1/2)$ *representation*

We begin with the spin 1/2 case. From the Lagrangian (12) we can extract Feynman rules of first and second order on the charge of the particle, while the propagator can be read form the inverse of equation (7). The cross section of the Compton scattering process is found in terms of the g_S parameter, the incident photon energy $\eta = \omega/m$ and the scattering angle in the Lab system $x = \cos\theta$. In the low energy limit we get:

$$\left[\frac{d\sigma_{1/2}(g_S, x, \eta)}{d\Omega}\right]_{\eta\to 0} = \frac{1}{2}(x^2+1)r_0^2, \tag{41}$$

where $r_0^2 = e^2/(4\pi m) = \alpha/m$ is the so called classical radius of the particle. The latter expression is in accordance with the Thompson limit of the scattering. In forward direction ($x = 1$) the result is energy-independent,

$$\left[\frac{d\sigma_{1/2}(g_S, x, \eta)}{d\Omega}\right]_{\theta=0} = r_0^2. \tag{42}$$

Therefore, at this stage, we cannot say much about the gyromagnetic factor since the forward cross section is a well behaved quantity at all energies independently of g. However, the cross section under investigation depends on g in other directions. Indeed, for the total cross section at high energies one finds

$$\sigma_{1/2}(g_S, \eta)|_{\eta\to\infty} = \frac{3}{128}(g_S-2)^2 g_S^2 \sigma_T, \tag{43}$$

where $\sigma_T = (8/3)\pi r_0^2$ denotes the Thompson cross section. Then, for the formalism to describe properly the EM interaction of the electron, one has to choose $g_S = 2$ with the aim to guarantee a vanishing cross section at high energies and in accordance with the experimental observations. In this way, the EM moments from the second order formalism (36) reproduce the Dirac results when we consider the Compton scattering process. In addition it has been shown [11] that the general result for an arbitrary g factor can be used to approximately describe a fermion with $g \neq 2$, namely the proton with $g_P \approx 5.58$, in which case the cross section at a fixed angle will look as presented in Figure 1.

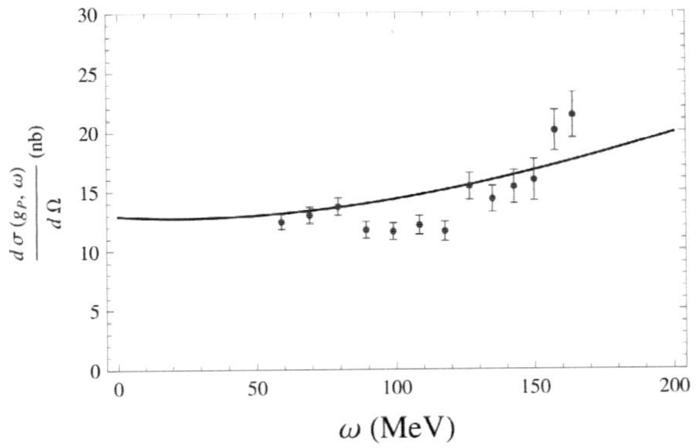

Figure 1. Differential cross section as a function of the photon energy at the fixed angle $\theta = 107^o$. The solid curve is the prediction of the second order formalism, the experimental data obtained by the TAPS experiment is extracted from Ref. [12].

4.2. Spin 1 particles in the $(1/2, 1/2)$ *representation*

The Compton scattering process can be handled in parallel to the case of spin 1/2 fermions. Also here we are going to search for additional information about the gyromagnetic factor from the behavior of the Compton cross section in the ultraviolet. Also in this case we recover the Thompson result in the low energy limit,

$$\left[\frac{d\sigma_1(g_V, x, \eta)}{d\Omega}\right]_{\eta\to 0} = \frac{1}{2}(x^2+1)r_0^2. \tag{44}$$

The more interesting part is the forward cross section since it depends on the g-value:

$$\left[\frac{d\sigma_1(g_V, x, \eta)}{d\Omega}\right]_{\theta=0} = \left(\frac{1}{24}(g_V-2)^4\eta^2+1\right)r_0^2, \tag{45}$$

with $\eta = \omega/m$ as before. We notice that, for an arbitrary g_V parameter, the differential cross section will increase with the energy, as a consequence the total cross section will be undetermined in the high energy limit unless we have $g_V = 2$, in which case we obtain an energy independent forward cross section and a total cross section looks as displayed in Figure 2. Then

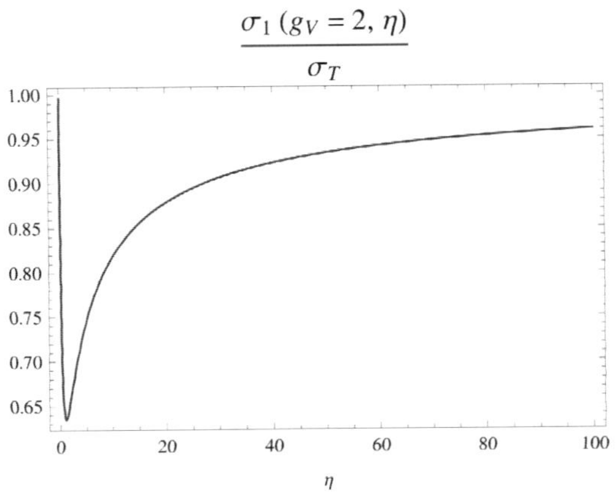

Figure 2. Result from the covariant projector formalism for the cross section of the Compton scattering process off spin 1 particles normalized to the Thompson cross section.

the forward Compton scattering process restricts the value of the gyromagnetic factor, and the related EM moments match the moments the W boson. In this way we relate the requirement on finite forward Compton scattering cross sections with the natural multipole moments of the particle.

Journal of Physics: Conference Series **378** (2012) 012033 doi:10.1088/1742-6596/378/1/012033

4.3. Spin 3/2 particles in the $(1/2, 1/2) \otimes [(1/2, 0) \oplus (0, 1/2)]$ *representation*

The Compton scattering process at low energies is insensitive to the g parameter where the classical limit is again recovered

$$\left[\frac{d\sigma_{3/2}(g,x,\eta)}{d\Omega}\right]_{\eta\to 0} = \frac{1}{2}(x^2+1)r_0^2, \tag{46}$$

the dependence on g is again present in the forward direction

$$\left[\frac{d\sigma_{3/2}(g,x,\eta)}{d\Omega}\right]_{\theta=0} = r_0^2 + \frac{r_0^2}{81}(g-2)(5(g-2)^3 - 18(g-2) - 36)\eta^2 + \frac{8r_0^2}{81}g^2(g-2)\eta^4. \tag{47}$$

Similarly to spin 1 case, the scattering process requires $g = 2$ for an energy independent forward cross section. The result for the cross section however is more complex than that of the vector case, for a spin 3/2 particle it is found that the choice of the g parameter can prevent the differential cross section to grow with the energy in the forward direction, but will not do so in other directions. At any rate, the forward scattering constraint still provides a tool to restrict the g value to $g = 2$. Same parametrization has been independently found as the only choice capable to ensure causality of the theory [3]. Notice that within the RS formalism the cross section steadily increases with the energy in all directions, including at $\theta = 0$, [13, 4]

$$\left[\frac{d\sigma_{RS}(g,x,\eta)}{d\Omega}\right]_{\theta=0} = \frac{1}{81}(8\eta^4 + 56\eta^2 + 81)r_0^2, \tag{48}$$

and in disadvantage with respect to the Covariant projector result shown in Figure 3. Inspection

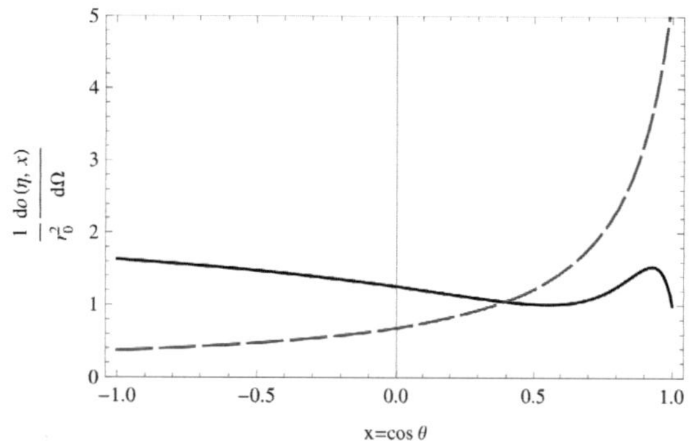

Figure 3. Differential cross section for spin 3/2 at $\eta = 1.5$ as a function of $\cos\theta$. The second order result with $g = 2$ is the thick line, the RS result is the dashed line.

of the latter Figure shows that the dashed curve (RS prediction) increases in the forward direction, while the thick one (Covariant projection prediction with $g = 2$) approaches the constant value of r_0^2. Choosing $g = 2$ furthermore would help fixing the EM moments.

5. Conclusions

- The covariant projector formalism allows for a complete description of the electromagnetic interactions of a particle of spin s embedded in a given representation of the Poincaré group.
- We have shown that the Dirac equation is not the only way to properly describe spin 1/2 fermions.
- By considering a general antisymmetric part in the equation of motion, the covariant projector formalism in the vector case is free of the problems exhibited by the Proca equation in an EM environment.

- Rarita-Scwhinger equation fails to properly describe the complete structure of a spin 3/2 particle in the $(1/2, 1/2) \otimes [(1/2, 0) \oplus (0, 1/2)]$ representation of the HLG, this can be seen in the Gordon-like decomposition of the electromagnetic current, as a consequence the EM moments of a RS particle include a dipole magnetic moment with $g = 2/3$ instead of the expected universal value $g = 2$.
- We have found that Compton scattering off a spin 3/2 particle from a second order formalism exhibits a cross section that grows with the energy except in the forward direction when $g = 2$, a condition also required for the causality of the theory, for this reason we regard the covariant projector formalism as a good candidate to describe the natural EM moments of a spin 3/2 particle.

References

[1] Proca A 1937 *Compt. Rend.* **190** 1377
[2] Rarita W and Schwinger J 1941 *Phys.Rev.* **60** 61
[3] Napsuciale M, Kirchbach M and Rodriguez S 2006 *Eur.Phys.J.* **A29** 289–306 (*Preprint* `hep-ph/0606308`)
[4] Delgado-Acosta E and Napsuciale M 2009 *Phys.Rev.* **D80** 054002 (*Preprint* `0907.1124`)
[5] Kleefeld F 2000 *J.Phys.G* (*Preprint* `nucl-th/0012076`)
[6] Foldy L L 1952 *Phys. Rev.* **87**(5) 688
[7] Aitchison I and Hey A 2003 *Gauge Theories in Particle Physics, Vol. 1* 3rd ed (Taylor & Francis)
[8] Lee T 1981 *Particle Physics and Introduction to Field Theory* 1st ed (Harwood Academic)
[9] Honzawa N and Ishida S 1992 *Phys.Rev.* **C45** 47
[10] Holstein B R 2006 *Am. J. Phys.* **74** 1104–1111 (*Preprint* `hep-ph/0607187`)
[11] Delgado-Acosta E, Napsuciale M and Rodriguez S 2011 *Phys.Rev.* **D83** 073001 (*Preprint* `1012.4130`)
[12] Olmos de León V *et al* 2001 *The European Physical Journal A - Hadrons and Nuclei* **10**(2) 207
[13] Dormuth K and Teshima R 1970 *Lett.Nuovo Cim.* **4S1** 796

Journal of Physics: Conference Series **378** (2012) 012034 doi:10.1088/1742-6596/378/1/012034

Baryon chiral perturbation theory transferred to hole-doped antiferromagnets on the honeycomb lattice

F-J Jiang[1], F Kämpfer[2], B Bessire[3], M Wirz[4], C P Hofmann[5], and U-J Wiese[6]

[1] Department of Physics, National Taiwan Normal University, 88, Sec. 4, Ting-Chou Rd. Taipei 116, Taiwan
[2] BKW FMB Energy Ltd, Energy Trading Unit, 3000 Bern, Switzerland
[3] Institute of Applied Physics, Bern University, Sidlerstrasse 5, CH-3012 Bern, Switzerland
[4] Mathematical Institute, Bern University, Sidlerstrasse 5, CH-3012 Bern, Switzerland
[5] Facultad de Ciencias, Universidad de Colima, Bernal Díaz del Castillo 340, Colima C.P. 28045, Mexico
[6] Institute for Theoretical Physics, Albert Einstein Center for Fundamental Physics, Bern University, Sidlerstrasse 5, CH-3012 Bern, Switzerland

E-mail: `christoph@ucol.mx`

Abstract. A systematic low-energy effective field theory for hole-doped antiferromagnets on the honeycomb lattice is constructed. The formalism is then used to investigate spiral phases in the staggered magnetization as well as the formation of two-hole bound states.

1. Motivation

Although the phenomenon of high-temperature superconductivity was discovered more than twenty-five years ago [1], the dynamical mechanism behind it still remains a mystery today. The reason may be readily identified: Due to the nonperturbative nature of the problem, analytic studies usually suffer from uncontrolled approximations, while numerical simulations of the microscopic Hubbard and t-J-type models suffer from a severe fermion sign problem away from half-filling.

In the present work, we will address the physics of the antiferromagnetic precursors of high-temperature superconductivity from a universal and model-independent point of view, based on the method of effective Lagrangians. The effective Lagrangian method is very well-established in particle physics, where chiral perturbation theory (CHPT) [2, 3] represents the effective theory of quantum chromodynamics (QCD). It operates in the Goldstone boson sector of QCD, taking into account the octet of the pseudoscalar mesons that emerge due to the spontaneously broken chiral symmetry. Baryon chiral perturbation theory [4, 5, 6, 7] is an extension of CHPT as it also includes heavy particles such as the baryon octet or decuplet, i.e. those degrees of freedom that remain massive in the chiral limit.

Spontaneous symmetry breaking is also a very common phenomenon in condensed matter physics: Magnons e.g. are the Goldstone bosons of a spontaneously broken spin symmetry $SU(2)_s \rightarrow U(1)_s$, while phonons emerge due to a spontaneously broken translation symmetry.

Journal of Physics: Conference Series **378** (2012) 012034 doi:10.1088/1742-6596/378/1/012034

For these systems effective Lagrangians have been constructed in Refs. [8, 9, 10, 11] and various applications have been considered in Refs. [12, 13, 14, 15, 16, 17, 18].

However, the incorporation of heavy degrees of freedom in the condensed matter domain – in a transparent and systematic manner – has only been achieved recently. An effective theory for weakly doped antiferromagnets, the analog of baryon chiral perturbation theory, has been constructed for hole- and electron-doped antiferromagnets on the square lattice in Refs.[19, 20, 21] and for hole-doped antiferromagnets on the honeycomb lattice in Ref.[22]. Further applications of the formalism were discussed in Refs.[23, 24, 25], demonstrating that the effective Lagrangian method allows one to gain insight into the physics of these insulating precursors of high-temperature superconductors in a systematic and unambiguous manner.

In this presentation we will focus on hole-doped antiferromagnets on a honeycomb lattice. We first review the systematic construction of the effective Lagrangian for magnons and holes for these systems and will then consider two applications of the formalism: The emergence of spiral phases in the staggered magnetization order parameter as well as the formation of two-hole bound states mediated by magnon exchange.

We would also like to point out that in a recent article on an analytically solvable microscopic model for a hole-doped ferromagnet in 1+1 dimensions [26], the correctness of the effective field theory approach was demonstrated by comparing the effective theory predictions with the microscopic calculation. Likewise, in a series of high-accuracy investigations of the antiferromagnetic spin-$\frac{1}{2}$ quantum Heisenberg model on a square lattice using the loop-cluster algorithm [27, 28, 29, 30], the Monte Carlo data were confronted with the analytic predictions of magnon chiral perturbation theory and the low-energy constants were extracted with permille accuracy. All these tests unambiguously demonstrate that the effective Lagrangian approach provides a rigorous and systematic derivative expansion for both ferromagnetic and antiferromagnetic systems.

2. Construction of the effective field theory for holes and magnons

The effective Lagrangian approach is based on a symmetry analysis of the underlying theory, i.e., in the present case the Hubbard or t-J-type models which are believed to be minimal models for high-temperature superconductors [31]. We thus first perform a rigorous symmetry analysis of these models in order to then construct the effective Lagrangian for holes and magnons.

2.1. Symmetry analysis

Let $c_{xs}^\dagger$ denote the operator which creates a fermion with spin $s \in \{\uparrow, \downarrow\}$ on a lattice site $x = (x_1, x_2)$. The corresponding annihilation operator is c_{xs}. These fermion operators obey the canonical anticommutation relations

$$\{c_{xs}^\dagger, c_{ys'}\} = \delta_{xy}\delta_{ss'}, \ \{c_{xs}, c_{ys'}\} = \{c_{xs}^\dagger, c_{ys'}^\dagger\} = 0. \tag{1}$$

The second quantized Hubbard Hamiltonian is defined by

$$H = -t \sum_{\langle x,y\rangle;\, s=\uparrow,\downarrow} (c_{xs}^\dagger c_{ys} + c_{ys}^\dagger c_{xs}) + U \sum_x c_{x\uparrow}^\dagger c_{x\uparrow} c_{x\downarrow}^\dagger c_{x\downarrow} - \mu' \sum_{x;\, s=\uparrow,\downarrow} c_{xs}^\dagger c_{xs}, \tag{2}$$

where $\langle x, y\rangle$ indicates summation over nearest neighbors, t is the hopping parameter, and the parameter $U > 0$ fixes the strength of the Coulomb repulsion between two fermions located on the same lattice site. The parameter μ' denotes the chemical potential.

Using the fermion creation and annihilation operators, we define the following $SU(2)_s$ Pauli spinors

$$c_x^\dagger = \left(c_{x\uparrow}^\dagger, c_{x\downarrow}^\dagger\right), \qquad c_x = \left(\begin{array}{c} c_{x\uparrow} \\ c_{x\downarrow} \end{array}\right). \tag{3}$$

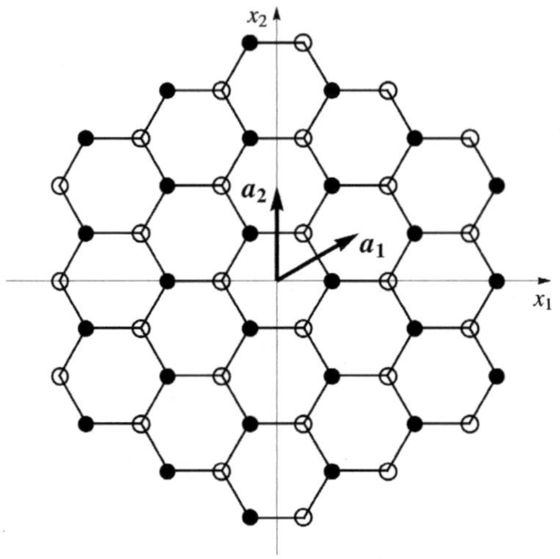

Figure 1. *Bipartite non-Bravais honeycomb lattice consisting of two triangular Bravais sublattices. The translation vectors are* a_1 *and* a_2*.*

In terms of these operators, the Hubbard model can be reformulated as

$$H = -t \sum_{\langle xy \rangle} (c_x^\dagger c_y + c_y^\dagger c_x) + \frac{U}{2} \sum_x (c_x^\dagger c_x - 1)^2 - \mu \sum_x (c_x^\dagger c_x - 1). \tag{4}$$

The parameter $\mu = \mu' - \frac{U}{2}$ controls doping where the fermions are counted with respect to half-filling.

Since all terms in the effective Lagrangian must be invariant under all symmetries of the Hubbard model, a careful symmetry analysis of Eq.(4) is needed. Let us divide the symmetries of the Hubbard model into two categories: Continuous symmetries, which are internal symmetries of Eq.(4), and discrete symmetries, which are symmetry transformations of the underlying honeycomb lattice depicted in figure 1.

The continuous symmetries are the $SU(2)_s$ spin rotation symmetry, the $U(1)_Q$ fermion number symmetry and its non-Abelian extension $SU(2)_Q$. The discrete symmetries include translations (D_i) along the two primitive lattice vectors of the honeycomb lattice, rotations (O) by 60 degrees around the center of a hexagon, and reflections (R) at the x_1-axis going through the center of the hexagon. There is also time reversal which is implemented by an anti-unitary operator T. This symmetry will be considered later on in the effective field theory framework. Furthermore, in the construction of the effective field theory for magnons and holes, it will turn out to be useful to incorporate the combined symmetry O' consisting of a spatial rotation O and a global $SU(2)_s$ spin rotation $g = i\sigma_2$. Likewise, we define the composed transformation T', consisting of a regular time-reversal T and the specific spin rotation $g = i\sigma_2$.

In [32, 33], Yang and Zhang proved the existence of a non-Abelian extension of the $U(1)_Q$ fermion number symmetry in the half-filled Hubbard model. This pseudospin symmetry contains $U(1)_Q$ as a subgroup. The $SU(2)_Q$ symmetry is realized on the square as well as on the honeycomb lattice and is generated by the three operators

$$Q^+ = \sum_x (-1)^x c_{x\uparrow}^\dagger c_{x\downarrow}^\dagger, \qquad Q^- = \sum_x (-1)^x c_{x\downarrow} c_{x\uparrow},$$

$$Q^3 = \sum_x \frac{1}{2}(c_{x\uparrow}^\dagger c_{x\uparrow} + c_{x\downarrow}^\dagger c_{x\downarrow} - 1) = \frac{1}{2}Q. \quad (5)$$

The factor $(-1)^x$ distinguishes between the two triangular sublattices A and B of the honeycomb lattice. Defining Q^1 and Q^2 through $Q^\pm = Q^1 \pm iQ^2$, one readily shows that the $SU(2)_Q$ Lie-algebra $[Q^a, Q^b] = i\varepsilon_{abc}Q^c$, with $a, b, c \in \{1, 2, 3\}$, indeed is satisfied and that $[H, \vec{Q}] = 0$ with $\vec{Q} = (Q^1, Q^2, Q^3)$ for the Hubbard Hamiltonian with $\mu = 0$.

In order to write the Hubbard Hamiltonian Eq.(2) or Eq.(4) in a manifestly invariant form under $SU(2)_s \times SU(2)_Q$, we arrange the fermion operators in a 2×2 matrix-valued operator, arriving at the fermion representation

$$C_x = \begin{pmatrix} c_{x\uparrow} & (-1)^x\, c_{x\downarrow}^\dagger \\ c_{x\downarrow} & -(-1)^x c_{x\uparrow}^\dagger \end{pmatrix}. \quad (6)$$

This allows us to write down the Hubbard Hamiltonian in the manifestly $SU(2)_s$, $U(1)_Q$, D_i, O, O' and R invariant form

$$H = -\frac{t}{2}\sum_{x,i} \mathrm{Tr}[C_x^\dagger C_{x+\hat{i}} + C_{x+\hat{i}}^\dagger C_x] + \frac{U}{12}\sum_x \mathrm{Tr}[C_x^\dagger C_x C_x^\dagger C_x] - \frac{\mu}{2}\sum_x \mathrm{Tr}[C_x^\dagger C_x \sigma_3]. \quad (7)$$

The σ_3 Pauli matrix in the chemical potential term prevents the Hubbard Hamiltonian from being invariant under $SU(2)_Q$ away from half-filling. For $\mu \neq 0$, $SU(2)_Q$ is explicitly broken down to its subgroup $U(1)_Q$. In addition, the pseudospin symmetry is realized in Eq.(7) only for nearest-neighbor hopping. As soon as next-to-nearest-neighbor hopping is included, the $SU(2)_Q$ invariance gets lost even for $\mu = 0$. The continuous $SU(2)_Q$ symmetry contains a discrete particle-hole symmetry. Although this pseudospin symmetry is not present in real materials, it will play an important role in the construction of the effective field theory. The identification of the final effective fields for holes will lead us to explicitly break the $SU(2)_Q$ symmetry in subsection 2.3.

2.2. Effective field theory for magnons

In this section we review the construction of the effective theory for antiferromagnetic magnons on the honeycomb lattice. The basic object in the effective theory is the staggered magnetization order parameter of the antiferromagnet, which is described by a unit-vector field

$$\vec{e}(x) = (\sin\theta(x)\cos\varphi(x), \sin\theta(x)\sin\varphi(x), \cos\theta(x)), \quad (8)$$

in the coset space $SU(2)_s/U(1)_s = S^2$, where $x = (x_1, x_2, t)$ denotes a point in (2+1)-dimensional space-time. A key ingredient for constructing the effective field theory is the nonlinear realization of the global $SU(2)_s$ spin symmetry which is spontaneously broken down to its $U(1)_s$ subgroup, which is discussed in detail in Ref. [19].

It turns out that the above parametrization of the magnon degrees of freedom through the vector $\vec{e}(x)$ is not appropriate for this construction. Rather, one uses composite magnon fields $v_\mu(x)$ whose components will be used to couple the magnons to the fermions. The composite magnon field is defined by

$$v_\mu(x) = u(x)\partial_\mu u(x)^\dagger, \quad (9)$$

where the matrix $u(x)$ is related to the original magnetization vector $\vec{e}(x) = (e_1(x), e_2(x), e_3(x))$ by

$$\begin{aligned} u(x) &= \frac{1}{\sqrt{2(1+e_3(x))}}\begin{pmatrix} 1+e_3(x) & e_1(x) - ie_2(x) \\ -e_1(x) - ie_2(x) & 1+e_3(x) \end{pmatrix} \\ &= \begin{pmatrix} \cos\left(\frac{\theta(x)}{2}\right) & \sin\left(\frac{\theta(x)}{2}\right)\exp(-i\varphi(x)) \\ -\sin\left(\frac{\theta(x)}{2}\right)\exp(i\varphi(x)) & \cos\left(\frac{\theta(x)}{2}\right) \end{pmatrix}. \end{aligned} \quad (10)$$

The coupling of magnons to holes is then realized through the matrix-valued anti-Hermitean field

$$v_\mu(x) = iv_\mu^a(x)\sigma_a, \qquad v_\mu^\pm(x) = v_\mu^1(x) \mp iv_\mu^2(x), \tag{11}$$

which decomposes into an Abelian "gauge" field $v_\mu^3(x)$ and two vector fields $v_\mu^\pm(x)$ "charged" under the unbroken subgroup $U(1)_s$. Here $\vec{\sigma}$ are the Pauli matrices. These fields have a well-defined transformation behavior under the symmetries which the effective theory inherits from the underlying microscopic t-J model:

$$\begin{aligned}
SU(2)_s: \quad & v_\mu(x)' = h(x)(v_\mu(x) + \partial_\mu)h(x)^\dagger, \\
D_i: \quad & {}^{D_i}v_\mu(x) = v_\mu(x), \\
O: \quad & {}^{O}v_1(x) = \tau(Ox)\Big\{\tfrac{1}{2}v_1(Ox) + \tfrac{\sqrt{3}}{2}v_2(Ox) + \tfrac{1}{2}\partial_1 + \tfrac{\sqrt{3}}{2}\partial_2\Big\}\tau(Ox)^\dagger, \\
& {}^{O}v_2(x) = \tau(Ox)\Big\{-\tfrac{\sqrt{3}}{2}v_1(Ox) + \tfrac{1}{2}v_2(Ox) - \tfrac{\sqrt{3}}{2}\partial_1 + \tfrac{1}{2}\partial_2\Big\}\tau(Ox)^\dagger, \\
& {}^{O}v_t(x) = \tau(Ox)(v_t(Ox) + \partial_t)\tau(Ox)^\dagger, \\
R: \quad & {}^{R}v_1(x) = v_1(Rx), \quad {}^{R}v_2(x) = -v_2(Rx), \\
& {}^{R}v_t(x) = v_t(Rx), \\
T: \quad & {}^{T}v_i(x) = \tau(Tx)(v_i(Tx) + \partial_i)\tau(Tx)^\dagger, \\
& {}^{T}v_t(x) = -\tau(Tx)(v_t(Tx) + \partial_t)\tau(Tx)^\dagger.
\end{aligned} \tag{12}$$

In the above symmetry transformations, we have introduced the matrix $\tau(x)$,

$$\tau(x) = \begin{pmatrix} 0 & -\exp(-i\varphi(x)) \\ \exp(i\varphi(x)) & 0 \end{pmatrix}. \tag{13}$$

Finally, the Abelian "gauge" transformation

$$h(x) = \exp(i\alpha(x)\sigma_3) \tag{14}$$

belongs to the unbroken $U(1)_s$ subgroup of $SU(2)_s$ and acts on the composite vector fields as

$$\begin{aligned}
v_\mu^3(x)' &= v_\mu^3(x) - \partial_\mu\alpha(x), \\
v_\mu^\pm(x)' &= v_\mu^\pm(x)\exp(\pm 2i\alpha(x)).
\end{aligned} \tag{15}$$

The magnon action, in terms of the composite magnon field $v_\mu(x)$, can now be expressed as

$$S[v_\mu^\pm] = \int d^2x\, dt\; 2\rho_s \left(v_i^+ v_i^- + \frac{1}{c^2} v_t^+ v_t^-\right). \tag{16}$$

Although the expression $v_\mu^+ v_\mu^-$ looks like a mass term of a charged vector field, it is just the kinetic term of a massless Goldstone boson, since it contains derivatives acting on $u(x)$.

2.3. Effective field theory for holes and magnons

Analytic calculations as well as Monte Carlo simulations in t-J-like models on the honeycomb lattice have revealed that at small doping holes occur in pockets centered at lattice momenta $k^\alpha = -k^\beta = (0, \frac{4\pi}{3\sqrt{3}a})$, and their copies in the periodic Brillouin zone [34, 35]. The honeycomb lattice, illustrated in figure 1, is a bipartite non-Bravais lattice which consists of two triangular Bravais sublattices. The corresponding Brillouin zone and the location of the corresponding hole pockets are shown in figure 2. The single-hole dispersion relation for the t-J model on the honeycomb lattice is illustrated in figure 3.

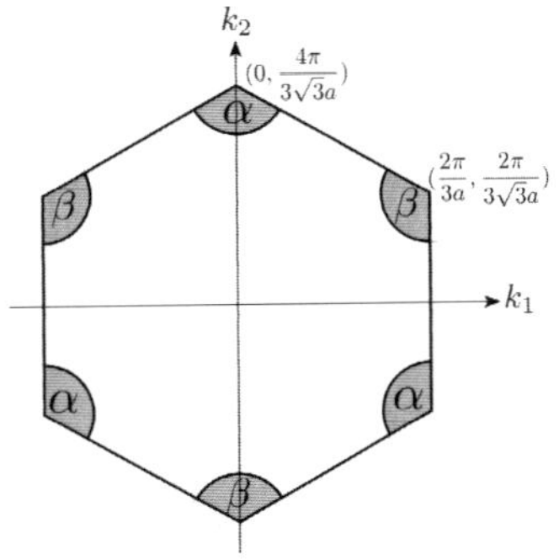

Figure 2. *Brillouin zone of the honeycomb lattice with corresponding hole pockets.*

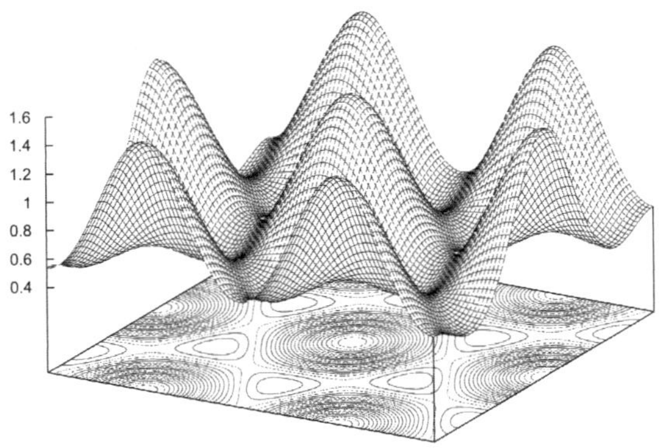

Figure 3. *Energy-momentum dispersion relation $E_h(k)/t$ for a single hole in the t-J model on the honeycomb lattice for $J/t = 2$.*

The effective field theory is defined in the space-time continuum and the holes are described by Grassmann-valued fields $\psi_s^f(x)$ carrying a "flavor" index $f = \alpha, \beta$ that characterizes the corresponding hole pocket. The index $s = \pm$ denotes spin parallel (+) or antiparallel (−) to the local staggered magnetization. Under the various symmetry operations the hole fields transform as

$$
\begin{aligned}
SU(2)_s: &\quad \psi_\pm^f(x)' = \exp(\pm i\alpha(x))\psi_\pm^f(x), \\
U(1)_Q: &\quad {}^Q\psi_\pm^f(x) = \exp(i\omega)\psi_\pm^f(x), \\
D_i: &\quad {}^{D_i}\psi_\pm^f(x) = \exp(ik_i^f a_i)\psi_\pm^f(x), \\
O: &\quad {}^O\psi_\pm^\alpha(x) = \mp\exp(\mp i\varphi(Ox) \pm i\frac{2\pi}{3})\psi_\mp^\beta(Ox), \\
&\quad {}^O\psi_\pm^\beta(x) = \mp\exp(\mp i\varphi(Ox) \mp i\frac{2\pi}{3})\psi_\mp^\alpha(Ox), \\
R: &\quad {}^R\psi_\pm^\alpha(x) = \psi_\pm^\beta(Rx), \qquad {}^R\psi_\pm^\beta(x) = \psi_\pm^\alpha(Rx),
\end{aligned}
$$

Journal of Physics: Conference Series **378** (2012) 012034 doi:10.1088/1742-6596/378/1/012034

$$
\begin{aligned}
T: \quad & {}^T\psi^{\alpha}_{\pm}(x) = \exp(\mp i\varphi(Tx))\psi^{\beta\dagger}_{\pm}(Tx), \\
& {}^T\psi^{\beta}_{\pm}(x) = \exp(\mp i\varphi(Tx))\psi^{\alpha\dagger}_{\pm}(Tx), \\
& {}^T\psi^{\alpha\dagger}_{\pm}(x) = -\exp(\pm i\varphi(Tx))\psi^{\beta}_{\pm}(Tx), \\
& {}^T\psi^{\beta\dagger}_{\pm}(x) = -\exp(\pm i\varphi(Tx))\psi^{\alpha}_{\pm}(Tx).
\end{aligned} \tag{17}
$$

Here $U(1)_Q$ is the fermion number symmetry of the holes. Interestingly, in the effective continuum theory the location of holes in lattice momentum space manifests itself as a "charge" k_i^f under the displacement symmetry D_i.

Now that the relevant low-energy degrees of freedom have been identified and the transformation rules of the corresponding fields have been worked out, the construction of the effective action is uniquely determined. The low-energy effective action of magnons and holes is constructed as a derivative expansion. At low energies, terms with a small number of derivatives dominate the dynamics. Since the holes are heavy nonrelativistic fermions, one time-derivative counts like two spatial derivatives. Here we limit ourselves to terms with at most one temporal or two spatial derivatives. One then constructs all terms consistent with the symmetries listed above. The effective action can be written as

$$
S\left[\psi^{f\dagger}_{\pm}, \psi^{f}_{\pm}, v^{\pm}_{\mu}, v^{3}_{\mu}\right] = \int d^2x \; dt \sum_{n_\psi} \mathcal{L}_{n_\psi}, \tag{18}
$$

where n_ψ denotes the number of fermion fields that the various terms contain. The leading terms in the pure magnon sector take the form

$$
\mathcal{L}_0 = 2\rho_s \left(v_i^+ v_i^- + \frac{1}{c^2} v_t^+ v_t^- \right). \tag{19}
$$

The leading terms with two fermion fields (containing at most one temporal or two spatial derivatives), describing the propagation of holes as well as their couplings to magnons, are given by

$$
\begin{aligned}
\mathcal{L}_2 = \sum_{f=\alpha,\beta;\; s=+,-} \Big[& M\psi^{f\dagger}_s \psi^f_s + \psi^{f\dagger}_s D_t \psi^f_s + \tfrac{1}{2M'} D_i \psi^{f\dagger}_s D_i \psi^f_s + \Lambda \psi^{f\dagger}_s (isv_1^s + \sigma_f v_2^s)\psi^f_{-s} \\
& + iK[(D_1 + is\sigma_f D_2)\psi^{f\dagger}_s (v_1^s + is\sigma_f v_2^s)\psi^f_{-s} \\
& - (v_1^s + is\sigma_f v_2^s)\psi^{f\dagger}_s (D_1 + is\sigma_f D_2)\psi^f_{-s}] \\
& + \sigma_f L \psi^{f\dagger}_s \epsilon_{ij} f^3_{ij} \psi^f_s + N_1 \psi^{f\dagger}_s v_i^s v_i^{-s} \psi^f_s \\
& + is\sigma_f N_2 (\psi^{f\dagger}_s v_1^s v_2^{-s} \psi^f_s - \psi^{f\dagger}_s v_2^s v_1^{-s} \psi^f_s) \Big].
\end{aligned} \tag{20}
$$

Here M is the rest mass and M' is the kinetic mass of a hole, Λ and K are hole-one-magnon couplings, while L, N_1, and N_2 are hole-two-magnon couplings. Note that all low-energy constants are real-valued. The sign σ_f is $+$ for α and $-$ for β. We have introduced the field strength tensor of the composite Abelian "gauge" field

$$
f^3_{ij}(x) = \partial_i v^3_j(x) - \partial_j v^3_i(x), \tag{21}
$$

and the covariant derivatives D_t and D_i acting on $\psi^f_{\pm}(x)$ as

$$
\begin{aligned}
D_t \psi^f_{\pm}(x) &= \left[\partial_t \pm i v^3_t(x) - \mu\right] \psi^f_{\pm}(x), \\
D_i \psi^f_{\pm}(x) &= \left[\partial_i \pm i v^3_i(x)\right] \psi^f_{\pm}(x).
\end{aligned} \tag{22}
$$

Journal of Physics: Conference Series **378** (2012) 012034 doi:10.1088/1742-6596/378/1/012034

The chemical potential μ enters the covariant time-derivative like an imaginary constant vector potential for the fermion number symmetry $U(1)_Q$. It is remarkable that the term proportional to Λ with just a single (uncontracted) spatial derivative satisfies all symmetries. Due to the small number of derivatives it contains, this term dominates the low-energy dynamics of a lightly hole-doped antiferromagnet on the honeycomb lattice. Interestingly, for antiferromagnets on the square lattice, a corresponding term, which was first identified by Shraiman and Siggia, is also present in the hole-doped case [20]. On the other hand, a similar term is forbidden by symmetry reasons in the electron-doped case [21]. For the honeycomb geometry we even identify a second hole-one-magnon coupling, K, whose contribution, however, is sub-leading. Interestingly, hole- or electron-doped antiferromagnets on the square lattice do not allow terms containing the field-strength tensor f_{ij} in $\mathcal{L}_2$.

Finally, the leading terms without derivatives and with four fermion fields are given by

$$\begin{aligned}\mathcal{L}_4 &= \sum_{s=+,-} \Big\{ \frac{G_1}{2} (\psi_s^{\alpha\dagger}\psi_s^{\alpha}\psi_{-s}^{\alpha\dagger}\psi_{-s}^{\alpha} + \psi_s^{\beta\dagger}\psi_s^{\beta}\psi_{-s}^{\beta\dagger}\psi_{-s}^{\beta}) \\ &+ G_2\, \psi_s^{\alpha\dagger}\psi_s^{\alpha}\psi_s^{\beta\dagger}\psi_s^{\beta} + G_3\, \psi_s^{\alpha\dagger}\psi_s^{\alpha}\psi_{-s}^{\beta\dagger}\psi_{-s}^{\beta} \Big\}. \end{aligned} \tag{23}$$

The low-energy four-fermion coupling constants G_1, G_2, and G_3 are also real-valued. Although potentially invariant under all symmetries, terms with two identical hole fields vanish due to the Pauli principle.

It is important to note that in the above construction of the effective Lagrangian for the t-J model which contains holes only, a crucial step was to identify the degrees of freedom that correspond to the holes. In order to remove the electron degrees of freedom one has to explicitly break the particle-hole $SU(2)_Q$ symmetry, leaving the ordinary fermion number symmetry $U(1)_Q$ intact. This task can be achieved by constructing all possible fermionic mass terms that are invariant under the various symmetries. Picking the eigenvectors which correspond to the lowest eigenvalues of the mass matrices then allows one to separate electrons from holes. This procedure is described in detail in Ref. [22]

We would like to point out that the leading order terms in the effective Lagrangian for magnons and holes constructed above exhibit two accidental global symmetries. First, we notice that for $c \to \infty$ and without the term proportional to iK in $\mathcal{L}_2$, Eqs.(19) and (20), have an accidental Galilean boost symmetry. Although the Galilean boost symmetry is explicitly broken at higher orders of the derivative expansion, this symmetry has physical implications: The leading one-magnon exchange between two holes, to be discussed in the next section, can be investigated in their rest frame without loss of generality.

In addition, we notice an accidental global rotation symmetry $O(\gamma)$. Except for the term proportional to iK, $\mathcal{L}_2$ of Eq.(20) is invariant under a continuous spatial rotation by an angle γ. This symmetry is not present in the Λ-term of the square lattice. The $O(\gamma)$ invariance has some interesting implications for the spiral phases in a lightly doped antiferromagnet on the honeycomb lattice, as we will see in the next section.

3. Applications

In this section we will consider two nontrivial applications of the effective Lagrangian constructed above: The emergence of spiral phases in the staggered magnetization vector order parameter and the formation of two-hole bound states.

3.1. Spiral phases

Let us first investigate whether so-called spiral phases in the staggered magnetization can occur upon hole–doping. This problem has been studied before using microscopic theories

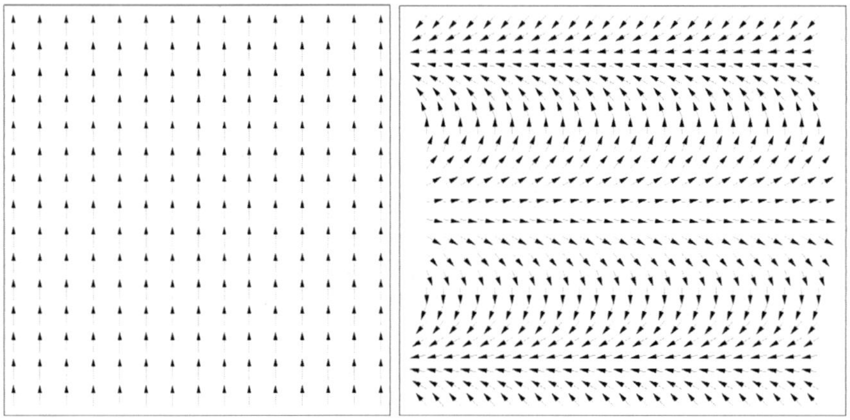

Figure 4. Left panel: Homogeneous phase with constant staggered magnetization. Right panel: Spiral phase with helical structure in the staggered magnetization.

[36, 37, 38, 39]. In Ref. [24] the problem was studied systematically in the case of a square lattice using effective Lagrangians. Here we consider the honeycomb lattice – details of the calculation can be found in Ref. [25].

In our study we assume that the holes are doped into the system homogeneously and thus limit ourselves to configurations of the staggered magnetization that are either homogeneous themselves or generate a constant background field $v_i(x)$ for the charge carriers. As was shown in [24], the most general configuration of this kind represents a spiral in the staggered magnetization. We also assume that four-fermion contact interactions are weak and take them into account perturbatively – again, for details see Refs.[21, 24].

When one populates the various hole pockets, one must distinguish a total of four different cases. First we have α-pockets and β-pockets. Moreover, the energy eigenstates of a hole pocket f can acquire two values, E_-^f and E_+^f, where the former corresponds to the lower energy. One thus starts considering the case of populating all four hole pockets, i.e. with both flavors $f = \alpha, \beta$ and with both energy indices $\pm$. One then proceeds with populating only three, only two and, finally, only one hole pocket. This gives rise to various phases of the staggered magnetization, which can be either homogeneous or a spiral phase as depicted in figure 4.

More precisely, the energy densities of the various phases take the form

$$\epsilon_i = \epsilon_0 + Mn + \frac{1}{2}\kappa_i n^2. \tag{24}$$

Here ϵ_0 is the energy density of the system at half-filling and n is the total density of holes. The index i refers to the number of hole pockets that are populated in the corresponding phase. The compressibilities κ_i are given by

$$\begin{aligned}
\kappa_1 &= \frac{2\pi}{M'} - \frac{\Lambda^2}{4\rho_s}, \\
\kappa_2 &= \frac{\pi}{M'} - \frac{\Lambda^2}{4\rho_s} + \frac{1}{4}(G_2 + G_3), \\
\kappa_3 &= \frac{2\pi}{3M'}\left(1 - \frac{1}{8}\frac{M'\Lambda^2}{3\pi\rho_s - M'\Lambda^2}\right)
\end{aligned}$$

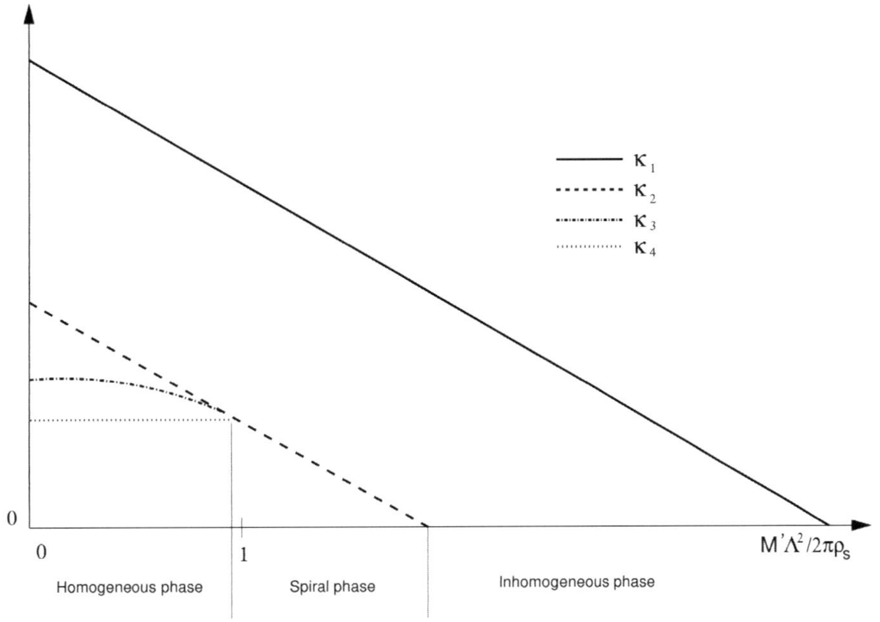

Figure 5. *The compressibilities κ_i as functions of $M'\Lambda^2/2\pi\rho_s$ determine the stability ranges of the various phases. A homogeneous phase, a spiral, or an inhomogeneous phase are energetically favorable, for large, intermediate, and small values of ρ_s, respectively.*

$$
\begin{aligned}
&+ \frac{4\pi\rho_s - M'\Lambda^2}{(3\pi\rho_s - M'\Lambda^2)^2}\frac{1}{16}\Big[8(G_1 + G_2 + G_3)\pi\rho_s \\
&- (4G_1 + 3G_2 + 3G_3)M'\Lambda^2\Big], \\
\kappa_4 &= \frac{\pi}{2M'} + \frac{1}{4}(G_1 + G_2 + G_3), \qquad (25)
\end{aligned}
$$

and shown in figure 5 as functions of $M'\Lambda^2/2\pi\rho_s$. For large values of ρ_s, spiral phases cost a large amount of magnetic energy and the homogeneous phase is more stable. To be more precise, in this regime one has $\kappa_4 < \kappa_3 < \kappa_2 < \kappa_1$. Notice that κ_1 is always larger than κ_2 for any value of ρ_s. As ρ_s decreases and reaches the value

$$\rho_s = \frac{M'\Lambda^2}{2\pi} + \frac{(M')^2\Lambda^2 G_1}{4\pi^2}, \qquad (26)$$

at leading order in the 4-fermi couplings one finds $\kappa_2 = \kappa_3 = \kappa_4$. For smaller values of ρ_s, the two-pocket spiral is energetically favored until κ_2 becomes negative and the system becomes unstable against the formation of spatial inhomogeneities of a yet undetermined type.

Interestingly, unlike in the square lattice case, due to the accidental continuous $O(\gamma)$ spatial rotation symmetry, at leading order a spiral does not have an a priori preferred spatial direction. It is instructive to compare the results presented here with the results obtained in the square lattice case [24]. Qualitatively the stability ranges of various phases are the same for both lattice geometries except that the one-pocket spiral is never energetically favored on the honeycomb lattice while it is favorable in a small parameter regime on the square lattice.

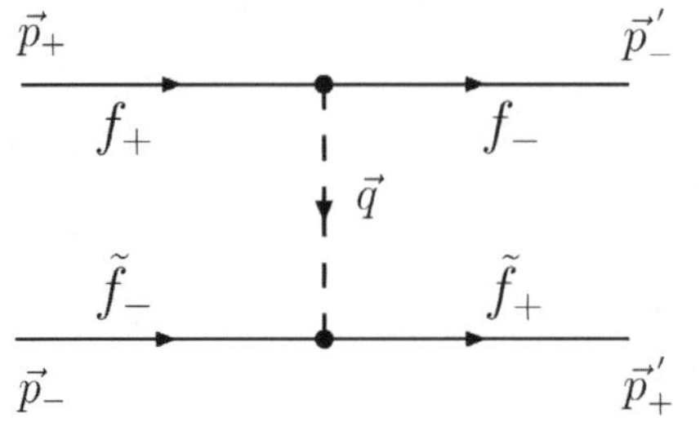

Figure 6. *Tree-level Feynman diagram for one-magnon exchange between two holes.*

3.2. Two-hole bound states

In the effective theory framework, at low energies, holes interact with each other via magnon exchange. Since the long-range dynamics is dominated by one-magnon exchange, we will calculate the one-magnon exchange potentials between two holes of the same flavor α and β and of different flavor, and then address the question regarding the formation of two-hole bound states. The same problem was also considered in Refs. [40, 41, 42] using microscopic theories.

We expand in the magnon fluctuations $m_1(x)$ and $m_2(x)$ around the ordered staggered magnetization

$$\vec{e}(x) = \left(\frac{m_1(x)}{\sqrt{\rho_s}}, \frac{m_2(x)}{\sqrt{\rho_s}}, 1\right) + \mathcal{O}\left(m^2\right). \tag{27}$$

For the composite magnon fields this implies

$$\begin{aligned} v_\mu^\pm(x) &= \frac{1}{2\sqrt{\rho_s}}\partial_\mu[m_2(x) \pm i m_1(x)] + \mathcal{O}\left(m^3\right), \\ v_\mu^3(x) &= \frac{1}{4\rho_s}[m_1(x)\partial_\mu m_2(x) - m_2(x)\partial_\mu m_1(x)] + \mathcal{O}\left(m^4\right). \end{aligned} \tag{28}$$

Vertices with $v_\mu^3(x)$ involve at least two magnons, such that one-magnon exchange results from vertices with $v_\mu^\pm(x)$ only. As a consequence, two holes can exchange a single magnon only if they have anti-parallel spins (+ and $-$), which are both flipped in the magnon-exchange process. We denote the momenta of the incoming and outgoing holes by $\vec{p}_\pm$ and $\vec{p}_\pm{}'$, respectively. The momentum carried by the exchanged magnon is denoted by $\vec{q}$. The incoming and outgoing holes are asymptotic free particles with momentum $\vec{p} = (p_1, p_2)$ and energy $E(\vec{p}) = M + p_i^2/2M'$. One-magnon exchange between two holes is associated with the Feynman diagram in figure 6. Evaluating these Feynman diagrams, one finds the resulting potentials for the various combinations of flavors $f, \tilde{f} \in \{\alpha, \beta\}$ and couplings $F, \tilde{F} \in \{\Lambda, K\}$.

The leading contribution to the low-energy physics, as we noted earlier, comes from the Λ vertex. From here on, we therefore concentrate on the potential with two Λ vertices only. In coordinate space the $\Lambda\Lambda$-potentials are given by

$$\langle \vec{r}_+{}'\vec{r}_-{}'|V_{\Lambda\Lambda}^{f\tilde{f}}|\vec{r}_+\vec{r}_-\rangle = V_{\Lambda\Lambda}^{f\tilde{f}}(\vec{r})\ \delta(\vec{r}_+ - \vec{r}_-{}')\ \delta(\vec{r}_- - \vec{r}_+{}'), \tag{29}$$

with

$$V_{\Lambda\Lambda}^{ff}(\vec{r}) = -\frac{\Lambda^2}{2\rho_s}\delta^{(2)}(\vec{r}), \qquad V_{\Lambda\Lambda}^{ff'}(\vec{r}) = \frac{\Lambda^2}{2\pi\rho_s\vec{r}^{\,2}}\exp(2i\sigma_f\varphi). \tag{30}$$

Here $\vec{r} = \vec{r}_+ - \vec{r}_-$ denotes the distance vector between the two holes and φ is the angle between $\vec{r}$ and the x_1-axis. The δ-functions in Eq.(29) ensure that the holes do not change their position during the magnon exchange. Note that the one-magnon exchange potentials are instantaneous although magnons travel with the finite speed c. Retardation effects occur only at higher orders.

Interestingly, in the $\Lambda\Lambda$ channel, one-magnon exchange over long distances between two holes can only happen for holes of opposite flavor. For two holes of the same flavor, one-magnon exchange acts as a contact interaction. Here we concentrate on the long-range physics of weakly bound states of holes and therefore will only consider the binding of holes of different flavor.

Let us therefore investigate the Schrödinger equation for the relative motion of two holes with flavors α and β. In the following, we will take care of short distance interactions by imposing a hard-core boundary condition on the pair's wave function. Due to the accidental Galilean boost invariance, without loss of generality, we can consider the hole pair in its rest frame. The total kinetic energy of the two holes is given by

$$T = \sum_{f=\alpha,\beta} T^f = \sum_{f=\alpha,\beta} \frac{p_i^2}{2M'} = \frac{p_i^2}{M'}. \tag{31}$$

We now introduce the two probability amplitudes $\Psi_1(\vec{r})$ and $\Psi_2(\vec{r})$ which represent the two flavor-spin combinations $\alpha_+\beta_-$ and $\alpha_-\beta_+$, respectively, with the distance vector $\vec{r}$ to point from the β to the α hole. Because the holes undergo a spin flip during the magnon exchange, the two probability amplitudes are coupled through the magnon exchange potentials and the Schrödinger equation describing the relative motion of the hole pair is a two-component equation. Using the explicit form of the potentials, it takes the form

$$\begin{pmatrix} -\frac{1}{M'}\Delta & \gamma\frac{1}{r^2}\exp(-2i\varphi) \\ \gamma\frac{1}{r^2}\exp(2i\varphi) & -\frac{1}{M'}\Delta \end{pmatrix} \begin{pmatrix} \Psi_1(\vec{r}) \\ \Psi_2(\vec{r}) \end{pmatrix} = E \begin{pmatrix} \Psi_1(\vec{r}) \\ \Psi_2(\vec{r}) \end{pmatrix}, \tag{32}$$

with

$$\gamma = \frac{\Lambda^2}{2\pi\rho_s}. \tag{33}$$

As it turns out, magnon-mediated forces can lead to bound states only if the low-energy constant Λ is larger than a critical value given by

$$\Lambda_c = \sqrt{\frac{2\pi\rho_s}{M'}}. \tag{34}$$

Interestingly, the same critical value also arose in the investigation of spiral phases in a lightly doped antiferromagnet on the honeycomb lattice in Eq.(26). There it marked the point where spiral phases become energetically favorable compared to the homogeneous phase.

With the separation ansatz

$$\Psi_1(r,\varphi) = R_1(r)\exp(im_1\varphi), \qquad \Psi_2(r,\varphi) = R_2(r)\exp(im_2\varphi), \tag{35}$$

the relevant equation which is associated with an attractive potential and thus potentially leads to the formation of bound states amounts to

$$\left[-\left(\frac{d^2}{dr^2} + \frac{1}{r}\frac{d}{dr}\right) + (1 - \gamma M')\frac{1}{r^2}\right] R(r) = -M'|E|R(r), \tag{36}$$

with $E = -|E|$ and $R(r) = R_1(r) - R_2(r)$. It has to be pointed out that this equation refers to the case $m_1 = -1$ and $m_2 = 1$ where the system (32) can be decoupled (see [22]).

The same equation also occurred in the square lattice case [20, 23] and can be solved along the same lines. As it stands, the equation is ill-defined because the $1/r^2$ potential is too singular at the origin. However, we have not yet included the contact interaction proportional to the 4-fermion coupling G_3. Here, in order to keep the calculation analytically feasible, we model the short-range repulsion by a hard core radius r_0, i.e. we require $R(r_0) = 0$ for $r \leq r_0$. Eq.(36) is solved by a modified Bessel function

$$R(r) = AK_\nu(\sqrt{M'|E|}r), \qquad \nu = i\sqrt{\gamma M' - 1}, \tag{37}$$

with A being a normalization constant. Demanding that the wave function vanishes at the hard core radius gives a quantization condition for the bound state energy. The quantum number n then labels the n-th excited state. For large n, the binding energy is given by

$$E_n \sim -\frac{1}{M' r_0^2} \exp\left(\frac{-2\pi n}{\sqrt{\gamma M' - 1}}\right). \tag{38}$$

The binding is exponentially small in n and there are infinitely many bound states. While the highly excited states have exponentially small energy, for sufficiently small r_0 the ground state could have a small size and be strongly bound. However, for short-distance physics the effective theory should not be trusted quantitatively. Still, as long as the binding energy is small compared to the relevant high-energy scales, our result is valid and receives only small corrections from higher-order effects such as two-magnon exchange.

We now turn to the discussion of the angular part of the wave equation. The ansatz (35) leads to the following solution for the ground state wave function

$$\Psi(r,\varphi) = \begin{pmatrix} \Psi_1(\vec{r}) \\ \Psi_2(\vec{r}) \end{pmatrix} = R(r) \begin{pmatrix} \exp(-i\varphi) \\ -\exp(i\varphi) \end{pmatrix}. \tag{39}$$

Under the 60 degrees rotation O, using the transformation rules of Eq.(17), one obtains

$$^O\Psi(r,\varphi) = -\,\Psi(r,\varphi). \tag{40}$$

Interestingly, the wave function for the ground state of two holes of flavors α and β thus exhibits f-wave symmetry.[1] The corresponding probability distribution depicted in figure 7, on the other hand, seems to show s-wave symmetry. However, the relevant phase information is not visible in this picture, because only the probability density is shown. Interestingly, for two-hole bound states on the square lattice, the wave function for the the ground state of two holes of flavors α and β shows p-wave symmetry, while the corresponding probability distribution (which again does not contain the relevant phase information) resembles $d_{x^2-y^2}$ symmetry [20]. Remarkably, the ground state wave function (39) of a bound hole pair on the honeycomb lattice remains invariant under the reflection symmetry R, the shift symmetries D_i, as well as under the accidental continuous rotation symmetry $O(\gamma)$.

We find it quite remarkable that the f-wave character of the two-hole bound state on the honeycomb lattice is an immediate consequence of our systematic effective field theory analysis. The question regarding the true symmetry of the pairing state realized in the dehydrated version of $Na_2CoO_2 \times y\, H_2O$, still seems to be controversial [43]. Still, a careful analysis of the available experimental data for this compound suggests that the pairing symmetry indeed is f-wave [44].

4. Conclusions

The effective theory for the insulating antiferromagnetic precursors of high-temperature superconductors is the condensed matter analog of baryon chiral perturbation theory: Magnons

[1] Strictly speaking, the continuum classification scheme of angular momentum eigenstates does not apply here, since we are not dealing with a continuous rotation symmetry.

Journal of Physics: Conference Series **378** (2012) 012034 doi:10.1088/1742-6596/378/1/012034

Figure 7. *Probability distribution for the ground state of two holes of flavors α and β.*

play the role of the pseudoscalar mesons, while the holes and electrons are the analog of the baryon octet. We have analyzed the symmetries of the underlying Hubbard and $t-J$-type models in detail and reviewed the systematic construction of the effective low-energy field theory for weakly hole-doped antiferromagnets on the honeycomb lattice. The procedure is fully systematic order by order in a derivative expansion of the magnon and hole fields. As two nontrivial applications we have considered the existence of spiral phases in the staggered magnetization order parameter and the formation of two-hole bound states mediated by magnon exchange.

Regarding spiral phases, we have limited ourselves to constant composite vector fields $v_i(x)'$ which implies that the fermions experience a constant background field. Unlike in the square lattice case, due to the accidental continuous $O(\gamma)$ spatial rotation symmetry, at leading order a spiral does not have an a priori preferred spatial direction. However, since the $O(\gamma)$ symmetry is broken explicitly by the higher-order terms, once such terms are included, one expects the spiral to align with a lattice direction. We also investigated the stability of spiral phases in the presence of 4-fermion couplings. If these couplings can be treated perturbatively, for sufficiently large values of ρ_s, the homogeneous phase is energetically favored. With decreasing ρ_s, a two-pocket spiral becomes energetically more favorable. In contrast to the square lattice case, the one-pocket spiral is never favored. For small values of ρ_s the two-pocket spiral becomes unstable against the formation of inhomogeneities of a yet undetermined type.

Regarding the formation of two-hole bound states mediated by magnon-exchange, we have studied the effect of the magnon-hole vertex. Again in contrast to the square lattice case, it turned out that the magnon-hole coupling constant Λ must exceed a critical value in order to obtain two-hole bound states. Our analysis implies that the wave function for the ground state of two holes of flavors α and β exhibits f-wave symmetry (while the corresponding probability distribution seems to suggest s-wave symmetry). This is quite different from the square lattice case, where the wave function for the ground state of two holes of flavors α and β exhibits p-wave symmetry (while the corresponding probability distribution resembles $d_{x^2-y^2}$ symmetry).

It is quite remarkable that all these results unambiguously follow from the very few basic assumptions of the systematic low-energy effective theory, such as symmetry, locality, and unitarity. The effective theory provides a theoretical framework in which the low-energy dynamics of lightly hole-doped antiferromagnets can be investigated in a systematic manner. In particular, after the low-energy parameters have been adjusted appropriately, the resulting low-energy physics is completely equivalent to the one of the Hubbard or t-J model.

Acknowledgments

C.P.H. would like to thank the members of the Institute for Theoretical Physics at Bern University for their warm hospitality during a visit at which this project was completed. Support by CONACYT grant No. 50744-F is gratefully acknowledged.

References

[1] Bednorz J C and Müller K A 1986 *Z. Phys.* B **64** 189
[2] Weinberg S 1979 *Physica* **96 A** 327
[3] Gasser J and Leutwyler H 1985 *Nucl. Phys.* B **250** 465
[4] Gasser J, Sainio M E and Švarc A 1988 *Nucl. Phys.* B **307** 779
[5] Jenkins E and Manohar A 1991 *Phys. Lett.* B **255** 558
[6] Bernard V , Kaiser N, Kambor J, and Meissner U-G 1992 *Nucl. Phys.* B **388** 315
[7] Becher T and Leutwyler H 1999 *Eur. Phys. J.* C **9** 643
[8] Hasenfratz P and Niedermayer F 1993 *Z. Phys.* B **92** 91
[9] Leutwyler H 1994 *Phys. Rev.* D **49** 3033
[10] Leutwyler H 1997 *Helv. Phys. Acta* **70** 275
[11] Román J M and Soto J 1999 *Int. J. Mod. Phys.* B **13** 755
[12] Hofmann C P 1999 *Phys. Rev.* B **60** 388
[13] Hofmann C P 1999 *Phys. Rev.* B **60** 406
[14] Román J M and Soto J 1999 *Ann. Phys.* **273** 37
[15] Román J M and Soto J 2000 *Phys. Rev.* B **62** 3300
[16] Hofmann C P 2002 *Phys. Rev.* B **65** 094430
[17] Hofmann C P 2010 *Phys. Rev.* B **81** 014416
[18] Hofmann C P 2011 *Phys. Rev.* B **84** 064414
[19] Kämpfer F, Moser M and Wiese U-J 2005 *Nucl. Phys.* B **729** 317
[20] Brügger C, Kämpfer F, Moser M, Pepe M and Wiese U-J 2006 *Phys. Rev.* B **74** 224432
[21] Brügger C, Hofmann C P, Kämpfer F, Moser M, Pepe M and Wiese U-J 2007 *Phys. Rev.* B **75** 214405
[22] Kämpfer F, Bessire, B, Wirz, M, Hofmann C P, Jiang F-J and Wiese 2012 to appear in *Phys. Rev.* B
[23] Brügger C, Kämpfer F, Pepe M and Wiese U-J 2006 *Eur. Phys. J.* B **53** 433
[24] Brügger C, Hofmann C P, Kämpfer F, Pepe M and Wiese U-J 2007 *Phys. Rev.* B **75** 014421
[25] Jiang F-J, Kämpfer F, Hofmann C P, and Wiese U-J 2009 *Eur. Phys. J.* B **69** 473
[26] Gerber U, Hofmann C P, Kämpfer F and Wiese U-J 2010 *Phys. Rev.* B **81** 064414
[27] Wiese U-J and Ying H P 1994 *Z. Phys.* B **93** 147
[28] Gerber U, Hofmann C P, Jiang F-J, Nyfeler M and Wiese U-J 2009 *J. Stat. Mech.: Theory Exp.* P03021
[29] Jiang F-J and Wiese U-J 2011 *Phys. Rev.* B **83** 155120
[30] Gerber U, Hofmann C P, Jiang F-J, Palma G, Stebler P and Wiese U-J 2011 *J. Stat. Mech.: Theory Exp.* P06002
[31] Anderson P W 1987 *Science* **235** 1196
[32] Zhang S 1990 *Phys. Rev. Lett.* **65** 120
[33] Yang C N and Zhang S C 1990 *Mod. Phys. Lett.* B **4** 759
[34] Lüscher A , Läuchli A, Zheng W, and Sushkov O 2006 *Phys. Rev.* B **73** 155118
[35] Jiang F-J, Kämpfer F, Nyfeler M and Wiese U-J 2008 *Phys. Rev.* B **78** 214406
[36] Shraiman B I and Siggia E D 1988 *Phys. Rev. Lett.* **60** 740; **61** 467; **62** 1564
[37] Shraiman B I and Siggia E D 1992 *Phys. Rev.* B **46** 8305
[38] Sushkov O P and Kotov V N 2004 *Phys. Rev.* B **70** 024503
[39] Kotov V N and Sushkov O P 2004 *Phys. Rev.* B **70** 195105
[40] Sushkov O P and Flambaum V V 1993 *Physica* C **206** 269
[41] Kuchiev M J and Sushkov O P 1993 *Physica* C **218** 197
[42] Flambaum V V, Kuchiev M Y and Sushkov O P 1994 *Physica* C **227** 267
[43] Ivanova N B, Ovchinnikov S G, Korshunov M M, Eremin I M and Kazak N V 2009 *Phys. Usp.* **52** 789
[44] Mazin I I and Johannes M D 2005 *Nature Physics* **1** 91

Journal of Physics: Conference Series **378** (2012) 012035 doi:10.1088/1742-6596/378/1/012035

$q - \bar{q}$, tetraquark and mixing in a dynamical model of the strong interaction

I. Toledano Juárez and G. Toledo Sánchez

Instituto de Física, Universidad Nacional Autónoma de México, AP 20-364, México D.F. 01000, México

E-mail: toledo@fisica.unam.mx

Abstract. We perform a variational Monte Carlo simulation of a system consisting of two quarks and two antiquarks. Using a dynamical model of the strong interaction (string-flip model) we explore the formation of mesons and/or a tetraquark for a set of different density conditions. We characterize the single properties of these states and the modifications upon the mixing. The competition among the different configurations and its implications on several observables are exhibited.

1. Introduction

The Standard Model of Particle Physics tell us that hadronic matter is composed of quarks, and that these bound states must be color-singlets. The known states that can be color-singlets are the mesons ($q - \bar{q}$), baryons ($3q$) and their corresponding antiparticles. However, glueballs, hybrids and multiquark states are also color-singlets and are predicted as bound states in models inspired by Quantum Chromodynamics (QCD). In particular, mesons with exotic quantum numbers are non-($q - \bar{q}$) objects. The multiquark ($q - q - \bar{q} - \bar{q}$) or tetraquark state is the simplest on its kind and has been under study theoretically and experimentally, in order to find clear signatures of its existence [1, 2, 3]. In this work we present a preliminary study on the formation of this state from the presence of two initially uncorrelated q-qbar states, which are externally forced to approach to each other. We analyze the system using a QCD-inspired model (*String-Flip*)[4], incorporating a many-body quark potential able to confine quarks within color-singlet clusters, quantum correlations between quarks and a 3-Dimensional evolution of the system. At low densities the model describes a system of quarks as isolated hadrons, whereas at high densities, the system resembles a free Fermi gas of quarks. In the present case we consider all the quarks (antiquarks) as made of a single light flavor and all with the same color (anticolor) quantum number.

2. Variational wave function

The system wave function requires to account for the correlations between identical particles and the clustering formation in a color singlet way. These can be achieved by considering a variational wave function of the following form:

$$\Psi_\lambda(\vec{r}) = e^{-\lambda V(\vec{r})}\Phi_{FG}, \tag{1}$$

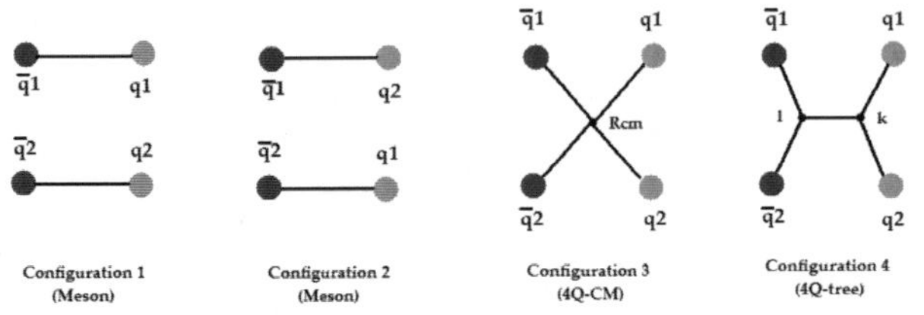

Figure 1. Possible configurations of the system. Two meson systems and two tetraquark systems

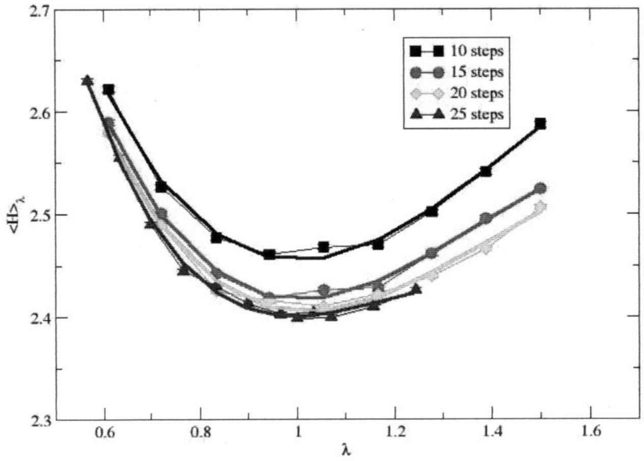

Figure 2. Example of the minimization of the energy as a function of the variational parameter and the efficiency of the $4Q - tree$ potencial algorithm for a given number or steps. We can observe the convergence on the system lowest energy as the number of steps increases.

where λ is the single variational parameter, $V(\vec{r})$ is a many-body potential which will be described in the next section, and Φ_{FG} is the Fermi gas wave function given by a product of Slater determinants, one for each color-flavor combination of quarks, whose single wave functions correspond to that of a particle in a box. The structure of the wave function allows some simplifications in the calculations. After an integration by parts, the expectation value of the Hamiltonian operator for a given particle density can be evaluated as:

$$\langle H \rangle_\lambda = T_{FG} + 2\lambda^2 \langle W \rangle_\lambda + \langle V \rangle_\lambda , \tag{2}$$

where W is a kinetic term induced by the interaction and T_{FG} is the kinetic energy of a free Fermi gas.

3. Many body potential

The many-body potential is defined as the optimal clustering of quarks into color-singlets objects, where by optimal we mean the configuration that minimizes the potential energy. By this, we distinguish four possible configurations in the system:

- *Meson-like*: In this case we distinguish a pairing between color and anticolors, building two mesons. The system allows two possible pairings (configurations 1 & 2, Fig. 1) and the

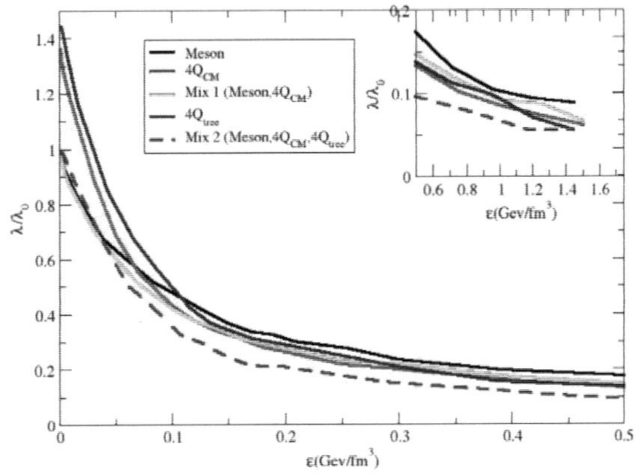

Figure 3. Evolution of the optimal variational parameter as a function of the energy density. The variational parameter is normalized to the theoretical solution of a harmonically bound quark-antiquark pair (isolated meson). We can observe that for the meson case the ratio converges to one as expected, while for the tetraquark cases this is up to 1.4 times larger and rapidly decreasing as the energy density increases. In the mixed cases the parameter is dominated by the meson configuration and thus the variational parameter resembles that of a mesonic system. All of them become very alike at high energy densities.

many-body potential made up of mesons is then given by:

$$V_{q\bar{q}} = \min_{P} \sum_{i=1}^{2} v[r_{q_i}, r_{\bar{q}_i}], \tag{3}$$

where $v[r_i, r_j] = \frac{1}{2}k(r_i - r_j)^2$, we use a harmonic potential to describe the interaction between particles. The minimum takes into account both permutations of the system.

- *Tetraquark-like*: In this case, we consider two different potentials in which we can confine four quarks into a color-singlet, denoted by $4Q - cm$ (configuration 3, Fig. 1) as linked through their center of mass. This is described by the following potential:

$$V_{4Q-cm} = \sum_{i=1}^{2} V\left[\vec{R}_{cm}, \vec{r_{q_i}}\right] + \sum_{i=1}^{2} V\left[\vec{R}_{cm}, \vec{r_{\bar{q}_i}}\right] \tag{4}$$

where $\vec{R}_{cm}$ is the vector position of the center of mass. Another possibility to link four quarks is by using the so called steiner-tree configuration, denoted by $4Q - Tree$ (configuration 4, Fig. 1). This is the most efficient way to confine four particles into a cluster [5, 6]. The potential has the following form:

$$V_{4Q-tree} = \sum_{i=1}^{2} V\left[\vec{k}, \vec{r_{q_i}}\right] + \sum_{i=1}^{2} V\left[\vec{l}, \vec{r_{\bar{q}_i}}\right] + V[\vec{k}, \vec{l}] \tag{5}$$

The implementation of such configuration requires an algorithm to search over the $\vec{k}$ and $\vec{l}$ vectors providing the minimal length of the links.

Journal of Physics: Conference Series **378** (2012) 012035 doi:10.1088/1742-6596/378/1/012035

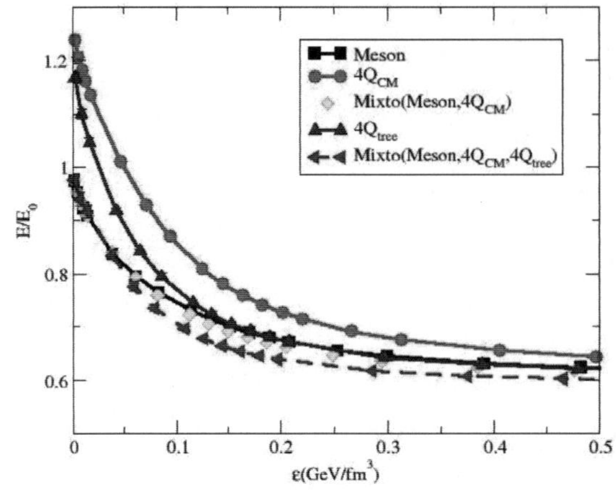

Figure 4. Evolution of the Energy per particle as a function of energy density. The energy is normalized to the theoretical solution of a harmonically bound quark-antiquark pair (isolated meson). We can observe that for the meson case the ratio converges to one as expected, while for the tetraquark cases this is around 1.2 times larger and rapidly decreasing as the energy density increases. In the mixed cases the energy is dominated by the meson configuration and thus the mixed system resembles that of a mesonic system. Note that the energy for the tetraquark is lower when considering the steiner tree configuration than when considering the center of mass configuration. This fact makes a substantial difference in the mixed case, showing that in the presence of the steiner tree configuration the mixed system is more stable than the pure mesonic state. All of them become very alike at high energy densities.

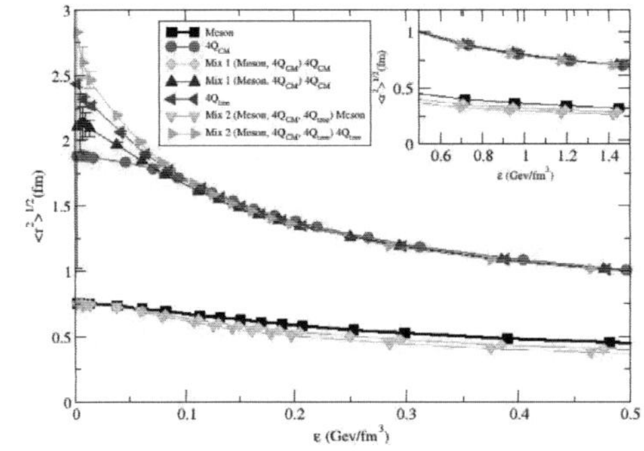

Figure 5. Evolution of the mean square radius as a function of the energy density. We can identify two typical structures, one for the meson and another for the tetraquark. They are well separated from each other regardless of the kind of potentials involved. The meson square radius is less sensitive to the energy density than that of the tetraquark, becoming more alike at high energy densities but still substantial differences remain

4. Results

To determine the variational parameter (λ) as a function of particle density (ρ) we proceed as follows: first we select the value of ρ in the box that confines the quarks, which, for a fixed

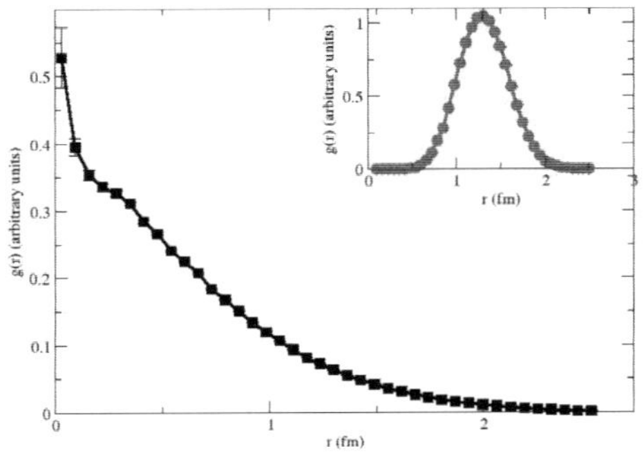

Figure 6. We show the $q - \bar{q}$ (meson-meson) correlation function (upper panel) for an energy density of 0.1074 GeV/fm^3. The quark-antiquark correlation is large at short distances, as any quark can find an antiquark close to each other within a meson structure. The distance for the vanishing of the correlation can be understood then as the radial size of quark cloud of the meson. In the upper panel the meson-meson correlation takes its maximum around 1.3 fm. This can be seen as the typical distance among mesons at that energy density. Comparing with the quark-antiquark correlation function we can conclude that there is an small overlap between mesons.

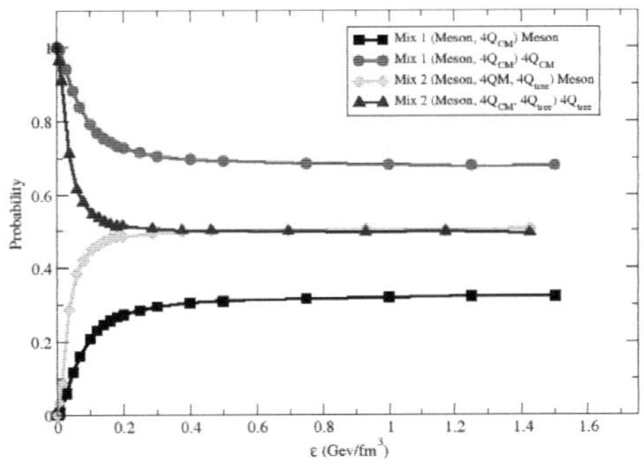

Figure 7. Probability to form mesons or a tetraquark as a function of the energy density. When the tetraquark is built up using a center of mass potential the probability of appearance increases from zero at low energy densities to around 30% at high energy densities. In the case that the tetraquark is formed by a steiner tree, its probability rapidly converges to 50%, making a perfect mixing with the mesonic state.

number of particles, means changing the box size. Then, we compute the energy of the system as a function of λ. The minimum of the energy determines the optimal λ.

$$\frac{\partial \langle \Psi_\lambda | H | \Psi_\lambda \rangle}{\partial \lambda} = \frac{\partial E_\lambda}{\partial \lambda} = 0 \tag{6}$$

For the $4Q-Tree$ configuration, we compute the $\vec{k}$ and $\vec{l}$ vectors using an algorithm that explores their magnitude and direction for a set of steps in spherical coordinates, taking $\vec{Rcm}$ as a starting point. The algorithm then takes the ones producing the lowest potential energy. In Fig. 2 we show an example of the minimization of E_λ and exhibit the efficiency of the algorithm when including 10, 15, 20 and 25 steps. It is clear the energy convergence and the almost independence of the optimal λ parameter on them.

The optimal value for λ was determined for a set of densities. In Figure 3 we show the result as a function of the energy density. The energy density is used here after as a parameter to exhibit the evolution on a set of observables of the system: The total energy (Fig. 4), the square root of the mean square radius (Fig. 5), and the correlation function between particles (Fig. 6). In addition, we show the probability of having a meson or a tetraquark in the simulation (Fig. 7).

5. Conclusions

We have presented a preliminary study on the possible formation of a tetraquark state as produced when two mesons (namely two quarks and two antiquarks) are forced to approach to each other. Using a dynamical approximation which allows the competition among meson-like and tetraquark-like configurations. We have characterized each state and their modifications upon the mixing. We find that the total energy, mean square radius, and correlation function of a meson are susceptive to the effects of the mixing with a tetraquark state. The tetraquark state properties are also modified by the mixing, in particular the mean square radius. A substantial reduction on the ground state energy of the mixed system is achieved when including the steiner-tree configuration, making this smaller than the pure mesonic state as the energy density increases.

We have implemented a Monte Carlo simulation to study the tetraquark formation in a systematic way, we expect to explore the effects when considering different flavors, in particular it will be interesting to study the heavy-quark regime. An extended analysis, discussion and final results will be presented elsewhere.

Acknowledgments

We acknowledge the support of CONACyT, Mexico.

References

[1] J. Vijande, A. Valcarce, J. M. Richard, Phys. Rev. D 76, 114013 (2007)
[2] P. Bicudo, M. Cardoso. Phys. Rev. D 83, 094010 (2011)
[3] Chenrong Deng, Jialun Ping, Fan Wang, T. Goldman. Phys. Rev. D 82, 074001 (2010)
[4] G. Toledo Sánchez and J. Piekarewicz Phys. Rev. C 65, 045208 (2001)
[5] Fumiko Okiharu, Hideo Suganuma, Toru T. Takahashi, Physical Review D 72, 014505 (2005) (and references therein)
[6] Okiharu, F ; Doi, T ; Ichie, H ; Iida, H ; Ishii, N ; Oka, M ; Suganuma, H. arXiv:hep-ph/0507187 (2005)

XIII Mexican Workshop on Particles and Fields IOP Publishing
Journal of Physics: Conference Series **378** (2012) 012036 doi:10.1088/1742-6596/378/1/012036

Conformal symmetry breaking and degeneracy of high-lying unflavored mesons

Mariana Kirchbach[1], **Adrian Pallares-Rivera**[1], **Cliffor Compean**[2] **and Alfredo Raya**[2]

[1]Institute of Physics, Autonomous University of San Luis Potosi, Manuel Nava 6, San Luis Potosi, SLP 98290, Mexico

[2]Institute of Physics and Mathematics, University of Michoacan San Nicolas de Hidalgo, Building C-3, University City, Morelia Michoacan 58040, Mexico

E-mail: mariana@ifisica.uaslp.mx, pallares@ifisica.uaslp.mx, cliffor@ifm.imich.mx, raya@ifm.umich.mx

Abstract. We show that though conformal symmetry can be broken by the dilaton, such can happen without breaking the conformal degeneracy patterns in the spectra. Our argumentation goes as follows: We departure from the gauge-gravity duality which predicts on the boundaries of the AdS_5 geometry a conformal theory, associated with QCD at high temperatures, and consider $S^1 \times S^3$ slicing. The inverse radius, R, of S^3 relates to the temperature of the deconfinement phase transition and has to satisfy, $\hbar c/R \gg \Lambda_{QCD}$. On S^3, whose isometry group is $SO(4)$, we then focus on the eigenvalue problem of the conformal Laplacian there, given by $\frac{1}{R^2}\left(\mathcal{K}^2+1\right)$, with $\mathcal{K}^2$ standing for the Casimir invariant of the $so(4)$ algebra. This eigenvalue problem describes the spectrum of a scalar particle, to be associated with a $q\bar{q}$ system. Such a spectrum is characterized by a $(K+1)^2$-fold degeneracy of its levels, with $K \in [0,\infty)$. We then break the conformal S^3 metric, $ds^2 = d\chi^2 + \sin^2\chi(d\theta^2 + \sin^2\theta d\varphi^2)$ -in polar χ, θ, and azimuthal φ coordinates- according to, $\widetilde{ds^2} = e^{-b\chi}\left((1+b^2/4)d\chi^2 + \sin^2\chi(d\theta^2+\sin^2\theta d\varphi^2)\right)$, and attribute the symmetry breaking scale $b\hbar^2c^2/R^2$ to the dilaton. Next we show that the above metric deformation is equivalent to a breaking of the conformal curvature of S^3 by a term proportional to $b\cot\chi$, and that the perturbed conformal Laplacian is equivalent to $\left(\widetilde{\mathcal{K}}^2 + c_K\right)$, with c_K a representation constant, and $\widetilde{\mathcal{K}}^2$ being again an $so(4)$ Casimir invariant, but this time in a representation unitarily inequivalent to the 4D rotational one. As long as the spectra before and after the symmetry breaking happen to be determined each by eigenvalues of a Casimir invariant of an $so(4)$, no matter whether or not in a representation that generates the orthogonal group $SO(4)$ as a subgroup of the conformal group $SO(2,4)$, the degeneracy patterns remain unaltered though the conformal symmetry breaks at the level of the representation of the algebra. We fit the S^3 radius and the $\hbar^2c^2b/R^2$ scale to the high-lying excitations in the spectra of the unflavored mesons, as reported by the Crystal Barrel collaboration, and observe the correct tendency of the $\hbar c/R = 373$ MeV value to notably exceed Λ_{QCD}. The size of the symmetry breaking scale is calculated as $\hbar c\sqrt{b}/R = 673.7$ MeV.

1. Introduction and background

The five-dimensional Anti-de Sitter space, AdS_5, is presently one of the most intriguing geometries in field theoretical studies due to its relevance in establishing the link between brane theory and QCD. The fact is that according to the gauge-gravity duality conjecture [1], a string theory at the boundary of AdS_5 appears dual to a conformal field theory in (1+3) dimensions, associated with QCD [2]. The AdS_5 geometry is defined as a surface embedded in a (2+4)-dimensional flat ambient space of two time-like (X_0, X_5) and four space-like (X_1, X_2, X_3, X_4) dimensions, according to [3]

$$\begin{aligned} X_0^2 + X_5^2 - X_1^2 - X_2^2 - X_3^2 - X_4^2 &= \\ (X_0+X_4)(X_0-X_4) + (X_5+X_3)(X_5-X_3) &- (X_1 - iX_2)(X_1+iX_2) = L^2, \end{aligned} \tag{1}$$

where L^2 is a scale. In changing variables to

$$X_0 + X_4 = e^v, \qquad X_\mu = e^v x_\mu, \quad \mu = 5,1,2,3, \tag{2}$$

the equation (1) becomes

$$e^v(X_0 - X_4) + e^{2v}(x_5^2 - x_1^2 - x_2^2 - x_3^2) = L^2. \tag{3}$$

This equation shows that the AdS_5 space contains a flat Minkowski space-time, warped by the exponential factor of $\exp(2v)$. Dividing now by that very warp factor, and taking the $v \to \infty$ limit, one encounters the Minkowskian light-cone, $\mathcal{C}^{1+3}$ as,

$$\mathcal{C}^{1+3}: \quad \lim_{v\to\infty}\left[\frac{X_0 - X_4}{e^v} + x_5^2 - x_1^2 - x_2^2 - x_3^2\right] = \lim_{v\to\infty}\frac{L^2}{e^{2v}} \to 0. \tag{4}$$

The v variable is known as the *holographic* variable. Therefore, the intersections of the AdS_5 null-ray cone with $X_0 + X_4 = e^v$ hyperplanes, produce slices, Π_v, called branes, which are copies of flat Minkowski space-times in the x_μ coordinates with $\mu = 5,1,2,3$. In terms of the holographic variable, the conformal AdS_5 boundary corresponds to the $v \to \infty$ limit, in which case the warp factor blows up. QCD on $\mathcal{C}^{1+3}$ as its emerges at the AdS_5 boundary, is known as holographic QCD and has produced a variety of interesting insights into the properties of hadrons (see ref. [4] for a recent review).

However, flat Minkowski space-time, in combination with an interaction potential of an infinite range, represent a significant obstacle toward the formulation of a finite-temperature field theory and the description of the deconfinement phase transition. The reason is that the absence of a scale presents a serious difficulty in defining the QCD chemical potential. This difficulty has been circumvented [2], [5], [6] in replacing the flat Minkowski space-time by the compactified one, $S^1 \times S^3$, which equally well describes the conformal AdS_5 boundary, as visible from casting eq. (1) into the form [7]

$$\begin{aligned} &\det U - \det g = L^2, \\ &U = \begin{pmatrix} X_5 + iX_0 & 0 \\ 0 & X_5 - iX_0 \end{pmatrix}, \qquad g = \begin{pmatrix} X_4 + iX_3 & iX_1 - X_2 \\ iX_1 + X_2 & X_4 - iX_3 \end{pmatrix}, \\ &X_0 + iX_5 = R_0 e^{i\eta}, \qquad X_1 + iX_2 = R\sin\chi\sin\theta e^{i\varphi}, \\ &X_3 = R\sin\chi\cos\theta, \qquad X_4 = R\cos\chi, \quad R_0^2 - R^2 = L^2. \end{aligned} \tag{5}$$

The null-ray AdS_5 cone can now be approached for $R_0 \to R$. The proof that $S^1 \times S^3$ is a conformally compactified flat Minkowski space-time, $\mathbf{R}^{1+3}$, has been elaborated in the literature in great detail [8], [9], and the explicit map that takes the $\mathbf{R}^{1+3}$ metric to the metric of $S^1 \times S^3$ can be found for example in [7]. The line interval on the unit hypersphere reads,

$$ds^2 = d\chi^2 + \sin^2\chi(d\theta^2 + \sin^2\theta d\varphi^2). \tag{6}$$

Particle motions on this space have been equally well studied, the ref. [7] providing valuable insights in that regard. In particular, it has been found that the free geodesic motion of a scalar particle is described in terms of the eigenvalue problem of the conformal Laplacian, $\mathbf{L}_{S^3}$,

$$\mathbf{L}_{S^3} = -\frac{1}{R^2}\Delta_{S^3} + \mathcal{R}_{S^3}, \qquad -\frac{1}{R^2}\Delta_{S^3} = -\frac{1}{R^2}\left(\frac{\mathrm{d}^2}{\mathrm{d}\chi^2} + 2\cot\chi\frac{\mathrm{d}}{\mathrm{d}\chi} - \frac{\mathbf{L}^2}{\sin^2\chi}\right) = \frac{1}{R^2}\mathcal{K}^2, \tag{7}$$

where Δ_{S^3} is the Laplace-Beltrami operator on S^3, $\mathcal{R}_{S^3}$ stands for the conformal curvature, and $\mathcal{K}^2$ is the operator of the squared four-dimensional (4D) angular momentum which acts as a Casimir invariant of the isometry algebra $so(4)$ of S^3. The $so(4)$ algebra in the representation chosen in (7) has the property to generate the orthogonal group $SO(4)$. On the unit hypersphere, the spectral problem of $\mathbf{L}_{S^3}$ reads,

$$\begin{aligned}
\mathbf{L}_{S^3}Y_{Klm}(\vec{\Omega}) = (-\Delta_{S^3}+1)Y_{Klm}(\vec{\Omega}) &= \left(\mathcal{K}^2+1\right)Y_{Klm}(\vec{\Omega}) = (K+1)^2Y_{Klm}(\vec{\Omega}), \quad \vec{\Omega}=(\chi,\theta,\varphi)\\
Y_{Klm}(\vec{\Omega}) = \mathcal{S}^l_K(\chi)Y^m_l(\theta,\varphi), \quad & \mathcal{S}^l_K(\chi) = \sin^l\chi\,\mathcal{G}^{l+1}_{K-l}(\cos\chi),\\
K\in[0,\infty), \quad l\in[0,K], \quad & m\in[-l,+l].
\end{aligned} \tag{8}$$

Here, $Y_{Klm}(\vec{\Omega})$ are the hyper-spherical harmonics, $\mathcal{G}^{\alpha}_n(\cos\chi)$ denote the Gegenbauer polynomials, and $Y^m_l(\theta,\varphi)$ stand for the ordinary three-dimensional spherical harmonics. The χ dependent part of the hyperspherical harmonics, i.e. the $\mathcal{S}^l_K(\chi)$ function, is sometimes referred to as "quasi-radial" function [10], a notation which we occasionally shall make use of. It is obvious, that the spectrum in (8) is characterized by a $(K+1)^2$ fold degeneracy of the levels. Such a spectrum as whole would fit into an infinite unitary representation of the conformal group, $SO(2,4)$, where the group $SO(4)$ appears in the reduction chain,

$$SO(2,4)\subset SO(4)\subset SO(3)\subset SO(2). \tag{9}$$

It is within this context that $(K+1)^2$-fold degeneracy is associated with conformal symmetry, in parallel to the H atom where $(K+1)$ acquires the meaning of the principal quantum number of the Coulomb potential problem. However, conformal symmetry can be at most an approximate symmetry of QCD because it requires

- massless quarks,
- scale independent strong coupling, and
- a massless dilaton.

While the first two conditions seem to be reasonably respected within the unflavored sector of QCD, in which the u and d quark masses can be considered sufficiently small at the scale of the nucleon excitations, which start around 1500 MeV, and in the infrared regime, where the running coupling constant starts walking toward a fixed value [11], the dilaton mass can not be disregarded so far. The breaking of the conformal symmetry is frequently treated in terms of breaking the metric of the Minkowski space-time encountered at the AdS_5 boundary, i.e. by deforming the warp factor (soft-walls) [4]. Notice, the dilaton can be introduced in the action, $\mathcal{S}$, as a background field in an overall exponential according to,

$$\mathcal{S} = \int \mathrm{d}^DXe^{-\Phi(z)}\sqrt{-g}\mathcal{L}, \quad \Phi(z) = (\mu z)^\nu, \quad z = e^{-v}, \tag{10}$$

where μ sets a mass scale, usually taken as Λ_{QCD} for simplicity. As shown in [12], [13], the dilaton modifies the particle motion and leaves a print in the meson spectra.

We here adapt the essentials of the above scheme to the geometry of the compactified Minkowski space-time, and design a pragmatic quark model of conformal symmetry breaking on $S^1 \times S^3$ through global S^3 metric deformation by an exponential factor of the second polar angle. We then study the impact of this deformation on the spectra of the high-lying unflavored mesons reported by the Cristal Barrel collaboration in the compilation of [14]. Our aim is to reveal a possibility to deform the conformal curvature in (6) by a scale, and without removing the conformal degeneracy patterns from the spectra.

The outline of the paper is as follows. In the next section we present the mechanism of conformal symmetry breaking without removing the conformal degeneracy patterns in the spectra. In section 3 we analyze the Crystal-Barrel data [14] on the high-lying unflavored meson spectra with the aim to obtain an estimate for the S^3 radius and the order of the scale of the conformal symmetry breaking which we attribute to the dilaton. The paper ends with concise conclusions.

2. Breaking the conformal curvature of S^3 by an interaction

Any perturbation of the quantum mechanical geodesic motion on S^3 by an interaction potential, $V(\vec{\Omega})$, breaks the conformal curvature as,

$$\hbar^2c^2\mathcal{R}_{S^3} \longrightarrow \hbar^2c^2\widetilde{\mathcal{R}}_{S^3} = \frac{\hbar^2c^2}{R^2} + V(\vec{\Omega}), \tag{11}$$

and the perturbed Laplacian is no longer conformal, neither is its spectral problem necessarily exactly solvable. Not so for the potential, given by,

$$V(\chi) = -2B\cot\chi, \tag{12}$$

with $B = b\hbar^2c^2/R^2$, and a dimensionless b, which gives rise to an exactly solvable $\left(\hbar^2c^2\mathbf{L}_{S^3} + V(\chi)\right)$ spectral problem, and one which moreover happens to carry same $(K+1)^2$-fold degeneracy patterns as the spectral problem of the intact conformal Laplacian in (8). The $\cot\chi$ potential is special insofar as it is a harmonic function on S^3, just as is the Coulomb potential in E_3, and can be treated along the line of harmonic analysis and potential theory. Using harmonic functions as potentials brings the advantage of minimizing the Dirichlet functional integral. Occasionally, the cotangent potential on S^3 is referred to as "curved" Coulomb potential. In the present section we wish to take a closer look on the quantum mechanical cotangent potential problem on S^3 from the perspective of conformal Laplacians [15, 16]. The cotangent potential problem on S^3, first considered by Schrödinger [17, 18], is equivalent to the relativistic four-dimensional rigid-rotator problem perturbed by a $\cot\chi$ interaction,

$$\frac{\hbar^2c^2}{R^2}\left[-\Delta_{S^3} + 1 - 2b\cot\chi\right]\Psi_{K\widetilde{l}\widetilde{m}}(\vec{\Omega}) = \frac{\hbar^2c^2}{R^2}\left[(K+1)^2 - \frac{b^2}{(K+1)^2}\right]\Psi_{K\widetilde{l}\widetilde{m}}(\vec{\Omega}). \tag{13}$$

The solutions of equation (13) are well studied and known. Specifically in [19], they have been expressed in terms of real non-classical Romanovski polynomials, $R_n^{\alpha,\beta}(\cot\chi)$ (reviewed in [20]), and in units of $\hbar = 1 = c$, $R = 1$, read,

$$\begin{aligned}
\Psi_{K\widetilde{l}\widetilde{m}}(\vec{\Omega}) &= e^{\frac{\alpha_K\chi}{2}}\psi_K^{\widetilde{l}}(\chi)Y_{\widetilde{l}}^{\widetilde{m}}(\theta,\varphi),\\
\psi_K^{\widetilde{l}}(\chi) &= \sin^K\chi R_{K-\widetilde{l}}^{\alpha_K,\beta_K-1}(\cot\chi),\\
\alpha_K = -\frac{2b}{K+1}, &\qquad \beta_K = -K.
\end{aligned} \tag{14}$$

Journal of Physics: Conference Series **378** (2012) 012036 doi:10.1088/1742-6596/378/1/012036

The Romanovski polynomials satisfy the following differential hypergeometric equation,

$$(1+x^2)\frac{\mathrm{d}^2R_n^{\alpha,\beta}}{\mathrm{d}x^2}+2\left(\frac{\alpha}{2}+\beta x\right)\frac{\mathrm{d}R_n^{\alpha,\beta}}{\mathrm{d}x}-n(2\beta+n-1)R_n^{\alpha,\beta}=0. \tag{15}$$

They are obtained from the following weight function,

$$\omega^{\alpha,\beta}(x)=(1+x^2)^{\beta-1}\exp(-\alpha\cot^{-1}x), \tag{16}$$

by means of the Rodrigues formula,

$$R_n^{\alpha,\beta}(x)=\frac{1}{\omega^{\alpha,\beta}(x)}\frac{\mathrm{d}^n}{\mathrm{d}x^n}\left[(1+x^2)^n\omega^{\alpha,\beta}(x)\right]. \tag{17}$$

Recently, it has been shown in [19] that the $\psi_K^{\widetilde{l}}(\chi)$ functions allow for finite decompositions in the $SO(4)$ basis of the quasi-radial functions $S_K^l(\chi)$ in (8). Namely, the following relation has been found to hold valid,

$$\psi_K^{\widetilde{l}}(\chi)=\sum_{\widetilde{l}}^{K}C_l\ S_K^l(\chi). \tag{18}$$

The explicit expressions for the coefficients C_l defining the full similarity transformation between the invariant spaces of the perturbed and intact Laplacians on S^3 can be consulted in [19]. Upon introducing the notations,

$$\omega_K(\chi)=\frac{\alpha_K\chi}{2}\equiv-\frac{b\chi}{K+1},\qquad b=-\omega_K'(\chi)(K+1), \tag{19}$$

and insertion of (18) into (14), allows to rewrite (13) to,

$$\begin{aligned}\left[\mathbf{L}_{S^3}+2\omega_K'(\chi)(K+1)\cot\chi\right]e^{\omega_K(\chi)}\psi_K^{\widetilde{l}}(\chi) &=\\ &e^{\omega_K(\chi)}\sum_{\widetilde{l}}^{K}C_l\left[(K+1)^2-\omega_K'(\chi)^2\right]S_K^l(\chi).\end{aligned} \tag{20}$$

Dragging now the exponential factor in the l.h.s. in (20) from the right to the very left, recalling that $\mathbf{L}_{S^3}=\mathcal{K}^2+1$, and making use of (18), amounts to

$$\begin{aligned}e^{\omega_K(\chi)}\left[\mathcal{K}^2+1-\omega_K'(\chi)^2-2\omega_K'(\chi)\mathcal{D}_{\mathbf{K}}\right]\sum_{\widetilde{l}}C_l\ S_K^l(\chi) &=\\ &e^{\omega_K(\chi)}\sum_{\widetilde{l}}^{K}C_l\left[(K+1)^2-\omega_K'(\chi)^2\right]S_K^l(\chi),\\ \mathcal{D}_{\mathbf{K}} &= \frac{\mathrm{d}}{\mathrm{d}\chi}-K\cot\chi.\end{aligned} \tag{21}$$

Using in the r.h.s. of (21) the relationship, $(\mathcal{K}^2+1)\,S_K^l(\chi)=(K+1)^2S_K^l(\chi)$, known from (7), allows to equivalently cast eq. (21) as,

$$\begin{aligned}&\left(\widetilde{\mathcal{K}}^2+1-\omega_K'(\chi)^2\ -\ 2\omega_K'(\chi)e^{\omega_K(\chi)}\mathcal{D}_{\mathbf{K}}e^{-\omega_K(\chi)}\right)e^{\omega_K(\chi)}\sum_{\widetilde{l}}^{K}C_l\ S_K^l(\chi)\\ &=\sum_{\widetilde{l}}^{K}\left(\widetilde{\mathcal{K}}^2+1-\omega'(\chi)^2\right)e^{\omega_K(\chi)}C_l\ S_K^l(\chi).\end{aligned} \tag{22}$$

This equality implies the following important property of the $\mathcal{D}_K$ operator,

$$\begin{aligned}\left(\widetilde{\mathcal{K}}^2 - 2\omega_K'(\chi)e^{\omega_K(\chi)}\mathcal{D}_{\mathbf{K}}e^{-\omega_K(\chi)}\right)e^{\omega_K(\chi)}\sum_{\widetilde{l}}^{K} C_l\, \mathcal{S}_K^l(\chi) &= \sum_{\widetilde{l}}^{K}\widetilde{\mathcal{K}}^2\, e^{\omega_K(\chi)}C_l\, \mathcal{S}_K^l(\chi),\\ \widetilde{\mathcal{K}}^2 &= e^{\omega_K(\chi)}\mathcal{K}^2 e^{-\omega_K(\chi)}.\end{aligned} \tag{23}$$

Now, the equivalence of eqs. (20)–(21), and (22) allows to draw the final conclusion as

$$\begin{aligned}\left[\mathcal{K}^2 + 1 + 2\omega_K'(\chi)(K+1)\cot\chi\right]e^{\omega(\chi)}\psi_K^{\widetilde{l}}(\chi) &= \sum_{\widetilde{l}}^{K} C_l\left[\widetilde{\mathcal{K}}^2 + 1 - \omega'(\chi)^2\right]e^{\omega_K(\chi)}\mathcal{S}_K^l(\chi)\\ &= \left[(K+1)^2 - \omega_K'(\chi)^2\right]e^{\omega(\chi)}\psi_K^{\widetilde{l}}(\chi),\end{aligned} \tag{24}$$

with $\psi_K^{\widetilde{l}}$ from (18), and $\omega_K'(\chi)$ being a representation constant. In this fashion, the eigenvalue problem of the cotangent-broken conformal Laplacian on S^3 presents itself as the eigenvalue problem of a Casimir invariant of the $so(4)$ algebra in a representation unitarily inequivalent to the 4D-rotational. As long as the eigenvalues of any Casimir invariant remain unaltered under any well defined similarity transformations, no matter unitary or not, the degeneracy patterns in the spectral problems of the conformal and the cotangent-broken Lapalacians result same. The Killing vectors of this new $so(4)$ algebra,

$$\begin{aligned}\widetilde{J}_{ik} = e^{\omega_K(\chi)}J_{ik}e^{-\omega_K(\chi)} &= e^{\omega_K(\chi)}\left(X_j\frac{\partial}{\partial X_k} - X_k\frac{\partial}{\partial X_j}\right)e^{-\omega_K(\chi)},\\ \left[\widetilde{J}_{ik}, \widetilde{J}_{kj}\right] = \widetilde{J}_{ji}, &\qquad \chi = \cot^{-1}\frac{X_4}{\sqrt{X_1^2 + X_2^2 + X_3^2}},\end{aligned} \tag{25}$$

refer to a surface (call it $\widetilde{S}^3$) which will differ in shape from S^3. In this fashion, the breaking of the conformal symmetry reveals itself through metric deformation. Equivalently, one can say that it reveals itself at the level of the representation functions, in reference to the difference between the hyper-spherical harmonics, $Y_{Klm}(\vec{\Omega})$ in (8), and $\Psi_{K\widetilde{lm}}(\vec{\Omega})$ in (14). Indeed, the global metric of the S^3 ball is incorporated by the scalar of the 4D rotational isometry algebra $so(4)$, which, in being the lowest $\mathbf{L}_{S^3}$ eigenstate, is the hyper-spherical harmonic, $Y_{000}(\vec{\Omega})$,

$$S^3: \quad Y_{000}(\vec{\Omega}) = \sum_{i=1}^{4} X_i^2 = \cos^2\chi + \sin^2\chi\left(\cos^2\theta + \sin^2\theta\left(\cos^2\varphi + \sin^2\varphi\right)\right) = 1. \tag{26}$$

Correspondingly, the global metric of the $\widetilde{S}^3$ surface will be implemented by the scalar of the non-unitary $so(4)$ algebra in (23), (25) and is given by the exponentially rescaled hyper-sphere,

$$\widetilde{S}^3: \quad e^{-b\chi}Y_{000}(\vec{\Omega}) = e^{-b\chi}\left(\cos^2\chi + \sin^2\chi\left(\cos^2\theta + \sin^2\theta\left(\cos^2\varphi + \sin^2\varphi\right)\right)\right) = e^{-b\chi}. \tag{27}$$

The line element on the deformed surface $\widetilde{S}^3$ is then easily calculated as,

$$\mathrm{d}^2\widetilde{s} = e^{-b\chi}\left((1 + b^2/4)\mathrm{d}^2\chi + \sin^2\chi\left(\mathrm{d}\theta^2 + \sin^2\theta\mathrm{d}^2\varphi\right)\right). \tag{28}$$

In the following we conjecture that the $\hbar^2c^2b/R^2$ scale of conformal symmetry breaking can be attributed to the dilaton, the only significant source of such a breaking.

Notice that the perturbance of the conformal Laplacian on S^3 by a cotangent interaction does not even produce a conformal map [15], [16], as it could have happened by a different, in general, gradient potential.

Journal of Physics: Conference Series **378** (2012) 012036 doi:10.1088/1742-6596/378/1/012036

In summary, a subtle mode of symmetry breaking has been identified in which the symmetry breaking happens at the level of the representation functions of the isometry algebra of a given geometric manifold, and reveals itself through a metric deformation, while the spectral problems of the intact and broken Laplacians are characterized by identical degeneracy patterns.

It seems opportune to refer to this symmetry breaking mechanism as "symmetry breaking camouflaged by degeneracy", or, shortly, "camouflaged symmetry breaking".

3. Degeneracy of high-lying unflavored mesons

In the following we employ the cotangent-broken conformal Laplacian on S^3 in (13) in the description of the spectra of the high-lying unflavored mesons as reported by the Crystal Barrel collaboration (data compilation from [14]) and adjust the R and b potential parameters. The cotangent potential derives its utility for employment in the description of hadron spectra from its close relationship to the Cornell potential, predicted by Lattice QCD [21], as visible from its Taylor series decomposition,

$$-2b\cot\frac{\widehat{r}}{R} = -\frac{2bR}{\widehat{r}} + \frac{4b}{3R}\,\widehat{r} + ..., \quad \text{for} \quad \chi = \frac{\widehat{r}}{R}, \tag{29}$$

where $\widehat{r}$ denotes the geodesic distance on S^3, and R was the S^3 radius.

The Crystal Barrel data on the spectra of the high-lying unflavored mesons show well pronounced conformal degeneracy patterns in the region above $\approx$ 1200 MeV. Our case is that

- these patterns emerge in consequence of the breaking of the conformal $SO(2,4)$ group symmetry at the level of the representation of its algebra, and in accordance with eqs. (24), (28),
- such a breaking may be attributed to the dilaton.

In Figs. (1–4) we present the spectra of the mesons of interest. Tables 1 and 2 contain the results of the least square parameter fit.

Table 1. The potential parameters from the least square fit to the meson spectra by means of eq. (13).

Isospin	Lowest spin	R [fm]	$(\hbar^2c^2b)/R^2[\text{GeV}^2]$
I=0	$J_{\text{min}}=0^-$	0.5157279	0.459666
I=0	$J_{\text{min}}=1^-$	0.5293829	0.425489
I=1	$J_{\text{min}}=0^-$	0.5443879	0.522783
I=1	$J_{\text{min}}=1^-$	0.5268850	0.411833

Table 2. The meson masses from the least square fit to the spectra by means of eq. (13) (last column). The label K stands for the four-dimensional angular momentum characterizing a degenerate level in accordance with eq. (13). The fourth column contains the averaged mass of the experimentally observed states as displayed in the respective plots on the Figs. 1-4.

I	J_{min}	K	experiment [MeV]	fit [MeV]
0	0^-	1	1278.33	1349.60
		2	1692.5	1659.48
		3	1970.66	1961.83
		4	2274.22	2280.1
1	0^-	1	1328.33	1404.89
		2	1696.66	1708.67
		3	2031	1985.04
		4	2253.75	2273.69
0	1^-	1	1325	1416.19
		2	1653.33	1697.43
		3	1994.14	1979.19
		4	2269.77	2279.69
1	1^-	1	1343.33	1385.85
		2	1682.5	1670.28
		3	1825.85	1957.16
		4	2256.66	2262.75

4. Discussion and conclusions

The Tables 1 and 2 show that our least square fits to the Crystal Barrel data on the high-lying unflavored mesons by means of the cotangent broken conformal Laplacian on S^3 in eq. (13) provide quite a reasonable description of the observed degeneracies in the spectra under investigation and in support of our hypothesis that the breaking of the conformal symmetry can occur at the level of the representation of the conformal algebra and without affecting the degeneracies. The b parameter is obtained as

$$b = 3.2793 \pm 0.0697. \tag{30}$$

We furthermore observe that to a very good approximation, both the S^3 radius and the $\hbar^2 c^2 b/R^2$ scale are isospin and parity independent, as should be for a conformal symmetry breaking scale due to the dilaton. The mean value of the inverse radius obtained from our fits corresponds to a temperature of $T = \hbar c/R$= 373 MeV which reasonably fulfills the requirement to be notably larger than $\Lambda_{QCD} = 175$ MeV. In order to extract the scale parameter of the conformal symmetry breaking within our approach, we make use of eq. (29), i.e. of $\chi = \frac{\widehat{r}}{R}$, and rewrite the $\exp(-b\chi)$ factor equivalently as,

$$\exp(-b\chi) = \exp\left(-\left[\left(\frac{\sqrt{b}}{R}\right)^2 \left(\sqrt{\widehat{r}\, R}\right)^2\right]\right) \equiv \exp(\mu r),$$
$$\mu = \frac{\sqrt{b}}{R}, \qquad r = \sqrt{\widehat{r}\, R}. \tag{31}$$

Evaluating this scale with the numbers listed in Table 1 amounts to a mean value of $\mu = 673.7$ MeV, which is closer to the dilaton mass than to $\Lambda_{QCD} = 175$ MeV in (10). We conclude

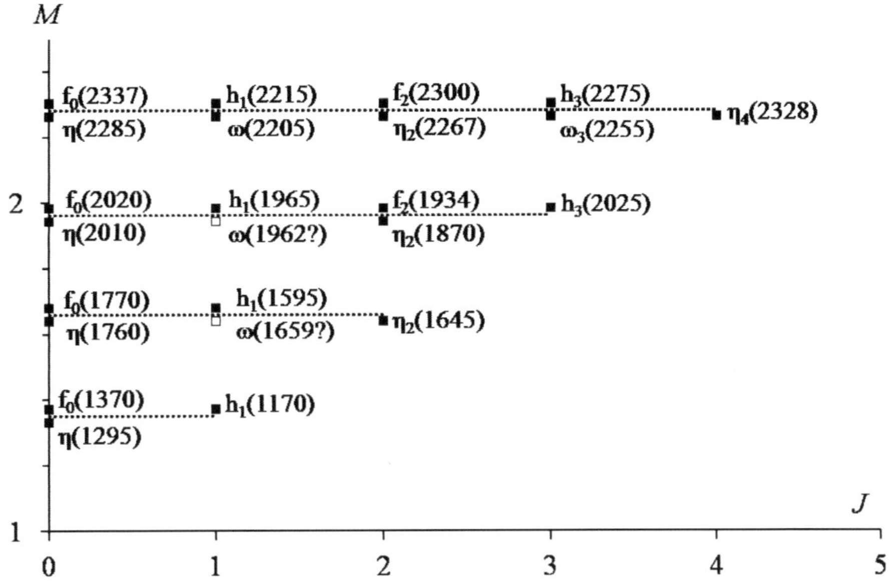

Figure 1. High-lying excitations of the isosinglet pseudoscalar η meson according to the Crystal Barrel measurements in the compilation of [14]. On the one side, the quantum numbers of the mesons in the four groups displayed here fit well into the respective $K = 1, 2, 3$, and 4 eigenstates of the conformal Laplacian on S^3 in (8). At the same time, they fit equally well into the eigenstates of the broken Laplacian in (24), which shows that the breaking of the conformal group symmetry at the level of the representation of its algebra by the $\hbar^2 c^2 b/R^2$ mass scale does not leave a print in the degeneracies in the spectrum. Empty squares denote "missing" resonances. These are additionally marked by a question-mark inside the round brackets, while the accompanying number is our prediction for the respective mass. Notice that not only the conformal degeneracy patterns are well pronounced, also practically all the states with spins lower than the maximal, in a level, ($J_{\max} = K$), appear parity duplicated, a phenomenon that is indicative of a manifest realization of the chiral symmetry in the Wigner-Wyle mode at that scale.

that the Crystal Barrel data on high lying unflavored mesons support our model of conformal symmetry breaking by a mass scale at the level of the representation of the underlying $so(4)$ algebra, and without loosing the degeneracies in the spectra.

Journal of Physics: Conference Series **378** (2012) 012036 doi:10.1088/1742-6596/378/1/012036

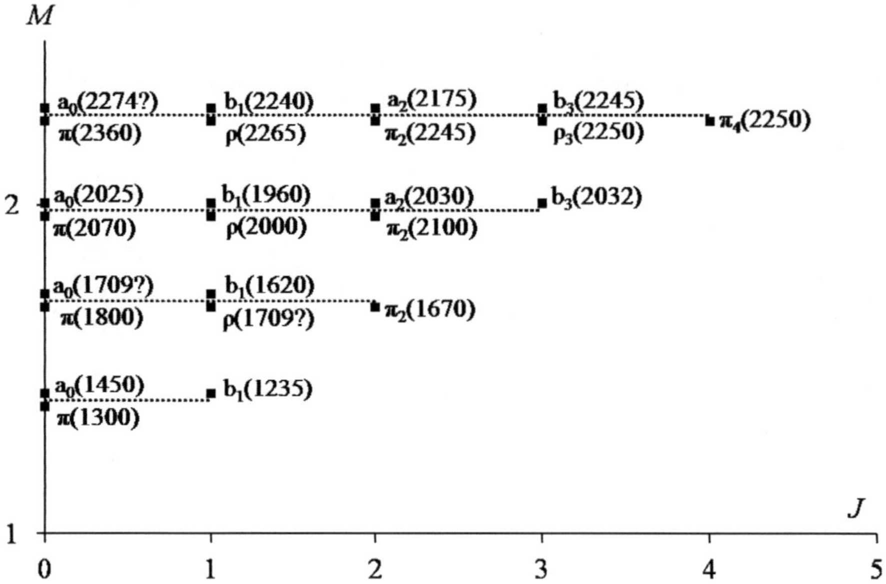

Figure 2. High-lying excitations of the isotriplet pseudoscalar π meson according to the Crystal Barrel measurements in the compilation of [14]. For notations and clarifying comments see Fig. 1.

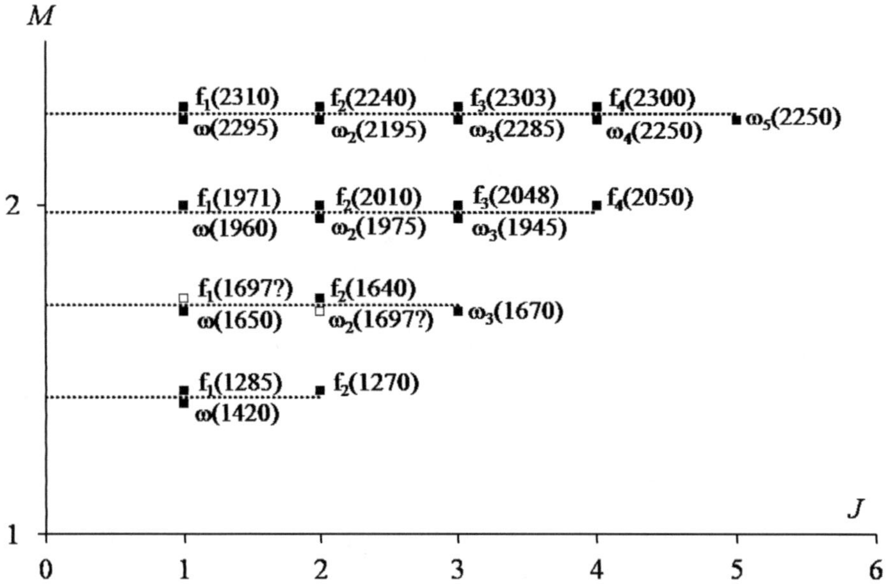

Figure 3. High-lying excitations of the isosinglet vector ω meson according to the Crystal Barrel measurements in the compilation of [14]. For notations and clarifying comments see Fig. 1.

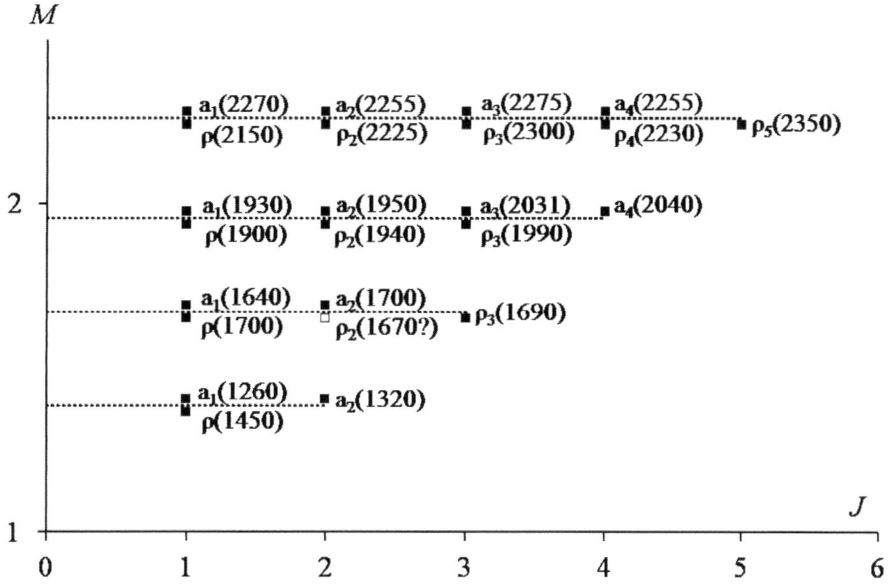

Figure 4. High-lying excitations of the isotriplet vector ρ meson according to the Crystal Barrel measurements in the compilation of [14]. For notations and clarifying comments see Fig. 1.

References

[1] Maldacena J 1998, *Phys. Rev. Lett.* **80**, 4859.
[2] Witten E 1998, *Adv. Theor. Math. Phys.* **2**, 233.
[3] Moschella U 2005, *Séminaire Poincaré* **1**, 1.
[4] de Teramond G F and Brodsky S J 2011, *J. Phys. Conf. Ser.* **287**, 012007.
[5] de Boer J 2002, *Proceedings of the 10 Int. Conf. on Supersymmetry and Unification of Fund. Interactions (SUSY02), Hamburg, Vol.* **1**, 512.
[6] Hands S, Hollowood T J and Myers J C 2010, *JHEP* **1007** 086.
[7] Gibbons G W and Steif A R 1995, *Phys. Lett. B* **346**, 255.
[8] Lüscher M and Mack G 1975, *Commun. Math. Phys.* **41**, 203.
[9] Gibbons G W 2011 Anti-de-Sitter space and its uses, *Preprint* arxiv:1110.1206v1 [hep-th].
[10] Pervushin V N, Pogosyan G S, Sissakian A N and Vinitsky S I 1993, *Phys. Atom. Nucl.* **5b**, 1027.
[11] Deur A, Burkert V, Chen J P and Korsch W 2008, *Phys. Lett.* **665**, 349.
[12] Colangelo P, De Fazio F, Gianuzzi F, Jugeau F and Nicotri S 2008, *Phys. Rev. D* **78**, 055009.
[13] Kelley T M, Bartz S B and Kapusta J I, *Phys. Rev. D* **83**, 016002 (2011).
[14] Afonin S S 2006, *Eur. Phys. J. A* **29**, 327.
[15] Rosenberg S 1998, *The Laplacian on a Riemannian manifold* (London: London Mathematical Society, Student Text 31, Cambridge Univerisity Press).
[16] Gurski M J 2006, *Conformal invariants and non-linear elliptic equations, Proc. of the Int. Congress of Mathematicians, Madrid, Spain, pp. 203-212.*
[17] Schrödinger E 1940, *Proc. Roy. Irish Acad. A* **46**, 9.
[18] Schrödinger E 1941, *Proc. Roy. Irish Acad.* **46**, 183.
[19] Pallares-Rivera A and Kirchbach M 2011, *J. Phys. A:Math.Theor.* **44**, 445302.
[20] Raposo A, Weber H J, Alvarez-Castillo D E and Kirchbach M 2007, *C. Eur. J. Phys.* **5**, 253.
[21] Takahashi T T, Suganuma H, Nemoto Y and Matsufuru H 2002, *Phys. Rev. D* **65**, 114509.

Journal of Physics: Conference Series **378** (2012) 012037 doi:10.1088/1742-6596/378/1/012037

Dyson-Schwinger equations for IR Yang-Mills theory in different dimensions.

Pietro Dall'Olio

Instituto de Fisica y Matematicas, Universidad Michoacana de San Nicolas de Hidalgo, Edificio C-3, Ciudad Universitaria, A. Postal 2-82, 58040. Morelia, Michoacan, Mexico.

E-mail: pietro@ifm.umich.mx

Abstract. A numerical solution of the coupled Dyson-Schwinger equations for the ghost and gluon propagators in Yang-Mills theory is presented in Landau gauge. Aimed at investigating the infrared behavior of the propagators, the equations are simplified by neglecting the gluon loops, according to the ghost dominance hypothesis motivated by the Gribov-Zwanziger scenario. The equations are solved with an iterative method, eliminating the ultraviolet divergence through a continuous regulator function depending on the cut off scale. The solutions, derived for different values of the Euclidean space-time dimension, present scaling (the infrared exponents are obtained) or decoupling behavior, depending on whether the horizon condition is or not implemented. Moreover, it is shown that the running coupling constant approaches a constant value in the IR, corresponding to an attractive fixed point.

1. Introduction

In the last few decades much effort has been devoted to determine the low-energy (or long distance) properties of Green's functions in QCD. In fact, despite of their gauge dependence, the infrared (IR) anlaysis of correlation functions is believed to be able to shed some light on the understanding of some mechanisms such as confinement, dynamical chiral symmetry breaking, etc.

The main difficulty of this task is the necessity to appeal to some non-perturbative technique, since the usual perturbative renormalization scheme brings to the generation of a IR Landau pole that jeopardizes the whole perturbative approach at low energies (see however [1; 2]).

Calculations on the Lattice probably represent the most reliable source of information to probe the IR sector of QCD. However, formulations on a continuum, where renormalization procedures are demanded, are employed as independent tools and in order to check the Lattice results, whose extrapolations at zero momentum present some difficulties, due to the finite nature of the Lattice.

Functional methods, in particular, like Dyson-Schwinger equations (DSEs) or Functional Renormalization Group equations (FRGs) have been extensively used to calculate (analytically or numerically) 2 or 3 point correlation functions.

These approaches have the advantage, over the "hard" numerical Lattice formulation, to better enlighten the analytical structure of the theory in the IR. However, even in the simpler pure Yang-Mills sector that is considered here, the analysis is far to be conclusive, since some truncations and approximations are required.

A quite long debate, especially, has been pursued, on which of two qualitatively different solutions, in the Landau gauge, for the gluon and the ghost propagators are realized in the deep IR. These are known as the *scaling* solution, that accounts for a vanishing gluon propagator and an enhanced ghost dressing function (the ghost propagator diverges faster than its tree-level counterpart at low momentum), and the *decoupling* solution, where both the ghost dressing function and the gluon propagator approach finite values at vanishing momentum, with the gluon presenting therefore a massive behavior.

While the former solution is certainly more desirable, since it is compatible with the maintaining of the BRST symmetry at the quantum level and fulfills the Kugo-Ojima criterion for confinement [3; 4], it is the latter one that seems to better match the latest lattice results [5; 6]. As it has been pointed out recently [7], there is also a theoretically motivation for being more inclined towards the decoupling solution and it has to do with taking seriously into account the problem of Gribov copies [8]: the dynamics of the auxiliary fields introduced to localize the *horizon* function, in the Gribov-Zwanziger framework [9; 10] that constraints the configuration space of the gauge field into the first Gribov region (and that explicitly breaks the BRST symmetry), would generate condensates that affect the gluon propagator behavior, forcing it to a finite value at vanishing momentum.

Nevertheless, both solutions found for the gluon propagator account for the violation of reflection positivity [11], which is interpreted as a clue for gluon confinement.

In this contribution numerical solutions of the coupled DSEs for the gluon and the ghost 2-point functions in Landau gauge are presented in a simple truncation scheme that relies on the hypothesis of ghost dominance at low energies, that is supported by both the Gribov-Zwanziger scenario and its refined version, and is therefore supposed to give confident results at least in the deep IR region. Despite of this simplification, no particular ansatz is made regarding the functional dependence on the momentum of the propagators, since the equations are solved by iteration according to the prescription described in [12].

The solutions are calculated for different values of Euclidean space-time dimension, in order to compare them with other numerical and analytical results. Eventually, both type of solutions are found, depending on whether a boundary condition for the ghost, the horizon condition, is imposed or not. Besides, for the scaling solutions, the same IR exponents are found, in the 4 dimensional case, that were found in previous works [13; 14], with more refined truncations, and for the decoupling solution the same deep IR behavior that has been observed in the Lattice and calculated analytically [1; 2] is encountered. This suggesting that this minimal truncation scheme for the DSEs is really the only relevant for the deep IR behavior of the correlation functions.

2. Dyson-Schwinger equations in Yang-Mills theory

DSEs are a powerful tool to extract non-perturbative information about the correlation functions of a quantum field theory. They correspond to the quantum version of the classical equations of motion, since they enclose both the stationary action principle and the equal time commutation relations between the fields. In a functional formulation of quantum field theory, their derivation depends only on the assumption of the existence of a well defined integration measure, since they are based on considering as vanishing the total functional derivative inside the integral of the generating functional $Z[j]$. In a compact notation:

$$0 = \int \mathcal{D}\varphi \frac{\delta}{\delta\varphi} e^{-S[\varphi]+j\cdot\varphi} = \left\{ -\frac{\delta S}{\delta\varphi}\left[\frac{\delta}{\delta j}\right] + j \right\} Z[j], \tag{1}$$

that for the effective action $\Gamma[\phi]$, the generator of 1-PI diagrams, yields:

$$\frac{\delta\Gamma[\phi]}{\delta\phi} - \frac{\delta S}{\delta\phi}\left[\phi + \left(\frac{\delta^2\Gamma}{\delta\phi\delta\phi}\right)^{-1}\frac{\delta}{\delta\phi}\right] = 0. \tag{2}$$

Taking several derivates w.r.t. the fields, one gets exact equations for the proper vertices generated by the effective action (for a complete discussion see [11]). In particular, here the Yang-Mills action in Euclidean space-time, in Landau gauge, is considered:

$$S = \int dx^d \left[\frac{1}{4}F^a_{\mu\nu}(x)F^a_{\mu\nu}(x) + \partial_\mu \bar{c}^a(x) D^{ab}_\mu c^b(x) + i\partial_\mu A^a_\mu(x) B^a(x)\right], \tag{3}$$

with $F^a_{\mu\nu}(x) = \partial_\mu A^a_\nu(x) - \partial_\nu A^a_\mu(x) + g f^{abc} A^b_\mu(x) A^c_\nu(x)$, $\quad D^{ab}_\mu = \delta^{ab}\partial_\mu + g f^{acb} A^c_\mu(x)$,

where the ghost term reproduce the Faddeev-Popov determinant in the gauge fixing process, and the Nakanishi-Lautrup field $B^a(x)$ is introduced in order to constrain the vector field in the Landau gauge $(\partial_\mu A_\mu(x) = 0)$.

When the DSE for the proper 2-point function of the gluon is derived, pure gluonic diagrams involving 3 and 4 gluon proper vertices appear (plus a tadpole diagram that contributes only with an infinite constant that can be absorbed in the renormalization process). In previous works these vertices have been parametrized with tensor structures, whose kernels were built by using additional physical information that could come by studying their own DSEs (that of course involve higher order vertices and must be truncated at some point) or by constraining the vertices to satisfy the Slavnov-Taylor identities (STI) that come from the BRST symmetry.

In this work such diagrams are just ignored, since it is believed, as pointed out by Gribov and Zwanziger, that the restriction of the functional integral to the region where the Fadeev-Popov operator is positive definite (the first Gribov region), which should be implemented in order to get rid of the extra gauge copies left by the Fadeev-Popov prescription, make the ghost propagator more divergent than the gluon one at zero momentum. This is also confirmed in the aforementioned refined Gribov-Zwanziger scenario [7], although the ghost dominance becomes weaker.

Moreover, from the point of view of the Wilsonian Renormalization Group (RG), as it is explained in [2], if the BRST symmetry is broken in the IR by the implementation of the horizon function, a mass term for the gluon would not be forbidden anymore, and this would change the scaling dimension of the gluon, in a way that the gluon vertices would become irrelevant operators near the IR fixed point.

It therefore seems reasonable to conjecture that these diagrams which contain only gluonic internal legs, will not give any relevant contribution to the propagators in the extreme IR. Besides, if at low energies BRST symmetry breaking has to be taken seriously, as it seems to be the case, an attempt to construct the gluon proper vertices through the analysis of the STI would not be valid anymore. More care should be taken in this way, maybe replacing the usual BRST symmetry with a more general symmetry shown by Curci-Ferrari [15] for their model that accounts for a gluon mass term.

It is noteworthy that the functional restriction to the Gribov region does not affect the form of the DSEs (that is why the horizon function was not added in (3)), since an eventual change would come from the boundary term in (1), but this is zero due to the fact that the Fadeev-Popov operator has zero eigenvalue at the boundary (the Gribov horizon). The effect of the Gribov restriction only reveals itself in some proper boundary conditions for the ghost propagator, that at this stage are already implemented through the choice of this truncation scheme.

Finally, the truncated DSEs for the gluon and the ghost 2-point function reduce themselves to the coupled equations that are shown diagrammatically in Fig. 1. Although in the original equations referring to these diagrams, one of the ghost-gluon vertex is dressed, this has been

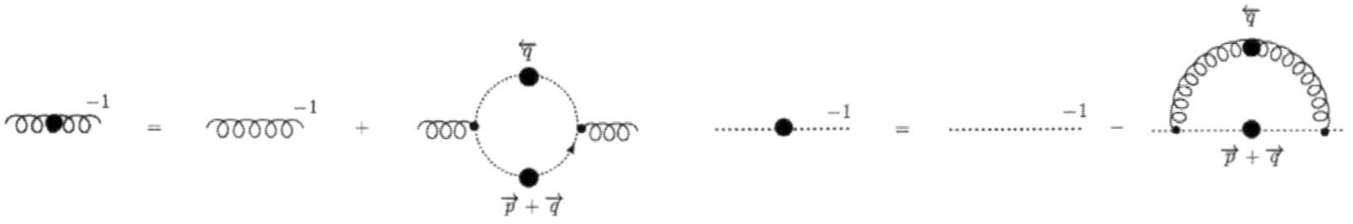

Figure 1. Truncated DSEs for the gluon and the ghost 2-point functions. The large black circles refer to the dressed propagator, while the small ones to the bare vertices.

replaced with the bare one. Such approximation is not so drastic, since from Taylor's argument [16] it is understood that, in Landau gauge, perturbative corrections at all orders to the proper ghost-gluon vertex are finite and scale-independent, at the symmetrical point, making the full vertex differ from the bare one only by a finite constant factor that can be absorbed in the renormalized coupling constant. Extending the approximation to the IR scales does not seem to be bold either, as it shown by the Lattice results in that region [17].

Besides, Taylor's argument would be strictly valid only for the transversal (in the gluon momentum) part of the vertex, which is the only part that will be considered here. That is because, although the BRST breaking gives rise to a longitudinal part for the self energy of the gluon 2-point function, the gluon connected propagator continues to be transversal in Landau gauge, which is guaranteed by the DSE of the Nakanishi-Lautrup field. This means that the gluon DSE can be projected on the transversal part of the external momentum without losing any valuable information.

With this last approximation at hand, the building blocks of the diagrams are given, in momentum space, by:

$$\begin{aligned}
\langle A_\mu^a(p)A_\nu^b(-q)\rangle &= \frac{P_{\mu\nu}^T(p)}{\omega(p)}\delta^{ab}(2\pi)^d\delta^d(p-q), \qquad \text{with} \quad P_{\mu\nu}^T(p)=\delta_{\mu\nu}-\frac{p_\mu p_\nu}{p^2}. \\
\langle c^a(p)\bar{c}^b(-q)\rangle &= -\frac{d(p)}{p^2}\delta^{ab}(2\pi)^d\delta^d(p-q), \\
(2\pi)^{3d}\frac{\delta^3\Gamma}{\delta\bar{c}^a(p)\delta c^b(q)\delta A_\nu^c(k)} &= P_{\mu\nu}^T(k)\, igf^{abc}p_\nu(2\pi)^d\delta^d(p+q+k),
\end{aligned} \tag{4}$$

where a transversal projector is included in the ghost-gluon vertex, for the reason just explained. The functions $d(p)$ and $\omega(p)$ represent respectively the ghost dressing function and the inverse gluon propagator, and are the unknown functions that will be determined by numerically solving the closed system of DSEs, which are given by the following expressions:

$$\begin{aligned}
\omega(p) &= Z_3 p^2 + g^2\frac{N_c}{d-1}\int\frac{d^dq}{(2\pi)^d}\frac{d(q)d(|\,p+q\,|)(1-(\hat{p}\cdot\hat{q})^2)}{(p+q)^2}, \\
d^{-1}(p) &= \tilde{Z}_3 - g^2N_c\int\frac{d^dq}{(2\pi)^d}\frac{d(|\,p+q\,|)(1-(\hat{p}\cdot\hat{q})^2)}{\omega(q)(p+q)^2},
\end{aligned} \tag{5}$$

where N_c is the number of colors, and the following relations have been used:

Journal of Physics: Conference Series **378** (2012) 012037 doi:10.1088/1742-6596/378/1/012037

$$f^{acd} f^{bcd} = N_c \delta^{ab}, \qquad P^T_{\mu\mu}(p) = d - 1, \tag{6}$$

the second one resulting from projecting the gluon equation on the transversal part of the external momentum. Z_3 and $\tilde{Z}_3$ are the usual wave function renormalization constants of the gluon and the ghost respectively. Note that, according to Taylor's argument, the vertex renormalization constant $\tilde{Z}_1$ has been omitted.

3. Numerical solutions

The above DSEs are solved with a cut-off regularization scheme borrowed from FRG [12]. Inside the integrals (5) the following regulator function:

$$R(q, \Lambda) = q^2 \exp\left[\frac{\Lambda^2}{q^2} - \frac{q^2}{\Lambda^2}\right] \tag{7}$$

is inserted in a subtracting new term, in such a way that smoothly suppresses the ultraviolet (UV) modes larger than the cut-off scale Λ:

$$\begin{aligned} \omega(p) = Z_3 p^2 + g^2 \frac{N_c}{d-1} \int \frac{d^d q}{(2\pi)^d} (1 - (\hat{p}\cdot\hat{q})^2) \Bigg\{ \frac{d(q) d(\mid p+q \mid)}{(p+q)^2} \\ - \frac{q^2}{[q^2 d^{-1}(q) + R(q,\Lambda)][(p+q)^2 d^{-1}(\mid p+q \mid) + R(\mid p+q \mid)]} \Bigg\}, \\ d^{-1}(p) = \tilde{Z}_3 - g^2 N_c \int \frac{d^d q}{(2\pi)^d} (1 - (\hat{p}\cdot\hat{q})^2) \Bigg\{ \frac{d(\mid p+q \mid)}{\omega(q)(p+q)^2} \\ - \frac{1}{[\omega(q) + R(q,\Lambda)][(p+q)^2 d^{-1}(\mid p+q \mid) + R(\mid p+q \mid)]} \Bigg\}. \end{aligned} \tag{8}$$

It is clear that such insertion does not affect the integrands for values of internal momentum much smaller than Λ, because the exponential contained in the regulator function make these extra terms vanish. On the contrary, for values of momentum much larger than Λ, the regulator function goes to zero and the extra terms become equal to the original ones, so that the integrands vanish. In this way the regulator function acts like a smooth UV cut-off.

The above system of equations is solved numerically by iteration, using as a starting guess for $\omega(p)$ and $d(p)$ a constant function (other trials have been made starting with other simple functions, which have given the same results) and putting the outcomes back inside the integrals, until they converge to the solutions.

Since in any dimension the integration are reduced, in spherical coordinates, to the double integrals over the modulus of the loop momentum and over the angle that this forms with the external momentum, it is immediate to perform the calculations for different values (even non integer) of dimension. The integrals are numerically evaluated using the Gauss-Legendre algorithm, up to an accuracy of about 150 nodes for each integration. The unknown functions are determined for a range of external momentum belonging to $[10^{-7}\Lambda,\, 10\Lambda]$.

The values of the external momentum are chosen at the zeros of Chebyshev polynomials, so that, at each step of the iteration process, inside the integrals the functions can be evaluated also for the values of momentum interpolating the Chebyshev roots, through the Chebyshev approximation, for which 120 nodes are used.

Moreover, in order to improve the convergence of the iteration process, a relaxation method is employed: at each step, inside the integrals, a linear combination of the two foregoing results

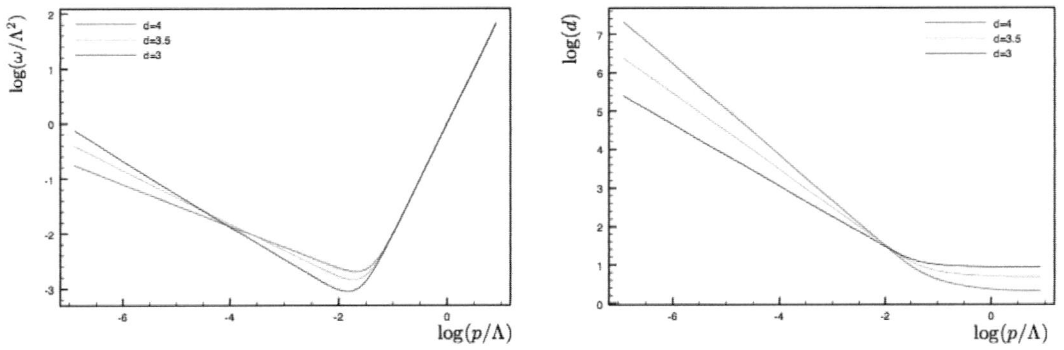

Figure 2. Scaling solutions for the gluon inverse propagator $\omega(p)$ and the ghost dressing function $d(p)$ at dimension $d = 3, 3.5, 4$.

is entered, *e.g.* $\omega = r * \omega_n + (1 - r) * \omega_{n-1}$, where ω_n is the outcome of the iteration coming before the running one, and $r \in [0, 1]$ is a parameter that is adjusted at each step (the best choice being found at $r \sim 0.6$).

It is worth to mention that an extrapolation is also needed, in the routine that evaluates the integrals, for the values of $d(p)$ and $\omega(p)$ at momenta that exceed the limits of the range defined for the Chebyshev approximation. In the high UV, for $p > \Lambda$, this is done by approximating the functions with their bare counterpart (*i.e.* $Z_3 p^2$ and $\tilde{Z}_3$), since the integrals in the right sides of the DSEs vanish at very large momenta ($p \gg \Lambda$) due to the presence of the regulator function. On the other hand, in the deep IR, the values are extrapolated by doing a power-like fit (in practice a linear fit in logarithmic scale) of the functions, since it is the behavior that is expected for the propagators at a fixed point.

3.1. Scaling solution

At each step of the iteration one has to fix the renormalization constants Z_3 and $\tilde{Z}_3$. For the ghost, this is done, in a first moment, by imposing the horizon condition, *i.e.* by imposing that the ghost dressing function $d(p)$ diverges at zero momentum. In practice this is done by subtracting the ghost DSE at zero external momentum. This is what Gribov had predicted from the effect of the restriction to the first Gribov region at the second perturbative order, and that Zwanziger confirmed at all orders.

For the gluon, the renormalization constant is fixed by imposing that in the UV region, for momenta $p \lesssim \Lambda$, the gluon propagator approaches its tree-level form, $\omega(p) = p^2$, as a zeroth order realization of *asymptotic freedom*, without taking into account any resummation of perturbative corrections. This is done, at each step, by performing a quadratic fit in UV that determines the constants used to parametrize the bare gluon term:

$$Z_3 p^2 = pa^2 + m^2, \tag{9}$$

where a mass counter term is considered as generated by the gauge symmetry breaking cut-off regularization. The exact scale at which the normalization condition is implemented is not fundamental for the qualitative behavior of the solution, since (see appendix B of [14]) this truncation scheme preserves multiplicative renormalization, so that imposing the normalization condition at a different scale would only change the final result by a multiplicative factor. Moreover, in the DSEs, $g = 1$ and $N_c = 3$ are settled.

Journal of Physics: Conference Series **378** (2012) 012037 doi:10.1088/1742-6596/378/1/012037

Table 1. IR exponents and running coupling constant at zero momentum.

dimension	α	β	$\alpha_R(0)$
3	0.60	0.80	0.76
3.5	0.48	0.99	1.52
4	0.38	1.19	2.97

The solutions with such renormalization scheme are depicted in Fig. 2 in a double logarithmic scale, for three different values of Euclidean dimension. As one can observe, while in the UV $d(p)$ and $\omega(p)$ present their classical behavior, coherently with asymptotic freedom, in the deep IR they exhibit a power-like dependence:

$$\omega(p)_{p\to 0} \sim p^{-\alpha}, \qquad d(p)_{p\to 0} \sim p^{-\beta}, \tag{10}$$

with the IR exponents α and β that satisfy the well known sum rule (or scaling relation):

$$\alpha = 2\beta + 2 - d, \tag{11}$$

that can be checked by looking at the numerical values in Table 1. The sum rule for the exponents is easily found, by simple power counting consistency, if one inserts in the truncated DSEs, with bare ghost-gluon vertices, power-like ansatze for $\omega(p)$ and $d(p)$. It is noteworthy that doing that, two different solutions for the IR exponents are found, analytically, both satisfying the sum rule [18; 19], given by the following expressions:

$$\beta = d - 2, \qquad \beta \approx \frac{2d-2}{5}, \tag{12}$$

where the second one is an approximated formula within 2% error, and corresponds to the solutions that are found here. In fact, as pointed out in [20], this is the only solution that is compatible with both DSEs and FRGs.

Besides, by taking an ϵ-expansion around the IR upper critical dimension ($d = 2$) of a Yang-Mills theory that accounts for a gluon mass term [2], one can see that the former solution of (12) corresponds to an IR repulsive fixed point, while the latter corresponds to an IR attractive fixed point (although for this last one, a fine-tuning of an irrelevant ghost operator is required, which is the equivalent of performing the horizon condition), *i.e.* is the only one physical relevant in the IR. Through that analysis, it is also interesting to note that such scaling solution is therefore consistent with the BRST symmetry breaking and the generation of a gluon mass term.

Taking advantage of the finite renormalization of the ghost-gluon vertex it is possible to define a dimensionless running coupling constant depending only on $\omega(p)$ and $d(p)$:

$$g_R(p) = p^{d/2-1}\frac{d(p)}{\omega(p)^{1/2}}. \tag{13}$$

As it can be seen in Fig. 3 the running coupling constant approaches finite constant values at low momenta, for the scaling solutions, pointing out the reaching of an IR fixed point.

As before mentioned, this simple truncation scheme is thought to give confident results only in the deep IR, since in the mid-momentum region and in the UV the gluon vertices should contribute and resummed perturbative corrections should be taken into account. Nevertheless, it is comforting to recover asymptotic freedom, at least qualitatively, as it can be seen by the dumping of the running coupling at high momentum.

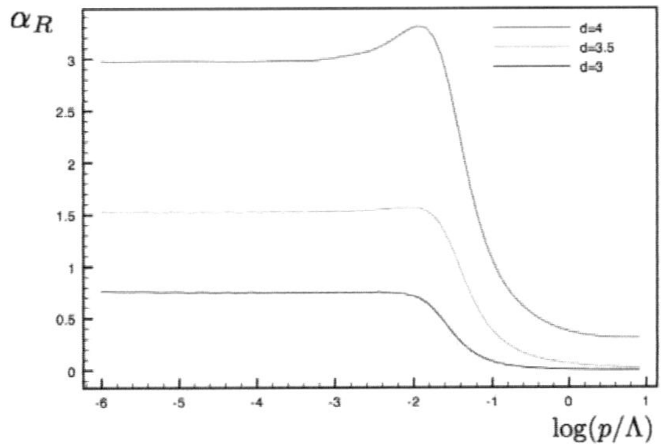

Figure 3. Running coupling constant $\alpha_R = g_R^2/4\pi$ for the scaling solutions at dimension $d = 3, 3.5, 4$.

3.2. Decoupling solution
If the condition on the IR ghost enhancing is loosen up, *i.e.* the horizon condition is not explicitly implemented, a different kind of solution is numerically found, the aforementioned decoupling solution, for which the scaling relation (11) is not satisfied. The ghost wave function renormalization constant $\tilde{Z}_3$ is now fixed, similarly to the gluon normalization condition, by fixing the ghost dressing function to its tree-level form, *i.e.* $d(p) = 1$, at a scale near below the cut-off scale. This is achieved by performing, at each step of the iteration, a numerical fit for a constant function in this UV region.

With this renormalization scheme at hand, obviously at high momenta the same behavior as before is obtained, with the gluon propagator and the ghost dressing function approaching their bare values. However, $\omega(p)$ and $d(p)$ now do not diverge in the IR, but both approach finite values. It is therefore worthy to represent them only in a narrow range of deep IR momenta, to highlight in what way these functions flow towards the zero momentum point.

In Figs. 4,5 the gluon propagator $\omega(p)^{-1}$ is depicted, in order to compare it with other numerical and analytical results. In the 3 dimensional case, a linear rise of the gluon propagator from a constant value at zero momentum can be observed. This is also seen in the lattice [5] and predicted in the ϵ-expension formulation near the IR fixed point [2].

This last one, ignoring the gluon kinetic term in the action, also predicts a quadratically rise in the deep IR for the gluon propagator in 4 dimension, which cannot be seen directly in Fig. 5, because it is hidden by the p^2 contribution coming from the UV fixed point. However, if the kinetic term is subtracted from the gluon propagator, the quadratically rise can be approximately observed (Fig. 6, where p^2 is put in the x-axis to make the quadratical dependence more evident).

Lastly, the running coupling constant can be seen to vanish (Fig. 7) at zero momentum, being the trivial point the only IR stable fixed point in this case [2]. As predicted in [2], α_R grows from the fixed point linearly in 3 dimension and quadratically in 4 dimension.

4. Conclusions
In this work it has been argued how a minimal truncation scheme for the DSEs of the ghost and the gluon 2-point functions, where only diagrams with internal ghost legs are included, is sufficient to get the most relevant information about their behavior in the deep IR. This is justified by the IR ghost dominance effect, predicted, in Landau gauge, by the Gribov-Zwanziger scenario and its refined version.

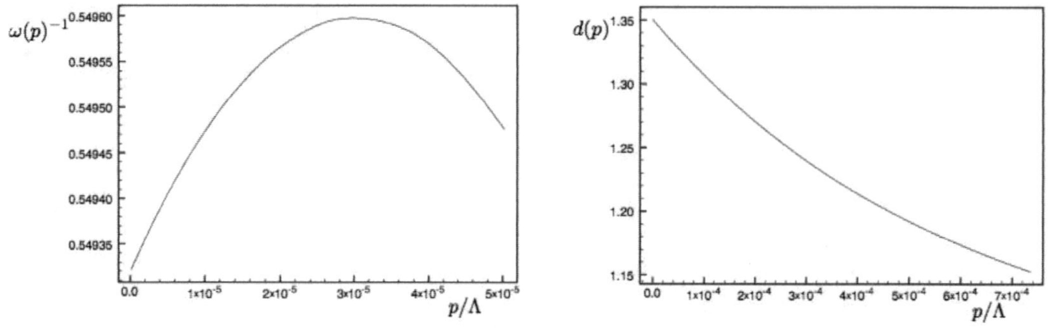

Figure 4. Decoupling solutions for the gluon propagator $\omega(p)^{-1}$ and the ghost dressing function $d(p)$ in 3 dimension.

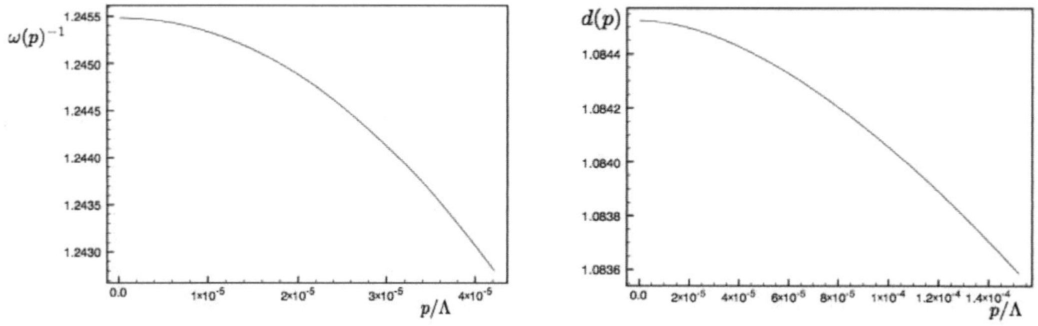

Figure 5. Decoupling solutions for the gluon propagator $\omega(p)^{-1}$ and the ghost dressing function $d(p)$ in 4 dimension.

Within such truncation scheme, where Taylor's argument has been used to approximate the ghost-gluon vertex, two kind of solutions have been found for different dimensions: the scaling solution, where the gluon propagator and the ghost dressing function show an IR power-like behavior with exponents different form zero, and the decoupling solution, where both approach finite values at vanishing momenta.

The discriminating factor for finding one solution rather than the other resulted to be the horizon condition, *i.e.* imposing explicitly the divergency of the ghost dressing function at zero momentum, whether implementing or not this condition corresponds to two inequivalent renormalization schemes.

It is noteworthy that the DSEs have been solved by iteration, without any external input concerning the unknown propagators, nor without performing any angular approximation inside the loop integrals.

For the scaling solutions, the IR exponents have been evaluated and found to correspond, at least in 4 dimensions, to the same values encountered in previous works. Moreover, a running coupling constant, depending only on the 2-point functions, can be defined, and is found to approach a constant value in the IR, corresponding to a stable fixed point.

For the decoupling solution of the gluon propagator, a linear and a quadratically (if the

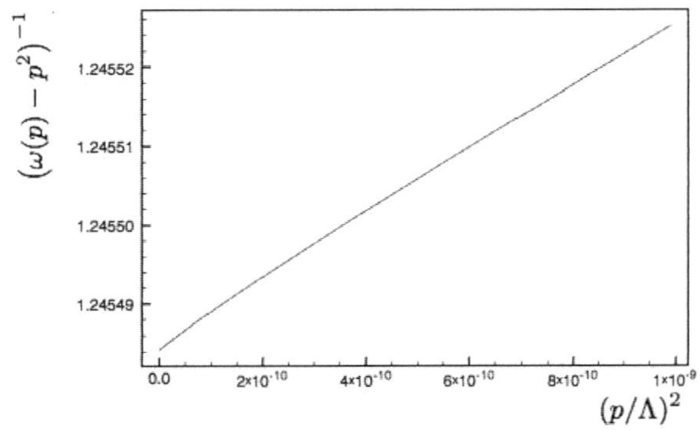

Figure 6. Decoupling solution for the gluon propagator in 4 dimension with the term p^2 subtracted.

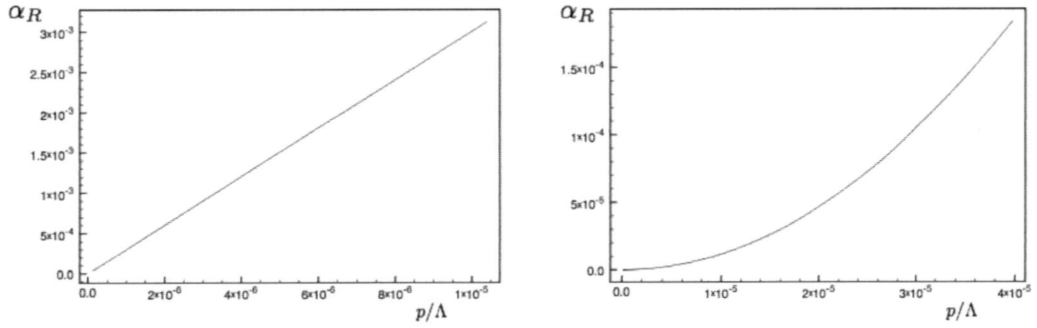

Figure 7. Running coupling constant $\alpha_R = g_R^2/4\pi$ for the decoupling solution in 3 (left) and 4 (right) dimension.

kinetic term is subtracted) behavior has been found, in the deep IR, in 3 and 4 dimensions respectively, in agreement with the latest Lattice calculations (the quadratically behavior has been observed for the Lattice parameter $\beta = 0$) and with the ϵ-expansion IR results. For these solutions, the running coupling constant vanishes at zero momentum, linearly in 3 dimensions, and quadratically in 4 dimensions.

References

[1] Tissier M and Wschebor N 2011 *Phys.Rev.* **D84** 045018 (*Preprint* 1105.2475)

[2] Weber A 2011 (*Preprint* arXiv/1112.1157)

[3] Kugo T 1995 *Talk given at International Symposium on BRS Symmetry on the Occasion of Its 20th Anniversary, Kyoto, Japan, 18-22 Sep 1995. Published in *Kyoto 1995, BRS symmetry* 107-119.*

[4] Kugo T and Ojima I 1979 *Progress of Theoretical Physics Supplement* **66** 1–130

[5] Cucchieri A and Mendes T 2009 *PoS* **QCD-TNT09** 026 (*Preprint* 1001.2584)

[6] Cucchieri A and Mendes T 2010 *Phys. Rev. D* **81**(1) 016005

[7] Dudal D, Gracey J A, Sorella S P, Vandersickel N and Verschelde H 2008 *Phys. Rev.* **D78** 065047 (*Preprint* 0806.4348)

[8] Gribov V N 1978 *Nucl. Phys.* **B139** 1

[9] Zwanziger D 1989 *Nucl.Phys.* **B323** 513–544

[10] Zwanziger D 1993 *Nucl.Phys.* **B399** 477–513

[11] Alkofer R and von Smekal L 2001 *Phys. Rept.* **353** 281 (*Preprint* hep-ph/0007355)

[12] Leder M, Pawlowski J M, Reinhardt H and Weber A 2011 *Phys. Rev.* **D83** 025010 (*Preprint* 1006.5710)

[13] Fischer C and Alkofer R 2002 *Phys.Lett.* **B536** 177–184 (*Preprint* hep-ph/0202202)

[14] Fischer C S, Maas A and Pawlowski J M 2009 *Annals Phys.* **324** 2408–2437 (*Preprint* 0810.1987)

[15] Curci G and Ferrari R 1976 *Nuovo Cim.* **A32** 151–168

[16] Taylor J C 1971 *Nucl. Phys.* **B33** 436–444

[17] Cucchieri A, Mendes T and Mihara A 2004 *JHEP* **12** 012 (*Preprint* hep-lat/0408034)

[18] Zwanziger D 2002 *Phys. Rev. D* **65**(9) 094039

[19] Lerche C and von Smekal L 2002 *Phys. Rev. D 65* **D 65** 125006 (*Preprint* hep-ph/0202194)

[20] Fischer C S and Pawlowski J M 2009 *Phys. Rev.* **D80** 025023 (*Preprint* 0903.2193)

XIII Mexican Workshop on Particles and Fields IOP Publishing
Journal of Physics: Conference Series **378** (2012) 012038 doi:10.1088/1742-6596/378/1/012038

Meson-loop contributions in the quark model

Roelof Bijker

Instituto de Ciencias Nucleares, Universidad Nacional Autónoma de México, A.P. 70-543, 04510 México, D.F., México

E-mail: bijker@nucleares.unam.mx

Elena Santopinto

I.N.F.N., Sezione di Genova, via Dodecaneso 33, I-16145, Italy

E-mail: santopinto@ge.infn.it

Abstract. We present a review of some recent results obtained in the unquenched quark model for baryons. The effects of sea quarks are taken into account in an explicit form via a QCD-inspired creation mechanism of the quark-antiquark pairs. In this approach, the contribution of the quark-antiquark pairs can be studied for any inital baryon and for any flavor of the $q\bar{q}$ pairs ($u\bar{u}$, $d\bar{d}$ and $s\bar{s}$). It is shown that, whereas the inclusion of $q\bar{q}$ pairs (or meson loops) does not affect the baryon magnetic moments, it immediately leads to an excess of $\bar{d}$ over $\bar{u}$ in the proton and introduces a sizeable contribution of orbital angular momentum to the spin of the proton. The contribution of $s\bar{s}$ pairs to the magnetic moment and the radius of the proton is found to be small, in agreement with the latest experimental results and recent lattice calculations.

1. Introduction

In the constituent quark model (CQM) hadrons are described as a system of constituent (or valence) quarks and antiquarks, qqq for baryons and $q\bar{q}$ for mesons. Despite the success of the quark model there is strong evidence for the existence of exotic degrees of freedom (other than valence quarks) in hadrons. Common features of constituent quark models for baryons are the effective degrees of freedom of three constituent quarks (qqq configurations), the $SU(6)$ spin-flavor symmetry and a long-range confining potential. In general, CQMs reproduce the mass spectrum of baryon resonances reasonably well, but at the same time, they show very similar deviations for other observables, such as photocouplings, helicity amplitudes and strong decays. Since the photocouplings depend mostly on the spin-flavor structure, all models that have the same $SU(6)$ structure in common, show the same behavior, *e.g.* the photocouplings for the $\Delta(1232)$ are underpredicted by a large amount, even though their ratio is reproduced correctly. In general, the helicity amplitudes (or transition form factors) show deviations from CQM calculations at low values of Q^2 [1, 2, 3]. The problem of missing strength at low Q^2 indicates that some fundamental mechanism is lacking in the dynamical description of hadronic structure. This mechanism can be identified with the production of quark-antiquark pairs, which becomes more important in the outer region of the nucleon.

Additional evidence for such higher Fock components in the baryon wave function ($qqq - q\bar{q}$ configurations) comes from CQM studies of strong decay widths of baryon resonances that are on average underpredicted by CQMs [4, 5, 6], the spin-orbit splitting of $\Lambda(1405)$ and $\Lambda(1520)$,

Journal of Physics: Conference Series **378** (2012) 012038 doi:10.1088/1742-6596/378/1/012038

and the large η decay widths of $N(1535)$, $\Lambda(1670)$ and $\Sigma(1750)$. More direct evidence for the importance of quark-antiquark components in the proton comes from measurements of the $\bar{d}/\bar{u}$ asymmetry in the nucleon sea [7, 8], the proton spin crisis [9, 10] and parity-violating electron scattering experiments which report a nonvanishing strange quark contribution, albeit small, to the charge and magnetization distributions [11, 12].

The aim of this contribution is to present a review of some results recently obtained for the unquenched quark model, in which the effects of quark-antiquark pair creation ($u\bar{u}$, $d\bar{d}$ and $s\bar{s}$) are taken into account in an explicit form via a $^3P_0$ coupling mechanism [13, 14, 15]. In order to test the consistency of the formalism we first calculate the baryon magnetic moments [13] which constitute one of the early successes of the CQM. Next we discuss the spin and flavor content of the proton. Whereas the flavor asymmetry and the orbital angular momentum are dominated by pion loops, the contribution of the sea quark spins to the proton spin arises almost entirely from (excited) vector meson loops [16]. Finally, we study the strange magnetic moment and the strange radius of the proton [17].

2. Unquenched quark model

The role of higher Fock components in baryon wave functions has been studied by many authors in the context of meson cloud models, such as the cloudy bag model, meson convolution models and chiral models [7, 18]. In these models, the flavor asymmetry of the proton can be understood in terms of couplings to the pion cloud. There have also been several attempts to study the importance of higher Fock components in the context of the constituent quark model. In this respect we mention the work by Riska and coworkers who introduce a small number of selected higher Fock components which are then fitted to reproduce the experimental data [19]. However, these studies lack an explicit model or mechanism for the mixing between the valence and sea quarks. The Rome group studied the pion and nucleon electromagnetic form factors in a Bethe-Salpeter approach, mainly thanks to the dressing of photon vertex by means of a vector-meson dominance parametrization [20]. Koniuk and Guiasu used a convolution model with CQM wave functions and an elementary emission model for the coupling to the pion cloud to calculate the magnetic moments and the helicity amplitudes from the nucleon to the Δ resonance [21]. It was found that the nucleon magnetic moments were unchanged after renomalization of the parameters, but that the missing strength in the helicity amplitudes of the Δ could not be explained with pions only.

The impact of $q\bar{q}$ pairs in hadron spectroscopy was originally studied by Törnqvist and Zenczykowski in a quark model extended by the $^3P_0$ model [22]. Even though their model only includes a sum over ground state baryons and ground state mesons, the basic idea of the importance to carry out a sum over a complete set of intermediate states was proposed in there. Subsequently, the effects of hadron loops in mesons was studied by Geiger and Isgur in a flux-tube breaking model in which the $q\bar{q}$ pairs are created in the $^3P_0$ state with the quantum numbers of the vacuum [23, 24, 25]. In this approach, the quark potential model arises from an adiabatic approximation to the gluonic degrees of freedom embodied in the flux-tube [26]. It was shown that cancellations between apparently uncorrelated sets of intermediate states occur in such a way that the modification in the linear potential can be reabsorbed, after renormalization, in the new strength of the linear potential [24]. In addition, the quark-antiquark pairs do not destroy the good CQM results for the mesons [24] and preserve the OZI hierarchy [25] provided that the sum be carried out over a large tower of intermediate states. A first application of this procedure to baryons was presented in [27] in which the importance of $s\bar{s}$ loops in the proton were studied by taking into account the contribution of the six diagrams of Fig. 1 in combination with harmonic oscillator wave functions for the baryons and mesons and a $^3P_0$ pair creation mechanism. This approach has the advantage that the effects of quark-antiquark pairs are introduced explicitly via a QCD-inspired pair-creation mechanism, which opens the

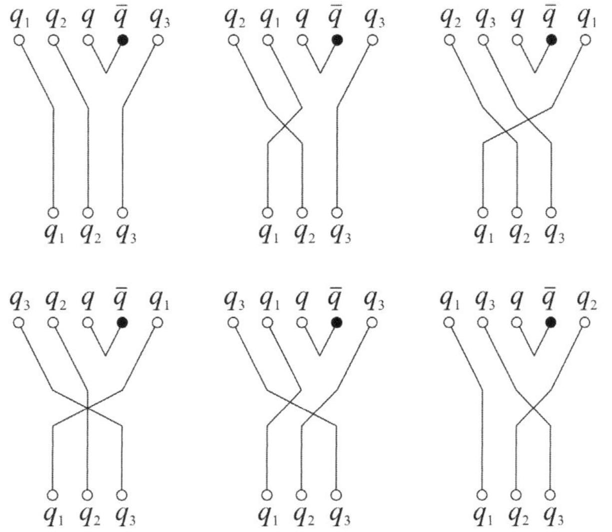

Figure 1. Quark line diagrams for $A \rightarrow BC$ with $q\bar{q} = s\bar{s}$ and $q_1 q_2 q_3 = uud$.

possibility to study the importance of $q\bar{q}$ pairs in baryons and mesons in a systematic and unified way.

The present approach is motivated by these earlier studies on extensions of the quark model to include the effects of $q\bar{q}$ pairs [22, 27]. Our approach is based on a CQM to which the quark-antiquark pairs with vacuum quantum numbers are added as a perturbation employing a $^3P_0$ model for the $q\bar{q}$ pair creation [13, 14, 27]. The pair-creation mechanism is inserted at the quark level and the one-loop diagrams are calculated by summing over a complete set of intermediate baryon-meson states. Under these assumptions, the baryon wave function consists of a zeroth order three-quark configuration $\mid A\rangle$ plus a sum over all possible higher Fock components due to the creation of $^3P_0$ quark-antiquark pairs

$$\mid \psi_A\rangle \;=\; \mathcal{N}\left[\mid A\rangle + \sum_{BCl}\int d\vec{k} \mid BC\vec{k}\,lJ\rangle \frac{\langle BC\vec{k}\,lJ \mid T^\dagger \mid A\rangle}{M_A - E_B - E_C}\right] . \qquad (1)$$

Here A denotes the initial baryon, B and C represent the intermediate baryon and meson, and M_A, E_B and E_C are their respective energies, $\vec{k}$ and l the relative radial momentum and orbital angular momentum of B and C, and J is the total angular momentum $\vec{J} = \vec{J}_B + \vec{J}_C + \vec{l}$. The operator $T^\dagger$ creates a quark-antiquark pair in the $^3P_0$ state with the quantum numbers of the vacuum: $L = S = 1$ and $J = 0$ [13, 14, 28]

$$\begin{aligned} T^\dagger \;=\; & -3\sum_{ij}\int d\vec{p}_i\, d\vec{p}_j\, \delta(\vec{p}_i + \vec{p}_j)\, C_{ij}\, F_{ij}\, \Gamma(\vec{p}_i - \vec{p}_j) \\ & [\chi_{ij} \times \mathcal{Y}_1(\vec{p}_i - \vec{p}_j)]^{(0)}\, b_i^\dagger(\vec{p}_i)\, d_j^\dagger(\vec{p}_j) . \end{aligned} \qquad (2)$$

Here, $b_i^\dagger(\vec{p}_i)$ and $d_j^\dagger(\vec{p}_j)$ are the creation operators for a quark and antiquark with momenta $\vec{p}_i$ and $\vec{p}_j$, respectively. The quark pair is characterized by a color singlet wave function C_{ij}, a flavor singlet wave function F_{ij} and a spin triplet wave function χ_{ij} with spin $S = 1$. The solid harmonic $\mathcal{Y}_1(\vec{p}_i - \vec{p}_j)$ indicates that the quark and antiquark are in a relative P wave. The $SU(3)$ flavor symmetry of the valence quark configuration $\mid A\rangle$ is broken by the quark-antiquark pairs via the energy denominator, but the $SU(2)$ isospin symmetry is still preserved. In the special case of the closure limit in which the energy denominator of Eq. (1) is a constant, the flavor symmetry of the valence quark configuration is recovered.

Since the operator $T^\dagger$ creates a pair of constituent quarks, a Gaussian quark-antiquark creation vertex function was introduced by which the pair is created as a finite object with an effective size, rather than as a pointlike object. In momentum space it is given by

$$\Gamma(\vec{p}_i - \vec{p}_j) \;=\; \gamma_0 \, e^{-r_q^2(\vec{p}_i-\vec{p}_j)^2/6} \;. \tag{3}$$

The width has been determined from meson decays to be approximately 0.25−0.35 fm [25, 27, 29]. Here we take the average value, $r_q = 0.30$ fm. Finally, the dimensionless constant γ_0 is the intrinsic pair creation strength which can be determined from the strong decays of baryons [4].

The strong coupling vertex

$$\langle BC\vec{k}\, lJ \mid T^\dagger \mid A \rangle \;, \tag{4}$$

was derived in explicit form in the harmonic oscillator basis [28]. In the present calculations, we use harmonic oscillator wave functions in which there is a single oscillator parameter for the baryons and another one for the mesons which, following [27], are taken to be $\beta_{\rm baryon} = 0.32$ GeV [30] and $\beta_{\rm meson} = 0.40$ GeV [23], respectively.

In order to calculate the effects of quark-antiquark pairs on an observable, one has to evaluate the contribution of all possible intermediate states. By using a combination of group theoretical and computational techniques, the sum over intermediate states is carried out up to saturation and not only for the first few shells as in previous studies [22, 27]. Not only does this have a significant impact on the numerical result, but it is necessary for consistency with the OZI-rule and the success of CQMs in hadron spectroscopy. In addition, the contributions of quark-antiquark pairs can be evaluated for any initial baryon (ground state or resonance) and for any flavor of the $q\bar{q}$ pair (not only $s\bar{s}$ as in [27], but also $u\bar{u}$ and $d\bar{d}$), and for any model of baryons and mesons, as long as their wave functions are expressed in the basis of harmonic oscillator wave functions [13, 14].

In the calculations presented in this contribution, we use harmonic oscillator wave functions up to five oscillator shells for the intermediate baryons and mesons. All parameters were taken from the literature without attempting to optimize their values in order to improve the agreement with experimental data [13, 14].

2.1. Closure limit

Before discussing an application of the unquenched model to baryon magnetic moments and spins, we study the so-called closure limit in which the intermediate states appearing in Eq. (1) are degenerate in energy and hence the energy denominator becomes a constant independent of the quantum numbers of the intermediate states. In the closure limit, the evaluation of the contribution of the quark-antiquark pairs (or the higher Fock components) simplifies considerably, since the sum over intermediate states can be solved by closure and the contribution of the quark-antiquark pairs to the matrix element reduces to

$$\mathcal{O}_{\rm sea} \;\propto\; \langle A \mid T\,\hat{\mathcal{O}}\,T^\dagger \mid A \rangle \;. \tag{5}$$

Since the $^3P_0$ pair-creation operator of Eq. (2) is a flavor singlet and the energy denominator in Eq. (1) is reduced to a constant in the closure limit, the higher Fock components of the baryon wave function have the same $SU(3)$ flavor symmetry as the valence quark configuration $\mid A \rangle$.

At a qualitative level, the closure limit helps to explain the phenomenological success of the CQM because the $SU(3)$ flavor symmetry of the baryon wave function is preserved. As an example, the strange content of the proton vanishes in the closure limit due to many cancelling contributions in the sum over intermediate states in Eq. (1). Away from the closure limit, the strangeness content of the proton is expected to be small, in agreement with the experimental data from parity-violating electron scattering (for some recent data see [11, 12]). Even though in this case the cancellations are no longer exact, many intermediate states contribute with opposite signs, and the net result is nonzero, but small. This means that even if the flavor symmetry of the CQM is broken by the higher Fock components, the net results are still to a large extent determined by the flavor symmetry of the valence quark configuration. Similar arguments were applied to the preservation of the OZI hierarchy in the context of the flux-tube breaking model [25]. Therefore, the closure limit not only provides simple expressions for the relative flavor content of physical observables, but also gives further insight into the origin of cancellations between the contributions from different intermediate states.

In addition, the closure limit imposes very stringent conditions on the numerical calculations, since it involves the sum over all possible intermediate states. Therefore, the closure limit provides a highly nontrivial test of the computer codes which involves both the spin-flavor sector, the permutation symmetry, the construction of a complete set of intermediate states in spin-flavor space for each radial excitation and the implementation of the sum over all of these states.

In the following, we study the effects of quark-antiquark pairs on the magnetic moments, radii, and the flavor and spin content of baryons in the general case, *i.e.* beyond the closure limit.

2.2. Magnetic moments

The unquenching of the quark model has to be carried out in such a way as to preserve the phenomenological successes of the constituent quark model. In applications to mesons, it was shown that the inclusion of quark-antiquark pairs does not destroy the good CQM results [24] and preserves the OZI hierarchy [25]. In a similar fashion, in this Section we will show that the CQM results for the magnetic moments of the octet baryons also hold in the unquenched CQM [13].

It is well known that the CQM gives a good description of the baryon magnetic moments, even in its simplest form in which the baryons are treated in terms of three constituent quarks in a relative S-wave. The quark magnetic moments are determined by fitting the magnetic moments of the proton, neutron and Λ hyperon to give $\mu_u = 1.852$, $\mu_d = -0.972$ and $\mu_s = -0.613$ μ_N [31].

In the unquenched CQM the baryon magnetic moments also receive contributions from the quark spins of the pairs and the orbital motion of the quarks

$$\vec{\mu} = \sum_q \mu_q \left[2\vec{s}(q) + \vec{l}(q) - 2\vec{s}(\bar{q}) - \vec{l}(\bar{q}) \right] , \tag{6}$$

where $\mu_q = e_q\hbar/2m_qc$ is the quark magnetic moment. In Fig. 2 we show a comparison between the experimental values of the magnetic moments of the octet baryons (circles) and the theoretical values obtained in the CQM (squares) and in the unquenched quark model (triangles). The results for the unquenched quark model were obtained in a calculation involving a sum over intermediate states up to five oscillator shells for both baryons and mesons. The results obtained in the unquenched quark model are practically identical to the ones in the CQM, which shows that the addition of the quark-antiquark pairs preserves the good CQM results for the baryon magnetic moments. The effect of the $q\bar{q}$ pairs could be absorbed into renormalized values of the quark magnetic moments to $\mu_u = 2.066$, $\mu_d = -1.110$ and $\mu_s = -0.633$ μ_N, A similar feature was found in the context of the flux-tube breaking model for mesons in which it was shown that the inclusion of quark-antiquark pairs preserved the linear behavior of the confining potential as well as the OZI hierarchy [25]. The change in the linear potential caused

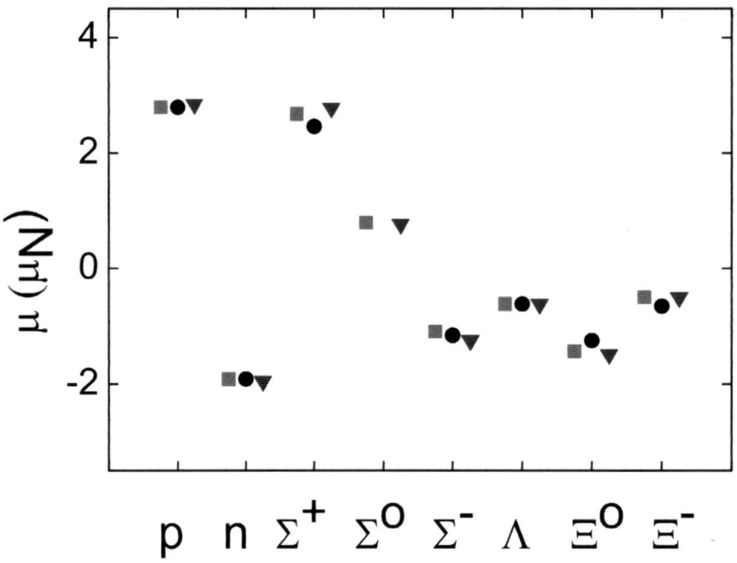

Figure 2. Magnetic moments of octet baryons: experimental values from PDG [31] (circles), CQM (squares) and unquenched quark model (triangles).

by the bubbling of the pairs in the string could be absorbed into a renormalized strength of the linear potential. The largest difference is observed for the charged Σ hyperons, but the relation between the magnetic moments of Σ hyperons $\mu(\Sigma^0) = [\mu(\Sigma^+) + \mu(\Sigma^-)]/2$ [32] is preserved in the unquenched calculation as a consequence of isospin symmetry.

The results for the magnetic moments can be understood qualitatively in the closure limit in which the relative contribution of the quark spins from the quark-antiquark pairs is the same as that from the valence quarks [13]. Moreover, since in the closure limit the contribution of the orbital angular momentum is small in comparison to that of the quark spins, the results for the baryon magnetic moments are almost indistinguishable from those of the CQM. Away from the closure limit, even though the relations between the different contributions no longer hold exactly, they are still valid approximately. In addition, there is now a contribution from the orbital part (at the level of ~ 5 %) which is mainly due to the baryon-pion channel.

In summary, the inclusion of the effects of quark-antiquark pairs preserves, after renormalization, the good results of the CQM for the magnetic moments of the octet baryons.

2.3. Flavor content

The flavor asymmetry of the proton $\mathcal{A}(p)$ is related to the Gottfried integral S_G for the difference of the proton and neutron electromagnetic structure functions as

$$S_G = \int_0^1 \frac{F_2^p(x) - F_2^n(x)}{x} dx = \frac{1}{3} - \frac{2}{3} \int_0^1 [\bar{d}_p(x) - \bar{u}_p(x)]\, dx = \frac{1}{3}[1 - 2\mathcal{A}(p)] \, . \tag{7}$$

Under the assumption of a flavor symmetric (or rather flavor independent) sea one obtains the Gottfried sum rule $S_G = 1/3$ [8, 33], whereas any deviation from this value is an indication of the $\bar{d}/\bar{u}$ asymmetry of the nucleon sea, thus providing evidence of the existence of higher

Table 1. Contributions to the flavor asymmetry of the proton [14].

	Unquenched QM	
	0-4 $\hbar\omega$	0 $\hbar\omega$
$N\pi$	0.195	0.177
$\Delta\pi$	−0.016	−0.010
$N\pi\eta_8\eta_1$	−0.028	−0.018
$N\rho$	0.050	0.012
$\Delta\rho$	−0.017	−0.003
$N\rho\omega_8\omega_1$	−0.033	−0.010
Total	0.151	0.147

Fock components (such as $qqq - q\bar{q}$ configurations) in the proton wave function. The first clear evidence of a violation of the Gottfried sum rule came from the New Muon Collaboration (NMC) [37] which was later confirmed by Drell-Yan experiments [34, 35] and a measurement of semi-inclusive deep-inelastic scattering [36]. All experiments show evidence that there are more $\bar{d}$ quarks in the proton than there are $\bar{u}$ quarks [8]. The final NMC value is 0.2281 ± 0.0065 at $Q^2 = 4$ (GeV/c)^2 for the Gottfried integral over the range $0.004 \leq x \leq 0.8$ [37], which implies a flavor asymmetric sea. The observed flavor asymmetry is far too large to be accounted for by processes that can be described by QCD in perturbative regime and therefore has to attributed to non-perturbative QCD mechanisms. It was shown in the framework of the meson-cloud model, that the coupling of the nucleon to the pion cloud provides a mechanism that is able to produce a flavor asymmetry due to the dominance of $n\pi^+$ among the virtual configurations [38].

In the unquenched quark model, the flavor asymmetry of the proton can be calculated directly from the difference of the number of $\bar{d}$ and $\bar{u}$ sea quarks in the proton, even in the absence of explicit information on the (anti)quark distribution functions. Table 1 shows that the flavor asymmetry for the proton in the UCQM is 0.151 [14] which corresponds to a value of the Gottfried integral of 0.232, remarkably close to the experimental value. The main contribution to the flavor asymmetry of the proton is due to the pion loops, especially the $n\pi^+$ intermediate state, thus confirming in an explicit calculation the explanation given in Ref. [38] in the context of the meson-cloud model. In addition, we find that there are important contributions from the $\Delta\pi$ channel and, especially, from the off-diagonal terms $p\pi^0$-$p\eta_8$ and $p\pi^0$-$p\eta_1$ which together are of the order of 15-20 % of that of the $N\pi$ channel, but with the opposite sign (see Table 1). The contribution of the intermediate vector mesons is very small due to a cancelation between the $n\rho^+$ and the $\Delta\rho$ channels and the cross terms $p\rho^0$-$p\omega_8$ and $p\rho^0$-$p\omega_1$. Kaon loops do not contribute to the proton flavor asymmetry. Table 1 shows that the full four-shell calculation is dominated by the contribution of the ground state intermediate baryons and mesons (0 $\hbar\omega$). Both columns show the same qualitative behavior: dominance of the pion loops with a small negative correction of the order of 10-15 % due to the off-diagonal terms involving π and η pseudoscalar mesons and an almost vanishing contribution from the vector mesons.

Since the unquenched quark model is valid not only for the proton, but for all baryons (ground state or resonance), it is straightforward to calculate the flavor asymmetries of the other octet baryons. For the Σ^+ hyperon and the Ξ^0 cascade particle we find $\mathcal{A}(\Sigma^+) = 0.126$ and $\mathcal{A}(\Xi^0) = -0.001$ [14], respectively. The flavor asymmetries of the remaining octet baryons can be obtained by using the isospin symmetry of the unquenched quark model [14]. For example, the excess of $\bar{d}$ over $\bar{u}$ in the proton is related to the excess of $\bar{u}$ over $\bar{d}$ in the neutron, $\mathcal{A}(p) = -\mathcal{A}(n)$. Similar relations hold for the other octet baryons: $\mathcal{A}(\Sigma^+) = -\mathcal{A}(\Sigma^-)$, $\mathcal{A}(\Xi^0) = -\mathcal{A}(\Xi^-)$ and

Journal of Physics: Conference Series **378** (2012) 012038 doi:10.1088/1742-6596/378/1/012038

Table 2. Relative flavor asymmetries of octet baryons

Model	$\mathcal{A}(\Sigma^+)/\mathcal{A}(p)$	$\mathcal{A}(\Xi^0)/\mathcal{A}(p)$	Ref.
Unquenched CQM	0.833	–0.005	[14]
Octet couplings	0.353	–0.647	[39]
Chiral QM	2	1	[40]
Balance Model	3.083	2.075	[41]

$\mathcal{A}(\Lambda) = \mathcal{A}(\Sigma^0) = 0$. Just as for the proton, the flavor asymmetry of the other octet baryons is expected to be dominated by pion loops, whereas the other contributions are suppressed by the energy denominator in Eq. (1). For the Σ hyperon this is indeed the case, but for the cascade particles the pion loops are suppressed by the value of the $SU(3)$ flavor coupling which is a factor of 5 smaller than that for the proton. Hence for the Ξ hyperons there is no dominant contribution. Since for the Ξ hyperon all contributions are roughly of the same order and small, and moreover some with a positive and others with a negative sign, the value of the flavor asymmetry of the cascade particles is calculated to be small [14].

In Table 2, we show a comparison of some predictions for the flavor asymmetry of the Σ^+ and Ξ^0 hyperons relative to that of the proton. In the unquenched quark model, the flavor asymmetry of the proton is predicted to be of the same order as that of the Σ^+ hyperon and much larger than that of the cascade particle

$$\mathcal{A}(p) \sim \mathcal{A}(\Sigma^+) \gg |\mathcal{A}(\Xi^0)| \ . \tag{8}$$

This behavior is very different from that obtained in the chiral quark model $\mathcal{A}(\Sigma^+) = 2\mathcal{A}(p) = 2\mathcal{A}(\Xi^0)$ [40], the balance model $\mathcal{A}(\Sigma^+) > \mathcal{A}(\Xi^0) > \mathcal{A}(p)$ [41], and the octet model $\mathcal{A}(p) > |\mathcal{A}(\Xi^0)| > \mathcal{A}(\Sigma^+)$ [39]. The values for the chiral quark model and the balance model were taken from [42].

In order to distinguish between the predictions of the different models and to obtain a better understanding of the non-perturbative structure of QCD, new experiments are needed to measure the flavor asymmetry of hyperons. In particular, the flavor asymmetry of charged Σ hyperons can obtained from Drell-Yan experiments using charged hyperon beams on the proton [39] or by means of backward $K^\pm$ electroproduction [43].

2.4. Spin content

The contribution of the quark spins to the spin of the proton can be obtained from the proton spin structure function g_1^p in combination with the neutron and hyperon semileptonic decays [10]. The observation by the European Muon Collaboration that the total quark spin constitutes only a small fraction of the spin of the nucleon [9] sparked an enormous interest in the spin structure of the proton [10]. Recent experiments show that approximately one third of the proton spin is carried by quarks [44, 45], and that the gluon contribution is rather small (either positive or negative) and compatible with zero [46]. This rules out the possibility that most of the missing spin be carried by the gluon and indicates that the origen of the missing spin of the proton has to be attributed to other mechanisms.

In the unquenched quark model, the effect of hadron loops on the fraction of the proton spin carried by the quark (antiquark) spins and orbital angular momentum can be studied in an explicit way [13]. As in other effective models [10], gluonic effects associated with the axial anomaly are not included, and therefore the contribution from the gluons is missing from the outset. The total spin of the proton can then be written as the sum of the contributions from

Journal of Physics: Conference Series **378** (2012) 012038 doi:10.1088/1742-6596/378/1/012038

Table 3. Contribution of quark spins $\Delta\Sigma$ and orbital angular momentum ΔL to the spin of the proton and the Λ hyperon

		CQM	Unquenched QM		
			Valence	Sea	Total
p	$\Delta\Sigma$	1	0.378	0.298	0.676
	$2\Delta L$	0	0.000	0.324	0.324
	$2\Delta J$	1	0.378	0.622	1.000
Λ	$\Delta\Sigma$	1	0.422	0.429	0.851
	$2\Delta L$	0	0.000	0.149	0.149
	$2\Delta J$	1	0.422	0.578	1.000

the quark (and antiquark) spins and orbital angular momentum

$$1 = 2\Delta J = \Delta\Sigma + 2\Delta L . \tag{9}$$

Table 3 shows that the inclusion of the quark-antiquark pairs has a dramatic effect on the spin content of the proton. Whereas in the CQM the proton spin is carried entirely by the (valence) quarks, in the unquenched calculation the contributions of the valence quark spins, the sea quark spins and the orbital angular momentum to the proton spin are comparable in size and equal to approximately 38, 30 and 32 %, respectively. The importance of orbital angular momentum to the proton spin was discussed many years ago by Sehgal [47] and Ratcliffe [48] in the context of the quark-parton model and, more recently, by Myhrer and Thomas in framework of the bag model [49].

The large contribution of orbital angular momentum is a consequence of the fact that the effects of pion loops for the proton flavor asymmetry and the contribution of orbital angular momentum to the proton spin are identical $\mathcal{A}(p) = \Delta L$ due to the spin and isospin properties [50]. Since in the unquenched calculations both the flavor asymmetry and the orbital angular momentum are dominated by pion loops, this relation is to a good approximation still valid in the UCQM, $\mathcal{A}(p) = 0.151$ and $\Delta L = 0.162$, respectively.

The situation for the quark spins is completely different. In the unquenched calculations, the contributions of valence and sea quarks are given by $\Delta\Sigma_{\rm val} = 0.378$ and $\Delta\Sigma_{\rm sea} = 0.298$, respectively. While the orbital angular momentum arises almost entirely from the $N\pi$ channel, the sea quark spins are dominated by the intermediate vector mesons with a relatively small contribution from the pseudoscalar mesons [17]. Since the convergence of the sum over the contribution of intermediate vector mesons is much slower, the sum was carried out over five complete oscillator shells for both the intermediate baryons and mesons [13].

The experimental data on the spin structure of the proton have raised many questions about the contributions of valence and sea quarks, gluons and orbital angular momentum to the proton spin. In this respect it is of interest to investigate the spin structure of other octet baryons, in particular the Λ hyperon. In most studies, additional assumptions had to be made about the sea quarks in order to get an estimate of its spin content. For example, the assumption that both valence and sea quarks are related by $SU(3)$ flavor symmetry, allows to express the spin content of the Λ hyperon in terms of that of the proton [51, 52, 53] and gives rise to equal contributions of the quark spins $(\Delta\Sigma)_\Lambda = (\Delta\Sigma)_p$. In the unquenched quark model there is no need to make additional assumptions about the nature of the sea. Table 3 shows that the contribution of quark spins for the Λ is larger than that for the proton, $(\Delta\Sigma)_\Lambda > (\Delta\Sigma)_p$, which is a result of $SU(3)$ flavor breaking by the sea quarks.

Journal of Physics: Conference Series **378** (2012) 012038 doi:10.1088/1742-6596/378/1/012038

2.5. Strangeness

Even though the nucleon carries no net strangeness, it may have a nonvanishing distribution of strangeness. In 1987 Kaplan and Manohar [54] observed that neutral current experiments could provide information on the strange matrix elements of the nucleon. The parity-violating elastic scattering (PVES) of electrons on nucleon, as suggested by Beck and McKeown in 1988 [55], has demonstrated to provide a powerful tool to determine the strangeness contribution to by combining the weak form factors $G^{Z,p}$ of the proton with the electromagnetic form factors of the nucleon $G^{\gamma,p}$ and $G^{\gamma,n}$ [56, 57, 58, 59, 60, 61]. The strangeness contribution may be extracted by performing a flavor decomposition [55, 62]

$$\begin{aligned}
G^{\gamma,p}_{E,M} &= \frac{2}{3}G^{u}_{E,M} - \frac{1}{3}\left(G^{d}_{E,M} + G^{s}_{E,M}\right) , \\
G^{\gamma,n}_{E,M} &= \frac{2}{3}G^{d}_{E,M} - \frac{1}{3}\left(G^{u}_{E,M} + G^{s}_{E,M}\right) , \\
G^{Z,p}_{E,M} &= (1 - \frac{8}{3}\sin^2\theta_W)G^{u}_{E,M} + (-1 + \frac{4}{3}\sin^2\theta_W)\left(G^{d}_{E,M} + G^{s}_{E,M}\right) . \qquad (10)
\end{aligned}$$

Even though the first measurements indicated large and positive values of G^s_M, *e.g.* the SAMPLE collaboration in 1999 found $G^s_M = 0.61 \pm 0.17 \pm 0.21 \pm 0.19$ at $Q^2 = 0.1$ (GeV/c)^2 [63], the more recent values obtained by the HAPPEX, A4, G0 collaborations for G^s_M and G^s_E are much smaller. The most recent measurements show that the contribution of strange quarks to the electric and magnetic form factors is compatible with zero within the experimental errors [58, 59, 60]. Another possibility to determine the strangeness in the proton was suggested by Pate in 2004 [64] by combining experimental data on neutrino scattering with the electromagnetic form factors of the nucleon.

Theoretically, the strangeness of the proton has been addressed in terms of two static observables, the strange magnetic moment and the strange radius. They can be extracted from the behavior of the strange form factors of the nucleon near the origin $Q^2 = 0$ as

$$\begin{aligned}
\mu_s &= e_s\, G^s_M(0) , \\
R^2_s &= -6e_s \left.\frac{dG^s_E}{dQ^2}\right|_{Q^2=0} . \qquad (11)
\end{aligned}$$

We note, that this definition differs from the one in [62] by the electric charge of the strange quark. The values of μ_s and R^2_s can be extracted from the experimental data as follows. In Ref. [56], after measuring G^s_M at $Q^2 = 0.1$ (GeV/c)^2, the strange magnetic moment μ_s is extrapolated by considering the momentum dependence of $G^s_M(Q^2)$ taken from Ref. [65]. The resulting value is $\mu_s = -0.003 \pm 0.097 \pm 0.103 \pm 0.023\ \mu_N$. In Ref. [66] the strange magnetic moment of the proton was obtained from a fit to the complete set of parity violating scattering experimental data to give $\mu_s = -0.04 \pm 0.18 \pm 0.02\ \mu_N$.

Since R^2_s is proportional to the slope of G^s_E in the origen, the evaluation of the strange radius R^2_s requires a measurement of the strange electric form factor at a small value of Q^2 in combination with $G^s_E(0) = 0$ (the nucleon carries no net strangeness). The values of the strange radius obtained in two global analysis of the available experimental data are $R^2_s = 0.002 \pm 0.014$ fm^2 [66] and $R^2_s = -0.006 \pm 0.013$ fm^2 [67].

The first theoretical calculation of the strange form factors of the nucleon, performed by Jaffe in 1989 [68], reported quite large results for G^s_E and G^s_M, thus triggering the interest for this kind of observables. Subsequent theoretical calculations of μ_s and R^2_s, obtained through lattice QCD calculations, hadronic models and effective hadronic theory, vary widely both in absolute value and in the predicted sign $\mu_s(p)$ and $R^2_s(p)$ [62, 69]. The results from recent lattice QCD calculations show small values for both $\mu_s(p)$ and $R^2_s(p)$ [70, 71].

Journal of Physics: Conference Series **378** (2012) 012038 doi:10.1088/1742-6596/378/1/012038

Finally, we discuss the calculation of the strange magnetic moment and radius of the proton in the unquenched CQM. The results are obtained in a calculation involving a sum over intermediate states up to four oscillator shells for both baryons and mesons. In the UCQM formalism, the strange magnetic moment of the proton is given by the expectation value of the operator

$$\vec{\mu}_s = \sum_i \mu_{i,s} \left[2\vec{s}(q_i) + \vec{\ell}(q_i) - 2\vec{s}(\bar{q}_i) - \vec{\ell}(\bar{q}_i) \right] . \qquad (12)$$

Here $\mu_{i,s}$ is the magnetic moment of the quark i times a projector. In the UCQM the strange magnetic moment of the proton arises from the sea quarks. There is a contribution from the quark spins of the $s\bar{s}$ pair $-0.0004\ \mu_N$, as well as from its orbital motion $-0.0002\ \mu_N$. Both contributions are small and give a total strange magnetic moment $\mu_s = -0.0006\ \mu_N$ [17].

Similarly, the strange radius of the proton is calculated as the expectation value of the operator

$$R_s^2 = \sum_{i=1}^{5} e_{i,s} \left(\vec{r}_i - \vec{R}_{CM} \right)^2 , \qquad (13)$$

where $e_{i,s}$ is the electric charge of the quark i times a projector on strangeness, and $\vec{r}_i$ and $\vec{R}_{CM}$ are the coordinates of the quark i and the center of mass, respectively. The strange radius of the proton is calculated to be -0.004 fm^2 [17].

In conclusion, the effects of the higher Fock components on the strange magnetic moment and the strange radius of the proton are found to be negligible. Our results are compatible with the latest experimental data and recent lattice calculations.

3. Summary and conclusion

In this contribution, we presented a short review of some general features of an unquenched quark model for baryons in which the effects of sea quarks are taken into account in an explicit form via a $^3P_0$ creation mechanism of the quark-antiquark pairs ($u\bar{u}$, $d\bar{d}$ and $s\bar{s}$). This provides the possibility to address many open problems in baryon structure and spectroscopy. It was shown that, whereas the inclusion of $q\bar{q}$ pairs (or meson loops) does not affect the baryon magnetic moments, it immediately leads to an excess of $\bar{d}$ over $\bar{u}$ in the proton and introduces a sizeable contribution of orbital angular momentum to the spin of the proton. In addition, the contribution of $s\bar{s}$ pairs to the magnetic moment and the radius of the proton was found to be small, in agreement with the latest experimental results and recent lattice calculations.

Even though different models of hadron structure may show similar results for the properties of the proton, often their predictions for the other octet baryons exhibit large variations. Therefore, in order to be able to distinguish between the predictions of different models of hadron structure and to obtain a better understanding of the non-perturbative structure of QCD, new experiments are needed to measure the properties of other octet baryons, such as the Σ and Λ hyperons. In particular, the flavor asymmetry of charged Σ hyperons can obtained from Drell-Yan experiments using charged hyperon beams on the proton [39] or by means of backward $K^{\pm}$ electroproduction [43].

The results for the magnetic moments, the spin and flavor content of octet baryons are very promising and encouraging. The inclusion of the effects of quark-antiquark pairs in a general and consistent way, as suggested here, may provide a major improvement to the constituent quark model which increases considerably its range of applicability.

Acknowledgments

This work was supported in part by CONACyT (grant 78833) and PAPIIT-UNAM (grant IN113711), Mexico and INFN, Italy.

Journal of Physics: Conference Series **378** (2012) 012038 doi:10.1088/1742-6596/378/1/012038

References

[1] Aznauryan I G *et al* 2008 *Phys. Rev.* C **78** 045209
Aznauryan I G *et al* 2009 *Phys. Rev.* C **80** 055203
Aznauryan I G, Burkert V D, Lee T S H and Mokeev V 2011 *J. Phys.: Conf. Ser.* **299** 012008
[2] Bijker R, Iachello F and Leviatan A 1994 *Ann. Phys. (N.Y.)* **236** 69
[3] Bijker R, Iachello F and Leviatan A 1996 *Phys. Rev.* C **54** 1935
[4] Capstick S and Roberts W 1994 *Phys. Rev.* D **49** 4570
[5] Bijker R, Iachello F and Leviatan A 2000 *Ann. Phys. (N.Y.)* **284** 89
[6] Bijker R, Iachello F and Leviatan A 1997 *Phys. Rev.* D **55** 2862
[7] Kumano S 1998 *Phys. Rep.* **303** 183
[8] Garvey G T and Peng J C 2001 *Prog. Part. Nucl. Phys.* **47** 203
[9] Ashman J *et al* 1988 *Phys. Lett.* B **206** 364
[10] Bass S D 2008 *The spin structure of the proton* (Singarpore: World Scientific)
Thomas A W 2009 *Int. J. Mod. Phys.* E **18** 1116
Kuhn S E, Chen J P and Leader E 2009 *Progr. Part. Nucl. Phys.* **63** 1
[11] Acha A *et al* 2007 *Phys. Rev. Lett.* **98** 032301
[12] Baunack S *et al* 2009 *Phys. Rev. Lett.* **102** 151803
[13] Bijker R and Santopinto E 2009 *Phys. Rev.* C **80** 065210
[14] Santopinto E and Bijker R 2010 *Phys. Rev.* C **82** 062202(R)
[15] Bijker R and Santopinto E 2010 *J. Phys.: Conf. Ser.* **239** 012009
[16] Bijker R and Santopinto E 2011 *J. Phys.: Conf. Ser.* **322** 012014
[17] Bijker R, Ferretti J and Santopinto E 2012 *Phys. Rev.* C **85** in press
[18] Speth J and Thomas A W 1998 *Adv. Nucl. Phys.* **24** 83
[19] Zou B S and Riska D O 2005 *Phys. Rev. Lett.* **95** 072001
An C S, Riska D O and Zou B S 2006 *Phys. Rev.* C **73** 035207
Riska D O and Zou B S 2006 *Phys. Lett.* B **636** 265
Li Q B and Riska D O 2007 *Nucl. Phys.* A **791** 406
[20] De Melo J P B C, Frederico T, Pace E and Salmé G 2006 *Phys. Rev.* D **73** 074013
De Melo J P B C, Frederico T, Pace E, Pisano S and Salmé G 2007 *Nucl. Phys.* A **782** 69c
De Melo J P B C, Frederico T, Pace E, Pisano S and Salmé G 2009 *Phys. Lett.* B **671** 153
[21] Guiasu I and Koniuk R 1987 *Phys. Rev.* D **36** 2757
[22] Törnqvist N A and Zenczykowski P 1984 *Phys. Rev.* D **29** 2139
Törnqvist N A 1985 *Acta Phys. Polon.* B **16** 503 and 683
Zenczykowski P 1986 *Ann. Phys. (N.Y.)* **169** 453
[23] Kokoski R and Isgur N 1987 *Phys. Rev.* D **35** 907
[24] Geiger P and Isgur N 1990 *Phys. Rev.* D **41** 1595
[25] Geiger P and Isgur N 1991 *Phys. Rev. Lett.* **67** 1066
Geiger P and Isgur N 1991 *Phys. Rev.* D **44** 799
Geiger P and Isgur N 1993 *Phys. Rev.* D **47**, 5050
[26] Isgur N and Paton J 1983 *Phys. Lett.* B **124** 247
Isgur N and Paton J 1985 *Phys. Rev.* D **31** 2910
[27] Geiger P and Isgur N 1997 *Phys. Rev.* D **55** 299
[28] Roberts W and Silvestre-Brac B 1992 *Few-Body Systems* **11** 171
[29] Silvestre-Brac B and Gignoux C 1991 *Phys. Rev.* D **43** 3699
[30] Isgur N and Karl G 1978 *Phys. Rev.* D **18** 4187
Isgur N and Karl G 1979 *Phys. Rev.* D **19** 2653
Isgur N and Karl G 1979 *Phys. Rev.* D **20** 1191
[31] Nakamura K *et al* 2010 *J. Phys. G: Nucl. Part. Phys.* **37** 075021
[32] Marshak R, Okubo S and Sudarshan G 1957 *Phys. Rev.* **106** 599
Coleman S and Glashow S L 1961 *Phys. Rev. Lett.* **6** 423
[33] Gottfried K 1967 *Phys. Rev. Lett.* **18** 1174
[34] Baldit A *et al* 1994 *Phys. Lett.* B **332** 244
[35] Towell R S *et al* 2001 *Phys. Rev.* D **64** 052002
[36] Ackerstaff K *et al* 1998 *Phys. Rev. Lett.* **81** 5519
[37] Amaudruz P *et al* 1991 *Phys. Rev. Lett.* **66** 2712
Arneodo M *et al* 1997 *Nucl. Phys.* B **487** 3
[38] Thomas A W 1983 *Phys. Lett.* B **126** 97 (1983)
Henley E M and Miller G A 1990 *Phys. Lett.* B **251** 453
[39] Alberg M, Henley E M, Ji X and Thomas A W 1996 *Phys. Lett.* B **389** 367

Alberg M, Falter T and Henley E M 1998 *Nucl. Phys.* A **644** 93
[40] Eichten E J, Hinchcliffe I and Quigg C 1992 *Phys. Rev.* D **45** 2269
[41] Zhang Y J, Deng W Z and Ma B Q 2002 *Phys. Rev.* D **65** 114005
[42] Shao L, Zhang Y J and Ma B Q 2010 *Phys. Lett.* B **686** 136
[43] Osipenko R, private communication.
[44] Airapetian A *et al* 2007 *Phys. Rev.* D **75** 012007
[45] Alexakhin V Y *et al* 2007 *Phys. Lett.* B **647** 8
[46] Bradamante F 2008 *Prog. Part. Nucl. Phys.* **61** 229
[47] Sehgal L M 1974 *Phys. Rev.* D **10** 1663
[48] Ratcliffe P G 1987 *Phys. Lett.* B **192** 180
[49] Myhrer F and Thomas A W 2008 *Phys. Lett.* B **663** 302
[50] Garvey G T 2010 *Phys. Rev.* C **81** 055212
[51] Jaffe R L 1996 *Phys. Rev.* D **54** R6581
[52] Boros C and Zuo-tang L 1998 *Phys. Rev.* D **57** 4491
[53] Ashery D and Lipkin H J 1999 *Phys. Lett.* B **469** 263
[54] Kaplan D B and Manohar A 1988 *Nucl. Phys.* B **310** 527
[55] McKeown R D 1989 Phys. Lett. B **219** 140
Beck D H 1989 *Phys. Rev.* D **39** 3248
[56] Hasty R *et al* 2000 Science **290** 2117
[57] Ito T M *et al* 2004 *Phys. Rev. Lett.* **92** 102003
Beise E J *et al* 2005 *Prog. Part. Nucl. Phys.* **54** 289
[58] Acha A *et al* 2007 *Phys. Rev. Lett.* **98** 032301
[59] Baunack S *et al* 2009 *Phys. Rev. Lett.* **102** 151803
[60] Androic D *et al* 2010 *Phys. Rev. Lett.* **104** 012001
[61] Armstrong D S *et al* 2005 *Phys. Rev. Lett.* **95** 092001
[62] Beck D H and McKeown R D 2001 *Ann. Rev. Nucl. Part. Sci.* **51** 189
Beck D H and Holstein B R 2001 *Int. J. Mod. Phys.* E **10** 1
[63] Spayde D T *et al* 2000 *Phys. Rev. Lett.* **84** 1106
[64] Pate S F 2004 *Phys. Rev. Lett.* **92** 082002
Pate S F, McKee D W and Papavassiliou V 2008 *Phys. Rev.* C **78** 015207
[65] Hemmert T R, Meissner U G and Steininger S 1998 *Phys. Lett.* B **437** 184
[66] Young R D *et al* 2006 *Phys. Rev. Lett.* **97** 102002
[67] Liu J *et al* 2007 *Phys. Rev.* C **76** 025202
[68] Jaffe R L 1989 *Phys. Lett.* B **229** 275
[69] Bijker R 2006 *J. Phys. G: Nucl. Part. Phys.* **32** L49
[70] Leinweber D B *et al* 2005 *Phys. Rev. Lett.* **94** 212001
Leinweber D B *et al* 2006 *Phys. Rev. Lett.* **97** 022001
[71] Doi T *et al* 2009 *Phys. Rev.* D **80** 094503 (2009)

Journal of Physics: Conference Series **378** (2012) 012039 doi:10.1088/1742-6596/378/1/012039

Analysis of baryon properties in large-N_c chiral perturbation theory

Rubén Flores-Mendieta[1], María A Hernández-Ruiz[1,2] and Christoph P Hofmann[3]

[1] Instituto de Física, Universidad Autónoma de San Luis Potosí, Álvaro Obregón 64, Zona Centro, San Luis Potosí, S.L.P. 78000, México

[3] Facultad de Ciencias, Universidad de Colima, Bernal Díaz del Castillo 340, Colima, Colima 28045, México

Abstract. We discuss how to compute corrections to one-loop order to the baryon axial-vector current and magnetic moments in heavy baryon chiral perturbation theory in the large-N_c limit, where N_c is the number of colors. Loop graphs with octet and decuplet intermediate states cancel to various orders in N_c as a consequence of the large-N_c spin-flavor symmetry of QCD baryons. These cancellations are explicitly shown for the general case of N_f flavors of light quarks. We also indicate how to compare our resultant expressions with the corresponding ones obtained within conventional heavy baryon chiral perturbation theory at the physical values $N_c = 3, N_f = 3$. We observe an excellent agreement order by order.

1. Introductory remarks

Nowadays it is commonly accepted that QCD is the theory of the strong interactions. However, the analytic calculations of the spectrum and properties of hadrons are not possible within QCD because the theory is strongly coupled at low energies. In this respect, chiral perturbation theory and the $1/N_c$ expansion (where N_c is the number of colors) have shed much light on the subject.

Chiral perturbation theory exploits the symmetry of the QCD Lagrangian under $SU(3)_L \times SU(3)_R \times U(1)_V$ transformations of the three flavors of light quarks in the limit $m_q \to 0$. Chiral symmetry is spontaneously broken by the QCD vacuum to the subgroup $SU(3)_V \times U(1)_V$, giving rise to an octet of Goldstone bosons. Physical observables can be expanded order by order in powers of p^2/Λ_χ^2 and m_Π^2/Λ_χ^2, or equivalently, m_q/Λ_χ, where p is the meson momentum, m_Π is the mass of the Goldstone boson and Λ_χ is the scale of chiral symmetry breaking. When baryons are accounted for in the theory it is convenient to introduce velocity-dependent baryon fields, so that the expansion of the baryon chiral Lagrangian in powers of m_q and $1/M_B$ (where M_B is the baryon mass) is manifest [1, 2]. The resultant theory, referred to as *heavy baryon chiral perturbation theory*, was first used to compute the chiral logarithmic corrections to the baryon axial vector current for baryon semileptonic decays due to meson loops [1, 2]. While these corrections are large when only octet baryon intermediate states are kept [1], the inclusion of decuplet baryon intermediate states yields sizable cancellations between one-loop corrections [2].

[2] Present address: Facultad de Ciencias Químicas, Universidad Autónoma de Zacatecas, Apartado Postal 585, Zacatecas, Zacatecas 98060, México

The $1/N_c$ expansion, on the other hand, has led to a considerable theoretical progress in understanding the spin-flavor structure of QCD baryons [3, 4]. In the large-N_c limit there is an exact contracted $SU(2N_f)$ spin-flavor symmetry (where N_f is the number of light quark flavors) [5], which can be used to classify large-N_c baryon states and matrix elements. Physical quantities can be analyzed in the large-N_c limit, where corrections arise at relative orders $1/N_c$, $1/N_c^2$ and so on; this is the origin of the $1/N_c$ expansion. Various static properties of baryons have been computed within this formalism, namely, masses [3, 4, 6], vector and axial couplings [3, 4, 7, 8] and magnetic moments [7, 9, 10], to name but a few.

In the present paper, we use a combined expansion in m_q and $1/N_c$ to show how to compute one-loop corrections specifically to the baryon axial-vector current and magnetic moment. The groundwork material on this topic has already been settled down in Refs. [11, 12, 13] so we limit ourselves to review some basic material. The $1/N_c$ chiral effective Lagrangian for the lowest-lying baryons was constructed in Ref. [14] and describes the interactions of the spin-$\frac{1}{2}$ baryon octet and the spin-$\frac{3}{2}$ baryon decuplet with the pion nonet. Within this framework it is possible to analyze the renormalization of the baryon axial vector current at the one-loop level [12]. As already pointed out in Refs. [3, 4, 5], there are large-N_c cancellations between individual Feynman diagrams, provided one sums over all baryon states in a complete multiplet of the large-N_c $SU(6)$ spin-flavor symmetry and uses axial coupling ratios given by the large-N_c spin-flavor symmetry. In Ref. [11] the general structure of the various large-N_c cancellations was analyzed. Here we briefly outline the evaluation of the corresponding operator expressions that involve complicated products of commutators and/or anticommutators of $SU(6)$ spin-flavor operators. The reduction of these operator products to a physical operator basis, although tedious, is nevertheless doable. The final expressions explicitly exhibit how these large-N_c cancellations occur.

Also, it is instructive to perform a comparison of the results obtained within the combined formalism with the corresponding ones obtained in conventional heavy baryon chiral perturbation theory (including both octet and decuplet baryons), where no $1/N_c$ expansion is involved. Both approaches agree – the large-N_c cancellations are guaranteed to occur as a consequence of the contracted $SU(6)$ spin-flavor symmetry present in the limit $N_c \to \infty$: No large numerical cancellations between loop diagrams with intermediate octet states and low-energy constants of the next-to-leading order effective Lagrangian, containing the effects of decuplet states, arise.

The present paper is organized as follows. In Sec. 2 we provide a brief review on heavy baryon chiral perturbation theory and introduce our notation and conventions; we also present some expressions for the one-loop corrections to the baryon axial-vector current and magnetic moments within this formalism. In Sec. 3 we give an overview of the $1/N_c$ chiral effective Lagrangian for the lowest-lying baryons. In Secs. 3.1 and 3.2 we discuss, in the combined approach, the one-loop corrections to the baryon-axial vector current and magnetic moments, respectively, and indicate how to compare our expressions with the ones discussed in Sec. 2. We close this paper with some concluding remarks.

2. Baryons in chiral perturbation theory

The heavy baryon chiral Lagrangian as first introduced in Refs. [1, 2] was constructed in terms of the pion field

$$\pi = \frac{1}{\sqrt{2}} \begin{pmatrix} \frac{1}{\sqrt{2}}\pi^0 + \frac{1}{\sqrt{6}}\eta & \pi^+ & K^+ \\ \pi^- & -\frac{1}{\sqrt{2}}\pi^0 + \frac{1}{\sqrt{6}}\eta & K^0 \\ K^- & \bar{K}^0 & -\frac{2}{\sqrt{6}}\eta \end{pmatrix}, \tag{1}$$

the baryon octet field

$$B_v = \begin{pmatrix} \frac{1}{\sqrt{2}}\Sigma_v^0 + \frac{1}{\sqrt{6}}\Lambda_v & \Sigma_v^+ & p_v \\ \Sigma_v^- & -\frac{1}{\sqrt{2}}\Sigma_v^0 + \frac{1}{\sqrt{6}}\Lambda_v & n_v \\ \Xi_v^- & \Xi_v^0 & -\frac{2}{\sqrt{6}}\Lambda_v \end{pmatrix}, \tag{2}$$

and the baryon decuplet field

$$T^\mu_{abc}, \tag{3}$$

where T^μ_{abc} is a Rarita-Schwinger field which contains both spin-1/2 and spin-3/2 pieces and obeys the constraint $\gamma^\mu T_\mu = 0$.

The coupling of the pseudoscalar pion field with the baryon matter fields occurs through the vector and axial vector combinations

$$V^\mu = \frac{1}{2}(\xi\partial^\mu\xi^\dagger + \xi^\dagger\partial^\mu\xi), \qquad A^\mu = \frac{i}{2}(\xi\partial^\mu\xi^\dagger - \xi^\dagger\partial^\mu\xi), \tag{4}$$

where

$$\xi = e^{i\pi/f}, \qquad \Sigma = \xi^2 = e^{2i\pi/f}, \tag{5}$$

and $f \approx 93$ MeV is the pion decay constant. Under $SU(3)_L \times SU(3)_R$ chiral symmetry, the above fields transform as

$$\begin{aligned} &\xi \to L\xi U^\dagger = U\xi R^\dagger, && \Sigma \to L\Sigma R^\dagger, \\ &B \to UBU^\dagger, && T^\mu_{abc} \to U_a^d U_b^e U_c^f T^\mu_{def}, \end{aligned} \tag{6}$$

where U is defined implicitly by the transformation of ξ in Eq. (6).

The most general Lagrangian at lowest order is

$$\begin{aligned} \mathcal{L}_{\text{baryon}} &= i\,\mathrm{Tr}\,\bar{B}_v(v\cdot\mathcal{D})B_v - i\,\bar{T}_v^\mu(v\cdot\mathcal{D})T_{v\mu} + \Delta\,\bar{T}_v^\mu T_{v\mu} + 2D\,\mathrm{Tr}\,\bar{B}_v S_v^\mu\{A_\mu, B_v\} \\ &\quad + 2F\,\mathrm{Tr}\,\bar{B}_v S_v^\mu[A_\mu, B_v] + \mathcal{C}\,(\bar{T}_v^\mu A_\mu B_v + \bar{B}_v A_\mu T_v^\mu) + 2\,\mathcal{H}\,\bar{T}_v^\mu S_v^\nu A_\nu T_{v\mu}, \end{aligned} \tag{7}$$

where D, F, $\mathcal{C}$, and $\mathcal{H}$ are the baryon-pion couplings and $\Delta = m_T - m_B$ is the decuplet-octet mass difference.

On the other hand, to first order in the quark mass matrix, $\mathcal{M} = \mathrm{diag}(m_u, m_d, m_s)$, the chiral Lagrangian is expressed as

$$\begin{aligned} \mathcal{L}^{\mathcal{M}}_{\text{baryon}} &= \sigma\,\mathrm{Tr}\,[\mathcal{M}(\Sigma+\Sigma^\dagger)]\,\mathrm{Tr}(\bar{B}_v B_v) - \tilde{\sigma}\,\mathrm{Tr}\,[\mathcal{M}(\Sigma+\Sigma^\dagger)]\,\bar{T}_v^\mu T_{v\mu} \\ &\quad + b_D\,\mathrm{Tr}\,\bar{B}_v\{\xi^\dagger\mathcal{M}\xi^\dagger + \xi\mathcal{M}\xi, B_v\} + b_F\,\mathrm{Tr}\,\bar{B}_v[\xi^\dagger\mathcal{M}\xi^\dagger + \xi\mathcal{M}\xi, B_v] \\ &\quad + c\,\bar{T}_v^\mu\left(\xi^\dagger\mathcal{M}\xi^\dagger + \xi\mathcal{M}\xi\right)T_{v\mu}\,. \end{aligned} \tag{8}$$

Here σ and $\tilde{\sigma}$ are the quark mass parameters of the octet and decuplet, respectively. The flavor octet quark mass splittings are described by the parameters b_D and b_F for the baryon octet, and by the parameter c for the baryon decuplet.

Journal of Physics: Conference Series **378** (2012) 012039 doi:10.1088/1742-6596/378/1/012039

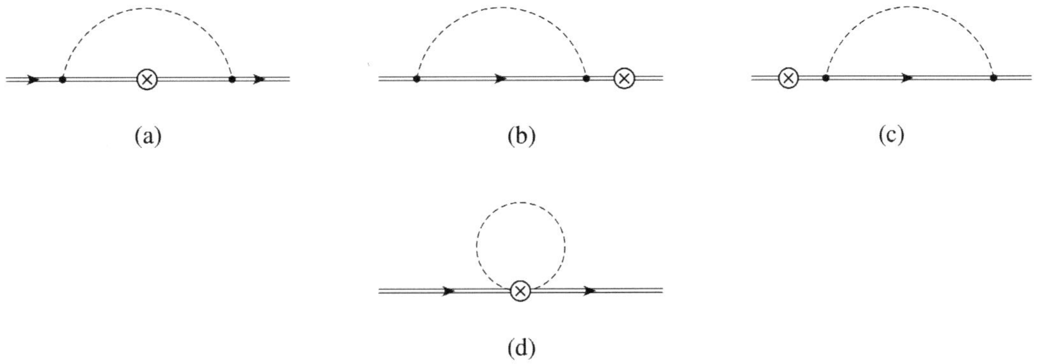

Figure 1. Feyman diagrams that yield one-loop corrections to the baryon axial-vector current. The dashed lines denote mesons and the solid lines denote baryons. The crossed circle represents the insertion of an axial-vector current.

2.1. Chiral corrections to the baryon axial current

The renormalization of the baryon axial current can now be readily obtained by using the effective Lagrangian (7). The leading nonanalytic correction is a chiral logarithm produced by the one-loop Feynman graphs displayed in Fig. 1. The renormalized baryon axial current can be written as [1, 2, 12]

$$\langle B_i|J_\mu^A|B_j\rangle = \left[\alpha_{ij}^A + \sum_\Pi \left(\bar{\beta}_{ij}^{A,\Pi} - \bar{\lambda}_{ij}^\Pi \alpha_{ij}^A\right) \frac{m_\Pi^2}{16\pi^2 f^2} \ln\left(\frac{m_\Pi^2}{\mu^2}\right)\right] \bar{u}_{B_i}\gamma_\mu\gamma_5 u_{B_j}, \tag{9}$$

where α_{ij}^A is the lowest order result, $\bar{\lambda}_{ij}^\Pi = \lambda_{ij}^\Pi + \lambda'^\Pi_{ij}$ is the one-loop corrections due to wavefunction renormalization

$$\sqrt{Z_i Z_j} = 1 + \sum_\Pi \bar{\lambda}_{ij}^\Pi \frac{m_\Pi^2}{16\pi^2 f^2} \ln\left(\frac{m_\Pi^2}{\mu^2}\right), \qquad \bar{\lambda}_{ij}^\Pi = \frac{1}{2}(\bar{\lambda}_i^\Pi + \bar{\lambda}_j^\Pi), \tag{10}$$

$\bar{\beta}_{ij}^{A,\Pi} = \beta_{ij}^{A,\Pi} + \beta'^{A,\Pi}_{ij}$ and Π stands for π, K, and η mesons. The unprimed and primed quantities are contributions with intermediate octet and decuplet baryons, respectively. The explicit formulas for the chiral coefficients α_{ij}^A, $\bar{\beta}_{ij}^{A,\Pi}$, and $\bar{\lambda}_{ij}^\Pi$ can be found in Appendix A of Ref. [12].

The correction to the axial current can be expressed in a more compact form as [12]

$$\delta\langle B_i|J_\mu^A|B_j\rangle = \left(\sum_\Pi a_{ij}^{A,\Pi} F(m_\Pi)\right) \bar{u}_{B_i}\gamma_\mu\gamma_5 u_{B_j}, \tag{11}$$

where

$$a_{ij}^{A,\Pi} = \bar{\beta}_{ij}^{A,\Pi} - \bar{\lambda}_{ij}^\Pi \alpha_{ij}^A, \tag{12}$$

and

$$F(m_\Pi) = -\frac{m_\Pi^2}{16\pi^2 f^2} \ln\left(\frac{m_\Pi^2}{\mu^2}\right), \tag{13}$$

where $F(m_\Pi)$ stands for the integration over the loop and only the non-analytic pieces have been retained. Equation (11) can be split into flavor singlet, flavor octet and flavor-**27** pieces in terms of flavor singlet, octet, and **27** linear combinations of $F(\pi)$, $F(K)$, and $F(\eta)$ [14], namely,

$$\delta\langle B_i|J_\mu^A|B_j\rangle = \left[\frac{1}{8}\left[3F(\pi)+4F(K)+F(\eta)\right]b_{ij,\mathbf{1}}^A + \frac{2\sqrt{3}}{5}\left(\frac{3}{2}F(\pi)-F(K)-\frac{1}{2}F(\eta)\right)b_{ij,\mathbf{8}}^A + \left(\frac{1}{3}F(\pi)-\frac{4}{3}F(K)+F(\eta)\right)b_{ij,\mathbf{27}}^A\right]\bar{u}_{B_i}\gamma_\mu\gamma_5 u_{B_j}, \quad (14)$$

where

$$b_{ij,\mathbf{1}}^A = -a_{ij}^{A,\pi} + a_{ij}^{A,K} + a_{ij}^{A,\eta}, \quad (15)$$

$$b_{ij,\mathbf{8}}^A = -\frac{1}{\sqrt{3}}\left(a_{ij}^{A,\pi} - \frac{1}{2}a_{ij}^{A,K} - a_{ij}^{A,\eta}\right), \quad (16)$$

$$b_{ij,\mathbf{27}}^A = -\frac{3}{40}\left(a_{ij}^{A,\pi} - 3a_{ij}^{A,K} + 9a_{ij}^{A,\eta}\right), \quad (17)$$

In Eq. (13) the mass of the meson is denoted by its particle label.

2.2. Chiral corrections to baryon magnetic moments

The chiral Lagrangian with electromagnetism incorporated was derived in Ref. [15]. The one-loop diagrams that contribute to baryon magnetic moments are displayed in Figs. 2 and 3. Corrections arising from these loop graphs are order $\mathcal{O}(m_q^{1/2})$ and $\mathcal{O}(m_q \ln m_q)$, respectively.

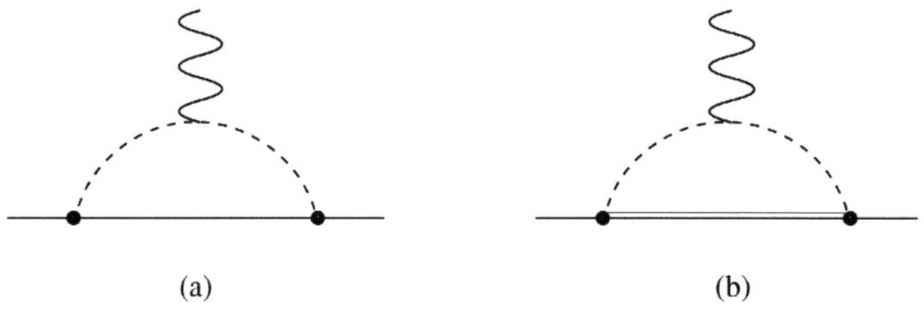

Figure 2. Feynman diagrams which yield nonanalytic $m_q^{1/2}$ corrections to the magnetic moments of octet baryons. Dashed lines denote mesons and single and double solid lines denote octet and decuplet baryons, respectively. For decuplet baryons and decuplet-octet transitions the diagrams are similar.

For octet baryons, the magnetic moments computed in Ref. [15] can be rewritten as

$$\mu_i = \alpha_i + \sum_{X=\pi,K}\beta_i^{(X)}I(m_X,0) + \sum_{X=\pi,K}\beta_i'^{(X)}I(m_X,\Delta) + \sum_{X=\pi,K,\eta}\frac{1}{32\pi^2 f^2}(\overline{\gamma}_i^{(X)} - 2\overline{\lambda}_i^{(X)}\alpha_i)m_X^2\ln\frac{m_X^2}{\mu^2}, \quad (18)$$

where α_i corresponds to the tree-level value of baryon i, $\beta_i^{(X)}$ and $\beta_i'^{(X)}$ are the contributions arising from loop graphs of Fig. 2 and the remaining coefficients come from loop graphs of Fig. 3. For μ_{Σ^-}, for instance, the corresponding chiral coefficients listed in Ref. [15] read

$$\beta_{\Sigma^-}^{(\pi)} = \frac{2}{3}D^2 + 2F^2, \qquad \beta_{\Sigma^-}^{(K)} = (D-F)^2, \qquad \beta_{\Sigma^-}'^{(\pi)} = -\frac{1}{18}\mathcal{C}^2, \qquad \beta_{\Sigma^-}'^{(K)} = -\frac{1}{9}\mathcal{C}^2. \quad (19)$$

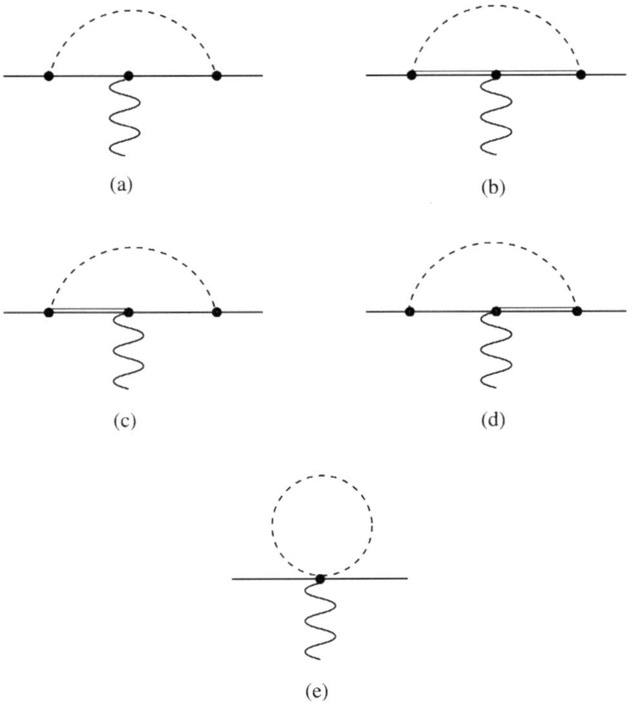

Figure 3. Feynman diagrams which yield nonanalytic $m_q \ln m_q$ corrections to the magnetic moments of octet baryons. Dashed lines denote mesons and single and double solid lines denote octet and decuplet baryons, respectively. The wavefunction renormalization graphs are omitted in the figure but are nevertheless considered in the analysis.

3. Baryons in large N_c chiral perturbation theory

The $1/N_c$ baryon chiral Lagrangian obtained in Ref. [14] is expressed as

$$\mathcal{L}_{\rm baryon} = i\mathcal{D}^0 - M_{\rm hyperfine} + {\rm Tr}\left(\mathcal{A}^i\lambda^a\right)A^{ia} + {\rm Tr}\left(\mathcal{A}^i\frac{2I}{\sqrt{6}}\right)A^i + \ldots, \tag{20}$$

with

$$\mathcal{D}^0 = \partial^0 I + {\rm Tr}\left(\mathcal{V}^0\lambda^a\right)T^a + \frac{1}{3}{\rm Tr}\left(\mathcal{V}^0 I\right)N_c I. \tag{21}$$

Each term in Eqs. (20) and (21) represents itself a baryon operator which can be expressed in terms of polynomials in the spin-flavor generators J^i, T^a, and G^{ia} [4]. Here $M_{\rm hyperfine}$ denotes the spin splittings of the tower of baryon states with spins $\frac{1}{2}, \ldots, \frac{N_c}{2}$ in the flavor representations. The vector and axial vector combinations of the pion fields are given by

$$\mathcal{V}^0 = \frac{1}{2}\left(\xi\partial^0\xi^\dagger + \xi^\dagger\partial^0\xi\right), \qquad \mathcal{A}^i = \frac{i}{2}\left(\xi\nabla^i\xi^\dagger - \xi^\dagger\nabla^i\xi\right), \tag{22}$$

which couple to vector and axial vector baryon currents, respectively. The $\ell = 1$ flavor octet axial vector pion combination couples to the flavor octet baryon axial vector current A^{ia}. It is spin-1 object, an octet under $SU(3)$ and its $1/N_c$ expansion is [4]

$$A^{ia} = a_1 G^{ia} + \sum_{n=2,3}^{N_c} b_n \frac{1}{N_c^{n-1}}\mathcal{D}_n^{ia} + \sum_{n=3,5}^{N_c} c_n \frac{1}{N_c^{n-1}}\mathcal{O}_n^{ia}, \tag{23}$$

where $\mathcal{D}_n^{ia}$ are diagonal operators with nonzero matrix elements only between states with the same spin, whereas $\mathcal{O}_n^{ia}$ are purely off-diagonal operators with nonzero matrix elements only

Journal of Physics: Conference Series **378** (2012) 012039 doi:10.1088/1742-6596/378/1/012039

between states of different spin. The explicit forms for these operators can be found in Ref. [4]. At the physical value $N_c = 3$, Eq. (23) can be truncated after the fourth term

$$A^{ia} = a_1 G^{ia} + b_2 \frac{1}{N_c} J^i T^a + b_3 \frac{1}{N_c^2} \mathcal{D}_3^{ia} + c_3 \frac{1}{N_c^2} \mathcal{O}_3^{ia}, \qquad (24)$$

where

$$\mathcal{D}_3^{ia} = \{J^i, \{J^j, G^{ja}\}\}, \qquad (25)$$

$$\mathcal{O}_3^{ia} = \{J^2, G^{ia}\} - \frac{1}{2}\{J^i, \{J^j, G^{ja}\}\}. \qquad (26)$$

On the other hand, in the limit of exact $SU(3)$ flavor symmetry, the baryon mass operator is defined by

$$M = \langle B' \left| \mathcal{H}_{\text{QCD}} \right| B \rangle, \qquad (27)$$

where $\mathcal{H}_{\text{QCD}}$ is the QCD Hamiltonian in the chiral limit $m_i \to 0$. M transforms as a $(0, 1)$ under $SU(2) \times SU(3)$ symmetry and its $1/N_c$ expansion is [14]

$$M = m_{(0)}^{0,1} N_c I + \sum_{n=2,4}^{N_c - 1} m_{(n)}^{0,1} \frac{1}{N_c^{n-1}} J^n. \qquad (28)$$

The first term on the right-hand side of Eq. (28) is the overall spin-independent mass of the baryon multiplet and is removed from the chiral Lagrangian by the heavy baryon field redefinition [1]. The remaining terms are spin-dependent so they define $M_{\text{hyperfine}}$. For $N_c = 3$, the hyperfine mass expansion reduces to the operator

$$M = m_{(2)}^{0,1} \frac{1}{N_c} J^2. \qquad (29)$$

3.1. Corrections to the baryon axial-vector current in large-N_c chiral perturbation theory

The leading nonanalytic corrections to the baryon axial current is of the form $m^2 \ln m^2$ arising from the loop diagrams shown in Fig. 1 The normalized axial current, in the degeneracy limit $\Delta \to 0$, can be expressed as

$$A^{kc} + \delta A^{kc}, \qquad (30)$$

where the correction δA^{kc} reads [11, 12]

$$\delta A^{kc} = \frac{1}{2}[A^{ia}, [A^{ib}, A^{kc}]]\Pi^{ab} - \frac{1}{2}[T^a, [T^b, A^{kc}]]\Pi^{ab}. \qquad (31)$$

The first term on the right-hand side of Eq. (31) comes from three-vertex diagrams, Figs. 1(a-c) whereas the second term is the contribution of the one-vertex diagram, Fig. 1(d). Here Π^{ab} is a symmetric tensor which contains meson loop integrals with the exchange of a single meson. Π^{ab} decomposes into flavor singlet, flavor **8** and flavor **27** representations as [14]

$$\begin{aligned} \Pi^{ab} &= \frac{1}{8}\left[3F(\pi) + 4F(K) + F(\eta)\right]\delta^{ab} + \frac{2\sqrt{3}}{5}\left(\frac{3}{2}F(\pi) - F(K) - \frac{1}{2}F(\eta)\right) d^{ab8} \\ &+ \left(\frac{1}{3}F(\pi) - \frac{4}{3}F(K) + F(\eta)\right)\left(\delta^{a8}\delta^{b8} - \frac{1}{8}\delta^{ab} - \frac{3}{5}d^{ab8}d^{888}\right), \end{aligned} \qquad (32)$$

where $F(m)$ is a function of the meson mass m and is defined in this formalism by the integral

$$\delta^{ij} F(m) = \frac{1}{f^2} \int \frac{d^4k}{(2\pi)^4} \frac{i^3(\mathbf{k}^i)(-\mathbf{k}^j)}{(k^2-m^2)(k\cdot v)^2}, \tag{33}$$

for the first term in Eq. (31), and by the integral

$$F(m) = \frac{1}{f^2} \int \frac{d^4k}{(2\pi)^4} \frac{i}{k^2-m^2}, \tag{34}$$

for the second term in Eq. (31).

The correction to the axial current can be organized as follows

$$\delta A^{kc} = \frac{1}{2}\left([A^{ia},[A^{ib},A^{kc}]] - [T^a,[T^b,A^{kc}]]\right)\left(\Pi_{\mathbf{1}}^{ab} + \Pi_{\mathbf{8}}^{ab} + \Pi_{\mathbf{27}}^{ab}\right). \tag{35}$$

For instance, for the singlet contribution we get

$$[A^{ia},[A^{ib},A^{kc}]]\Pi_{\mathbf{1}}^{ab} = \frac{1}{8}\left[3F(\pi) + 4F(K) + F(\eta)\right][A^{ia},[A^{ia},A^{kc}]], \tag{36}$$

and to relative order in $1/N_c^2$ in the $1/N_c$ expansion one has [12]

$$\begin{aligned}
&[A^{ia},[A^{ia},A^{kc}]] = \\
&\quad \Bigg[\left(\frac{3}{4}N_F - \frac{1}{N_F}\right)a_1^3 - 2\left(\frac{1}{N_F}+\frac{1}{N_c}\right)a_1^2b_2 + \left(\frac{1}{2} - \frac{1}{N_F} + \frac{N_F-2}{N_c} - 3\frac{N_F}{N_c^2}\right)a_1b_2^2 \\
&\qquad - \left(1 + 2\frac{N_F}{N_c} - 2\frac{N_F-4}{N_c^2}\right)a_1^2b_3 - \left(1 + 2\frac{N_F}{N_c} - \frac{N_F}{N_c^2}\right)a_1^2c_3\Bigg] G^{kc} \\
&\quad + \frac{1}{N_c}\Bigg[\left(\frac{9}{4}N_F - \frac{1}{N_F} + 2\right)a_1^2b_2 + 2\left(1+\frac{N_F}{N_c}\right)\left(1-\frac{1}{N_F}\right)a_1b_2^2 \\
&\qquad -3\left(1+\frac{N_F}{N_c}\right)a_1^2b_3 - \frac{1}{2}\left(1+\frac{N_F}{N_c}\right)a_1^2c_3\Bigg] J^kT^c \\
&\quad + \frac{1}{N_c^2}\Bigg[\left(\frac{13}{4}N_F - \frac{3}{N_F} + 4\right)a_1^2b_3 + \frac{1}{2}(1+N_F)a_1^2c_3\Bigg]\mathcal{D}_3^{kc} \\
&\quad + \frac{1}{N_c^2}\Bigg[\left(N_F - \frac{8}{N_F} + 2\right)a_1^2b_3 + \left(\frac{15}{4}N_F - \frac{1}{N_F} + 3\right)a_1^2c_3\Bigg]\mathcal{O}_3^{kc} + \mathcal{O}\left(\frac{1}{N_c^3}\right).
\end{aligned} \tag{37}$$

Similar expressions can be obtained for the octet and **27** contributions; they can be found in Ref. [12] and will not be repeated here.

Let us remark that there is a nontrivial N_c dependence of the matrix elements of the generators J^i, T^a, and G^{ia} in the weight diagrams for the $SU(3)$ flavor representations of the spin-$\frac{1}{2}$ and spin-$\frac{3}{2}$ baryons [4]. In other words, factors of T^a/N_c and G^{ia}/N_c are of order 1 somewhere in the weight diagram, whereas factors of J^i/N_c are of order $1/N_c$ everywhere. If we restrict ourselves to baryons with spins of order unity, the N_c counting rules can be summarized as [11]

$$T^a \sim N_c, \qquad G^{ia} \sim N_c, \qquad J^i \sim 1. \tag{38}$$

Under these assumptions, one naively could assert that the left-hand-side of Eq. (37) is order N_c^3: a factor of N_c for each axial-vector current operator. However, after explicit computation, we notice that the right-hand-side of Eq. (37) is at most of order N_c, according to counting rules (38), i.e., there are large-N_c cancellations in the evaluation of the double commutator provided one sums over all baryon states in a complete multiplet of the large-N_c $SU(6)$ spin-flavor symmetry, i.e., over both the octet and decuplet, and uses axial coupling ratios given by the large-N_c spin-flavor symmetry

Similarly, for the contribution of the Feynman diagram Fig. 1(d) one finds [12]

$$\delta^{ab}[T^a,[T^b,A^{kc}]] = N_f A^{kc}, \tag{39}$$

$$d^{ab8}[T^a,[T^b,A^{kc}]] = \frac{N_f}{2} d^{c8e} A^{ke}, \tag{40}$$

$$\delta^{a8}\delta^{b8}[T^a,[T^b,A^{kc}]] = f^{c8d} f^{8de} A^{ke}. \tag{41}$$

In the double commutator involved in Eqs. (39)-(41) we can observe that there are again large-N_c cancellations in the structure.

3.1.1. Comparison with heavy baryon chiral perturbation theory. Equation (35) can be straightforwardly compared with the analogous expression obtained in conventional chiral perturbation theory, Eq. (15)-(17), by making use of the identifications [14]

$$\begin{aligned} D &= \frac{1}{2}a_1 + \frac{1}{6}b_3, \\ F &= \frac{1}{3}a_1 + \frac{1}{6}b_2 + \frac{1}{9}b_3, \\ \mathcal{C} &= -a_1 - \frac{1}{2}c_3, \\ \mathcal{H} &= -\frac{3}{2}a_1 - \frac{3}{2}b_2 - \frac{5}{2}b_3. \end{aligned} \tag{42}$$

Both expressions agree order by order [12].

3.2. Baryon magnetic moments in large-N_c chiral perturbation theory

In the large-N_c limit, the baryon magnetic moments possess the same kinematical properties as the baryon axial-vector couplings; as a result, the operators used to describe these quantities are practically identical [7]. The baryon magnetic moment operator is also a spin-1 object and an octet under $SU(3)$. For definiteness, in analogy with Eq. (24), the $1/N_c$ expansion of the operator which yields the baryon magnetic moment is [13]

$$M^{kc} = m_1 G^{kc} + m_2 \frac{1}{N_c} \mathcal{D}_2^{kc} + m_3 \frac{1}{N_c^2} \mathcal{D}_3^{kc} + m_4 \frac{1}{N_c^2} \mathcal{O}_3^{kc}, \tag{43}$$

where we have truncated the series at the physical value $N_c = 3$. If we assume $SU(3)$ symmetry, the unknown coefficients m_i are independent of k so they are unrelated to the ones of expansion (24) in this limit. The magnetic moments are proportional to the quark charge matrix $\mathcal{Q} = \mathrm{diag}(2/3, -1/3, -1/3)$, so they can be separated into isovector and isoscalar components, M^{k3} and M^{k8}, respectively. Accordingly, from Eq. (43), we define the baryon magnetic moment operator as

$$M^k = M^{kQ} \equiv M^{k3} + \frac{1}{\sqrt{3}} M^{k8}. \tag{44}$$

Hereafter the spin index k of M^k will be set to 3 whereas the flavor index Q will stand for $Q = 3 + (1/\sqrt{3})8$ so any operator of the form X^Q should be understood as $X^3 + (1/\sqrt{3})X^8$.

Now, one-loop corrections to baryon magnetic moments emerge from the Feynman graphs depicted in Fig. 2 and 3, which contribute to orders $\mathcal{O}(m_q^{1/2})$ and $m_q \ln m_q$, respectively.

Graphs from Fig. 2 involve π and K emission and reabsorption only (the η meson does not contribute). For degenerate heavy baryons interacting with mesons, these diagrams depend on a function $I(m_\Pi)$ of the meson mass m_Π, which is obtained by performing the Feynman loop integration. Thus, in the $\Delta \to 0$ limit, we have

$$M_{\rm fig2}^k = \epsilon^{ijk} A^{ia} A^{jb} \Gamma^{ab}. \tag{45}$$

Here Γ^{ab} is an antisymmetric tensor which contains the integrals over the loops and is given in Ref. [7]. $i\Gamma^{ab}$ can be represented by a Hermitian matrix which is diagonal in a basis corresponding to particles of definite quantum numbers. This matrix has eight eigenvalues: four of them are zero and correspond to the four neutral mesons, two of them are equal and opposite eigenvalues $\pm I(m_K)$ corresponding to $K^\pm$, respectively, and the remaining two are equal and opposite eigenvalues $\pm I(m_\pi)$ corresponding to $\pi^\pm$, respectively. Thus [7]

$$\Gamma^{ab} = A_0 \Gamma_0^{ab} + A_1 \Gamma_1^{ab} + A_2 \Gamma_2^{ab}, \tag{46}$$

where the coefficients A_i are linear combinations of the functions $I(m_\pi)$ and $I(m_K)$ and read

$$A_0 = \frac{1}{3}[I(m_\pi) + 2I(m_K)], \tag{47}$$

$$A_1 = \frac{1}{3}[I(m_\pi) - I(m_K)], \tag{48}$$

$$A_2 = \frac{1}{\sqrt{3}}[I(m_\pi) - I(m_K)], \tag{49}$$

and the tensors Γ_i^{ab} are written as

$$\Gamma_0^{ab} = f^{abQ}, \tag{50}$$

$$\Gamma_1^{ab} = f^{ab\overline{Q}}, \tag{51}$$

$$\Gamma_2^{ab} = f^{aeQ} d^{be8} - f^{beQ} d^{ae8} - f^{abe} d^{eQ8}. \tag{52}$$

Although Γ_0^{ab} and Γ_1^{ab} are both $SU(3)$ octets, they have quite different physical interpretations. The former transforms as the electric charge whereas the latter also transforms as the electric charge but rotated by π in isospin space [7]. In what follows any operator of the form $X^{\overline{Q}}$ should be understood as $X^3 - (1/\sqrt{3})X^8$. Notice also that X^Q and $X^{\overline{Q}}$ fall into different octet representations. On the other hand, Γ_2^{ab} breaks $SU(3)$ as $\mathbf{10} + \overline{\mathbf{10}}$ [7].

In the degeneracy limit $\Delta \to 0$ and retaining only the nonanalytic pieces in m_q, the integral over the loop, which comprises the proper factors to give the correct dimensions, can be expressed as [15]

$$I(m_\Pi) = \frac{1}{8\pi^2 f^2} M_N m_\Pi, \tag{53}$$

where f is the pion decay constant and M_N and m_Π denote the nucleon and the meson masses, respectively.

Thus, the one-loop correction arising from Fig. 2 can be decomposed into the pieces emerging from the $\mathbf{8}$ and $\mathbf{10} + \overline{\mathbf{10}}$ representations as follows,

$$M_{\rm fig2}^k = A_0 M_{\mathbf{8},\rm fig2}^{kQ} + A_1 M_{\mathbf{8},\rm fig2}^{k\overline{Q}} + A_2 M_{\mathbf{10}+\overline{\mathbf{10}},\rm fig2}^{kQ}, \tag{54}$$

where the different contributions read

$$M_{\mathbf{8},\rm fig2}^{kc} = \epsilon^{ijk} f^{abc} A^{ia} A^{jb}, \tag{55}$$

and

$$M^{kc}_{\mathbf{10}+\overline{\mathbf{10}},\text{fig2}} = \epsilon^{ijk}(f^{aec}d^{be8} - f^{bec}d^{ae8} - f^{abe}d^{ec8})A^{ia}A^{jb}. \tag{56}$$

The reduction of the operator products contained in $M^{kc}_{\mathbf{8}}$ and $M^{kc}_{\mathbf{10}+\overline{\mathbf{10}}}$ to the order considered here are given in Ref. [13]. Gathering together partial results, we have for example, for the octet contribution,

$$\begin{aligned} M^{kc}_{\mathbf{8},\text{fig2}} = {} & \\ & \left[-\frac{N_c+3}{2}a_1^2 - \frac{3}{N_c}a_1b_2 - \frac{2(N_c+3)}{N_c^2}a_1b_3\right]G^{kc} + \frac{1}{2}\left[a_1^2 - \frac{1}{N_c^2}(2a_1b_3 - 9a_1c_3 + 3b_2^2)\right]\mathcal{D}_2^{kc} \\ & - \frac{1}{N_c^2}\left[\frac{N_c+3}{2}a_1c_3 + \frac{3}{N_c}b_2b_3\right]\mathcal{D}_3^{kc} - \frac{1}{N_c^2}\left[N_ca_1b_2 + (N_c+3)a_1b_3 + \frac{N_c+3}{2}a_1c_3 + \frac{3}{N_c}b_2c_3\right] \\ & \times \mathcal{O}_3^{kc} + \frac{1}{N_c^2}a_1c_3\mathcal{D}_4^{kc} - \frac{1}{N_c^3}b_2c_3\mathcal{O}_5^{kc} + \ldots \end{aligned} \tag{57}$$

For the other flavor representation the corresponding expression can be found in Ref. [13]. Equation (57) has been rearranged to exhibit explicitly leading and subleading terms in $1/N_c$. It is now evident that this expression yields matrix elements at most of order $\mathcal{O}(N_c^2)$, according to the N_c-counting rules discussed above. In addition, f and M_N are $\mathcal{O}(\sqrt{N_c})$ and $\mathcal{O}(N_c)$, respectively, so the one-loop contribution M^k_{fig2}, Eq. (54), is order $\mathcal{O}(N_c)$. In the limit of small m_s, the symmetry breaking part of M^k_{fig2} is $\mathcal{O}(m_s^{1/2})$ so the overall contribution of Eq. (54) to baryon magnetic moments is $\mathcal{O}(m_s^{1/2}N_c)$.

As for nonanalytic corrections of order $m_q \ln m_q$, we can separate the different contribution as those coming from Figs. 3(a-d) and that from Fig. 3(e). For the first kind we have

$$M^k_{\text{fig3a--d}} = \left[A^{ja},\left[A^{jb},M^k\right]\right]\Pi^{ab}. \tag{58}$$

Here, Π^{ab} is again a symmetric tensor which contains meson loop integrals and has the same structure as (32) except that this time

$$F(m_\Pi,\mu) = \frac{m_\Pi^2}{32\pi^2 f^2}\ln\frac{m_\Pi^2}{\mu^2}, \tag{59}$$

where μ is the scale of dimensional regularization and only nonanalytic terms in m_q have been kept.

Now, in the operator reduction of the structure (58) some subtleties arise. The appearance of the new parameters m_i makes unfeasible the direct application of Eqs. (37), for instance, to obtain the corresponding loop contribution M^k_{fig3ad}. Indeed, new terms need be calculated. We remark that, because the operator basis is complete, the reduction is doable. In Ref. [13] the full calculation is presented and will not be repeated here. What is important to emphasize is the large-N_c cancallations observed is the structure of this contribution, as expected.

On the other hand, the contribution to the magnetic moments of the Feynman diagram displayed in Fig. 3(e) possesses the structure

$$M^k_{\text{fig3e}} = \left[T^a,\left[T^b,M^k\right]\right]\Pi^{ab}, \tag{60}$$

where Π^{ab} is the symmetric tensor introduced in Eq. (32). Equation (60) can also be separated into flavor singlet, flavor octet, and flavor **27** pieces as

$$M^k_{\text{fig3e}} = F_{\mathbf{1}}M^{kQ}_{\mathbf{1},\text{fig3e}} + F_{\mathbf{8}}M^{kQ}_{\mathbf{8},\text{fig3e}} + F_{\mathbf{27}}M^{kQ}_{\mathbf{27},\text{fig3e}}, \tag{61}$$

Journal of Physics: Conference Series **378** (2012) 012039 doi:10.1088/1742-6596/378/1/012039

where this time the group structures of the double commutator read

Flavor singlet contribution

$$\begin{aligned} M^{kc}_{\mathbf{1},\text{fig3e}} &= [T^a,[T^a,M^{kc}]] \\ &= N_f M^{kc}. \end{aligned} \tag{62}$$

Flavor octet contribution

$$\begin{aligned} M^{kc}_{\mathbf{8},\text{fig3e}} &= d^{ab8}[T^a,[T^b,M^{kc}]] \\ &= \frac{N_f}{2} d^{c8e} M^{ke}. \end{aligned} \tag{63}$$

Flavor **27** *contribution*

$$\begin{aligned} M^{kc}_{\mathbf{27},\text{fig3e}} &= [T^8,[T^8,M^{kc}]] \\ &= f^{c8e} f^{8eg} M^{kg}. \end{aligned} \tag{64}$$

Let us notice that, in order for $M^{kc}_{\mathbf{27},\text{fig3e}}$ to be a truly **27** contribution, singlet and octet pieces must be subtracted off. In the above equations, the free flavor index c will be set to $Q = 3 + (1/\sqrt{3})8$. By doing this, expression (64) as it stands, will vanish.

The total correction arising from Fig. 3 is then given by

$$M^k_{\text{fig3}} = M^k_{\text{fig3a-d}} + M^k_{\text{fig3e}}. \tag{65}$$

Corrections to the baryon magnetic moments are then obtained by computing the matrix elements of operator M^k_{fig3} between $SU(6)$ baryon states, namely,

$$\mu_B^{(\text{fig3})} = \langle B|M^3_{\text{fig3}}|B\rangle, \tag{66}$$

where B stands for either an octet or a decuplet baryon. For decuplet to octet transition magnetic moments, we also have

$$\mu_{TB}^{(\text{fig3})} = \langle T|M^3_{\text{fig3}}|B\rangle. \tag{67}$$

The singlet piece of M^k_{fig3} yields magnetic moments that satisfy the Coleman-Glashow relations whereas violations to them are due to the **8** and **27** pieces.

3.2.1. Total one-loop corrections to baryon magnetic moments. At this point, we can summarize our findings and provide analytic results. Thus, the final expression of the magnetic moment of baryon B up to one-loop order can be organized as

$$\mu_B = \mu_B^{(0)} + \mu_B^{(\text{fig2})} + \mu_B^{(\text{fig3})}. \tag{68}$$

Applications of this expression are given in Ref. [13] and will not be repeated here.

3.3. Comparison with conventional heavy baryon chiral perturbation theory.

It is instructive to compare our computation of baryon magnetic moments at the physical value $N_c = 3$ with the one obtained in the framework of conventional heavy baryon chiral perturbation theory, *i.e.*, the effective field theory with no $1/N_c$ expansion. In Ref. [14] it has been shown that there is a one-to-one correspondence between the parameters of the $1/N_c$ baryon chiral Lagrangian at $N_c = 3$ and the octet and decuplet chiral Lagrangian. The baryon-pion couplings are related to the coefficients of the $1/N_c$ expansion of A^{ia}, Eq. (24), at $N_c = 3$ by Eq. (42). On the other hand, the magnetic moments in conventional heavy baryon chiral perturbation theory are parametrized by four $SU(3)$ invariants μ_D, μ_F, μ_C and μ_T [15] while in the present analysis they are parametrized in terms of m_i, with $i = 1, \ldots, 4$, introduced in Eq. (43). The relations between these parameters read

$$\begin{aligned} \mu_D &= \frac{1}{2}m_1 + \frac{1}{6}m_3, \\ \mu_F &= \frac{1}{3}m_1 + \frac{1}{6}m_2 + \frac{1}{9}m_3, \\ \mu_C &= \frac{1}{2}m_1 + \frac{1}{2}m_2 + \frac{5}{6}m_3, \\ \mu_T &= -2m_1 - m_4. \end{aligned} \tag{69}$$

In the literature, there are some analyses of baryon magnetic moments within heavy baryon chiral perturbation theory which allow us to carry out a comparison of our respective results in the limit $\Delta \to 0$, where Δ is the decuplet-octet mass difference. The work by Jenkins *et. al.* [15] about octet baryons allows a full comparison between one-loop corrections whereas the papers by Geng *et. al.* [17] for decuplet baryons, and Arndt and Tiburzi [18] for decuplet-octet transitions only allow partial comparisons of contributions emerging from loop graphs of Fig. 2.

For octet baryons, we obtain a remarkable agreement with the theoretical expressions displayed in Eq. (16) of Ref. [15], except for the global factor $-5/2$ missing in the loop contributions of Fig. 2(c) of this reference [which corresponds to Fig. 3(b) in the present paper]. When fixing this omission, the agreement is achieved for all nine observables.

As for decuplet baryons, starting from Eq. (17) of Ref. [17] and working in the limit $\Delta \to 0$, we find a very good agreement in both calculations.

Finally, Arndt and Tiburzi [18] present the calculation of baryon decuplet to octet electromagnetic transition form factors in quenched and partially quenched chiral perturbation theory, and provide the corresponding $SU(3)$ coefficients emerging from these schemes. They also present the counterparts of such coefficients that appear in chiral perturbation theory. These are precisely the coefficients we need to compare with. We find that, except for for transitions $\Sigma^{*+}\Sigma^+$ and $\Xi^{*0}\Xi^0$, the theoretical expressions differ by a global sign.

We would like to close this section by stating that, to the order of approximation implemented here, both approaches lead to the same results. This fact causes no surprise. Previous works have shown this matching for baryon masses [14] and axial-vector couplings [19] in a systematic way.

4. Summary and conclusions

In this paper we show how to compute corrections to some baryon properties to one-loop order within heavy baryon chiral perturbation theory in the large-N_c limit. We specialize our analysis to the computation of the baryon axial-vector current and magnetic moments and contrast them with their corresponding counterparts obtained in the frame of conventional heavy baryon chiral perturbation theory. Thus, in Sec. 2 we introduce some basic material on heavy baryon chiral perturbation theory and then we proceed with a review of some basic material on large-N_c chiral perturbation theory in Sec. 3.

As a starting point, we use the fact that in the large-N_c limit both the baryon axial-vector couplings and the baryon magnetic moments share the same kinematical properties so they can be analyzed in terms of the same operators. Hence, in Sec. 3.1, we construct the $1/N_c$ expansion of baryon axial-vector current operator and compute the corrections which emerge from Feynman diagrams at one-loop order. In Sec. 3.2 we thus give the baryon magnetic moment operator based on the expansion deduced for the axial-vector current operator. We also analyze one-loop corrections in the degeneracy limit $\Delta \equiv M_T - M_B \to 0$, where M_T and M_B are the $SU(3)$ invariant masses of the decuplet and octet baryon multiplets, respectively.

The final analytic expressions can be crosschecked with other calculations and with experiment. Existing analytic results in conventional heavy baryon chiral perturbation theory allow a detailed comparison at one-loop order. Barring a few exceptions (global multiplicative factors and/or opposite signs), the comparison with existing analytic results is a successful one. The advantage of our approach is that one only needs to construct a universal operator $A^{kc} + \delta A^{kc}$, evaluate the matrix elements of this operator between $SU(6)$ baryon states and compute the integrals over the loops. Here we perform the analysis to relative order $\mathcal{O}(1/N_c^2)$. It is now clear that if we had involved ourselves in evaluating higher order contributions, we would have faced a much more complicated computation, perhaps not yet needed.

Returning to the main discussion about the comparison of this approach with conventional heavy baryon chiral perturbation theory, it should be emphasized that these two formulations yield to identical results. Nonetheless, in a given context, one or the other might be more inviting for computational ease.

This work has been partially supported by Consejo Nacional de Ciencia y Tecnología (México).

References

[1] Jenkins E and Manohar A V 1991 *Phys. Lett.* B **255** 558
[2] Jenkins E and Manohar A V 1991 *Phys. Lett.* B **259** 353
[3] Dashen R F, Jenkins E and Manohar A V 1994 *Phys. Rev.* D **49** 4713
[4] Dashen R F, Jenkins E and Manohar A V 1995 *Phys. Rev.* D **51** 3697
[5] Dashen R F and Manohar A V 1993 *Phys. Lett.* B **315** 425, 438
[6] Jenkins E and Lebed R F 1995 *Phys. Rev.* D **52** 282
[7] Dai J, Dashen R F, Jenkins E and Manohar A V 1996 *Phys. Rev.* D **53** 273
[8] Flores-Mendieta R, Jenkins E and Manohar A V 1998 *Phys. Rev.* D **58** 094028
[9] Lebed R F and Martin D R 2004 *Phys. Rev.* D **70** 016008
[10] Jenkins E 2011 (Preprint arXiv:1111.2055 [hep-ph])
[11] Flores-Mendieta R, Hofmann C P, Jenkins E and Manohar A V 2000 *Phys. Rev.* D **62** 034001
[12] Flores-Mendieta R and Hofmann C P 2006 *Phys. Rev.* D **74** 094001
[13] Flores-Mendieta R 2009 *Phys. Rev.* D **80** 094014
[14] Jenkins E 1996 *Phys. Rev.* D **53** 2625
[15] Jenkins E, Luke M E, Manohar A V and Savage M J 1993 *Phys. Lett.* B **302** 482
[16] Geng L S, Camalich J M, Alvarez-Ruso L and Vacas M J 2008 *Phys. Rev. Lett.* **101** 222002
[17] Geng L S, Camalich J M and Vicente Vacas M J 2009 *Phys. Rev.* D **80** 034027
[18] D. Arndt D and Tiburzi B C 2004 *Phys. Rev.* D **69** 014501
[19] Flores-Mendieta R 2004 *Phys. Rev.* D **70** 114036

Journal of Physics: Conference Series **378** (2012) 012040 doi:10.1088/1742-6596/378/1/012040

The $\omega\rho\pi$ coupling and the influence of heavier resonances

D. García Gudiño and G. Toledo Sánchez

Instituto de Física, Universidad Nacional Autónoma de México, AP 20-364, México D.F. 01000, México

E-mail: toledo@fisica.unam.mx

Abstract. We determine the value of the $\omega - \rho - \pi$ mesons coupling ($g_{\omega\rho\pi}$), in the context of the vector meson dominance model, from radiative decays, the $\omega \to 3\pi$ decay width and the $e^+e^- \to 3\pi$ cross section. For the last two observables we consider the effect of either a heavier resonance ($\rho'(1450)$) or a contact term . A weighted average of the results from the set of observables yields $g_{\omega\rho\pi} = 13.9 \pm 0.1$ GeV^{-1} in absence of those contributions, and $g_{\omega\rho\pi} = 11.9 \pm 0.2$ GeV^{-1} or $g_{\omega\rho\pi} = 11.4 \pm 0.1$ GeV^{-1} when including the ρ' or contact term respectively. Improved measurements of these observables and the $\rho'(1450)$ meson parameters are needed to give a definite answer on the pertinence of the inclusion of this last one in the considered processes.

1. Introduction

The high precision experiments have become a very powerful tool to test the standard model and to look for physics beyond it. For example, the measurement of the muon magnetic moment performed by the BNL-E821 [1] experiment is able to scrutiny contributions from different natures: The electromagnetic, weak and strong contributions simultaneously. While the contributions from the first two are known within the current experimental precision, a lack of knowledge of the strong interaction do not allow to draw definite conclusion on the observation of new physics [2]. The strong interaction regime involved here is not other but the hadronic one. The experimental data on hadronic production from electron-positron annihilation and tau decays [3, 4, 5] have increasing accuracy, thus providing a rich information that can lead to a deeper understanding of the strong interaction.

In particular, 2π 3π and 4π decays are the dominant decay modes. These are mainly driven by the low mass mesons interactions. Here we are interested in the strong interaction between the ω, ρ and π mesons which can be encoded in a single parameter, denoted by $g_{\omega\rho\pi}$.

The direct determination of the magnitude of $g_{\omega\rho\pi}$ would require the observation of the ω decaying into the others, which is not allowed, since there is not enough phase space for the three particles to be on the mass shell. Therefore, it must be extracted by indirect means. For example, from the above mentioned annihilation process and tau decays. A theoretical framework is required to describe such processes and link the parameters to the physical states, a Chiral approach based on the low energy symmetries of Quantum Chromodynamics (QCD) and the so called vector meson dominance model (VMD) are able to describe them. Although they have different spirit, they both manage to resume the low energy manifestation of the strong interaction.

Decay	$\lvert g_{\omega\rho\pi}\rvert$ [GeV^{-1}]
$\rho^- \to \pi^-\gamma$	11.3 ± 0.9
$\rho^0 \to \pi^0\gamma$	13.1 ± 0.9
$\omega \to \pi^0\gamma$	11.4 ± 0.2
$\pi^0 \to \gamma\gamma$	12.8 ± 0.3
Weighted Average	11.9 ± 0.2

Table 1. Determination of $|g_{\omega\rho\pi}|$ from radiative decays.

In this work we determine the value of $g_{\omega\rho\pi}$, in the context of VMD, from several processes: i) radiative decays of the form $V \to \pi\gamma$ (V: vector meson), ii) the $\omega \to 3\pi$ decay and iii) the $e^+e^- \to 3\pi$ cross section. These two last are known to be dominated by the $\rho\pi \to 3\pi$ channel [5, 6], here we also take into account the heavier resonance $\rho'(1450)$ and a $\omega \to 3\pi$ contact term. We find the effective value of $g_{\omega\rho\pi}$ which is consistent with all the considered observables.

The VMD Lagrangian including the ρ, π and ω mesons can be set as:

$$\begin{aligned}\mathcal{L} &= g_{\rho\pi\pi}\epsilon_{abc}\rho^a_\mu\pi^b\partial^\mu\pi^c + g_{\omega\rho\pi}\delta_{ab}\epsilon^{\mu\nu\lambda\sigma}\partial_\mu\omega_\nu\partial_\lambda\rho^a_\sigma\pi^b \\ &+ g_{3\pi}\epsilon_{abc}\epsilon^{\mu\nu\lambda\sigma}\omega_\mu\partial_\nu\pi^a\partial_\lambda\pi^b\partial_\sigma\pi^c + \frac{em_V^2}{g_V}V_\mu A^\mu,\end{aligned} \tag{1}$$

where we have made explicit the notation regarding the couplings and the corresponding fields and, in the last term, V refers in general to vector mesons and A^μ refers to the photon field. The couplings in this context are free parameters to be determined from experiment. Relations between them and those coming from low energy theorems can be drawn by building models which incorporate vector mesons into the chiral symmetric lagrangians [7, 8, 9, 10, 11]. In the following we will determine such couplings from the experimental data and draw some comparisons whenever possible.

2. Radiative decays

The $g_{\omega\rho\pi}$ coupling can be obtained from vector mesons radiative decays considering that the photon emission is mediated by a neutral vector meson [12]:

- $\omega \to \pi\rho \to \pi\gamma$ decay
- $\rho \to \pi(\omega,\phi) \to \pi\gamma$ decay
- $\pi^0 \to (\omega,\phi)\rho \to \gamma\gamma$ decay.

In Table 1, we exhibit the numerical values obtained for the above processes. Neglecting small correlations induced by the VMD vector mesons parameters, we can compute a weighted average, which yields 11.9 ± 0.2 with the error dominated by the most precise $\omega \to \pi^0\gamma$ channel. A standard average gives $g_{\omega\rho\pi} = 12.2 \pm 1.3$ GeV^{-1}, with errors added in quadratures and is dominated by the uncertainty in the $\rho^0 \to \pi^0\gamma$ decay width. In the following we will refer to the weighted average from radiative decays as VMDr.

As a way of comparison, let us get the expected value from the agreement between VMD and low energy theorems [13] for the $\pi \to \gamma\gamma$ decay [9]: Considering only the $\rho-\omega$ channel (the $\rho-\phi$ channel makes a small effect), it requires that

$$|g_{\omega\rho\pi}| = |g_\rho g_\omega/8\pi^2 f_\pi| = 11.5\ GeV^{-1}. \tag{2}$$

If, in addition, we impose universality ($g_\rho = g_{\rho\pi\pi}$) and SU(3) symmetry ($g_\rho = 3g_\omega$) then

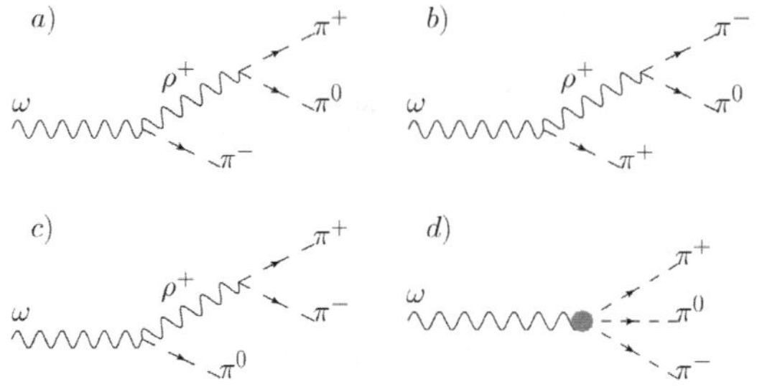

Figure 1. $\omega \to 3\pi$ process. The contribution from the ρ channel (a, b, c) and higher order contributions including the contact term (d).

$$|g_{\omega\rho\pi}| = |3g^2_{\rho\pi\pi}/8\pi^2 f_\pi| = 14.4\ GeV^{-1}. \tag{3}$$

By using the KSFR relation ($g_{\rho\pi\pi} = m_\rho/(\sqrt{2}f_\pi)$) [14] it takes the form

$$|g_{\omega\rho\pi}| = |3m^2_\rho/16\pi^2 f^3_\pi| = 14.2\ GeV^{-1}, \tag{4}$$

where $f_\pi = 0.093$ GeV. Therefore, at this stage, the value from radiative decays favors the one expected from the strict agreement between VMD and low energy theorems.

3. The $\omega \to 3\pi$ decay

Lets consider the process $\omega(\eta, q) \to \pi^+(p_1)\pi^-(p_2)\pi^0(p_3)$. Where the p_i are the corresponding 4-momenta and η is the ω meson polarization. The contributions to the amplitude from the ρ channel (Fig. 1a-c) and the contact term (Fig. 1d) can be set as follow:

$$\mathcal{M}_D = \imath\epsilon_{\mu\alpha\beta\gamma}\eta^\mu p_1^\alpha p_2^\beta p_3^\gamma \mathcal{A}, \tag{5}$$

where $\mathcal{A}$ is given by:

$$\mathcal{A} = 6g_{3\pi} + 2g_{\omega\rho\pi}g_{\rho\pi\pi}\left(D^{-1}[\rho^0, p_1 + p_2] + D^{-1}[\rho^+, p_1 + p_3] + D^{-1}[\rho^-, p_2 + p_3]\right), \tag{6}$$

where, $D[\rho, Q] = Q^2 - m^2_\rho + im_\rho\Gamma_\rho$ and the factors of 6 and 2 in $\mathcal{A}$ come from the cyclic permutations and momentum conservation used to bring the amplitude into the current form. The coupling $g_{\rho\pi\pi} = 5.95 \pm 0.02$ is fixed by the decay width of $\rho \to \pi\pi$, $\Gamma_\rho = 149.1 \pm 0.8$ MeV. Using these values and $g_{\omega\rho\pi}$ from radiative decays we can check that the prediction for the width, without taking into account the contact term ($g_{3\pi} = 0$) , is $\Gamma^\rho_{\omega\to3\pi} = 4.4 \pm 0.2$ MeV, which is 58% of the experimental value ($\Gamma^{exp}_{\omega\to3\pi} = 7.56 \pm 0.13$ MeV [5]). The correction by using an energy dependent width of the ρ is negligible compared with the error bars and the radiative corrections have been also estimated to be negligible [15].

In order to reach the 100% of the experimental width it may be necessary either to increase the coupling constant value up to 15.7 GeV^{-1} or to keep the value from radiative decays and include additional contributions like the contact term. Let us make a blind inclusion of this last in order to have an idea of its expected strength, needed to account for the observed decay width, this yields a quadratic equation for the coupling strength whose solutions are: $g_{3\pi} = -62 \pm 7$ and $+409 \pm 10$ GeV^{-3}.

3.1. The contact contribution

The contact term is not really a free parameter, it is fixed from the axial anomaly [16]. Let us recover some previous results, obtained by linking the low energy theorems to the corresponding amplitudes in VMD, while respecting the KSFR relation. It has been found [17, 18] that when the vector channel is saturated by the ρ meson, this contribute with

$$A^{\rho} = e^2 \frac{6}{g_\omega} \frac{g_{\omega\rho\pi} g_{\rho\pi\pi}}{m_\rho^2} = \frac{3}{2} \frac{e^2}{4\pi^2 f_\pi^3} = \frac{3}{2} A^{anomaly} \tag{7}$$

that is, three halves of the total amplitude as obtained from the Chiral anomaly [16]. Therefore, the contact term contributes with minus one half of the total amplitude. This was found by Rudaz [17] as a consistency requirement and by Cohen [18] as the one which satisfies axial Ward identities.

$$A^{contact} = e^2 \frac{6}{g_\omega} g_{3\pi} = \frac{-1}{2} A^{anomaly} \tag{8}$$

this fixes the corresponding coupling to be:

$$\rightarrow g_{3\pi} = -\frac{g_{\rho\pi\pi}}{16\pi^2 f_\pi^3} = -47\ GeV^{-3}. \tag{9}$$

where we have made use of the relationship among the couplings as discussed in the previous section. The result is close to one of the above solutions obtained to account for the experimental decay width. Considering this value we get $\Gamma_{\omega\to 3\pi} = 6.8 \pm 0.2$ MeV.

3.2. The ρ' channel

So far, we have considered the ρ channel and a Wess-Zumino fixed contact term as the only ways the decay can go through. However, decays via radial excitations may be also important, provided the mass suppression factor is not extremely large compared to the energies involved in the process. The $\rho'(1450)$ meson ($m_{\rho'} = 1465$ MeV and $\Gamma_{\rho'} = 400 \pm 60$ MeV), satisfies this condition for the ω decay regime. Let us explore the role of such contribution: The heavy mass of the ρ' allows to simplify its propagator leading to identify the global coupling as an effective contact term (Fig.2), in full analogy to Eqn. (1):

$$|g'_{3\pi}| \approx \frac{g_{\omega\rho'\pi} g_{\rho'\pi\pi}}{m_{\rho'}^2}. \tag{10}$$

The couplings involved in the right hand side are not settled, neither in the theoretical side nor experimentally [4, 19]. In order to make an estimate of their magnitudes, we assume that

$$g_{\omega\rho'\pi}/g_{\rho'\pi\pi} = g_{\omega\rho\pi}/g_{\rho\pi\pi} = 2. \tag{11}$$

Studies on the value of $|g_{\omega\rho'\pi}|$ have found it to lay in the interval 10-18 GeV^{-1} [19]. Under these assumptions we get $|g'_{3\pi}| \approx 46 \pm 23$ GeV^{-3}. We have evaluated the deviations from this value in Eqn. (10) due to momentum and width dependence of the propagators, which combined produce an increase of 6%. Thus, our estimate for the coupling is

$$|g'_{3\pi}| = 49 \pm 24\ GeV^{-3} \tag{12}$$

This is also close to the smallest value of the contact coupling obtained by requiring to get the 100% of the width, and to the result in Eqn. (9). In accordance, we choose its sign to be negative.

Journal of Physics: Conference Series **378** (2012) 012040 doi:10.1088/1742-6596/378/1/012040

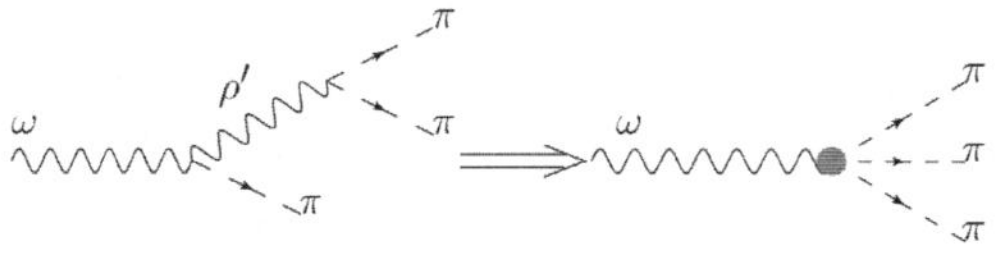

Figure 2. $\rho'(1450)$ as an effective contact term.

Reference	$\lvert g_{3\pi}\rvert$ [GeV^{-3}]
Rudaz, Cohen [17, 18]	47
Dominguez [20]	29 ± 3
Kuraev [7]	123
Kaymakcalan [8]	37
This work from $\Gamma(\omega \to 3\pi)$	65 ± 7
This work from ρ'	49 ± 24

Table 2. Determination of $|g_{3\pi}|$-like terms from several approaches. See text for details.

In Table 2, we have collected a set of values for the contact coupling computed in the literature from different approaches. [17] uses the VMD approach being consistent with low-energy theorems. [20] Extended the previous idea by including an infinite number of radial excitations. [7] works within a framework of a chiral effective langrangian. [8] uses an extension to the chiral lagrangian adding spin-1 fields, and our results obtained from different approaches.

It can be argued that given the different nature of these approaches a direct comparison between them is meaningless. However, we consider that it is interesting to quote their magnitudes as a way to exhibit that, besides the model dependence, there exist a regularity on these contributions.

To estimate the decay width we will not take the ρ' contribution simultaneously with the contact term, since its inclusion in the current form breaks the consistency between low energy theorems and VMD, which was already achieved by considering only the ρ meson [21].

4. The $e^+e^- \to \omega \to 3\pi$ cross section

Now, we explore the implications of the contact term or the ρ' in the $e^+e^- \to 3\pi$ cross section. Following the same notation as in the previous section, we can write the amplitude for the ω channel as follows:

$$\mathcal{M} = \frac{e^2 {m_\omega}^2}{g_\omega} \frac{\bar{v}\gamma^\mu u \epsilon_{\mu\alpha\beta\gamma} {p_1}^\alpha {p_2}^\beta {p_3}^\gamma}{q^2 \left(q^2 - m_\omega + i m_\omega \Gamma_\omega\right)} \mathcal{A} \tag{13}$$

where e is the positron electric charge, m_ω and Γ_ω are the mass and total width of the ω meson. In Figure 3 we have plotted the cross section as function of the center of mass energy for several cases: The experimental data from CMD2 [3] (circle symbols), the VMD prediction when the ω decay proceeds via the ρ meson for three cases: **i)** Using the central value for $g_{\omega\rho\pi}$ coupling determined from the radiative decays (VMD$_r$, solid line), **ii)** Including the ρ' contribution (decreased by 1 standard deviation, long dashed line) and **iii)** Including the contact contribution (dotted line). It is clear that the inclusion of either the ρ' or the contact term increases the central theoretical prediction based only on the ρ channel. The uncertainties on the ρ' allows to bring it down to the experimental data within one standard deviation. For the fixed contact term data can be only approached using the VMD$_r$ own error bars (not displayed).

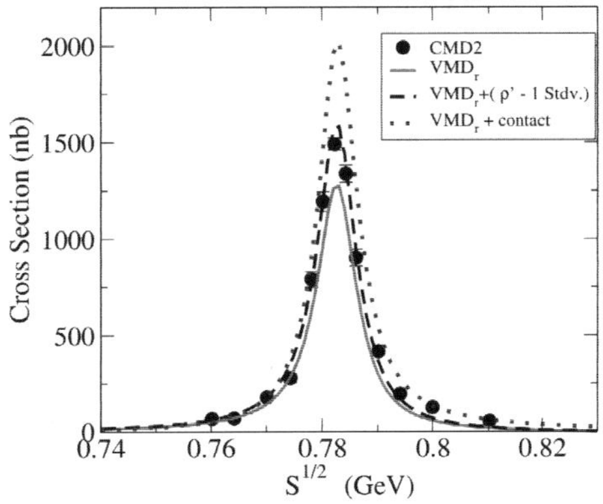

Figure 3. $e^+e^- \to \omega \to 3\pi$ cross section. Experimental data [3] (circle symbols) and the prediction by VMD using $g_{\omega\rho\pi}$ from radiative decays (solid line), and including either the ρ' term (decreased by 1stdv., dashed line) or the contact term (dot-dashed line).

A fit to the CMD2 data considering only the ρ channel, leaving $g_{\omega\rho\pi}$ as a free parameter leads to $|g_{\omega\rho\pi}| = 13.1 \pm 0.1$ GeV^{-1}.

5. Discussion

We have performed an analysis to obtain the $g_{\omega\rho\pi}$ coupling in the VMD approach, considering radiative decays, the $\omega \to 3\pi$ decay width and the $e^+e^- \to \omega \to 3\pi$ cross section. Our global result from all these observables are presented within two possible scenarios (See Table 3): i)Without contact nor ρ' term: In this case we average (weighted) the value predicted from radiative processes, the value which reproduces the experimental decay width for $\omega \to 3\pi$ and the value for the coupling obtained to fit the experimental $e^+e^- \to \omega \to 3\pi$ cross section. ii)With either ρ' or contact term: In this case we average the value predicted from radiative processes, the values from the decay width for $\omega \to 3\pi$ and the $e^+e^- \to \omega \to 3\pi$ cross section considering the existence of a contact term or the effect produced by the ρ' meson.

There is no ambiguity on the ρ' resonance contributions, besides the uncertainties associated to its corresponding mass, decay width and couplings. The inclusion of the contact term and the ρ channel alone in the $\omega \to 3\pi$ decay is consistent with the low energy theorems while fulfilling the KSFR relation. At this stage, we have not considered the contact and ρ' contributions simultaneously to avoid possible inconsistencies.

The value obtained for $g_{\omega\rho\pi}$ is sensitive to the inclusion of other contributions, in the context discussed above. Estimates, based on different approaches, yield similar values [17, 20, 22, 23], pointing out to the small model dependence of our results. Improved measurements of these observables and the $\rho'(1450)$ meson parameters are needed to settle the issues above mentioned.

Acknowledgments

We acknowledge the support of CONACyT, Mexico.

References

[1] G.W. Bennett et al., Phys. Rev. Lett. 89, 101804 (2002); Erratum ibid. Phys. Rev. Lett. 89, 129903 (2002); G.W. Bennett et al., Phys. Rev. Lett. 92, 161802 (2004); G.W. Bennett et al., Phys. Rev. D73, 072003 (2006).

	ρ channel	ρ + contact	$\rho + \rho'$
VMD$_r$	11.9 ± 0.2	11.9 ± 0.2	11.9 ± 0.2
$\Gamma(\omega \to 3\pi)$	15.7 ± 0.1	12.8 ± 0.1	12.6 ± 1.3
$\sigma(e^+e^- \to 3\pi)$	13.1 ± 0.1	9.8 ± 0.1	9.7 ± 1.4
Average	13.9 ± 0.1	11.4 ± 0.1	11.9 ± 0.2

Table 3. The $\omega\rho\pi$ coupling (GeV^{-1}) from different scenarios.

[2] M. Benayoun, P. David, L. DelBuono, O. Leitner 2010 Eur. Phys. Jour C **65** 211 ; ibid 2010 **68** 355 .
[3] R.R. Akhmetshin et al. 2004 Phys. Lett. B **578** 285.
[4] K. W. Edwards et al. 2000 Phys. Rev. D **61** 072003 .
[5] K. Nakamura et al. (Particle Data Group), 2010 J. Phys. G **37**, 075021
[6] M. Gell-Mann, D. Sharp, W. G. Wagner, 1962 Phys. Rev. Lett. **8** 460.
[7] E. A. Kuraev, Z. Silagadze, 1995 Phys. Atom. Nucl. **58** 1589.
[8] Kaymakcalan, S. Rajeev, J. Schechter, 1984 Phys. Rev. D **30** 594.
[9] T. Fujiwara et al. , 1985 Prog. Theor. Phys. **73** 92.
[10] E.A. Kuraev, Z.K. Silagadze, 1992 Phys. Lett. B**292** 377.
[11] J. Gasser, G. Ecker, A. Pich and E. de Rafael, 1988 Nucl. Phys. B **321** 311.
[12] M. Gell-Mann and F. Zachariasen, 1961 Phys. Rev. **124** 953.
[13] S.Adler, B.W.Lee, S.Treiman, A.Zee, 1971 Phys. Rev. D**4** 3497; M.V. Terentev, 1971 JETP Lett. **14** 40; M.V. Terentev, 1972 Phys. Lett. **38B** 419.
[14] K.Kawarayabashi and M.Suzuki, 1966 Phys. Rev. Lett. **16** 255; Riazuddin and Fayyazuddin, 1966 Phys. Rev. **147** 1071.
[15] See for example R.R. Akhmetshin et al. 2000 Phys. Lett. B **476** 33.
[16] J. Wess , B. Zumino, 1971 Phys. Lett. B**37** 95; E. Witten, 1983 Nucl. Phys. B**223** 422.
[17] S. Rudaz, 1984 Phys. Lett. B **45** 281. See references therein for earlier works.
[18] T. D. Cohen, 1989 Phys. Lett. B **233** 467.
[19] N. N. Achasov and A. A. Kozhevnikov, 2000 Phys. Rev. D **62** 117503 and references therein.
[20] C. A. Dominguez, 1987 Mod. Phys. Lett. A **12** 983.
[21] A detailed analysis in this regard will be considered elsewhere. Arxiv:1106.1467[hep-ph]
[22] P.G.O. Freund and S. Nandi, 1974 Phys. Rev. Lett. **32** 181.
[23] J. L. Lucio et al. 2000 Phys. Rev. D **61** 034013 and references therein.

Journal of Physics: Conference Series **378** (2012) 012041 doi:10.1088/1742-6596/378/1/012041

Topological Summation in Lattice Gauge Theory

Wolfgang Bietenholz[1] **and Ivan Hip**[2]

[1] Instituto de Ciencias Nucleares, Universidad Nacional Autónoma de México
A.P. 70-543, C.P. 04510 Distrito Federal, Mexico

[2] Faculty of Geotechnical Engineering, University of Zagreb
Hallerova aleja 7, 42000 Varaždin, Croatia

E-mail: wolbi@nucleares.unam.mx

Abstract. In gauge theories the field configurations often occur in distinct topological sectors. In a lattice regularised system with chiral fermions, these sectors can be defined by referring to the Atiyah-Singer Index Theorem. However, if such a model is simulated with local updates of the lattice gauge configuration, the Monte Carlo history tends to get stuck in one sector for many steps, in particular on fine lattices. Then expectation values can be measured only within specific sectors. Here we present a pilot study in the 2-flavour Schwinger model which explores methods of approximating the complete result for an observable — corresponding to a suitable sum over all sectors — based on numerical measurements in a few specific topological sectors. We also probe various procedures for an indirect evaluation of the topological susceptibility, starting from such topologically restricted measurements.

1. Topological sectors in gauge theories

Our general framework in this article is the functional integral formulation of quantum physics in Euclidean space. In this setting, the set of configurations may occur in disjoint subsets, so that all continuously deformed configurations belong to the same subset. Such subsets are known as *topological sectors*. Continuous deformations capture all configurations in one topological sector, but none of any different sector (general aspects are discussed *e.g.* in Refs. [1]).

The simplest example where this situation occurs is a quantum mechanical scalar particle moving on the circle S^1, with periodic boundary conditions in Euclidean time. The expectation value of some observable $\mathcal{O}$ in this system is given by

$$\langle \mathcal{O} \rangle = \frac{1}{Z} \int \mathcal{D}\varphi \ \mathcal{O}[\phi] \exp(-S[\varphi]) \ , \quad \text{where} \quad Z = \int \mathcal{D}\varphi \ \exp(-S[\varphi]) \tag{1}$$

is the partition function, and $\mathcal{D}\varphi$ is the sum over all closed paths $\varphi(t) \in S^1$ in some period T, *i.e.* $t \in [0, T]$ and $\varphi(0) = \varphi(T)$. The set of all these paths is naturally divided into disjoint subsets, which are characterised by the winding number

$$Q = \frac{1}{2\pi} \int_0^T dt \ \dot{\varphi} \in \mathbb{Z} \ , \tag{2}$$

which represents in this case the *topological charge*. Continuous path deformations cannot change Q, hence these subsets are indeed topological sectors.

Topological sectors also occur in a variety of gauge theories [1]. Let us consider gauge configurations in a Euclidean space with periodic boundary conditions (a torus). If they are split into topological sectors, the characteristic topological charge Q is also denoted as the Pontryagin index. Two examples are

$$\begin{aligned} 2\mathrm{d}\ U(1) \quad &: \quad Q = \frac{1}{\pi} \int d^2x\, \epsilon_{\mu\nu} F_{\mu\nu}(x)\ , \\ 4\mathrm{d}\ SU(3) \quad &: \quad Q = \frac{1}{32\pi^2} \mathrm{Tr} \int d^4x\, F_{\mu\nu}(x)\, \tilde{F}_{\mu\nu}(x)\ , \end{aligned} \tag{3}$$

where $F_{\mu\nu}$ is the field strength tensor, and $\tilde{F}_{\mu\nu} := \epsilon_{\mu\nu\rho\sigma} F_{\rho\sigma}$. Gauge configurations can be continuously deformed only within a fixed topological sector, hence the functional integral splits into separate integrals for each $Q \in \mathbb{Z}$.

Let us now address such a gauge theory in the presence of *chiral fermions, i.e.* massless fermions with a Dirac operator D that anti-commutes with γ_5, $D\gamma_5 + \gamma_5 D = 0$. In this case the zero modes of the Dirac operator have a definite chirality ± 1.

For such a Dirac operator, in a given gauge background, we denote the number of zero modes with chirality $+1$ (-1) as n_+ (n_-). Their difference is the *fermion index*

$$\nu := n_- - n_+ \in \mathbb{Z}\ . \tag{4}$$

The *Atiyah-Singer Index Theorem* [2] states that for any gauge configuration, this index coincides with the topological charge

$$Q \stackrel{!}{=} \nu\ . \tag{5}$$

2. Lattice regularisation

The lattice regularisation reduces the (Euclidean) space to discrete sites x, which are separated by some finite lattice spacing a. The latter implies an UV regularisation of the corresponding quantum field theory. Matter field variables are now defined on each lattice site, *e.g.* $\bar{\Psi}_x$, Ψ_x for fermion fields, while gauge fields can be formulated as *compact link variables* $U_{x,\mu} \in \{\,\text{gauge group}\,\}$. It is a great virtue that this formulation is *gauge invariant* even on the regularised level, so in this approach no gauge fixing is needed.

A priori there are *no* topological sectors anymore in the lattice regularised system; all configurations can now be continuously deformed into each other. Still, the desired connection to the continuum theory motivates the attempt to introduce somehow (the analogue of) topological sectors also on the lattice. A number of suggestions appeared in the literature, often with a somewhat questionable conceptual basis. A clean formulation emerged only at the very end of the last century, based on chiral lattice fermions. Their lattice Dirac operator D cannot simply anti-commute with γ_5 due to the notorious doubling problem of lattice fermions [3], but it may obey the *Ginsparg-Wilson Relation* (GWR), which reads (in its simplest form)

$$D\gamma_5 + \gamma_5 D = aD\gamma_5 D\ . \tag{6}$$

This still guarantees a lattice deformed — but exact — version of the chiral symmetry [4]. The latter also implies that the corresponding lattice Dirac operator has exact zero modes with a

Journal of Physics: Conference Series **378** (2012) 012041 doi:10.1088/1742-6596/378/1/012041

definite chirality, as in the continuum. Hence we can adopt the Index Theorem [5] and define the topological charge of a lattice gauge configuration as $Q := \nu$.

We remark that random lattice gauge configurations always occur with $n_+ = 0$ or $n_- = 0$; configurations with a cancellation in the lattice fermion index also exist (the free fermion is an example), but their probability measure seems to vanish.[1]

3. Monte Carlo simulation

Observables in quantum gauge theory can be evaluated beyond perturbation theory by means of Monte Carlo simulations. The idea is to use a sizeable set of gauge configurations $[U]$ (consisting of link variables all over the lattice volume), which are generated randomly with the probability distribution

$$p[U] = \det D[U] \ \exp(-S_{\rm gauge}[U]) \ . \tag{7}$$

Here we assume a fermion action, which is bilinear in the Grassmann valued spinor fields $\bar{\Psi}$, Ψ. Their functional integration $\mathcal{D}\bar{\Psi}\mathcal{D}\Psi$ is carried out already, giving rise to the fermion determinant $\det D[U]$.

The summation over this set of configurations yields a numerical measurement of expectation values $\langle \dots \rangle$, in particular of n-point functions. These results obviously come with some statistical error (since the available set of configurations is finite), and a systematic error (*e.g.* due to the finite lattice spacing a, which usually requires a continuum extrapolation). Both can be estimated and reduced if necessary by extended simulations. On the other hand, we stress again that the result is fully *non-perturbative;* we deal with the complete action in the exponent, *i.e.* we capture directly the given model at finite interaction strength.

Practical algorithms for the generation of gauge configurations (with the given probability distribution) perform local updates, *i.e.* in one step a configuration is modified just locally. Iterating such steps many times leads to a (quasi-)independent new configuration, to be used for the next measurement. Changing a gauge configuration drastically in a single step is also conceivable in principle, but in practice such algorithms tend to be inefficient.[2]

A problem with a sequence of local updates is, however, that it hardly ever changes the topological sector — although one should do so frequently in order to sample correctly the entire space of configurations. This problem is particularly striking in the attempts to simulate QCD with chiral quarks; the JLQCD Collaboration performed very extensive 2-flavour QCD simulations of this kind [6] — which led to interesting results — but the Monte Carlo histories were always confined to the trivial topological sector of charge $Q = 0$.

Most QCD simulations with dynamical quarks involve a non-chiral lattice quark formulation, since Ginsparg-Wilson fermions are tedious to simulate. In particular Wilson fermions (and variants thereof) have the disadvantage of additive mass renormalisation, but the problem with sampling the topological sectors is less severe so far. However, that property depends on the lattice spacing; typical values that have been used in the past are $a \approx 0.05 \dots 0.1$ fm. Once one tries to proceed to even finer lattices, the problem of confinement of the Monte Carlo history to a single topological sector is expected to show up also in this formulation [7].[3]

[1] The same holds for the specific lattice field configurations which are exactly *on* a topological boundary, so we can ignore them.

[2] Cluster algorithms are a counter example for certain spin models, but no efficient application to lattice gauge theories is known so far.

[3] To be more explicit: any algorithm has to obey "detailed balance", *i.e.* the transition probabilities of some configuration C_1 to C_2 and vice versa have to match the probability ratio for these configurations to occur

Journal of Physics: Conference Series **378** (2012) 012041 doi:10.1088/1742-6596/378/1/012041

So we have to address the question how to handle Monte Carlo simulations if the history tends to be trapped for a very long (computing) time, *i.e.* for many, many update steps, in *one* topological sector. What are then the prospects for measuring some n-point function, or the topological susceptibility

$$\chi_t := \frac{1}{V}\Big(\langle Q^2\rangle - \langle Q\rangle^2\Big) \qquad (V = \text{volume})\ , \tag{8}$$

which actually require the summation over a variety of topological sectors, with suitable statistical weights?

This is a delicate and highly relevant issue. Here we address it in a toy model study of the 2-flavour Schwinger model, which we simulated [8] with *dynamical overlap hypercube fermions;* this is one version of chiral lattice fermions [9], with a Dirac operator that solves the GWR (6).[4] We designed and applied a variant of the Hybrid Monte Carlo algorithm, which is particularly suitable for this type of lattice fermions [8].

4. The Schwinger model

The Schwinger model [11] represents Quantum Electrodynamics on a plane (QED_2). It is a popular toy model; in particular it shares with QCD the property of fermion confinement [12] (although the gauge group is Abelian) and the presence of topological sectors, see eq. (3). On the other hand there are qualitative differences, such as the absence of a running gauge coupling in the Schwinger model. In the continuum its Lagrangian can be written as

$$\mathcal{L}(\bar\Psi,\Psi,A_\mu) = \bar\Psi(x)\Big[\gamma_\mu(\mathrm{i}\,\partial_\mu + gA_\mu(x)) + m\Big]\Psi(x) + \frac{1}{2}F_{\mu\nu}(x)F_{\mu\nu}(x)\ . \tag{9}$$

We are interested in the case of $N_f = 2$ degenerate fermion flavours of mass $m \ll g$, where Ref. [13] made the following predictions:

$$\text{chiral condensate} \qquad \Sigma := -\langle\bar\Psi\Psi\rangle = 0.388\ldots(mg^2)^{1/3}\ , \tag{10}$$

$$\text{pion mass} \qquad M_\pi = 2.008\ldots(m^2 g)^{1/3}\ . \tag{11}$$

As in 2-flavour QCD, a "meson" singlet and a triplet emerge, the former (latter) being massive (massless) in the chiral limit $m \to 0$, cf. eq. (11). Referring to this analogy, and in agreement with the literature, we denote the triplet as "pions". Its emergence in 2 dimensions might appear somewhat surprising; the theoretical background of these "quasi-Nambu-Goldstone bosons" was first discussed in Ref. [14].

5. Numerical measurement at fixed topology

We simulated the 2-flavour Schwinger model at $\beta := 1/g^2 = 5$ [8]. This implies smooth gauge configurations (mean plaquette value $\simeq 0.9$). Also the "meson" dispersion relations confirm that lattice artifacts are tiny [8], hence we can confront our results directly with the continuum predictions (10), (11), without really needing a continuum extrapolation. On the other hand, finite size effects are significant, and they are in fact necessary for our discussion of topology dependent observables. We simulated on $L \times L$ lattices of sizes L = 16, 20, 24, 28, 32 with

(Boltzmann weights), $p(C_1 \to C_2)/p(C_2 \to C_1) = \exp(S[C_1] - S[C_2])$. The boundaries between topological sectors are surrounded by zones of high action, *i.e.* low probability. As the lattice spacing a is reduced, their weight $p(C)$ decreases with a high power of a [7]. Hence a sequence of small update steps will rarely tunnel through such a boundary.

[4] In this formulation we insert an improved kernel into the overlap formula, instead of the Wilson kernel of the standard overlap operator [10]. The virtues include an improved locality and scaling behaviour, and approximate rotation symmetry [9].

L	m	number of configurations $\nu = 0$	$\lvert\nu\rvert = 1$	$\lvert\nu\rvert = 2$	total	topological transitions
16	0.01	2428	307		2735	7
16	0.03	1070	508		1578	2
16	0.06	741	660		1401	7
16	0.09	919	587	1	1507	7
16	0.12	664	501	248	1413	8
16	0.18	791	563	50	1404	15
16	0.24	576	978	56	1637	17
L	m	number of configurations $\nu = 0$	$\lvert\nu\rvert = 1$	$\lvert\nu\rvert = 2$	$\lvert\nu\rvert = 3$	total
20	0.01	435	304			739
24	0.01			278	273	551
28	0.01		240		180	420
32	0.01	138	98	82		318
32	0.06	91	293			384

Table 1. Statistics for lattice size $L = 16$ (above) and $L > 16$ (below) and various fermion masses m, in distinct topological sectors (ν is the fermion index). Multiple starts of Monte Carlo histories were necessary to get access to various topological sectors, since topological transitions were very rare.

fermion masses in the range $m = 0.01 \dots 0.24$ (in lattice units). Our statistics is displayed in Table 1.

Let us first address the *Dirac spectrum*. All the eigenvalues of a lattice Dirac operator (before adding the mass), which obeys the GWR (6), are located on the circle in the complex plane with centre and radius $1/a$, as illustrated in Fig. 1. This confirms that the zero modes are exact,

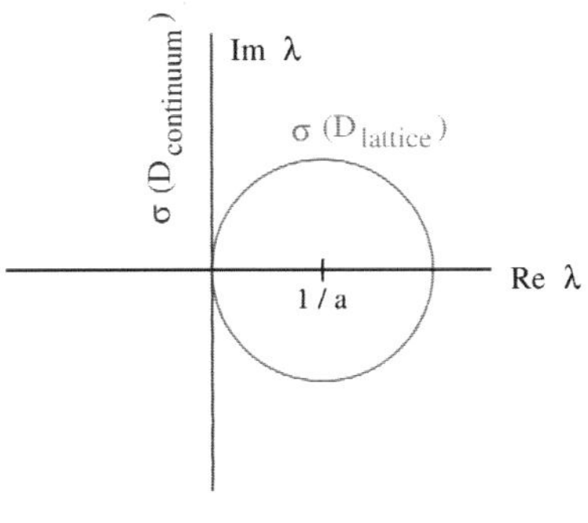

Figure 1. The spectrum of a lattice Dirac operator that fulfils the GWR (6) is located on a circle in the complex plane. In the continuum limit $a \to 0$ it turns into the imaginary axis.

and we have mentioned before that their fermion index is identified with the topological charge, $Q = \nu = n_- - n_+$.[5]

In this study we could evaluate the complete Dirac spectrum for our lattice configurations

[5] Actually throughout this study only the absolute value $|\nu|$ matters.

(which is not feasible in 4 dimensions, except for tiny lattices). To make this spectrum compatible with the continuum formulation, we map it stereographically onto the imaginary axis [15]. Based on the eigenvalues λ_i that we obtain after this mapping, we obtain the *chiral condensate*

$$\Sigma := -\langle\bar{\Psi}\Psi\rangle = \frac{1}{V}\left\langle \sum_i \frac{1}{|\lambda_i| + m} \right\rangle . \tag{12}$$

The sum can be computed for each configuration, but expectation values can only be measured within fixed topological sectors. Table 1 shows that topological transitions are indeed so rare that the entire space of configurations is not well sampled, but specific sectors are explored well. Hence we measure results for the expectation values of the chiral condensate at specific values of $|\nu|$,

$$\Sigma_{|\nu|} = -\langle\bar{\Psi}\Psi\rangle_{|\nu|} = \frac{1}{V}\left\langle \sum_i \frac{1}{|\lambda_i| + m} \right\rangle_{|\nu|} := \frac{|\nu|}{mV} + \varepsilon_{|\nu|} . \tag{13}$$

In the last expression we split off the zero mode contribution to $\Sigma_{|\nu|}$, which dominates at small mass m (and $\nu \neq 0$), and we denote the rest as $\varepsilon_{|\nu|}$. Numerical results are shown in Fig. 2.

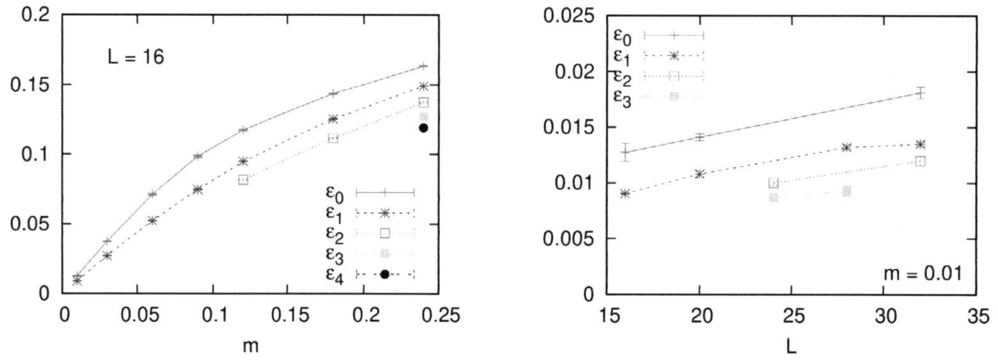

Figure 2. Numerical results for the chiral condensate at |topological charge| = $|\nu|$, after subtracting the zero mode contribution, cf. eq. (13), at $L = 16$ (on the left) and at $m = 0.01$ (on the right).

It is a generic property of stochastic Hermitian matrices (such as $\gamma_5 D$) that zero eigenvalues repel the low-lying non-zero modes. This suggests the inequality

$$\varepsilon_0 > \varepsilon_1 > \varepsilon_2 \ldots \tag{14}$$

at fixed m and V, which is confirmed consistently by the plot in Fig. 2 on the left. Moreover the plot on the right shows that

$$\varepsilon_i(V_1) > \varepsilon_i(V_2) \quad \text{for} \quad V_1 > V_2 , \tag{15}$$

which is less obvious: in a larger volume more eigenvalues cluster near zero, which supersedes the pre-factor $1/V$.

The rest of this article is devoted to tests of three different methods for approximately extracting "physical" quantities (*i.e.* quantities which are properly summed over all topological sectors), based on measurements in a few specific sectors.

Journal of Physics: Conference Series **378** (2012) 012041 doi:10.1088/1742-6596/378/1/012041

6. Gaussian evaluation of the topological susceptibility

We first assume a Gaussian distribution of the topological charges — this is certainly reasonable, for instance precision tests in $SU(3)$ pure gauge theory revealed at most tiny deviations from this behaviour [16]. It implies that the chiral condensate is composed as

$$\Sigma = \sum_{\nu=-\infty}^{\infty} p(|\nu|)\, \Sigma_{|\nu|} \quad , \qquad p(|\nu|) = \frac{\exp\{-\nu^2/(2V\chi_t)\}}{\sum_\nu \exp\{-\nu^2/(2V\chi_t)\}} \ . \tag{16}$$

Parity symmetry assures that $\langle \nu \rangle = 0$, hence the topological susceptibility simplifies to

$$\chi_t = \frac{\langle \nu^2 \rangle}{V} \ . \tag{17}$$

In most volumes we have data for $\Sigma_0 \dots \Sigma_Q$, *i.e.* up to some maximal topological charge Q. Thanks to inequality (14) all the higher charge contributions — for $|\nu| > Q$ — are bounded as

$$\frac{|\nu|}{mV} < \Sigma_{|\nu|} < \frac{|\nu|}{mV} + \varepsilon_Q \ . \tag{18}$$

Hence for a given value of the susceptibility χ_t the sum in eq. (16) can be performed, up to a uncertainty which affects Σ only mildly, since $\Sigma_{|\nu|}$ for high charges contribute only little.

In two volumes, $L = 24$ and 28, some $\Sigma_{|\nu|}$ data are missing for $|\nu| < Q$ (see Table 1); in these cases we can again fix a minimal and a maximal value for $\varepsilon_{|\nu|}$, this time based on inequality (15) and the results in the next smaller and next larger volume.

So we can probe any ansatz for χ_t and compute the corresponding value of Σ up to a modest uncertainty. We require the result to agree (within errors) with the prediction (10). In this way we determine χ_t. Fig. 3 shows the results for $L = 16$ and $m = 0.01,\ 0.03,\ 0.06$ — for higher masses the assumption $m \ll g \simeq 0.45$, which is needed for the prediction (10), seems to fail. Since the theory refers to infinite volume, we expect the result to improve for increasing m (*i.e.* for shorter correlation length) within the allowed range.

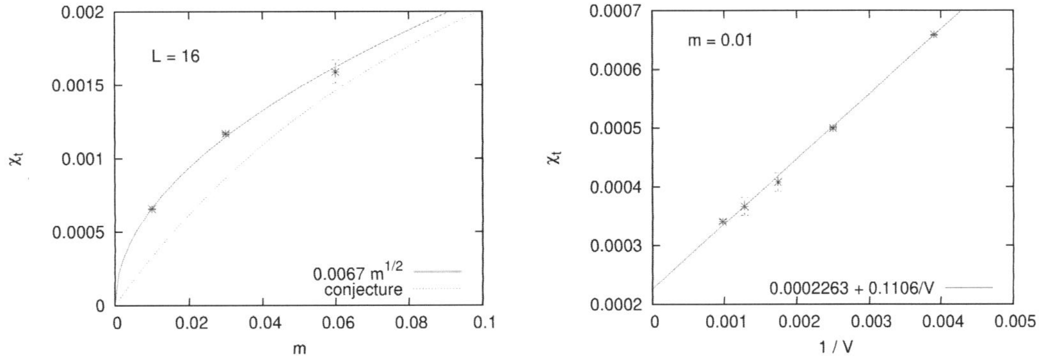

Figure 3. Our results for the topological susceptibility χ_t based on Gaussian summation. On the left: in a fixed volume $V = 16^2$ our data are compatible with an interpolation $\chi_t(m) \propto \sqrt{m}$, and for increasing mass (*i.e.* reduced finite size effects) we approach the conjecture (19). On the right: at fixed $m = 0.01$ we obtain results for χ_t, which seem to follow a behaviour linear in $1/V$.

Journal of Physics: Conference Series **378** (2012) 012041 doi:10.1088/1742-6596/378/1/012041

The result is compared to a QCD-inspired conjectured of Ref. [17] (for N_f flavours, in a large volume),

$$\frac{1}{\chi_t} = \frac{N_f}{m}\,\Sigma(N_f = 1, m = 0) + \frac{1}{\chi_t(N_f = 0)} \ . \tag{19}$$

The first ingredient has been computed analytically, $\Sigma(N_f = 1, m = 0) \simeq 0.16\,g$ [14], and the quenched susceptibility $\chi_t(N_f = 0) \simeq 0.000332$ has been measured numerically [18]. Fig. 3 confirms that the corresponding curve approaches the fit through our values for increasing m.

Alternatively we fix the mass $m = 0.01$ and compare the results for $L = 16 \dots 32$ (plot in Fig. 3 on the right). In our largest volume, $V = 32^2$, we obtained $\chi_t = 0.000341(4)$, which is close to the value given by conjecture (19), $\chi_t = 0.000332$. An infinite volume extrapolation of our data, however, leads to a smaller susceptibility of $\chi_t = 0.000226(5)$.

7. Correlation of the topological charge density

A drawback of the method in Section 6 is that a known reference quantity is needed (here it was Σ), and results in various topological sectors are required. This is not the case for an approach suggested in Ref. [19], which derived a "model independent formula" for the *correlation of the topological charge density* ρ_t in one sector,

$$\lim_{|x| \to \infty} \langle \rho_t(x) \rho_t(0) \rangle_{|\nu|} \simeq -\frac{1}{V}\chi_t + \frac{\nu^2}{V^2} + O(V^{-3}) \ . \tag{20}$$

For tests in 2-flavour QCD we refer to Ref. [20]. (The original formula even includes a correction for a possible deviation from a Gaussian distribution of the topological charges, which we neglect.) In order to justify the assumptions in the derivation of this formula, we have to assume a large expectation value $\langle \nu^2 \rangle = V\chi_t$, and a small ratio $|\nu|/\langle \nu^2 \rangle$.

As an example, we show in Fig. 4 the corresponding correlation at $L = 16$, $\nu = 0$ and various masses. Numerically the density was computed from the simplest lattice version of $\rho_t = \epsilon_{12} F_{12}$ (this is not problematic in the current setting, where we are always dealing with smooth configurations).

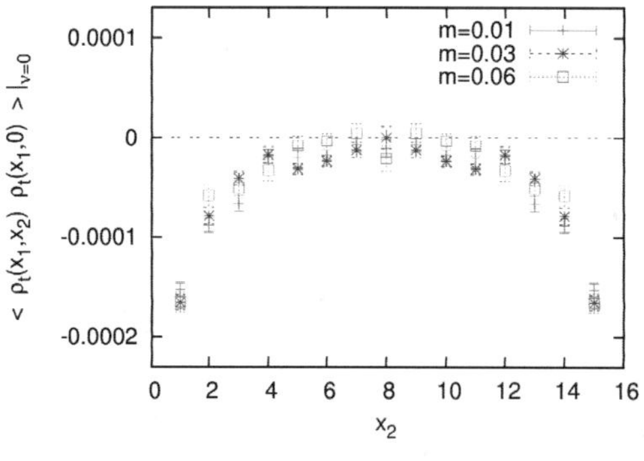

Figure 4. The correlation of the topological charge density for $L = 16$, $\nu = 0$ and $m = 0.01$, 0.03, 0.06. The statistical noise does not allow us to resolve a plateau value at large distances — overcoming this problem would require a huge statistics.

At large distances one should find a plateau value, which would then yield χ_t. In particular for $m = 0.06$, where the maximal distance might be sufficient to see the asymptotic behaviour, we expect (based on the data and the conjectured formula in Section 6) a plateau value of $\chi_t/V \approx -6 \cdot 10^{-6}$. However, our statistical errors are of $O(10^{-5})$, so in order to clearly resolve this plateau we would need about 50 000 to 100 000 configurations (cf. Table 1). We conclude that the applicability of this method requires unfortunately a very large statistics.

8. Approximate topological summation of observables

We now proceed to the main approach in this study. It is a method that does not require a known input observable either (as in Section 7), but measurements in various topological sectors and volumes are needed. In fact this is the input which is usually accessible. Then one tries to extract a (topologically summed) observable $\langle O \rangle$ by employing the approximation formula

$$\langle O \rangle_{|\nu|} \approx \langle O \rangle + \frac{c}{V\chi_t}\left(1 - \frac{\nu^2}{V\chi_t}\right) \qquad (c = \text{const.}) \ . \tag{21}$$

This formula has been derived first for the pion mass in QCD [21], but it applies generally to observables in a field theory with topology [8]. As in Section 7 one assumes a Gaussian distribution of the topological charges, and a large value of $V\chi_t$, as well as a small ratio $|\nu|/\langle \nu^2 \rangle$, are favourable for the validity of the approximations involved in the derivation. This approximation formula could be truly powerful in QCD and elsewhere, but it has never been tested before.

8.1. Application to the chiral condensate

Let us apply formula (21) to the chiral condensate. It is convenient to modify the notation,

$$\Sigma_\nu \approx \Sigma - \frac{A}{V} + \nu^2 \frac{B}{V^2} \ , \qquad A = -\frac{c}{\chi_t} \ , \quad B = -\frac{c}{\chi_t^2} \ . \tag{22}$$

The unknown quantities are Σ, A and B, and we are ultimately interested in Σ and $\chi_t = A/B$. They can be determined (in the framework of this approximation) by numerical results for some Σ_ν:

- At fixed m and V, we can determine B, for instance from Σ_0 and Σ_1.
- If we keep m fixed but consider two volumes, $V_1 \neq V_2$, we can further determine A, *e.g.* based on Σ_0.

In total, it takes (at least) three Σ_ν values, involving two volumes, to obtain results for Σ and χ_t.

We follow this sequence of steps and start with the determination of B. If we use as our input the measurements in the topological sectors with $|\nu| = k,\ \ell$ (at fixed m and V), we denote the result as $B_{k,\ell}$,

$$\frac{1}{V} B_{k,\ell} = V \frac{\Sigma_k - \Sigma_\ell}{k^2 - \ell^2} = \frac{1}{m(k+\ell)} + V \frac{\varepsilon_k - \varepsilon_\ell}{k^2 - \ell^2} \ . \tag{23}$$

The semi-classical term, $1/(m(k+\ell))$, tends to vary strongly for different choices of k and ℓ. Ideally the quantum effects should render the results for $B_{k,\ell}$ similar again. As an example, we show in Fig. 5 results for $B_{k,0}$ at $L = 16$, $m = 24$. In fact the non-perturbative results are much more stable in k than the semi-classical contributions alone. Hence the first consistency test is passed well.

We proceed to the determination of A, and therefore of Σ, based on Σ_0 measurements in two volumes with sizes (L_1, L_2). Here we consider $m = 0.01$ and we give two examples:

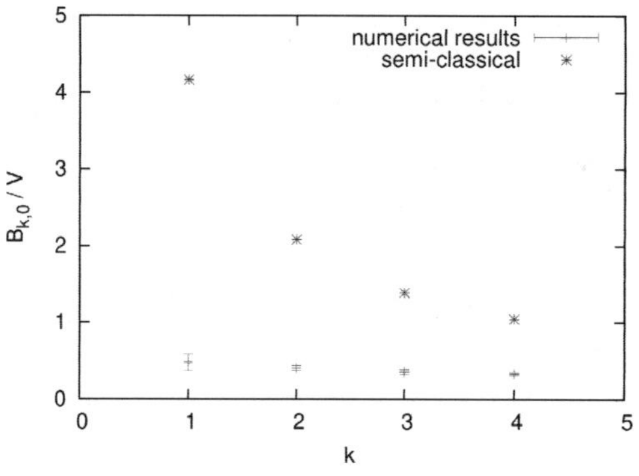

Figure 5. Results for the auxiliary variables $B_{k,0}$ (cf. eqs. (22), (23)) at $L = 16$, $m = 0.24$, for $k = 1, \ldots, 4$. We see that the numerical results are quite consistent, in contrast to the semi-classical values.

- The Σ_0 values in $(L_1, L_2) = (16, 32)$ yield $\Sigma = 0.0199(7)$.
- The Σ_0 values in $(L_1, L_2) = (20, 32)$ yield $\Sigma = 0.0207(12)$.

Thus the consistency looks fine again, but these results are far below the prediction (10), $\Sigma = 0.04888$ (in an infinite volume). In this case, our results are strongly affected by finite size effects, which is not surprising: for the given fermion mass, the correlation length in infinite volume (given by eq. (11)) would be $\xi = 1/M_{\pi,V=\infty} \simeq 14$. The relatively small boxes enhance the Dirac eigenvalues $|\lambda_i|$, such that Σ decreases.

So the mass $m = 0.06$ should be more promising, where theory predicts $\xi \simeq 4.3$. Here we only have data in $(L_1, L_2) = (16, 32)$, for $|\nu| = 0,\ 1$, so we cannot repeat the above consistency tests. Nevertheless we can evaluate $\chi_t = 0.00118(30)$ (which is just compatible with the conjecture (19), $\chi_t \simeq 0.00146$). We further insert our most reliable result for B, namely $B_{1,0}$ measured in $L = 32$, and arrive at a result for Σ, which is indeed close to the theoretical prediction (11),

$$\Sigma_{\rm numerical} = 0.0883(69) \quad , \quad \Sigma_{\rm theory} = 0.0888 \ . \tag{24}$$

8.2. Application to the pion mass

Let us also test the approximate summation formula (21) by applying it to the pion mass. As we mentioned before, this was the original idea of Ref. [21] (though that work referred to QCD). We re-write approximation (21) in the notation analogous to (22),

$$M_{\pi,|\nu|} \approx M_\pi - \frac{A}{V} + \frac{B}{V^2}\nu^2 \ . \tag{25}$$

However, we now adopt a strategy which differs from the previous consideration of Σ: at fixed m we determine the three unknown parameters A, B, M_π directly by a least-square fit for some set of numerical $M_{\pi,|\nu|}$ values.

- For $m = 0.01$ we have in total 11 measurements of $M_{\pi,|\nu|}$ (see Table 1), and we include the most promising ones. We need at least two volumes, so we take the largest two with

Journal of Physics: Conference Series **378** (2012) 012041 doi:10.1088/1742-6596/378/1/012041

$(L_1, L_2) = (28, 32)$. Moreover we only include the topological sectors with $|\nu| \leq 1$, which are favourable for the condition that $|\nu|/\langle\nu^2\rangle$ should be small. This leads to

$$\left.\begin{array}{rcc} & M_{\pi,0} & M_{\pi,1} \\ L=28: & & 0.146(4) \\ L=32: & 0.05(1) & 0.160(8) \end{array}\right\} \stackrel{\text{fit}}{\Longrightarrow} M_\pi = 0.073(25)\ , \qquad (26)$$

which matches well the theoretical prediction, $M_\pi = 0.071$ (albeit with a large error).

• We proceed to $m = 0.06$, where we only have data for $(L_1, L_2) = (16, 32)$. Hence we have less choice in this case, but the finite size effects are less severe. Again we include the results for $|\nu| \leq 1$, which corresponds to four input measurements this time, and we arrive at

$$\left.\begin{array}{rcc} & M_{\pi,0} & M_{\pi,1} \\ L=16: & 0.041(1) & 0.271(4) \\ L=32: & 0.23(1) & 0.232(7) \end{array}\right\} \stackrel{\text{fit}}{\Longrightarrow} M_\pi = 0.233(8)\ . \qquad (27)$$

Also this result agrees well with the theoretical pion mass, $M_\pi = 0.235$, and this time also the uncertainty is modest.

9. Conclusions

We have addressed a quite generic problem of lattice simulations in gauge theories with dynamical (quasi-)chiral fermions. The Monte Carlo histories of such simulations tend to get trapped in *one* topological sector for a very long (simulation) time, *i.e.* over many update steps of the lattice gauge configuration. A conceptual issue that one has to address in this situation is *ergodicity*, a property which is compulsory for a correct algorithm. Here we studied a more practical question: how can we evaluate the expectation value of some observable $\langle O\rangle$, when only numerical measurements restricted to a few topological sectors, $\langle O\rangle_\nu$, are available?

The dominant subject in contemporary lattice simulations is QCD with dynamical quarks. Here the problem of topological restriction is most striking when one deals with chiral lattice quarks (of overlap [10] or Domain Wall [22] type), which solve the Ginsparg-Wilson Relation (eq. (6) or generalisations thereof). The use of Wilson type quarks is more widespread because they are much faster to simulate, though plagued by additive mass renormalisation and problems related to operator mixing.[6] [7]

Here the aforementioned topological problem is less severe so far, but it is expected to show up as well when simulations will be carried out on finer and finer lattices, say with lattice spacing $a < 0.05$ fm. This renders the lattice QCD formulation more and more continuum-like, which is in general welcome, but it also makes it more difficult to change the topological sector.

This problem is not manifest in a very large volume, where $\langle O\rangle_\nu$ is the same for all indices ν (this property agrees with approximations (20), (21)). However, to suppress the topological dependence and other finite size effects, the volume has to be large compared to the correlation

[6] Also that problem is avoided by the use of Ginsparg-Wilson fermions [23].

[7] For completeness we add that "staggered fermions" are widespread as well in lattice QCD. They are also quick to simulate, and they do not suffer from additive mass renormalisation, but the number of flavours is not flexible. Therefore it is now popular to take the fourth root of the fermion determinant (cf. eq. (7)), which formally corresponds to a single flavour, but this is harmful for locality, which is conceptually important. The question if this is a reason to worry in practice is highly controversial. In any case, neither Wilson nor staggered fermions do provide a sound definition of the topological charge since there is no well-defined fermion index, in contrast to Ginsparg-Wilson fermions [5]. Hence one has to refer to some rather hand-waving definition in these cases.

length, which is given by the inverse pion mass, $\xi_{\rm QCD} \approx 1.4$ fm $\ll La$. But when a is very small, this requires a huge lattice size L, which makes simulations again very tedious.

As a way out, the use of open boundary conditions in the Euclidean time direction has recently been advocated, so that topological charge can gradually flow in or out of the volume during a simulation [24]. In our study, however, we stay with periodic boundary conditions for the gauge fields, which guarantee that the topological charge is always integer, along with (discrete) translation invariance. As a toy model we considered the Schwinger model with two light, degenerate flavours, which were represented on the lattice by dynamical overlap hypercube fermions. In a set of small or moderate volumes, this only enabled measurements inside some specific topological sectors. In order to establish a link to the "physical" quantities, we tested three methods to approximate the topological summation:

- The confrontation of a Gaussian summation with a known observable allows us to fix the topological susceptibility χ_t. This method is robust, but it requires a known input quantity. This is available in the 2-flavour Schwinger model [13] (we used the chiral condensate), but not in general.

- Next we tested a method to evaluate χ_t based on the correlation function of the topological charge density [19]. More precisely, one searches for an asymptotic plateau of this correlation at large distances, which should amount to $-\chi_t/V$ (at $\nu = 0$). Unfortunately this value tends to be tiny for realistic settings, hence its resolution requires a very large statistics.

- Our main goal was the test of an approximate summation formula given in Ref. [21], which could provide a "physical" result $\langle O \rangle$, using only measurements of some topologically restricted observables $\langle O \rangle_\nu$ as an input — for various values of $|\nu|$, in at least two volumes. This method is potentially powerful, but it has never been tested before.
Our results suggest that it may work, if the assumptions used in the derivation of this formula are reasonably well justified. In particular, $V\chi_t = \langle \nu^2 \rangle$ should be "large", but it is difficult to predict explicitly what this means. In our settings this quantity was always below 0.5, but nevertheless we found decent (though not very precise) results for the topologically summed chiral condensate and pion mass. This observation is encouraging for applications in QCD simulations with dynamical quarks.

Acknowledgements: Stanislav Shcheredin and Jan Volkholz have contributed to this work at an early stage. We also thank Poul Damgaard, Stephan Dürr, Hidenori Fukaya and Jim Hetrick for helpful comments. This work was supported by the Croatian Ministry of Science, Education and Sports (project 0160013) and by the Deutsche Forschungsgemeinschaft through Sonderforschungsbereich Transregio 55 (SFB/TR55) "Hadron Physics from Lattice QCD".

References

[1] Polyakov A M 1987 *Gauge Fields and Strings* (Harwood Academic Publishers)
Rajaraman R 1987 *Solitons and Instantons* (North-Holland Personal Library)
Coleman S 1988 *Aspects of Symmetry* (Cambridge University Press)
[2] Atiyah M F and Singer I M 1968 *Ann. Math.* **87** 484
[3] Nielsen H B and Ninomiya M 1981 *Nucl. Phys.* B **185** 20
[4] Lüscher M 1998 *Phys. Lett.* B **428** 342
[5] Hasenfratz P, Laliena V and Niedermayer F 1998 *Phys. Lett.* B **427** 125
[6] Fukaya H *et al.* 2007 *Phys. Rev. Lett.* **98** 172001; 2007 *Phys. Rev.* D **76** 054503
[7] Lüscher M 2010 *JHEP* **1008** 071; 2010 *PoS(LATTICE2010)* 015

[8] Bietenholz W, Hip I, Shcheredin S and Volkholz J 2011 *arXiv:1109.2649 [hep-lat]*
[9] Bietenholz W 1999 *Eur. Phys. J.* C **6** 537; Bietenholz W and Hip I 2000 *Nucl. Phys.* B **570** 423; Bietenholz W 2002 *Nucl. Phys.* B **644** 223; Bietenholz W and Shcheredin S 2006 *Nucl. Phys.* B **754** 17
[10] Neuberger H 1998 *Phys. Lett.* B **417** 141; 1998 *Phys. Lett.* B **427** 353
[11] Schwinger J 1962 *Phys. Rev.* **128** 2425
[12] Coleman S R, Jackiw R and Susskind L 1975 *Annals Phys.* **93** 267
[13] Smilga A V 1997 *Phys. Rev.* D **55** 443
[14] Coleman S R 1976 *Annals Phys.* **101** 239
[15] Bietenholz W, Jansen K and Shcheredin S 2003 *JHEP* **0307** 033
[16] Del Debbio L, Giusti L and Pica C 2005 *Phys. Rev. Lett.* **94** 032003
Dürr S, Fodor Z, Hoelbling C and Kurth T 2007 *JHEP* **0704** 055
[17] Dürr S 2001 *Nucl. Phys.* B **611** 281
[18] Dürr S and Hoelbling C 2005 *Phys. Rev.* D **71** 054501
[19] Aoki S, Fukaya H, Hashimoto S and Onogi T 2007 *Phys. Rev.* D **76** 054508
[20] Aoki S *et al.* (JLQCD and TWQCD Collaborations) 2008 *Phys. Lett.* B **665** 294
[21] Brower R, Chandrasekharan S, Negele J W and Wiese U-J 2003 *Phys. Lett.* B **560** 64
[22] Kaplan D B 1992 *Phys. Lett.* B **288** 342
[23] Hasenfratz P 1998 *Nucl. Phys.* B **525** 401
[24] Lüscher M and Schaefer S 2011 *JHEP* **1107** 036

Journal of Physics: Conference Series **378** (2012) 012042 doi:10.1088/1742-6596/378/1/012042

The infrared fixed point of Landau gauge Yang-Mills theory: A renormalization group analysis

Axel Weber

Instituto de Física y Matemáticas, Universidad Michoacana de San Nicolás de Hidalgo, Edificio C-3, Ciudad Universitaria, A. Postal 2-82, 58040 Morelia, Michoacán, Mexico

E-mail: axel@ifm.umich.mx

Abstract. The infrared behavior of gluon and ghost propagators in Landau gauge Yang-Mills theory has been at the center of an intense debate over the last decade. Different solutions of the Dyson-Schwinger equations have a different behavior of the propagators in the infrared: in the so-called scaling solutions both propagators follow a power law, while in the decoupling solutions the gluon propagator shows a massive behavior. The latest lattice results favor the decoupling solutions. In this contribution, after giving a brief overview of the present status of analytical and semi-analytical approaches to the infrared regime of Landau gauge Yang-Mills theory, we will show how Callan-Symanzik renormalization group equations in an epsilon expansion reproduce both types of solutions and single out the decoupling solutions as the infrared-stable ones for space-time dimensions greater than two, in agreement with the lattice calculations.

1. Introduction

The infrared regime of quantum chromodynamics is one of the prime examples of non-perturbative physics giving rise to such fascinating phenomena as the confinement of quarks and the dynamical breaking of chiral symmetry. The only method so far to obtain quantitatively reliable results directly from the underlying theory has been lattice gauge theory, and it has done so only in a limited fashion. For a long time, it was generally believed that analytical tools could not be of much help in the quantitative description of the phenomena mentioned before. This point of view is now slowly changing. In this contribution, before describing our own approach to the subject to which the title refers, we will try to give an overview, from our perspective, of what has been achieved so far in (the pure gauge sector of) infrared quantum chromodynamics with analytical and semi-analytical methods. Needless to say, progress with these methods has also been limited, but there have been some recent very promising developments.

The story begins in 1978 with the publication of Gribov's seminal work on the quantization of non-abelian gauge theories [1]. It was clear by that time that for the quantization of a gauge theory in the continuum the gauge needs to be fixed, and that, at least for perturbation theory, covariant gauges are particularly convenient. Specializing to the Landau gauge, the functional integral over the gauge field then includes two characteristic pieces: a functional delta distribution that implements the gauge condition $\partial_\mu A_\mu^a = 0$, and a functional determinant (Jacobian) that arises when one changes the integration variables from the general gauge fields A_μ^a to gauge equivalent fields $\bar{A}_\mu^a$ that fulfill the gauge condition, and the corresponding gauge transformations U, $A^a_\mu = (\bar{A}^U)^a_\mu$. The delta distribution can conveniently be written with the help of a bosonic auxiliary field called the Nakanishi-Lautrup field,

$$\delta(\partial_\mu A^a_\mu) = \int D[B] \exp\left(-i \int d^D x\, B^a \partial_\mu A^a_\mu\right), \tag{1}$$

while the Jacobian is represented in a local way as a functional integral over (fermionic scalar) ghost and antighost fields,

$$\det(-\partial_\mu D^{ab}_\mu) = \int D[c, \bar{c}] \exp\left(\int d^D x\, \bar{c}^a \partial_\mu D^{ab}_\mu c^b\right). \tag{2}$$

We are generally working in D-dimensional Euclidean space-time, and the covariant derivative (in the adjoint representation) is defined as

$$D^{ab}_\mu = \delta^{ab}\partial_\mu - g f^{abc} A^c_\mu . \tag{3}$$

In the functional integral of the theory (for gauge invariant operators) the integral over the gauge transformations U then factorizes and can consistently be omitted, and one is left in the pure gauge sector with an integral over A^a_μ, B^a, c^a and $\bar{c}^a$, and the (negative of the) Faddeev-Popov action

$$S_{\mathrm{FP}} = \int d^D x \left(\frac{1}{4} F^a_{\mu\nu} F^a_{\mu\nu} + i B^a \partial_\mu A^a_\mu + \partial_\mu \bar{c}^a D^{ab}_\mu c^b\right) \tag{4}$$

in the exponent.

As it turns out, the beta function for this theory is negative giving rise to the celebrated asymptotic freedom of non-abelian gauge theories and the applicability of perturbation theory in the ultraviolet regime. In the language of renormalization group theory, non-abelian gauge theories have a trivial fixed point (with vanishing renormalized coupling constant $g_R(\mu)$) which is ultraviolet-attractive. On the other hand, the renormalized coupling constant runs into a Landau pole at the characteristic scale Λ_{QCD}, and perturbation theory (around the trivial fixed point) becomes useless in the infrared, i.e., for energy-momentum scales of the order of and below Λ_{QCD}.

2. Gauge copies

In 1978, Gribov [1] observed that the gauge fixing procedure leading to the Faddeev-Popov action as described above is actually not consistent in non-abelian theories. The problem arises in the change of variables in the functional integral over the gauge field: the choice of a gauge-fixed field $\bar{A}^a_\mu$, $\partial_\mu \bar{A}^a_\mu = 0$, that lies on the same gauge orbit as the initial field A^a_μ, $A^a_\mu = (\bar{A}^U)^a_\mu$ for an appropriate gauge transformation U, is not unique. In other words, there exist different gauge fields, called gauge copies or Gribov copies, that fulfill the gauge condition and are, nevertheless, connected by a gauge transformation.

In particular, if we look at an infinitesimal gauge transformation generated by ω^a, we have

$$\partial_\mu (A^\omega)^a_\mu = \partial_\mu A^a_\mu + \partial_\mu D^{ab}_\mu \omega^b . \tag{5}$$

Then it is possible to have both, A^a_μ and $(A^\omega)^a_\mu$, fulfill the gauge condition if ω^a is a zero mode of the Faddeev-Popov operator $(-\partial_\mu D^{ab}_\mu)$. Hence Gribov's proposal to restrict the integration over the gauge field to the region where the Faddeev-Popov operator is positive definite, now called the (first) Gribov region and denoted by Ω. Note that at the boundary of the Gribov region, the (first) Gribov horizon $\partial\Omega$, we have that $\det(-\partial_\mu D^{ab}_\mu) = 0$ because the Faddeev-Popov operator develops a zero mode there.

Although gauge copies which are related by an infinitesimal gauge transformation are excluded by the restriction to Ω, it is not clear that finitely related gauge copies cannot exist within Ω. In fact, it is now known that such copies do exist in Ω, and the consequent thing to do would be to restrict the integration over the gauge fields further to a smaller region that is called the fundamental modular region Λ. However, the restriction to Λ is very hard to implement, in numerical calculations (lattice gauge theory with Landau gauge fixing) and even more so in analytical (or semi-analytical) calculations. Fortunately, according to an argument by Zwanziger [2], the difference between the restrictions to Ω or to Λ has no effect on the correlation functions.

For the rest of this contribution, we will hence consider the restriction of the integral over the gauge field to the Gribov region Ω. One of the important implications of this restriction as compared to an unrestricted integral over the gauge field, according to an argument by Gribov [1], is the impossibility of a Landau pole for the renormalized coupling constant to occur (at a momentum scale different from zero). Realizing the restriction to Ω in an approximate way, Gribov was furthermore able to predict the infrared behavior of the gluon and ghost propagators. Here we will present an exact way to implement the restriction of the integral which is due to Zwanziger [3] and based on Gribov's ideas, and come back to the propagators later.

Zwanziger found a way to realize the (Heavyside) theta functional for the restriction of the integral over the gluon field to the Gribov region as a Gaussian (in the thermodynamic limit):

$$\int_\Omega D[A] = \int D[A]\,\theta(-\partial_\mu D_\mu^{ab}) = \int D[A]\,\exp\left(-\gamma^4 \int d^D x\, h(x)\right), \tag{6}$$

with the horizon function $h(x)$ defined as

$$h(x) = g^2 f^{abc} f^{cde} A_\nu^b \left[(-\partial_\mu D_\mu)^{-1}\right]^{ad} A_\nu^e \,. \tag{7}$$

For the correct normalization, the horizon condition

$$\langle h(x) \rangle = 4(N^2-1) \tag{8}$$

has to be fulfilled (in an SU(N) gauge theory). This condition fixes the new parameter γ with the dimension of mass (in terms of Λ_{QCD}, or vice versa).

Now $h(x)$ is a complicated nonlocal functional because it contains the inverse of the Faddeev-Popov operator. A local form of the theory, necessary for the application of the usual quantum field theoretical machinery, can be achieved by introducing two additional auxiliary vector fields (and the corresponding antifields), one of them bosonic ϕ_μ^{ab} and the other fermionic ω_μ^{ab}, each carrying two color indices (in the adjoint representation). The gauge sector of the theory including the restriction to the first Gribov region is then described by a functional integral over A_μ^a, B^a, c^a, $\bar{c}^a$, ϕ_μ^{ab}, $\bar{\phi}_\mu^{ab}$, ω_μ^{ab} and $\bar{\omega}_\mu^{ab}$, and the exponential of the (negative of the) Gribov-Zwanziger action [4]

$$\begin{aligned} S_{\mathrm{GZ}} = S_{\mathrm{FP}} &- \int d^D x \left(\partial_\mu \bar{\phi}_\nu^{ac} D_\mu^{ab} \phi_\nu^{bc} - \partial_\mu \bar{\omega}_\nu^{ac} D_\mu^{ab} \omega_\nu^{bc} + g f^{abc} \partial_\mu \bar{\omega}_\nu^{ad} D_\mu^{be} c^e \phi_\nu^{cd}\right) \\ &- \gamma^2 \int d^D x \left(g f^{abc} A_\mu^a \phi_\mu^{bc} + g f^{abc} A_\mu^a \bar{\phi}_\mu^{bc} + 4(N^2-1)\gamma^2\right), \end{aligned} \tag{9}$$

see eq. (4). The horizon condition can be rewritten in terms of the auxiliary fields as

$$\langle g f^{abc} A_\mu^a (\phi_\mu^{bc} + \bar{\phi}_\mu^{bc}) \rangle = 8(N^2-1)\gamma^2 \,. \tag{10}$$

The Gribov-Zwanziger action leads to the tree-level gluon propagator (in momentum space)

$$\langle A_\mu^a(p) A_\nu^b(-q) \rangle = \frac{p^2}{p^4+\lambda^4} \left(\delta_{\mu\nu} - \frac{p_\mu p_\nu}{p^2}\right) \delta^{ab} (2\pi)^D \delta(p-q) \,, \tag{11}$$

with the characteristic scale λ defined as

$$\lambda^4 = 2g^2 N \gamma^4 \,. \tag{12}$$

For $p^2 \gg \lambda^2$, one recovers the usual (tree-level) ultraviolet behavior, while the propagator tends to zero in the limit $p^2 \to 0$. The tree-level ghost propagator, on the other hand, behaves like

$$\langle c^a(p) \bar{c}^b(-q) \rangle \propto \frac{1}{p^4} \, \delta^{ab} (2\pi)^D \delta(p-q) \tag{13}$$

in the infrared regime $p^2 \ll \lambda^2$. The divergence in the limit $p^2 \to 0$ is much stronger than for the tree-level ghost propagator in Faddeev-Popov theory that does not take the restriction to the Gribov region into account. It is clear, then, that the effect of the restriction to the Gribov region is rather drastic for the infrared behavior of the propagators. The forms (11) and (13) of the propagators have already been found by Gribov in his approximate description [1].

An important aspect of the Gribov-Zwanziger framework is the breakdown of BRST invariance, a global fermionic symmetry of the Faddeev-Popov action that plays a fundamental role in the perturbative renormalization of the theory and the construction of the physical Hilbert space [5]. There is a simple intuitive argument that shows how the breaking of BRST invariance is an inevitable consequence of the restriction to the Gribov region [6]: the BRST transformation of the gluon field has the form of an infinitesimal gauge transformation, but close to the Gribov horizon, an infinitesimal gauge transformation can relate gauge copies, see eq. (5), and hence the transformed field falls outside the Gribov region.

3. Dyson-Schwinger equations

In an initially independent development, the Dyson-Schwinger equations of quantum chromodynamics were approximately solved in the infrared regime. As in the previous section, we here focus on the pure gauge sector, i.e., Yang-Mills theory. The Dyson-Schwinger equations are derived from the Faddeev-Popov action (4), and the restriction to the Gribov region is not implemented in any way. While this is clear from the independence of this development, the reader is justified to wonder why it would make any sense to present the solution of Dyson-Schwinger equations that ignore the restriction to the Gribov region, when we have argued at the end of the foregoing section that the restriction to the Gribov region has a severe effect on the infrared behavior of the propagators.

However, as Zwanziger has clarified [7], the restriction to the Gribov region actually leaves the Dyson-Schwinger equations unchanged: Dyson-Schwinger equations are derived from the fact that the functional integral over a (functional) derivative vanishes. When one restricts the functional integral over the gauge field in the Faddeev-Popov formulation to the Gribov region Ω, contributions from the boundary $\partial\Omega$ are to be expected. However, there are no such contributions in this case since the integrand vanishes on $\partial\Omega$ due to the presence of the Faddeev-Popov determinant as remarked earlier, and the resulting Dyson-Schwinger equations are the same as for the unrestricted functional integral.

Solutions to the Dyson-Schwinger equations for the gluon and ghost propagators have been found, in a certain approximation, first in refs. [8, 9]. The results are an infrared suppressed gluon propagator and an infrared enhanced ghost propagator as compared to the tree-level propagators of Faddeev-Popov theory, which is qualitatively consistent with the results (11) and (13) in the Gribov-Zwanziger framework. Then, with hindsight, we can simplify the Dyson-Schwinger equations in the infrared regime by omitting the diagrams that contain gluon loops, or in general, in the Dyson-Schwinger equation for a given correlation function only keep the diagrams with the biggest number of ghost loops since the latter dominate the infrared behavior. This is sometimes referred to as ghost dominance.

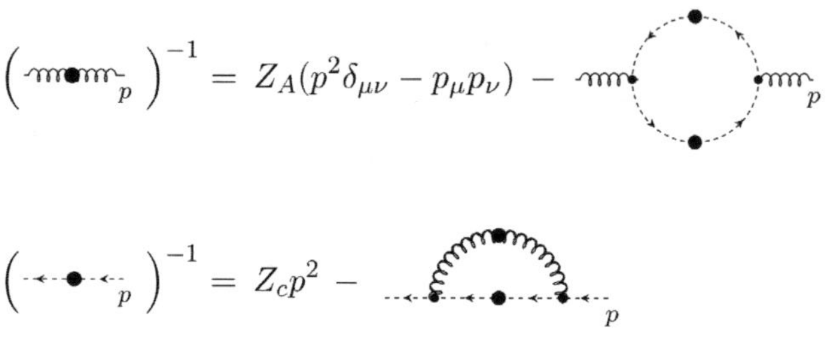

Figure 1. Simplified Dyson-Schwinger equations for the gluon and ghost propagators containing the dominant contributions in the infrared regime.

Furthermore, Taylor's non-renormalization theorem [10] allows us to replace the full ghost-gluon vertex by the bare one. That this replacement is valid to a good approximation even in the non-perturbative regime has been checked in lattice calculations [11]. With these simplifications, the Dyson-Schwinger equations for the gluon and ghost propagators take the form depicted in fig. 1. This simplified system of equations not only correctly describes the extreme infrared limit of the theory, but it also gives a qualitative description of the extreme ultraviolet, asymptotically free regime, and it can thus serve as a useful laboratory for the study of several properties of the complete theory [12]. For a quantitatively correct description of the ultraviolet regime, in particular the correct value of the (one-loop) beta function, however, one needs to take additional contributions into account in the first equation of fig. 1, and more sophisticated approximation schemes have been devised to this end [13, 14].

If one is exclusively interested in the extreme infrared regime, the simplified Dyson-Schwinger equations can even be solved analytically. In order to describe the solutions, we will introduce some notation: we write the propagators in terms of their dressing functions $G(p^2)$ and $F(p^2)$,

$$\begin{aligned}\langle A^a_\mu(p) A^b_\nu(-q)\rangle &= \frac{G(p^2)}{p^2}\left(\delta_{\mu\nu} - \frac{p_\mu p_\nu}{p^2}\right)\delta^{ab}(2\pi)^D\delta(p-q)\,,\\ \langle c^a(p)\bar{c}^b(-q)\rangle &= \frac{F(p^2)}{p^2}\,\delta^{ab}(2\pi)^D\delta(p-q)\,.\end{aligned} \tag{14}$$

For the analytical solutions, we make power-like ansaetze for the dressing functions:

$$G(p^2) \propto (p^2)^{-\alpha_G}\,, \qquad F(p^2) \propto (p^2)^{-\alpha_F}\,. \tag{15}$$

Inserting the ansaetze (15) in the system of approximate Dyson-Schwinger equations in fig. 1, one obtains, for any dimension D between two and four, *two* consistent solutions known as scaling solutions [7, 15]. One of these solutions is

$$\alpha_F(D) = \frac{D-2}{2}\,, \qquad \alpha_G(D) = -\frac{D}{2} \tag{16}$$

in dimension D, for $2 < D < 4$ (this solution is not completely consistent in dimensions two and four due to logarithmic corrections). The infrared exponents of the other scaling solution have to be obtained, for general D, from the solution of a transcendental equation. A good approximation (within 2% error for any D with $2 \leq D \leq 4$) is given by the simple formulae

$$\alpha_F(D) = \frac{D-1}{5}\,, \qquad \alpha_G(D) = -\frac{16-D}{10}\,. \tag{17}$$

For both solutions, the infrared exponents fulfill the sum rule

$$\alpha_G + 2\alpha_F = \frac{D-4}{2} \tag{18}$$

in any dimension D. The solutions also have in common that $\alpha_F > 0$ and $1 + \alpha_G < 0$, implying that the ghost propagator diverges more strongly in the infrared than the (Faddeev-Popov) tree-level propagator, while the gluon propagator tends to zero in the limit $p^2 \to 0$, just like the tree-level propagators in the Gribov-Zwanziger framework.

Taking into account that the Dyson-Schwinger equations are not altered by restricting the functional integral to the Gribov region, we interpret the solutions (16) and (17) as (approximately) incorporating the quantum corrections to the tree-level results (11) and (13) presented in the previous section. It is also clear that the power-like behavior (15) of the solutions can only be obtained by summing up the quantum corrections (in some approximation) to *any* loop order. On the other hand, it is a priori not so clear which of the two solutions (16) and (17) is the physically relevant one for general dimension D. Also, one would like to have a systematic scheme to improve on the approximations employed in the derivation of the Dyson-Schwinger equations in fig. 1, and it has turned out to be very difficult to formulate such a scheme.

An important consequence of Taylor's non-renormalization theorem is the natural definition of a running coupling constant as

$$g_R^2(p^2) = (p^2)^{(D-4)/2}\, G(p^2)\, F^2(p^2)\, g^2\,. \tag{19}$$

The factor $(p^2)^{(D-4)/2}$ compensates for the dimension of the bare coupling constant g^2, so that $g_R^2(p^2)$ is a dimensionless quantity. Then it is clear from the sum rule (18) that the running coupling constant tends to a finite fixed point value in the infrared limit. This value can be determined analytically for the two scaling solutions (16) and (17), and it turns out to vary continuously with the dimension D.

One of the characteristic features of the scaling solutions (including their tree-level forms (11), (13)) is the infrared enhancement of the ghost propagator. In fact, it may be shown [4] that the horizon condition (8) is equivalent to

$$\lim_{p^2 \to 0} F^{-1}(p^2) = 0\,, \tag{20}$$

the infrared divergence of the ghost dressing function. It is well understood by now that implementing the condition (20) necessarily leads to the scaling solutions, be it in the tree-level form (11) and (13) or in the "resummed" form (16) or (17).

4. Decoupling solutions

With the analytical and semi-analytical predictions of the propagators in the infrared regime at hand, it seemed natural to corroborate these predictions with the help of completely non-perturbative lattice calculations. While it is not necessary (nor natural) to fix the gauge in these calculations, it is also not too difficult to implement the Landau gauge on the lattice: for a given gauge field configuration A_μ^a, consider the functional

$$\int d^D x\, (A^U)_\mu^a (A^U)_\mu^a \tag{21}$$

for all possible gauge transformations U, i.e., along the gauge orbit that contains A_μ^a. Then the extrema (or saddle points) of this functional along the orbit correspond to the fields $\bar{A}_\mu^a = (A^U)_\mu^a$ that fulfill the gauge condition $\partial_\mu \bar{A}_\mu^a = 0$, the local minima correspond to fields $\bar{A}_\mu^a$ in the first

Gribov region Ω, and the absolute minima to fields in the fundamental modular region Λ. For the huge lattices that are necessary to measure the infrared behavior of the propagators, it is extremely difficult (numerically costly) to determine the absolute minimum of the functional (21) along a given gauge orbit, however, it is relatively easy to find a local minimum and hence to restrict the gauge fields to the Gribov region.

The first lattice results for the propagators in the Landau gauge (with the restriction to the Gribov region Ω) show a clear increase of the ghost dressing function $F(p^2)$ towards the infrared, and a maximum and subsequent marked decrease of the gluon dressing function $G(p^2)$ as one follows p^2 from the ultraviolet to the infrared, in agreement with the scaling solutions of the Dyson-Schwinger equations. However, it was difficult to discern for a long time whether the ghost dressing function would actually diverge in the limit $p^2 \to 0$ and whether the gluon propagator itself would vanish in the same limit.

Meanwhile, it was discovered that the Dyson-Schwinger equations of the theory permit another type of solutions now known as decoupling solutions [16, 17, 18] (see also [14]). In these solutions, the ghost dressing functions tends toward a finite constant in the limit $p^2 \to 0$, while the gluon propagator has a finite (nonzero) limiting value similarly to the propagator of a massive particle. Then one still has ghost dominance in the sense that the gluon propagator is infrared suppressed in comparison with the ghost propagator, and the reduction of the full Dyson-Schwinger equations to the approximation represented in fig. 1 is justified in the infrared regime.

In terms of the infrared exponents defined in eq. (15), one has for the decoupling solutions that

$$\alpha_F = 0\,, \qquad \alpha_G = -1 \tag{22}$$

in any dimension D. In particular, the decoupling solutions do not respect the sum rule (18) for the exponents (except at $D = 2$). Furthermore, the running coupling constant defined in eq. (19) tends to zero in the limit $p^2 \to 0$ for the decoupling solutions (for dimensions $D > 2$).

The discovery of the decoupling solutions of the Dyson-Schwinger equations initiated a long and often passionate discussion in the community about the true infrared behavior of the Landau gauge propagators. In 2007, at the lattice conference in Regensburg, three lattice groups presented their results on the infrared behavior of the propagators in Landau gauge Yang-Mills theory on very large lattices [19, 20, 21]. The majority of the community has interpreted these results as confirming the *decoupling solutions* of the Dyson-Schwinger equations in three and four space-time dimensions. In the same year, lattice calculations in *two* dimensions [22] confirmed the realization of the second scaling solution (17).

At first sight, the decoupling solutions are at odds with the tree-level propagators (11) and (13) and the horizon condition in the form (20) that result in the Gribov-Zwanziger scenario. However, indications were found in ref. [6] (see also [23]) that the auxiliary fields form a condensate $\langle \bar{\phi}^{ab}_\mu \phi^{ab}_\mu \rangle$ (among other possible condensates). As a consequence, the horizon condition (10) does *not* imply the infrared divergence (20) of the ghost dressing function any more, and through the linear coupling of the fields ϕ^{ab}_μ and $\bar{\phi}^{ab}_\mu$ to the gluon field, see eq. (9), the condensate leads to a mass term for the gluons.

In the simplest case [6] in this so-called refined Gribov-Zwanziger framework, i.e., with the condensates taken into account, the tree-level propagators are of the form

$$\langle A^a_\mu(p) A^b_\nu(-q) \rangle = \frac{p^2 + M^2}{p^4 + M^2 p^2 + \lambda^4} \left(\delta_{\mu\nu} - \frac{p_\mu p_\nu}{p^2} \right) \delta^{ab} (2\pi)^D \delta(p - q) \tag{23}$$

(see eq. (12) for the definition of the characteristic scale λ), and

$$\langle c^a(p) \bar{c}^b(-q) \rangle \propto \frac{1}{p^2} \delta^{ab} (2\pi)^D \delta(p - q) \tag{24}$$

in the infrared regime. The gluon propagator can also be written in terms of an effective momentum-dependent mass $M_{\text{eff}}(p^2)$,

$$\frac{p^2 + M^2}{p^4 + M^2 p^2 + \lambda^4} = \frac{1}{p^2 + M_{\text{eff}}(p^2)}, \qquad M_{\text{eff}}(p^2) = \frac{\lambda^4}{p^2 + M^2}. \tag{25}$$

In four space-time dimensions, a successfull fit of the tree-level gluon propagator in the refined Gribov-Zwanziger framework with an additional $\langle A^a_\mu A^a_\mu \rangle$-condensate to the lattice data in the infrared and intermediate momentum regimes has been performed in ref. [24] (see also [25]).

At this point, from the perspective of the Dyson-Schwinger equations, the decoupling solution as well as both of the scaling solutions are consistent and valid solutions, and from these equations alone there is no reason to prefer one of the solutions over the others. In the refined Gribov-Zwanziger framework, on the other hand, the decoupling solution with the tree-level propagators (23) and (24) is preferred according to the lowest-order calculation of the effective potential for various local composite operators [23]. The problem with this latter approach is rather of a technical nature: it appears extremely cumbersome to carry the calculations to higher orders.

5. Callan-Symanzik equations

5.1. A gluon mass term

For the rest of this contribution, we will describe a recent approach to the subject from a renormalization group perspective [26]. We start from the Dyson-Schwinger equations of Faddeev-Popov theory which, as discussed before, do not change when one restricts the functional integral over the gluon field to the Gribov region Ω. Since the iterative solution of the Dyson-Schwinger equations generates the perturbative expansion, the same is true for the perturbative series. On the other hand, the (re)normalization conditions that define the quantum theory can certainly change as a result of the restriction.

One of the important consequences of the restriction to the Gribov region, as mentioned at the end of section 2, is the breaking of BRST invariance. The resulting violation of the Slavnov-Taylor identites leads, almost inevitably, to the generation of a gluon mass term as the only possible properly relevant term in the (effective) action with the field content of Faddeev-Popov theory, i.e., without the additional auxiliary fields of the Gribov-Zwanziger action. Note that a ghost mass term cannot be induced for the same kinematical reasons that lead to Taylor's non-renormalization theorem. We will, hence, include a gluon mass term

$$\int d^D x \, \frac{1}{2} A^a_\mu m^2 A^a_\mu \tag{26}$$

in the Faddeev-Popov action.

Such an action has been considered first, as a special case among a much more general class of actions, in ref. [27] where the extension of the BRST transformation and the Slavnov-Taylor identities to the massive case is discussed (however, the Nakanishi-Lautrup field is not introduced in ref. [27]). The Faddeev-Popov action with a gluon mass term has also been used in a recent perturbative calculation [28, 29] to which we come back later. Since the mass term originates from the breaking of the BRST symmetry as a consequence of the restriction to the Gribov region, we expect the (renormalized) mass parameter to be of the order of the characteristic scale Λ_{QCD}. It is then clear that the introduction of the mass term has no important effect in the ultraviolet regime, i.e., for momenta $p^2 \gg \Lambda^2_{\text{QCD}}$. In particular, the value of the ultraviolet beta function is unchanged and, of course, the theory is asymptotically free. This is consistent with the fact that the restriction to the Gribov region is irrelevant for the perturbative expansion in the ultraviolet regime where only "small" fields A^a_μ contribute.

Journal of Physics: Conference Series **378** (2012) 012042 doi:10.1088/1742-6596/378/1/012042

Now, we would at least naively expect that in the extreme infrared regime $p^2 \ll m^2$ the mass term (26) dominates over the other term in the action that is quadratic in A_μ^a,

$$\int \frac{d^D p}{(2\pi)^D} A_\mu^a(-p) \left(p^2 \delta_{\mu\nu} - p_\mu p_\nu\right) A_\nu^a(p) \,. \tag{27}$$

In fact, this is true also for the loop corrections as we will now illustrate with the help of a relevant example: consider the one-loop ghost self-energy

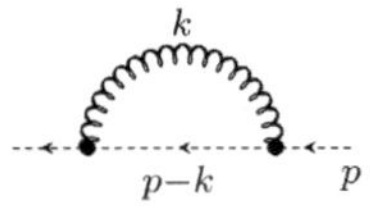

$$= N g^2 \delta^{ab} \int \frac{d^D k}{(2\pi)^D} p_\mu \frac{1}{(p-k)^2} (p_\nu - k_\nu) \frac{1}{k^2 + m^2} \left(\delta_{\mu\nu} - \frac{k_\mu k_\nu}{k^2}\right) . \tag{28}$$

It is worth mentioning that the tree-level gluon propagator in Landau gauge is exactly transverse in momentum space even though the mass term (26) has a longitudinal part. The diagram (28) can be evaluated for arbitrary dimension D, e.g., by introducing Feynman parameters. We will here concentrate on $D = 2$, which will turn out to play a particularly important role, and comment on other dimensions later. Then, evaluating (28) in $D = 2$ and expanding the result in powers of (p^2/m^2) for the extreme infrared behavior gives

$$-\frac{1}{2} \frac{N g^2}{4\pi} \delta^{ab} \frac{p^2}{m^2} \left[\ln \frac{p^2}{m^2} - 1 - \frac{1}{2} \frac{p^2}{m^2} + \mathcal{O}\left(\left(p^2/m^2\right)^2\right)\right] . \tag{29}$$

On the other hand, neglecting the term (27) in the action corresponds to the replacement

$$\frac{1}{m^2 + k^2} \left(\delta_{\mu\nu} - \frac{k_\mu k_\nu}{k^2}\right) \longrightarrow \frac{1}{m^2} \left(\delta_{\mu\nu} - \frac{k_\mu k_\nu}{k^2}\right) \tag{30}$$

for the tree-level gluon propagator. Making this replacement in the loop diagram (28) leads to the result

$$-\frac{1}{2} \frac{N g^2}{4\pi} \delta^{ab} \frac{p^2}{m^2} \left(\frac{2}{\epsilon} + \gamma_E - \ln(4\pi) + \ln \frac{p^2}{\kappa^2}\right) \tag{31}$$

(plus terms of order ϵ) in $D = 2 + \epsilon$ dimensions, with an arbitrary unit of mass κ. The appearance of an ultraviolet divergency in two dimensions is due to the radical change in the ultraviolet behavior of the gluon propagator. To be precise, the symbol g in eq. (31) stands for the bare coupling constant multiplied with a power of κ such that g has the dimension of mass, and it coincides with the coupling constant appearing in eq. (29) only in the limit $D \to 2$.

The results (29) and (31) are obviously different. However, if we implement an appropriate normalization condition for the term proportional to p^2 in the proper ghost two-point function, then *after renormalization* the two results will only differ by contributions of the order (p^2/m^2) relative to the leading term, since renormalization takes care of the constants multiplying p^2. Expressions (29) and (31) are then equivalent in the infrared limit, and in this sense the replacement (30) of the tree-level gluon propagator is justified. The argument can easily be extended to any dimension D in the range $2 \leq D \leq 4$.

We can, hence, neglect the term (27) in the action as long as we are only interested in describing the extreme infrared regime of the theory. These modifications, the introduction of a mass term for the gluons and the subsequent omission of the term (27), have profound consequences for the renormalization group anaylisis of the theory. We begin this analysis by considering the free ($g = 0$) part of the action which is now given by

$$S_{\mathrm{IR}}^0 = \int d^D x \left(\frac{1}{2} A_\mu^a m^2 A_\mu^a + i B^a \partial_\mu A_\mu^a + \partial_\mu \bar{c}^a \partial_\mu c^a\right) . \tag{32}$$

Journal of Physics: Conference Series **378** (2012) 012042 doi:10.1088/1742-6596/378/1/012042

Performing a Wilsonian renormalization group transformation on this free field theory (see, e.g., ref. [30]), the only nontrivial step is the rescaling of the space-time coordinates x and the fields. Writing the rescaling of the coordinates as $x \to x/s$ with a parameter $s > 1$, the action (32) is invariant under the renormalization group transformation if we rescale the gluon field as

$$A_\mu^a(x) \to s^{D/2} A_\mu^a(sx) \,. \tag{33}$$

This behavior under rescaling defines the canonical dimension of the gluon field which is hence $D/2$ and manifestly differs from its usual canonical dimension $(D-2)/2$. The reason is that the action (32) corresponds to the high-temperature fixed point rather than the usually considered critical Gaussian fixed point [30], as far as the gluon field is concerned. The appearance of the high-temperature fixed point naturally connects to the model of the stochastic vacuum. The canonical dimension of the ghost fields, on the other hand, is the ordinary $(D-2)/2$.

Reintroducing the coupling terms of Faddeev-Popov theory into the action, the scaling dimensions around the infrared fixed point action (32) turn out to be different for the different couplings. Concretely, both the three-gluon and the four-gluon couplings become perturbatively *irrelevant*, while the ghost-gluon coupling is relevant only for dimensions $D < 2$ (and marginal at $D = 2$). In other words, the upper critical dimension of the theory is two. This can also be concluded from an infrared power counting analysis to all orders of perturbation theory [29].

Hence, for an infrared analysis of the theory in dimensions $D > 2$, we would be justified to work with the free action (32) alone, which obviously directly leads to the decoupling solution for the propagators (in the extreme infrared limit). It will turn out to be very interesting, however, to include the coupling that is marginal at $D = 2$ dimensions and to treat dimensions above two in an epsilon expansion. We will hence add the coupling term

$$S_{\rm IR}^1 = m\, \kappa^{(2-D)/2}\, \bar{g} f^{abc} \int d^D x\, \partial_\mu \bar{c}^a A_\mu^b\, c^c \tag{34}$$

to the free action (32), where we have introduced the dimensionless coupling constant $\bar{g}$. Incidentally, keeping the ghost-gluon coupling while neglecting the three- and four-gluon couplings is precisely equivalent to keeping only the diagrams with the biggest number of (internal) ghost propagators, that is, ghost dominance.

5.2. *Infrared fixed point*

In the theory defined by $S_{\rm IR}^0 + S_{\rm IR}^1$, the one-loop expressions for the propagators are precisely those represented in fig. 1, except that the propagators in the loop diagrams should be replaced with the tree-level propagators. We have given the result for the ghost self-energy in $D = 2 + \epsilon$ dimensions already in eq. (31). For the gluon self-energy, we get in $D = 2 + \epsilon$ dimensions

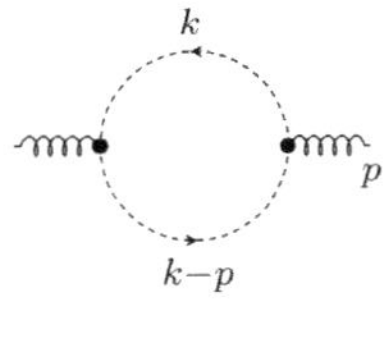

$$= -N m^2 (\kappa^2)^{(2-D)/2}\, \bar{g}^2 \delta^{ab} \int \frac{d^D k}{(2\pi)^D} (k_\mu - p_\mu) \frac{1}{k^2} k_\nu \frac{1}{(k-p)^2}$$

$$= \frac{1}{2} \frac{N \bar{g}^2}{4\pi} m^2 \delta^{ab} \left[\left(\frac{2}{\epsilon} + \gamma_E - \ln(4\pi) + \ln \frac{p^2}{\kappa^2} - 2 \right) \delta_{\mu\nu} + 2 \frac{p_\mu p_\nu}{p^2} \right] . \tag{35}$$

It is important for the consistency of the renormalization procedure that the counterterm can be chosen local, in this case proportional to $m^2 \delta_{\mu\nu}$. The gluon self-energy is *not* transverse in momentum, which is not a surprise since BRST invariance is broken. Only the transverse part, given by the coefficient of $\delta_{\mu\nu}$ in the expression (35), enters the result for the gluon propagator.

For the renormalization of the theory, we introduce renormalized fields as $A^a_\mu = Z_A^{1/2} A^a_{R,\mu}$ and $c^a = Z_c^{1/2} c^a_R$ and normalize the renormalized propagators to

$$\left\langle A^a_{R,\rho}(p) A^b_{R,\sigma}(-q) \right\rangle \Big|_{p^2=\mu^2} = \frac{1}{m^2} \left(\delta_{\rho\sigma} - \frac{p_\rho p_\sigma}{p^2} \right) \delta^{ab} (2\pi)^D \delta(p-q) ,$$

$$\left\langle c^a_R(p) \bar{c}^b_R(-q) \right\rangle \Big|_{p^2=\mu^2} = \frac{1}{\mu^2} \delta^{ab} (2\pi)^D \delta(p-q) , \tag{36}$$

at the renormalization scale μ. Using the expressions (31) and (35) for the self-energies, the normalization conditions (36) define the field renormalization constants Z_A and Z_c as functions of the scale μ. The one-loop anomalous dimensions are then easily calculated to be

$$\gamma_A = \mu^2 \frac{d}{d\mu^2} \ln Z_A = \frac{1}{2} \frac{N\bar{g}^2}{4\pi} , \qquad \gamma_c = \mu^2 \frac{d}{d\mu^2} \ln Z_c = -\frac{1}{2} \frac{N\bar{g}^2}{4\pi} . \tag{37}$$

We define the dimensionless renormalized coupling constant as usual from the renormalized vertex function at the symmetric point,

$$\begin{aligned} \Gamma_{R,\bar{c}Ac}(p,q,r)\big|_{p^2=q^2=r^2=\mu^2} &= Z_A^{1/2}(\mu) Z_c(\mu)\, \Gamma_{\bar{c}Ac}(p,q,r)\big|_{p^2=q^2=r^2=\mu^2} \\ &= m\, \mu^{-\epsilon/2}\, \bar{g}_R(\mu) f^{abc} i p_\rho\, (2\pi)^D \delta(p+q+r) \end{aligned} \tag{38}$$

(cf. eq. (34)), where we have omitted the Lorentz and color indices on the symbol $\Gamma_{\bar{c}Ac}$ of the vertex function. We can now calculate the beta function. Taylor's non-renormalization theorem implies that the bare vertex function at the symmetric point is independent of the scale μ, so that the one-loop beta function becomes

$$\beta(\epsilon, \bar{g}_R) = \mu^2 \frac{d}{d\mu^2} \bar{g}_R = \frac{1}{2} \bar{g}_R \left(\frac{\epsilon}{2} + \gamma_A + 2\gamma_c \right) = \frac{1}{2} \bar{g}_R \left(\frac{\epsilon}{2} - \frac{1}{2} \frac{N\bar{g}_R^2}{4\pi} \right) . \tag{39}$$

From eq. (39) we read off the existence of two fixed points for $\epsilon > 0$ ($D > 2$): a trivial infrared-stable one, and an infrared-unstable one at

$$\frac{N\bar{g}_R^2}{4\pi} = \epsilon . \tag{40}$$

We begin with the discussion of the infrared-unstable fixed point. The μ-independence of the bare propagators leads to differential equations for the renormalized propagators: using the definition of the anomalous dimensions we obtain the Callan-Symanzik equations

$$\mu^2 \frac{d}{d\mu^2} \left\langle A^a_{R,\rho}(p) A^b_{R,\sigma}(-q) \right\rangle = -\gamma_A \left\langle A^a_{R,\rho}(p) A^b_{R,\sigma}(-q) \right\rangle ,$$

$$\mu^2 \frac{d}{d\mu^2} \left\langle c^a_R(p) \bar{c}^b_R(-q) \right\rangle = -\gamma_c \left\langle c^a_R(p) \bar{c}^b_R(-q) \right\rangle . \tag{41}$$

Inserting the fixed point value (40) of the coupling constant in the anomalous dimensions, the differential equations for the renormalized propagators may be integrated with the normalization conditions (36) as initial conditions. The results are

$$\left\langle A^a_{R,\rho}(p) A^b_{R,\sigma}(-q) \right\rangle = \frac{1}{m^2} \left(\frac{p^2}{\mu^2} \right)^{\epsilon/2} \left(\delta_{\rho\sigma} - \frac{p_\rho p_\sigma}{p^2} \right) \delta^{ab} (2\pi)^D \delta(p-q) ,$$

$$\left\langle c^a_R(p) \bar{c}^b_R(-q) \right\rangle = \frac{1}{p^2} \left(\frac{\mu^2}{p^2} \right)^{\epsilon/2} \delta^{ab} (2\pi)^D \delta(p-q) . \tag{42}$$

Journal of Physics: Conference Series **378** (2012) 012042 doi:10.1088/1742-6596/378/1/012042

This is *exactly* the infrared behavior of the first scaling solution (16). So the renormalization group is able to reproduce one of the scaling solutions in the infrared regime and furthermore demonstrates that it corresponds to an infrared-*unstable* fixed point.

Now we turn to the trivial infrared-stable fixed point. The fact that the stable fixed point is at $\bar{g}_R = 0$ implies that perturbation theory with the action $S^0_{\rm IR} + S^1_{\rm IR}$ should give a good description of the infrared regime. In fact, it was shown in refs. [28, 29] that one-loop perturbation theory reproduces the lattice results very well in four dimensions and at least qualitatively in three dimensions. The one-loop contributions that include the three- and four-gluon vertices were taken into account in this comparison, however, they are suppressed by powers of (p^2/m^2) for small momenta.

Actually, the renormalization group approach can be used to go beyond the mere fact that the decoupling solution in the extreme infrared limit corresponds to a stable fixed point. The way the coupling constant approaches the fixed point value (zero) can be obtained by integrating the differential equation (39) for $\bar{g}_R(\mu)$. The result is

$$\frac{N\bar{g}_R^2(\mu)}{4\pi} = \frac{(\mu^2/\Lambda^2)^{\epsilon/2}}{1+(\mu^2/\Lambda^2)^{\epsilon/2}}\,\epsilon\,, \tag{43}$$

where we have introduced a new characteristic scale Λ. It is defined as

$$\frac{N\bar{g}_R^2(\Lambda)}{4\pi} = \frac{\epsilon}{2} \tag{44}$$

and should eventually be related to $\Lambda_{\rm QCD}$ (this relation, however, is beyond the scope of the present approach). Substituting the result (43) for $\bar{g}$ in the anomalous dimensions (37), the integration of the differential equations (41) yields

$$\begin{aligned}\left\langle A^a_{R,\rho}(p)A^b_{R,\sigma}(-q)\right\rangle &= \frac{1}{m^2}\,\frac{1+(p^2/\Lambda^2)^{\epsilon/2}}{1+(\mu^2/\Lambda^2)^{\epsilon/2}}\left(\delta_{\rho\sigma}-\frac{p_\rho p_\sigma}{p^2}\right)\delta^{ab}(2\pi)^D\delta(p-q)\,,\\ \left\langle c^a_R(p)\bar{c}^b_R(-q)\right\rangle &= \frac{1}{p^2}\,\frac{1+(\mu^2/\Lambda^2)^{\epsilon/2}}{1+(p^2/\Lambda^2)^{\epsilon/2}}\,\delta^{ab}(2\pi)^D\delta(p-q)\,.\end{aligned} \tag{45}$$

The infrared behavior of the propagators predicted by eq. (45) is at least qualitatively confirmed by lattice calculations in $D = 3$ dimensions [31]. In $D = 4$ space-time dimensions ($\epsilon = 2$), the increase of the gluon propagator proportional to p^2 starting from a non-zero constant at $p^2 = 0$ as described by eq. (45) is actually seen [32] in lattice calculations at lattice parameter $\beta = 0$, which in continuum language means that the proper Yang-Mills action

$$\int d^Dx\,\frac{1}{4}F^a_{\mu\nu}F^a_{\mu\nu} \tag{46}$$

is omitted in the probability distribution, and only the delta functional (1) and the Faddeev-Popov determinant (2) (and the restriction to the Gribov region) are left.

In order to reproduce the $D = 4$ lattice results with the Yang-Mills action taken into account, we have to reintroduce the term (27) into the action $S^0_{\rm IR} + S^1_{\rm IR}$. The reason is that this term, although perturbatively irrelevant given the scaling behavior (33) of the gluon field, is of the same order in (p^2/m^2) as the correction to the tree-level gluon propagator obtained from the renormalization group improvement in eq. (45) for $\epsilon = 2$. Since at the one-loop level there are no quantum corrections to (27) considered as a composite operator, we can just add this term to the renormalization-group improved result (45) for the proper gluonic two-point function (inverse propagator for the transverse part), which results in

$$\left\langle A^a_{R,\rho}(p)A^b_{R,\sigma}(-q)\right\rangle = \left(p^2 + m^2\,\frac{1+\mu^2/\Lambda^2}{1+p^2/\Lambda^2}\right)^{-1}\left(\delta_{\rho\sigma}-\frac{p_\rho p_\sigma}{p^2}\right)\delta^{ab}(2\pi)^D\delta(p-q)\,. \tag{47}$$

This is precisely of the form (23) of the infrared propagator found in the refined Gribov-Zwanziger framework (without an additional $\langle A_\mu^a A_\mu^a \rangle$-condensate), with the effective momentum-dependent mass given by (cf. eq. (25))

$$M_{\text{eff}}^2(p^2) = \frac{m^2(\mu^2 + \Lambda^2)}{p^2 + \Lambda^2}\,. \tag{48}$$

In the special case of two space-time dimensions ($\epsilon = 0$), the two fixed points considered so far coalesce, and the result is one infrared-unstable trivial fixed point. The instability implies that the free action (32) cannot describe the infrared physics in two dimensions. Indeed, calculations on the lattice confirm that in two dimensions the decoupling solution corresponding to the action (32) is *not* realized [22]. Instead, apparently the second scaling solution (17) is found, and we have not yet succeeded in giving a description of the corresponding fixed point (see, however, below).

5.3. Lifshitz point

Up to now, we have not made any attempt to implement the horizon condition in the form (20). It is not difficult to do so: on the classical level, for a local (and covariant) formulation, we have to replace the ghost term in the action (32) with

$$\int d^D x\, \frac{1}{b^2} \partial_\mu \bar{c}^a (-\partial^2) \partial_\mu c^a\,, \tag{49}$$

where we have introduced an arbitrary dimensionful parameter b. The form (49) of the action is known in statistical physics as an isotropic Lifshitz point [33, 34].

Using the term (49) in the free action changes the canonical dimension of the ghost fields. Scale invariance of the free action now implies the scaling behavior

$$c^a(x) \to s^{(D-4)/2}\, c^a(sx)\,, \qquad \bar{c}^a(x) \to s^{(D-4)/2}\, \bar{c}^a(sx)\,. \tag{50}$$

Then, around the Lifshitz point, the three- and four-gluon couplings are perturbatively irrelevant just as before, while the ghost-gluon coupling becomes relevant for dimensions $D < 6$.

In order to describe the infrared behavior of the theory with the horizon condition (20) implemented in dimensions between two and four, we use an epsilon expansion around $D = 6$. We introduce the coupling term

$$S_{\text{IR}}^1 = \frac{m}{b^2}\, \kappa^{(6-D)/2}\, \bar{g} f^{abc} \int d^D x\, \partial_\mu \bar{c}^a A_\mu^b\, c^c \tag{51}$$

with the dimensionless coupling constant $\bar{g}$ (different from the coupling constant in eq. (34)), and follow the procedure applied before in close analogy, starting with the calculation of the ghost and gluon self-energies, now at dimension $D = 6 - \epsilon$ and with the tree-level ghost propagator proportional to b^2/k^4. We stress that a fine tuning of the tree-level ghost term that appears in eq. (32) is necessary in order to fulfill the horizon condition (20) to one-loop order, just like in ordinary ϕ^4-theory the tree-level mass term has to be fine-tuned to reach the critical point.

We use normalization conditions and a definition of the renormalized coupling constant $\bar{g}_R(\mu)$ in strict analogy to eqs. (36) and (38), adapted to the new form (49) of the ghost term and the coupling (51). Taylor's non-renormalization theorem continues to hold in the present situation, and the result for the one-loop beta function is

$$\beta(\epsilon, \bar{g}_R) = \mu^2 \frac{d}{d\mu^2}\, \bar{g}_R = -\frac{1}{2}\, \bar{g}_R \left(\frac{\epsilon}{2} - \frac{1}{2} \frac{N \bar{g}_R^2}{(4\pi)^3} \right)\,. \tag{52}$$

It thus turns out that the infrared-stable fixed point for dimensions $D < 6$ ($\epsilon > 0$) is the nontrivial one. The final result for the renormalization-group improved infrared propagators corresponding to the infrared-stable fixed point is

$$\begin{aligned}\langle A^a_{R,\rho}(p)A^b_{R,\sigma}(-q)\rangle &= \frac{1}{m^2}\left(\frac{p^2}{\mu^2}\right)^{\epsilon/12}\left(\delta_{\rho\sigma} - \frac{p_\rho p_\sigma}{p^2}\right)\delta^{ab}(2\pi)^D\delta(p-q)\,,\\ \langle c^a_R(p)\bar{c}^b_R(-q)\rangle &= \frac{b^2}{p^4}\left(\frac{p^2}{\mu^2}\right)^{5\epsilon/24}\delta^{ab}(2\pi)^D\delta(p-q)\,.\end{aligned} \tag{53}$$

In the notation of eq. (15), we obtain the following values for the infrared exponents:

$$\alpha_F(D) = \frac{5D-6}{24}\,, \qquad \alpha_G(D) = -\frac{18-D}{12}\,. \tag{54}$$

The exponents fulfill the sum rule (18). The values (54) should be compared to the (approximate) exponents of the second scaling solution (17),

$$\alpha_F(D) = \frac{5D-5}{25}\,, \qquad \alpha_G(D) = -\frac{16-D}{10}\,. \tag{55}$$

The two results coincide (exactly) at $D = 6$, and in the range $2 \leq D \leq 4$ they differ the most at $D = 2$ (where the approximation (55) turns out to be exact): $\alpha_F = 1/6$ vs. $\alpha_F = 1/5$ and $\alpha_G = -8/6$ vs. $\alpha_G = -7/5$. However, compared to our experience with the epsilon expansion in statistical physics, the coincidence of the infrared exponents (54) and (55) is unexpectedly good, and we consider them to be approximations to the same solution of the exact Dyson-Schwinger equations of the theory.

5.4. Discussion

We have reproduced the two scaling solutions and the decoupling solution of the Dyson-Schwinger equations with the help of an analytical renormalization group analysis. Only the second scaling solution (17) and the decoupling solution correspond to infrared-stable fixed points, while the first scaling solution (16) corresponds to an infrared-unstable fixed point in dimensions D between two and four.

In fact, the Lifshitz point which leads to the second scaling solution is only stable if one implements the horizon condition (20). As remarked after eq. (51), the implementation of the horizon condition requires a fine tuning of the tree-level term

$$\int d^Dx\,\partial_\mu\bar{c}^a\partial_\mu c^a\,. \tag{56}$$

Consequently, the nontrivial fixed point in eq. (52) is infrared-*unstable* unless one insists in implementing the horizon condition.

We hence conclude that in dimensions $D > 2$ the physically realized solution according to lattice simulations and the refined Gribov-Zwanziger approach is the only one that corresponds to an infrared-stable fixed point. In other words, the only input in the Callan-Symanzik approach is the gluon mass term that is generated by the breaking of BRST invariance, then the renormalization group flow automatically leads to the decoupling solution without further assumptions. In two space-time dimensions, on the other hand, the trivial fixed point corresponding to the decoupling solution becomes unstable, and the lattice results point to the second scaling solution in this case. However, according to the explanation above the fixed point corresponding to this scaling solution should also be unstable. Indeed, calculating

Journal of Physics: Conference Series **378** (2012) 012042 doi:10.1088/1742-6596/378/1/012042

the one-loop corrections to (56) considered as a composite operator, the fixed point becomes even more strongly infrared-repulsive with decreasing dimension. Hence, unless there is some reason to implement the horizon condition precisely in two dimensions, we do not yet have a satisfactory understanding of the infrared behavior of the theory in two dimensions within the Callan-Symanzik approach.

In conclusion, we have presented a new approach to the infrared regime of Yang-Mills theory in Landau gauge that is based on a renormalization group analysis with Callan-Symanzik equations in an epsilon expansion. In contrast to the Dyson-Schwinger equations, the Callan-Symanzik equations are analytical and inherently systematic as an approximation scheme. Furthermore, they are technically relatively simple and economic as compared to the Gribov-Zwanziger approach with the additional auxiliary fields necessary for a local formulation. This should be an important advantage in the further elaboration of the theory, in particular the inclusion of condensates (as in the refined Gribov-Zwanziger framework) and dynamical quarks.

Acknowledgments

Support by CIC-UMSNH and Conacyt project CB-2009/131787 is gratefully acknowledged.

References

[1] Gribov V N 1978 *Nucl. Phys.* B **139** 1
[2] Zwanziger D 2004 *Phys. Rev.* D **69** 016002
[3] Zwanziger D 1989 *Nucl. Phys.* B **323** 513
[4] Zwanziger D 1993 *Nucl. Phys.* B **399** 477
[5] Kugo T and Ojima I 1979 *Prog. Theor. Phys. Suppl.* **66** 1
[6] Dudal D, Gracey J A, Sorella S P, Vandersickel N and Verschelde H 2008 *Phys. Rev.* D **78** 065047
[7] Zwanziger D 2002 *Phys. Rev.* D **65** 094039
[8] von Smekal L, Hauck A and Alkofer R 1997 *Phys. Rev. Lett.* **79** 3591
[9] von Smekal L, Hauck A and Alkofer R 1998 *Annals Phys.* **267** 1
[10] Taylor J C 1971 *Nucl. Phys.* B **33** 436
[11] Cucchieri A, Mendes T and Mihara A 2004 *JHEP* **12** 012
[12] Dall'Olio P 2011 *El Régimen Infrarrojo en Teorías de Yang-Mills* MSc thesis (Universidad Michoacana de San Nicolás de Hidalgo)
[13] Fischer C and Alkofer R 2002 *Phys. Lett.* B **536** 177
[14] Fischer C S, Maas A and Pawlowski J M 2009 *Annals Phys.* **324** 2408
[15] Lerche C and von Smekal L 2002 *Phys. Rev.* D **65** 125006
[16] Aguilar A C and Natale A A 2004 *JHEP* **0408** 057
[17] Boucaud P, Brüntjen T, Leroy J P, Le Yaouanc A, Lokhov A, Micheli J, Pène O and Rodríguez-Quintero J 2006 *JHEP* **0606** 001
[18] Frasca M 2008 *Phys. Lett.* B **670** 73
[19] Bogolubsky I L, Ilgenfritz E M, Müller-Preussker M and Sternbeck A 2007 *Proc. Sci.* LAT2007 290
[20] Cucchieri A and Mendes T 2007 *Proc. Sci.* LAT2007 297
[21] Sternbeck A, von Smekal L, Leinweber D B and Williams A G 2007 *Proc. Sci.* LAT2007 340
[22] Maas A 2007 *Phys. Rev.* D **75** 116004
[23] Dudal D, Sorella S P and Vandersickel N 2011 *Phys. Rev.* D **84** 065039
[24] Dudal D, Oliveira O and Vandersickel N 2010 *Phys. Rev.* D **81** 074505
[25] Cucchieri A, Dudal D, Mendes T and Vandersickel N 2011 *Preprint* arXiv:1111.2327 [hep-lat]
[26] Weber A 2011 *Preprint* arXiv:1112.1157 [hep-th]
[27] Curci G and Ferrari R 1976 *Nuovo Cim.* A **32** 151
[28] Tissier M and Wschebor N 2010 *Phys. Rev.* D **82** 101701
[29] Tissier M and Wschebor N 2011 *Phys. Rev.* D **84** 045018
[30] Le Bellac M 1991 *Quantum and Statistical Field Theory* (Oxford: Oxford University Press)
[31] Cucchieri A and Mendes T 2009 *Proc. Sci.* QCD-TNT09 026.
[32] Cucchieri A and Mendes T 2010 *Phys. Rev.* D **81** 016005
[33] Hornreich R M, Luban M, and Shtrikman S 1975 *Phys. Rev. Lett.* **35** 1678
[34] Mergulhão, Jr. C and Carneiro C E I 1998 *Phys. Rev.* B **58** 6047